50 Years
of ANDERSON
LOCALIZATION

50 Years of ANDERSON LOCALIZATION

Editor

Elihu Abrahams
University of California, Los Angeles, USA

World Scientific

NEW JERSEY · LONDON · SINGAPORE · BEIJING · SHANGHAI · HONG KONG · TAIPEI · CHENNAI

Published by

World Scientific Publishing Co. Pte. Ltd.

5 Toh Tuck Link, Singapore 596224

USA office: 27 Warren Street, Suite 401-402, Hackensack, NJ 07601

UK office: 57 Shelton Street, Covent Garden, London WC2H 9HE

British Library Cataloguing-in-Publication Data
A catalogue record for this book is available from the British Library.

Cover credit: Speckle pattern of transmission of red light from a helium neon laser through 60 glass slides. Sharp peaks represents regions in which light overlaps a localized mode. Courtesy of Azriel Genack.

50 YEARS OF ANDERSON LOCALIZATION

ISBN-13 978-981-4299-06-0
ISBN-10 981-4299-06-5
ISBN-13 978-981-4299-07-7 (pbk)
ISBN-10 981-4299-07-3 (pbk)

Desk Editor: Tan Hwee Yun

Printed by FuIsland Offset Printing (S) Pte Ltd. Singapore

INTRODUCTION

At the time of the publication of this volume, more than fifty years have passed since the appearance in *The Physical Review* of Philip W. Anderson's landmark paper titled *Absence of Diffusion in Certain Random Lattices.*[1] During the decades since, the phenomenon predicted and explained in that paper became known as "Anderson localization" and has been widely recognized as one of the fundamental concepts in the physics of condensed matter and disordered systems. Anderson's 1977 Nobel Prize, shared with Nevill Mott and John Van Vleck, is based in part on that seminal paper.

Anderson was initially motivated to understand the influence of disorder on spin diffusion and on electron transport. In the years since, the concepts and results that he created have found their way across a wide range of other topics. Among them are nano– and meso–scale technology, seismology, acoustic waves, quantum optics, ultracold atomic gases, localization of light.

The chapters contributed by Phil Anderson, David Thouless and T. V. Ramakrishnan explain clearly some of the early history of the understanding of the localization phenomenon. Earlier discussions of the background and content of Anderson's 1958 paper may be found in Thouless' 1970 review[2] and in Anderson's 1977 Nobel Lecture.[3]

In the area of electron transport, not much was done on the localization problem for more than a decade after the 1958 paper. What might be called the modern era of localization began in the 1970s, with the introduction of scaling ideas by Licciardello and Thouless,[4] Wegner[5] and Schuster.[6] As a matter of fact, it was the Schuster paper that set Phil Anderson, Don Licciardello, T. V. Ramakrishnan, and I thinking about the statistical mechanics analogy, one–parameter scaling and the beta function of scaling theory. The consequence was our 1979 Physical Review Letter,[7] often called the "gang of four" paper ("G4"). As is well known, we concluded that the metal–insulator transition is continuous, i.e. there is no minimum metallic conductivity and that all states in two dimensions are localized. The history of these developments is beautifully reviewed by David Thouless in his contribution to this volume.

There are a number of papers that are not often quoted now, although they made significant impact when they appeared. Here, I take this opportunity to mention some of them and to place them in historical context. The functional integral formulation for correlation functions of a disordered electron system and disorder averaging by means of the $n \to 0$ replica trick was developed by several people, notably Amnon Aharony and Yoseph Imry.[8] John Cardy, in 1978,[9] reformulated the functional integral representation and the n-replica method. He showed how to control the saddle point of the equivalent Ginzburg–Landau action and obtained power laws for the energy dependence of the density of states.

Following upon G4, Shinobu Hikami, Anatoly Larkin and Yosuke Nagaoka[10] incorporated scattering mechanisms with different symmetries (spin–orbit scattering, magnetic impurities), inelastic scattering, and crucially, magnetic field into the treatment of the crossed graphs of Langer and Neal,[11] which are the basis of the scaling behavior derived in G4. Here, the magnetoresistance was derived and this became, and remains, the diagnostic of choice for all subsequent experiments. In this connection, see the chapters of Bergmann, Dynes and Giordano in this book.

Around 1980, there were a number of discussions[12] of the equivalence of the localization problem and the matrix nonlinear σ model. An especially transparent derivation was given by Shinobu Hikami in 1981.[13] He showed how the systematic perturbative treatment of the relevant diffusion propagators in the particle–hole ("diffuson") and particle–particle ("cooperon") channels leads to an effective Hamiltonian of the nonlinear σ model. His analysis of the propagators and their interaction vertices became the standard basis for subsequent perturbative treatments of various effects, including in particular early analyses[14] of the effect of electron–electron interactions. The development of matrix nonlinear σ model methods is reviewed by several contributors to this volume: Efetov, Mirlin *et al.*, Pruisken.

The physics of Anderson localization has had a pervasive influence on a broad variety of fundamental concepts and phenomena, including the quantum Hall effect, quantum criticality, symmetry and random matrix theory, multifractality, electron–electron interaction in disordered metals. These and other issues are explored in many of the chapters of this book.

Some of the pioneers in the field of disordered systems, both theorists and experimentalists, have contributed to this volume. It is a mark of the vitality of Anderson localization physics (and indeed of the contributors) that while a few chapters are essentially historical, the others report results of current research. Unfortunately, space constraints have prevented a comprehensive survey of all the past and current developments. In spite of this limitation the hope is that the reader will acquire an appreciation of the history of the physics of localization and its current manifestations.

References

1. P. W. Anderson, *Phys. Rev.* **109**, 1492 (1958).
2. D. J. Thouless, *J. Phys. C: Solid State Phys.* **4**, 1559 (1970).
3. P. W. Anderson, *Rev. Mod. Phys.* **50**, 191 (1978).
4. D. C. Licciardello and D. J. Thouless, *Phys. Rev. Lett.* **35**, 1475 (1974).
5. F. J. Wegner, *Z. Physik B* **25**, 327 (1976).
6. H. G. Schuster, *Z. Physik B* **31**, 99 (1978).
7. E. Abrahams, P. W. Anderson, D. C. Licciardello and T. V. Ramakrishnan, *Phys. Rev. Lett.* **42**, 673 (1979).
8. A. Aharony and Y. Imry, *J. Phys. C: Solid State Phys.* **10**, L487 (1977).
9. J. L. Cardy, *J. Phys. C: Solid State Phys.* **11**, L321 (1978).
10. S. Hikami, A. I. Larkin and Y. Nagaoka, *Prog. Theor. Phys.* **63**, 707 (1980).
11. J. S. Langer and T. Neal, *Phys. Rev. Lett.* **16**, 984 (1966).
12. F. J Wegner, *Z. Phys. B* **35**, 207 (1979) and for example, K. B. Efetov, A. I. Larkin and D. E. Khmelnitskii, *JETP* **52**, 568 (1980); A. J. McKane and M. Stone, *Annals of Physics* **131**, 36 (1981); A. B. Harris and T. C. Lubensky, *Solid State Comm.* **34**, 343 (1980).
13. S. Hikami, *Phys. Rev. B* **24**, 2671 (1981).
14. See the review of P. A. Lee and T. V. Ramakrishnan, *Rev. Mod. Phys.* **57**, 287 (1985).

CONTENTS

Chapter 1

THOUGHTS ON LOCALIZATION

Philip Warren Anderson

Department of Physics, Princeton University, Princeton, NJ 08544, USA

The outlines of the history which led to the idea of localization are available in a number of places, including my Nobel lecture. It seems pointless to repeat those reminiscences; so instead I choose to set down here the answer to "what happened next?" which is also a source of some amusement and of some modern interest.

A second set of ideas about localization has come into my thinking recently and is, again, of some modern interest: a relation between the "transport channel" ideas which began with Landauer, and many-body theory.

I have several times described the series of experiments, mostly on phosphorus impurities doped into Si (Si-P), done by Feher in 1955–56 in the course of inventing his ENDOR method, where he studies the nuclear spins coupled via hyperfine interaction to a given electron spin, via the effect of a nuclear resonance RF signal on the EPR of the electron spin. My study of these led to what Mott called "the 1958 paper" but in fact there is tangible evidence that the crucial part of the work dated to 1956–7. I have referred to the published discussion by David Pines which immediately followed Mott's famous paper in Can. J. Phys., 1956, where he described the joint, inconclusive efforts of the little group of E. Abrahams, D. Pines and myself to understand Feher' s observations. I also actually broke into print, at least in the form of two abstracts of talks, during that crucial period, and I show here facsimiles of these two abstracts (Figs. 1.1 and 1.2). The first is for the talk I gave at the International Conference on Theoretical Physics, in Seattle, October 1956. That conference was dominated by the parity violation talks of Lee and Yang, and by the appearance of Bogoliubov leading a Russian delegation; only third came the magnificent talk by Feynman on superfluid He and his failure to solve superconductivity. (A memory — on

APPLICATIONS OF PARAMAGNETIC RESONANCE TO SEMICONDUCTOR PHYSICS

by

P.W. Anderson
Bell Telephone Laboratories, Inc., Murray Hill, N. J.

ABSTRACT

This talk will present a number of the cases in which information of importance in semiconductor physics is to be gained from paramagnetic resonance work. In particular, we discuss first the work of Shulman on the relaxation in Si_{29} nuclear resonance due to scattering of electrons by the hyperfine interaction, thus telling us the value of the free electron wave function at the nucleus, and a certain average value of the hole wave function. Second is the detailed experimental observation of the wave-functions of donors in Si made possible by Fehwer's "double-resonance" technique, which by comparison with Kohn's theory can in principle locate the k_0 of the many-valley model. Thirdly, the observation by Feher of several exchange effects in donor electron resonance in Si gives us further information both on donor wave-functions and on the interactions of donors.

Fig. 1.1.   Anderson abstract, Seattle, 1956.

a ferry crossing Puget Sound in thick fog with T. D. Lee explaining that we were being guided by phonon exchange beteen the foghorns!)

The second, which is a fortuitously preserved abstract of a talk I gave at the RCA laboratories in February 1957, makes it quite clear that by now the "fog" had cleared and I was willing to point out the absence of spin diffusion in these experiments. But actually, if my memory is no more fallible than usual, the first talk used the idea of localization without trumpeting that use, so it is essentially the first appearance of the localization scheme: a minor sideline to all the great things reported at Seattle.

But what I want to point out is how both George Feher and I missed the opportunity of a lifetime to jump five decades in time and invent modern quantum information theory in 1957. What my newly-fledged localization idea had left us with is the fact that suddenly the phosphorus spins had become true Qbits, independent sites containing a quantum entity with two

INTERACTIONS OF DONOR SPINS IN SILICON

P.W. ANDERSON
Bell Telephone Laboratories, Inc.
Murray Hill, New Jersey

ABSTRACT

In this talk, I will describe some of the magnetic resonance experiments of Feher on donor atoms in Si, and discuss in particular those effects which depend on the interactions of the different donors.

Some of the topics which come under this general heading are: (1) the satellite lines, explained by Slichter as being caused by clusters, and what they tell us about the exchange integrals and the distribution of donors. (2) The non-existence of rapid relaxation because of motion within the "impurity band". This has important quantitative implications on transport in the impurity band. (3) Measurements involving "burning a hole in the line" show that "spin diffusion" is absent, and we discuss this theoretically. (4) Some more complicated minor effects involving exchange are discussed.

Fig. 1.2.   Anderson abstract, RCA Laboratories, Princeton NJ, 1957.

states and an SU(2) state space which could, via the proper sequence of magnetic field variation, saturating monochromatic signals, and "π pulses", be run through almost any unitary transformation we liked. Localization provided them the appropriate isolation, the reassurance that at least for several seconds or minutes there was no loss of quantum coherence until we came back to them with RF excitation. We could even hope to label the spins by tickling them with the appropriate RF signals. In other words, we had available the program which Steve Lyon is working on at Princeton right now. Of course, the very words like "Qbit" were far in the future, but we certainly saw those spins do some very strange things, some of which I tried to explain in my talk at Seattle.

But what did happen? George was in the course of getting divorced, remarried, and moving to UCSD at La Jolla and into the field of photosynthesis. While doing this he passed the whole problem — which we called "passage effects" — on to his postdoc, Meir Weger. I, on the other hand, in the summer and fall of 1957 met the BCS theory and had an idea about it which eventually turned out to be the "Anderson–Higgs" mechanism, which was exciting and totally absorbing; I also did not find Meir as imaginative and as eager to hear my thoughts as George had been. So both of us shamelessly abandoned the subject — much, I suspect, to our eventual benefit. It really was too early and everything seemed — and was — too complicated for us.

Progress from 1958 to 1978 was slow; even my interest in localization was only kept alive for the first decade by Nevill Mott's persistence and his encouragement of the experimental groups of Ted Davis and Helmut Fritzsche to keep after the original impurity band system, where they confirmed that localization was not merely a Mott phenomenon in that case. I think the valuable outcome for me was that I began to realize that Mottness and localization were not inimical but friendly: my worries in the 1957 days, that interactions would spoil everything, had been unnecessary.

By 1971, when Mott published his book on "Electronic Processes in Non-Crystalline Materials", I had apparently begun to take an interest again. At least, in 1970 I had begun to answer[1] some of the many doubters of the correctness of my ideas. The masterly first chapter of that book, in which Mott makes the connection between localization and the Joffe–Regel idea of breakdown of transport theory when $k\ell \sim 1$, and between this limit and his "minimum metallic conductivity" (MMC) now gradually began to seep into my consciousness; but it was only much later that the third relevant connection, to Landauer's discussion of the connection between conductivity and the transmission of a single lossless channel, became clear to me. I felt, I think correctly, that Landauer's conclusion that 1D always localizes was a trivial corollary of my 1958 discussion; a great deal of further thought has to be put into the Landauer formalism before it is a theory of localization proper (see for instance, but not only, my paper of 1981[2]).

After finally understanding Mott's ideas I became, for nearly a decade, a stanch advocate of Mott's MMC. I remember with some embarrassment defending it strongly to Wegner just a few days before the Gang of Four (G-IV) paper broke into our consciousness. I liked quoting, in talks during that period, an antique 1914 article about Bi films by W. F. G. Swann, later to be Director of the Bartol Research Foundation, and E. O. Lawrence's thesis professor, which confirmed the 2D version of MMC experimentally without having, of course, the faintest idea that the number he measured was e^2/h. But as often is the case, I was holding two incompatible points of view in my head simultaneously, because I was very impressed by David Thouless's first tentative steps towards a scaling theory and, in fact, we hired at Princeton his collaborator in that work, Don Licciardello.

Again, the story, such as it was, of the genesis of the G-IV paper has been repeatedly narrated. More obscure are the origins of the understanding of conductivity fluctuations and of the relationships between Landauer's formula, localization and conductance quantization. Here, at least for me, Mark Az'bel played an important role. He was the first person to explain to me that a localized state could be part of a transmission channel which, at a sufficiently carefully chosen energy, would necessarily have transmission nearly unity. To my knowledge, he never published any way of deriving the universality and the divergent magnitude of conductivity fluctuations, but he had, or at least communicated to me, the crucial insight, quite early. (In the only paper on the subject I published, his contributions were referred to as "unpublished".) In any case, in the end I came to believe that the real nature of the localization phenomenon could be understood best, by me at least, by Landauer's formula

$$G = \frac{e^2}{h} \mathrm{Tr}[tt^*], \qquad (1.1)$$

where $t_{\alpha\beta}$ is the transmissivity between incoming channel α and outgoing channel β, on the energy shell. As I showed in that paper, the statistics of conductance fluctuations is innate in this formula; also conductance quantization in a single channel. (Though there are subtle corrections to the simple theory I gave there because of the statistics of eigenvalue spacing, as Muttalib and, using a different formalism, Altshuler showed.)

But what might be of modern interest is the "channel" concept which is so important in localization theory. The transport properties at low frequencies can be reduced to a sum over one-dimensional "channels". What this is reminiscent of is Haldane and Luther's tomographic bosonization of the Fermi system, where we see an analogy between the Fermi surface "patches" of Haldane and the channels of localization theory. Is it possible that a truly general bosonization of the Fermion system is possible in terms of density operators in a manifold of channels[3]?

One motivation for pursuing the consequences of such a bosonization is the failures — this is not too strong a word — of standard quantum computational methods to deal properly with the sign problem of strongly interacting Fermion systems. Quantum Monte Carlo and even very sophisticated generalizations of QMC fail completely to identify the Fermi surface singularities which are inevitable in such systems.[4] Perhaps a "bosons in channels" reformulation is called for.

References

1. P. W. Anderson, *J. Non-Cryst. Solids* **4**, 433 (1970).
2. P. W. Anderson, *Phys. Rev. B* **23**, 4828 (1981).
3. K. B. Efetov, C. Pépin and H. B. Meier, *Phys. Rev. Lett.* **103**, 186403 (2009).
4. P. W. Anderson, con-mat/0810.0279; (submitted to *Phys. Rev. Lett.*).

Chapter 2

ANDERSON LOCALIZATION IN THE SEVENTIES AND BEYOND

David Thouless

Department of Physics, University of Washington,
Seattle, WA 98195, USA
thouless@phys.washington.edu

Little attention was paid to Anderson's challenging paper on localization for the first ten years, but from 1968 onwards it generated a lot of interest. Around that time a number of important questions were raised by the community, on matters such as the existence of a sharp distinction between localized and extended states, or between conductors and insulators. For some of these questions the answers are unambiguous. There certainly are energy ranges in which states are exponentially localized, in the presence of a static disordered potential. In a weakly disordered one-dimensional potential, all states are localized. There is clear evidence, in three dimensions, for energy ranges in which states are extended, and ranges in which they are diffusive. Magnetic and spin-dependent interactions play an important part in reducing localization effects. For massive particles like electrons and atoms the lowest energy states are localized, but for massless particles like photons and acoustic phonons the lowest energy states are extended.

Uncertainties remain. Scaling theory suggests that in two-dimensional systems all states are weakly localized, and that there is no minimum metallic conductivity. The interplay between disorder and mutual interactions is still an area of uncertainty, which is very important for electronic systems. Optical and dilute atomic systems provide experimental tests which allow interaction to be much less important. The quantum Hall effect provided a system where states on the Fermi surface are localized, but non-dissipative currents flow in response to an electric field.

1. Introduction

Anderson's article on *Absence of Diffusion in Certain Random Lattices*[1] appeared in March 1958, but its importance was not widely understood at the time. Science Citation Index lists 23 citations of the paper up to the end of 1963, including 3 by Mott, and there were another 20 citations in the next five years, including 7 more by Mott. I could find only one citation by Anderson himself in this period. A bare count of the number of citations overestimates the speed with which the novel ideas in this paper spread through the community. Many of the earliest papers with citations I could find, including Anderson's own,[2] had only a peripheral connection with this work.

The original context of the work was an experiment by Feher and Gere[3] in which quantum diffusion of a spin excitation away from its initial site was expected, but in which, for low concentrations of spins, the excitation appeared to remain localized, and to diffuse only by thermally activated hopping between discrete localized states. Anderson refers even-handedly to both spin diffusion and electrical conduction. The opening sentence in the Abstract reads "This paper presents a simple model for such processes as spin diffusion or conduction in the 'impurity band'." However, most of the earliest papers that referred to this work were concerned with spin diffusion rather than charge diffusion. I will phrase the discussion in terms of charge transport rather than spin transport, since this is the side with which I am most familiar.

The model Anderson studied involved a lattice on which an electron could hop from one site to one of its neighbors, with a spread in the energies of the sites produced by some source of disorder, such as a random electrostatic potential. By means of a perturbation expansion in powers of the strength of the hopping matrix, Anderson showed that for strong disorder, or close to the band edges (in the Lifshitz tail[4]), the electron eigenstates are localized, falling off exponentially with distance from the site of maximum amplitude, while for weak disorder, away from the band edges, there is no reason to doubt the applicability of the usual theory of metals and of strongly doped semiconductors, in which the disorder is regarded as inducing transitions of the electrons between Bloch waves in an energy band. For a degenerate Fermi gas of electrons at low temperature, the conductivity is nonzero if the Fermi surface is in a region of such Bloch wave like states.

Over those ten years, there were a number of developments that brought this paper more into the main stream. Firstly, the work of Anderson, Kondo and others on localized magnetic moments in conductors raised some of the same issues in a different context. Kohn[5] had argued that the difference between metals and insulators was that the electron wave functions were exponentially localized in insulators, but extended in metals. Ovshinsky[6] was arguing for the technological importance of amorphous semiconductors.

Mott and his colleagues[7,8] were analysing the electrical properties of highly disordered and amorphous materials. Domb, Fisher, and their colleagues and competitors, such as Pokrovsky and Kadanoff, were making a general study of continuous (second order) phase transitions.[9]

It is clear to me that it was primarily the work of Mott that brought the ideas that Anderson had developed to the center of the discussion of the difference between metals and insulators. In a lengthy review article on the theory of impurity conduction in semiconductors, Mott and Twose[7] gave a real discussion of Anderson localization, which included an early proof of the result that, in one dimension, even the weakest disorder should localize all states. There are some odd remarks in this paper, which suggest that they had not read Anderson's paper particularly carefully. They say that Anderson was concerned with spin diffusion, and Twose extended his work to impurity conduction, despite the explicit mention in both abstract and text of impurity conduction in Anderson's paper. They also say that the concentration needed for delocalization has nothing to do with the concentration for metallic conduction, which they discuss in their own paper. In Mott's frequently cited 1967 review article on electrons in disordered structures[8] the main ideas of both Mott and Anderson are put together in a clear and coherent manner.

In 1969 there was a conference in Cambridge, England, on amorphous and liquid semiconductors, which featured an argument about the Anderson mechanism for localization. Here, arguments were brought by Ziman and Brouers in opposition to the orthodox views of Anderson and Mott.[10] I knew almost nothing about localization, although I already knew Anderson well, but in the course of an accidental meeting with Ziman in Bristol, I promised to look into this dispute. I was overwhelmed by the strength of the arguments in Anderson's 1958 paper, and offered no comfort to Ziman. There was little to revise in Anderson's work, so the paper I wrote[11] was basically a review of it, which I hoped would be more accessible than the original paper appeared to be. It also prepared me to use numerical techniques to resolve some of the questions which analytical work had left open.

The confusion that was displayed in the 1969 Cambridge meeting[10] is clearly shown in the article by Lloyd,[12] which gives an exact result for the density of states of a tight-binding model in which the electron energy levels on each site are distributed with a Cauchy distribution. The result is that the average of the Green's function $1/(E - H)$ is just what one would get by replacing the random, independently distributed, site energies by a complex constant $\epsilon_0 + i\gamma$, where γ is the width of the Cauchy distribution, and ϵ_0 is its center; this is true for any regular lattice and in any number of dimensions. This result shows that the density of energy levels is not only smooth, but analytic, for the Cauchy distribution. Nevertheless, it seems unlikely that in two- or three-dimensional spaces there would be significant changes in the localization properties of potential fluctuations with a Cauchy distribution, which has a divergent variance, and the uniform distribution used by Anderson or a normal distribution, which have finite variances, because in dimension greater than unity the wave functions can just avoid sites with energies far from the mean. This intuitive argument is strongly supported by calculations using finite sized scaling techniques which are described in Sec. 3.

It is indeed true that the average of the Green's function does not differentiate between localized and extended states, and one must look at least at the mean square of the Green's function to study electrical conductivity. Nevertheless, many of us continued to accept that disorder in a metal would destroy the long-range Ruderman–Kittel oscillations in the interaction between magnetic ions, until it was pointed out in 1986 by Zyuzin and Spivak[13] that the same process of averaging the square modulus shows that the oscillations persist in the presence of disorder.

Most of us found features of Anderson's theory that were unfamiliar and uncomfortable, because they challenged assumptions that we had taken for granted. Within a few years it moved from being a misunderstood set of ideas to being one of the core components of our understanding of disordered condensed matter. Unfortunately we do not yet, after fifty years, have even a sketch of the complete theory of electrical conduction in disordered systems we would like, which must surely include simultaneously both the interactions between electrons, and also the dynamics of the substrate of ions and neutral atoms through which the electrons move.

Although the wide publicity which localization received in the 1970s was mostly focused on electrons and holes as carriers of charge, there are many other entities to which these ideas apply. Anderson's original paper on the problem was not directed to electrons as charge carriers, but to electrons as spin carriers. Phonons or photons in a disordered lattice can also be localized. Relatively recent spectacular work on cold atoms has led to the study of cold trapped atoms localized in a disordered or quasiperiodic optical lattice. On a larger scale, both microwaves and ultrasound have been localized in a random array of nonoverlapping conducting spheres. These developments are discussed in Sec. 6.

2. Characteristics of Localized States

For a long time it has been thought that the states of electrons, or of other charge carriers, are qualitatively different in conductors and in insulators, at least at sufficiently low temperatures. In a nonmagnetic metal the conductivity rises to a finite limit when the temperature is lowered towards zero, while in many insulators the conductivity drops as the exponential of $-E_0/k_B T$. Kohn[5] argued that the nonzero limit of conductivity in a metal reflects the fact that electron wave functions are extended through the system, while the insulators have localized charge carriers, that must overcome energy barriers of order E_0 to move through the system. It seems that only in the zero temperature limit is there a sharp distinction between metals and insulators, and for that reason, the metal–insulator transition, like the quantum Hall transition, is regarded as a *quantum phase transition*.

In an energy region where states are localized, the Hamiltonian has a point spectrum, in which the wave functions have well-defined eigenvalues and eigenvectors which are insensitive to boundary conditions or to what is the state of the system at large distances from the particular eigenstates under consideration. Since they are localized, the eigenstates which are dominant in one region are negligible in distant regions. Localized wave functions can respond to an electric field by a limited displacement proportional to the applied field. Extended states are energy levels that depend sensitively on the boundary conditions, and these can respond to an applied electric field by forming complex eigenstates representing currents that can flow right across the system. The spectrum is continuous, and the local density of states deviates smoothly from the average density of states. Localized and extended states cannot coexist at the same energy, unless the two are separated by an impenetrable barrier, as in a percolation problem. The feature of sensitivity of localized eigenstates to boundary conditions is a useful diagnostic tool in numerical simulations for distinguishing between the two types of states. In regions of localized states, the average density of states can be smooth and analytic in the neighborhood of the Fermi energy, as the example studied by Lloyd[12] shows, but the local density of states at a point or region, in which each state's contribution to the density of states is weighted by the square modulus of that state's overlap with the point, is highly irregular. There may not be gaps in the limit of a system of infinite size, but the spectral weights vary from the order of unity to weights that are exponentially small. There is a good explanation of this, with experimental results from the work of Feher and Gere[3] illustrating it, in Anderson's Nobel Lecture.[14]

The absence of an energy gap at the Fermi energy is a distinctive feature of Anderson localization, and leads to a mechanism for conductivity at low temperatures which was called *variable range hopping conductivity* by Mott. As a result of the competition between the exponential fall-off of hopping matrix elements with the distance between two localized states, and the algebraic growth in the availability of such low-lying states with distance, Mott[15] predicted that the conductivity would decrease as $\exp[-(T_0/T)^{1/4}]$ with decreasing temperature, rather than the $\exp[-(T_0/T)]$ calculated by Miller and Abrahams.[16] This behavior was observed in numerous experiments. However, Efros and Shklovskii[17] later argued convincingly and persistently that when the Coulomb interaction between the localized electrons is taken into account, the exponent in the expression for conductivity is changed from 1/4 to 1/2. The exponent of one half was indeed found when experiments were continued to lower temperatures. There are various other phenomena which depend on this variable range hopping, that are discussed in the book by Mott and Davis.[18]

The occurrence of variable range hopping is a feature of Anderson localization that does not occur in other mechanisms for a transition between insulating and conducting states. The filled bands, which were put forward by Wilson[19] as an explanation for insulating behavior in many cases, are characterized by a definite activation energy, so that a plot of the logarithm of conductivity as a function of $1/T$ is a straight line in the low temperature limit. For a variable range hopping there is no minimum excitation energy, so the slope of such a plot gets less and less negative as the temperature is lowered. Similar behavior is expected when all the electrons are bound in covalent bonds, even in a glassy material or in a quantum fluid. For materials such as NiO, with one electron per atom, or atomic hydrogen at low densities, Mott argued that the electrons were bound in position by the long range attraction between the electron and the positive ion which would be left behind if the electron escaped.[20] Such materials are known as *Mott insulators*. Kohn has shown that the transition goes through a series of transitions between excitonic states before the electron becomes free.[21] In this case also there is a definite energy barrier that has to be overcome for electrical conduction at low temperatures.

Anderson's original argument for the occurrence of complete localization in the presence of sufficiently strong disorder[1] was basically a very appealing one, closely related to the underlying physics, but it was dressed up in a notation that was not widely understood at that time. If V is the magnitude of the hopping matrix element from each site to its neighbors, and $\Delta\epsilon$ is the width of the distribution of the site energies, measured from some arbitrary energy, caused by the disorder, a perturbation theory can be developed as a formal series in $V/\Delta\epsilon$. It is plausible to argue, as Anderson[1] did, that such a series for the self-energy will converge and lead to an exponential fall-off with distance. This argument is by no means straightforward, as denominators relating to distant sites can be arbitrarily small, but it is correct, and a proof was given of this result in the case of large disorder, or near the band edge, by Fröhlich *et al.*[22,23] Although a given energy can be matched arbitrarily well at sufficient distance, the expected mismatch in energy decreases with distance R like R^{-d}, where d is the number of dimensions, the coupling to a distant resonant state falls exponentially with distance, so such distant resonances have a vanishingly small probability to delocalize the state if the disorder is sufficiently large. In this paper Anderson also argued that in a simple one-dimensional chain arbitrarily weak disorder would localize all states, a result which was proved by a number of subsequent authors, including Mott and Twose.[7]

At the end of a dinner in the Andersons' house in Cambridge, Anderson told me of a new approach to localization, and I left their house with a large envelope, on which a few suggestive equations were scribbled. I recognized these as being analogous to the Bethe–Peierls equations in statistical mechanics,[24] which were known to be exact for a Bethe lattice, an infinite regular lattice with no loops. With the help of my student Ragi Abou-Chacra, we showed that these equations led to a set of nonlinear integral equations whose solutions could be found numerically, and which displayed a transition between extended and localized solutions.[25] This result was mentioned in Anderson's Nobel lecture.[14]

In the 1958 paper, Anderson did not consider the effect of a magnetic field, an issue that became very important once experimental tests were made. Altshuler *et al.*[26] argued that the amplitude of a Green's function is enhanced at its starting point because, in the absence of a magnetic field, a path around a closed loop interferes constructively with the time reversed path around the same loop, thus reducing the amplitude for escape from the neighborhood of the origin. When time reversal symmetry is broken by the presence of an external magnetic field, this constructive interference at the origin of the path is destroyed, escape of the particle becomes easier, and the electrical resistance is lowered. Thus negative magnetoresistance is a sign of Anderson localization. When electron–electron interactions become important, the situation is more complicated.

3. Scaling Properties

The success which scaling arguments had in explaining the critical equilibrium properties near a continuous phase transition such as the critical point of a gas–liquid transition, the normal to superfluid transition in helium, or a paramagnetic to ferromagnetic transition,[27] suggested that similar arguments might be used to describe the neighborhood of the localized to extended state transition in the Anderson model at zero temperature. One of the earliest ideas of the scaling properties of localization developed from the identification of the conductance (in units of the quantum conductance h/e) of a finite block of material as the ratio of the strength of coupling of eigenstates in neighboring blocks to the energy spacing between levels in the same block. This coupling between neighboring blocks ΔE_L was identified with $\hbar/\tau$. where τ is the diffusion time for the electron to cross the block.[28,29] In the zero-temperature limit of a metal it is expected that the coupling $\hbar/\tau$ will scale as $\hbar D/L^2$, while the spacing between energy levels scales as $1/n(E_F)LA$, where D is the diffusion constant for the electrons, L is the length of the sample along the direction in which the coupling is measured, n is the density of states per unit volume at the Fermi energy, and A is its cross-sectional area. The ratio of the coupling to the energy spacing therefore scales as

$$\hbar D n(E_F) A/L. \qquad (3.1)$$

For a three-dimensional sample with $A \approx L^2$ this ratio increases as the length scale L increases, so metallic (diffusive) behavior at small length scales (but larger than the mean free path) should get more metallic at longer length scales.

As was pointed out by Yuval,[30,31] in a one-dimensional system of cross-sectional area A the ratio decreases as L increases, so at sufficiently large L the coupling between blocks is too small to delocalize the electrons at the Fermi energy, and the behavior should switch from diffusive to localized. A mathematical argument to this effect was given at about the same time by Goldsheid.[32] Experimental evidence began to accumulate for a crossover from metallic behavior in thin wires as the temperature was lowered,[33–35] although it was clear that in these experiments a simple Anderson theory did not work, because interactions between electrons were playing a significant role. The one-dimensional case was worked out in more detail by Anderson *et al.*[36]

At this level, the scaling theory tells us little about what happens in two dimensions, since the dependence on L cancels out. This is a marginal case that requires more careful treatment. A fuller scaling theory was developed by Abrahams, Anderson, Licciardello and Ramakrishnan,[37] who combined scaling with a more accurate theory of the metallic regime based on multiple scattering theory. This theory is a one-parameter scaling theory in which the *dimensionless conductance*

$$g(L) = \Delta E_L n(E_F)), \tag{3.2}$$

is a universal function of L, where $n_L(E_F)$ is the density of energy levels at the Fermi energy for a sample of size L. Here *universal* means that, for large L, the β-function $\beta(L) = d\ln g/d\ln L$ depends on the dimensionality of the sample, the symmetry of the Hamiltonian, and so on, but not on the particular form of the sample or its Hamiltonian. One should note, however, that this discussion is probably restricted to systems described by a one-particle potential, where the interactions between electrons close to the Fermi energy are irrelevant. Much of the behavior of g can be deduced from its known behavior in the high conductance limit, where weak scattering perturbation theory is relevant, and from the strong scattering limit, where exponential localization is known to occur. In the weak scattering limit in d dimensions, large g implies $g \propto L^{d-2}$, so $\beta \to d - 2$, which implies that as L increases, g increases in three dimensions, but decreases in one dimension, eventually switching over to the strong scattering regime at large length scales. In the strong scattering regime there is exponential localization, so that $\ln g \to -L/L_{\text{loc}}$, and $\beta \to \ln g$. This is so far just a reframing of the arguments of the previous two paragraphs, with some refinements.

In two dimensions, the scaling theory predicts that the behavior is marginal, and depends sensitively on the conditions. For the standard Anderson model with no spin-dependent terms, the function β approaches 0 from below as $\ln g$ increases. Integration of the differential equation for $g(L)$ gives g decreasing as L, logarithmically increases the size of the system, until it crosses over to a regime in which the resistance grows exponentially with size. Experimental measurements and numerical simulations support this general behavior, although some doubts have been cast on the details of this one parameter scaling theory. When there is a significant spin-orbit potential the scaling behavior is quite different, so that β approaches zero from above as L increases, and this two-dimensional system has a metallic phase, with a finite resistance per square, as well as a localized phase.

A different approach to scaling theory based on a diagrammatic expansion was introduced by Vollhardt and Wölfle.[38,39] This produced rather similar results, but seems to be more flexible in some cases.

This scaling theory was in serious conflict with the idea of a "minimum metallic conductivity", for which Mott argued forcefully, and for which considerable experimental and numerical evidence was obtained.[40,41] The main argument was based on what is known as the *Ioffe–Regel criterion*, that the mean free path could not be less than the spacing between atoms, and therefore the minimum metallic conductivity in three dimensions is of order $e^2/3\hbar a_{\min}$, where $a_{\min}$ is the relevant length scale, and in two dimensions the minimum metallic conductance is of order e^2/h. The results of our group were among the numerical results quoted by Mott in support of a minimum metallic conductivity,[29,42] but, as we made simulations of larger systems, we began to think that the evidence we had was not supportive of Mott's conclusion.[43] From the point of view of scaling theory, one expects a lower conductivity to be possible, because when the mean free path is of the order of $a_{\min}$, quantum interference will still reduce the diffusion rate below its classical value.

The valuable technique of finite size scaling, introduced by Fisher and Barber for the study of critical phenomena,[44] was used by Mackinnon and Kramer[45,46] to do elegant scaling calculations for the Anderson model which confirmed the applicability of these ideas. In order to study the behavior of a simple two-dimensional layer, they studied strips of varying width M, for each value of M varying the length N of the strip until they could get a good estimate of the limiting behavior of the transmission from end to end of the strips, which should be an exponential function $\exp(-N\lambda_M)$ of the length N of the strip, since the strip is one-dimensional. The finite size scaling hypothesis is

$$\lambda_M/M = f_2(\xi/M)\,, \qquad (3.3)$$

where ξ is the two-dimensional localization length, dependent on energy and degree of disorder, and f_2 is a "universal" function for the symmetry class. In two dimensions, the function f can be continued to rather low disorder and large values of ξ, showing that there is no delocalized regime apparent. In three dimensions, the localization length ξ diverges for nonzero values of disorder, and for lower disorder a similar procedure yields a separate scaling function for the conductance. This work gives clear evidence that there is no minimum metallic conductivity of the magnitude predicted by the Ioffe–Regel criterion for a system of noninteracting electrons.

This method was used to study the critical behavior of the Anderson model with a Cauchy distribution of site energies, which was discussed in Ref. 12. In two dimensions it was found that the scaling results fall on the same curve as they do for the uniform distribution, in agreement with the arguments that, despite the different behavior of the averaged Green's function, there should be no major difference in the localization properties.

A recent review of the scaling approach to Anderson localization was given by Evers and Mirlin.[48] Among other topics dealt with in this paper are the multifractality of the critical wave function, the need to characterize probability distributions in the localized regime more carefully, more general universality classes for the localization transition, and localization for Dirac Hamiltonians.

4. Effects of Interactions

There is no question that interactions between electrons and electron–phonon interactions are important for all metals and semiconductors. The long-range part of the Coulomb potential is usually taken into account in terms of a self-consistent screening potential that is used in the definition of the electrochemical potential in a sample; it is such a screened potential that determines the existence of an inversion layer between two semiconductors, or between a semiconductor and an insulator. The interactions are also taken into account when considering how the effects of localization are modified by inelastic scattering at nonzero temperatures, as was discussed briefly in Sec. 2. What is lacking is a convincing account of how the satisfactory theory we have of the localization of noninteracting particles (or waves, as I discuss in Sec. 6) can be adapted when interactions have to be taken into account beyond the mean field or Boltzmann-like collision approximations.

For example, we know that in the absence of interactions between fermions, the Fermi surface may be in a region where all low-energy excitations are localized, or in a region where excitations have diffusive behavior. I am not aware of theoretical work which shows that this property is robust, and survives when interactions are taken into account, although I regard the experimental work done to confirm localization properties at low temperatures as giving us a strong signal that this is correct.[18,49–52] On the other hand, the discovery that many disordered thin films display a metal–insulator transition at low temperatures[53] suggests either than we are wrong about localization in two-dimensional systems of noninteracting electrons, or that interactions can alter the situation. There is also evidence from theoretical[54] and numerical[55] studies that sufficiently strong repulsive interactions can delocalize a one-dimensional disordered system.

A tentative result on the effect of interactions on Anderson localization was obtained by Fleishman and Anderson,[59] and recently developed much further by Basko, Aleiner and Altshuler.[60] These authors have shown that at low enough temperatures, in the absence of the coupling to an external heat bath which promotes hopping conductivity, the electrical conductivity is exactly zero, but at higher temperatures the resistivity is finite. There is therefore, in the absence of an external heat bath, a finite temperature transition between an insulating state and a conducting state.

An obvious way of including interactions between particles, which has been used for many years, is to use the Hartree–Fock mean field theory, sometimes modified by using a pseudopotential when there is a strong short-ranged force between the particles, or by the kind of local density approximation intorduced by Kohn and Sham.[56] In the case of interacting bosons with repulsive interactions between them, mean field theory is particularly easy to use at low temperatures, where the superfluid transition occurs.[57,58] The total energy is reduced if the condensate wave function is fairly uniform, and this smoothes out the effects of short-ranged disorder.

5. Quantum Hall Effect

The integer quantum Hall effect was discovered by von Klitzing and collaborators in 1980,[61] in the two-dimensional electron gas at the interface of p-type bulk silicon and the insulating SiO_2 layer on the surface of a high-purity metal–oxide–semiconductor field effect transistor (MOSFET). This two-dimensional electron gas is an n-type *inversion layer* at the interface, which is held at a low temperature and in a high magnetic field, usually perpendicular to the inversion layer. The Fermi energy of the electrons was was controlled by varying the gate voltage, which gives an electric field normal to the surface, just as it is when the device is used as a transistor. There was a current source on one edge of the rectangular region where the electron gas was confined, and a current drain on the opposite edge, while there were two voltage leads on each of the other edges. The voltage leads were arranged in opposing pairs, so that the transverse resistance could be determined from the voltage across one of the pairs, and the longitudinal resistance could be determined from the voltage between two leads on the same edge.

The crucial observation was that, when the gate voltage was varied at a sufficiently low temperature and high magnetic field, there was a series of very flat plateaus in the measured transverse (Hall) resistance. In between these plateaus, the Hall resistance decreased smoothly with increasing gate voltage. The longitudinal resistance was too small to measure at the plateaus of the Hall voltage, but rose smoothly to a peak value between each pair of adjacent plateaus. The most startling result in the paper was that the value of the Hall resistance in each of the plateaus was

$$r_H = \frac{h}{\nu e^2}, \tag{5.1}$$

where ν is an integer, with a precision that, in this initial paper, was better than one part in 10^5.

There is an obvious question of why this quantization of Hall conductance as integer multiples of e^2/h should be so precise, and a profound answer to this was given by Laughlin in the year following the experimental discovery.[62] The less obvious question is why there are conductance plateaus at all, and the answer to this question involves the important part played by Anderson localization between the plateaus. A simple, but misleading, explanation of the integer quantum Hall effect is that electrons in a filled Landau level, where there are perpendicular electric and magnetic fields, $\mathcal{E}$, $\mathbf{B}$, drift at a speed $\mathcal{E}/B$ in the direction perpendicular to both fields. The filling of a Landau level leads to the quantized Hall conductance, and it can be shown to be invariant under some perturbations.[63–65] This does not explain why the Hall conductance is constant, not only when the Fermi energy is varied, but also when the magnetic field is varied, as it was in the later work experiments of Tsui, Stormer and Gossard.[66]

There is general agreement, supported by detailed experimental work on the capacitance of the inversion layer, that, under the conditions that lead to a plateau of the Hall conductance, there is no gap in the density of states, but the Fermi energy lies in a region of localized states, which do not contribute either to Hall conductance nor to longitudinal (dissipative) dc conductance; they can, however, contribute a capacitative response to alternating currents. The nondissipative Hall current is carried by states well below the Fermi surface. The localized states close to the Fermi energy play a role in quantum Hall devices similar to that which they play in the impurity bands of semiconductors, where they pin the Fermi energy and stabilize the electrical properties of bulk semiconductors against the effects of random perturbations. A detailed description of a semiclassical model that leads to such a result was given by Kazarinov and Luryi,[67] who argued that, in an isotropically disordered potential for an annulus, classical percolation could only occur at a single energy for each cyclotron orbit, which would be spread into a narrow energy band by an electric field across the annulus. Other orbits are confined to a local region of space, and do not contribute either to the Hall current, nor to a dissipative dc transport current. This discussion can be generalized to deal with quantum delocalization near this energy band, and with Anderson localization away from the percolation energy. As the temperature is lowered, the Hall conductance plateaus get sharper and broader, and the longitudinal resistance between the plateaus gets larger, as one would expect from the theory of activated hopping between localized states. It should be noted that the nondissipative Hall current continues even when the Fermi energy is well away from the energy of extended states. There are energy bands or singular energies at which this current can flow along equipotentials that are connected around loops encircling the system, whatever is happening at the Fermi energy.

The work of Tsui, Stormer and Gossard[66] was carried out on the inversion layer formed on the interface between p-type GaAs and n-type $Ga_{1-x}Al_xAs$, which allows for a much longer electron mean free path, because SiO_2 is usually rather disordered, but the dopants in the $Ga_{1-x}Al_xAs$ materials can be kept far from the interface of the heterojunction. In this work, it was discovered that the plateaus for a relatively pure material could have fractional values of the quantized conductance, with fractions whose denominators are usually odd numbers. Since Laughlin's theory of the integer quantum Hall effect generalizes Bloch's unpublished theorem that the energy levels of a loop of electrons are periodic in the flux threading the loop, with period h/e, it does not allow a longer period such as $3h/e$, as is seen for the $\nu = 1/3$ plateau, unless this state is threefold degenerate.[68,69] It became clear that the appearance of fractional quantum numbers could not be explained entirely by the disordered one-particle substrate potential, but must be produced by the interaction between electrons, and such a theory was published by Laughlin in 1983,[70] although he was slow to recognize the degeneracy which was hidden by the elegance of his formalism and by his choice of boundary conditions. Laughlin's theory takes account of the interactions between the electrons, and agrees with remarkable accuracy with numerical simulations of small systems, but ignores the substrate disorder. It is true that fractional states with increasing denominator emerge as the disorder is reduced and the temperature is lowered, but in reality both disorder and electron–electron interaction play a part in all real quantum Hall systems.

6. Localization in Other Systems

The original paper on Anderson localization[1] was inspired by experimental work showing that spin diffusion did not always occur in solids,[3] but it was framed in a language that could be applied equally naturally to electrons in solids. The main framework of later discussions was on electrons in low-temperature solids, and I do not doubt that a major reason for this was the emphasis given to it by Mott.[8] Other sorts of waves are also subjected to Anderson localization, but in many cases there is an important difference that they are subjected to an attenuation of their intensity, unlike electrons satisfying the Schrödinger equation, or electron or nuclear spins in a solid, whose number is conserved. Numerical simulations of electron systems has given us useful confirmation and corrections of theoretical ideas about the effects of static disorder on such systems, but, as always, such simulations suffer from two major problems: the size of systems that can be handled on a computer is severely limited, and the number of time intervals for which the system can be followed is also limited. Both these shortcomings can be overcome by studying real physical analogs whose internal details can be followed more easily than those of electrons in solids. There is a third problem that can not be overcome so readily by using analogs, which is that probability distributions of relevant parameters need to be deduced from the properties of many independent samples.

In December 2008, a conference on Fifty years of Anderson localization was organized in Paris by Lagendijk, van Tigelen and Wiersma; I am grateful to have been invited to the meeting, and found it very stimulating. A summary of it by the organizers appears in *Physics Today*,[71] and in this article there are descriptions of various other systems in which localization has been studied. Studies of light waves in a scattering but non-absorbing medium were suggested by John,[72] but the right conditions for localization were hard to achieve, and more than ten years passed before strong localization of light without absorption was achieved by using finely powdered gallium arsenide.[73]

The group at Winnipeg and Grenoble used a random array of 4 mm aluminum beads held in position by brazing their contacts with one another to detect localization in ultrasound propagated through the air outside the beads.[74] The comparatively large scale of this system allows for more detailed observation of the waves than is possible in many systems. A similar situation exists for the scattering of microwaves by composite media, which has been studied by a number of groups, including Chabanov *et al.*[75]

Attempts to show localization in a three-dimensional photonic material have not yet, as far as I know, been successful, but Segev and his collaborators at the Technion in Israel have shown transverse localization in a photonic crystal, with a transition from gaussian spread of the wave function to exponential fall away from the axis.[76]

Another topic discussed in the *Physics Today* survey, and discussed in more detail in the following article by two of the major participants in this field,[77] is the study of localization and delocalization in a one-dimensional Bose condensate of noninteracting bosons at very low temperatures. Aspect heads a group in Paris, and Inguscio's group is centered on Florence, and detailed descriptions of the work of each group were published in *Nature* in 2008.[78,79] In both cases, the scattering length of the interatomic potential was reduced to zero by tuning the magnetic field. The system in each case was made one-dimensional by having a weak confining potential for the atoms in the longitudinal direction and a strong potential in each transverse direction. In the Paris work, disorder is added to the confining potential by using laser speckle, which has the advantages that the fluctuations in the potential are bounded, so that it cannot introduce impenetrable barriers in the potential, and that it can be varied in a reproducible manner by using different screens to produce the speckle. In the Florence work the potential used is not random but is quasiperiodic, the sum of two potentials with incommensurate periods. The theory of localization by such quasiperiodic potentials was worked out by Aubry and André many years ago.[80] These experiments show the expected transitions between diffusive behavior for weaker disorder to a confined core with an exponentially falling density in the tail for stronger disorder. It should not be a sharp transition in this one dimensional case, even for weak disorder, but the expected exponential localization length becomes too large to observe.

7. Summary

In this review, I have concentrated on the seed that Anderson sowed more than fifty years ago, and on the plant which bloomed so prolifically ten years later. I have tried to explain what I regard as the major developments in the second decade of the theory, and its major importance for understanding the quantum Hall effect. I have only touched on other more recent developments, both on the experimental side and on the theoretical side, except that I have given a fuller account of the influence of localization on the quantum Hall effect, which seems to me to be as profound as its influence on impurity bands in semiconductors.

It is not hard, even now, to see why the physics community was so slow to accept Anderson's ideas. An effective field, such as is used in the *coherent potential approximation*,[81] which replaces the disordered potential by an effective average potential, including a dissipative term, is much more comfortable to deal with than a probability distribution. However, when one starts asking about the transition between a weak scattering situation in a strongly doped semiconductor and a situation which is surely disordered enough to localize the electrons in certain energy ranges, as in the band edge of a weakly doped semiconductor, the need for care becomes apparent. A similar situation arose in the study of critical points in thermal physics, where the simplicity of the mean field pictures introduced by van der Waals and Landau seems to have overcome the early experimental evidence for a different behavior near the critical point.

Acknowledgments

This is an appropriate place to acknowledge the immense debt I owe to Phil Anderson for his friendship, guidance and encouragement over the past fifty years. I should also acknowledge that I would not have understood this area so well without Sir Nevill Mott's careful questioning of what he and other people believed about it. My other debts to members of the physics community who have contributed to my education in this area are too numerous to acknowledge here.

I wish to thank the Isaac Newton Institute in Cambridge for the hospitality of its program on Fifty Years of Anderson Localization, in the second half of 2008.

References

1. P. W. Anderson, *Phys. Rev.* **109**, 1492 (1958).
2. P. W. Anderson, *Phys. Rev.* **114**, 1002 (1959).
3. G. Feher and E. A. Gere, *Phys. Rev.* **114**, 1245 (1959).
4. I. M. Lifshitz, *Adv. Phys.* **13**, 483 (1964).
5. W. Kohn, *Phys. Rev. A* **133**, 171 (1964).
6. S. R. Ovshinsky, *Phys. Rev. Lett.* **21**, 1450 (1964).
7. N. F. Mott and W. D. Twose, *Adv. Phys.* **10**, 107 (1961).
8. N. F. Mott, *Adv. Phys.* **16**, 49 (1967).
9. M. E. Fisher, *Rept. Progr. Phys.* **30**, 615 (1967).
10. J. M. Ziman, F. Brouers and P. W. Anderson, in *Proceedings of 3rd International Conference*, ed. N. F. Mott (North-Holland Pub. Co., Amsterdam, 1970), pp. 426–435.
11. D. J. Thouless, *J. Phys. C: Solid State Phys.* **4**, 1559 (1970).
12. P. Lloyd, *J. Phys. C: Solid State Phys.* **2**, 1717 (1969).
13. A. Yu. Zyuzin and B. Z. Spivak, *JETP Lett.* **43**, 234 (1986).

14. P. W. Anderson, *Rev. Mod. Phys.* **50**, 191 (1978)
15. N. F. Mott, *Phil. Mag.* **19**, 835 (1969).
16. A. Miller and E. Abrahams, *Phys. Rev.* **120**, 745 (1960).
17. A. L. Efros and B. I. Shklovskii, *J. Phys. C: Solid State Phys.* **8**, L49 (1975).
18. N. F. Mott and E. A. Davis, *Electronic Processes in Non-Crystalline Materials* (Oxford University Press, 1979).
19. A. H. Wilson, *Proc. R. Soc. London, Ser. A* **133**, 458 (1931).
20. N. F. Mott, *Proc. Phys. Soc. London, Ser. A* **62**, 416 (1949).
21. W. Kohn, *Phys. Rev. Lett.* **119**, 789 (1967).
22. J. Fröhlich and T. Spencer, *Phys. Rep.* **103**, 9 (1984).
23. J. Fröhlich, F. Martinelli, E. Scoppol and T. Spencer, *Commun. Math. Phys.* **101**, 22 (1985).
24. H. A. Bethe, *Proc. R. Soc. London, Ser. A* **150**, 552 (1935).
25. R. Abou-Chacra, P. W. Anderson and D. J. Thouless, *J. Phys. C: Solid State Phys.* **7**, 1734 (1974).
26. B. L. Altshuler, A. G. Aronov, A. I. Larkin and D. E. Khmelnitskii, *Zh. Éksp. Teor. Fiz.* **81**, 768 (1981), translation in *Soviet Physics JETP*.
27. M. E. Fisher and K. F. Wilson, *Phys. Rev. Lett.* **28**, 548 (1972).
28. J. T. Edwards and D. J. Thouless, *J. Phys. C: Solid State Phys.* **5**, 807 (1972).
29. D. C. Licciardello and D. J. Thouless, *J. Phys. C: Solid State Phys.* **8**, 4157 (1975).
30. G. Yuval, *Phys. Lett. A* **53**, 136 (1975).
31. D. J. Thouless, *Phys. Rev. Lett.* **39**, 1167 (1977).
32. I. Ya. Goldsheid, *Dokl. Akad. Nauk. SSSR* **224**, 1248 (1975).
33. G. J. Dolan and D. D. Osheroff, *Phys. Rev. Lett.* **43**, 721 (1979).
34. N. Giordano, W. Gilson and D. E. Prober, *Phys. Rev. Lett.* **43**, 725 (1979).
35. P. Chaudhari and H.-U. Habermeier, *Phys. Rev. Lett.* **44**, 40 (1980).
36. P. W. Anderson, D. J. Thouless, E. Abrahams and D. S. Fisher, *Phys. Rev. B* **22**, 3519 (1980).
37. E. Abrahams, P. W. Anderson, D. C. Licciardello and T. V. Ramakrishnan, *Phys. Rev. Lett.* **42**, 673 (1979).
38. D. Vollhardt and P. Wölfle, *Phys. Rev. Lett.* **45**, 842 (1980).
39. D. Vollhardt and P. Wölfle, *Phys. Rev. Lett.* **48**, 699 (1982).
40. N. F. Mott, *Phil. Mag.* **26**, 1015 (1972).
41. N. F. Mott, M. Pepper, S. Pollitt, R. H. Wallis and C. J. Adkins, *Proc. R. Soc. London, Ser. A* **345**, 169 (1975).
42. D. C. Licciardello and D. J. Thouless, *Phys. Rev. Lett.* **35**, 1475 (1975).
43. D. C. Licciardello and D. J. Thouless, *J. Phys. C: Solid State Phys.* **11**, 925 (1978).
44. M. E. Fisher and M. N. Barber, *Phys. Rev. Lett.* **28**, 1516 (1972).
45. A. MacKinnon and B. Kramer, *Phys. Rev. Lett.* **47**, 1546 (1981).
46. A. MacKinnon and B. Kramer, *Z. Phys. B* **53**, 1 (1983).
47. A. MacKinnon, *J. Phys. C: Solid State Phys.* **17**, L289 (1984).
48. F. Evers and A. D. Mirlin, *Rev. Mod. Phys.* **80**, 1355 (2008).
49. M. A. Paalanen, T. F. Rosenbaum, G. A. Thomas and R. N. Bhatt, *Phys. Rev. Lett.* **48**, 1284 (1982).

50. T. F. Rosenbaum, R. F. Milligan, M. A. Paalanen, G. A. Thomas, R. N. Bhatt and W. Lin, *Phys. Rev. B* **27**, 7509 (1983).
51. G. A. Thomas, M. A. Paalanen and T. F. Rosenbaum, *Phys. Rev. B* **27**, 3897 (1983).
52. S. Waffenschmidt, C. Pfleiderer and H. von Löhneysen, *Phys. Rev. Lett.* **83**, 3005 (1999).
53. E. Abrahams, S. V. Kravchenko and M. P. Sarachik, *Rev. Mod. Phys.* **73**, 251 (2001).
54. T. Giamarchi and H. J. Schulz, *Phys. Rev. B* **37**, 325 (1988).
55. P. Schmitteckert, T. Schulze, C. Schuster, P. Schwab and U. Eckern, *Phys. Rev. Lett.* **80**, 560 (1998).
56. W. Kohn and L. J. Sham, *Phys. Rev. A* **140**, 1133 (1965).
57. J. R. Ensher, D. S. Jin, M. R. Matthews, C. E. Wieman and E. A. Cornell, *Phys. Rev. Lett.* **77**, 4984 (1996).
58. J. R. Abo-Shaeer, C. Raman, J. M. Vogels and W. Ketterle, *Science* **292**, 476 (2001).
59. L. Fleishman and P. W. Anderson, *Phys. Rev. B* **21**, 2366 (1980).
60. D. M. Basko, I. L. Aleiner and B. L. Altshuler, *Annals of Phys.* **321**, 1126 (2006).
61. K. von Klitzing, G. Dorda and M. Pepper, *Phys. Rev. Lett.* **45**, 494 (1980).
62. R. B. Laughlin, *Phys. Rev. B* **23**, 5632 (1981).
63. H. Aoki and T. Ando, *Solid State Commun.* **38**, 1079 (1981).
64. R. E. Prange, *Phys. Rev. B* **23**, 4802 (1981).
65. D. J. Thouless, *J. Phys. C: Solid State Phys.* **14**, 3475 (1981).
66. D. C. Tsui, H. L. Stormer and A. C. Gossard, *Phys. Rev. Lett.* **48**, 1559 (1982).
67. R. F. Kazarinov and S. Luryi, *Phys. Rev. B* **25**, 7626 (1982).
68. P. W. Anderson, *Phys. Rev. B* **28**, 2264 (1983).
69. D. J. Thouless and Y. Gefen, *Phys. Rev. Lett.* **66**, 806 (1991).
70. R. B. Laughlin, *Phys. Rev. Lett.* **50**, 141 (1983).
71. A. Lagendijk, B. A. van Tiggelen and D. S. Wiersma, *Phys. Today* **62**(8), 24 (2008).
72. S. John, *Phys. Rev. Lett.* **53**, 2169 (1984).
73. D. S. Wiersma, P. Bartolini, A. Lagendijk and R. Righini, *Nature* **390**, 671 (1997).
74. H. Hu, A. Strybulevlch, J. H. Page, S. E. Skipetrov and B. A. van Tiggelen, *Nat. Phys.* **4**, 945 (2008).
75. A. A. Chabanov, M. Stoytchev and A. Z. Genack, *Nature* **404**, 850 (2000).
76. T. Schwartz, G. Bartal, S. Fishman and M. I. Segev, *Nature* **446**, 52 (2007).
77. A. Aspect and M. Inguscio, *Phys. Today* **62**(8), 30 (2008).
78. J. Billy, V. Josse, Z. Zuo, A. Bernard, B. Hambrecht, P. Lugan, D. Clément, L. Sanchez-Palencia, P. Bouyer and A. Aspect, *Nature* **453**, 891 (2008).
79. G. Roati, C. D'Errico, L. Fallani, M. Fattori, C. Fort, M. Zaccanti, G. Modugno, M. Modugno and M. Inguscio, *Nature* **453**, 895 (2008).
80. S. Aubry and G. André, *Ann. Israel Phys. Soc.* **3**, 133 (1980).
81. F. Yonezawa and K. Morigaki, *Prog. Theor. Phys. Suppl.* **53**, 1 (1972).

Chapter 3

INTRINSIC ELECTRON LOCALIZATION IN MANGANITES

T. V. Ramakrishnan*

*Department of Physics, Indian Institute of Science,
Bangalore 560012, India*

We mention here an unusual disorder effect in manganites, namely the ubiquitous hopping behavior for electron transport observed in them over a wide range of doping. We argue that the implied Anderson localization is **intrinsic** to manganites, because of the existence of polarons in them which are spatially localized, generally at random sites (unless there is polaron ordering). We have developed a microscopic two fluid lb model for manganites, where l denotes lattice site localized l polarons, and b denotes band electrons. Using this, and the self-consistent theory of localization, we show that the occupied b states are Anderson localized in a large range of doping due to the scattering of b electrons from l polarons. Numerical simulations which further include the effect of long range Coulomb interactions support this, as well the existence of a novel polaronic Coulomb glass. A consequence is the inevitable hopping behaviour for electron transport observed in doped insulating manganites.

1. Introduction

It is a great pleasure to participate in this celebration of fifty years of one of the great and enduring organizing principles of modern condensed matter physics, namely Anderson localization. I thought that I would describe some work we have been doing recently on Anderson localization in manganites, which seems to us to be inevitable for this family of compounds. It is an example of the ways in which the principle is relevant for significant observed behavior of real systems.

Manganites (namely rare earth manganites substitutionally doped with alkaline earths, $Re_{1-x}Ak_xMnO_3$) show a rich and unusual variety of poorly understood phenomena[1,2] including colossal magnetoresistance. Here I

---

*Also at: Department of Physics, Banaras Hindu University, Varanasi 221005, India.

concentrate on one of them, namely the fact that in the doping range $0.2 \lesssim x \lesssim 0.5$ in which most manganites have a paramagnetic insulator to ferromagnetic metal transition on lowering the temperature and also show no orbital long range order, the electrical transport in the paramagnetic insulating state is **always** characterized by hopping behavior, namely the electrical resistivity depends on temperature T as $\exp(T/T_0)^\alpha$ where α ranges from $(1/4)$ to $(1/2)$ to 1!. I argue here that this arises from the fact that manganites are intrinsically disordered electronically, with some electrons existing as small polarons (l), and others as band electrons (b). This leads to two broad reasons for hopping behavior. First, the scattering of the b electrons from the randomly located l polarons leads to their Anderson localization. This is inevitable unless the polarons move fast or the system has long range order. Thus Anderson localization is a generic feature of manganite physics. I describe below our theory[3,4] for the effect; before this, I summarize some features of manganites. Details are found in numerous reviews e.g. Refs. 1 and 2. I also summarize a microscopic two fluid (lb) model for them proposed by us (Sec. 2), with some applications.[5–8] Second, in doped manganites, the long range Coulomb interaction has qualitative effects, describable e.g. as quantum Coulomb glass formation of the l polarons. This has important consequences for the transport behavior, in particular Efros–Shklovskii hopping involving l polarons.

The parent compound ($x = 0$) is an insulator, with an antiferromagnetic ground state. This ground state continues till about $x \lesssim 0.1$, beyond which it is generally a ferromagnetic insulator. The insulating ground state gives way to a metallic ground state (both ferromagnetic) for $x \gtrsim 0.2$. The existence of a ferromagnetic insulating ground state is a challenge for a simple double exchange (Refs. 1,2 and 9,10) theory of manganites, since in such an approach, ferromagnetism and metallicity necessarily go together. A strange feature of the ferromagnetic insulating state ($0.1 < x < 0.2$) is the Efros–Shklovskii hopping transport in single crystals of them.[11] Between $x \sim 0.2$ and $x \sim 0.5$, the manganite in general has no orbital or charge long range order. The ground state is metallic and ferromagnetic, and on heating above the Curie temperature T_c, it becomes a paramagnetic insulator. Beyond $x \sim 0.5$, one seems to have, on lowering the temperature, charge/orbital order whose period depends on x, as well as an antiferromagnetic ground state. A general "phase diagram" is shown in Fig. 3.1. (In reality there are many exceptions, e.g. $La_{1-x}Sr_xMnO_3$ is a metal for all $x > 0.1$, while $Pr_{1-x}Ca_xMnO_3$ is a bad metal/insulator for all x). Here I concentrate on the transport properties

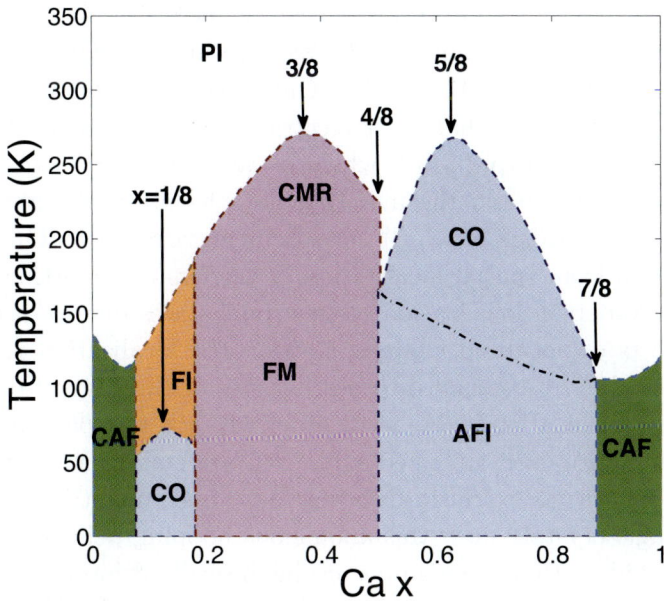

Fig. 3.1. Idealized phase diagram for doped manganites (close to the real phase diagram of $La_{1-x}Ca_xMnO_3$). The hole doping is the x axis and the temperature is the y axis. With increasing hole doping, one has C type antiferro insulating, ferromagnetic insulating, charge ordered insulating, ferro metallic, antiferro/charge ordered insulating, and C type antiferro insulating phases (CAF, FI, CO, FM, CO/AFI, CAF) phases as indicated. The high temperature phase is a paramagnetic insulator. Special commensurate densities are indicated. The CMR region is near the para insulator/ferro metal boundary.

of the doped manganites in the doping regime $0.1 \lesssim x \lesssim 0.5$, where the system is an insulator (ferromagnetic at low temperatures) for all temperatures when $0.1 \lesssim x \lesssim 0.2$, and is a paramagnetic insulator above T_c for $0.2 \lesssim x \lesssim 0.5$. One observes localization related hopping transport in all manganites in the above doping regime.

The manganites are in an orbital fluid/glass state e.g. for doping $x \sim 0.3$ somewhere in the middle of this regime, namely there is no long range orbital order. There is a transition identifiable clearly as paramagnetic to ferromagnetic, the evidence for the latter being from magnetic hysteresis measurements. This transition is nearly coincident with the insulator metal transition. The Curie temperature varies from about 360 to about 110 K; for a given system, (chemical constitution and hole density) it seems to decrease with increasing disorder. (The correlation is clearly brought out in the detailed work of Attfield and coworkers who explored the connection between ferromagnetic T_c and disorder; see e.g. Ref. 12).

The manganites are actually insulating over much of the doping (or x) range for most members of the family. One of the common features of the hopping transport[13–16] is that their electrical resistivity depends on temperature as $\rho(T) \sim \rho_0 \exp(T/T_0)^\alpha$ where α varies from $(1/4)$ to $(1/2)$ to 1 depending on doping x and the material, both its physical state and even on the temperature range in which measurements are made. This behavior arises in general from spatial localization of carriers transporting electrical current, e.g. Anderson localization of electronic states due to disorder and Coulomb effects on localized charged states. The localization is generally assumed to be due to external disorder because of the general observation that Anderson localization is due to disorder which is **not** part of the clean, perfect system. However, unexpectedly, even systems which are clean i.e. are known to have no extrinsic disorder, e.g. high quality ferromagnetic insulating single crystals show this anomalous hopping behavior.[11] We develop here the ideas that manganites are intrinsically disordered because of the l polaron (above a low crossover temperature T^*, see below), and the localization of states (Anderson localization in the case of b electrons and the localization of l polarons) is the underlying cause of hopping behavior[17]; the Coulomb interaction which is inevitably present has an effect on the hopping exponent.

We have developed a microscopic two fluid model for manganites[5,6] and have applied it to describe a number of properties, including intrinsic nanoscopic inhomogeneities due in our view to inevitable Coulomb interactions.[7,8] In this model (outlined below in Sec. 2), the low energy electronic states of a manganite are argued to arise from two Fermi fluids (interpenetrating at the atomic level), namely an immobile small Jahn–Teller lattice polaron fluid l and a band fluid b. Over a wide range of x and T (e.g. for $0.20 \lesssim x \lesssim 0.5$ and all T in this range of x), the l fluid has no spatial long range order, and thus the b electrons move in the random medium of the l polarons, effectively avoiding them because of the large mutual repulsion, described by the Mott–Hubbard U. Disorder in manganites is thus, finally, an inevitable consequence of strong local interactions: the large electron lattice coupling leads to small Jahn–Teller polarons effectively site localized at random sites, and then the Mott–Hubbard U can be thought of as an effective disorder potential for the b electrons. We calculate (in Sec. 3) the Anderson localization effects due to this disorder in a novel approach for strongly correlated systems (implemented via the self consistent theory of localization[18]), and show that the b electron states are localized, the mobility edge being below the chemical potential for a wide range of doping.[19] We have shown earlier[7,8] that long range Coulomb interactions result in strong nanoscopic electronic inhomogeneities with b electron puddles in an l polaron background. This is a Coulomb glass with Efros–Shklovskii hopping transport. Within this modeling of manganites, the b electron states are shown there to be localized in about the same range of x, through a calculation of the participation ratio of their eigenstates. Thus the electron hopping transport universally observed in insulating manganites (e.g. variable range hopping or VRH,[20] or Coulomb interaction dominated Efros–Shklovskii[21] hopping) is inevitable; external disorder, e.g. due to ion size mismatch, random strains or grain boundaries contributes further to localization effects. In such an intrinsically inhomogeneous two electron fluid/glass model as above for manganites, hopping transport cannot be quantitatively described by the simple homogeneously disordered one component model[22,23] which has been used to fit data. We discuss our results for manganites and juxtapose them with the experimental situation in Sec. 4.

2. A Two Fluid Model for Manganites

The two fluid model for manganites describes it as atomically interpenetrating polaronic and band electronic fluids, symbolized by l (localized, polaronic) and b (broad band). It takes into account, approximately, all the largest local interactions, namely the electron phonon coupling leading to polaron formation, the Mott–Hubbard U, and the Hund's rule coupling J_H. In a lattice (tight binding) description, we notice that at each Mn site the five fold degeneracy of the Mn d electron orbitals is lifted in the octahedrally symmetric crystal field of the O ions surrounding each Mn ion, and one has t_{2g} and e_g levels, the former lying lower. Since for hole doping x, the nominal ionic configuration is $\{(1-x)Mn^{3+}, xMn^{4+}\}$ with four electrons in the former configuration and three electrons in the latter (both in 3d orbitals), one can describe the system as having three t_{2g} electrons on each Mn site, and one e_g electron on a fraction $(1-x)$ of the sites. The former three together can be thought of as constituting a spin $S = 3/2$ because of large positive J_H, and the e_g electron as forming a Jahn–Teller polaron by locally distorting the lattice and removing the original two fold degeneracy of the e_g state. The binding energy of the polaron is denoted by E_{JT}. We assume it to be site localized. Since the e_g state is doubly degenerate, at each site there is necessarily another "antipolaron" state. Thus at the single site level, there is a spin $S = 3/2$ which can be thought of as a classical spin for most purposes, a Jahn–Teller polaron (called l) with energy E_{JT} which we assume to be small and site localized, and an antipolaronic state with site energy E_{JT} (The latter has an undiminished hopping amplitude t_{ij} and forms a broad band; we therefore denote it as a b electron).

The bare amplitude for electron hopping to the nearest neighbour is t_{ij}; Its effective size (and sign) depends on the actual d orbitals involved in the hopping. We simplify it in the following ways. For the polaron, there is a Huang–Rhys reduction factor $\exp(-2E_{JT}/\hbar\omega_0)$ at temperatures smaller than $\hbar\omega_0$. This factor is smaller than $(1/200)$ for values of E_{JT} and $\hbar\omega_0$ typical in manganites, so that the effective hopping amplitude is very small, the characteristic energy being $\lesssim 150$ K $\sim T^*$. To the zeroth approximation, and for investigation of effects at temperatures higher than this value, we simply ignore it, i.e. assume that the polaron does not hop. In this sense, our model differs essentially from that of Littlewood, Shraiman and Millis[24] who were the first to emphasize the crucial effect of polarons in the CMR problem. These authors assumed that the l polarons hop from site to site with undiminished amplitude. Below this temperature T^*, polaron hopping leads to a coherent single quantum fluid; polarons as well defined entities disappear, for example. Our theory which ignores lb hopping is not applicable in this regime.

The dependence of the nearest neighbour hopping amplitude t_{ij} on the orbital degrees of freedom means that the overlap of the b states on nearest neighbour sites depends on the orbital degrees of freedom; at each site, the orbital degrees of freedom of the b electron are orthogonal to the polaronic state. This leads to a fairly complicated b band, and to orbital correlations. We simplify the model by averaging over orbital degrees of freedom and denoting the orbitally averaged hopping amplitude by a single number t_{ij}.

The lb Hamiltonian is thus written as

$$H_{lb} = (-E_{JT} - \mu) \sum_{i,\sigma} n_{li\sigma} - \mu \sum_{i,\sigma} n_{bi\sigma} + U_{dd} \sum_{i,\sigma} n_{li\sigma} n_{bi\sigma}$$
$$- t \sum_{<ij>,\sigma} (b^\dagger_{i\sigma} b_{j\sigma} + h.c.) - J_H \sum_i (\vec{\sigma}_{li} + \vec{\sigma}_{bi}).\vec{S}_i - J_F \sum_{<ij>} \vec{S}_i.\vec{S}_j$$

$$(2.1)$$

where the chemical potential μ is chosen to be the same for l polarons and b electrons, and is such that the number of electrons per site at $T = 0$ has the average value $(1 - x)$.

The *lb* model (Eq. (2.1)) is most naturally viewed as the strong coupling, high temperature theory for the orbital fluid regime of manganites, ie the large x and T domain where there is no long range orbital/charge order, but the two fluids are in chemical equilibrium, namely have the same chemical potential. There is quantum mechanical hybridization between l and b orbitals on the nearest neighbour sites which leads to a quantum coherent single Fermi fluid on an energy scale lower than the hybridization energy $T^*(\sim 150\mathrm{K})$, namely at rather low temperatures. The DMFT calculations in Refs. 5 and 6 (which are identical to the single site CPA for a two component alloy), assume a homogeneous interpenetrating quantum incoherent two fluid system. The Hamiltonian Eq. (2.1) cannot be solved exactly. We solve it in the single site DMFT, which method has been used for this problem in Ref. 24 for example. We assume the spin $S = 3/2$ to be classical, and take J_H and U to be ∞ (they are large, much larger than t, and the results for the low energy behavior do not depend much on their actual large values, so we take the above limit). The basic energy scale is $t_{ij} = t$, and the dimensionless ratios of interest are (E_{JT}/t), $(k_B T/t)$, and (J_F/t). In the single site DMFT, the b electron band has a semicircular density of states, and its width depends on the average hopping amplitude of the b electron, which increases with x and is determined self consistently. It also increases with decreasing T, because of the double exchange effect. The electron density for different values of the temperature (indicated) are shown in Fig. 3.2 (this is exactly the same as Fig. 3.3 of arXiv:cond-mat/0308396). For the parameters chosen, $T_c =208\mathrm{K}$ and coincides with the metal–insulator transition temperature. The polaronic level E_{JT} appears as a sharp peak in $\rho(\omega)$; the two fluid nature of the system, and origin of the CMR in the strong magnetic field dependence of the polaron b electron hopping energy are clear (The b bandwidth decreases most with field near T_c, so that the magnetoresistance is negative and colossal there).

Such a model system will phase separate into macroscopic l and b domains in the absence of long range charge interactions, since this necessarily leads to greater kinetic energy decrease for the b electrons. Electrostatic Coulomb interaction between the cations, anions and electrons in the manganite mutes this phase separation into nanoscopic electronic inhomogeneity, in which b electron puddles of various sizes are in equilibrium with the surrounding l polarons. An extended *lb* model, with additional terms describing the Coulomb interactions has the Hamiltonian Eq. (2.2)

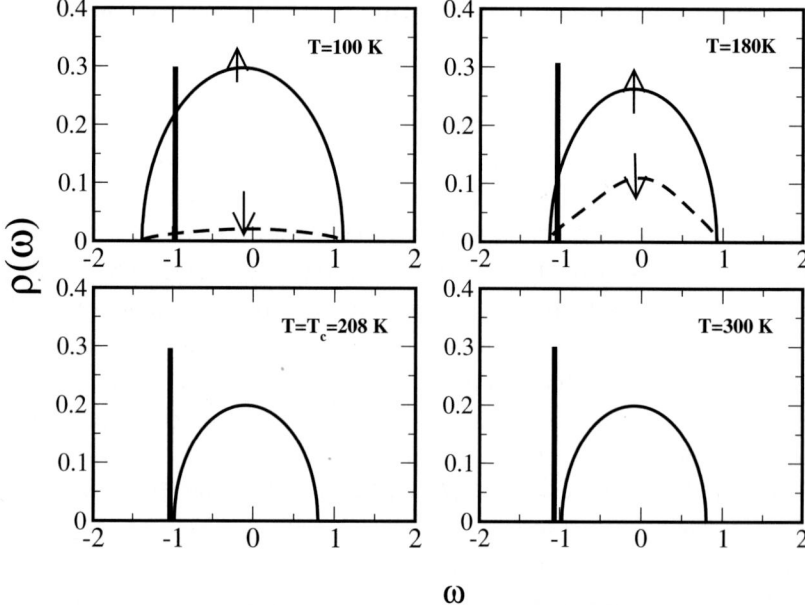

Fig. 3.2.   Evolution of the spectral function $\rho_b(\omega)$ for $E_{JT} = 1$ (in units of t), with J_F such that $T_c = 208$ (appropriate for LSMO), and $J_H = \infty, U = \infty$. Shown is the evolution of the spectral density and effective bandwidth as a function of temperature for $T = 100, 180, 208$ (T_c), and 300 K. The doping $x = 0.175$. Thick lines represent the effective l level, and up and down arrows indicate up and down spin spectral functions.

$$H_{lb}^{ext} = E_{Ak} + \sum_i \Phi_i \hat{q}_i + V_0 \sum_{\{ij\}} \frac{1}{r_{ij}} \hat{q}_i \hat{q}_j$$

$$- t \sum_{<ij>} (b_i^\dagger b_j + b_j^\dagger b_i) + E_{JT} \sum_i h_i^\dagger h_i. \qquad (2.2)$$

This has been extensively investigated numerically in Refs. 7 and 8. The Coulomb interaction is approximated by a Hartree term which is known to correctly describe screening. The l polarons occupy random sites on the lattice, and the exact b eigenstates are determined in this background. The whole problem is solved self consistently. Section 3 mentions some of the results that relate to Anderson localization (calculated via participation ratios). We find that while in the absence of Coulomb interaction, the system will phase separate, Coulomb interaction mutes it into nanoscopic regions essentially of l and b ; this is the scale of charge inhomogeneity essentially determined by the screening of Coulomb interactions. Refs. 7 and 8 also show that a simple, homogeneous, annealed alloy model such as the basic lb

model of Refs. 5 and 6 gives good results for the nanoscopically inhomogeneous system.

3. Localization in the Manganites

In the single site DMFT used above, while the strong correlation effects are known to be treated successfully,[25] Anderson localization is missing, as is well known. For example, the DMFT is equivalent to a $d = \infty$ or CPA theory, where there is no localization. Localization, for large d, is a $(1/d)$ effect. However, in the absence of a systematic $(1/d)$ approach to localization for large d, we adopt the following hybrid approach[3,4] to investigate Anderson localization in manganites. We calculate the mean free path for the b electrons from the DMFT or the CPA as above. We then use this mean free path to estimate the localization behavior from the self-consistent theory of localization.[18] As is well known, the latter is a development of the theory of weak localization[26] in which the frequency dependent conductivity and thence the localization length are calculated self-consistently. It takes account of the consequences of disorder and long distance quantum interference effects (in low spatial dimensions!) and enables one to find the localization length as a function of electron energy, starting from a given mean free path at short length scales. The former is determined from DMFT, and incorporates strong correlation effects nonperturbatively.

The self-consistent equation for ac diffusivity $D(\omega)$ (related to conductivity via the Einstein relation $\sigma(\omega) = 2e^2 \rho D(\omega)$) is given by

$$D(\omega) = D_0 - \frac{1}{\rho \pi \hbar (2\pi)^d} \int_{1/l_{\rm el}}^{1/l_{\rm inel}} \frac{d\vec{k}}{k^2 - \frac{i\omega}{D(\omega)}} \tag{3.1}$$

where $l_{\rm el}$ and $l_{\rm inel}$ are the elastic and inelastic mean free paths respectively. For states localized near the Fermi energy with localization length λ, the leading behavior of $(-i\omega/D(\omega))$ is $\sim \lambda^{-2}$. Using this equation, Economou and Soukoulis[27] developed a potential well analogy, namely an analogy where λ is connected with the bound state size in a potential well, the dc conductivity is related to the depth V_0 of the potential well and the self-consistent equation (Eq. (3.1)) with the appropriate Schrödinger equation for the bound state in the potential well. The equation for $\omega \to 0$ is

$$\frac{1}{\Omega |V_0(E)|} = \frac{2m^*}{(2\pi)^d \hbar^2} \int \frac{d^d k}{k^2 + k_b^2}. \tag{3.2}$$

The upper and lower limits of k in Eqs. (3.1) and (3.2) are the same. For localized states, $D(\omega) \to 0$ as $\omega \to 0$ and is therefore omitted in Eq. (3.2).

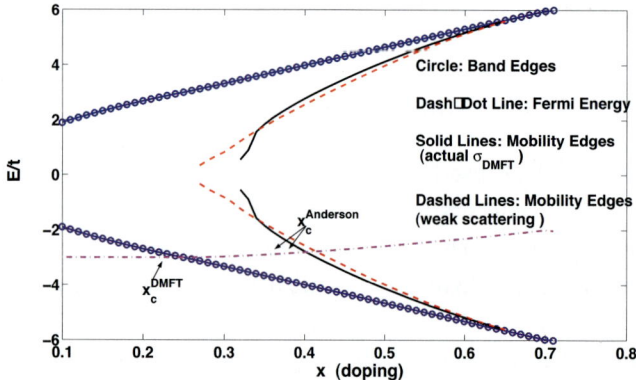

Fig. 3.3. Mobility edge (ϵ^*) trajectory vs. doping from the self consistent theory of localization (solid line: from actual σ_{DMFT}; dashed line: weak scattering). Also shown are the trajectories of the Fermi energy (ϵ_F) and the effective b band bottom (x_c^{DMFT}). The Jahn–Teller binding energy $E_{\text{JT}} = 3$ in units of t. The temperature $T = 0$, and the system is assumed ferromagnetic.

Here k_b is the inverse localization length λ^{-1}. We insert the dc conductivity σ_0 and the mean free path l_{el} calculated in the single site DMFT in Eq. (3.1), assume l_{inel} to be infinity, and then calculate the value k^* for which Eq. (3.1) is satisfied, and for which k_b is zero. Then the mobility edge energy is $E = \epsilon^*$. We determine the mobility edge this way for different values of x and E_{JT} (which affect σ_0 and l_{el}). For example, we plot (in Fig. 3.3) the mobility edge trajectory for $E_{\text{JT}} = 3t$ and $T = 0$ (ferromagnetic ground state). (The lattice is simple cubic, so that the bare band bottom is at $6t$). Also shown are the Fermi energy and the effective band bottom (of the semicircular b electron density of states). A general result is that the Fermi energy increases as x increases, while the effective band bottom and the mobility edge energies decrease as x increases. We notice that for $x < x_c = 0.25$, the band bottom is above the Fermi level. The system is a band insulator, with only site localized polaronic states occupied. For $x > x_c$, the b band states are occupied, but according to our calculation of the mobility edge, the latter is below the Fermi energy and above the effective band bottom for $0.25 < x < 0.40$, so that in this regime of doping, the b electron states relevant for transport (near the Fermi energy) are localized. Quite generally, states near the band edge are expected to be Anderson localized. For $x > 0.40$, the states near the Fermi energy are extended. The hole concentration at which the mobility edge intersects the Fermi energy depends on (E_{JT}/t), decreasing with the latter. We have also calculated the localization length as a function of energy.

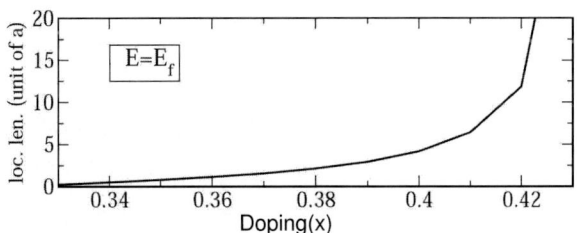

Fig. 3.4. Doping variation of localization length evaluated at the Fermi energy, in units of the lattice constant, for the same parameters as in Fig. 3.3.

Its value at the Fermi energy is plotted for different values of x in Fig. 3.4. It is seen to be small, of the order of a lattice constant, over much of the doping range over which the b states are occupied.

Coulomb interactions, in addition to being crucially the cause of nanoscopic inhomogeneities, have several important effects on transport and localization. First, it was shown by Shenoy *et al.*[7,8] that the system is a Coulomb insulator, with a typical Coulomb pseudogap whose density of states goes as the square of the excitation energy with respect to the Fermi energy, and that the numerically obtained b bandwidth (without approximations) follows quite accurately the DMFT estimate.[7,8] The corresponding Efros–Shklovskii resistivity goes as $\rho(T) \sim \rho_0 \exp(T/T_0)^{1/2}$. Second, in this model, there are several hole doping concentrations that are of relevance. The system consists of puddles of b electrons in immobile l polarons. At a concentration x_{c1}, the Fermi energy is such that these puddles begin to be occupied by electrons (in analogy with x_c of the DMFT, to which it is very close, as seen from simulations including Coulomb interactions[7,8]). At a somewhat higher concentration x_{c2}, the nanoscopic puddles "percolate". Figure 3.5 shows the computed E_{JT} vs. doping curve, and exhibits the critical hole doping x_c in the DMFT, the density x_{c1} at which the puddle b states begin to be occupied, and the density x_{c2} at which they percolate. We notice that x_c, which is calculated using a microscopically homogeneous approximation such as the DMFT, and x_{c1}, which manifestly has inhomogeneities and Coulomb interactions, are very close to each other. This implies that it is not bad to ignore the Coulomb interaction as well as nanoscopic inhomogeneities, and use the DMFT. The system is still not a metal, since the occupied b electronic states are Anderson localized, as evidenced by their inverse participation ratios. From the latter, the mobility edge has been calculated. Its trajectory as a function of x is

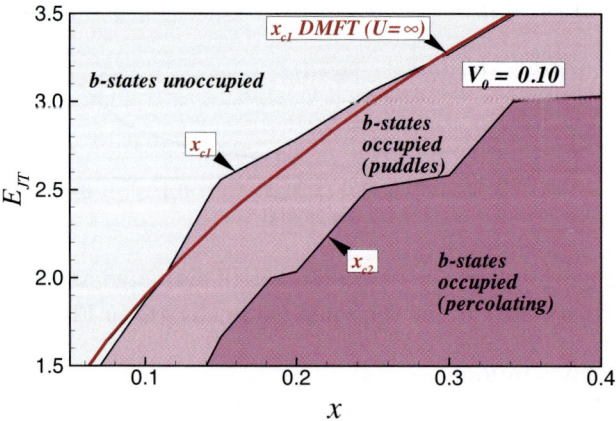

Fig. 3.5. E_{JT} vs. x showing various phase boundaries, along with the DMFT parabola for comparison. The Coulomb energy V_0 is 0.01 in units of the strength of the bare Coulomb interaction for the nearest neighbours, and unit electric charge.

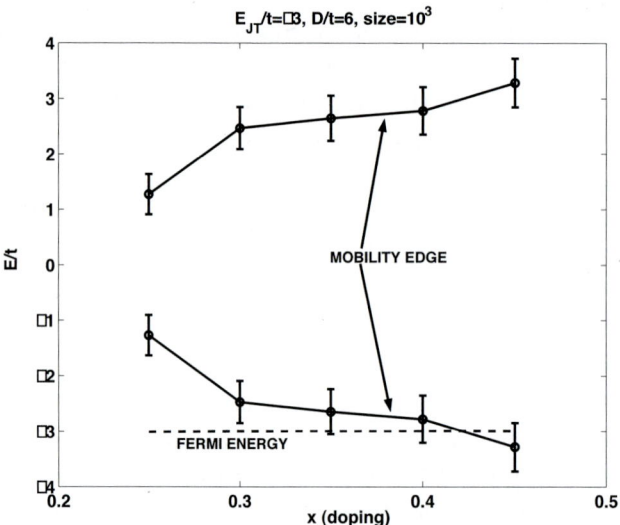

Fig. 3.6. Mobility edge trajectory vs. doping, calculated from inverse participation ratio, obtained from real space simulations of the extended lb Hamiltonian Eq. (2.2).

shown in Fig. 3.6. It is encouragingly similar to that of the mobility edge calculated above (Fig. 3.3) in a completely different way, using a particular approximation for localization. At a slightly higher higher concentration x_{c3}, the geometrically percolating b states are extended, and the system is a metal.

4. Comparison with Experiments

In this section, we confront our calculations with experiment. We deliberately do not compare specific experimental results with specific predictions of our calculations. The reason essentially is that our predictions are for the $T = 0$ ferromagnetic phase, while the experiments are for a **variety** of systems (difficult to specify in terms of our model parameters) at $T \neq 0$; these systems can be ferromagnetic or paramagnetic at that temperature. However, we point out here ways in which our calculations strongly suggest hopping transport under very different conditions (schematically indicated in Fig. 3.7 in the *lb* model), in the light of results mentioned above in Sec. 3.

We first argue that our conclusions on Anderson localization of b electron eigenstates continue to be valid for $T \neq 0$, and in the paramagnetic state. At $T \neq 0$, l_{inel} is finite, so that in the integral Eq. (3.2) has a nonzero lower limit. Since l_{inel} is quite large, the lower limit is still close to zero. Further, since the elastic mean free path is still quite short (the system is a dirty metal in the *lb* model, with hardly any temperature dependence in the

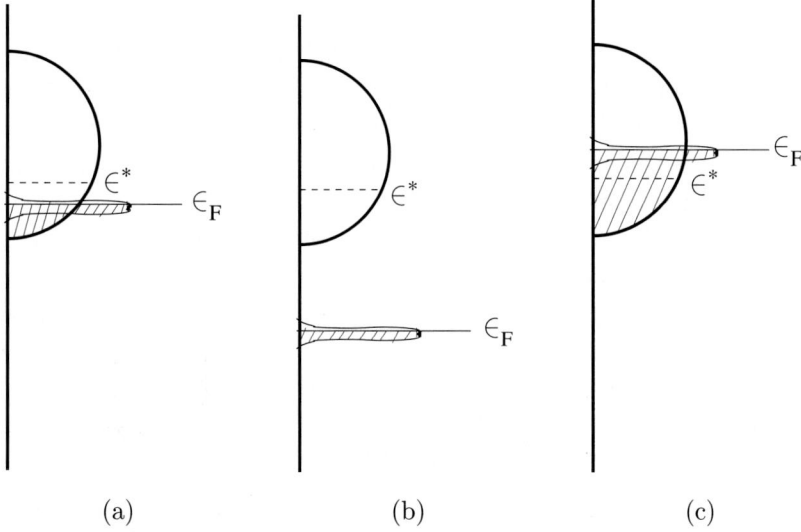

(a)   (b)   (c)

Fig. 3.7. Schematic diagrams of the polaronic (l) and broad band (b) states in the extended *lb* model with Coulomb interactions. The l state is broadened into a narrow band as shown, due to the Coulomb interactions. The b band has a semicircular density of states. The mobility edge energy ϵ^* and the Fermi energy ϵ_F are shown. Occupied states are shown by shaded lines. The system can be paramagnetic or ferromagnetic; it is an insulator in (a) with essentially Mott variable range hopping; and in (b) with predominantly Efros–Shklovskii hopping. It is a band metal in (c).

"metallic regime"), the upper limit of the integral hardly changes. So, the integral does not change much, nor does the mobility edge determined from it. Further, the disorder which determines localization of b electrons is the $U n_b n_l$ term involving the b electrons and the l polarons on site, independent of spin. Thus the localization effect is independent of the spin orientations of the polarons and the b electrons, i.e. whether the system is paramagnetic or ferromagnetic.

We show, in Figs. 3.7(a)–3.7(c), schematically, three possibilities for the relative placements of the Coulomb broadened l level, the b band, the Fermi level, and the mobility edge (drawn on the basis of our results mentioned earlier). In Figs. 3.7(a) and 3.7(b), the system described is an insulator, which can be ferromagnetic or paramagnetic. In Fig. 3.7(c), one has a metal, which can again be paramagnetic or ferromagnetic. In Fig. 3.7(a), the conduction will lead to Mott variable range hopping behavior with presumably overwhelms the Efros–Shklovskii contribution. (The hopping is exponentially reduced further because of the Coulomb energy caused landscape through which the localized states have to tunnel). In plain DMFT, the system will be a metal, with a small number of carriers in the b band. The closest experimental analogy is to the $T = 180$ K spectral density feature in Fig. 3.2. If localization effects are included, this system is expected to be insulating with variable range hopping. It is thus quite likely that the ferromagnetic metal phase, on increasing the temperature, becomes first an Anderson insulator (paramagnetic or ferromagnetic) and then a band like insulator. Another kind of hopping transport is schematically implied in Fig. 3.7(b). This describes, in the DMFT, and on ignoring the Coulomb interaction, an insulating phase close to the band insulator appropriate to the $T = 300$ K diagram of Fig. 3.2. The sharp polaron level broadens into a softgapped polaron band. The hopping transport in this case definitely has an Efros–Shklovskii form. In addition, one might expect activated hopping of the l polaron to an unoccupied b electron state. This can lead to a peculiar kind of hopping behavior, e.g. carriers are created in the otherwise unoccupied b band by the thermal promotion of the l polaron, and these carriers if localized, show variable range hopping, and if extended, plain Drude conductivity. Thus the temperature dependence of resistivity is quite complicated; while there is no single exponent α, the dominant term is perhaps Efros–Shklovskii hopping with $\alpha = 1/2$. It seems plausible that in the doping regime $0.1 \lesssim x \lesssim 0.2$ for which the system is a ferromagnetic insulator, the scenario of Fig. 3.7b is appropriate, so that one expects (and finds[11]) Efros–Shklovskii hopping. For higher doping, in the range $0.2 \lesssim x \lesssim 0.5$, it seems natural from the above that the index α changes from $(1/4)$ at lower temperatures to $(1/2)$ at higher temperatures (e.g. Figs. 3.7(a) and 3.7(b)). This indeed happens. We thus believe that in the *lb* model, the existence of small polarons l (which are localized anyway, and which are present at random sites if there is no long range polaronic order), their forming a Coulomb glass, the inevitable Anderson localization of b electrons in the random field of these l polarons, all combine to produce the ubiquitous hopping transport behavior observed in them.

Acknowledgments

I am thankful to the Department of Science and Technology, New Delhi, for research support through the Ramanna Fellowship. I would also like to thank H. R. Krishnamurthy, V. B. Shenoy and P. Sanyal for discussion and collaboration, and Sumilan Banerjee for help with the manuscript.

References

1. M. B. Salamon and M. Jaime, *Rev. Mod. Phys.* **73**, 583 (2001).
2. T. Chatterji (ed.), *Colossal Magnetoresistive Manganites* (Kluwer Academic Publishers, Dordrecht, 2004).
3. P. Sanyal, H. R. Krishnamurthy, V. B. Shenoy and T. V. Ramakrishnan, to be published (2009).
4. P. Sanyal, Ph.D thesis, Indian Institute of Science, unpublished, 2007.
5. T. V. Ramakrishnan, H. R. Krishnamurthy, S. R. Hassan and G. V. Pai, *Phys. Rev. Lett.* **92**, 157203 (2004). Also, Chapter 10 of Ref. 2.
6. G. V. Pai, S. R. Hassan, H. R. Krishnamurthy and T. V. Ramakrishnan, *Europhys. Lett.* **64**, 696 (2003).
7. V. B. Shenoy, T. Gupta, H. R. Krishnamurthy and T. V. Ramakrishnan, *Phys. Rev. Lett.* **98**, 097201 (2007).
8. V. B. Shenoy, T. Gupta, H. R. Krishnamurty and T. V. Ramakrishnan, *Phys. Rev. B* **80**, 125121 (2009).
9. C. Zener, *Phys. Rev.* **52**, 403 (1951).
10. P. W. Anderson and H. Hasegawa, *Phys. Rev.* **100**, 675 (1955).
11. H. Jain and A. K. Raychaudhuri, [cond-mat.str-el] arXiv:0804.4646v1.
12. L. M. Rodriguez and J. P. Attfield, *Phys. Rev. B* **54**, R15622 (1996).
13. M. Viret, L. Rannoo and J. M. D. Coey, *Phys. Rev. B* **55**, 8067 (1997).
14. C. Ciu, T. A. Tyson, Z. Zhang, J. P. Carlo and Y. Qin, *Phys. Rev. B* **67**, 104107 (2003).
15. E. Bose, S. Karmakar, B. K. Chaudhuri, S. Pal, C. Martin, S. Herbert and A. Maignan, *J. Phys. Cond. Mat.* **19**, 266218 (2007).
16. L. Sudheendra and C. N. R. Rao, *J. Phys. Cond. Mat.* **15**, 3029 (2003).

17. In reality, the l polarons, unless they form a glass, do move slowly, because of the exponentially reduced Huang–Rhys intersite hopping amplitude; though here we assume the l polarons to be static. This neglect is a very good approximation especially in the temperature range here since the slow motion of polarons does not affect electron localization at temperatures of interest to us where the system is an incoherent two component quantum fluid; somewhat like in an insulating fluid in which electron states continue to be localized despite the slow motion of the ions.

18. D. Vollhardt and P. Wölfle, *Phys. Rev. Lett.* **45**, 482 (1980); *Phys. Rev. B* **22**, 4666 (1980).

19. This intrinsic disorder is qualitatively different from that proposed to exist in models of manganites with only double exchange; in such a model it has been argued that the thermal disorder of t_{2g} spins leading to e.g. electron hopping disorder can localize a sizeable fraction of e.g. states (for example C. M. Varma, *Phys. Rev. B* **54**, 7328, (1996)). However, finite size scaling calculations (Qiming Li, Jun Zhang, A. R. Bishop and C. M. Soukoulis, *Phys. Rev. B* **56**, R4541, (1997) show that the fraction of states localized by this process is extremely small, and is therefore not relevant to the presence of an insulating phase above the Curie temperature T_c. This remark also applies, I believe, to the magnetic localization mechanism of Coey *et al.* (see Ref. 13) in which it is argued that the Hund's rule enslavement of the hopping spin, and the incomplete ferromagnetic order in the fixed spins will tend to localize electronic states.

20. N. F. Mott, *J. Non-Cryst. Solids* **1**, 1 (1968).

21. A. L. Efros and B. I. Shklovskii, *J. Phys. C: Solid State Phys.* **8**, L49 (1975).

22. N. F. Mott and E. A. Davies, *Electronic Processes in Noncrystalline Solids*, 2nd edn. (Oxford University Press, New York, 1979).

23. A. G. Zabrodskii, *Phil. Mag.* **81**, 1131 (2001).

24. A. J. Millis, P. B. Littlewood and B. I. Shraiman, *Phys. Rev. Lett.* **77**, 175 (1996); A. J. Millis, R. Mueller and B. Shraiman, *Phys. Rev. B* **54**, 5389 (1996); *Phys. Rev. B* **54**, 5405 (1996).

25. A. Georges, G. Kotliar, W. Krauth and M. J. Rosenberg, *Rev. Mod. Phys.* **68**, 13 (1996).

26. E. Abrahams, P. W. Anderson, D. C. Licciardello and T. V. Ramakrishnan, *Phys. Rev. Lett.* **42**, 673 (1979).

27. E. N. Economou and C. M. Soukoulis, *Phys. Rev. B* **28**, 1093 (1983); E. N. Economou, C. M. Soukoulis and A. D. Zdetesis, *Phys. Rev. B* **30**, 1686 (1984).

Chapter 4

SELF-CONSISTENT THEORY OF ANDERSON LOCALIZATION: GENERAL FORMALISM AND APPLICATIONS

P. Wölfle* and D. Vollhardt†

*Institute for Condensed Matter Theory,
Institute for Nanotechnology and DFG-Center for Functional Nanostructures,
Karlsruhe Institute of Technology, D-76131 Karlsruhe, Germany*

†*Theoretical Physics III, Center for Electronic Correlations and Magnetism,
Institute for Physics, University of Augsburg, D-86135 Augsburg, Germany*

The self-consistent theory of Anderson localization of quantum particles or classical waves in disordered media is reviewed. After presenting the basic concepts of the theory of Anderson localization in the case of electrons in disordered solids, the regimes of weak and strong localization are discussed. Then the scaling theory of the Anderson localization transition is reviewed. The renormalization group theory is introduced and results and consequences are presented. It is shown how scale-dependent terms in the renormalized perturbation theory of the inverse diffusion coefficient lead in a natural way to a self-consistent equation for the diffusion coefficient. The latter accounts quantitatively for the static and dynamic transport properties except for a region near the critical point. Several recent applications and extensions of the self-consistent theory, in particular for classical waves, are discussed.

1. Introduction to Anderson Localization

1.1. *Brief historical review*

The localization of quantum particles by a static random potential, or of classical waves by random fluctuations of the medium, is one of the most intriguing phenomena in statistical physics. The key ingredient of localization, wave interference, was introduced in P. W. Anderson's seminal paper "Absence of Diffusion in Certain Random Lattices".[1] There it was shown that electrons may be localized by a random potential, so that diffusion is suppressed, even in a situation where classical particles would be delocalized. The fundamental reason for the localizing effect of a random potential on quantum particles or classical waves is the multiple interference of wave

43

components scattered by randomly positioned scattering centers. The interference effect takes place, as long as the propagation is coherent.

It is interesting to note that the first application of the idea of localization concerned the spin diffusion D of electrons and not the electrical conductivity σ. Anderson considered a tight-binding model of electrons on a crystal lattice, with energy levels at each site chosen from a random distribution.[1] The traditional view had been, that scattering by the random potential causes the Bloch waves to lose well-defined momentum on the length scale of the mean-free path ℓ. Nevertheless, the wavefunction was thought to remain extended throughout the sample. Anderson pointed out that if the disorder is sufficiently strong, the particles may become localized, in that the envelope of the wave function $\psi(\boldsymbol{r})$ decays exponentially from some point $\boldsymbol{r}_0$ in space:

$$|\psi(\boldsymbol{r})| \sim \exp(|\boldsymbol{r} - \boldsymbol{r}_0|/\xi), \tag{1.1}$$

where ξ is the localization length.

There exist a number of review articles on the Anderson localization problem. The most complete account of the early work was presented by Lee and Ramakrishnan.[2] The seminal early work on interaction affects is presented in Ref. 3. A complete account of the early numerical work can be found in Ref. 4. A path integral formulation of weak localization is presented in Ref. 5. Several more review articles and books are cited along the way. In the following, we will use units with Planck's constant $\hbar$ and Boltzmann's constant k_B equal to unity, unless stated otherwise.

1.2. *Electrons and classical waves in disordered systems*

The wavefunction $\psi(\boldsymbol{r})$ of a single electron of mass m in a random potential $V(\boldsymbol{r})$ obeys the stationary Schrödinger equation

$$\left(-\frac{\hbar^2}{2m}\boldsymbol{\nabla}^2 + V(\boldsymbol{r}) - E\right)\psi(\boldsymbol{r}) = 0. \tag{1.2}$$

In the simplest case $V(\boldsymbol{r})$ may be assumed to obey Gaussian statistics with $\langle V(\boldsymbol{r})V(\boldsymbol{r}')\rangle = \langle V^2\rangle\delta(\boldsymbol{r} - \boldsymbol{r}')$, but many of the results presented below are valid for a much wider class of models. Electrons propagating in the random potential $V(\boldsymbol{r})$ will be scattered on average after a time τ. For weak random potential the scattering rate is given by

$$\frac{1}{\tau} = \pi N(E)\,\langle V^2\rangle \tag{1.3}$$

where $N(E)$ is the density of states at the energy E of the electron. In a metal the electrons carrying the charge current are those at the Fermi energy $E = E_F$. Within the time τ the electron travels a distance $\ell = v_F \tau$, where v_F is its velocity.

In close analogy the wave amplitude $\psi(\boldsymbol{r})$ of a classical monochromatic wave of frequency ω obeys the wave equation

$$\left(\frac{\omega^2}{c^2(\boldsymbol{r})} + \boldsymbol{\nabla}^2 \right) \psi(\boldsymbol{r}) = 0. \tag{1.4}$$

Here $c(\boldsymbol{r})$ is the wave velocity at position $\boldsymbol{r}$ in an inhomogeneous medium, assumed to be a randomly fluctuating quantity. The main difference between the Schrödinger equation and the wave equation is that in the wave equation the "random potential" $1/c^2(\boldsymbol{r})$ is multiplied by ω^2, so that disorder is suppressed in the limit $\omega \to 0$. By contrast, in the quantum case disorder will be dominant in the limit of low energy E. A further difference may arise if the wave amplitude is a vector quantity as, e.g., in the case of electromagnetic waves in $d = 3$ dimensions.

In real systems particles or wave packets are not independent, but interact. Electrons are coupled by the Coulomb interaction, leading to important effects that go much beyond the single particle model. Similarly, wave packets interact via nonlinear polarization of the medium. Apart from these complications, the physics of electronic wave packets and classical wave packets is quite similar. In the following, we will present most of the discussion in the language of electronic wave packets.

1.3. *Weak localization*

The all-important effect of wave interference is most clearly seen in the limit of weak scattering, where it already may cause localization, but only in reduced dimensions. While it is difficult to observe full localization at finite temperature T, on account of the effect of interactions limiting the phase coherence, the dramatic signatures of localization are visible at finite T in the form of "weak localization".[6,7]

An electron or a wave packet moving through a disordered medium will be scattered by the random potential on the average after propagating a distance ℓ, the mean-free path. On larger length scales the propagation is diffusive. Weak localization is a consequence of destructive interference of two wave components starting at some point and returning to the same point after traversing time-reversed paths. Let the probability amplitudes for the wave packet to move from point $\boldsymbol{r}_0$ along some path C_1 back to $\boldsymbol{r}_0$ be A_1

and along a different path C_2 be A_2, then the transition probability for the particle to move either along C_1 or along C_2 will be

$$w = |A_1 + A_2|^2 = w_{cl} + w_{int},   (1.5)$$

where $w_{cl} = |A_1|^2 + |A_2|^2$ and $w_{int} = 2Re(A_1^* A_2)$. For any two paths the interference term w_{int} may be positive or negative, and thus averages to zero. However, if $A_2 = A_r$ is the amplitude of the time-reverse of path $A_1 = A$ and if time reversal invariance holds, then $A = A_r$, i.e., the probability of return w is enhanced by a factor of two compared to the probability w_{cl} of a classical system:

$$w = 4|A|^2 = 2w_{cl}.   (1.6)$$

In that case the probability for transmission is reduced, which leads to a reduced diffusion coefficient and a reduced conductivity. One may estimate the correction to the conductivity in the following qualitative way. The relative change of the conductivity σ by the above interference effect is equal to the probability of interference of two wave packets of extension λ, the wavelength, after returning to the starting point. The infinitesimal probability of return to the origin in time t of a particle diffusing in d dimension is given by $(4\pi Dt)^{-d/2} d^3r$ where D is the diffusion coefficient. Since the volume of interference in the time interval $[t, t+dt]$ is $\lambda^{d-1} v dt$, where v is the velocity of the wave packet, one finds the quantum correction to the conductivity $\delta\sigma$ as[6,7]

$$\frac{\delta\sigma}{\sigma_0} \approx -\int_\tau^{\tau_\phi} \frac{v\lambda^{d-1} dt}{(4\pi Dt)^{d/2}} = \begin{cases} -c_3 \dfrac{\lambda^2}{\ell^2}\left(1 - \dfrac{\tau}{\tau_\phi}\right), & d = 3 \\[2mm] -c_2 \dfrac{\lambda}{\ell} \ln(\tau_\phi/\tau), & d = 2 \\[2mm] -c_1 \left(\sqrt{\dfrac{\tau_\phi}{\tau}} - 1\right), & d = 1. \end{cases} \qquad (1.7)$$

Here $D = \frac{1}{d}v^2\tau$ is the diffusion constant, $\ell = v\tau$ is the mean-free path, τ is the mean time between successive elastic collisions, $\sigma_0 = e^2 n\tau/m$ is the Drude conductivity with n as the particle density, and c_i are constants of order unity. The upper limit of the integral is the phase relaxation time τ_ϕ, i.e., the average time after which phase coherence is lost due to inelastic or other phase-shifting processes. For weak localization processes to exist at all, the inequality $\tau_\phi \gg \tau$ must hold. We note that the correction in three and two dimensions depends on the ratio of wavelength λ to mean-free path ℓ, and gets smaller in the limit of weak disorder, where $\lambda/\ell \ll 1$. In two and one dimension the correction grows large in the limit $\tau/\tau_\phi \to 0$ since one expects the phase relaxation rate $1/\tau_\phi$ for a system in thermodynamic equilibrium to go to zero for $T \to 0$. By contrast, in some cases a plateau behavior of $1/\tau_\phi$ as a function of temperature has been found experimentally, which gave rise to the speculation that the zero point fluctuations may cause decoherence. However, given a unique ground state, it is difficult to understand how a particle in the system may loose its phase coherence. Several physical mechanisms that may lead to a plateau of $1/\tau_\phi$ have been identified. For a recent discussion of these issues see Ref. 8.

With $\tau/\tau_\phi \to 0$ for $T \to 0$ the weak localization quantum correction will be large in any system in $d = 1, 2$, no matter how weak the disorder. As we will see, this behavior signals the fact that there are no extended states in $d = 1, 2$ dimensions. The characteristic length L_ϕ over which a wave packet retains phase coherence is related to τ_ϕ by the diffusion coefficient $L_\phi = \sqrt{D\tau_\phi}$. In systems of restricted dimension, e.g., films of thickness a or wires of diameter a, the effective dimensionality of the system with respect to localization is determined by the ratio L_ϕ/a. Namely, for $L_\phi \ll a$ the system is three-dimensional ($3d$), while for $L_\phi \gg a$ diffusion over time τ_ϕ takes place in the restricted geometry of the film or wire, and the effective dimension is therefore 2 or 1.

1.4. *Strong localization and the Anderson transition*

The appearance of localized states is easily understood in the limit of very strong disorder: localized orbitals will then exist at positions where the random potential forms a deep well. The admixture of adjacent orbitals by the hopping amplitudes will only cause a perturbation that does not delocalize the particle. The reason for this is that nearby orbitals will have sufficiently different energies so that the amount of admixture is small. On the other hand, orbitals close in energy will in general be spatially far apart, so that their overlap is exponentially small. Thus, we can expect the wave functions in strongly disordered systems to be exponentially localized. Whether the particles become delocalized when the disorder strength is reduced, is a much more complex question. In one dimension it can be shown rigorously that all states are localized, no matter how weak the disorder.[9–11]

In three dimensions, the accepted view is that the particles are delocalized for weak disorder. In general, localized and extended states of the same energy do not coexist, since in a typical situation any small perturbation would lead to hybridization and thus to the delocalization of a localized

state. We can therefore assume that the localized and extended states of
a given energy are separated. For increasing disorder strength η there will
then be a sharp transition from delocalized to localized states at a critical
disorder strength η_c. A qualitative criterion as to when an Anderson transition is expected in $3d$ systems has been proposed by Ioffe and Regel.[12] It
states that as the mean free path ℓ becomes shorter with increasing disorder,
the Anderson transition occurs when ℓ is of the order of the wavelength λ
of the particle (which amounts to the condition $k_F \ell \sim 1$ in metals, where
k_F is the Fermi wave number). As we will see later, in $1d$ or $2d$ systems, ℓ
may be much longer than the wavelength and the particles are nonetheless
localized. In fact, the relevant mean free path here is the one with respect
to momentum transfer. A similar situation exists when we fix the disorder strength, but vary the energy E. Electrons in states near the bottom
of the energy band are expected to be localized even by a weakly disordered potential, whereas electrons in states near the band center (in $d = 3$)
will be delocalized, provided the disorder is not too strong. Thus there
exists a critical energy E_c separating localized from delocalized states, the
so-called mobility edge.[13,14] The electron mobility as a function of energy
is identically zero on the localized side (at zero temperature), and increases
continuously with energy separation $|E - E_c|$ in the delocalized, or metallic,
phase. The continuous character of this quantum phase transition, termed
Anderson transition, is a consequence of the scaling theory to be presented
below.

Historically the continuous nature of the metal–insulator transition in
disordered solids has been a point of controversy for many years. According
to an earlier theory by Mott[13,14] the conductivity changes discontinuously
at the transition, such that a "minimum metallic conductivity" exists on the
metallic side of the transition. Numerical simulations[4] have shown beyond
doubt that the transition is instead continuous, at least in the absence of
interactions.

In the much more complex situation of interacting electrons one finds for
the Hubbard model without disorder, using the Dynamical Mean-Field Theory (DMFT), that the Mott–Hubbard metal–insulator transition is discontinuous at finite temperatures, and that it becomes continuous in the limit
$T \to 0$.[15,16] For the Hubbard model in the presence of disorder ("Anderson–
Hubbard model") at $T = 0$ the situation is similar: the Mott–Hubbard
metal–insulator transition is discontinuous for finite disorder and becomes
continuous in the limit of vanishing disorder.[17,18]

2. Fundamental Theoretical Concepts of Anderson Localization

The Anderson localization transition is a quantum phase transition, i.e., it is a transition at zero temperature tuned by a control parameter, e.g., the disorder strength, particle energy, or wave frequency. Unlike other quantum phase transitions, the Anderson transition does not have an obvious order parameter. Nonetheless, there exists a dynamically generated length scale, the localization or correlation length ξ, which tends to infinity as the transition is approached. Therefore, by drawing an analogy with magnetic phase transitions, Wegner early on proposed scaling properties.[19] Later, he formulated a field-theoretic description of the Anderson transition in the form of a non-linear sigma model (NLσM) of interacting matrices (rather than vectors, as for magnetic systems).[20] The NLσM was later formulated in the mathematically more tractable supersymmetric form.[21]

2.1. *Scaling theory of the conductance*

Wegner[19] argued that the Anderson localization transition should be described in the language of critical phenomena of continuous (quantum) phase transitions. This requires the assumption of a correlation length ξ diverging as a function of disorder strength η at the critical point

$$\xi(\eta) \sim |\eta - \eta_c|^{-\nu}. \tag{2.1}$$

The conductivity is then expected to obey the scaling law

$$\sigma(\eta) \sim \xi^{2-d} \sim (\eta_c - \eta)^s; \quad \eta < \eta_c, \quad d > 2. \tag{2.2}$$

This follows from the fact that σ, written in units of $e^2/(2\pi\hbar)$, has dimension $(1/\text{length})^{d-2}$, and the only characteristic length near the transition is the correlation length ξ. By comparing the conductivity exponent s with the exponent of ξ one finds

$$s = \nu(d - 2). \tag{2.3}$$

On the other hand, the conductance g of a d-dimensional cube of length L, which for a good metal of conductivity σ is given by $g(L) = \sigma L^{d-2}$, must obey the scaling property

$$g(\eta; L) = \Phi(L/\xi). \tag{2.4}$$

This means that g is a function of a single parameter L/ξ, so that each value of L/ξ corresponds to a value g.

2.2. Renormalization group equation

It then follows that $g(L)$ obeys the renormalization group (RG) equation

$$\frac{d\ln g}{d\ln L} = \beta(g), \tag{2.5}$$

where $\beta(g)$ is a function of g only, and does not depend on disorder. In a landmark paper, Abrahams, Anderson, Licciardello and Ramakrishnan[6] proposed the above equation and calculated the β-function in the limits of weak and strong disorder. A confirmation of the assumption of scaling was obtained from a calculation of the next-order term.[7]

At strong disorder one expects all states to be localized, with average localization length ξ. It then follows that $g(L)$ is an exponentially decreasing function of L:

$$g(L) \sim \exp(-L/\xi). \tag{2.6}$$

In comparison with the ohmic dependence $g \sim L^{d-2}$, this is a very non-ohmic behavior. The β-function is then given by

$$\beta(g) \sim \ln(g/g_c) < 0. \tag{2.7}$$

At weak disorder one finds from $g \sim L^{d-2}$ that

$$\beta(g) = d - 2. \tag{2.8}$$

The important question of whether the system is delocalized (metal) or localized (insulator) may be answered by integrating the RG equation from some starting point L_0, where $g(L_0)$ is known. Depending on whether $\beta(g)$ is positive or negative along the integration path, the conductance will scale to infinity or to zero, as L goes to infinity.

In $d = 3$ dimensions one has $\beta(g) > 0$ at large g, but $\beta(g) < 0$ at small g. Thus, there exists a critical point at $g = g_c$, where $\beta(g_c) = 0$, separating localized and delocalized behavior.

On the other hand, in $d = 1$ dimension one has $\beta(g) < 0$ at large and small g, and by interpolation also for intermediate values of g, so that there is no transition in this case and all states are localized.

The dimension $d = 2$ apparently plays a special role, as in this case $\beta(g) \rightarrow 0$ for $g \rightarrow \infty$. In order to determine whether $\beta > 0$ or < 0 for large g one has to calculate the scale dependent (i.e., L-dependent) corrections to the Drude result at large g. This is precisely the weak localization correction already mentioned above. For a system of finite length $L < L_\phi$ we should replace $\frac{1}{\tau_\phi} = DL_\phi^{-2}$ in Eq. (1.7) by DL^{-2}, leading to

$$g(L) = \sigma_0 - a \ \ln\left(\frac{L}{\ell}\right), \tag{2.9}$$

where a diagrammatic calculation[6] gives $a = 2/\pi$ and $\sigma_0 = \ell/\lambda_F$ (in units of $e^2/\hbar$; λ_F is the Fermi wave length) has been used. It follows that

$$\beta(g) = -\frac{a}{g}, \quad d = 2, \tag{2.10}$$

so that we can expect $\beta(g) < 0$ for all g, implying that again all states are localized. This result is valid for the "usual" type of disorder, i.e., in case all symmetries, in particular time reversal symmetry (required for the weak localization correction to be present) are preserved. If time-reversal invariance is broken, e.g., by spin-flip scattering at magnetic impurities, the weak localization effect is somewhat reduced in dimensions $d = 2 + \epsilon, \epsilon \ll 1$, but is not completely removed. The first correction term in the β-function is then proportional to $-1/g^2$ (see, e.g., Ref. 21) implying that all states are still localized (in $d = 3$ dimensions the leading correction term is again $\sim 1/g$; see Ref. 22). In the presence of a magnetic field the situation is more complex, since the scaling of the Hall conductance is coupled to the scaling of g. As a result, one finds exactly one extended state per Landau energy level, which then gives rise to the quantum Hall effect.[23] On the other hand, if spin-rotation invariance is broken, but time-reversal invariance is preserved, as is the case of spin-orbit scattering, the correction term is proportional to $+1/g$, i.e., it is *anti-localizing*. In this case the β-function in $d = 2$ dimensions has a zero, implying the existence of an Anderson transition.[24]

2.3. *Critical exponents*

In the neighborhood of the critical point at $g = g_c$ in $d = 3$ we may expand the β-function as

$$\beta(g) = \frac{1}{y}\left[\frac{g - g_c}{g_c}\right], \quad |g - g_c| \ll g_c. \tag{2.11}$$

Integrating the RG equation for $g > g_c$ from $g(\ell) = g_0$ to $\beta \to 1$ at large L we find $g(L) = \sigma L$, where

$$\sigma \sim \frac{1}{\ell}(g(\ell) - g_c)^y. \tag{2.12}$$

Since $[g(\ell) - g_c] \propto (\eta_c - \eta)$, we conclude that the inverse of the slope of the β-function, y, is equal to the conductivity exponent $s = y$.

Similarly, one finds on the localized side $(g < g_c)$

$$g(L) \sim g_c \exp[-c(g_c - g(\ell))^y L/\ell] \sim g_c \exp(-L/\xi), \qquad (2.13)$$

from which the localization length follows as

$$\xi \sim \ell |\eta - \eta_c|^{-y}. \qquad (2.14)$$

The critical exponent ν governing the localization length is therefore $\nu = y = s$ in $d = 3$ dimensions.

Since the critical conductance $g_c = O(1)$ in $d = 3$, there exist no analytical methods to calculate the β-function in the critical region in a quantitative way. A perturbative expansion in $2+\epsilon$ dimensions, where $g_c \gg 1$, is possible, but the expansion in ϵ is not well-behaved, so that it cannot be used to obtain quantitative results for s and ν in $d = 3$. There exist, however, reliable results on ν from numerical studies, according to which $s = \nu = 1.58 \pm 0.02$.[4,25]

2.4. *Dynamical scaling*

The dynamical conductivity $\sigma(\omega)$, i.e., the a.c. conductivity at frequency ω, in the thermodynamic limit in $d = 3$ obeys the scaling law[26,27]

$$\sigma(\omega; \eta) = \frac{1}{\xi}\Phi(L_\omega/\xi), \qquad (2.15)$$

where the scaling function Φ has been introduced in Eq. (2.4). Here L_ω is the typical length which a wave packet travels in the time of one cycle, $1/\omega$. Since the motion is diffusive it obeys $L_\omega = \sqrt{D(\omega)/\omega}$. It is important to note that the diffusion coefficient $D(\omega)$ is energy scale dependent and is related to the conductivity via the Einstein relation

$$\sigma(\omega) = \hbar N(E)D(\omega), \qquad (2.16)$$

where $N(E)$ is the density of states at the particle energy E.

At the Anderson transition, where $\xi \to \infty$, we expect $\sigma(\omega)$ to be finite. It follows that $\lim_{\xi \to \infty} \Phi(L_\omega/\xi) \sim \xi/L_\omega$ and consequently

$$\sigma(\omega; \eta) \sim \frac{1}{L_\omega}, \quad \eta = \eta_c. \qquad (2.17)$$

This is a self-consistent equation for $\sigma(\omega)$, with solution

$$\sigma(\omega) \sim \omega^{1/3}, \quad \eta = \eta_c. \qquad (2.18)$$

More precisely, in the above expressions ω should be replaced by the imaginary frequency $-i\omega$, such that $\sigma(\omega)$ is a complex-valued quantity.

In a more general notation, introducing the dynamical critical exponent z by $\sigma(\omega) \sim \omega^{1/z}$, we conclude that $z = 3$. The dynamical scaling is valid in a wide neighborhood of the critical point, defined by $\omega > \frac{1}{\tau}(\xi/\ell)^{-z} \sim |\eta - \eta_c|^{\nu z}$, where $\nu z \approx 4.8$. This scaling regime is accessible in experiment, not only by measuring the dynamical conductivity directly, but also by observing that at finite temperature the scaling in ω is cut off by the phase relaxation rate $1/\tau_\phi$.[27] Therefore, assuming a single temperature power law $1/\tau_\phi \sim T^p$, one finds the following scaling law for the temperature dependent d.c. conductivity

$$\sigma(T;\eta) \sim T^{p/3}\Phi_T(\xi T^{p/3}). \tag{2.19}$$

Using this scaling law, one may in principle determine the critical exponent ν from the temperature dependence of the conductivity in the vicinity of the critical point. In the case of disordered metals or semiconductors, where studies of this type have been performed, the effect of electron-electron interaction has to be taken into account. One major modification in the above is that the Einstein relation is changed. Namely, the single-particle density of states (which is not critical) is replaced by the compressibility $\partial n/\partial \mu$, with n as the density and μ as the chemical potential, which in the presence of the long-range Coulomb interaction is expected to vanish at the transition, i.e., the system becomes incompressible. Another change is that the frequency cutoff is given by the temperature. The critical exponents determined from experiment vary widely, from $s = 0.5$ (Ref. 28) and $s = 1$ (Ref. 29) to $s = 1.6$ (Refs. 30 and 31), and from $z = 2$ (Ref. 30) to $z = 2.94$ (Ref. 31).

3. Renormalized Perturbation Theory of Quantum Transport in Disordered Media

The field-theoretic description in terms of the nonlinear σ model (NLσM) mentioned in the beginning of Sec. 2 is believed to be an exact framework within which the critical properties of the Anderson transition may be, in principle, calculated exactly. The mapping of the initial microscopic model onto the NLσM requires a number of simplifications, so that the noncritical properties like the critical disorder η_c, the behavior in anisotropic systems, or systems of finite extension are no longer well represented by this model. In addition, it is not known how to solve the NLσM in cases of major interest, such as in $d = 3$ dimensions.

It is therefore useful to consider approximation schemes, which on one hand keep the information about the specific properties of the system and on the other hand account approximately for the critical properties at the transition. Such a scheme, the self-consistent theory of Anderson localization, is available at least for the orthogonal ensemble (in which both, time reversal and spin rotation symmetry are conserved). This approach has been developed by us in Refs. 32 and 33 and was reviewed in Ref. 34. It may be termed "self-consistent one-loop approximation" in the language of renormalization group theory but has, in fact, been derived following a somewhat different logic as will be discussed below.

The appropriate language to formulate a microscopic theory of quantum transport or wave transport in disordered media is a renormalized perturbation theory in the disorder potential. The building blocks of this theory for the model defined by Eq. (2) are (i) the renormalized one-particle retarded (advanced) Green's functions averaged over disorder

$$G_{\boldsymbol{k}}^{R,A}(E) = [E - k^2/2m - \Sigma_{\boldsymbol{k}}^{R,A}(E)]^{-1}, \tag{3.1}$$

where $\Sigma_{\boldsymbol{k}}^{R}(E) = (\Sigma_{\boldsymbol{k}}^{A}(E))^*$ is the self-energy, and (ii) the random potential correlator $\langle V^2 \rangle$. The self-energy Σ is a non-critical quantity and can be approximated by $\Sigma_{\boldsymbol{k}}^{R}(E) \simeq -i/2\tau$, where $1/\tau$ is the momentum relaxation rate entering the Drude formula of the conductivity and isotropic scattering is assumed.

The quantity of central interest here is the diffusion coefficient D. It follows from very general considerations[35] that the density-response function describing the change in density caused by an external space and time dependent chemical potential is given by

$$\chi(\boldsymbol{q},\omega) = \frac{D(\boldsymbol{q},\omega)q^2}{-i\omega + D(\boldsymbol{q},\omega)q^2}\chi_0, \tag{3.2}$$

where $D(\boldsymbol{q},\omega)$ is a generalized diffusion coefficient. The static susceptibility (which is non-critical in the model of non-interacting particles) is given by $\chi_0 = N_F$, where N_F is the density of states at the Fermi level. The form of χ is dictated by particle number conservation and may be expressed in terms of $G^{R,A}$ as

$$\chi(\boldsymbol{q},\omega) = -\frac{\omega}{2\pi i} \sum_{\boldsymbol{k},\boldsymbol{k}'} \Phi_{\boldsymbol{k}\boldsymbol{k}'}(\boldsymbol{q},\omega) + \chi_0. \tag{3.3}$$

The two-particle quantity

$$\Phi_{\boldsymbol{k}\boldsymbol{k}'}(\boldsymbol{q},\omega) = \left\langle G_{k_+,k'_+}^{R} G_{k_-,k'_-}^{A} \right\rangle, \tag{3.4}$$

where $G_{k,k'}^{R,A}$ are non-averaged single-particle Green's functions, $k_{\pm} = (\boldsymbol{k} \pm \boldsymbol{q}/2, E \pm \omega/2)$, and the angular brackets denote averaging over disorder, may be written in terms of the irreducible vertex function U as

$$\Phi_{\boldsymbol{k}\boldsymbol{k}'}(\boldsymbol{q},\omega) = G_{k_+}^{R} G_{k_-}^{A} \left[\delta_{\boldsymbol{k},\boldsymbol{k}'} + \sum_{\boldsymbol{k}''} U_{\boldsymbol{k}\boldsymbol{k}''}(\boldsymbol{q},\omega)\Phi_{\boldsymbol{k}''\boldsymbol{k}'}(\boldsymbol{q},\omega) \right]. \tag{3.5}$$

In a diagrammatic formulation the vertex function U is given by the sum of all particle-hole irreducible diagrams of the four-point vertex function. By expressing $G^R G^A$ as

$$G^R_{k_+} G^A_{k_-} = \frac{\Delta G_k}{\omega - k \cdot q/m - \Delta \Sigma_k}, \tag{3.6}$$

where $\Delta G_k = G^R_{k_+} - G^A_{k_-}$ and $\Delta \Sigma_k = \Sigma^R_{k_+} - \Sigma^A_{k_-}$ one may rewrite Eq. (3.5) in the form of a kinetic equation

$$\left(\omega - \frac{k \cdot q}{m} - \Delta \Sigma_k\right) \Phi_{kk'} = -\Delta G_k \left[\delta_{kk'} + \sum_{k''} U_{kk''} \Phi_{k''k'}\right]. \tag{3.7}$$

By summing Eq. (3.7) over k, k' one finds the continuity equation

$$\omega \Phi(q, \omega) - q \Phi_j(q, \omega) = 2\pi i N_F \tag{3.8}$$

with the density-relaxation function

$$\Phi(q, \omega) = \sum_{k,k'} \Phi_{kk'}(q, \omega), \tag{3.9}$$

and the current-density relaxation function

$$\Phi_j(q, \omega) = \sum_{k,k'} \frac{k \cdot \hat{q}}{m} \Phi_{kk'}(q, \omega), \tag{3.10}$$

where $\hat{q} = q/\mid q \mid$. Here the Ward identity $\Delta \Sigma_k = \sum_{k'} U_{kk'} \Delta G_{k'} = \sum_{k'} U_{k'k} \Delta G_{k'}$ has been used.[33] Since the Ward identity plays a central role in the derivation of the self-consistent equation, we provide a short proof which does not rely on the perturbation expansion employed in Ref. 33. Instead the proof follows the derivation of a similar Ward identity in the case of wave propagation in disordered media.[36] Starting from the equations of motion of the single particle Green's function before impurity averaging

$$\left[E + \frac{\omega}{2} + i0 + \frac{1}{2m}\nabla^2_{r_1} - V(r_1)\right] G^R\left(r_1, r_2; E + \frac{\omega}{2}\right) = \delta(r_1 - r_2), \tag{3.11}$$

$$\left[E - \frac{\omega}{2} - i0 + \frac{1}{2m}\nabla^2_{r_3} - V(r_3)\right] G^A\left(r_3, r_4; E - \frac{\omega}{2}\right) = \delta(r_3 - r_4) \tag{3.12}$$

we multiply the first of these equations by $G^A(r_3, r_4; E - \frac{\omega}{2})$ and the second by $G^R(r_1, r_2; E + \frac{\omega}{2})$ and take the difference. We now perform the limit $r_1 \to r_3$, upon which the terms containing the disorder potential $V(r_i)$, $i = 1, 3$, cancels out. Finally, the disorder average is taken and the result is Fourier transformed into momentum space, with the result

$$\sum_{k}(\omega - \frac{k \cdot \hat{q}}{m})\Phi_{kk'}(q,\omega) = G^R_{k'_+} - G^A_{k'_-} \tag{3.13}$$

Comparing with Eq. (3.7) it is seen that the Ward identity indeed holds.

In the hydrodynamic limit, i.e., $\omega\tau \ll 1$, $q\ell \ll 1$, the current density is proportional to the gradient of the density, which is expressed in Fourier space by

$$\Phi_j + iqD(q,\omega)\Phi = 0. \tag{3.14}$$

In fact, multiplying Eq. (3.7) by $k \cdot \hat{q}/m$ and summing over k and k', one may derive relation (3.14) and by comparison finds

$$D_0/D(q,\omega) = 1 - \eta\frac{2E}{mn}\sum_{k,k'}(k\cdot\hat{q})G^R_{k_+} G^A_{k_-} U_{kk'}(q,\omega)G^R_{k'_+} G^A_{k'_-}(k'\cdot\hat{q}), \tag{3.15}$$

where $\eta = \pi N_F\langle V^2\rangle = \frac{1}{2\pi E\tau}$ is the disorder parameter, and $D_0 = \frac{1}{d}v^2\tau$ is the bare diffusion constant.

As the Anderson transition is approached the left-hand-side of Eq. (3.15) will diverge for $q, \omega \to 0$, and therefore the irreducible vertex U has to diverge, too. The leading divergent contribution to U is given by the set of diagrams obtained by using the following property of the full vertex function Γ (the sum of all four-point vertex diagrams) in the presence of time-reversal symmetry[32,33]:

$$\Gamma_{kk'}(q,\omega) = \Gamma_{(k-k'+q)/2,(k'-k+q)/2}(k+k',\omega). \tag{3.16}$$

This relation follows if one twists the particle-hole (p-h) diagrams of Γ such that the lower line has its direction reversed, i.e., the diagram becomes a particle-particle (p-p) diagram. Now, if time-reversal symmetry holds, one may reverse the arrow on the lower Green's function lines if one lets $k \to -k$ at the same time. This operation transforms p-p-diagrams back into p-h diagrams, so that an identity is established relating each diagram of Γ to its transformed diagram Γ^T, which yields the above relation.

The leading singular diagrams of Γ give rise to the diffusion pole

$$\Gamma_D = \frac{1}{2\pi N_F\tau^2}\frac{1}{-i\omega + Dq^2}, \tag{3.17}$$

where D is the renormalized diffusion coefficient. These diagrams are of the ladder-type and therefore reducible. Their transformed counterparts

Γ_D^T are, however, irreducible and thus contribute to U. We may therefore approximate the singular part of U by

$$U_{\boldsymbol{kk'}}^{\text{sing}} = \frac{1}{2\pi N_F \tau^2} \frac{1}{-i\omega + D(\boldsymbol{k} + \boldsymbol{k'})^2}. \tag{3.18}$$

In low-order perturbation theory U^{sing} is given by the "maximally crossed diagrams", which when summed up give a result $U^{\text{sing},0}$ similar to Eq. (3.18), with D replaced by the diffusion constant D_0. When $U^{\text{sing},0}$ is substituted as a vertex correction into the conductivity diagram, the result is exactly the weak-localization correction discussed in Sec. 1.3. The structure of the kernel $U_{\boldsymbol{kk'}}$ has been analyzed from a general viewpoint in Ref. 37. The importance of the diffusion pole for the Anderson localization problem was discussed in Refs. 38 and 39 in connection with the derivation of mean-field theories for disordered systems in the limit of high spatial dimensions.

4. Self-Consistent Theory of Anderson Localization

It follows from Eq. (3.15) that for $d \leq 2$ even the lowest-order correction in the disorder parameter η to the inverse diffusion coefficient (obtained by replacing $U_{\boldsymbol{kk'}}$ by $U^{\text{sing},0}$) yields a contribution which, in principle, diverges in the limit $\omega \to 0$. This infrared divergence depends crucially on the dimension d and leads to a breakdown of perturbation theory in dimensions $d \leq 2$. In higher dimensions the divergence takes place at finite disorder strength. Since the fundamental reason for the divergence of $D_0/D(0,0)$, Eq. (3.15), is the presence of diffusion poles in the kernel $U_{\boldsymbol{kk'}}$, and since these diffusion poles depend on the renormalized diffusion coefficient, Vollhardt and Wölfle[32,33] interpreted Eq. (3.15) as a self-consistent equation for the diffusion coefficient.

By construction Eq. (3.15) is in agreement with perturbation theory. An earlier attempt to set up a self-consistent equation in the spirit of mode-mode coupling theory[40] failed to reproduce the weak localization results, as it did not account for quantum interference effects. A later *ad hoc* modification of the latter theory led to a self-consistency scheme[41,42] which is in partial agreement with the one presented here, the main difference being that an additional classical (i.e., not interference related) mechanism of localization is included.

When U^{sing} from Eq. (3.18) is substituted for U, Eq. (3.15) for the diffusion coefficient $D(\omega)$ (i.e., in the limit $q \to 0$) leads to the following self-consistent equation for the frequency-dependent diffusion coefficient

$D(\omega)$[32,33]:

$$\frac{D_0}{D(\omega)} = 1 + \frac{k_F^{2-d}}{\pi m} \int_0^{1/\ell} dQ \frac{Q^{d-1}}{-i\omega + D(\omega)Q^2}. \tag{4.1}$$

Here we assumed that a finite limit $\lim_{q\to 0} D(q,\omega) = D(\omega)$ exists, and that Q is limited to $1/\ell$ in the diffusive regime.

Equation (4.1) may be re-expressed as

$$\frac{D(\omega)}{D_0} = 1 - \eta d k_F^{2-d} \int_0^{1/\ell} dQ \frac{Q^{d-1}}{-i\omega/D(\omega) + Q^2}. \tag{4.2}$$

4.1. Results of the self-consistent theory of Anderson localization

In $d = 3$, Eq. (3.14) has a solution in the limit $\omega \to 0$ up to a critical disorder strength η_c

$$D = D_0(1 - \frac{\eta}{\eta_c}), \quad \eta < \eta_c = \frac{1}{\sqrt{3\pi}}, \tag{4.3}$$

which implies the critical exponent of the conductivity $s = 1$. The ω-dependence of $D(\omega)$ at the critical point is obtained as[43]

$$D(\omega) = D_0(\omega\tau)^{1/3}, \quad \eta = \eta_c, \tag{4.4}$$

implying a dynamical critical exponent $z = 3$ in agreement with the exact result of Wegner.[19]

At stronger disorder, $\eta > \eta_c$, all states are found to be localized. The localization length ξ, defined by $\xi^{-2} = \lim_{\omega\to 0}(-i\omega/D(\omega))$, is found as

$$\xi = \frac{\sqrt{\pi}}{2}\ell \left|1 - \frac{\eta}{\eta_c}\right|^{-1}, \tag{4.5}$$

i.e., the exponent is $\nu = 1$. For general d in the interval $2 < d < 4$ one finds Wegner scaling, $s = \nu(d - 2)$. An extension of the self-consistent theory with respect to the momentum dependence of the renormalized diffusion coefficient near the Anderson transition has been proposed in Ref. 44. It leads to a modified critical exponent of the localization exponent, $\nu = 1/(d-2) + 1/2$, which is in much better agreement with numerical results in $d = 3$. The conductivity exponent is found to be unchanged ($s = 1$), i.e., Wegner scaling is no longer obeyed.

In dimensions $d \leq 2$, there is no metallic-type solution. The localization length is found as

$$\xi = \ell \left[\exp \frac{1}{\eta} - 1 \right]^{1/2}, \quad d = 2$$
$$\xi \cong c_1 \ell, \quad d = 1 \tag{4.6}$$

where the coefficient $c_1 \approx 2.6$, while the exact result is $c_1 = 4$.[11]

The β-function has been derived from the self-consistent equation for the length-dependent diffusion coefficient, where a lower cutoff $1/L$ has been applied to the Q-integral in Eq. (4.2). The result[44] for $d = 3$ dimensions in the metallic regime is given by

$$\beta(g) = \frac{g - g_c}{g}, \quad y > y_c = \frac{1}{\pi^2}, \tag{4.7}$$

and in the localized regime by

$$\beta(g) = 1 - \frac{1}{\pi^2 g} \frac{1+x}{1+x^2} e^{-x} - \frac{x^2}{1+x}, \quad g < g_c. \tag{4.8}$$

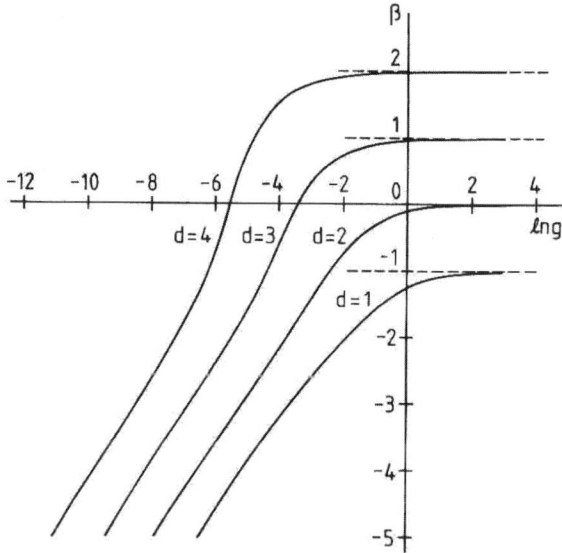

Fig. 4.1.   Renormalization group β-function in dimensions $d = 1, 2, 3$ for the orthogonal ensemble, as obtained from the self-consistent theory.[44]

Here $x = x(g)$ is the inverse function of

$$g = \frac{1}{\pi^2}(1+x)e^{-x}\left(1 - x\arctan\frac{1}{x}\right). \tag{4.9}$$

The β-functions in $d = 1, 2, 3$ obtained in this way are shown in Fig. 4.1.

The phase boundary separating localized and extended states in a disordered three-dimensional system may be determined approximately by a variety of methods. For electrons on a cubic lattice with nearest-neighbor hopping and one orbital per site with random energy ϵ_i chosen from a box distribution in the interval $[-W/2, W/2]$, the phase diagram has been determined by numerical simulations[45] as shown in Fig. 4.2.

Also shown is the result of an analytic expression obtained from the self-consistent theory[46] applied to a tight-binding model, where the coherent potential approximation (CPA) was used to evaluate the single-particle properties; no adjustable parameters enter. The agreement is seen to be very good.

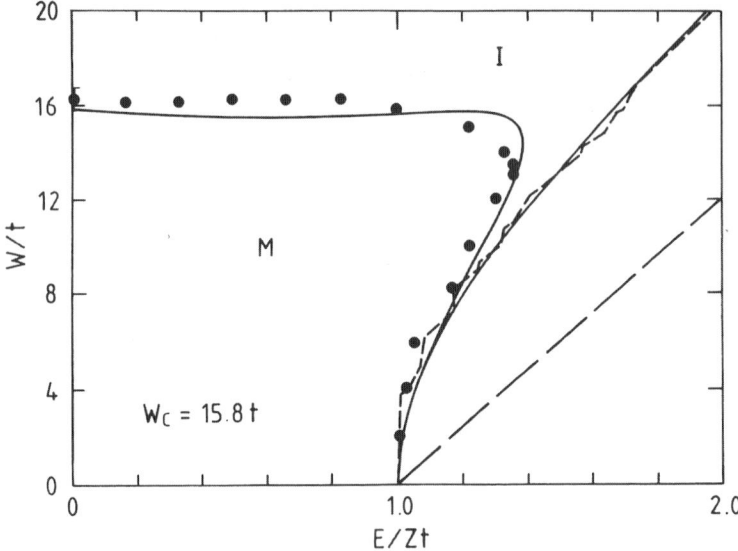

Fig. 4.2. Phase diagram showing metallic (M) and insulating (I) regions of the tight-binding model with site-diagonal disorder (box distribution of width W). Dots: numerical study;[4] solid line: self-consistent theory.[46] The remaining lines are bounds on the energy spectrum; see Ref. 34.

5. Applications of the Self-Consistent Theory of Anderson Localization

The self-consistent theory of Anderson localization proposed by us in 1980[32,33] was applied and extended to account for many of the salient features of disordered systems. Here we briefly review the more recent developments, not yet described in our review.[34] While initially the main interest had focussed on disordered electronic systems, in recent years the interest shifted to localization of classical waves and even more recently, to ultracold atom systems. We first review an extension of the self-consistent theory to the case of weak applied magnetic and electric fields.

5.1. *Effect of static magnetic and electric fields*

5.1.1. *Magnetic fields*

One of the limitations of the self-consistent theory has been the difficulty to treat scale dependent contributions to the conductivity in the presence of a magnetic field in perturbation theory. As explained above, a magnetic field induces a "mass" in the Cooperon propagator and therefore removes the localizing interference effect leading to localization of all states in $d \leq 2$ dimensions. On the other hand, studies of the nonlinear σ-model show that in higher (two-loop) order scale dependent terms appear which are generated solely by diffusion propagators (diffusons). There is, however, a general theorem of perturbation theory, related to gauge invariance, stating that the singular contribution of any diagram with one diffuson and an arbitrary decoration with additional impurity lines cancels within a group of related diagrams.[33] The way out of this apparent contradiction has only been found very recently.[47]

A satisfactory generalized self-consistent theory for the case of unitary symmetry, including the two-loop and higher contributions has not been formulated yet. Nonetheless there is a parameter regime of weak magnetic field B ($\omega_c = eB/mc$) and moderately strong disorder, $\omega_c \tau << 1/\epsilon_F \tau \lesssim 1$ in which the one-loop contributions still dominate over the two-loop contributions and a generalized self-consistent theory may be formulated. The most complete discussion of this approach was given by Bryksin and Kleinert,[48] who proposed a set of two coupled self-consistent equations for the diffusion coefficients D_{ph} in the particle-hole channel (diffuson) and D_{pp} in the particle-particle channel (Cooperon) of a $2d$ system:

$$\frac{D_{pp}}{D_0} = 1 - g\left[\psi\left(\frac{1}{2} + \ell_B^2\kappa^2 + \frac{\ell_B^2}{4\tau_\phi D_{ph}}\right) - \psi\left(\frac{1}{2} + \frac{\ell_B^2}{4\tau_\phi D_{ph}}\right)\right], \quad (5.1)$$

$$\frac{D_{ph}}{D_0} = 1 - g\ln\left(1 + \frac{\tau_\phi}{\tau}\frac{D_{pp}}{D_0}\right). \quad (5.2)$$

Here $\psi(z)$ is the digamma function, $g = 2/(\pi k_F \ell)$ is the coupling constant, $\ell_B = (c/eB)^{1/2}$ is the magnetic length, and $\kappa = 1/(\sqrt{2}\ell)$. The solution of these equations allows one to extend the results of weak localization theory, e.g., for the negative magnetoresistance, to the regime of moderately strong disorder, leading to renormalized values of the parameters of weak localization theory. Good agreement has been found with experimental data in that range.[48]

5.1.2. *Electric fields*

An applied static electric field $\boldsymbol{E}$ affects the localization physics in the following way: electrons drifting under the influence of $\boldsymbol{E}$ experience a reduced probability of return, weakening the localization effect provided by interference of return paths. This effect is incorporated into the Cooperon dynamics, leading to a new term in the diffusion pole

$$\Gamma_D = \frac{1}{2\pi N_F \tau^2}\frac{1}{-i\omega + Dq^2 + i\mu_d \boldsymbol{q}\cdot\boldsymbol{E}}, \quad (5.3)$$

where $\mu_d = e/(m\tau)$ is the mobility. The electric field term leads to the appearance of a localization transition even in dimension $d = 1, 2$. Near the transition in $d = 1$ the diffusion coefficient is found as[48]

$$D(E) = \begin{cases} D_0(1 - E_0/E), & \text{for } E > E_0 \\ 0 & \text{for } E_0 < E, \end{cases} \quad (5.4)$$

in agreement with the exact result in Ref. 49, where $E_0 = (2n/(e\pi N_F^2 D_0)$ is the threshold field. In $d = 2$ dimensions, the behavior above threshold is logarithmic:

$$D(E) = \begin{cases} D_0\ln(E/E_0), & \text{for } E > E_0 \\ 0 & \text{for } E_0 < E, \end{cases} \quad (5.5)$$

where $E_0 = (4\epsilon_F/e\kappa)\exp(-\pi k_F\ell/2)$. The relaxation of the charge current following a sudden switch on of the electric field has been considered in Ref. 50. There it was found that the current has a long time tail $\propto t^{-1/2}$ as a consequence of the infrared singular behavior of the Cooperon pole.

The way in which electric and magnetic fields affect transport near the localization transition in anisotropic systems was studied in Refs. 51 and 52.

5.2. *Anisotropic systems, films and wires*

The question of how the scaling properties of the conductance are modified in anisotropic systems was first addressed in Ref. 53. There it was established that even in the presence of an anisotropic electronic band structure and an anisotropic impurity-scattering cross section, the one-parameter scaling theory holds. The ratios of the components of the conductivity tensor are invariant under scaling, implying that the geometric mean of the conductivity components plays the role of the scaling quantity. This feature is preserved by the self-consistent theory. Numerical studies of anisotropic systems[54,55] appeared to cast doubts on the one-parameter scaling hypothesis. However, a later more careful study of the problem in $d = 2$ dimensions showed that indeed one-parameter scaling is obeyed[56]: the ratio of the localization lengths (in the direction of the principal axes) turns out to be proportional to the square root of the ratio of the conductivities. A comparison with the self-consistent theory in the somewhat simpler form of the "potential-well analogy"[57] showed again qualitative agreement. Localization in anisotropic systems has also been considered in a model with anisotropic random potential correlations, and the phase diagram has been mapped out within an extension of the self-consistent theory.[58] The same authors explored the consequences of finite-range correlations of the random potential within a generalization of the self-consistent theory.[59]

A somewhat different but related question is the behavior of the conductance of a film of finite thickness, or a wire of finite diameter. There is no doubt that in the thermodynamic limit these systems behave like true $2d$ or $1d$ systems. It is, however, interesting to understand how this behavior is approached. Numerical studies of metallic disordered films as a function of film thickness seemed to indicate a localization transition as a function of thickness,[60] in contradiction to the results of the self-consistent theory applied to this system. A further study by the same authors[61] on systems of finite thickness in a magnetic field explored the possibility of a delocalization transition controlled by both the thickness and the magnetic field. The transitions obtained are pseudo-transitions marking a crossover from strong to weak localization, as confirmed in a later more accurate numerical study.[62]

5.3. *Anderson localization of classical waves*

The concept of the self-consistent theory of localization can be carried over to the case of propagation of classical waves in disordered media. Here we sketch the formulation following the presentation of Kroha, Soukoulis, and Wölfle.[63] For scalar waves propagating in a medium of randomly positioned point scatterers of density $n_I = a^{-3}$, modelled by spheres of volume V_s, the average phase velocity is given by $c_{ph} = c_0[1 + (V_s/a^3)\Delta\epsilon]^{-1/2}$, where c_0 is the bare phase velocity and $\Delta\epsilon$ characterizes the strength of the scattering ("dielectric contrast"). The Green's function of the wave equation is defined as

$$G_{\boldsymbol{k}}(\omega) = [G_0^{-1}(\omega) - \Sigma_{\boldsymbol{k}}(\omega)], \qquad (5.6)$$

where $G_0^{-1}(\omega) = \omega^2 - c_0^2 k^2$. The self-energy Σ may be determined within the CPA (see, e.g., Ref. 64) provided it is independent of k. Then the bare diffusion constant is found as

$$D_0 = 2c(\omega)\frac{c_0}{\omega}G_0^{-1}(\omega)\sum_{\boldsymbol{k}}(\boldsymbol{k}\cdot\widehat{\boldsymbol{q}})^2(\mathrm{Im}G_{\boldsymbol{k}}^A)^2. \qquad (5.7)$$

The renormalized diffusion coefficient may be shown to satisfy the self-consistency equation

$$D(\Omega) = D_0 - 2\left[c(\omega)\frac{c_0}{\omega}\right]^2 \frac{\mathrm{Im}\Sigma}{(\mathrm{Im}G_0)^2}\frac{D(\Omega)}{D_0}\sum_{\boldsymbol{k},\boldsymbol{k}'}(\boldsymbol{k}\cdot\widehat{\boldsymbol{q}})\frac{\mathrm{Im}G_{\boldsymbol{k}}(\mathrm{Im}G_{\boldsymbol{k}'})^2}{-i\Omega + D(\Omega)(\boldsymbol{k}+\boldsymbol{k}')^2}(\boldsymbol{k}'\cdot\widehat{\boldsymbol{q}}).$$
$$(5.8)$$

Here Ω is the external frequency while ω is the frequency of the waves which enter in one-particle quantities. This equation can be solved in the limit $\Omega \to 0$ to obtain the diffusion coefficient in the delocalized phase and the localization length $\xi = \lim_{\Omega\to 0}[D(\Omega)/(-i\Omega)]^{1/2}$ in the localized phase. One finds that it is much harder to localize classical waves as compared to electrons, and there is only a narrow region of the phase diagram (at reasonable contrast $\Delta\epsilon$) where localization is found.[63]

A more realistic theory of the propagation of electromagnetic waves in disordered materials with loss or gain mechanisms keeping the vector character of the fields has been worked out by Lubatsch, Kroha, and Busch.[65] We briefly sketch the main results here. The electric field amplitude $\boldsymbol{E}_\omega(\boldsymbol{r})$ of an electromagnetic wave of frequency ω in a medium with random dielectric constant $\epsilon(\boldsymbol{r};\omega) = \bar{\epsilon}(\omega) + \Delta\epsilon(\boldsymbol{r};\omega)$, obeys the wave equation

$$\nabla \times (\nabla \times \boldsymbol{E}_\omega(\boldsymbol{r})) - \frac{\omega^2}{c^2}\epsilon(\boldsymbol{r};\omega)\boldsymbol{E}_\omega(\boldsymbol{r}) = \omega\boldsymbol{J}_\omega(\boldsymbol{r}), \qquad (5.9)$$

where $\langle\Delta\epsilon(\boldsymbol{r};\omega)\rangle = 0$. In the following, the random part of the dielectric function will be modelled as $\Delta\epsilon(\boldsymbol{r};\omega) = -(c^2/\omega^2)h(\omega)V(\boldsymbol{r})$. The Green's functions of the wave equation, after disorder averaging, are defined as

$$G_{\bm{k}}^{R,A}(\omega) = \left[\left(\frac{\omega^2}{c^2} \bar{\epsilon} - k^2 \right) \bm{P} - \bm{\Sigma}_{\bm{k}}^{R,A}(\omega) \right]^{-1}. \tag{5.10}$$

Here $\bm{G}$ and the self-energy $\bm{\Sigma}$ are (3×3) tensors and $\bm{P} = \bm{1} - \hat{\bm{k}} \otimes \hat{\bm{k}}$ is the projector onto the transverse subspace (here and in the following the hat symbol denotes a unit vector). Transport properties are contained in the two-particle correlation function (a tensor of rank four)

$$\bm{\Phi}_{\bm{kk'}}(\bm{q}, \Omega) = \left\langle \bm{G}_{k_+, k'_+}^{R} \otimes \bm{G}_{k_-, k'_-}^{A} \right\rangle, \tag{5.11}$$

where $k_\pm = (\bm{k} \pm \bm{q}/2, \omega \pm \Omega/2)$, etc., which obeys the Bethe–Salpeter equation

$$\bm{\Phi}_{\bm{kk'}}(\bm{q}, \Omega) = \bm{G}_{k_+}^{R} \otimes \bm{G}_{k_-}^{A} \left[\delta_{\bm{kk'}} + \sum_{\bm{k''}} \bm{U}_{\bm{kk''}}(\bm{q}, \Omega) \bm{\Phi}_{\bm{k''k'}}(\bm{q}, \Omega) \right]. \tag{5.12}$$

As in the case of electrons in a random potential considered above, the Bethe–Salpeter equation may be converted into a kinetic equation for the integrated intensity correlation tensor $\bm{\Phi}_{\bm{k}}(\bm{q}, \Omega) = \sum_{\bm{k'}} \bm{\Phi}_{\bm{kk'}}(\bm{q}, \Omega)$ of the form

$$\left(\Delta \bm{G}_{\bm{k},0}^{-1}(\omega) - \Delta \bm{\Sigma}_{\bm{k}} \right) \bm{\Phi}_{\bm{k}} = -\Delta \bm{G}_{\bm{k}} \left[\bm{1} \otimes \bm{1} + \sum_{\bm{k''}} \bm{U}_{\bm{kk''}} \bm{\Phi}_{\bm{k''}} \right], \tag{5.13}$$

where $\Delta \bm{G}_{\bm{k},0}^{-1}(\omega) = [\bm{G}_{k_+,0}^{R}]^{-1} \otimes \bm{1} - \bm{1} \otimes [\bm{G}_{k_-,0}^{A}]^{-1}$, $\Delta \bm{\Sigma}_{\bm{k}} = \bm{\Sigma}_{k_+}^{R} \otimes \bm{1} - \bm{1} \otimes \bm{\Sigma}_{k_-}^{A}$, and $\Delta \bm{G}_{\bm{k}} = \bm{G}_{k_+}^{R} \otimes \bm{1} - \bm{1} \otimes \bm{G}_{k_-}^{A}$.

The kinetic equation serves to derive the energy conservation equation and the equivalent of Fick's law:

$$\left[\Omega + \frac{i}{\tau_L(\Omega)} \right] P_E(\bm{q}, \Omega) + \bm{q} \cdot \bm{J}_E(\bm{q}, \Omega) = S(\bm{q}, \Omega) \tag{5.14}$$

$$\bm{J}_E(\bm{q}, \Omega) = i P_E(\bm{q}, \Omega) \bm{D}(\Omega) \cdot \bm{q}. \tag{5.15}$$

Here

$$P_E(\bm{q}, \Omega) = (\omega/c_p)^2 \sum_{\bm{k}} \bm{\Phi}_{\bm{k}}(\bm{q}, \Omega) \tag{5.16}$$

is the energy-density relaxation function, with c_p as the renormalized phase velocity, and

$$\bm{J}_E(\bm{q}, \Omega) = (\omega/c_p) v_E(\omega) \sum_{\bm{k}} (\bm{k} \cdot \hat{\bm{q}}) \bm{\Phi}_{\bm{k}}(\bm{q}, \Omega) \tag{5.17}$$

is the energy-current density relaxation function, with $\bm{v}_E(\omega)$ as the energy transport velocity; for the definitions of c_p and $\bm{v}_E(\omega)$ we refer the reader to

Ref. 65. When energy absorption by the medium is taken into account (as expressed by the imaginary part of the dielectric function), or conversely, if a medium with gain is considered, energy is not conserved, as expressed by the loss/gain rate $\frac{1}{\tau_L(\Omega)}$. The energy-diffusion coefficient tensor $\boldsymbol{D}(\Omega)$ is found as

$$\boldsymbol{D}(\Omega) = \frac{1}{3}v_E(\omega)\boldsymbol{l}_T, \qquad (5.18)$$

where the tensor of transport mean free path is given by

$$\boldsymbol{l}_T = \frac{c_p}{\omega}(A + \kappa)\boldsymbol{l}. \qquad (5.19)$$

Here the main contribution to $\boldsymbol{l}$ has a form which is analogous to Eq. (3.15):

$$\boldsymbol{l}^{-1} = a_1^{-1}\sum_{\boldsymbol{k},\boldsymbol{k}'}(\boldsymbol{k}\cdot\hat{\boldsymbol{q}})\Delta G_{\boldsymbol{k}}U_{\boldsymbol{k}\boldsymbol{k}''}(0,\Omega)\Delta G_{\boldsymbol{k}'}(\boldsymbol{k}'\cdot\hat{\boldsymbol{q}}), \qquad (5.20)$$

and a_1 and A are defined in Ref. 65. The quantity κ describes scattering caused by a mismatch of absorption/gain between the scattering objects and the medium.

The energy density propagator $P_E(\boldsymbol{q},\Omega)$ in the limit of small $\boldsymbol{q},\Omega$ follows from Eqs. (5.14), (5.15) as

$$P_E(\boldsymbol{q},\Omega) = \left[\Omega + \frac{i}{\tau_L(\Omega)} + i\boldsymbol{q}\cdot\boldsymbol{D}(\Omega)\cdot\boldsymbol{q}\right]^{-1}S(\boldsymbol{q},\Omega). \qquad (5.21)$$

Replacing $U_{\boldsymbol{k}\boldsymbol{k}''}(0,\Omega)$ by its singular part proportional to the diffusion propagator $P_E(\boldsymbol{q},\Omega)$, one arrives at a self-consistent equation for the diffusion coefficient tensor. The latter provides a framework for the description of the interplay between localization and stimulated emission in materials with gain, i.e., the problem of the random laser.[66]

The predictions of the self-consistent theory have also been probed by comparison with numerical results[67] for transmission of waves in unbounded 1d and 2d systems and through strips of finite width. Good overall agreement is found.

The localization of phonons and the ultrasound attenuation in layered crystals with random impurities has been studied within the self-consistent theory in Ref. 68.

5.4. *Transport through open interfaces*

Most of the discussion so far considered transport in infinitely extended systems, with the exception of the scaling theory for systems of length L. In some cases, however, transport through plate-shaped systems in the direction perpendicular to the plate surface is of interest. As pointed out by van Tiggelen and collaborators,[69,70] the weak localization physics changes near an open boundary, as the finite probability of escape through the interface diminishes the return probability necessary for interference. In the framework of the self-consistent theory, this effect may be taken into account quantitatively. To this end it is useful to express Eq. (4.1) in position-energy space as

$$\frac{D_0}{D(\omega)} = 1 + 2\pi \frac{k_F^{2-d}}{m} C(r, r'), \qquad (5.22)$$

where $C(r, r')$ is a solution of the diffusion equation, and a cut-off $Q < 1/\ell$ to the momentum was applied to the spectrum of the Q-modes in Eq. (4.1)

$$[-i\omega + D(\omega)\nabla^2]C(r, r') = \delta(r - r'). \qquad (5.23)$$

The above formulation now allows one to describe position-dependent diffusion processes, as they appear near the sample surface in a confined geometry, e.g., transmission through a slab. In that case the diffusion coefficient may be assumed to be position dependent, $D = D(r, \omega)$. Then $C(r, r\prime)$ obeys the modified diffusion equation[71]

$$[-i\omega + \nabla D(r, \omega)\nabla]C(r, r') = \delta(r - r'). \qquad (5.24)$$

The solution is subject to an appropriate boundary condition at the surface of the sample. A microscopic derivation of the above equation in diagrammatic language was given in Ref. 72. Further confirmation of the theory was obtained in Ref. 73, where the above equations was derived within the nonlinear σ-model framework in the weak coupling limit. The theory accounts very well for the localization properties of accoustic waves transmitted through a strongly scattering plate.[71]

It is natural to ask whether a position dependent diffusion coefficient will change the critical behavior obtained from the scaling properties of the conductance of finite size samples. This question was addressed in Ref. 74 with the result that the critical exponents are unchanged and the β-function is hardly modified by the improved approximation. The scaling of the transmission coefficient for classical waves through a disordered madium near the Anderson transition was considered within the position dependent self-consistent theory in Ref. 75.

The transmission of microwave pulses through quasi one-dimensional samples has been measured recently and was analyzed in terms of the self-consistent theory.[76] It was found that while the self-consistent theory can account very well for the propagation at intermediate times, it fails at longer times when the transport occurs by hopping between localized regions.

Anderson localization of atoms in a Bose–Einstein condensate released from a trap and subject to a random potential has been considered in the framework of the self-consistent theory in Ref. 77. The authors show that the scaling properties govern the dynamical behavior of the expanding atom cloud, so that the critical exponents determine the power law in time obeyed by the expanding cloud size.

6. Conclusion

Anderson localization in disordered systems continues to be a very lively field of research. Current investigations do not concentrate so much on disordered electrons but on classical waves (light, electromagnetic microwaves, acoustic waves), or ultracold atoms in the presence of disorder. Although the fundamental concepts of Anderson localization are well understood by now, there still remain a number of open questions. Some of them are related to the analytical theory of critical properties near the Anderson transition. Others concern the quantitative description of realistic materials, e.g., the question under which conditions light or acoustic waves become localized. The self-consistent theory of Anderson localization has been, and will continue to be, a versatile tool for the investigation of these problems. It allows one to incorporate the detailed characteristics of the system such as the energy dispersion relation, the particular form of disorder, the shape of the sample, and loss or gain mechanisms in an efficient way. The self-consistent theory is not only applicable to stationary transport problems, but also to dynamical situations such as pulse propagation or the behavior after a sudden switch-on.

As Anderson localization is a wave-interference phenomenon, the limitations of phase coherence are an important subject of study in this context. By now Anderson localization has been observed in many different systems beyond doubt. On the other hand, the observation of the Anderson transition itself is a much more challenging task. Here the recent investigations of classical waves and atomic matter waves offer fascinating, new perspectives which will undoubtedly lead to a deeper understanding of the localization phenomenon.

Acknowledgments

We thank Vaclav Janiš, Hans Kroha, Khandker Muttalib, and Costas Soukoulis for many fruitful discussions. Financial support by the TTR 80 of the Deutsche Forschungsgemeinschaft is gratefully acknowledged.

References

1. P. W. Anderson, *Phys. Rev.* **109**, 1492 (1958).
2. P. A. Lee and T. V. Ramakrishnan, *Rev. Mod. Phys.* **57**, 287 (1985).
3. B. L. Altshuler and A. G. Aronov, in *Electron-Electron Interactions in Disordered Systems*, eds. A. L. Efros and M. Pollak, Vol. 10 of the series *Modern Problems in Condensed Matter Sciences* (North-Holland, Amsterdam, 1985).
4. B. Kramer and A. MacKinnon, *Rep. Prog. Phys.* **56**, 1469 (1993).
5. S. Chakravarty and A. Schmid, *Phys. Rep.* **140**, 193 (1986).
6. E. Abrahams, P. W. Anderson, D. Licciardello and T. V. Ramakrishnan, *Phys. Rev. Lett.* **42**, 673 (1979).
7. L. P. Gorkov, A. I. Larkin and D. E. Khmelnitskii, *JETP Lett.* **30**, 228 (1979).
8. J. van Delft, *Int. J. Mod. Phys. B* **22**, 727 (2008).
9. N. F. Mott and W. D. Twose, *Adv. Phys.* **10**, 107 (1961).
10. R. E. Borland, *Proc. R. Soc. London, Ser. A* **274**, 529 (1963).
11. V. L. Berezinskii, *Sov. Phys.-JETP* **38**, 620 (1974).
12. A. F. Ioffe and A. R. Regel, *Prog. Semicond.* **4**, 237 (1960).
13. N. F. Mott, *Metal-Insulator Transitions* (Taylor and Francis, London, 1974).
14. N. F. Mott and E. A. Davis, *Electronic Processes in Non-Crystalline Materials* (Clarendon Press, Oxford, 1979).
15. A. Georges, G. Kotliar, W. Krauth and M. J. Rozenberg, *Rev. Mod. Phys.* **68**, 13 (1996).
16. F. Kotliar and D. Vollhardt, *Phys. Today* **57**, 53 (2004).
17. K. Byczuk, W. Hofstetter and D. Vollhardt, *Phys. Rev. Lett.* **94**, 056404 (2005).
18. K. Byczuk, W. Hofstetter and D. Vollhardt, to be published in *Fifty Years of Anderson Localization*, ed. E. Abrahams (World Scientific, Singapore, 2010), preprint arXiv:1002.3696.
19. F. Wegner, *Z. Phys. B* **25**, 327 (1976).
20. F. Wegner, *Z. Phys. B* **35**, 207 (1979).
21. K. B. Efetov, *Supersymmetry in Disorder and Chaos* (Cambridge University Press, Cambridge, 1997).
22. T. Nakayama, K. A. Muttalib and P. Wölfle, unpublished.
23. A. M. M. Pruisken, *The Quantum Hall Effect*, eds. S. Prange and S. Girvin (Springer, Berlin, 1987).
24. S. Hikami, A. I. Larkin and Y. Nagaoka, *Progr. Theor. Phys.* **63**, 707 (19810).
25. K. Slevin and T. Ohtsuki, *Phys. Rev. Lett.* **82**, 382 (1999).
26. B. Shapiro and E. Abrahams, *Phys. Rev. B* **24**, 4889 (1981).
27. Y. Imry, Y. Gefen and D. Bergman, *Phys. Rev. B* **26**, 3436 (1982).

28. M. A. Paalanen, T. F. Rosenbaum, G. A. Thomas and R. N. Bhatt, *Phys. Rev. Lett.* **48**, 1284 (1982).
29. S. B. Field and T. F. Rosenbaum, *Phys. Rev. Lett.* **55**, 522 (1985).
30. S. Bogdanovich, M. P. Sarachik and R. N. Bhatt, *Phys. Rev. Lett.* **82**, 137 (1999).
31. S. Waffenschmidt, C. Pfleiderer and H. V. Löhneysen, *Phys. Rev. Lett.* **83**, 3005 (1999).
32. D. Vollhardt and P. Wölfle, *Phys. Rev. Lett.* **45**, 482 (1980).
33. D. Vollhardt and P. Wölfle, *Phys. Rev. B* **22**, 4666 (1980).
34. D. Vollhardt and P. Wölfle, in *Electronic Phase Transitions*, eds. W. Hanke and Yu. V. Kopaev, Vol. 32 of the series *Modern Problems in Condensed Matter Sciences* (North-Holland, Amsterdam, 1992), p. 1.
35. D. Forster, *Hydrodynamic Fluctuations, Broken Symmetry and Correlation Functions* (Benjamin, Reading, 1975).
36. Y. N. Barabanenkov, L. M. Zurk and M. Y. Barabanenkov, *J. Electromagn. Waves Appl.* **9**, 1393 (1995).
37. I. M. Suslov, *JETP Lett.* **105**, 1198 (2007).
38. V. Janiš and J. Kolorenč, *Phys. Rev. B* **71**, 033103 (2005).
39. V. Janiš and J. Kolorenč, *Phys. Rev. B* **71**, 245106 (2005).
40. W. Götze, P. Prelovsek and P. Wölfle, *Solid State Commun.* **29**, 369 (1979).
41. P. Prelovsek, *Phys. Rev. B* **23**, 1304 (1981).
42. D. Belitz, A. Gold and W. Götze, *Z. Physik B* **44**, 273 (1981).
43. B. Shapiro, *Phys. Rev. B* **25**, 4266 (1982).
44. A. M. García-García, *Phys. Rev. Lett.* **100**, 076404 (2008).
45. B. Bulka, M. Schreiber and B. Kramer, *Z. Phys. B* **66**, 21 (1987).
46. J. Kroha, T. Kopp and P. Wölfle, *Phys. Rev. B* **41**, 888 (1990).
47. P. Ostrovskii, unpublished.
48. V. V. Bryksin, H. Schlegel and P. Kleinert, *Phys. Rev. B* **49**, 13697 (1994).
49. V. N. Prigodin, *JETP Lett.* **52**, 1185 (1980).
50. V. V. Bryksin and P. Kleinert, *J. Phys. C* **6**, 7879 (1994).
51. P. Kleinert and V. V. Bryksin, *Phys. Rev. B* **52**, 1649 (1995).
52. V. V. Bryksin and P. Kleinert, *Z. Phys. B* **101**, 91 (1996).
53. P. Wölfle and R. N. Bhatt, *Phys. Rev. B* **30**, R3542 (1984).
54. I. Zambetaki, Q. Li, E. N. Economou and C. M. Soukoulis, *Phys. Rev. Lett.* **76**, 3614 (1996).
55. I. Zambetaki, Q. Li, E. N. Economou and C. M. Soukoulis, *Phys. Rev.* **54**, 12221 (1997).
56. Q. Li, S. Katsoprinakis, E. N. Economou and C. M. Soukoulis, *Phys. Rev.* **56**, R4297 (1997).
57. E. N. Economou and C. M. Soukoulis, *Phys. Rev. B* **28**, 1093 (1983).
58. Q.-J. Chu and Z.-Q. Zhang, *Phys. Rev. B* **48**, 10761 (1993).
59. Q.-J. Chu and Z.-Q. Zhang, *Phys. Rev. B* **39**, 7120 (1989).
60. R. K. B. Singh and B. Kumar, *Phys. Rev. B* **66**, 075123 (2002).
61. R. K. B. Singh and B. Kumar, *Phys. Rev. B* **69**, 115420 (2004).
62. V. Z. Cerovski, R. K. B. Singh and M. Schreiber, *J. Phys.: Cond. Matt.* **18**, 7155 (2006).

63. J. Kroha, C. M. Soukoulis and P. Wölfle, *Phys. Rev. B* **47**, 11093 (1993).
64. C. M. Soukoulis, E. N. Economou, G. S. Grest and M. H. Cohen, *Phys. Rev. Lett.* **62**, 575 (1989).
65. A. Lubatsch, J. Kroha and K. Busch, *Phys. Rev. B* **71**, 184201 (2005).
66. R. Frank, A. Lubatsch and J. Kroha, *Phys. Rev.* **73**, 245107 (2006).
67. O. I. Lobkis and W. Weaver, *Phys. Rev. E* **71**, 011112 (2005).
68. E. P. Chulkin, A. P. Zhernov and T. N. Kalugina, *Low Temp. Phys.* **25**, 912 (1999).
69. S. E. Skipetrov and B. A. van Tiggelen, *Phys. Rev. Lett.* **96**, 043602 (2006).
70. B. A. van Tiggelen, A. Lagendijk and D. S. Wiersma, *Phys. Rev. Lett.* **84**, 4333 (2000).
71. H. Hu, A. Strybulevych, J. H. Page, S. E. Skipetrov and B. A. van Tiggelen, *Nature Phys.* **4**, 945 (2008).
72. N. Cherroret and S. E. Skipetrov, *Phys. Rev. E* **77**, 046608 (2008).
73. C. Tian, *Phys. Rev. B* **77**, 064205 (2008).
74. N. Cherroret, Ph.D. Thesis, Université Joseph Fourier (Grenoble, 2009).
75. N. Cherroret, S. E. Skipetrov and B. van Tiggelen, *Phys. Rev. B* **80**, 037101 (2009).
76. Z. Q. Zhang, A. A. Chabanov, S. K. Cheung, C. H. Wong and A. Z. Genack, *Phys. Rev. B* **79**, 037101 (2009).
77. S. E. Skipetrov, A. Minguzzi, B. A. van Tiggelen and B. Shapiro, *Phys. Rev. Lett.* **100**, 165301 (2008).

Chapter 5

ANDERSON LOCALIZATION AND SUPERSYMMETRY

K. B. Efetov

Theoretische Physik III, Ruhr-Universität Bochum
44780 Bochum, Germany
efetov@tp3.rub.de

The supersymmetry method for study of disordered systems is shortly reviewed. The discussion starts with a historical introduction followed by an explanation of the idea of using Grassmann anticommuting variables for investigating disordered metals. After that the nonlinear supermatrix σ-model is derived. Solution of several problems obtained with the help of the σ-model is presented. This includes the problem of the level statistics in small metal grains, localization in wires and films, and Anderson metal–insulator transition. Calculational schemes developed for studying these problems form the basis of subsequent applications of the supersymmetry approach.

1. Introduction

The prediction of the new phenomenon of the Anderson localization[1] has strongly stimulated both theoretical and experimental study of disordered materials. This work demonstrates the extraordinary intuition of the author that allowed him to make outstanding predictions. At the same time, one could see from that work that quantitative description of the disordered systems was not a simple task and many conclusions were based on semi-qualitative arguments. Although many interesting effects have been predicted in this way, development of theoretical methods for quantitative study of quantum effects in disordered systems was clearly very demanding.

The most straightforward way to take into account disorder is using perturbation theory in the strength of the disorder potential.[2] However, the phenomenon of the localization is not easily seen within this method and the conventional classical Drude formula for conductivity was considered in Ref. 2 as the final result for the dimensionality $d > 1$. This result is obtained after summation of diagrams without intersection of impurity lines. Diagrams with intersection of the impurity lines give a small contribution if the disorder potential is not strong, so that $\varepsilon_0\tau \gg 1$, where ε_0 is the energy of the particles (Fermi energy in metals) and τ in the elastic scattering time.

Although there was a clear understanding that the diagrams with the intersection of the impurity lines were not small for one-dimensional chains, $d = 1$, performing explicit calculations for those systems was difficult. This step has been done considerably later by Berezinsky[3] who demonstrated localization of all states in $1D$ chains by summing complicated series of the perturbation theory. This result confirmed the conclusion of Mott and Twose[4] about the localization in such systems made previously. As concerns the higher dimensional systems, $d > 1$, the Anderson transition was expected at a strong disorder but it was clear that the perturbation theory could not be applied in that case.

So, the classical Drude theory was considered as a justified way of the description of disordered metals in $d > 1$ and $\varepsilon_0\tau \gg 1$. At the same time, several results for disordered systems could not be understood within this simple generally accepted picture.

In 1965, Gorkov and Eliashberg[5] suggested a description of level statistics in small disordered metal particles using the random matrix theory (RMT) of Wigner–Dyson.[6,7] At first glance, the diagrammatic method of Ref. 2 had to work for such a system but one could not see any indication on how the formulae of RMT could be obtained diagrammatically. Of course, the description of Ref. 5 was merely a hypothesis and the RMT had not been used in the condensed matter before but nowadays it looks rather strange that this problem did not attract an attention.

The prediction of localization in thick wires for any disorder made by Thouless[8] could not be understood in terms of the traditional summing of the diagrams either but, again, there was no attempt to clarify this disagreement. Apparently, the diagrammatic methods were not very widely used in that time and therefore not so many people were interested in resolving such problems.

Actually, the discrepancies were not discussed in the literature until 1979, the year when the celebrated work by Abrahams *et al.*[9] appeared. In this work, localization of all states for any disorder already in $2D$ was predicted. This striking result has attracted so much attention that it was simply unavoidable that people started thinking about how to confirm it diagrammatically. The only possibility could be that there were some diverging quantum

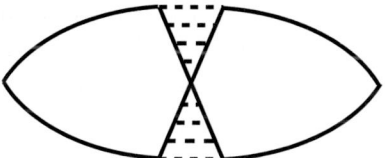

Fig. 5.1. Diverging contribution to conductivity (cooperon).

corrections to the classical conductivity and soon the mechanism of such divergencies has been discovered.[10-12]

It turns out that the sum of a certain class of the diagrams with intersecting impurity lines diverges in the limit of small frequencies $\omega \to 0$ in a low dimension $d \leq 2$. This happens for any weak disorder and is a general phenomenon. The corresponding contribution is represented in Fig. 5.1. The ladder in this diagram can be considered as an effective mode known now as "cooperon". This mode has a form of the diffusion propagator and its contribution to the conductivity $\sigma(\omega)$ can be written in the form

$$\sigma(\omega) = \sigma_0 \left(1 - \frac{1}{\pi\nu} \int \frac{1}{D_0 \mathbf{k}^2 - i\omega} \frac{d^d \mathbf{k}}{(2\pi)^d} \right), \qquad (1.1)$$

where $D_0 = v_0^2 \tau / 3$ is the classical diffusion coefficient and $\sigma_0 = 2e^2 \nu D_0$ is the classical conductivity. The parameters v_0 and ν are the Fermi velocity and density of states on the Fermi surface.

Similar contributions arise also in other quantities. Equation (1.1) demonstrates that in the dimensions $d = 0, 1, 2$ the correction to conductivity diverges in the limit $\omega \to 0$. It is very important that the dimension is determined by the geometry of the sample. In this sense, small disordered particles correspond to zero dimensionality, $d = 0$, and wires to $d = 1$.

The contribution coming from the diffusion mode, Eq. (1.1), is conceptually very important because it demonstrates that the traditional summation of the diagrams without the intersection of the impurity lines is not necessarily applicable in low dimensionality. One can see that most important contributions come from the diffusion modes that are obtained by summation of infinite series of diagrams containing electron Green functions.

The cooperon contribution, Eq. (1.1), has a simple physical meaning. It is proportional to the probability for a scattered electron wave to come back and interfere with itself.[13] The interference implies the quantum coherence and this condition is achieved at low temperatures. There are many interesting effects related to this phenomenon but discussion of these effects and experiments is beyond the scope of this chapter.

It is also relevant to mention that the cooperon contribution is cut by an external magnetic field, which leads to a negative magnetoresistance.[14] At the same time, higher order contributions can still diverge in the limit $\omega \to 0$ and these divergencies are not avoidable provided the coherence is not lost due to, e.g., inelastic processes.

In this way, one can reconcile the hypothesis about the Wigner–Dyson level statistics in disordered metal particles and assertion about the localization in thick wires and $2D$ films with the perturbation theory in the disorder potential. The divergences due to the contribution of the diffusion modes make the perturbation theory inapplicable in the limit $\omega \to 0$ and therefore one does not obtain just the classical conductivity using this approach. Of course, summing the divergent quantum corrections is not sufficient to prove the localization in the low dimensional systems and one should use additional assumptions in order to confirm the statements. Usually, the perturbation theory is supplemented by the scaling hypothesis[9] in order to make such far going conclusions.

At the same time, the divergence of the quantum corrections to the conductivity makes the direct analytical consideration very difficult for small ω because even the summation of all orders of the perturbation theory does not necessarily lead to the correct result. For example, the formulae for the level–level correlation functions[6,7] contain oscillating parts that cannot be obtained in any order of the perturbation theory.

All this meant that a better tool had to be invented for studying the localization phenomena and quantum level statistics. Analyzing the perturbation theory, one could guess that a low energy theory explicitly describing the diffusion modes rather than single electrons might be an adequate method.

The first formulation of such a theory was proposed by Wegner[15] (actually, almost simultaneously with Ref. 10). He expressed the electron Green functions in terms of functional integrals over conventional complex numbers $S(r)$, where r is the coordinate, and averaged over the disorder using the replica trick. Then, decoupling the effective interaction by an auxiliary matrix field Q, he was able to integrate over the field $S(r)$ and represent physical quantities of interest in terms of a functional integral over the $N \times N$ matrices Q, where N is the number of replicas that had to be put to zero at the end of the calculations. Assuming that the disorder is weak, the integral over the eigenvalues of the matrix Q was calculated using the saddle point approximation.

As a result, a field theory in a form of a so called σ-model was obtained. Working with this model one has to integrate over $N \times N$ matrices Q obeying the constraint $Q^2 = 1$. The σ-model is renormalizable and renormalization group equations were written in Ref. 15. These equations agreed with the perturbation theory of Eq. (1.1) and with the scaling hypothesis of Ref. 9.

However, the saddle point approximation was not carefully worked out in Ref. 15 because the saddle points were in the complex plane, while the original integration had to be done over the real axis. This question was addressed in the subsequent publications.[16,17]

In the work of Ref. 16, the initial derivation of Ref. 15 was done more carefully shifting the contours of the integration into the complex plane properly. In this way, one could reach the saddle point and integrate over the eigenvalues of matrix Q coming to the constraint $Q^2 = 1$. After calculating this integral, one is left with the integration over Q that can be written as

$$Q = U \Lambda U^{-1}, \quad \Lambda = \begin{pmatrix} 1 & 0 \\ 0 & -1 \end{pmatrix}, \tag{1.2}$$

where U is an $2N \times 2N$ pseudo-orthogonal or pseudo-unitary matrix. This matrices vary on a hyperboloid, which corresponds to a noncompact group of the rotations. This group is quite unusual for statistical physics.

In contrast, the method of Ref. 17 was based on representing the electron Green functions in a form of functional integrals over anticommuting Grassmann variables and the use of the replica trick. One could average over the disorder as well and further decouple the effective interaction by a gaussian integration over Q. The integration over the anticommuting variables leads to an integral over Q. The integral over the eigenvalues of Q can be calculated using, again, the saddle point method, while the saddle points are now on the real axis. As a result, one comes to a σ-model with Q-fields of the form of Eq. (1.2). However, now one obtains $2N \times 2N$ matrices U varying on a sphere and the group of the rotations is compact.

The difference in the symmetry groups of the matrices Q of these two approaches looked rather unusual and one could only hope that in the limit $N = 0$ imposed by the replica method the results would have agree with each other.

This is really so for the results obtained in Refs. 16 and 17 by using the renormalization group method or perturbation theory. The compact replica σ-model of Ref. 17 has later been extended by Finkelstein[21] to interacting electron systems. An additional topological term was added to this model by Pruisken[22] for studying the integer quantum Hall effect. So, one could hope that the replica σ-models would help to solve many problems in the localization theory.

However, everything turned out to be considerably more complicated for non-perturbative calculations. Desperate attempts[18] to study the level–level statistics in a limited volume and localization in disordered wires lead the present author to the conclusion that the replica σ-model of Ref. 17 could not give any reasonable formulae. Calculation of the level–level correlation function using both the compact and noncompact replica σ-models was discussed later by Verbaarschot and Zirnbauer[19] with a similar result. [Recently, formulae for several correlation functions for the unitary ensemble ($\beta = 2$) have nevertheless been obtained[20] from the replica σ-models by viewing the replica partition function as Toda Lattice and using links with Panleve equations.]

The failure in performing non-perturbative calculations with the replica σ-models lead the present author to constructing another type of the σ-model that was not based on the replica trick. This method was called supersymmetry method, although the word "supersymmetry" is often used in field theory in a more narrow sense. The field theory derived for the disordered systems using this approach has the same form of the σ-model as the one obtained with the replica trick and all perturbative calculations are similar.[23]

An attempt to calculate the level–level correlation function lead to a real surprise: the method worked[24] leading in a rather simple way to the famous formulae for the level–level correlation functions known in the Wigner–Dyson theory,[6,7] thus establishing the relevance of the latter to the disordered systems. Since then, one could use the RMT for calculations of various physical quantities in mesoscopic systems or calculate directly using the zero-dimensional supermatrix σ-model.

The calculation of the level correlations in small disordered systems followed by the full solution of the localization problem in wires,[25] on the Bethe lattice and in high dimensionality.[26–30] These works have demonstrated that the supersymmetry technique was really an efficient tool suitable for solving various problems of theory of disordered metals.

By now, several reviews and a book have been published[31–36] where numerous problems of disordered, mesoscopic and ballistic chaotic system are considered and solved using the supersymmetry method. The interested reader can find all necessary references in those publications.

The present paper is not a complete review of all the works done using the supersymmetry method. Instead, I describe here the main steps leading to the supermatrix σ-model and first problems solved using this approach. I will try to summarize at the end what has become clear in the last almost 30 years of the development and what problems await their resolution.

2. Supermatrix Non-Linear σ-Model

The supersymmetry method is based on using both integrals over conventional complex numbers S_i and anticommuting Grassmann variables χ_i obeying the anticommutation relations

$$\chi_i \chi_j + \chi_j \chi_i = 0. \tag{2.1}$$

The integrals over the Grassmann variables are used following the definition given by Berezin[37]

$$\int d\chi_i = 0, \quad \int \chi_i d\chi_i = 1. \tag{2.2}$$

With this definition one can write the Gaussian integral I_A over the Grassmann variables as

$$I_A = \int \exp\left(-\chi^+ A\chi\right) \prod_{i=1}^{N} d\chi_i^* d\chi_i = \det A, \tag{2.3}$$

which is different from the corresponding integral over complex numbers by presence of $\det A$ instead of $(\det A)^{-1}$ in the R.H.S. In Eq. (2.3), χ is a vector having as components the anticommuting variables χ_i (χ^+ is its transpose with components χ_i^*) and A is an $N \times N$ matrix.

One can introduce supervectors Φ with the components Φ_i,

$$\Phi_i = \begin{pmatrix} \chi_i \\ S_i \end{pmatrix} \tag{2.4}$$

and write Gaussian integrals for these quantities

$$I_S = \pi^{-N} \int \exp\left(-\Phi^+ F\Phi\right) \prod_i d\chi_i^* d\chi_i dS^* dS = SDet\, F. \tag{2.5}$$

In Eq. (2.5), F is a supermatrix with block elements of the form

$$F_{ik} = \begin{pmatrix} a_{ik} & \sigma_{ik} \\ \rho_{ik} & b_{ik} \end{pmatrix}, \tag{2.6}$$

where a_{ik} and b_{ik} are complex numbers and σ_{ik}, ρ_{ik} are Grassmann variables. The superdeterminant (Berezinian) $SDet\, F$ in Eq. (2.5) has the form

$$SDet\, F = \det\left(a - \sigma b^{-1}\rho\right) \det b^{-1}. \tag{2.7}$$

Another important operation is supertrace STr

$$STrF = Tra - Trb. \tag{2.8}$$

Using these definitions, one can operate with supermatrices in the same way as with conventional matrices. Note a very important consequence of Eq. (2.5) for supermatrices F_0 that do not contain the anticommuting variables and are equal to unity in the superblocks F_{ik} in Eq. (2.6) ($a_{ik} = b_{ik}$). In this case one obtains

$$I_S[F_0] = 1. \tag{2.9}$$

For such supermatrices, one can write a relation that is the basis of the supersymmetry method in disordered metals

$$F_{0ik}^{-1} = \int \Phi_i \Phi_k^+ \exp\left(-\Phi^+ F \Phi\right) d\Phi, \tag{2.10}$$

where $d\Phi = \pi^{-N} \prod_i^N d\chi_i^* d\chi_i dS^* dS$.

The weight denominator in the integral in Eq. (2.10) is absent and this form is analogous to what one has using the replica trick. Applying this representation to correlation functions describing disordered systems, one can average over the disorder just in the beginning before making approximations. This is what is done when deriving the supermatrix σ-model and let me sketch this derivation.

Many quantities of interest can be expressed in terms of products of retarded G_ε^R and advanced G_ε^A Green functions of the Schrodinger equation. Using Eq. (2.10) one can write these functions as integrals over supervectors Φ (see Refs. 31 and 33)

$$G_\varepsilon^{R,A}\left(y, y'\right) = \mp i \int \Phi_\alpha\left(y\right) \Phi_\alpha^+\left(y'\right)$$

$$\times \exp\left[i \int \Phi^+\left(x\right) \left(\pm\left(\varepsilon - H\right) + i\delta\right) \Phi\left(x\right) dx\right] D\Phi^+ D\Phi, \tag{2.11}$$

where x and y stand for both the space and spin variables.

The Hamiltonian H in Eq. (2.11) consists of the regular H_0 and random H_1 parts

$$H = H_0 + H_1, \quad \langle H_1 \rangle = 0, \tag{2.12}$$

where the angular brackets $\langle ... \rangle$ stand for the averaging over the disorder.

The most important contribution to such quantities as conductivity and density-density correlation function is expressed in terms of a product

$$K_\omega\left(\mathbf{r}\right) = 2 \left\langle G_{\varepsilon-\omega}^A\left(\mathbf{r}, 0\right) G_\varepsilon^R\left(0, \mathbf{r}\right) \right\rangle, \tag{2.13}$$

where $\mathbf{r}$ is a coordinate and ω is the frequency of the external electric field.

In order to express the function $K_\omega\left(\mathbf{r}\right)$ in terms of an integral over supervectors, one should double the size of the supervectors. Introducing such supervectors ψ one represents the function $K_\omega\left(\mathbf{r}\right)$ in terms of a Gaussian integral without a weight denominator. This allows one to average immediately this function over the random part. In the case of impurities described by a white noise disorder potential $u(\mathbf{r})$, one comes after averaging to the following expression

$$K_\omega(\mathbf{r}) = 2 \int \psi_\alpha^1(0)\,\psi_\alpha^1(\mathbf{r})\,\psi_\beta^2(\mathbf{r})\,\psi_\beta^2(0)\exp(-L)\,D\psi, \qquad (2.14)$$

where

$$L = \int \left[i\bar{\psi}(\varepsilon - H_0)\psi + \frac{1}{4\pi\nu\tau}(\bar{\psi}\psi)^2 - \frac{i(\omega + i\delta)}{2}\bar{\psi}\Lambda\psi \right] d\mathbf{r}. \qquad (2.15)$$

Equation (2.15) was obtained assuming the averages

$$\langle u(\mathbf{r})\,u(\mathbf{r}')\rangle = \frac{1}{2\pi\nu\tau}\delta(\mathbf{r} - \mathbf{r}'), \qquad \langle u(\mathbf{r})\rangle = 0, \qquad (2.16)$$

where ν is the density of states and τ is the elastic scattering time.

The fields $\bar{\psi}$ in Eqs. (2.14) and (2.15) are conjugate to ψ, the matrix Λ is in the space of the retarded-advanced Green functions and equals

$$\Lambda = \begin{pmatrix} 1 & 0 \\ 0 & -1 \end{pmatrix}. \qquad (2.17)$$

The infinitesimal $\delta \to +0$ is added to guarantee the convergence of the integrals over the commuting components S of the supervectors ψ.

The Lagrangian L, Eq. (2.15) has a form corresponding to a field theory of interacting particles. Of course, physically this interaction is fictitious but this formal analogy helps one to use approximations standard for many body theories.

The first approximation done in the supersymmetry method is singling out slowly varying pairs in the interaction term. This is done writing it as

$$L_{int} = \frac{1}{4\pi\nu\tau}\int (\bar{\psi}\psi)^2\,d\mathbf{r} = \frac{1}{4\pi\nu\tau}\sum_{\mathbf{p}_1+\mathbf{p}_2+\mathbf{p}_3+\mathbf{p}_4=0}(\bar{\psi}_{\mathbf{p}_1}\psi_{\mathbf{p}_2})(\bar{\psi}_{\mathbf{p}_3}\psi_{\mathbf{p}_4})$$

$$\approx \frac{1}{4\pi\nu\tau}\sum_{p_1,p_2,q<q_0}[(\bar{\psi}_{\mathbf{p}_1}\psi_{-\mathbf{p}_1+\mathbf{q}})(\bar{\psi}_{\mathbf{p}_2}\psi_{-\mathbf{p}_2-\mathbf{q}})$$

$$+ (\bar{\psi}_{\mathbf{p}_1}\psi_{\mathbf{p}_2})(\bar{\psi}_{-\mathbf{p}_2-\mathbf{q}}\psi_{-\mathbf{p}_1+\mathbf{q}}) + (\bar{\psi}_{\mathbf{p}_1}\psi_{\mathbf{p}_2})(\bar{\psi}_{-\mathbf{p}_1+\mathbf{q}}\psi_{-\mathbf{p}_2-\mathbf{q}})], \qquad (2.18)$$

where q_0 is a cutoff parameter, $q_0 < 1/l$, where l is the mean free path.

The next step is making a Hubbard–Stratonovich transformation decoupling the products of slowly varying pairs by auxiliary slowly varying fields. The term in the second line in Eq. (2.18) is not important and the terms in the third line are equal to each other provided one uses the form of the supervectors ψ of Refs. 31 and 33.

After decoupling, one obtains an effective Lagrangian quadratic in the fields $\psi, \bar{\psi}$ and one can integrate out the fields $\psi, \bar{\psi}$ in Eq. (2.14) and obtain a functional integral over the supermatrix field $Q(\mathbf{r})$. The corresponding free energy functional $F[Q]$ takes the form

$$F[Q] = \int \left[-\frac{1}{2} STr \ln \left(\varepsilon - H_0 - \frac{(\omega + i\delta)}{2} \Lambda - \frac{iQ(\mathbf{r})}{2\tau} \right) + \frac{\pi\nu}{8\tau} STr Q^2 \right] d\mathbf{r}$$
(2.19)

and physical quantities should be obtained integrating correlation functions containing Q over Q with the weight $\exp(-F[Q])$.

The integrals with $F[Q]$ can be simplified using the saddle point approximation. The position of the minimum of $F[Q]$ is found in the limit $\omega \to 0$ by solving the equation

$$Q = \frac{i}{\pi\nu} \left[\left(H_0 + \frac{i}{2\tau} Q(\mathbf{r}) \right)^{-1} \right]_{\mathbf{r},\mathbf{r}}.$$
(2.20)

One can find rather easily a coordinate independent solution of Eq. (2.20). Writing H_0 in a general form as

$$H_0 = \varepsilon(-i\nabla_r) - \varepsilon_0$$
(2.21)

and Fourier transforming the latter, one should calculate the integral over the momenta $\mathbf{p}$. In the limit $\varepsilon_0 \tau \gg 1$ one comes to the general solution

$$Q^2 = 1.$$
(2.22)

Although the supermatrix Q^2 is fixed by Eq. (2.22), the supermatrix Q is not. Supermatrices Q of the form of Eq. (1.2) are solutions for any 8×8 supermatrices U satisfying the condition $U\bar{U} = 1$. With this constraint they are neither unitary nor pseudo-unitary as it was in Refs. 16 and 17. Actually, they consist of both unitary and pseudo-unitary sectors "glued" by the anticommuting variables. This unique symmetry is extremely important for basic properties of many physical quantities.

The degeneracy of the minimum of the free energy functional $F[Q]$ results in the existence of gapless in the limit $\omega \to 0$ excitations (Goldstone modes). This are diffusion modes: so called "cooperons" and "diffusons". These modes formally originate from fluctuating Q obeying the constraint (2.22).

In order to write the free energy functional describing the fluctuations, we assume that supermatrices $Q(\mathbf{r})$ obeying Eq. (2.22) slowly vary in space. Assuming that ω is small, $\omega\tau \ll 1$, but finite and expanding $F[Q]$ in this quantity and gradients of Q, one comes to the supermatrix σ-model

$$F[Q] = \frac{\pi\nu}{8} \int STr \left[D_0 (\nabla Q)^2 + 2i(\omega + i\delta)\Lambda Q \right] d\mathbf{r},$$
(2.23)

where $D_0 = v_0^2 \tau / d$ is the classical diffusion coefficient (v_0 is the Fermi velocity and d is the dimensionality of the sample) and the 8×8 supermatrix Q obeys the constraint (2.22).

Calculation of, e.g., the function $K_\omega(\mathbf{r})$, Eq. (2.14), reduces to calculation of a functional integral over Q

$$K_\omega(\mathbf{r}) = 2 \int Q_{\alpha\beta}^{12}(0) Q_{\beta\alpha}^{21}(\mathbf{r}) \exp\left(-F[Q]\right) DQ. \qquad (2.24)$$

Equations (2.23) and (2.24) are reformulations of the initial problem of disordered metal in terms of a field theory that does not contain disorder because the averaging over the initial disorder has already been carried out. The latter enters the theory through the classical diffusion coefficient D_0. The supermatrix σ-model, described by Eq. (2.23) resembles σ-models used for calculating contributions of spin waves for magnetic materials. At the same time, the noncompactness of the symmetry group of the supermatrices Q makes this σ-model unique.

In order to obtain classical formulae and first quantum corrections one can parametrize the supermatrix Q as

$$Q = W + \Lambda \left(1 - W^2\right)^{1/2}, \quad W = \begin{pmatrix} 0 & Q^{12} \\ Q^{21} & 0 \end{pmatrix} \qquad (2.25)$$

and make an expansion in W in Eqs. (2.23) and (2.24). Keeping quadratic in W terms both in $F[Q]$ and in the pre-exponential in Eq. (2.24), one has to compute Gaussian integrals over W. Fourier transforming the function K_ω, one obtains

$$K_\omega(\mathbf{k}) = \frac{4\pi\nu}{D_0 \mathbf{k}^2 - i\omega}. \qquad (2.26)$$

Equation (2.26) is the classical diffusion propagator. Taking into account higher orders in W, one can compute weak localization corrections to the diffusion coefficient. The first order correction is written in Eq. (1.1).

The precise symmetry of Q depends on the presence of magnetic or spin-orbit interactions. In analogy with symmetries of random matrix ensembles in the Wigner–Dyson theory,[6,7] one distinguishes between the orthogonal ensemble (both magnetic and spin orbit interactions are absent), unitary (magnetic interactions are present) and symplectic (spin-orbit interactions are present but magnetic interactions are absent).

Actually, more symmetry classes are possible. They are fully classified by Altland and Zirnbauer.[38] In the next sections, solutions of several important problems solved with the help of the σ-model, Eq. (2.23), will be presented.

3. Level Statistics in Small Metal Particles

The first non-trivial problem solved with the supermatrix σ-model was the problem of describing the level statistics in small disordered metal particles. At first glance, this problem is not related to the Anderson localization. However, in the language of the σ-model, the solutions of these problems is study of the field theory, Eq. (2.23), in different dimensions. The localization can be obtained in the dimensions $d = 1, 2$ and 3, while the Wigner–Dyson level statistics can be obtained for the zero-dimensional version of the σ-model.

What is the zero dimensionality of the free energy functional $F[Q]$, Eq. (2.23), can easily be understood. In a finite volume, the space harmonics are quantized. The lowest harmonics corresponds to the homogeneous in the space supermatrix Q. The energy of the first excited harmonics E_1 can be estimated as

$$E_1 = E_c/\Delta, \tag{3.1}$$

where energy E_c,

$$E_c = \pi^2 D_0/L^2 \tag{3.2}$$

is usually called the Thouless energy.

The other energy scale Δ,

$$\Delta = (\nu V)^{-1}, \tag{3.3}$$

where V is the volume, is the mean level spacing.

It is clear from Eqs. (2.23) and (2.24) that in the limit

$$E_c \gg \Delta, \omega \tag{3.4}$$

one may keep in these equations only the zero space harmonics of Q, so that this supermatrix does not depend on the coordinates. One can interpret this limit as zero-dimensional one and replace the functional $F[Q]$ by the function $F_0[Q]$,

$$F_0[Q] = \frac{i\pi(\omega + i\delta)}{4\Delta} STr(\Lambda Q). \tag{3.5}$$

The function $R(\omega)$ that determines the correlation between the energy levels is introduced as

$$R(\omega) = \left\langle \frac{\Delta^2}{\omega} \sum_{k,m} (n(\varepsilon_k) - n(\varepsilon_m)) \delta(\omega - \varepsilon_m + \varepsilon_k) \right\rangle. \tag{3.6}$$

It is proportional to the probability of finding two levels at a distance ω.

Using the supersymmetry approach, one can represent the functions $R(\omega)$ in terms of a definite integral over the supermatrices Q

$$R(\omega) = \frac{1}{2} - \frac{1}{2} Re \int Q_{11}^{11} Q_{11}^{22} \exp\left(-F_0[Q]\right) dQ. \tag{3.7}$$

In order to calculate the integral in Eq. (3.7), one should choose a certain parametrization for the supermatrix Q.

It is convenient to write the supermatrix Q in the form

$$Q = U Q_0 \bar{U}, \quad Q_0 = \begin{pmatrix} \cos\hat{\theta} & i\sin\hat{\theta} \\ -i\sin\hat{\theta} & -\cos\hat{\theta} \end{pmatrix}, \quad U = \begin{pmatrix} u & 0 \\ 0 & v \end{pmatrix}, \tag{3.8}$$

where all anticommuting variables are packed in the supermatrix blocks u and v. It is clear that the (pseudo) unitary supermatrix U commutes with Λ, which drastically simplifies the integrand in Eq. (3.7).

Instead of the integration in Eq. (3.7) over the elements of the supermatrix Q with the constraint (2.22), one can integrate over the elements of the matrix $\hat{\theta}$ and the matrices u and v. Of course, it is necessary to write a proper Jacobian (Berezinian) of the transformation to these variables. The latter depends only on the elements of $\hat{\theta}$ and therefore the elements of u and v appear only in the pre-exponential in Eq. (3.7). The integration over the supermatrices u and v is quite simple and one comes to definite integrals over the elements of $\hat{\theta}$.

The number of the independent variables in the blocks $\hat{\theta}$ depends on the ensemble considered. The supermatrices Q written for the unitary ensemble have the simplest structure and the blocks $\hat{\theta}$ contains only 2 variables $0 < \theta < \pi$ and $0 < \theta_1 < \infty$. The corresponding blocks $\hat{\theta}$ for the orthogonal and symplectic ensembles contain 3 independent variables. All the transformations are described in details in Refs. 31–33.

In order to get an idea about what one obtains after the integration over u and v in Eq. (3.8), I write here an expression for the unitary ensemble only

$$R(\omega) = 1 + \frac{1}{2} Re \int_1^\infty \int_{-1}^1 \exp\left[i(x+i\delta)(\lambda_1 - \lambda)\right] d\lambda_1 d\lambda, \tag{3.9}$$

where $x = \pi\omega/\Delta$, $\lambda_1 = \cosh\theta_1$, and $\lambda = \cos\theta$.

So, the calculation of the level-level correlation function is reduced to an integral over 2 or 3 variables depending on the ensemble considered. The final result for the orthogonal $R_{\text{orth}}(\omega)$, unitary $R_{\text{unit}}(\omega)$, and symplectic $R_{\text{sympl}}(\omega)$ ensembles calculated using Eq. (3.7) takes the following form

$$R_{\text{orth}}(\omega) = 1 - \frac{\sin^2 x}{x^2} - \frac{d}{dx}\left(\frac{\sin x}{x}\right) \int_1^\infty \frac{\sin xt}{t} dt, \tag{3.10}$$

$$R_{\text{unit}}(\omega) = 1 - \frac{\sin^2 x}{x^2}, \tag{3.11}$$

$$R_{\text{sympl}}(\omega) = 1 - \frac{\sin^2 x}{x^2} + \frac{d}{dx}\left(\frac{\sin x}{x}\right)\int_0^1 \frac{\sin xt}{t}dt. \tag{3.12}$$

Equations (3.10)–(3.12) first obtained for the disordered metal particles[24,31] identically agree with the corresponding formulae of the Wigner–Dyson theory[6,7] obtained from the ensembles of random matrices. This agreement justified the application of the RMT for small disordered particles suggested in Ref. 5.

Actually, to the best of my knowledge, this was the first explicit demonstration that RMT could correspond to a real physical system. Its original application to nuclear physics was in that time phenomenological and confirmed by neither analytical nor numerical calculations.

A direct derivation of Eqs. (3.10)–(3.12) from Gaussian ensembles of the random matrices using the supermatrix approach was done in the review.[32] This allowed the authors to compute certain average compound-nucleus cross sections that could not be calculated using the standard RMT route.

The proof of the applicability of the RMT to the disordered systems was followed by the conjecture of Bohigas, Giannonni and Schmid[39] about the possibility of describing by RMT the level statistics in classically chaotic clean billiards. Combination of the results for clean and disordered small systems (billiards) has established the validity of the use of RMT in mesoscopic systems. Some researches use for explicit calculations methods of RMT but many others use the supermatrix zero-dimensional σ-model (for review see, e.g. Refs. 34, 40 and 41). At the same time, the σ-model is applicable to a broader class of systems than the Wigner–Dyson RMT because it can be used in higher dimensions as well. Actually, one can easily go beyond the zero dimensionality taking higher space harmonics in $F[Q]$, Eq. (2.23). In this case, the universality of Eqs. (3.10)–(3.11) is violated. One can also study this limit for $\omega \gg \Delta$ using the standard diagrammatic expansions of Ref. 2 and this was done in Ref. 42.

The other versions of the σ-model (based on the replica trick and Keldysh Green functions) have not shown a comparable efficiency for studying the mesoscopic systems, although the formula for the unitary ensemble, Eq. (3.11), has been obtained by these approaches.[20,43]

The results reviewed in this section demonstrate that the development of the theory of the energy level statistics in small systems and of related phenomena in mesoscopic systems have been tremendously influenced by the ideas of the Anderson localization because important results have been obtained by methods developed for studying the latter.

4. Anderson Localization in Quantum Wires

The one-dimensional σ-model corresponds to quantum wires. These objects are long samples with a finite cross-section S that should be sufficiently large,

$$Sp_0^2 \gg 1, \qquad (4.1)$$

where p_0 is the Fermi momentum. In other words, the number of transversal channels should be large. This condition allows one to neglect non-homogeneous in the transversal direction variations of Q. Of course, the inequality $\varepsilon_0 \tau \gg 1$ should be fulfilled as before.

Then, the σ-model can be written in the form

$$F\left[Q\right] = \frac{\pi \tilde{\nu}}{8} \int \left[D_0 \left(\frac{dQ}{dx}\right)^2 + 2i\omega\Lambda Q \right] dx, \qquad (4.2)$$

where $\tilde{\nu} = \nu S$.

Again, depending of the presence of magnetic and/or spin-orbit interactions the model has different symmetries (orthogonal, unitary and symplectic). It is important to emphasize that Eq. (4.2) is not applicable for disordered chains or thin wires where the inequality (4.1) is not fulfilled. However, the explicit solutions show that the low frequency behavior of all these systems is the same.

Computation of the correlation function $K_\omega\left(x\right)$, Eq. (2.24), with the one-dimensional σ-model can be performed using the transfer matrix technique. Following this method, one reduces the calculation of the functional integral in Eq. (2.24) to solving an effective Schrödinger equation in the space of the elements of the supermatrix Q and calculating matrix elements of Q entering the pre-exponential in Eq. (2.24). This has been done in Ref. 44 and presented also in the subsequent publications.[31,33]

At first glance, this procedure looks very complicated due to a large number of the elements in the supermatrices Q. Fortunately, the symmetries of the free energy functional $F\left[Q\right]$ in Eq. (4.2) help one again to simplify the calculations.

In order to derive the transfer matrix equations, one should subdivide the wire into small slices and write recursive equations taking at the end the continuous limit. Instead of this artificial subdivision, it is more instructive to consider a realistic model of a chain of grains coupled by tunnelling. The free energy functional $F_J[Q]$ for such a chain can be written in the form

$$F_J[Q] = STr\left(-\sum_{i,j} J_{ij}Q_iQ_j + \frac{i(\omega + i\delta)\pi}{4\Delta}\sum_i \Lambda Q_i\right), \qquad (4.3)$$

where $J_{ij} = J$ for nearest neighbors and $J_{ij} = 0$ otherwise. The summation runs in Eq. (4.3) over the grains. The coupling constant J can be expressed in terms of the matrix elements of the tunnelling from grain to grain T_{ij} but at the moment this explicit relation is not important.

In the limit $J \gg 1$, only small variations of the supermatrix Q in space are important and the functional $F_J[Q]$, Eq. (4.3), can be approximated by $F[Q]$, Eq. (4.2). The classical diffusion coefficient D_0 corresponding to Eq. (4.3) takes the form

$$D_0 = \frac{4\Delta}{\pi}\sum_i J_{ij}(r_i - r_j)^2. \qquad (4.4)$$

The correlation function K_ω, Eq. (2.24), should also be taken at the discrete coordinates r_i numerating the grains. Then, it can be re-written identically in the form

$$K_\omega(r_1, r_2) = 2\pi^2\nu\tilde{\nu}\int \Psi(Q_1)(Q_1)_{11}^{12}$$

$$\times \Gamma(r_1, r_2; Q_1, Q_2)(Q_2)_{11}^{21}\Psi(Q_2)\,dQ_1dQ_2, \qquad (4.5)$$

where the kernel $\Gamma(r_1, r_2; Q_1, Q_2)$ is the partition function of the segment between the points r_1 and r_2. It is assumed that integration for this kernel is performed over all Q except Q_1 and Q_2 at the points r_1 and r_2. So the kernel $\Gamma(r_1, r_2; Q_1, Q_2)$ depends on supermatrices Q_1, Q_2 and distances $r_2 - r_1$ (the point r_2 is to the right of the point r_1). The function $\Psi(Q)$ is the partition function of the parts of the wire located to the right of the point r_2 and to the left of the point r_1. This function depends only on the supermatrix Q at the end points r_1 or r_2.

Comparing the functions $\Psi(Q)$ at neighboring grains, one comes to the following equation

$$\Psi(Q) = \int N(Q, Q')Z_0(Q')\Psi(Q')\,dQ', \qquad (4.6)$$

where

$$N(Q, Q') = \exp\left(\frac{\alpha}{4}STrQQ'\right), \quad \alpha = 8J$$

$$Z_0(Q) = \exp\left(\frac{\beta}{4}STr\Lambda Q\right), \quad \beta = \frac{-i(\omega + i\delta)\pi}{\Delta}. \qquad (4.7)$$

A similar equation can be written for the kernel $\Gamma(r_1, r_2; Q_1, Q_2)$. Comparing this function at the neighboring points r and $r+1$, one obtains the recurrence equation

$$
\Gamma(r, r'; Q, Q') - \int N(Q, Q'') \mathcal{Z}_0(Q'') \Gamma(r+1, r'; Q'', Q') dQ''
$$
$$
= \delta_{rr'} \delta(Q - Q'). \tag{4.8}
$$

The δ-function entering Eq. (4.8) satisfies the usual equality

$$
\int f(Q') \delta(Q - Q') dQ' = f(Q). \tag{4.9}
$$

Equations (4.5)–(4.9) reduce the problem of calculation of a functional integral over $Q(r)$ to solving the integral equations and calculation of the integrals with their solutions. In the limit $J \gg 1$, the integral equations can be reduced to differential ones. Their solution can be sought using again the parametrization (3.8). The function $\Psi(Q)$ is assumed to be a function of the elements of the block $\hat{\theta}$. Then, one obtains the differential equation for Ψ in the form

$$
\mathcal{H}_0 \Psi = 0. \tag{4.10}
$$

The explicit form of the operator $\mathcal{H}_0$ depends on the ensemble considered. The simplest equation is obtained for the unitary ensemble for which the operator $\mathcal{H}_0$ takes the form

$$
\mathcal{H}_0 = -\frac{1}{2\pi\tilde{\nu}D_0} \left[\frac{1}{J_\lambda} \frac{\partial}{\partial\lambda} J_\lambda \frac{\partial}{\partial\lambda} + \frac{1}{J_\lambda} \frac{\partial}{\partial\lambda_1} J_\lambda \frac{\partial}{\partial\lambda_1} \right] - i(\omega + i\delta) \pi\tilde{\nu}(\lambda_1 - \lambda),
$$
$$
\tag{4.11}
$$

where

$$
J_\lambda = (\lambda_1 - \lambda)^{-2}.
$$

Similar equations can be written for the central part entering Eq. (4.5). Solving these equations and substituting the solutions into Eq. (4.5), one can determine (at least numerically) the frequency dependence of the function $K_\omega(r_1, r_2)$ and, hence, of the conductivity for all frequencies in the region $\omega\tau \ll 1$ and distances $|r_1 - r_2| p_0 \gg 1$.

The calculation becomes considerably simpler in the most interesting case of low frequencies $\omega \ll (\tilde{\nu}^2 D_0)^{-1}$. In this limit, the main contribution into the integral in Eq. (4.5) comes from large $\lambda_1 \gg 1$ and the solution Ψ of Eq. (4.10) is a function of only this variable.

Introducing a new variable

$$
z = -i\omega 2\pi^2 \tilde{\nu}^2 D_0 \lambda_1, \tag{4.12}
$$

one can reduce Eq. (4.11) to the form

$$-z \frac{d^2 \Psi(z)}{dz^2} + \Psi(z) = 0, \tag{4.13}$$

with the boundary condition

$$\Psi(0) = 1. \tag{4.14}$$

The Fourier transformed function $K_\omega(k)$ takes the form

$$K_\omega(k) = \frac{4\pi\nu A(k)}{-i\omega}, \qquad A(k) = \int_0^\infty (\Phi_k(z) + \Phi_{-k}(z)) \Psi(z) \, dz, \tag{4.15}$$

where the function $\Phi_k(z)$ satisfies the following equation

$$-\frac{d}{dz}\left(z^2 \frac{d\Phi_k(z)}{dz}\right) + ikL_c\Phi_k(z) + z\Phi_k(z) = \Psi(z), \tag{4.16}$$

with the length L_c equal to

$$L_c = 2\pi\nu SD_0. \tag{4.17}$$

The length L_c is actually the localization length, which will be seen from the final result. Equations (4.13)–(4.16) can also be obtained for the orthogonal and symplectic ensembles but with different localization lengths L_c. The result can be written as

$$L_c^{\text{symplectic}} = 2L_c^{\text{unitary}} = 4L_c^{\text{orthogonal}}. \tag{4.18}$$

The residue of the function K_ω is proportional to the function $p_\infty(r, r', \varepsilon)$ introduced by Anderson,[1]

$$p_\infty(r, r', \varepsilon) = \sum_k |\phi_k(r)|^2 |\phi_k(r')|^2 \delta(\varepsilon - \varepsilon_k), \tag{4.19}$$

where $\phi_k(r)$ are exact eigenfunctions.

Equations (4.13)–(4.16) exactly coincide with the low frequency limit of equations derived by Berezinsky[3] provided the length L_c is replaced by the mean free path l, which shows that the low frequency limit of the one dimensional systems is universal.

The exact solution of Eqs. (4.13)–(4.16) leads to the following expression

$$p_\infty(x) = \frac{\pi^2\nu}{16L_c} \int_0^\infty \left(\frac{1+y^2}{1+\cosh\pi y}\right)^2 \exp\left(-\frac{1+y^2}{4L_c}|x|\right) y \sinh \pi y \, dy. \tag{4.20}$$

In the limit $x \gg L_c$, Eq. (4.20) reduces to a simpler form

$$p_\infty(x) \approx \frac{\nu}{4\sqrt{\pi}L_c}\left(\frac{\pi^2}{8}\right)^2 \left(\frac{4L_c}{|x|}\right)^{3/2} \exp\left(-\frac{|x|}{4L_c}\right). \tag{4.21}$$

The exponential form of $p_\infty(x)$ proves the localization of the wave functions and shows that the length L_c is the localization length. Note, however, the presence of the pre-exponential $|x|^{-3/2}$. Due to the factor the integral over x of $p_\infty(x)$ remains finite even in the limit $L_c \to \infty$. Actually, one obtains

$$\int_{-\infty}^{\infty} p_\infty(x)\, dx = \nu, \tag{4.22}$$

which proves the localization of all states.

At small $k \ll L_c^{-1}$, the function $A(k)$ in Eq. (4.15) takes the form

$$A(k) = 1 - 4\zeta(3) k^2 L_c^2$$

and the static dielectric permeability ϵ equals

$$\epsilon = -4\pi e^2 \nu \left. \frac{d^2 A(k)}{dk^2} \right|_{k=0} = 32\zeta(3) e^2 \nu L_c^2, \tag{4.23}$$

where $\zeta(x)$ is the Riemann ζ-function.

All these calculations have been performed for a finite frequency ω and the infinite length of the sample. One can also consider the case of the zero frequency and a finite length L. A full analysis of this limit has been presented by Zirnbauer[45] who calculated the average conductivity as a function of L.

There is another Fokker–Planck approach to study transport of disordered wires developed by Dorokhov, Mello, Pereyra, and Kumar[46,47] (DMPK method). It can be applied also to thin wires with a small number of channels. At the same time, this method cannot be used for finite frequencies. In the case of thick wires with a large number of the channels and zero frequencies, the equivalence of the supersymmetry to the DMPK method has been demonstrated by Brouwer and Frahm.[48]

Many interesting problems of banded random matrices[49] and quantum chaos (like kicked rotor[50]) can be mapped onto the $1D$ supermatrix σ-model. However, a detailed review of these interesting directions of research is beyond the scope of this paper.

5. Anderson Localization in 2 and $2 + \epsilon$ Dimensions

Study of localization in 2 and $2 + \epsilon$ using the replica σ-model was started by Wegner[15] using a renormalization group (RG) technique. He was able to write the RG equations for the orthogonal and unitary ensembles that could be used in 2 dimensions and extended into $2 + \epsilon$ dimensions for $\epsilon \ll 1$. The latter was done with a hope that putting $\epsilon = 1$ at the end of the calculations one could extract at least qualitatively an information about the Anderson metal–insulator transition in 3 dimensions. Based on this calculation, a conclusion about the localization at any weak disorder in $2D$ was made. As concerns $2 + \epsilon$, an unstable fixed point was found, which following the standard arguments by Polyakov[51] signaled the existence of the metal–insulator transition.

The symplectic case was considered within the compact replica σ-model in Ref. 17 using the same method of RG and it was shown that the resistivity had to vanish in the limit of $\omega \to 0$. The difference between the replica σ-models used in Refs. 15 and 16 (noncompact) and Ref. 17 (compact) is not essential when applying the RG scheme.

Exactly the same results are obtained with the supermatrix σ-model using the RG technique[31,33,52] and let me sketch the derivation here. As usual in the RG method, one introduces a running cutoff parameter and coupling constants depending on this cutoff. The σ-model for such couplings can be written as

$$F = \frac{1}{t} \int STr \left[(\nabla Q)^2 + 2i\tilde{\omega}\Lambda Q \right] dr, \qquad (5.1)$$

where $\tilde{\omega} = \omega/D_0$. The bare value of t equals $t = 8 \left(\pi\nu D_0 \right)^{-1}$ (c.f. Eq. (2.23)).

The σ-model looks similar to classical spin σ-models considered in Ref. 51 and one can follow the RG procedure suggested in that work. Using the constraint (2.22), one can write the supermatrix Q in the form

$$Q = V\Lambda\bar{V}, \qquad (5.2)$$

where $V\bar{V} = 1$ so that V is a pseudo-unitary supermatrix.

In order to integrate over a momentum shell one can represent the supermatrix V in the form

$$V(r) = \tilde{V}(r) V_0(r), \qquad (5.3)$$

where V_0 is a supermatrix fast varying in space and $\tilde{V}$ is slowly varying one. These supermatrices have the same symmetry as the supermatrix V.

Substituting Eq. (5.3) into Eq. (5.1) one can write the free energy functional $F[Q]$ in the form

$$F = \frac{1}{t} \int STr \left[(\nabla Q_0)^2 + 2 [Q_0, \nabla Q_0] \Phi + [Q_0, \Phi]^2 + 2i\tilde{\omega}\overline{\tilde{V}}\Lambda\tilde{V}Q_0 \right] dr, \quad (5.4)$$

$$Q_0 = V_0\Lambda\bar{V}_0, \qquad \Phi = \overline{\tilde{V}}\nabla\tilde{V} = -\bar{\Phi}.$$

The next step of the RG procedure is to integrate over the fast varying matrices Q_0 and reduce to a functional containing only slowly varying variables

V. After this integration, the free energy F in Eq. (5.4) should be replaced by energy $\tilde{F}$ describing the slow fluctuations

$$\tilde{F} = -\ln \int \exp\left(-F\right) DQ_0. \tag{5.5}$$

The integration over the supermatrix Q_0 can be done using a parametrization (2.25) or a more convenient parametrization

$$Q_0 = \Lambda\left(1 + P\right)\left(1 - P\right)^{-1}, \qquad P\Lambda + \Lambda P = 0. \tag{5.6}$$

Integration over the fast variation means that one integrates over Fourier transformed P_k with $\lambda k_0 < k < k_0$, where k_0 is the upper cutoff and $\lambda < 1$. As a result of the integration, one comes to the same form of the functional F as in Eq. (5.1). The constant $\tilde{\omega}$ does not change under the renormalization but the new coupling constant $\tilde{t}$ can be written as

$$\tilde{t}^{-1} = t^{-1}\left(1 + \frac{\alpha t}{8} \int_{\lambda k_0}^{k_0} \frac{d^d k}{k^2\left(2\pi\right)^d}\right). \tag{5.7}$$

The correction to the coupling constant t, Eq. (5.7), is written in the first order in t. The parameter α depends on the ensemble and equals

$$\alpha = \begin{cases} -1, & \text{orthogonal} \\ 0, & \text{unitary} \\ 1 & \text{symplectic} \end{cases}. \tag{5.8}$$

Stretching the coordinates in the standard way and changing the notation for the coupling constant $t \to 2^{d+1}\pi d\Gamma\left(d/2\right)t$, where Γ is the Euler Γ-function one obtains the RG equation for t

$$\beta\left(t\right) = \frac{dt}{d\ln\lambda} = \left(d - 2\right)t + \alpha t^2, \tag{5.9}$$

where $\beta\left(t\right)$ means the Gell–Mann–Low function.

In 2D, the solution of this equation for the coupling constant t (proportional to resistivity) takes the form

$$t\left(\omega\right) = \frac{t_0}{1 + \alpha t_0 \ln\left(1/\omega\tau\right)}. \tag{5.10}$$

For sufficiently high frequencies ω, the resistivity and the diffusion coefficient $D\left(\omega\right)$ proportional to $t^{-1}\left(\omega\right)$ coincide with their bare values.

Decreasing the frequency ω results in growing the resistivity for the orthogonal ensemble until the coupling constant $t\left(\omega\right)$ becomes of the order 1. Then, the RG scheme is no longer valid because the expansion in t in the R.H.S. of Eq. (5.9) is applicable only for $t \ll 1$. However, it is generally believed that t diverges in the limit $\omega \to 0$ and this should mean the localization of all states with an exponentially large localization length

$$L_c \propto \exp\left(1/t_0\right). \tag{5.11}$$

In the symplectic ensemble, the resistivity $t\left(\omega\right)$ decreases with decreasing the frequency ω. This interesting result was obtained in the first order in t_0 by Hikami, Larkin and Nagaoka.[53] However, Eq. (5.10) has a greater meaning.[17] If the bare t_0 is small, $t_0 \ll 1$, the effective resistivity $t\left(\omega\right)$ decays down to zero in the limit $\omega \to 0$. In this case the constant $t\left(\omega\right)$ is small for any frequency and the one loop approximation used in the derivation of Eq. (5.9) is valid for all frequencies. So, the solution for the symplectic ensembles, when used for the low frequencies, is the most reliable one obtained with the RG method.

As concerns the unitary ensemble, the first order contribution vanishes and one should calculate corrections of the second order. As a result, one comes to the following dependence of $t\left(\omega\right)$ on the frequency

$$t\left(\omega\right) = \frac{t_0}{\left(1 - t_0^2 \ln\left(1/\omega\tau\right)\right)^{1/2}}. \tag{5.12}$$

One can see from Eq. (5.12) that the resistivity $t\left(\omega\right)$ grows, as in the orthogonal ensemble, until it becomes of order 1. Again, this behavior is interpreted as localization for any disorder. The conclusions about the localization in $2D$ for the orthogonal and unitary ensembles were made first in Ref. 15 and this agreed with the results based on using the scaling hypothesis.[9]

Wegner developed also theory of the Anderson metal–insulator transition in the dimensionality $2+\epsilon$ for $\epsilon \ll 1$.[54] One can see that the RG Eq. (5.9) has a fixed point $t_c = \epsilon$, at which the Gell–Mann–Low function vanishes. At this point, the total resistance of the sample does not depend on the sample size and this point should correspond to the Anderson metal–insulator transition.

Linearizing function $\beta\left(t\right)$ near the fixed point t_c, one can solve Eq. (5.9). As a result one can find a characteristic (correlation) length ξ near the fixed point

$$\xi \sim \xi_0 \left(\frac{t_c - t_0}{t_c}\right)^{-1/y}, \quad y = -\beta'\left(t_c\right), \tag{5.13}$$

where ξ_0 is the size of a sample having the entire resistance t_0. Assuming that the length ξ is the only characteristic length in the system and that the conductivity σ is proportional to $t_c^{-1}\xi^{2-d}$, one can write the equation for the conductivity in the following form

$$\sigma = A\frac{e^2}{\xi_0^{d-2}t_c}\left(\frac{t_c - t}{t_c}\right)^s, \qquad s = \frac{d-2}{y}. \qquad (5.14)$$

The explicit values of the critical resistance t_c and the exponent s for the orthogonal and unitary ensembles equals

$$\tilde{t}_c = \begin{cases} d-2, & \text{orthogonal} \\ (2\,(d-2))^{1/2}, & \text{unitary} \end{cases} \qquad (5.15)$$

and

$$s = \begin{cases} 1 \\ 1/2 \end{cases}. \qquad (5.16)$$

Equations (5.13)–(5.16) demonstrate that the metal–insulator transition exists in any dimensionality $d > 2$ and the conductivity near the transition obeys a power law. Of course, this consideration is restricted by small $\epsilon = d - 2$ and one can use the result in $3D$ only qualitatively.

The scaling approach developed for small ϵ is similar to the one developed for conventional phase transitions in, e.g., spin models where one can also write σ-models. This method is not sensitive to whether the symmetry of the supermatrices Q is compact or noncompact. Using this approach, one comes to the conclusion that the Anderson metal–insulator transition is very similar to standard second order phase transitions.

In the next section, the same problem will be considered on the Bethe lattice or in a high dimensionality. Surprisingly, the result will be very different and the peculiarity of the solution originates from the noncompactness of the group of the symmetry of the supermatrices Q.

6. Anderson Metal–Insulator Transition on the Bethe Lattice or in a High Dimensionality

It is generally difficult to find the critical point for a transition between different states and describe the critical behavior in its vicinity. The Anderson metal–insulator transition is definitely not an exception in this respect. Usually, identifying a proper order parameter, one can get an idea about a transition using a mean field approximation. As concerns the Anderson transition, this is not possible. Although the σ-model, Eq. (2.23), looks very similar to spin models in a magnetic field, one cannot take an average of Q with the free energy $F[Q]$, Eq. (2.23), as the order parameter because it determines the average density of states and is not related to the Anderson transition.

At the same time, the mean field approximation works very well in high dimensionality or on special structures like the Bethe lattice.

The Anderson model of the Bethe lattice was studied for the first time by Abou-Chacra, Anderson and Thouless,[55] who proved the existence of the metal–insulator transition and found the position of the mobility edge. With the development of the supersymmetry technique it became possible to describe the critical behavior both in the metallic and insulating regime. Considering a granular model one could obtain results for the orthogonal, unitary and symplectic ensembles. Later the Anderson model has also been described.

It turned out that in all the cases the critical behavior was the same, which contrasts the results obtained within the $2+\epsilon$ expansion. This could not be a big surprise because for most phase transitions the high dimensional results are more "universal" than those obtained in lower dimensions. However, the results for the metallic and insulating regimes did not obey the conventional scaling and this was completely unexpected.

The first attempt to solve the granular version of the supermatrix σ-model on the Bethe lattice has been undertaken in Ref. 26. In this work, correct integral equation have been written for description of critical behavior near the metal–insulator transition and the position of the mobility edge has been found. However, attempts to find a solution of this equation related to scaling properties of the $2+\epsilon$ limit were not successful, which lead to wrong conclusions.

Studying numerically the integral equation derived in Ref. 26, Zirnbauer[27] found a very unusual behavior near the critical point and presented formal reasons explaining this behavior. Finally, the density–density correlation function has been calculated for the unitary[28] and orthogonal and symplectic ensembles.[29] This determined the diffusion coefficient in the metallic region and localization length and dielectric permeability in the insulating one.

The form of the density–density correlation function on the Bethe lattice differs from the one on conventional lattices. Therefore the problem of the Anderson localization has been considered on such lattices in an effective medium approximation.[30] The latter becomes exact on the real lattices in a high dimensionality $d \gg 1$ and the basic equations and results are similar. The derivation of the equations and the final results are shortly displayed below. A detailed discussion can be found in Ref. 33.

The scheme of the derivation of the equations is similar to the one presented in Sec. 4 for one-dimensional structures consisting of the grains. We start with Eq. (4.3) written on a d-dimensional lattice with $d \gg 1$ or on the Bethe lattice. Denoting by $\Psi(Q)$ the partition function of a branch of the tree structure with a fixed value Q at the base and comparing it with the partition function on the neighboring site, one comes to a non-linear integral equation

$$\Psi(Q) = \int N(Q, Q') Z_0(Q') \Psi^m(Q') dQ', \qquad (6.1)$$

where $m = 2d - 1$ for a d-dimensional lattice and is the branching number on the Bethe lattice. The functions $N(Q, Q')$ and $Z_0(Q)$ have been introduced in Eq. (4.7).

The case $m = 1$ corresponds to the one-dimensional chains of the grains and Eq. (6.1) coincides with Eq. (4.6) in this limit. In this particular case, Eq. (6.1) is linear and, as we have seen in Sec. 4, all states are localized for any disorder. However, at $m > 1$ the integral Eq. (6.1) is non-linear and has a bifurcation at a critical α_c corresponding to the Anderson metal–insulator transition.

The density-density correlation function K_ω, Eq. (2.24), can be written in the form

$$K_\omega(r_1, r_2) = -2\pi^2 \nu \tilde{\nu} \int Q_{33}^{12} P_{33}(r, Q) Z(Q) \Psi(Q) dQ, \qquad (6.2)$$

where the function $P(r, Q)$ satisfies for the high dimensional lattices the following equation

$$P(r, Q) - \sum_{r'} W(r - r') \int N(Q, Q') P(r', Q') Z(Q') dQ'$$

$$+ m \int N_2(Q, Q') P(r, Q') Z(Q') dQ' = \delta(r) Q^{21} \Psi(Q). \qquad (6.3)$$

In Eq. (4.4) the function $N_2(Q, Q')$ is equal to

$$N_2(Q, Q') = \int N(Q, Q'') N(Q'', Q) Z(Q'') dQ''$$

and

$$W(r - r') = \begin{cases} 1, & |r - r'| = 1 \\ 0, & |r - r'| \neq 1 \end{cases}.$$

The third term in the L.H.S. of Eq. (6.3) takes into account the fact that two segments of a broken line cannot coincide. Equations (6.2) and (6.3) are very similar to Eqs. (4.5) and (4.8) written for the $1D$ case. This is natural because in both cases loops are absent. Their solution for a function $\Psi(Q)$ found from Eq. (6.1) can be obtained making a spectral expansion of $P(r, Q)$ in eigenfunctions of the integral operators entering the L.H.S. of Eq. (6.3).

In principle, this procedure is straightforward. However, solving the integral Eq. (6.1) is not simple because it contains a large number of the elements of the supermatrix Q.

Fortunately, Eqs. (6.1)–(6.3) drastically simplify in the metallic regime near the metal–insulator transition and everywhere in the insulating regime, provided one considers the low frequency limit $\omega \to 0$. The formal reason for this simplification is that the main contribution into the correlation functions comes in these cases from the region of very large values of the variables $\lambda_1 \gtrsim \Delta/\omega \gg 1$. The same simplification has helped one to solve the problem of the localization in wires in Sec. 4.

Nevertheless, the full analysis is quite involved even for small ω. Details can be found again in Ref. 33 and here I display only the final results.

In the insulating regime, $\alpha < \alpha_c$, only $\Psi = 1$ is the solution of Eq. (6.1) in the limit $\omega = 0$. This solution of the simplified equation persists for all α but another solution appears in the region $\alpha > \alpha_c$. The latter solution considered as a function of $\theta_1 = \ln(2\lambda_1)$ has a form of a kink moving to infinity as $\alpha \to \alpha_c$. The position θ_{1c} of the kink depends on the distance from the critical point α_c as

$$\theta_{1c} = s\,(\alpha - \alpha_c)^{-1/2}, \tag{6.4}$$

where s is a number of order 1. The dependence of $\Psi(\lambda_1)$ is represented in Fig. 5.2.

Only this solution should be used for $\alpha > \alpha_c$ and this leads to a very non-trivial critical behavior of the diffusion coefficient.

The position of the critical point α_c and the critical behavior have been calculated for all 3 ensembles. For large m, the value α_c for the orthogonal and unitary ensembles is determined by the following equations

$$\begin{aligned}
\frac{2^{3/2}}{\pi}\left(\frac{\alpha_c}{2\pi}\right)^{1/2} m \ln \frac{\gamma}{\alpha_c} &= 1, \quad \text{orthogonal} \\
\left(\frac{\alpha_c}{2\pi}\right)^{1/2} m \ln \frac{2}{\alpha_c} &= 1, \quad \text{unitary}
\end{aligned} \tag{6.5}$$

One can see from Eq. (6.5) that the metallic region is broader for systems with the broken time reversal invariance. In other words, applying a magnetic field shifts the metal-insulating transition to larger values of α_c. This result correlates with the one, Eq. (5.15), obtained in $2 + \epsilon$ dimensions.

Although the position of the Anderson transition depends on the ensemble considered, the form of the correlation functions is the same.

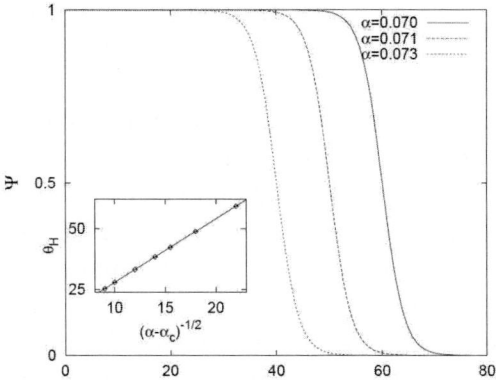

Fig. 5.2. Numerical solution $\Psi(\lambda_1)$ for $m = 2$ and different hopping amplitudes in the critical metallic regime. The inset shows θ_H defined by $\Psi(\cosh\theta_H) = 0.5$ as a function of $(\alpha - \alpha_c)^{-1/2}$. Reprinted (Fig. 2) with permission from *Phys. Rev. B* **45**, 11546 (1992). © American Physical Society.

In the insulating regime, the function $p_\infty(r)$, Eq. (4.19), takes for $r \gg L_c$ the following form

$$p_\infty(r) = \text{const } r^{-(d+2)/2} L_c^{-d/2} \exp\left(-\frac{r}{4L_c}\right), \tag{6.6}$$

where L_c is the localization length.

Near the transition the localization length L_c grows in a power law

$$L_c = \frac{\text{const}}{(\alpha_c - \alpha)^{1/2}}. \tag{6.7}$$

In this regime there is another interesting region of $1 \ll r \ll L_c$ where the function $p_\infty(r)$ decays in a power law

$$p_\infty(r) = \text{const } r^{-d-1}. \tag{6.8}$$

Remarkably, Eq. (6.6) obtained for $d \gg 1$ properly describes also the one-dimensional wires (c.f. Eq. (4.21)).

The integral of $p_\infty(r)$ over the volume is convergent for all $\alpha \leq \alpha_c$ and remains finite in the limit $\alpha \to \alpha_c$ indicating that the wave functions at the transition point decay rather fast. At the same time, all moments of this quantity diverge in this limit. The second moment determines the electric susceptibility κ,

$$\kappa\delta_{\alpha\beta} = e^2 \int r_\alpha r_\beta p_\infty(r)\, d^d r. \tag{6.9}$$

Near the transition calculation of the integral in Eq. (6.9) leads to the result

$$\kappa = 4\pi^2 \nu c L_c, \tag{6.10}$$

where c is a coefficient.

This equation shows that the susceptibility in the critical region is proportional to the localization length L_c and not to L_c^2 as it would follow from the one-parameter scaling[9] and results obtained in $2 + \epsilon$ dimensions. The unusual dependence of the susceptibility κ on L_c in Eq. (6.10) arises formally from the anomalous exponent $(d+2)/2$ in the power law behavior of the pre-exponent in Eq. (6.6).

As concerns the metallic regime, one comes for the real lattices to the diffusion propagator, Eq. (2.26), at all $\alpha > \alpha_c$. However, except the limit $\alpha \gg 1$, the diffusion coefficient D obtained now is different form the classical diffusion coefficient D_0. Its behavior in the critical region $\alpha - \alpha_c \ll \alpha_c$ is especially interesting. This is not a power law behavior as one could expect from the one parameter scaling. Instead, the diffusion coefficient decays near the transition exponentially

$$D = \mathrm{const} \frac{\exp\left[-s\left(\alpha - \alpha_c\right)^{-1/2}\right]}{\left(\alpha - \alpha_c\right)^{3/2}}. \tag{6.11}$$

This is a very unusual behavior. Formally, it follows from the non-compact symmetry of the supermatrices Q. For any compact symmetry, one would obtain in the same approximation a power law dependence of the diffusion coefficient on $\alpha - \alpha_c$. The exponential decay of the diffusion coefficient D, Eq. (6.11), follows from the shape of the function Ψ (see Fig. 5.2). The position of the kink $\lambda_H = \cosh\theta_H$, Eq. (6.7), goes to infinity as $\lambda_H \propto \exp\left[s\left(\alpha - \alpha_c\right)^{1/2}\right]$ and this results in the form (6.11) of the diffusion coefficient.

The same results, Eqs. (6.6)–(6.11), have been obtained later[57] for the Anderson model on the Bethe lattice and this completed the study of this model started in Ref. 55. The agreement of the results obtained for the Anderson model and granulated σ-model is, of course, not accidental because, the critical behavior is formed by long time correlations and the result should not be sensitive to short distance structures. As it has been discussed previously, the low frequency behavior of wires and strictly one-dimensional chains is also described by identical equations.

The exponential decay of the diffusion coefficient was interpreted in Ref. 33 in terms of tunnelling between quasi-localized states. This may happen provided the wave function is concentrated in centers with a large distance

$$\zeta \propto (\alpha - \alpha_c)^{-1/2} \qquad (6.12)$$

between them. The decay of the amplitudes of the wave functions in the single center is fast as it can be seen from the fast decay in Eq. (6.8). Then, the tunnelling leads to an overlap between the wave functions of the different centers and to formation of a conduction band with an effective bandwidth Γ,

$$\Gamma \propto \Delta \exp(-a\zeta), \qquad (6.13)$$

where a is a coefficient.

The exponential decay of the diffusion coefficient D, Eq. (6.11), can follow quite naturally from such a picture. Of course, the picture implies the existence of weakly overlapping centers of the localization near this transition. A well established multifractality of wave functions at the transition (for a recent review, see, e.g. Ref. 36) may point out on this strong inhomogeneity near the transition.

Another indication in favor of the presented picture comes from the fact that the solution Ψ of Eq. (6.1) looses the sensitivity to the existence of the transition at frequencies $\omega \gtrsim \Gamma$. This can be seen from a more detailed analysis of Eq. (6.1). The interpretation in terms of formation of a very narrow conduction band near the transition with the bandwidth Γ is consistent with this property of the solution Ψ.

The fixed point found in $2 + \epsilon$ for small ϵ corresponds to a weak disorder and the strong inhomogeneities are not seen in this approach. One cannot speak of a narrow conduction band near the transition in $2 + \epsilon$ dimensions within this picture.

In principle, centers of (quasi) localization exist in $2D$ and can be described in the framework of σ-model (for a review, see Refs. 33 and 35). However, the idea about these centers of the (quasi) localizations is not incorporated in the conventional $2 + \epsilon$ scheme. So, the standard continuation of the results obtained for small ϵ to $\epsilon = 1$ may result in loosing an important information.

The non-trivial form of the function Ψ (see e.g. Fig. 5.2) has lead the present author to the idea[30] that this function might play the role of an order parameter for the Anderson transition. It was guessed that a Laplace transform of this function could be related to a conductance distribution. This idea has been further developed in Ref. 56, where a functional in an extended space was constructed such that its minimum was reached at the function $\Psi(Q)$. This resembles the Landau theory of phase transitions but the role of the order parameter is played a by a function.

The concept of the function order parameter was also discussed in later works on the Bethe lattice.[57]

7. Discussion

In this paper, the basics of the supersymmetry method has been presented. It is explained how the non-linear supermatrix σ-model is derived and it is shown how one can calculate within this model. It is demonstrated how one comes to the Wigner–Dyson statistics in a limited volume and how one obtains Anderson localization in disordered wires. Renormalization group scheme is explained in 2 and $2 + \epsilon$ dimensions for small ϵ, renormalization group equations are written and solved. It is shown how one solves the problem of the Anderson metal–insulator transition on the Bethe lattice and high dimensionality.

From the technical point of view, all this was a demonstration how one can calculate in the dimensions $d = 0$, $d = 1$, $d = 2$, and $d \gg 1$. Due to the lack of the space, the present paper is not a complete review of the application of the supersymmetry technique and many interesting works are not mentioned. However, the calculational schemes presented here have been used in most of the subsequent works. So, having read this paper one can get an idea on how one can work in all situations where the supersymmetry method is useful.

This is a chapter in the book devoted to 50 years of the Anderson localization and I tried to describe shortly how one of the directions of the field was developing in 1980s after the second most important work on the Anderson localization[9] has been published. Many of the authors of the present volume entered this field motivated by this publication. I hope that the development of the supersymmetry method has been useful in solving several interesting problems of the Anderson localization.

Although the supersymmetry method proved to be an adequate method for studying disordered systems (at least, without electron–electron interaction), several very important problems have not been solved so far. In spite of the common belief that all states are localized in disordered films (orthogonal and unitary ensembles), the solution for the two-dimensional σ-model has not been found in the limit of low frequencies. The problem of the integer quantum Hall effect has not been solved either, although the idea about instantons[22] was very useful for the understanding of this phenomenon. The problem of describing the critical behavior near the transition between the Hall plateaus still awaits its resolution. One more interesting problem is to understand the critical behavior near the Anderson transition.

Of course, a lot of information comes from numerical simulations but solving the $2D$ problem analytically would be really a great achievement. As concerns the Anderson transition in $3D$, the hope to solve it exactly is not realistic because even a simpler Ising model has not been solved in spite of numerous attempts. However, in the conventional theory of phase transition one can start with a mean field theory justifiable in high dimensions, determine the upper critical dimension and then make an expansion near this dimensionality.

Unfortunately, until now a similar procedure has not been found for the Anderson transition, although the supermatrix σ-model resembles spin models for which this procedure is standard. This concerns also the $2D$ case, where conventional spin σ-models are solvable. However, the well developed methods like the Bethe Ansatz or methods of the conformal field theory do not work here.

The formal reason of the failure of these approaches for studying the supermatrix σ-model is that the group of the symmetry of the supermatrices Q is not compact. These supermatrices consist of a block varying on a sphere and another one with elements on the hyperboloid. The latter part of Q is formally responsible for the localization but its presence leads to difficulties when applying the well developed methods. It is clear that the importance of the noncompact symmetry is not fully appreciated.

I can only express my hope that these problems will be resolved in the next 50 years and the book devoted to 100 years of the Anderson localization will contain the complete theory of this phenomenon.

Acknowledgments

Financial support from the Transregio 12 Physics "Symmetries and Universality in Mesoscopic of the German Research society is gratefully acknowledged".

References

1. P. W. Anderson, *Phys. Rev.* **109**, 1492 (1958).
2. A. A. Abrikosov, L. P. Gorkov and I. E. Dzyaloshinskii, *Methods of Quantum Field Theory in Statistical Physics* (Prentice Hall, New York, 1963).

3. V. L. Berezinsky, *Zh. Exp. Teor. Phys.* **65**, 1251 (1973) [*Sov. Phys. JETP* **38**, 620 (1974)].

4. N. F. Mott and W. D. Twose, *Adv. Phys.* **10**, 107 (1961).

5. L. P. Gorkov and G. M. Eliashberg, *Zh. Eksp. Teor. Fiz.* **48**, 1407 (1965) [*Sov. Phys. JETP* **21**, 1940 (1965)].

6. E. Wigner, *Ann. Math.* **53**, 36 (1951); *ibid.* **67**, 325 (1958).

7. F. J. Dyson, *J. Math. Phys.* **3**, 140, 157, 166, 1199 (1962).

8. D. J. Thouless, *Phys. Rep.* **13**, 93 (1974); *J. Phys. C* **8**, 1803 (1975); *Phys. Rev. Lett.* **39**, 1167 (1977).

9. E. Abrahams, P. W. Anderson, D. C. Licciardello and T. V. Ramakrishnan, *Phys. Rev. Lett.* **42**, 673 (1979).

10. L. P. Gorkov, A. I. Larkin and D. E. Khmelnitskii, *Pis'ma Zh. Eksp. Teor. Fiz.* **30**, 248 (1979) [*Sov. Phys. JETP Lett.* **30**, 228 (1979)].

11. E. Abrahams, P. W. Anderson and T. V. Ramakrishnan, *Phil. Mag.* **42**, 827 (1980).

12. E. Abrahams and T. V. Ramakrishnan, *J. Non-Cryst. Sol.* **35**, 15 (1980).

13. D. E. Kmelnitskii, unpublished (1979).

14. B. L. Altshuler, D. E. Khmelnitskii, A. I. Larkin and P. A. Lee, *Phys. Rev. B* **20**, 5142 (1980).

15. F. Wegner, *Z. Phys. B* **35**, 207 (1979).

16. L. Schäfer and F. Wegner, *Z. Phys. B* **38**, 113 (1980).

17. K. B. Efetov, A. I. Larkin and D. E. Khmelnitskii, *Zh. Eksp. Teor. Fiz.* **79**, 1120 (1980) [*Sov. Phys. JETP* **52**, 568 (1980)].

18. K. B. Efetov, unpublished (1981).

19. J. J. M. Verbaarschot and M. R. Zirnbauer, *J. Phys. A* **18**, 1093 (1995).

20. E. Kanzieper, *Phys. Rev. Lett.* **89**, 205201 (2002).

21. A. M. Finkelstein, *Zh. Eksp. Teor. Fiz.* **84**, 168 (1983) [*Sov. Phys. JETP* **57**, 97 (1983)].

22. A. M. M. Pruisken, *Nucl. Phys. B* **235**, 277 (1984).

23. K. B. Efetov, *Zh. Eksp. Teor. Fiz.* **82**, 872 (1982) [*Sov. Phys. JETP* **55**, 514 (1982)].

24. K. B. Efetov, *Zh. Eksp. Teor. Fiz.* **83**, 33 (1982) [*Sov. Phys. JETP* **56**, 467 (1982)]; *J. Phys. C* **15**, L909 (1982).

25. K. B. Efetov and A. I. Larkin, *Zh. Eksp. Teor. Fiz.* **85**, 764 (1983) [*Sov. Phys. JETP* **58**, 444 (1983).

26. K. B. Efetov, *Pis'ma Zh. Eksp. Teor. Fiz.* **40**, 17 (1984) [*Sov. Phys. JETP Lett.* **40**, 738 (1984)]; *Zh. Eksp. Teor. Fiz.* **88**, 1032 (1985) [*Sov. Phys. JETP* **61**, 606 (1985).

27. M. R. Zirnbauer, *Nucl. Phys. B* **265** [**FS15**], 375 (1986); *Phys. Rev. B* **34**, 6394 (1986).

28. K. B. Efetov, *Zh. Eksp. Teor. Fiz.* **92**, 638 (1987) [*Sov. Phys. JETP* **65**, 360 (1987)].

29. K. B. Efetov, *Zh. Eksp. Teor. Fiz.* **93**, 1125 (1987) [*Sov. Phys. JETP* **66**, 634 (1987)].

30. K. B. Efetov, *Zh. Eksp. Teor. Fiz.* **94**, 357 (1988) [*Sov. Phys. JETP* **67**, 199 (1988)].

31. K. B. Efetov, *Adv. Phys.* **32**, 53 (1983).
32. J. J. M. Verbaarschot, H. A. Weidenmüller and M. R. Zirnbauer, *Phys. Rep.* **129**, 367 (1985).
33. K. B. Efetov, *Supersymmetry in Disorder and Chaos* (Cambridge University Press, New York, 1997).
34. T. Guhr, A. Müller-Groeling and H. A. Weidenmüller, *Phys. Rep.* **299**, 190 (1998).
35. A. D. Mirlin, *Phys. Rep.* **326**, 259 (2000).
36. F. Evers and A. D. Mirlin, *Rev. Mod. Phys.* **80**, 1355 (2008).
37. F. A. Berezin, *The Method of Second Quantization* (Academic Press, New York, 1966); *Introduction to Superanalysis* (MPAM, Vol. 9, D. Reidel, Dordrecht, 1987).
38. A. Altland and M. R. Zirnbauer, *Phys. Rev. B* **55**, 1142 (1997).
39. O. Bohigas, M. J. Gianonni and C. Schmidt, *Phys. Rev. Lett.* **52**, 1 (1984); *J. Physique Lett.* **45**, L1615 (1984).
40. J. P. Keating, D. E. Khmelnitskii and I. V. Lerner (eds.), *Supersymmetry and Trace Formula*, NATO ASI Series B: Physics, Vol. 370 (Kluwer Academic, New York, 1999).
41. C. W. J. Beenakker, *Rev. Mod. Phys.* **69**, 733 (1997).
42. B. L. Altshuler and B. I. Shklovskii, *Zh. Eksp. Teor. Fiz.* **91**, 220 (1986) [*Sov. Phys. JETP* **64**, 127 (1986)].
43. A. Altland and A. Kamenev, *Phys. Rev. Lett.* **85**, 5615 (2000).
44. K. B. Efetov and A. I. Larkin, *Zh. Eksp. Teor. Fiz.* **85**, 764 (1983) [*Sov. Phys. JETP* **58**, 444 (1983)].
45. M. R. Zirnbauer, *Phys. Rev. Lett.* **69**, 1584 (1992).
46. O. N. Dorokhov, *Pis'ma Zh. Eksp. Teor. Fiz.* **36**, 259 (1982) [*JETP Lett.* **36**, 318 (1982)].
47. P. A. Mello, P. Pereyra and N. Kumar, *Ann. Phys. (N.Y.)* **181**, 290 (1988).
48. P. W. Brouwer and K. Frahm, *Phys. Rev. B* **53**, 1490 (1996).
49. Y. V. Fyodorov and A. D. Mirlin, *Phys. Rev. Lett.* **67**, 2405 (1991).
50. A. Altland and M. R. Zirnbauer, *Phys. Rev. Lett.* **77**, 4536 (1996).
51. A. M. Polyakov, *Phys. Lett. B* **59**, 79 (1975).
52. K. B. Efetov, *Zh. Eksp. Teor. Fiz.* **82**, 872 (1982) [*Sov. Phys. JETP* **55**, 514 (1982).
53. S. Hikami, A. I. Larkin and Y. Nagaoka, *Prog. Teor. Phys.* **63**, 707 (1980).
54. F. Wegner, *Z. Phys. B* **25**, 327 (1976).
55. R. Abou-Chacra, P. W. Anderson and D. J. Thouless, *J. Phys. C* **6**, 1734 (1973).
56. K. B. Efetov, *Physica A* **167**, 119 (1990).
57. A. D. Mirlin and Y. V. Fyodorov, *Nucl. Phys. B* **366**, 507 (1991); *Europhys. Lett.* **25**, 669 (1994); *J. Phys. (Paris)* **4**, 655 (1994).

Chapter 6

ANDERSON TRANSITIONS: CRITICALITY, SYMMETRIES AND TOPOLOGIES

A. D. Mirlin*, F. Evers, I. V. Gornyi[†] and P. M. Ostrovsky[‡]

Institut für Nanotechnologie
Karlsruhe Institute of Technology, 76021 Karlsruhe, Germany

The physics of Anderson transitions between localized and metallic phases in disordered systems is reviewed. We focus on the character of criticality as well as on underlying symmetries and topologies that are crucial for understanding phase diagrams and the critical behavior.

1. Introduction

Quantum interference can completely suppress the diffusion of a particle in random potential, a phenomenon known as *Anderson localization.*[1] For a given energy and disorder strength the quantum states are either all localized or all delocalized. This implies the existence of *Anderson transitions* between localized and metallic phases in disordered electronic systems. A great progress in understanding of the corresponding physics was achieved in the seventies and the eighties, due to the developments of scaling theory and field-theoretical approaches to localization, which demonstrated connections between the Anderson transition and conventional second-order phase transitions; see review articles[2-4] and the book.[5]

During the last 15 years, considerable progress in the field has been made in several research directions. This has strongly advanced the understanding of the physics of Anderson localization and associated quantum phase

---

*Also at Institut für Theorie der kondensierten Materie, Karlsruhe Institute of Technology, 76128 Karlsruhe, Germany and Petersburg Nuclear Physics Institute, 188300 St. Petersburg, Russia.
[†]Also at A. F. Ioffe Physico-Technical Institute, 194021 St. Petersburg, Russia.
[‡]Also at L. D. Landau Institute for Theoretical Physics RAS, 119334 Moscow, Russia.

transitions and allows us to view it nowadays in a considerably broader and more general context.[6]

First, the *symmetry* classification of disordered systems has been completed. It has been understood that a complete set of random matrix theories includes, in addition to the three Wigner–Dyson classes, three chiral ensembles and four Bogoliubov-de Gennes ensembles.[7] Zirnbauer has established a relation between random matrix theories, σ-models and Cartan's classification of symmetric spaces, which provides the mathematical basis for the statement of completeness of the classification.[8] The additional ensembles are characterized by one of the additional symmetries — the chiral or the particle-hole one. The field theories (σ-models) associated with these new symmetry classes have in fact been considered already in the eighties. However, it was only after their physical significance had been better understood that the new symmetry classes were studied systematically.

Second, the classification of fixed points governing the localization transitions in disordered metals was found to be much richer than that of symmetries of random matrix ensembles (or field theories). The first prominent example of this was in fact given 25 years ago by Pruisken[9] who showed that the quantum Hall transition is described by a σ-model with an additional, topological, term. However, it is only recently that the variety of types of criticality — and, in particular, the impact of *topology* — was fully appreciated. Recent experimental discoveries of graphene and topological insulators have greatly boosted the research activity in this direction.

Third, an important progress in understanding the statistics of wave functions at criticality has been made. Critical wave functions show very strong fluctuations and long-range correlations that are characterized by *multifractality*[4,6,10,12] implying the presence of infinitely many relevant operators. The spectrum of multifractal exponents constitutes a crucially important characteristics of the Anderson transition fixed point. The understanding of general properties of the statistics of critical wave functions and their multifractality was complemented by a detailed study — analytical and numerical — for a number of localization critical points, such as conventional Anderson transition in various dimensionalities, 2D Dirac fermions in a random vector potential, integer quantum Hall effect (IQHE), spin quantum Hall effect (SQHE), 2D symplectic-class Anderson transition, as well as the power-law random banded matrix (PRBM) model.

Fourth, for several types of Anderson transitions, very detailed studies using both analytical and numerical tools have been performed. As a result, a fairly comprehensive *quantitative understanding* of the localization critical phenomena has been achieved. In particular, the PRBM model, which can be viewed as a 1D system with long-range hopping, has been analytically solved on its critical line.[6,11,12] This allowed a detailed study of the statistics of wave functions and energy levels at criticality. The PRBM model serves at present as a "toy model" for the Anderson criticality. This model possesses a truly marginal coupling, thus yielding a line of critical points and allowing to study the evolution of critical properties in the whole range from weak- to strong-coupling fixed points. Further recent advances in quantitative understanding of the critical behavior of Anderson transitions are related to exploration of network models of IQHE and its "relatives" from other symmetry classes, development of theories of disordered Dirac fermions, as well as large progress in numerical simulations.

Finally, important advances have been achieved in understanding the impact of the *electron–electron interaction* on Anderson transitions. While this article mainly deals with non-interacting systems, we will discuss most prominent manifestations of the interaction in Secs. 5 and 6.2.

This article presents an overview of field with an emphasis on recent developments. The main focus is put on conceptual issues related to phase diagrams, the nature of criticality, and the role of underlying symmetries and topologies. For a more detailed exposition of the physics of particular Anderson transition points and an extended bibliography, the reader is referred to a recent review in Ref. 6.

2. Anderson Transitions in Conventional Symmetry Classes

2.1. *Scaling theory, observables and critical behavior*

When the energy or the disorder strength is varied, the system can undergo a transition from the metallic phase with delocalized eigenstates to the insulating phase, where eigenfunctions are exponentially localized,[1]

$$|\psi^2(\mathbf{r})| \sim \exp(-|\mathbf{r} - \mathbf{r_0}|/\xi),\qquad(2.1)$$

and ξ is the localization length. The character of this transition remained, however, unclear for roughly 20 years, until Wegner conjectured, developing earlier ideas of Thouless,[14] a close connection between the Anderson transition and the scaling theory of critical phenomena.[15] Three years later, Abrahams, Anderson, Licciardello and Ramakrishnan formulated a *scaling theory* of localization,[16] which describes the flow of the dimensionless conductance g with the system size L,

$$d\ln g/d\ln L = \beta(g).\qquad(2.2)$$

This phenomenological theory was put on a solid basis after Wegner's discovery of the field-theoretical description of the localization problem in terms of a non-linear σ-model,[17] Sec. 2.2. This paved the way for the resummation of singularities in perturbation theory at or near two dimensions[18,19] and allowed to cast the scaling in the systematic form of a field-theoretical renormalization group (RG). A microscopic derivation of the σ-model worked out in a number of papers[20–22] has completed a case for it as the field theory of the Anderson localization.

To analyze the transition, one starts from the Hamiltonian $\hat{H}$ consisting of the free part $\hat{H}_0$ and the disorder potential $U(\mathbf{r})$:

$$\hat{H} = \hat{H}_0 + U(\mathbf{r}) ; \qquad \hat{H}_0 = \hat{\mathbf{p}}^2/2m. \qquad (2.3)$$

The disorder is defined by the correlation function $\langle U(\mathbf{r})U(\mathbf{r}')\rangle$; we can assume it to be of the white-noise type for definiteness,

$$\langle U(\mathbf{r})U(\mathbf{r}')\rangle = (2\pi\rho\tau)^{-1}\delta(\mathbf{r} - \mathbf{r}'). \qquad (2.4)$$

Here ρ is the density of states, τ the mean free time and $\langle\ldots\rangle$ denote the disorder average. Models with finite-range and/or anisotropic disorder correlations are equivalent with respect to the long-time, long-distance behavior to the white noise model with renormalized parameters (tensor of diffusion coefficients).[23]

The physical observables whose scaling at the transition point is of primary importance is the localization length ξ on the insulating side (say, $E < E_c$) and the DC conductivity σ on the metallic side ($E > E_c$),

$$\xi \propto (E_c - E)^{-\nu}, \qquad \sigma \propto (E - E_c)^s. \qquad (2.5)$$

The corresponding critical indices ν and s satisfy the scaling relation[15] $s = \nu(d - 2)$.

On a technical level, the transition manifests itself in a change of the behavior of the diffusion propagator,

$$\Pi(\mathbf{r_1}, \mathbf{r_2}; \omega) = \langle G^R_{E+\omega/2}(\mathbf{r_1}, \mathbf{r_2})G^A_{E-\omega/2}(\mathbf{r_2}, \mathbf{r_1})\rangle, \qquad (2.6)$$

where G^R, G^A are retarded and advanced Green functions,

$$G^{R,A}_E(\mathbf{r}, \mathbf{r}') = \langle\mathbf{r}|(E - \hat{H} \pm i\eta)^{-1}|\mathbf{r}'\rangle, \qquad \eta \to +0. \qquad (2.7)$$

In the delocalized regime Π has the familiar diffusion form (in the momentum space),

$$\Pi(\mathbf{q}, \omega) = 2\pi\rho(E)/(Dq^2 - i\omega), \qquad (2.8)$$

where ρ is the density of states (DOS) and D is the diffusion constant, related to the conductivity via the Einstein relation $\sigma = e^2\rho D$. In the insulating phase, the propagator ceases to have the Goldstone form (2.8) and becomes massive,

$$\Pi(\mathbf{r_1}, \mathbf{r_2}; \omega) \simeq \frac{2\pi\rho}{-i\omega} \mathcal{F}(|\mathbf{r_1} - \mathbf{r_2}|/\xi), \qquad (2.9)$$

with the function $\mathcal{F}(\mathbf{r})$ decaying exponentially on the scale of the localization length, $\mathcal{F}(r/\xi) \sim \exp(-r/\xi)$. It is worth emphasizing that the localization length ξ obtained from the averaged correlation function $\Pi = \langle G^R G^A \rangle$, Eq. (2.6), is in general different from the one governing the exponential decay of the typical value $\Pi_{\text{typ}} = \exp\langle \ln G^R G^A \rangle$. For example, in quasi-1D systems the two lengths differ by a factor of four.[12] However, this is usually not important for the definition of the critical index ν. We will return to observables related to critical fluctuations of wave functions and discuss the corresponding family of critical exponents in Sec. 2.3.

2.2. *Field-theoretical description*

2.2.1. *Effective field theory: Non-linear σ-model*

In the original derivation of the σ-model,[17,20–22] the replica trick was used to perform the disorder averaging. Within this approach, n copies of the system are considered, with fields ϕ_α, $\alpha = 1, \dots, n$ describing the particles, and the replica limit $n \to 0$ is taken in the end. The resulting σ-model is defined on the $n \to 0$ limit of either non-compact or compact symmetric space, depending on whether the fields ϕ_α are considered as bosonic or fermionic. As an example, for the unitary symmetry class (A), which corresponds to a system with broken time-reversal invariance, the σ-model target manifold is $U(n,n)/U(n) \times U(n)$ in the first case and $U(2n)/U(n) \times U(n)$ in the second case, with $n \to 0$. A supersymmetric formulation given by Efetov[5] combines fermionic and bosonic degrees of freedom, with the field Φ becoming a supervector. The resulting σ-model is defined on a supersymmetric coset space, e.g. $U(1,1|2)/U(1|1) \times U(1|1)$ for the unitary class. This manifold combines compact and non-compact features and represents a product of the hyperboloid $H^2 = U(1,1)/U(1) \times U(1)$ and the sphere $S^2 = U(2)/U(1) \times U(1)$ "dressed" by anticommuting (Grassmannian) variables. While being equivalent to the replica version on the level of the perturbation theory (including its RG resummation), the supersymmetry formalism allows also for a nonperturbative treatment of the theory, which is particularly important for the analysis of the energy level and eigenfunction statistics, properties of quasi-1D systems, topological effects, etc.[5,12,24,25]

Focusing on the unitary symmetry class, the expression for the propagator Π, Eq. (2.6) is obtained as

$$\Pi(\mathbf{r_1}, \mathbf{r_2}; \omega) = \int DQ \, Q_{12}^{bb}(\mathbf{r_1}) Q_{21}^{bb}(\mathbf{r_2}) e^{-S[Q]}, \qquad (2.10)$$

where $S[Q]$ is the σ-model action

$$S[Q] = \frac{\pi\rho}{4} \int d^d\mathbf{r} \, \mathrm{Str} \left[-D(\nabla Q)^2 - 2i\omega\Lambda Q \right]. \qquad (2.11)$$

Here, $Q = T^{-1}\Lambda T$ is a 4×4 supermatrix that satisfies the condition $Q^2 = 1$ and belongs to the σ-model target space described above, $\Lambda = \mathrm{diag}\{1, 1, -1, -1\}$, and Str denotes the supertrace. The size 4 of the matrix is due to (i) two types of the Green functions (advanced and retarded), and (ii) necessity to introduce bosonic and fermionic degrees of freedom to represent these Green's function in terms of a functional integral. The matrix Q consists thus of four 2×2 blocks according to its advanced-retarded structure, each of them being a supermatrix in the boson-fermion space. In particular, Q_{12}^{bb} is the boson-boson element of the RA block, and so on. One can also consider an average of the product of n retarded and n advanced Green functions, which will generate a σ-model defined on a larger manifold, with the base being a product of $U(n, n)/U(n) \times U(n)$ and $U(2n)/U(n) \times U(n)$ (these are the same structures as in the replica formalism, but now *without* the $n \to 0$ limit).

For other symmetry classes, the symmetry of the σ-model is different but the general picture is the same. For example, for the orthogonal class (AI) the 8×8 Q-matrices span the manifold whose base is the product of the non-compact space $O(2, 2)/O(2) \times O(2)$ and the compact space $Sp(4)/Sp(2) \times Sp(2)$. The σ-model symmetric spaces for all the classes (Wigner–Dyson as well as unconventional) are listed in Sec. 3.

2.2.2. *RG in $2 + \epsilon$ dimensions; ϵ-expansion*

The σ-model is the effective low-momentum, low-frequency theory of the problem, describing the dynamics of interacting soft modes — diffusons and cooperons. Its RG treatment yields a flow equation of the form (2.2), thus justifying the scaling theory of localization. The β-function $\beta(t) \equiv -dt/d\ln L$ can be calculated perturbatively in the coupling constant t inversely proportional to the dimensionless conductance, $t = 1/2\pi g$.[a]

---

[a]For spinful systems, g here does not include summation over spin projections.

This allows one to get the ϵ-expansion for the critical exponents in $2 + \epsilon$ dimensions, where the transition takes place at $t_* \ll 1$. In particular, for the orthogonal symmetry class (AI) one finds[27]

$$\beta(t) = \epsilon t - 2t^2 - 12\zeta(3)t^5 + O(t^6). \tag{2.12}$$

The transition point t_* is given by the zero of the $\beta(t)$,

$$t_* = \epsilon/2 - (3/8)\,\zeta(3)\epsilon^4 + O(\epsilon^5). \tag{2.13}$$

The localization length exponent ν is determined by the derivative

$$\nu = -1/\beta'(t_*) = \epsilon^{-1} - (9/4)\,\zeta(3)\epsilon^2 + O(\epsilon^3), \tag{2.14}$$

and the conductivity exponent s is

$$s = \nu\epsilon = 1 - (9/4)\,\zeta(3)\epsilon^3 + O(\epsilon^4). \tag{2.15}$$

Numerical simulations of localization on fractals with dimensionality slightly above 2 give the behavior of ν that is in good agreement with Eq. (2.14).[28] For the unitary symmetry class (A), the corresponding results read

$$\beta(t) = \epsilon t - 2t^3 - 6t^5 + O(t^7); \tag{2.16}$$
$$t_* = (\epsilon/2)^{1/2} - (3/2)\,(\epsilon/2)^{3/2} + O(\epsilon^{5/2}); \tag{2.17}$$
$$\nu = 1/2\epsilon - 3/4 + O(\epsilon) ; \qquad s = 1/2 - (3/4)\epsilon + O(\epsilon^2). \tag{2.18}$$

In 2D ($\epsilon = 0$) the fixed point t_* in both cases becomes zero: $\beta(t)$ is negative for any $t > 0$, implying that all states are localized. The situation is qualitatively different for the third — symplectic — Wigner–Dyson class. The corresponding β-function is related to that for the orthogonal class via $\beta_{\mathrm{Sp}}(t) = -2\beta_{\mathrm{O}}(-t/2)$, yielding[b]

$$\beta(t) = \epsilon t + t^2 - (3/4)\,\zeta(3)t^5 + O(t^6). \tag{2.19}$$

In 2D, the β-function (2.19) is positive at sufficiently small t, implying the existence of a truly metallic phase at $t < t_*$, with an Anderson transition at certain $t_* \sim 1$. This peculiarity of the symplectic class represents one of mechanisms of the emergence of criticality in 2D, see Sec. 4.1. The β-functions of unconventional symmetry classes will be discussed in Sec. 3.5.

---

[b]Here $t = 1/\pi g$, where g is the total conductance of the spinful system.

2.3. Critical wave functions: Multifractality

2.3.1. Scaling of inverse participation ratios and correlations at criticality

Multifractality of wave functions, describing their strong fluctuations at criticality, is a striking feature of the Anderson transitions.[29,30] Multifractality as a concept has been introduced by Mandelbrot.[31] Multifractal structures are characterized by an infinite set of critical exponents describing the scaling of the moments of some distribution. This feature has been observed in various complex objects, such as the energy dissipating set in turbulence, strange attractors in chaotic dynamical systems, and the growth probability distribution in diffusion-limited aggregation. For the present problem, the underlying normalized measure is just $|\psi^2(\mathbf{r})|$ and the corresponding moments are the inverse participation ratios (IPR)[c]

$$P_q = \int d^d\mathbf{r}|\psi(\mathbf{r})|^{2q}. \qquad (2.20)$$

At criticality, P_q show an anomalous scaling with the system size L,

$$\langle P_q \rangle = L^d \langle |\psi(\mathbf{r})|^{2q} \rangle \sim L^{-\tau_q}, \qquad (2.21)$$

governed by a continuous set of exponents τ_q. One often introduces fractal dimensions D_q via $\tau_q = D_q(q-1)$. In a metal $D_q = d$, in an insulator $D_q = 0$, while at a critical point D_q is a non-trivial function of q, implying wave function multifractality. Splitting off the normal part, one defines the anomalous dimensions Δ_q,

$$\tau_q \equiv d(q-1) + \Delta_q, \qquad (2.22)$$

which distinguish the critical point from the metallic phase and determine the scale dependence of the wave function correlations. Among them, $\Delta_2 \equiv -\eta$ plays the most prominent role, governing the spatial correlations of the "intensity" $|\psi|^2$,

$$L^{2d} \langle |\psi^2(\mathbf{r})\psi^2(\mathbf{r}')| \rangle \sim (|\mathbf{r} - \mathbf{r}'|/L)^{-\eta}. \qquad (2.23)$$

---

[c]Strictly speaking, P_q as defined by Eq. (2.20), diverges for sufficiently negative q ($q \leq -1/2$ for real ψ and $q \leq -3/2$ for complex ψ), because of zeros of wave functions related to their oscillations on the scale of the wave length. To find τ_q for such negative q, one should first smooth $|\psi^2|$ by averaging over some microscopic volume (block of several neighboring sites in the discrete version).

Equation (2.23) can be obtained from (2.21) by using that the wave function amplitudes become essentially uncorrelated at $|\mathbf{r} - \mathbf{r}'| \sim L$. Scaling behavior of higher order correlations, $\langle |\psi^{2q_1}(\mathbf{r_1})\psi^{2q_2}(\mathbf{r_2}) \dots \psi^{2q_n}(\mathbf{r_n})| \rangle$, can be found in a similar way, e.g.

$$L^{d(q_1+q_2)}\langle |\psi^{2q_1}(\mathbf{r_1})\psi^{2q_2}(\mathbf{r_2})| \rangle \sim L^{-\Delta_{q_1}-\Delta_{q_2}}(|\mathbf{r_1} - \mathbf{r_2}|/L)^{\Delta_{q_1+q_2}-\Delta_{q_1}-\Delta_{q_2}}. \tag{2.24}$$

Correlations of different (close in energy) eigenfunctions exhibit the same scaling,[32]

$$\left. \begin{array}{l} L^{2d}\langle |\psi_i^2(\mathbf{r})\psi_j^2(\mathbf{r}')| \rangle \\ L^{2d}\langle \psi_i(\mathbf{r})\psi_j^*(\mathbf{r})\psi_i^*(\mathbf{r}')\psi_j(\mathbf{r}') \rangle \end{array} \right\} \sim \left(\frac{|\mathbf{r} - \mathbf{r}'|}{L_\omega} \right)^{-\eta}, \tag{2.25}$$

where $\omega = \epsilon_i - \epsilon_j$, $L_\omega \sim (\rho\omega)^{-1/d}$, ρ is the density of states, and $|\mathbf{r} - \mathbf{r}'| < L_\omega$. For conventional classes, where the DOS is uncritical, the diffusion propagator (2.6) scales in the same way.

In the field-theoretical language (Sec. 2.2), Δ_q are the leading anomalous dimensions of the operators $\text{Tr}(Q\Lambda)^q$ (or, more generally, $\text{Tr}(Q\Lambda)^{q_1} \dots \text{Tr}(Q\Lambda)^{q_m}$ with $q_1 + \dots + q_m = q$).[29] The strong multifractal fluctuations of wave functions at criticality are related to the fact that $\Delta_q < 0$ for $q > 1$, so that the corresponding operators increase under RG. In this formalism, the scaling of correlation functions [Eq. (2.23) and its generalizations] results from an operator product expansion.[33–35]

2.3.2. *Singularity spectrum* $f(\alpha)$

The average IPR $\langle P_q \rangle$ are (up to the normalization factor L^d) the moments of the distribution function $\mathcal{P}(|\psi|^2)$ of the eigenfunction intensities. The behavior (2.21) of the moments corresponds to the intensity distribution function of the form

$$\mathcal{P}(|\psi^2|) \sim \frac{1}{|\psi^2|} L^{-d+f(-\frac{\ln|\psi^2|}{\ln L})} \tag{2.26}$$

Indeed, calculating the moments $\langle |\psi^{2q}| \rangle$ with the distribution (2.26), one finds

$$\langle P_q \rangle = L^d \langle |\psi^{2q}| \rangle \sim \int d\alpha \, L^{-q\alpha+f(\alpha)}, \tag{2.27}$$

where we have introduced $\alpha = -\ln|\psi^2|/\ln L$. Evaluation of the integral by the saddle-point method (justified at large L) reproduces Eq. (2.21), with the exponent τ_q related to the singularity spectrum $f(\alpha)$ via the Legendre transformation,

$$\tau_q = q\alpha - f(\alpha), \qquad q = f'(\alpha), \qquad \alpha = \tau_q'. \tag{2.28}$$

The meaning of the function $f(\alpha)$ is as follows: it is the fractal dimension of the set of those points $\mathbf{r}$ where the eigenfunction intensity is $|\psi^2(\mathbf{r})| \sim L^{-\alpha}$. In other words, in a lattice version of the model the number of such points scales as $L^{f(\alpha)}$.[36]

General properties of τ_q and $f(\alpha)$ follow from their definitions and the wave function normalization:

(i) τ_q is a non-decreasing, convex function ($\tau'_q \geq 0$, $\tau''_q \leq 0$), with $\tau_0 = -d$, $\tau_1 = 0$;

(ii) $f(\alpha)$ is a convex function ($f''(\alpha) \leq 0$) defined on the semiaxis $\alpha \geq 0$ with a maximum at some point α_0 (corresponding to $q = 0$ under the Legendre transformation) and $f(\alpha_0) = d$. Further, for the point α_1 (corresponding to $q = 1$) we have $f(\alpha_1) = \alpha_1$ and $f'(\alpha_1) = 1$.

If one formally defines $f(\alpha)$ for a metal, it will be concentrated in a single point $\alpha = d$, with $f(d) = d$ and $f(\alpha) = -\infty$ otherwise. On the other hand, at criticality this "needle" broadens and the maximum shifts to a position $\alpha_0 > d$, see Fig. 6.1.

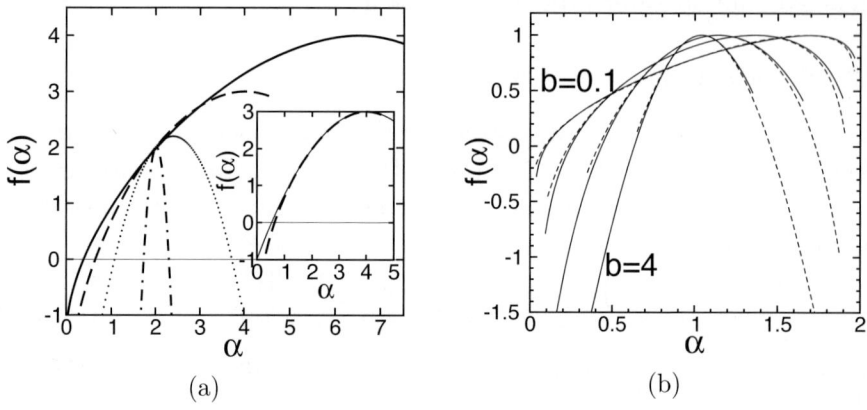

$$(a) \qquad\qquad (b)$$

Fig. 6.1.   Multifractality at Anderson transitions. (a) Singularity spectrum $f(\alpha)$ in $d = 2 + \epsilon$ for $\epsilon = 0.01$ and $\epsilon = 0.2$ (analytical), as well as 3D and 4D (numerical). With increasing d the spectrum gets broader, implying stronger multifractality. Inset: comparison between $f(\alpha)$ for 3D and the one-loop result of the $2+\epsilon$ expansion with $\epsilon = 1$ (solid).[41] (b) Singularity spectrum $f(\alpha)$ for the PRBM model. Evolution from weak to strong multifractality with decreasing parameter b is evident. Dashed lines represent $f(2 - \alpha) + \alpha - 1$, demonstrating the validity of Eq. (2.30).[38]

2.3.3. *Symmetry of the multifractal spectra*

As was recently shown,[37] the multifractal exponents for the Wigner–Dyson classes satisfy an exact symmetry relation

$$\Delta_q = \Delta_{1-q} , \qquad (2.29)$$

connecting exponents with $q < 1/2$ (in particular, with negative q) to those with $q > 1/2$. In terms of the singularity spectrum, this implies

$$f(2d - \alpha) = f(\alpha) + d - \alpha. \qquad (2.30)$$

The analytical derivation of Eqs. (2.29) and (2.30) is based on the super-symmetric σ-model; it has been confirmed by numerical simulations on the PRBM model[37,38] (see Fig. 6.1b) and 2D Anderson transition of the symplectic class.[39,40]

2.3.4. *Dimensionality dependence of multifractality*

Let us analyze the evolution[41] from the weak-multifractality regime in $d = 2 + \epsilon$ dimensions to the strong multifractality at $d \gg 1$.

In $2 + \epsilon$ dimensions with $\epsilon \ll 1$, the multifractality exponents can be obtained within the ϵ-expansion, Sec. 2.2.2. The 4-loop results for the orthogonal and unitary symmetry classes read[42]

$$\Delta_q^{(O)} = q(1 - q)\epsilon + \frac{\zeta(3)}{4}q(q - 1)(q^2 - q + 1)\epsilon^4 + O(\epsilon^5); \qquad (2.31)$$

$$\Delta_q^{(U)} = q(1 - q)(\epsilon/2)^{1/2} - \frac{3}{8}q^2(q - 1)^2\zeta(3)\epsilon^2 + O(\epsilon^{5/2}). \qquad (2.32)$$

Keeping only the leading term on the r.h.s. of Eqs. (2.31) and (2.32), we get the one-loop approximation for τ_q which is of parabolic form.

Numerical simulations[41] of the wave function statistics in 3D and 4D (see Fig. 6.1a) have shown a full qualitative agreement with analytical predictions, both in the form of multifractal spectra and in the shape of the IPR distribution. Moreover, the one-loop result of the $2 + \epsilon$ expansion with $\epsilon = 1$ describes the 3D singularity spectrum with a remarkable accuracy (though with detectable deviations). In particular, the position of the maximum, $\alpha_0 = 4.03 \pm 0.05$, is very close to its value $\alpha_0 = d + \epsilon$ implied by one-loop approximation. As expected, in 4D the deviations from parabolic shape are much more pronounced and $\alpha_0 = 6.5 \pm 0.2$ differs noticeably from 6.

The simulations[41] also show that fractal dimensions $D_q \equiv \tau_q/(q-1)$ with $q \gtrsim 1$ decrease with increasing d. As an example, for $q = 2$ we have $D_2 \simeq 2 - 2\epsilon$ in $2 + \epsilon$ dimensions, $D_2 = 1.3 \pm 0.05$ in 3D, and $D_2 = 0.9 \pm 0.15$ in 4D.

This confirms the expectation based on the Bethe-lattice results (Sec. 2.5) that $\tau_q \to 0$ at $d \to \infty$ for $q > 1/2$. Such a behavior of the multifractal exponents is a manifestation of a very sparse character of critical eigenstates at $d \gg 1$, formed by rare resonance spikes. In combination with the relation (2.29), this implies the limiting form of the multifractal spectrum at $d \to \infty$,

$$\tau_q = \begin{cases} 0 , & q \geq 1/2 \\ 2d(q - 1/2) , & q \leq 1/2 . \end{cases} \tag{2.33}$$

This corresponds to $f(\alpha)$ of the form

$$f(\alpha) = \alpha/2 , \qquad 0 < \alpha < 2d , \tag{2.34}$$

dropping to $-\infty$ at the boundaries of the interval $[0, 2d]$. It was argued[41] that the way the multifractality spectrum approaches this limiting form with increasing d is analogous to the behavior found[13] in the PRBM model with $b \ll 1$.

2.3.5. *Surface vs. bulk multifractality*

Recently, the concept of wave function multifractality was extended[43] to the surface of a system at an Anderson transition. It was shown that fluctuations of critical wave functions at the surface are characterized by a new set of exponent τ_q^{s} (or, equivalently, anomalous exponents Δ_q^{s}) independent from their bulk counterparts,

$$L^{d-1} \langle |\psi(\mathbf{r})|^{2q} \rangle \sim L^{-\tau_q^{\mathrm{s}}}, \tag{2.35}$$

$$\tau_q^{\mathrm{s}} = d(q - 1) + q\mu + 1 + \Delta_q^{\mathrm{s}}. \tag{2.36}$$

Here, μ is introduced for generality, in order to account for a possibility of non-trivial scaling of the average value, $\langle |\psi(\mathbf{r})|^2 \rangle \propto L^{-d-\mu}$, at the boundary in unconventional symmetry classes. For the Wigner–Dyson classes, $\mu = 0$. The normalization factor L^{d-1} is chosen such that Eq. (2.35) yields the contribution of the surface to the IPR $\langle P_q \rangle = \langle \int d^d\mathbf{r} |\psi(\mathbf{r})|^{2q} \rangle$. The exponents Δ_q^{s} as defined in Eq. (2.36) vanish in a metal and govern statistical fluctuations of wave functions at the boundary, $\langle |\psi(\mathbf{r})|^{2q} \rangle / \langle |\psi(\mathbf{r})|^2 \rangle^q \sim L^{-\Delta_q^{\mathrm{s}}}$, as well as their spatial correlations, e.g. $L^{2(d+\mu)} \langle |\psi^2(\mathbf{r})\psi^2(\mathbf{r}')| \rangle \sim (|\mathbf{r} - \mathbf{r}'|/L)^{\Delta_2^{\mathrm{s}}}$.

Wave function fluctuations are much stronger at the edge than in the bulk. As a result, surface exponents are important even if one performs a multifractal analysis for the whole sample, without separating it into "bulk" and "surface", despite the fact that the weight of surface points is down by a factor $1/L$.

The boundary multifractality was explicitly studied, analytically as well as numerically, for a variety of critical systems, including weak multifractality in 2D and $2 + \epsilon$ dimensions, the 2D spin quantum Hall transition,[43] the Anderson transition in a 2D system with spin-orbit coupling,[40] and the PRBM model.[38] The notion of surface multifractality was further generalized[40] to a corner of a critical system.

2.4. *Additional comments*

(i) For the lack of space we do not discuss the issues of IPR distributions at criticality and the role of ensemble averaging, as well as possible singularities in multifractal spectra; see Ref. 6.

(ii) Recently, an impressive progress was achieved in experimental studies of Anderson transitions in various systems.[44-48] The developed experimental techniques permit spatially resolved investigation of wave functions, thus paving the way to experimental study of multifractality. While the obtained multifractal spectra differ numerically from theoretical expectations (possibly because the systems were not exactly at criticality, or, in the case of electronic systems, pointing to importance of electron–electron interaction), the experimental advances seem very promising.

(iii) Recent theoretical work[49] explains the properties of a superconductor–insulator transition observed in a class of disordered films in terms of multifractality of electronic wave functions.

2.5. *Anderson transition in $d = \infty$: Bethe lattice*

The Bethe lattice (BL) is a tree-like lattice with a fixed coordination number. Since the number of sites at a distance r increases exponentially with r on the BL, it effectively corresponds to the limit of high dimensionality d. The BL models are the closest existing analogs of the mean-field theory for the case of the Anderson transition. The Anderson tight-binding model (lattice version of Eqs. (2.3) and (2.4)) on the BL was studied for the first time in Ref. 50, where the existence of the metal–insulator transition was proven and the position of the mobility edge was determined. Later, the BL versions of the σ-model (2.11)[51,52] and of the tight-binding model[53] were studied within the supersymmetry formalism, which allowed to determine the critical behavior. It was found that the localization length diverges in the way usual for BL models, $\xi \propto |E - E_c|^{-1}$, where E is a microscopic parameter driving the transition. When reinterpreted within the effective-medium approximation,[54,55] this yields the conventional mean-field value of the localization length exponent, $\nu = 1/2$. On the other hand, the critical behavior of other observables is very peculiar. The inverse participation ratios P_q with $q > 1/2$ have a finite limit at $E \to E_c$ when the critical point is approached from the localized phase and then jump to zero. By comparison with the scaling formula, $P_q \propto \xi^{-\tau_q}$, this can be interpreted as $\tau_q = 0$ for all $q \geq 1/2$. Further, in the delocalized phase the diffusion coefficient vanishes exponentially when the critical point is approached,

$$D \propto \Omega^{-1} \ln^3 \Omega \qquad \Omega \sim \exp\{\text{const } |E - E_c|^{-1/2}\}, \qquad (2.37)$$

which can be thought as corresponding to the infinite value, $s = \infty$, of the critical index s. The distribution of the LDOS $v \equiv \rho(r)/\langle\rho\rangle$ (normalized to its average value for convenience) was found to be of the form

$$\mathcal{P}(v) \propto \Omega^{-1/2} v^{-3/2}, \qquad \Omega^{-1} \ll v \ll \Omega, \qquad (2.38)$$

and exponentially small outside this range. Equation (2.38) implies for the moments of the LDOS:

$$\langle v^q \rangle \propto \Omega^{|q-1/2|-1/2}. \qquad (2.39)$$

The physical reason for the unconventional critical behavior was unraveled in Ref. 56. It was shown that the exponential largeness of Ω reflects the spatial structure of the BL: the "correlation volume" V_ξ (number of sites within a distance ξ from the given one) on such a lattice is exponentially large. On the other hand, for any finite dimensionality d the correlation volume has a power-law behavior, $V_d(\xi) \propto \xi^d \propto |E - E_c|^{\nu d}$, where $\nu \simeq 1/2$ at large d. Thus, the scale Ω cannot appear for finite d and, assuming some matching between the BL and large-d results, will be replaced by $V_d(\xi)$. Then Eq. (2.39) yields the following high-d behavior of the anomalous exponents Δ_q governing the scaling of the LDOS moments (Sec. 2.3),

$$\Delta_q \simeq d(1/2 - |q - 1/2|), \qquad (2.40)$$

or, equivalently, the results (2.33) and (2.34) for the multifractal spectra τ_q, $f(\alpha)$. These formulas describe the strongest possible multifractality.

The critical behavior of the conductivity, Eq. (2.37), is governed by the same exponentially large factor Ω. When it is replaced by the correlation volume $V_d(\xi)$, the power-law behavior at finite $d \gg 1$ is recovered, $\sigma \propto |E - E_c|^s$ with $s \simeq d/2$. The result for the exponent s agrees (within its accuracy, i.e. to the leading order in d) with the scaling relation $s = \nu(d-2)$.

3. Symmetries of Disordered Systems

In this section, we briefly review the symmetry classification of disordered systems based on the relation to the classical symmetric spaces, which was established in Refs. 7 and 8.

3.1. *Wigner–Dyson classes*

The random matrix theory (RMT) was introduced into physics by Wigner.[57] Developing Wigner's ideas, Dyson[58] put forward a classification scheme of ensembles of random Hamiltonians. This scheme takes into account the invariance of the system under time reversal and spin rotations, yielding three symmetry classes: unitary, orthogonal and symplectic. If the time-reversal invariance (T) is broken, the Hamiltonians are just arbitrary Hermitian matrices,

$$H = H^\dagger , \tag{3.1}$$

with no further constraints. This set of matrices is invariant with respect to rotations by unitary matrices; hence the name "unitary ensemble". In this situation, the presence or absence of spin rotation invariance (S) is not essential: if the spin is conserved, H is simply a spinless unitary-symmetry Hamiltonian times the unit matrix in the spin space. In the RMT one considers most frequently an ensemble of matrices with independent, Gaussian-distributed random entries — the Gaussian unitary ensemble (GUE). While disordered systems have much richer physics than the Gaussian ensembles, their symmetry classification is inherited from the RMT.

Let us now turn to the systems with preserved time-reversal invariance. The latter is represented by an antiunitary operator, $T = KC$, where C is the operator of complex conjugation and K is unitary. The time-reversal invariance thus implies $H = KH^T K^{-1}$ (we used the Hermiticity, $H^* = H^T$). Since acting twice with T should leave the physics unchanged, one infers that $K^* K = p$, where $p = \pm 1$. As was shown by Wigner, the two cases correspond to systems with integer ($p = +1$) and half-integer ($p = -1$) angular momentum. If $p = 1$, a representation can be chosen where $K = 1$, so that

$$H = H^T . \tag{3.2}$$

The set of Hamiltonians thus spans the space of real symmetric matrices in this case. This is the orthogonal symmetry class; its representative is the Gaussian orthogonal ensemble (GOE). For disordered electronic systems this class is realized when spin is conserved, as the Hamiltonian then reduces to that for spinless particles (times unit matrix in the spin space).

If T is preserved but S is broken, we have $p = -1$. In the standard representation, K is then realized by the second Pauli matrix, $K = i\sigma_y$, so that the Hamiltonian satisfies

$$H = \sigma_y H^{\mathrm{T}} \sigma_y. \tag{3.3}$$

It is convenient to split the $2N \times 2N$ Hamiltonian in 2×2 blocks (quaternions) in spin space. Each of them then is of the form $q = q_0\sigma_0 + iq_1\sigma_x + iq_2\sigma_y + iq_3\sigma_z$ (where σ_0 is the unit matrix and $\sigma_{x,y,z}$ the Pauli matrices), with real q_μ, which defines a real quaternion. This set of Hamiltonians is invariant with respect to the group of unitary transformations conserving σ_y, $U\sigma_y U^T = \sigma_y$, which is the symplectic group $\mathrm{Sp}(2N)$. The corresponding symmetry class is thus called symplectic, and its RMT representative is the Gaussian symplectic ensemble (GSE).

3.2. *Relation to symmetric spaces*

Before discussing the relation to the families of symmetric spaces, we briefly remind the reader how the latter are constructed.[59,60] Let G be one of the compact Lie groups $\mathrm{SU}(N)$, $\mathrm{SO}(N)$, $\mathrm{Sp}(2N)$, and $\mathfrak{g}$ the corresponding Lie algebra. Further, let θ be an involutive automorphism $\mathfrak{g} \to \mathfrak{g}$ such that $\theta^2 = 1$ but θ is not identically equal to unity. It is clear that θ splits $\mathfrak{g}$ in two complementary subspaces, $\mathfrak{g} = \mathfrak{K} \oplus \mathfrak{P}$, such that $\theta(X) = X$ for $X \in \mathfrak{K}$ and $\theta(X) = -X$ for $X \in \mathfrak{P}$. It is easy to see that the following Lie algebra multiplication relations holds:

$$[\mathfrak{K}, \mathfrak{K}] \subset \mathfrak{K}, \qquad [\mathfrak{K}, \mathfrak{P}] \subset \mathfrak{P}, \qquad [\mathfrak{P}, \mathfrak{P}] \subset \mathfrak{K}. \tag{3.4}$$

This implies, in particular, that $\mathfrak{K}$ is a subalgebra, whereas $\mathfrak{P}$ is not. The coset space G/K (where K is the Lie group corresponding to $\mathfrak{K}$) is then a compact symmetric space. The tangent space to G/K is $\mathfrak{P}$. One can also construct an associated non-compact space. For this purpose, one first defines the Lie algebra $\mathfrak{g}^* = \mathfrak{K} \oplus i\mathfrak{P}$, which differs from $\mathfrak{g}$ in that the elements in $\mathfrak{P}$ are multiplied by i. Going to the corresponding group and dividing K out, one gets a non-compact symmetric space G^*/K.

The groups G themselves are also symmetric spaces and can be viewed as coset spaces $G \times G/G$. The corresponding non-compact space is $G^{\mathbb{C}}/G$, where $G^{\mathbb{C}}$ is the complexification of G (which is obtained by taking the Lie algebra $\mathfrak{g}$, promoting it to the algebra over the field of complex numbers, and then exponentiating).

The connection with symmetric spaces is now established in the following way.[7,8] Consider first the unitary symmetry class. Multiplying a Hamiltonian matrix by i, we get an antihermitean matrix $X = iH$. Such matrices form the Lie algebra $\mathfrak{u}(N)$. Exponentiating it, one gets the Lie group $U(N)$, which is the compact symmetric space of class A in Cartan's classification. For the orthogonal class, $X = iH$ is purely imaginary and symmetric. The set of such matrices is a linear complement $\mathfrak{P}$ of the algebra $\mathfrak{K} = \mathfrak{o}(N)$ of imaginary antisymmetric matrices in the algebra $\mathfrak{g} = \mathfrak{u}(N)$ of antihermitean matrices. The corresponding symmetric space is $G/K = U(N)/O(N)$, which is termed AI in Cartan's classification. For the symplectic ensemble the same consideration leads to the symmetric space $U(N)/Sp(N)$, which is the compact space of the class AII. If we don't multiply H by i but instead proceed with H in the analogous way, we end up with associated non-compact spaces G^*/K. To summarize, the linear space $\mathfrak{P}$ of Hamiltonians can be considered as a tangent space to the compact G/K and non-compact G^*/K symmetric spaces of the appropriate symmetry class.

This correspondence is summarized in Table 6.1, where the first three rows correspond to the Wigner–Dyson classes, the next three to the chiral classes (Sec. 3.3) and last four to the Bogoliubov-de Gennes classes (Sec. 3.4). The last two columns of the table specify the symmetry of the corresponding σ-model. In the supersymmetric formulation, the base of the σ-model target space $\mathcal{M}_B \times \mathcal{M}_F$ is the product of a non-compact symmetric space $\mathcal{M}_B$ corresponding to the bosonic sector and a compact ("fermionic") symmetric space $\mathcal{M}_F$. (In the replica formulation, the space is $\mathcal{M}_B$ for bosonic or $\mathcal{M}_F$ for fermionic replicas, supplemented with the limit $n \to 0$.) The Cartan symbols for these symmetric spaces are given in the sixth column, and the compact components $\mathcal{M}_F$ are listed in the last column. It should be stressed that the symmetry classes of $\mathcal{M}_B$ and $\mathcal{M}_F$ are different from the symmetry class of the ensemble (i.e. of the Hamiltonian) and in most cases are also different from each other. Following the common convention, when we refer to a system as belonging to a particular class, we mean the symmetry class of the Hamiltonian.

It is also worth emphasizing that the orthogonal groups appearing in the expressions for $\mathcal{M}_F$ are $O(N)$ rather than $SO(N)$. This difference (which was irrelevant when we were discussing the symmetry of the Hamiltonians, as it does not affect the tangent space) is important here, since it influences topological properties of the manifold. As we will discuss in detail in Secs. 4 and 6, the topology of the σ-model target space often affects the localization properties of the theory in a crucial way.

Table 6.1. Symmetry classification of disordered systems. First column: symbol for the symmetry class of the Hamiltonian. Second column: names of the corresponding RMT. Third column: presence (+) or absence (−) of the time-reversal (T) and spin-rotation (S) invariance. Fourth and fifth columns: families of the compact and non-compact symmetric spaces of the corresponding symmetry class. The Hamiltonians span the tangent space to these symmetric spaces. Sixth column: symmetry class of the σ-model; the first symbol corresponds to the non-compact ("bosonic") and the second to the compact ("fermionic") sector of the σ-model manifold. The compact component $\mathcal{M}_F$ (which is particularly important for theories with non-trivial topological properties) is explicitly given in the last column. From Ref. 6.

Ham. class	RMT	T	S	Compact symmetric space	Non-Compact symmetric space	σ-model B|F	σ-model compact sector $\mathcal{M}_F$
Wigner–Dyson classes							
A	GUE	−	±	$U(N)\times U(N)/U(N) \equiv U(N)$	$GL(N,\mathbb{C})/U(N)$	AIII|AIII	$U(2n)/U(n)\times U(n)$
AI	GOE	+	+	$U(N)/O(N)$	$GL(N,\mathbb{R})/O(N)$	BDI|CII	$Sp(4n)/Sp(2n)\times Sp(2n)$
AII	GSE	+	−	$U(2N)/Sp(2N)$	$U^*(2N)/Sp(2N)$	CII|BDI	$O(2n)/O(n)\times O(n)$
chiral classes							
AIII	chGUE	−	±	$U(p+q)/U(p)\times U(q)$	$U(p,q)/U(p)\times U(q)$	A|A	$U(n)$
BDI	chGOE	+	+	$SO(p+q)/SO(p)\times SO(q)$	$SO(p,q)/SO(p)\times SO(q)$	AI|AII	$U(2n)/Sp(2n)$
CII	chGSE	+	−	$Sp(2p+2q)/Sp(2p)\times Sp(2q)$	$Sp(2p,2q)/Sp(2p)\times Sp(2q)$	AII|AI	$U(n)/O(n)$
Bogoliubov - de Gennes classes							
C		−	+	$Sp(2N)\times Sp(2N)/Sp(2N) \equiv Sp(2N)$	$Sp(2N,\mathbb{C})/Sp(2N)$	DIII|CI	$Sp(2n)/U(n)$
CI		+	+	$Sp(2N)/U(N)$	$Sp(2N,\mathbb{R})/U(N)$	D|C	$Sp(2n)$
BD		−	−	$SO(N)\times SO(N)/SO(N) \equiv SO(N)$	$SO(N,\mathbb{C})/SO(N)$	CI|DIII	$O(2n)/U(n)$
DIII		+	−	$SO(2N)/U(N)$	$SO^*(2N)/U(N)$	C|D	$O(n)$

3.3. *Chiral classes*

The Wigner–Dyson classes are the only allowed if one looks for a symmetry that is translationally invariant in energy, i.e. is not spoiled by adding a constant to the Hamiltonian. However, additional discrete symmetries may arise at some particular value of energy (which can be chosen to be zero without loss of generality), leading to novel symmetry classes. As the vicinity of a special point in the energy space governs the physics in many cases (i.e. the band center in lattice models at half filling, or zero energy in gapless superconductors), these ensembles are of large interest. They can be subdivided into two groups — chiral and Bogoliubov-de Gennes ensembles — considered here and in Sec. 3.4, respectively.

The chiral ensembles appeared in both contexts of particle physics and physics of disordered electronic systems about 15 years ago.[62–65] The corresponding Hamiltonians have the form

$$H = \begin{pmatrix} 0 & h \\ h^\dagger & 0 \end{pmatrix} , \tag{3.5}$$

i.e. they possess the symmetry

$$\tau_z H \tau_z = -H , \tag{3.6}$$

where τ_z is the third Pauli matrix in a certain "isospin" space. In the condensed matter context, such ensembles arise, in particular, when one considers a tight-binding model on a bipartite lattice with randomness in hopping matrix elements only. In this case, H has the block structure (3.5) in the sublattice space.

In addition to the chiral symmetry, a system may possess time reversal and/or spin-rotation invariance. In full analogy with the Wigner–Dyson classes, 3.1, one gets therefore three chiral classes (unitary, orthogonal, and symplectic). The corresponding symmetric spaces, the Cartan notations for symmetry classes, and the σ-model manifolds are given in the rows 4–6 of the Table 6.1.

3.4. *Bogoliubov-de Gennes classes*

The Wigner–Dyson and chiral classes do not exhaust all possible symmetries of disordered electronic systems.[7] The remaining four classes arise most naturally in superconducting systems. The quasiparticle dynamics in such systems can be described by the Bogoliubov-de Gennes Hamiltonian of the form

$$\hat{H} = \sum_{\alpha\beta}^{N} h_{\alpha\beta} c_\alpha^\dagger c_\beta + \frac{1}{2} \sum_{\alpha\beta}^{N} \left(\Delta_{\alpha\beta} c_\alpha^\dagger c_\beta^\dagger - \Delta_{\alpha\beta}^* c_\alpha c_\beta \right), \qquad (3.7)$$

where $c^\dagger$ and c are fermionic creation and annihilation operators, and the $N \times N$ matrices h, Δ satisfy $h = h^\dagger$ and $\Delta^T = -\Delta$, in view of hermiticity. Combining $c_\alpha^\dagger, c_\alpha$ in a spinor $\psi_\alpha^\dagger = (c_\alpha^\dagger, c_\alpha)$, one gets a matrix representation of the Hamiltonian, $\hat{H} = \psi^\dagger H \psi$, where

$$H = \begin{pmatrix} h & \Delta \\ -\Delta^* & -h^T \end{pmatrix}, \qquad h = h^\dagger, \quad \Delta = -\Delta^T. \qquad (3.8)$$

The minus signs in the definition of H result form the fermionic commutation relations between $c^\dagger$ and c. The Hamiltonian structure (3.8) corresponds to the condition

$$H = -\tau_x H^T \tau_x, \qquad (3.9)$$

(in addition to the Hermiticity $H = H^\dagger$), where τ_x is the Pauli matrix in the particle-hole space. Alternatively, one can perform a unitary rotation of the basis, defining $\tilde{H} = g^\dagger H g$ with $g = (1 + i\tau_x)/\sqrt{2}$. In this basis, the defining condition of class D becomes $\tilde{H} = -\tilde{H}^T$, so that $\tilde{H}$ is pure imaginary. The matrices $X = iH$ thus form the Lie algebra $\mathfrak{so}(2N)$, corresponding to the Cartan class D. This symmetry class described disordered superconducting systems in the absence of other symmetries.

Again, the symmetry class will be changed if the time reversal and/or spin rotation invariance are present. The difference with respect to the Wigner–Dyson and chiral classes is that now one gets four different classes rather than three. This is because the spin-rotation invariance has an impact even in the absence of time-reversal invariance, since it combines with the particle-hole symmetry in a non-trivial way. Indeed, if the spin is conserved, the Hamiltonian has the form

$$\hat{H} = \sum_{ij}^{N} \left[h_{ij} (c_{i\uparrow}^\dagger c_{j\uparrow} - c_{j\downarrow} c_{i,\downarrow}^\dagger) + \Delta_{ij} c_{i,\uparrow}^\dagger c_{j,\downarrow}^\dagger + \Delta_{ij}^* c_{i\downarrow} c_{j\uparrow} \right], \qquad (3.10)$$

where h and Δ are $N \times N$ matrices satisfying $h = h^\dagger$ and $\Delta = \Delta^T$. Similar to (3.8), we can introduce the spinors $\psi_i^\dagger = (c_{i\uparrow}^\dagger, c_{i\downarrow})$ and obtain the following matrix form of the Hamiltonian

$$H = \begin{pmatrix} h & \Delta \\ \Delta^* & -h^T \end{pmatrix}, \qquad h = h^\dagger, \quad \Delta = \Delta^T. \qquad (3.11)$$

It exhibits a symmetry property

$$H = -\tau_y H^T \tau_y. \tag{3.12}$$

The matrices $H = iX$ now form the Lie algebra $\mathfrak{sp}(2N)$, which is the symmetry class C.

If the time reversal invariance is present, one gets two more classes (CI and DIII). The symmetric spaces for the Hamiltonians and the σ-models corresponding to the Bogoliubov–de Gennes classes are given in the last four rows of the Table 6.1.

The following comment is in order here. Strictly speaking, one should distinguish between the orthogonal group SO(N) with even and odd N, which form different Cartan classes: SO($2N$) belongs to class D, while SO($2N$+1) to class B. In the conventional situation of a disordered superconductor, the matrix size is even due to the particle-hole space doubling, see Sec. 3.4. It was found, however, that the class B can arise in p-wave vortices.[75] In the same sense, the class DIII should be split in DIII-even and DIII-odd; the last one represented by the symmetric space SO($4N + 2$)/U($2N + 1$) can appear in vortices in the presence of time-reversal symmetry.

3.5. *Perturbative RG for σ-models of different symmetry classes*

Perturbative β-functions for σ-models on all the types of symmetric spaces were in fact calculated[26,27] long before the physical significance of the chiral and Bogoliubov-de Gennes classes has been fully appreciated. These results are important for understanding the behavior of systems of different symmetry classes in 2D. (We should emphasize once more, however, that this does not give a complete information about all possible types of criticality since the latter can be crucially affected by additional terms of topological character in the σ-model, see Secs. 4 and 6 below.)

One finds that in the classes A, AI, C and CI, the β-function is negative in 2D in the replica limit (at least, for small t). This indicates that normally all states are localized in such systems in 2D. (This conclusion can in fact be changed in the presence of topological or Wess–Zumino terms, Sec. 4.1.) Above 2D, these systems undergo the Anderson transition that can be studied within the $2 + \epsilon$ expansion, Sec. 2.2.2. For the classes AIII, BDI and CII (chiral unitary, orthogonal and symplectic classes, respectively), the $\beta(t) \equiv 0$ in 2D, implying a line of fixed points. Finally, in the classes AII, D and DIII, the β-function is positive at small t, implying the existence of a metal–insulator transition at strong coupling in 2D.

4. Criticality in 2D

4.1. *Mechanisms of criticality in 2D*

As was discussed in Sec. 2.2.2, conventional Anderson transitions in the orthogonal and unitary symmetry classes take place only if the dimensionality is $d > 2$, whereas in 2D all states are localized. It is, however, well understood by now that there is a rich variety of mechanisms that lead to emergence of criticality in 2D disordered systems.[61] Such 2D critical points have been found to exist for nine out of ten symmetry classes, namely, in all classes except for the orthogonal class AI. A remarkable peculiarity of 2D critical points is that the critical conductance g_* is at the same time the *critical conductivity*. We now list and briefly describe the mechanisms for the emergence of criticality.

4.1.1. *Broken spin-rotation invariance: Metallic phase*

We begin with the mechanism that has been already mentioned in Sec. 2.2.2 in the context of the Wigner–Dyson symplectic class (AII). In this case the β-function [(2.19) with $\epsilon = 0$] is positive for not too large t (i.e. sufficiently large conductance), so that the system is metallic (t scales to zero under RG). On the other hand, for strong disorder (low t) the system is an insulator, as usual, i.e. $\beta(t) < 0$. Thus, β-function crosses zero at some t_*, which is a point of the Anderson transition.

This mechanism (positive β-function and, thus, metallic phase at small t, with a transition at some t_*) is also realized in two of Bogoliubov-de Gennes classes — D and DIII. All these classes correspond to systems with broken spin-rotation invariance. The unconventional sign of the β-function in these classes, indicating weak antilocalization (rather then localization), is physically related to destructive interference of time reversed paths for particles with spin $s = 1/2$.

4.1.2. *Chiral classes: Vanishing β-function*

Another peculiarity of the perturbative β-function takes place for three chiral classes — AIII, BDI, ad CII. Specifically, for these classes $\beta(t) \equiv 0$ to all orders of the perturbation theory, as was first discovered by Gade and Wegner.[62,63] As a result, the conductance is not renormalized at all, serving as an exactly marginal coupling. There is thus a line of critical points for these models, labeled by the value of the conductance. In fact, the σ-models for these classes contain an additional term[62,63] that does not affect the absence of renormalization of the conductance but is crucial for the analysis of the behavior of the DOS.

4.1.3. *Broken time-reversal invariance: Topological θ-term and quantum Hall criticality*

For several classes, the σ-model action allows for inclusion of a topological term, which is invisible to any order of the perturbation theory. This is the case when the second homotopy group π_2 of the σ-model manifold $\mathcal{M}$ (a group of homotopy classes of maps of the sphere S^2 into $\mathcal{M}$) is non-trivial. From this point of view, only the compact sector $\mathcal{M}_F$ (originating from the fermionic part of the supervector field) of the manifold base matters. There are five classes, for which $\pi_2(\mathcal{M}_F)$ is non-trivial, namely A, C, D, AII and CII.

For the classes A, C and D, the homotopy group $\pi_2(\mathcal{M}_F) = \mathbb{Z}$. Therefore, the action $S[Q]$ may include the (imaginary) θ-term,

$$iS_{\text{top}}[Q] = i\theta N[Q] , \tag{4.1}$$

where an integer $N[Q]$ is the winding number of the field configuration $Q(\mathbf{r})$. Without loss of generality, θ can be restricted to the interval $[0, 2\pi]$, since the theory is periodic in θ with the period 2π.

The topological term (4.1) breaks the time reversal invariance, so it may only arise in the corresponding symmetry classes. The by far most famous case is the Wigner–Dyson unitary class (A). As was first understood by Pruisken,[9] the σ-model of this class with the topological term (4.1) describes the integer quantum Hall effect (IQHE), with the critical point of the plateau transition corresponding to θ=π. More recently, it was understood that counterparts of the IQHE exist also in the Bogoliubov-de Gennes classes with broken time-reversal invariance — classes C[66–70] and D.[71–74,79] They were called *spin* and *thermal* quantum Hall effects (SQHE and TQHE), respectively.

4.1.4. $\mathbb{Z}_2$ *topological term*

For two classes, AII and CII, the second homotopy group is $\pi_2(\mathcal{M}_F) = \mathbb{Z}_2$. This allows for the θ-term but θ can only take the values 0 and π. It has been recently shown[88] that the σ-model of the Wigner–Dyson symplectic class (AII) with a θ=π topological angle arises from a model of Dirac fermions with random scalar potential, which describes, in particular, graphene with long-range disorder. Like in the case of quantum-Hall systems, this topological term inhibits localization.

4.1.5. *Wess–Zumino term*

Finally, one more mechanism of emergence of criticality is the Wess–Zumino (WZ) term that may appear in σ-models of the classes AIII, CI and DIII. For these classes, the compact component $\mathcal{M}_F$ of the manifold is the group $H \times H/H = H$, where H is U(n), Sp($2n$) and O($2n$), respectively. The corresponding theories are called "principal chiral models". The WZ term has the following form:

$$iS_{\mathrm{WZ}}(g) = \frac{ik}{24\pi} \int d^2r \int_0^1 ds\, \epsilon_{\mu\nu\lambda} \mathrm{Str}(g^{-1}\partial_\mu g)(g^{-1}\partial_\nu g)(g^{-1}\partial_\lambda g), \quad (4.2)$$

where k is an integer called the level of the WZW model. The definition (4.2) of the WZ term requires an extension of the σ-model field $g(\mathbf{r}) \equiv g(x,y)$ to the third dimension, $0 \le s \le 1$, such that $g(\mathbf{r},0) = 1$ and $g(\mathbf{r},1) = g(\mathbf{r})$. Such an extension is always possible, since the second homotopy group is trivial, $\pi_2(H) = 0$, for all the three classes. Further, the value of the WZ term does not depend on the particular way the extension to the third dimension is performed. (This becomes explicit when one calculates the variation of the WZ term: it is expressed in terms of $g(\mathbf{r})$ only.) More precisely, there is the following topological ambiguity in the definition of $S_{\mathrm{WZ}}(g)$. Since the third homotopy group is non-trivial, $\pi_3(H) = \mathbb{Z}$, $S_{\mathrm{WZ}}(g)$ is defined up to an arbitrary additive integer n times $2\pi k$. This, however, does not affect any observables, since simply adds the phase $nk \times 2\pi i$ to the action.

The WZ term arises when one bosonizes certain models of Dirac fermions[76] and is a manifestation of the chiral anomaly. In particular, a σ-model for a system of the AIII (chiral unitary) class with the WZ term describes Dirac fermions in a random vector potential. In this case, the σ-model coupling constant is truly marginal (as is typical for chiral classes) and one finds a line of fixed points. On the other hand, for the class CI there is a single fixed point. The WZW models of these classes were encountered in the course of study of dirty d-wave superconductors[77,78] and, most recently, in the context of disordered graphene. We will discuss critical properties of these models in Sec. 4.2.3.

4.2. *Disordered Dirac Hamiltonians and graphene*

Localization and criticality in models of 2D Dirac fermions subjected to various types of disorder have been studied in a large number of papers and in a variety of contexts, including the random bond Ising model,[80] the quantum Hall effect,[83] dirty superconductors with unconventional pairing,[77-79] and some lattice models with chiral symmetry.[81] Recently, this class of problems has attracted a great deal of attention[82,84-89] in connection with its application to graphene.[91,92]

One of the most prominent experimentally discovered features of graphene is the "minimal conductivity" at the neutrality (Dirac) point. Specifically, the conductivity[93-95] of an undoped sample is close to e^2/h per spin per valley, remaining almost constant in a very broad temperature range — from room temperature down to 30 mK. This is in contrast with conventional 2D systems driven by Anderson localization into insulating state at low T and suggests that delocalization (and, possibly, quantum criticality) may emerge in a broad temperature range due to special character of disordered graphene Hamiltonian.

In the presence of different types of randomness, Dirac Hamiltonians realize all ten symmetry classes of disordered systems; see Ref. 90 for a detailed symmetry classification. Furthermore, in many cases the Dirac character of fermions induces non-trivial topological properties (θ-term or WZ term) of the corresponding field theory (σ-model). In Sec. 4.2.1, we review the classification of disorder in a two-flavor model of Dirac fermions describing the low-energy physics of graphene and types of criticality. The emergent critical theories will be discussed in Secs. 4.2.2–4.2.4.

4.2.1. *Symmetries of disorder and types of criticality*

The presentation below largely follows Refs. 87 and 89. We concentrate on a two-flavor model, which is in particular relevant to the description of electronic properties of graphene. Graphene is a semimetal; its valence and conduction bands touch each other in two conical points K and K' of the Brillouin zone. In the vicinity of these points, the electrons behave as massless relativistic (Dirac-like) particles. Therefore, the effective tight-binding low-energy Hamiltonian of clean graphene is a 4×4 matrix operating in the AB space of the two sublattices and in the K–K' space of the valleys:

$$H = v_0 \tau_3 \boldsymbol{\sigma} \mathbf{k}. \tag{4.3}$$

Here, τ_3 is the third Pauli matrix in the K–K' space, $\boldsymbol{\sigma} = \{\sigma_1, \sigma_2\}$ the two-dimensional vector of Pauli matrices in the AB space, and v_0 the velocity ($v_0 \simeq 10^8$ cm/s in graphene). It is worth emphasizing that the Dirac form of the Hamiltonian (4.3) does not rely on the tight-binding approximation but is protected by the symmetry of the honeycomb lattice which has two atoms in a unit cell.

Let us analyze the symmetries of the clean Hamiltonian (4.3) in the AB and KK' spaces. First, there exists an SU(2) symmetry group in the space of the valleys, with the generators[82]

$$\Lambda_x = \sigma_3\tau_1, \qquad \Lambda_y = \sigma_3\tau_2, \qquad \Lambda_z = \sigma_0\tau_3 , \qquad (4.4)$$

all of which commute with the Hamiltonian. Second, there are two more symmetries of the clean Hamiltonian, namely, time inversion operation (T_0) and chiral symmetry (C_0). Combining T_0, C_0, and isospin rotations $\Lambda_{0,x,y,z}$, one can construct twelve symmetry operations, out of which four (denoted as T_μ) are of time-reversal type, four (C_μ) of chiral type, and four (CT_μ) of Bogoliubov-de Gennes type:

$$
\begin{aligned}
T_0 &: \ A \mapsto \sigma_1\tau_1 A^T \sigma_1\tau_1, & C_0 &: \ A \mapsto -\sigma_3\tau_0 A \sigma_3\tau_0, & CT_0 &: \ A \mapsto -\sigma_2\tau_1 A^T \sigma_2\tau_1, \\
T_x &: \ A \mapsto \sigma_2\tau_0 A^T \sigma_2\tau_0, & C_x &: \ A \mapsto -\sigma_0\tau_1 A \sigma_0\tau_1, & CT_x &: \ A \mapsto -\sigma_1\tau_0 A^T \sigma_1\tau_0, \\
T_y &: \ A \mapsto \sigma_2\tau_3 A^T \sigma_2\tau_3, & C_y &: \ A \mapsto -\sigma_0\tau_2 A \sigma_0\tau_2, & CT_y &: \ A \mapsto -\sigma_1\tau_3 A^T \sigma_1\tau_3, \\
T_z &: \ A \mapsto \sigma_1\tau_2 A^T \sigma_1\tau_2, & C_z &: \ A \mapsto -\sigma_3\tau_3 A \sigma_3\tau_3, & CT_z &: \ A \mapsto -\sigma_2\tau_2 A^T \sigma_2\tau_2.
\end{aligned}
$$

It is worth recalling that the C and CT symmetries apply to the Dirac point ($E = 0$), i.e. to undoped graphene, and get broken by a non-zero energy E. We will assume the average isotropy of the disordered graphene, which implies that Λ_x and Λ_y symmetries of the Hamiltonian are present or absent simultaneously. They are thus combined into a single notation $\Lambda_\perp$; the same applies to $T_\perp$ and $C_\perp$. In Table 6.2, all possible matrix structures of disorder along with their symmetries are listed.

If all types of disorder are present (i.e. no symmetries is preserved), the RG flow is towards the conventional localization fixed point (unitary Wigner–Dyson class A). If the only preserved symmetry is the time reversal (T_0), again the conventional localization (orthogonal Wigner–Dyson class AI) takes place.[85] A non-trivial situation occurs if either (i) one of the chiral symmetries is preserved or (ii) the valleys remain decoupled. In Table 6.3, we list situations when symmetry prevents localization and leads to criticality and non-zero conductivity at $E=0$ (in the case of decoupled nodes — also at non-zero E). Models with decoupled nodes are analyzed in Sec. 4.2.2, and models with a chiral symmetry in Sec. 4.2.3 (C_0-chirality) and 4.2.4 (C_z-chirality).

Table 6.2.   Disorder symmetries in graphene. The first five rows represent disorders preserving the time reversal symmetry T_0; the last four — violating T_0. First column: structure of disorder in the sublattice (σ_μ) and valley (τ_ν) spaces. The remaining columns indicate which symmetries of the clean Hamiltonian are preserved by disorder.[87]

Structure	$\Lambda_\perp$	Λ_z	T_0	$T_\perp$	T_z	C_0	$C_\perp$	C_z	CT_0	$CT_\perp$	CT_z
$\sigma_0\tau_0$	$+$	$+$	$+$	$+$	$+$	$-$	$-$	$-$	$-$	$-$	$-$
$\sigma_{\{1,2\}}\tau_{\{1,2\}}$	$-$	$-$	$+$	$-$	$-$	$+$	$-$	$-$	$+$	$-$	$-$
$\sigma_{1,2}\tau_0$	$-$	$+$	$+$	$-$	$+$	$+$	$-$	$+$	$+$	$-$	$+$
$\sigma_0\tau_{1,2}$	$-$	$-$	$+$	$-$	$-$	$-$	$-$	$+$	$-$	$-$	$+$
$\sigma_3\tau_3$	$-$	$+$	$+$	$-$	$+$	$-$	$+$	$-$	$-$	$+$	$-$
$\sigma_3\tau_{1,2}$	$-$	$-$	$-$	$-$	$+$	$-$	$-$	$+$	$+$	$-$	$-$
$\sigma_0\tau_3$	$-$	$+$	$-$	$+$	$-$	$-$	$+$	$-$	$+$	$-$	$+$
$\sigma_{1,2}\tau_3$	$+$	$+$	$-$	$-$	$-$	$+$	$+$	$+$	$-$	$-$	$-$
$\sigma_3\tau_0$	$+$	$+$	$-$	$-$	$-$	$-$	$-$	$-$	$+$	$+$	$+$

4.2.2. *Decoupled nodes: Disordered single-flavor Dirac fermions and quantum-Hall-type criticality*

If the disorder is of long-range character, the valley mixing is absent due to the lack of scattering with large momentum transfer. For each of the nodes, the system can then be described in terms of a single-flavor Dirac Hamiltonian,

$$H = v_0[\boldsymbol{\sigma}\mathbf{k} + \sigma_\mu V_\mu(\mathbf{r})]. \tag{4.5}$$

Here, disorder includes random scalar (V_0) and vector ($V_{1,2}$) potentials and random mass (V_3). The clean single-valley Hamiltonian (4.5) obeys the effective time-reversal invariance $H = \sigma_2 H^T \sigma_2$. This symmetry ($T_\perp$) is not the physical time-reversal symmetry (T_0): the latter interchanges the nodes and is of no significance in the absence of inter-node scattering.

Remarkably, single-flavor Dirac fermions are never in the conventional localized phase! More specifically, depending on which of the disorders are present, four different types of criticality take place:

(i) The only disorder is the random vector potential ($V_{1,2}$). This is a special case of the symmetry class AIII. This problem is exactly solvable. It is characterized by a line of fixed points, all showing conductivity $4e^2/\pi h$, see Sec. 4.2.3.

Table 6.3. Possible types of disorder in graphene leading to criticality. The first three row correspond to C_z chiral symmetry leading to Gade–Wegner-type criticality, Sec. 4.2.4. The next three rows contain models with C_0 chiral symmetry (random gauge fields), inducing a WZ term in the σ-model action, Sec. 4.2.3. The last four rows correspond to the case of decoupled valleys (long-range disorder), see Sec. 4.2.2; in the last three cases the σ-model acquires a topological term with $\theta = \pi$. Adapted from Ref. 89.

Disorder	Symmetries	Class	Criticality	Conductivity
Vacancies, strong potential impurities	C_z, T_0	BDI	Gade	$\approx 4e^2/\pi h$
Vacancies + RMF	C_z	AIII	Gade	$\approx 4e^2/\pi h$
$\sigma_3\tau_{1,2}$ disorder	C_z, T_z	CII	Gade	$\approx 4e^2/\pi h$
Dislocations	C_0, T_0	CI	WZW	$4e^2/\pi h$
Dislocations + RMF	C_0	AIII	WZW	$4e^2/\pi h$
Ripples, RMF	Λ_z, C_0	2×AIII	WZW	$4e^2/\pi h$
Charged impurities	$\Lambda_z, T_\perp$	2×AII	$\theta = \pi$	$4\sigma_{Sp}^{**}$ or[a] $(4e^2/\pi h)\ln L$
Random Dirac mass: $\sigma_3\tau_0, \sigma_3\tau_3$	$\Lambda_z, CT_\perp$	2×D	$\theta = \pi$	$4e^2/\pi h$
Charged impurities + (RMF, ripples)	Λ_z	2×A	$\theta = \pi$	$4\sigma_U^*$

[a]Numerical simulations[96] reveal a flow towards the supermetal fixed point, $\sigma \simeq (4e^2/\pi h)\ln L \to \infty$.

(ii) Only random mass (V_3) is present. The system belongs then to class D. The random-mass disorder is marginally irrelevant, and the system flows under RG towards the clean fixed point, with the conductivity $4e^2/\pi h$.

(iii) The only disorder is random scalar potential (V_0). The system is then in the Wigner–Dyson symplectic (AII) symmetry class. As was found in Ref. 88, the corresponding σ-model contains a $\mathbb{Z}_2$ topological term with $\theta = \pi$ which protects the system from localization. The absence of localization in this model has been confirmed in numerical simulations.[96] The scaling function has been found in Ref. 96 to be strictly positive, implying a flow towards the "supermetal" fixed point.

(iv) At least two types of randomness are present. All symmetries are broken in this case and the model belongs to the Wigner–Dyson unitary class A. It was argued in Ref. 83 that it flows into the IQH transition fixed point. This is confirmed by the derivation of the corresponding σ-model,[78,88,89] which contains a topological term with $\theta = \pi$, i.e. is nothing but the Pruisken σ-model at criticality. A particular consequence of this is that the conductivity of graphene with this type of disorder is equal to the value σ_U^* of the longitudinal conductivity σ_{xx} at the critical point of the IQH transition multiplied by four (because of spin and valleys).

If a uniform transverse magnetic field is applied, the topological angle θ becomes energy-dependent. However, at the Dirac point ($E = 0$), where $\sigma_{xy} = 0$, its value remains unchanged, $\theta = \pi$. This implies the emergence of the half-integer quantum Hall effect, with a plateau transition point at $E = 0$.

4.2.3. *Preserved C_0 chirality: Random gauge fields*

Let us consider a type of disorder which preserves the C_0-chirality, $H = -\sigma_3 H \sigma_3$. This implies the disorder of the type $\sigma_{1,2}\tau_{0,1,2,3}$ being strictly off-diagonal in the σ space. Depending on further symmetries, three different C_0-chiral models arise:

(i) The only disorder present is $\sigma_{1,2}\tau_3$, which corresponds to the random abelian vector potential. In this case the nodes are decoupled, and the Hamiltonian decomposes in two copies of a model of the class AIII. This model characterized by a line of fixed points has already been mentioned in Sec. 4.2.2.

(ii) If the time-reversal symmetry T_0 is preserved, only the disorder of the type $\sigma_{1,2}\tau_{0,1,2}$ is allowed, and the problem is in the symmetry class CI. The model describes then fermions coupled to a SU(2) non-abelian gauge field, and is a particular case of analogous SU(N) models. This theory flows now into an isolated fixed point, which is a WZW theory on the level $k = -2N$.[35,77,97]

(iii) All C_0-invariant disorder structures are present. This describes Dirac fermions coupled to both abelian U(1) and non-abelian SU(2) gauge fields. This model is in the AIII symmetry class.

Remarkably, all these critical C_0-chiral models are exactly solvable. In particular, the critical conductivity can be calculated exactly and is independent on the disorder strength. A general proof of this statement based only on the gauge invariance is given in Ref. 87. (For particular cases it was earlier obtained in Refs. 83 and 98). The critical conductivity is thus the same as in clean graphene,

$$\sigma = 4e^2/\pi h. \qquad (4.6)$$

Spectra of multifractal exponents and the critical index of the DOS can also be calculated exactly, see Ref. 6.

4.2.4. Disorders preserving C_z chirality: Gade–Wegner criticality

Let us now turn to the disorder which preserves the C_z-chirality, $H = -\sigma_3\tau_3 H \sigma_3\tau_3$; according to Table 6.2, the corresponding disorder structure is $\sigma_{1,2}\tau_{0,3}$ and $\sigma_{0,3}\tau_{1,2}$. If no time-reversal symmetries are preserved, the system belongs to the chiral unitary (AIII) class. The combination of C_z-chirality and the time reversal invariance T_0 corresponds to the chiral orthogonal symmetry class BDI; this model has already been discussed in Sec. 4.1.2. Finally, the combination of C_z-chirality and T_z-symmetry falls into the chiral symplectic symmetry class CII. The RG flow and DOS in these models have been analyzed in Ref. 81. In all the cases, the resulting theory is of the Gade–Wegner type.[62,63] These theories are characterized by lines of fixed points, with non-universal conductivity. It was found[87,100] that for weak disorder, the conductivity takes approximately the universal value, $\sigma \simeq 4e^2/\pi h$. In contrast to the case of C_0 chirality, this result is, however, not exact. In particular, the leading correction to the clean conductivity is found in the second order in disorder strength.[87]

5. Electron–Electron Interaction Effects

Physically, the impact of interaction effects onto low-temperature transport and localization in disordered electronic systems can be subdivided into two distinct effects: (i) renormalization and (ii) dephasing.

Renormalization. The renormalization effects, which are governed by virtual processes, become increasingly more pronounced with lowering temperature. The importance of such effects in diffusive low-dimensional systems was demonstrated by Altshuler and Aronov, see Ref. 101. To resum the arising singular contributions, Finkelstein developed the RG approach based on the σ-model for an interacting system, see Ref. 102 for a review. This made possible an analysis of the critical behavior at the localization transition in $2 + \epsilon$ dimensions in the situations when spin-rotation invariance is broken (by spin-orbit scattering, magnetic field, or magnetic impurities). However, in the case of preserved spin-rotation symmetry, it was found that the strength of the interaction in spin-triplet channel scales to infinity at certain RG scale. This was interpreted as some kind of magnetic instability of the system; for a detailed exposition of proposed scenarios, see Ref. 103.

Recently, the problem has attracted a great deal of attention in connection with experiments on high-mobility low-density 2D electron structures (Si MOSFETs) giving an evidence in favor of a metal–insulator transition.[104] In Ref. 105, the RG for σ-model for interacting 2D electrons with a number of valleys $N > 1$ was analyzed on the two-loop level. It was shown that in the limit of large number of valleys N (in practice, $N = 2$ as in Si is already sufficient), the temperature of magnetic instability is suppressed down to unrealistically low temperatures and a metal–insulator transition emerges. The existence of interaction-induced metallic phase in 2D is due to the fact that, for a sufficiently strong interaction, its "delocalizing" effect overcomes the disorder-induced localization. Recent works[118,119] show that the RG theory describes well the experimental data up to lowest accessible temperatures. We will see in Sec. 6 that the Coulomb interaction may also lead to dramatic effects in the context of topological insulators.

The interaction-induced renormalization effects become extremely strong for correlated 1D systems (Luttinger liquids). While 1D systems provide a paradigmatic example of strong Anderson localization, a sufficiently strong attractive interaction can lead to delocalization in such systems. An RG treatment of the corresponding localization transition in a disordered interacting 1D systems was developed in Ref. 106, see also the book in Ref. 107. Recently, the interplay between Anderson localization,

Luttinger liquid renormalization, and dephasing has been studied in detail in Ref. 120.

Dephasing. We turn now to effects of dephasing governed by inelastic processes of electron–electron scattering at finite temperature T. The dephasing has been studied in great detail for metallic systems where it provides a cutoff for weak localization effects.[101] As to the Anderson transitions, they are quantum (zero-T) phase transitions, and dephasing contributes to their smearing at finite T. The dephasing-induced width of the transition scales as a power-law function of T. There is, however, an interesting situation when dephasing processes can create a localization transition. We mean the systems where all states are localized in the absence of interaction, such as wires or 2D systems. At high temperatures, when the dephasing is strong, so that the dephasing rate $\tau_\phi^{-1}(T)$ is larger than mean level spacing in the localization volume, the system is a good metal and its conductivity is given by the quasiclassical Drude conductivity with relatively small weak localization correction.[101] With lowering temperature, the dephasing gets progressively less efficient, the localization effects proliferate, and eventually the system becomes an Anderson insulator. What is the nature of this state? A natural question is whether the interaction of an electron with other electrons will be sufficient to provide a kind of thermal bath that would assist the variable-range hopping transport,[108] as it happens in the presence of a phonon bath. The answer to this question was given by Ref. 109, and it is negative. Fleishman and Anderson found that at low T, the interaction of a "short-range class" (which includes a finite-range interaction in any dimensionality d and Coulomb interaction in $d < 3$) is not sufficient to delocalize otherwise localized electrons, so that the conductivity remains strictly zero. In combination with the Drude conductivity at high-T, this implies the existence of transition at some temperature T_c.

This conclusion was recently corroborated by an analysis[110,111] in the framework of the idea of Anderson localization in Fock space.[112] In these works, the temperature dependence of conductivity $\sigma(T)$ in systems with localized states and weak electron–electron interaction was studied. It was found that with decreasing T the system first shows a crossover from the weak localization regime into that of "power-law hopping" over localized states (where σ is a power-law function of T), and then undergoes a localization transition. The transition is obtained both within a self-consistent Born approximation[111] and an approximate mapping onto a model on the Bethe lattice.[110] The latter yields also a critical behavior of $\sigma(T)$ above

T_c, which has a characteristic for the Bethe lattice non-power-law form $\ln \sigma(T) \sim (T - T_c)^{-\kappa}$ with $\kappa = 1/2$, see Sec. 2.5.

Up to now, this transition has not been observed in experiments[d], which indicate instead a smooth crossover from the metallic to the insulating phase with lowering T.[113-116] The reason for this discrepancy remains unclear. An attempt to detect the transition in numerical simulations also did not give a clear confirmation of the theory,[117] possibly because of strong restrictions on the size of an interacting system that can be numerically diagonalized. On the other hand, a very recent work[121] does report an evidence in favor of a transition of a Bethe-lattice character (though with different value of κ).

6. Topological Insulators

One of the most recent arenas where novel peculiar localization phenomena have been studied is physics of topological insulators.[122-129] Topological insulators are bulk insulators with delocalized (topologically protected) states on their surface.[e] As discussed above, the critical behavior of a system depends on the underlying topology. This is particularly relevant for topological insulators.

The famous example of a topological insulator is a two-dimensional (2D) system on one of quantum Hall plateaus in the integer quantum Hall effect. Such a system is characterized by an integer (Chern number) $n = \ldots, -2, -1, 0, 1, 2, \ldots$ which counts the edge states (here the sign determines the direction of chiral edge modes). The integer quantum Hall edge is thus a topologically protected one-dimensional (1D) conductor realizing the group $\mathbb{Z}$.

Another ($\mathbb{Z}_2$) class of topological insulators[122-124] can be realized in systems with strong spin-orbit interaction and without magnetic field (class AII) — and was discovered in 2D HgTe/HgCdTe structures in Ref. 125 (see also Ref. 127). A 3D $\mathbb{Z}_2$ topological insulator[126] has been found and investigated for the first time in $Bi_{1-x}Sb_x$ crystals. Both in 2D and 3D, $\mathbb{Z}_2$ topological insulators are band insulators with the following properties: (i) time reversal invariance is preserved (unlike ordinary quantum Hall systems); (ii)

---

[d]Of course, in a real system, phonons are always present and provide a bath necessary to support the hopping conductivity at low T, so that there is no true transition. However, when the coupling to phonons is weak, this hopping conductivity will have a small prefactor, yielding a "quasi-transition".

[e]Related topology-induced phenomena have been considered in Ref. 132 in the context of superfluid Helium-3 films.

Table 6.4.   Symmetry classes and "Periodic Table" of topological insulators.[134,135] The first column enumerates the symmetry classes of disordered systems which are defined as the symmetry classes H_p of the Hamiltonians (second column). The third column lists the symmetry classes of the classifying spaces (spaces of reduced Hamiltonians).[135] The fourth column represents the symmetry classes of a compact sector of the sigma-model manifold. The fifth column displays the zeroth homotopy group $\pi_0(R_p)$ of the classifying space. The last four columns show the possibility of existence of $\mathbb{Z}$ and $\mathbb{Z}_2$ topological insulators in each symmetry class in dimensions $d = 1, 2, 3, 4$. Adapted from Ref. 130.

	Symmetry classes				Topological insulators			
p	H_p	R_p	S_p	$\pi_0(R_p)$	d=1	d=2	d=3	d=4
0	AI	BDI	CII	$\mathbb{Z}$	0	0	0	$\mathbb{Z}$
1	BDI	BD	AII	$\mathbb{Z}_2$	$\mathbb{Z}$	0	0	0
2	BD	DIII	DIII	$\mathbb{Z}_2$	$\mathbb{Z}_2$	$\mathbb{Z}$	0	0
3	DIII	AII	BD	0	$\mathbb{Z}_2$	$\mathbb{Z}_2$	$\mathbb{Z}$	0
4	AII	CII	BDI	$\mathbb{Z}$	0	$\mathbb{Z}_2$	$\mathbb{Z}_2$	$\mathbb{Z}$
5	CII	C	AI	0	$\mathbb{Z}$	0	$\mathbb{Z}_2$	$\mathbb{Z}_2$
6	C	CI	CI	0	0	$\mathbb{Z}$	0	$\mathbb{Z}_2$
7	CI	AI	C	0	0	0	$\mathbb{Z}$	0
$0'$	A	AIII	AIII	$\mathbb{Z}$	0	$\mathbb{Z}$	0	$\mathbb{Z}$
$1'$	AIII	A	A	0	$\mathbb{Z}$	0	$\mathbb{Z}$	0

there exists a topological invariant, which is similar to the Chern number in QHE; (iii) this invariant belongs to the group $\mathbb{Z}_2$ and reflects the presence or absence of delocalized edge modes (Kramers pairs).[122]

Topological insulators exist in all ten symmetry classes in different dimensions, see Table 6.4. Very generally, the classification of topological insulators in d dimensions can be constructed by studying the Anderson localization problem in a $(d-1)$-dimensional disordered system.[134] Indeed, absence of localization of surface states due to the topological protection implies the topological character of the insulator.

In Sec. 6.1, we overview the full classification of topological insulators and superconductors.[134,135] In Sec. 6.2, we discuss $\mathbb{Z}_2$ topological insulators belonging to the symplectic symmetry class AII, characteristic to systems with strong spin-orbit interaction. Finally, in Sec. 6.3 we address, closely following Ref. 130, the interaction effects in $\mathbb{Z}_2$ topological insulators.

6.1. *Symmetry classification of topological insulators*

The full classification (periodic table) of topological insulators and super-conductors for all ten symmetry classes[7,8] was developed in Refs. 135 and 134. This classification determines whether the $\mathbb{Z}$ or $\mathbb{Z}_2$ topological insulator is possible in the d-dimensional system of a given symmetry class. In this section, we overview the classification of topological insulators closely following Refs. 135 and 134, and discuss the connection between the classification schemes of these papers.

All symmetry classes of disordered systems (see Sec. 3 and Table 6.1) can be divided into two groups: {A,AIII} and {all other}. The classes of the big group are labeled by $p = 0, 1, \ldots, 7$. Each class is characterized by (i) Hamiltonian symmetry class H_p; (ii) symmetry class R_p of the classifying space used by Kitaev;[135] (iii) symmetry class S_p of the compact sector $\mathcal{M}_F$ of the sigma-model manifold. The symmetry class R_p of the classifying space of reduced Hamiltonians characterizes the space of matrices obtained from the Hamiltonian by keeping all eigenvectors and replacing all positive eigenvalues by $+1$ and all negative by -1. Note that

$$R_p = H_{p+1}, \quad S_p = R_{4-p}. \tag{6.1}$$

Here and below cyclic definition of indices $\{0, 1, \ldots, 7\}$ (mod 8) and $\{0', 1'\}$ (mod 2) is assumed.

For the classification of topological insulators, it is important to know homotopy groups π_d for all symmetry classes. In Table 6.4 we list $\pi_0(R_p)$; other π_d are given by

$$\pi_d(R_p) = \pi_0(R_{p+d}). \tag{6.2}$$

The homotopy groups π_d have periodicity 8 (Bott periodicity).

There are two ways to detect topological insulators: by inspecting the topology of (i) classifying space R_p or of (ii) the sigma-model space S_p.

(i) Existence of topological insulator (TI) of class p in d dimensions is established by the homotopy group π_0 for the classifying space R_{p-d}:

$$\begin{cases} \text{TI of the type } \mathbb{Z} \\ \text{TI of the type } \mathbb{Z}_2 \end{cases} \iff \pi_0(R_{p-d}) = \begin{cases} \mathbb{Z} \\ \mathbb{Z}_2 \end{cases} \tag{6.3}$$

(ii) Alternatively, the existence of topological insulator of symmetry class p in d dimensions can be inferred from the homotopy groups of the sigma-model manifolds, as follows:

$$\begin{cases} \text{TI of the type } \mathbb{Z} \iff \pi_d(S_p) = \mathbb{Z} \\ \text{TI of the type } \mathbb{Z}_2 \iff \pi_{d-1}(S_p) = \mathbb{Z}_2 \end{cases} \quad (6.4)$$

The criterion (ii) is obtained if one requires existence of "non-localizable" boundary excitations. This may be guaranteed by either Wess-Zumino term in $d-1$ dimensions [which is equivalent to the $\mathbb{Z}$ topological term in d dimensions, i.e. $\pi_d(S_p) = \mathbb{Z}$] for a QHE-type topological insulator, or by the $\mathbb{Z}_2$ topological term in $d-1$ dimensions [i.e. $\pi_{d-1}(S_p) = \mathbb{Z}_2$] for a QSH-type topological insulator.

The above criteria (i) and (ii) are equivalent, since

$$\pi_d(S_p) = \pi_d(R_{4-p}) = \pi_0(R_{4-p+d}). \quad (6.5)$$

and

$$\pi_0(R_p) = \begin{cases} \mathbb{Z} & \text{for } p = 0, 4, \\ \mathbb{Z}_2 & \text{for } p = 1, 2. \end{cases} \quad (6.6)$$

Below we focus on 2D systems of symplectic (AII) symmetry class. One sees that this is the only symmetry class out of ten classes that supports the existence of $\mathbb{Z}_2$ topological insulators both in 2D and 3D.

6.2. $\mathbb{Z}_2$ topological insulators in 2D and 3D systems of class AII

A $\mathbb{Z}_2$ class of topological insulators belonging to the symmetry class AII was first realized in 2D HgTe/HgCdTe structures in Ref. 125. Such systems were found to possess two distinct insulating phases, both having a gap in the bulk electron spectrum but differing by edge properties. While the normal insulating phase has no edge states, the topologically non-trivial insulator is characterized by a pair of mutually time-reversed delocalized edge modes penetrating the bulk gap. Such state shows the quantum spin Hall (QSH) effect which was theoretically predicted in a model system of graphene with spin-orbit coupling.[122,131] The transition between the two topologically non-equivalent phases (ordinary and QSH insulators) is driven by inverting the band gap.[123] The $\mathbb{Z}_2$ topological order is robust with respect to disorder: since the time-reversal invariance forbids backscattering of edge states at the boundary of QSH insulators, these states are topologically protected from localization.

For clean 2D QSH systems with a bulk gap generated by spin-orbit interaction, the $\mathbb{Z}_2$ invariant can be constructed from the Bloch wave functions on the Brillouin zone[122] and is somewhat similar to the Chern number in the standard QHE. Formally, if the $\mathbb{Z}_2$ index is odd/even there is an odd/even number m of Kramers pairs of gapless edge states (here $m = 0$ is treated as even number). In the presence of disorder which generically back-scatters between different Kramers pairs, all the surface modes get localized if m was even in the clean system, while a single delocalized pair survives if m was odd.

Disorder was found to induce a metallic phase separating the two (QSH and ordinary) insulators.[136,137] The transition between metal and any of the two insulators occurs at the critical value of conductivity $g = g^* \approx 1.4$; both transitions are believed to belong to the same universality class, see Secs. 2.2.2 and 4.1.1. For $g < g^*$ all bulk states are eventually localized in the limit of large system, while for $g > g^*$ the weak antilocalization specific to the symplectic symmetry class drives the system to the "supermetallic" state, $g \to \infty$. The schematic phase diagram for the non-interacting case is shown in Fig. 6.2 (left panel).

A related three-dimensional (3D) $\mathbb{Z}_2$ topological insulator was discovered in Ref. 126 where crystals of $Bi_{1-x}Sb_x$ were investigated. The boundary in this case gives rise to a 2D topologically protected metal. Similarly to 2D topological insulators, the inversion of the 3D band gap induces an odd number of the surface 2D modes.[133,138] These states in BiSb have been studied experimentally in Refs. 126 and 128. Other examples of 3D topological insulators include BiTe and BiSe systems.[129] The effective 2D surface Hamiltonian has a Rashba form and describes a single species of 2D massless Dirac

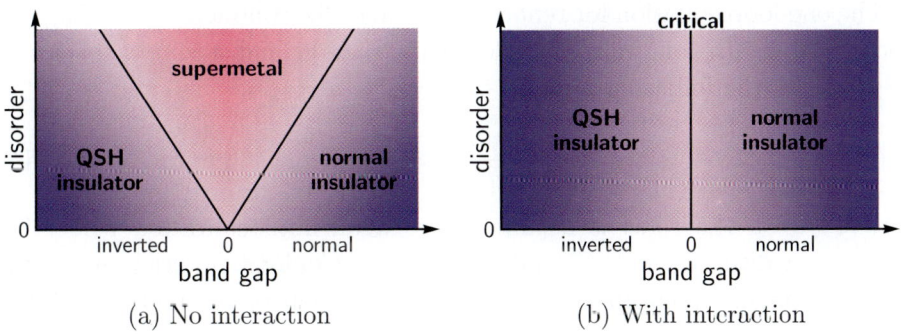

(a) No interaction   (b) With interaction

Fig. 6.2. The phase diagrams of a disordered 2D system demonstrating the QSH effect. Left: non-interacting case. Right: interacting case (with Coulomb interaction not screened by external gates). Interaction "kills" the supermetallic phase. As a result, the two insulating phases are separated by the critical line. Adapted from Ref. 130.

particles (cf. Ref. 139). It is thus analogous to the Hamiltonian of graphene with just a single valley. In the absence of interaction, the conductivity of the disordered surface of a 3D topological insulator therefore scales to infinity with increasing system size, see Sec. 4.2.2.

6.3. *Interaction effects on Z_2 topological insulators of class AII*

In this section, we overview the effect of Coulomb interaction between electrons in topological insulators.[130] Since a topological insulator is characterized by the presence of propagating surface modes, its robustness with respect to interactions means that interactions do not localize the boundary states. Indeed, arguments showing the stability of $\mathbb{Z}_2$ topological insulators with respect to interactions were given in Refs. 122,124,140 and 130. An additional argument in favor of persistence of topological protection in the presence of interaction is based on the replicated Matsubara sigma-model, in analogy with the ordinary QHE.[141] This theory possesses the same non-trivial topology as in the non-interacting case.

Can the topologically protected 2D state be a supermetal ($g \to \infty$) as in the non-interacting case? To answer this question, the perturbative RG applicable for large conductivity $g \gg 1$ has been employed in Ref. 130. It is well known that in a 2D diffusive system the interaction leads to logarithmic corrections to the conductivity,[101] see Sec. 5. These corrections (together with the interference-induced ones) can be summed up with the use of RG technique.[102,103]

The one-loop equation for renormalization of the conductivity in the symplectic class with long-range Coulomb interaction and a single species of particles has the following form:

$$\beta(g) = \frac{dg}{d\ln L} = -1/2. \tag{6.7}$$

Here $-1/2$ on the r.h.s. is a sum of the weak antilocalization correction $1/2$ due to disorder and -1 induced by the Coulomb interaction in the singlet channel. According to Eq. (6.7), the negative interaction-induced term in $\beta(g)$ dominates the scaling at large g. Therefore, for $g \gg 1$ the conductance decreases upon renormalization and the supermetal fixed point becomes repulsive.

Thus, on one hand, at $g \gg 1$ there is (i) scaling towards smaller g on the side of large g. On the other hand, surface states are topologically protected from localization, which yields (ii) scaling towards higher g on the side of small g. The combination of (i) and (ii) unavoidably leads to the conclusion that the system should scale to a critical state ($g \sim 1$). Indeed, there is no other way to continuously interpolate between negative (i) and positive (ii) beta functions: at some point $\beta(g)$ should cross zero. As a result, a critical point emerges due to the combined effect of interaction and topology.[130] In other words, if the system can flow neither towards a supermetal ($g \to \infty$) nor to an insulator ($g \to 0$), it must flow to an intermediate fixed point ($g \sim 1$). Remarkably, the critical state emerges on the surface of a 3D topological insulator without any adjustable parameters. This phenomenon can be thus called "self-organized quantum criticality".[130]

Let us now return to 2D $\mathbb{Z}_2$ topological insulators. The 2D disordered QSH system contains only a single flavor of particles, $N = 1$. Indeed, the spin-orbit interaction breaks the spin-rotational symmetry, whereas the valleys are mixed by disorder. As a result, the supermetal phase does not survive in the presence of Coulomb interaction: at $g \gg 1$ the interaction-induced localization wins. This is analogous to the case of the surface of a 3D topological insulator discussed above.

The edge of a 2D topological insulator is protected from the full localization.[122] This means that the topological distinction between the two insulating phases (ordinary and QSH insulator) is not destroyed by the interaction, whereas the supermetallic phase separating them disappears. Therefore, the transition between two insulators occurs through an interaction-induced critical state,[130] see Fig. 6.2 (right panel).

7. Summary

Despite its half-a-century age, Anderson localization remains a very actively developing field. In this article, we have reviewed some of recent theoretical advances in the physics of Anderson transitions, with an emphasis on manifestations of criticality and on the impact of underlying symmetries and topologies. The ongoing progress in experimental techniques allows one to explore these concepts in a variety of materials, including semiconductor structures, disordered superconductors, graphene, topological insulators, atomic systems, light and sound propagating in random media, etc.

We are very grateful to a great many of colleagues for fruitful collaboration and stimulating discussions over the years of research work in this remarkable field. The work was supported by the DFG — Center for Functional Nanostructures, by the EUROHORCS/ESF (IVG), and by Rosnauka grant 02.740.11.5072.

References

1. P. W. Anderson, *Phys. Rev.* **109**, 1492 (1958).
2. P. A. Lee and T. V. Ramakrishnan, *Rev. Mod. Phys.* **57**, 287 (1985).
3. B. Kramer and A. MacKinnon, *Rep. Prog. Phys.* **56**, 1469 (1993).
4. B. Huckestein, *Rev. Mod. Phys.* **67**, 357 (1995).
5. K. B. Efetov, *Adv. Phys.* **32**, 53 (1983); *Supersymmetry in Disorder and Chaos* (Cambridge University Press, 1997).
6. F. Evers and A. D. Mirlin, *Rev. Mod. Phys.* **80**, 1355 (2008).
7. A. Altland and M. R. Zirnbauer, *Phys. Rev. B* **55**, 1142 (1997).
8. M. R. Zirnbauer, *J. Math. Phys.* **37**, 4986 (1996); P. Heinzner, A. Huckleberry and M. R. Zirnbauer, *Commun. Math. Phys.* **257**, 725 (2005).
9. A. M. M. Pruisken, *Nucl. Phys. B* **235**, 277 (1984); in *The Quantum Hall Effect*, eds. R. E. Prange and S. M. Girvin (Springer, 1987), p. 117.
10. M. Janßen, *Int. J. Mod. Phys. B* **8**, 943 (1994).
11. A. D. Mirlin, Y. V. Fyodorov, F.-M. Dittes, J. Quezada and T. H. Seligman, *Phys. Rev. E* **54**, 3221 (1996).
12. A. D. Mirlin, *Phys. Rep.* **326**, 259 (2000).
13. F. Evers and A. D. Mirlin, *Phys. Rev. Lett.* **84**, 3690 (2000); A. D. Mirlin and F. Evers, *Phys. Rev. B* **62**, 7920 (2000).
14. D. Thouless, *Phys. Rep.* **13**, 93 (1974).
15. F. Wegner, *Z. Phys. B* **25**, 327 (1996).
16. E. Abrahams, P. W. Anderson, D. C. Licciardello and T. V. Ramakrishnan, *Phys. Rev. Lett.* **42**, 673 (1979).
17. F. Wegner, *Z. Phys. B* **35**, 207 (1979).
18. L. P. Gor'kov, A. I. Larkin and D. Khmel'nitskii, *JETP Lett.* **30**, 248 (1979).
19. D. Vollhardt and P. Wölfle, *Phys. Rev. B* **22**, 4666 (1980).
20. L. Schaefer and F. Wegner, *Z. Phys. B* **38**, 113 (1980).
21. K. Jüngling and R. Oppermann, *Z. Phys. B* **38**, 93 (1980).
22. K. B. Efetov, A. I. Larkin and D. E. Khmeniskii, *Sov. Phys. JETP* **52**, 568 (1980).
23. P. Wölfle and R. N. Bhatt, *Phys. Rev. B* **30**, 3542 (1984).
24. T. Guhr, A. Müller-Gröling and H. A. Weidenmüller, *Phys. Rep.* **299**, 189 (1998).
25. M. R. Zirnbauer, arXiv:math-ph/0404057 (2004).
26. S. Hikami, *Phys. Lett. B* **98**, 208 (1981).
27. F. Wegner, *Nucl. Phys. B* **316**, 663 (1989).
28. M. Schreiber and H. Grussbach, *Phys. Rev. Lett.* **76**, 1687 (1996).
29. F. Wegner, *Z. Phys. B* **36**, 209 (1980).
30. C. Castellani and L. Peliti, *J. Phys. A: Math. Gen.* **19**, L429 (1986).
31. B. B. Mandelbrot, *J. Fluid Mech.* **62**, 331 (1974).
32. J. T. Chalker, *Physica A* **167**, 253 (1990).
33. F. Wegner, in *Localisation and Metal Insulator Transitions*, eds. H. Fritzsche and D. Adler (Plenum, N. Y., 1985), p. 337.
34. B. Duplantier and A. W. W. Ludwig, *Phys. Rev. Lett.* **66**, 247 (1991).
35. C. Mudry, C. Chamon and X.-G. Wen, *Nucl. Phys. B* **466**, 383 (1996).

36. T. C. Halsey, M. H. Jensen, L. P. Kadanoff, I. Procaccia and B. I. Shraiman, *Phys. Rev. A* **33**, 1141 (1986).
37. A. D. Mirlin, Y. V. Fyodorov, A. Mildenberger and F. Evers, *Phys. Rev. Lett.* **97**, 046803 (2006).
38. A. Mildenberger, A. R. Subramaniam, R. Narayanan, F. Evers, I. A. Gruzberg and A. D. Mirlin, *Phys. Rev. B* **75**, 094204 (2007).
39. A. Mildenberger and F. Evers, *Phys. Rev. B* **75**, 041303(R) (2007).
40. H. Obuse, A. R. Subramaniam, A. Furusaki, I. A. Gruzberg and A. Ludwig, *Phys. Rev. Lett.* **98**, 156802 (2007).
41. A. Mildenberger, F. Evers and A. D. Mirlin, *Phys. Rev. B* **66**, 033109 (2002).
42. F. Wegner, *Nucl. Phys. B* **280**, 210 (1987).
43. A. R. Subramaniam, I. A. Gruzberg, A. W. W. Ludwig, F. Evers, A. Mildenberger and A. D. Mirlin, *Phys. Rev. Lett.* **96**, 126802 (2006).
44. M. Morgenstern, J. Klijn, Chr. Meyer and R. Wiesendanger, *Phys. Rev. Lett.* **90**, 056804 (2003); K. Hashimoto, C. Sohrmann, J. Wiebe, T. Inaoka, F. Meier, Y. Hirayama, R. A. Römer, R. Wiesendanger and M. Morgenstern, *Phys. Rev. Lett.* **101**, 256802 (2008).
45. J. Chabé, G. Lemarié, B. Grémaud, D. Delande, P. Szriftgiser and J. C. Garreau, *Phys. Rev. Lett.* **101**, 255702 (2008).
46. H. Hu, A. Strybulevych, J. H. Page, S. E. Skipetrov and B. A. van Tiggelen, *Nature Phys.* **4**, 945 (2008).
47. S. Faez, A. Strybulevych, J. H. Page, A. Lagendijk and B. A. van Tiggelen, *Phys. Rev. Lett.* **103**, 155703 (2009).
48. A. Richardella, P. Roushan, S. Mack, B. Zhou, D. A. Huse, D. D. Awschalom and A. Yazdani, *Science* **327**, 665 (2010).
49. M. V. Feigel'man, L. B. Ioffe, V. E. Kravtsov and E. Cuevas, arXiv:1002.0859.
50. R. Abou-Chacra, P. W. Anderson and D. J. Thouless, *J. Phys. C* **6**, 1734 (1973).
51. K. B. Efetov, *Sov. Phys. JETP* **61**, 606 (1985); *ibid.* **65**, 360 (1987).
52. M. R. Zirnbauer, *Phys. Rev. B* **34**, 6394 (1986); *Nucl. Phys. B* **265**, 375 (1986).
53. A. D. Mirlin and Y. V. Fyodorov, *Nucl. Phys. B* **366**, 507 (1991).
54. K. B. Efetov, *Physica A* **167**, 119 (1990).
55. Y. V. Fyodorov, A. D. Mirlin and H.-J. Sommers, *J. Phys. I France* **2**, 1571 (1992).
56. A. D. Mirlin and Y. V. Fyodorov, *Phys. Rev. Lett.* **72**, 526; *J. Phys. I France* **4**, 655.
57. E. P. Wigner, *Ann. Math.* **53**, 36 (1951).
58. F. J. Dyson, *J. Math. Phys.* **3**, 140, 1199 (1962).
59. S. Helgason, *Differential Geometry, Lie Groups, and Symmetric Spaces* (Academic Press, New York, 1978).
60. M. Caselle and U. Magnea, *Phys. Rep.* **394**, 41 (2004).
61. P. Fendley, arXiv:cond-mat/0006360v1 (2000).
62. R. Gade, *Nucl. Phys. B* **398**, 499 (1993).
63. R. Gade and F. Wegner, *Nucl. Phys. B* **360**, 213 (1991).
64. K. Slevin and T. Nagao, *Phys. Rev. Lett.* **70**, 635 (1993).
65. J. J. M. Verbaarschot and I. Zahed, *Phys. Rev. Lett.* **70**, 3852.

66. V. Kagalovsky, B. Horovitz, Y. Avishai and J. T. Chalker, *Phys. Rev. Lett.* **82**, 3516 (1999).

67. I. A. Gruzberg, A. W. W. Ludwig and N. Read, *Phys. Rev. Lett.* **82**, 4524 (1999).

68. T. Senthil, J. B. Marston and M. P. A. Fisher, *Phys. Rev. B* **60**, 4245 (1999).

69. E. J. Beamond, J. Cardy and J. T. Chalker, *Phys. Rev. B* **65**, 214301 (2002).

70. F. Evers, A. Mildenberger and A. D. Mirlin, *Phys. Rev. B* **67**, 041303(R) (2003); A. D. Mirlin, F. Evers and A. Mildenberger, *J. Phys. A: Math. Gen.* **36**, 3255 (2003).

71. S. Cho and M. P. A. Fisher, *Phys. Rev. B* **55**, 1025 (1997).

72. T. Senthil and M. P. A. Fisher, *Phys. Rev. B* **61**, 6966 (2000).

73. J. T. Chalker, N. Read, V. Kagalovsky, B. Horovitz, Y. Avishai and A. W. W. Ludwig, *Phys. Rev. B* **65**, 012506 (2002).

74. A. Mildenberger, F. Evers, A. D. Mirlin and J. T. Chalker, *Phys. Rev. B* **75**, 245321 (2007).

75. D. A. Ivanov, in *Vortices in Unconventional Superconductors and Superfluids*, eds. R. P. Huebener, N. Schopohl and G. E. Volovik (Springer, 2002), p. 253; *J. Math. Phys.* **43**, 126 (2002).

76. E. Witten, *Commun. Math. Phys.* **92**, 455 (1984).

77. A. A. Nersesyan, A. M. Tsvelik and F. Wenger, *Phys. Rev. Lett.* **72**, 2628 (1994); *Nucl. Phys. B* **438**, 561 (1995).

78. A. Altland, B. D. Simons and M. R. Zirnbauer, *Phys. Rep.* **359**, 283 (2002).

79. M. Bocquet, D. Serban and M. R. Zirnbauer, *Nucl. Phys. B* **578**, 628 (2000).

80. V. S. Dotsenko, *Adv. Phys.* **32**, 129 (1983).

81. S. Guruswamy, A. LeClair and A. W. W. Ludwig, *Nucl. Phys. B* **583**, 475 (2000).

82. E. McCann, K. Kechedzhi, V. I. Fal'ko, H. Suzuura, T. Ando and B. L. Altshuler, *Phys. Rev. Lett.* **97**, 146805 (2006).

83. A. W. W. Ludwig, M. P. A. Fisher, R. Shankar and G. Grinstein, *Phys. Rev. B* **50**, 7526 (1994)

84. D. V. Khveshchenko, *Phys. Rev. Lett.* **97**, 036802 (2006).

85. I. L. Aleiner and K. B. Efetov, *Phys. Rev. Lett.* **97**, 236801 (2006).

86. A. Altland, *Phys. Rev. Lett.* **97**, 236802 (2006).

87. P. M. Ostrovsky, I. V. Gornyi and A. D. Mirlin, *Phys. Rev B* **74**, 235443 (2006).

88. P. M. Ostrovsky, I. V. Gornyi and A. D. Mirlin, *Phys. Rev. Lett.* **98**, 256801 (2007).

89. P. M. Ostrovsky, I. V. Gornyi and A. D. Mirlin, *Eur. Phys. J. Special Topics* **148**, 63 (2007).

90. D. Bernard and A. LeClair, *J. Phys A* **35**, 2555 (2002).

91. A. K. Geim and K. S. Novoselov, *Nature Mater.* **6**, 183 (2007).

92. A. H. Castro Neto, F. Guinea, N. M. R. Peres, K. S. Novoselov and A. K. Geim, *Rev. Mod. Phys.* **81**, 109 (2009).

93. K. S. Novoselov, A. K. Geim, S. V. Morozov, D. Jiang, M. I. Katsnelson, I. V. Grigorieva, S. V. Dubonos and A. A. Firsov, *Nature (London)* **438**, 197 (2005).

94. Y. Zhang, Y.-W. Tan, H. L. Stormer and P. Kim, *Nature (London)* **438**, 201 (2005).
95. Y.-W. Tan, Y. Zhang, H. L. Stormer and P. Kim, *Eur. Phys. J. Special Topics* **148**, 15 (2007).
96. J. H. Bardarson *et al.*, *Phys. Rev. Lett.* **99**, 106801 (2007); K. Nomura *et al.*, *ibid.* **99**, 146806 (2007).
97. J.-S. Caux, *Phys. Rev. Lett.* **81**, 4196 (1998); J.-S. Caux, N. Taniguchi and A. M. Tsvelik, *Nucl. Phys. B* **525**, 621 (1998).
98. A. M. Tsvelik, *Phys. Rev. B* **51**, 9449 (1995).
99. J. Schwinger, *Phys. Rev.* **128**, 2425 (1962).
100. S. Ryu, C. Mudry, A. Furusaki and A. W. W. Ludwig, *Phys. Rev. B* **75**, 205344 (2007).
101. B. L. Altshuler and A. G. Aronov, in *Electron–Electron Interactions in Disordered Conductors*, eds. A. L. Efros and M. Pollak (Elsevier, 1985), p. 1.
102. A. M. Finkelstein, *Sov. Sci. Rev. A. Phys.* **14**, 1 (1990).
103. D. Belitz and T. R. Kirkpatrick, *Rev. Mod. Phys.* **66**, 261 (1994).
104. E. Abrahams, S. V. Kravchenko and M. P. Sarachik, *Rev. Mod. Phys.* **73**, 251 (2001).
105. A. Punnoose and A. M. Finkel'stein, *Phys. Rev. Lett.* **88**, 016802 (2001); *Science* **310**, 289 (2005).
106. T. Giamarchi and H. Schulz, *Phys. Rev. B* **37**, 325 (1988).
107. T. Giamarchi, *Quantum Physics in One Dimension* (Oxford University Press, 2004).
108. L. Fleishman, D. C. Licciardello and P. W. Anderson, *Phys. Rev. Lett.* **40**, 1340 (1978).
109. L. Fleishman and P. W. Anderson, *Phys. Rev. B* **21**, 2366 (1980).
110. I. V. Gornyi, A. D. Mirlin and D. G. Polyakov, *Phys. Rev. Lett.* **95**, 206603 (2005).
111. D. M. Basko, I. L. Aleiner and B. L. Altshuler, *Ann. Phys. (N.Y.)* **321**, 1126 (2006).
112. B. L. Altshuler, Y. Gefen, A. Kamenev and L. S. Levitov, *Phys. Rev. Lett.* **78**, 2803 (1997).
113. S.-Y. Hsu and J. M. Valles, Jr., *Phys. Rev. Lett.* **74**, 2331 (1995).
114. F. W. Van Keuls, H. Mathur, H. W. Jiang and A. J. Dahm, *Phys. Rev. B* **56**, 13263 (1997).
115. Y. B. Khavin, M. E. Gershenson and A. L. Bogdanov, *Phys. Rev. B* **58**, 8009 (1998).
116. G. M. Minkov, A. V. Germanenko, O. E. Rut, A. A. Sherstobitov and B. N. Zvonkov, *Phys. Rev. B* **75**, 235316 (2007).
117. V. Oganesyan and D. A. Huse, *Phys. Rev. B* **75**, 155111 (2007).
118. D. A. Knyazev, O. E. Omel'yanovskii, V. M. Pudalov and I. S. Burmistrov, *Phys. Rev. Lett.* **100**, 046405 (2008).
119. A. Punnoose, A. M. Finkel'stein, A. Mokashi and S. V. Kravchenko, arXiv:0910.5510.
120. I. V. Gornyi, A. D. Mirlin and D. G. Polyakov, *Phys. Rev. B* **75**, 085421 (2007).

121. C. Monthus and T. Garel, arXiv:1001.2984.
122. C. L. Kane and E. J. Mele, *Phys. Rev. Lett.* **95**, 146802 (2005); *ibid.* **95**, 226801 (2005).
123. B. A. Berncvig *et al.*, *Science* **314**, 1757 (2006); B. A. Bernevig and S.-C. Zhang, *Phys. Rev. Lett.* **96**, 106802 (2006).
124. X.-L. Qi *et al.*, *Phys. Rev. B* **78**, 195424 (2008).
125. M. König *et al.*, *Science* **318**, 766 (2007).
126. D. Hsieh *et al.*, *Nature* **452**, 970 (2008).
127. M. König *et al.*, *J. Phys. Soc. Jpn.* **77**, 031007 (2008); A. Roth *et al.*, *Science* **325**, 294 (2009).
128. D. Hsieh *et al.*, *Science* **323**, 919 (2008); P. Roushan *et al.*, *Nature* **460**, 1106 (2009).
129. D. Hsieh *et al.*, *Nature* **460**, 1101 (2009); Y. L. Chen *et al.*, *Science* **325**, 178 (2009); Y. Xia *et al.*, *Nature Phys.* **5**, 398 (2009); H. Zhang *et al.*, *ibid.* **5**, 438 (2009).
130. P. M. Ostrovsky, I. V. Gornyi and A. D. Mirlin, arXiv:0910.1338.
131. L. Sheng *et al.*, *Phys. Rev. Lett.* **95**, 136602 (2005).
132. G. E. Volovik, *JETP* **67**, 1804 (1988); G. E. Volovik and V. M. Yakovenko, *J. Phys.: Cond. Matt.* **1**, 5263 (1989).
133. L. Fu *et al.*, *Phys. Rev. Lett.* **98**, 106803 (2007); L. Fu and C. L. Kane, *Phys. Rev. B* **76**, 045302 (2007).
134. A. P. Schnyder *et al.*, *Phys. Rev. B* **78**, 195125 (2008); *AIP Conf. Proc.* **1134**, 10 (2009); S. Ryu *et al.*, arXiv:0912.2157.
135. A. Yu. Kitaev, *AIP Conf. Proc.* **1134**, 22 (2009).
136. M. Onoda *et al.*, *Phys. Rev. Lett.* **98**, 076802 (2007); S. Murakami *et al.*, *Phys. Rev. B* **76**, 205304 (2007).
137. H. Obuse *et al.*, *Phys. Rev. B* **76**, 075301 (2007); S. Ryu *et al.*, arXiv:0912.2158.
138. M. I. D'yakonov and A. V. Khaetskii, *JETP Lett.* **33**, 110 (1981).
139. V. A. Volkov and T. N. Pinsker, *Sov. Phys. Solid State* **23**, 1022 (1981); B. A. Volkov and O. A. Pankratov, *JETP Lett.* **42**, 178 (1985).
140. S.-S. Lee and S. Ryu, *Phys. Rev. Lett.* **100**, 186807 (2008).
141. A. M. M. Pruisken and I. S. Burmistrov, *Ann. Phys. (N.Y.)* **322**, 1265 (2007).

Chapter 7

SCALING OF VON NEUMANN ENTROPY AT THE ANDERSON TRANSITION

Sudip Chakravarty

Department of Physics and Astronomy, University of California Los Angeles,
Los Angeles, California 90024, USA
sudip@physics.ucla.edu

Extensive body of work has shown that for the model of a non-interacting electron in a random potential there is a quantum critical point for dimensions greater than two — a metal–insulator transition. This model also plays an important role in the plateau-to-plateau transition in the integer quantum Hall effect, which is also correctly captured by a scaling theory. Yet, in neither of these cases the ground state energy shows any non-analyticity as a function of a suitable tuning parameter, typically considered to be a hallmark of a quantum phase transition, similar to the non-analyticity of the free energy in a classical phase transition. Here we show that von Neumann entropy (entanglement entropy) is non-analytic at these phase transitions and can track the fundamental changes in the internal correlations of the ground state wave function. In particular, it summarizes the spatially wildly fluctuating intensities of the wave function close to the criticality of the Anderson transition. It is likely that all quantum phase transitions can be similarly described.

1. Introduction

Ever since Anderson's paper,[1] "Absence of Diffusion in Certain Random Lattices", it has been a theme in condensed matter physics to unravel the quantum phase transition between the itinerant and the localized electronic states.[2] The metal–insulator transition embodies the very basic concept of wave-particle complementarity in quantum mechanics. Itinerant states reflect the wave aspect, while the localized states reflect the particle aspect. In one-particle quantum mechanics without disorder, the wave and the particle descriptions are dual to each other. There is no fundamental distinction between them. Coherent superposition of waves are packets that act like spatially compact lumps of energy and momentum, or particles. In contrast, in a disordered medium the metallic state described by non-normalizable wave functions is separated by a quantum phase transition, the Anderson transition, when it exists, from the insulating state with normalizable wave functions. In the insulating state, particles are tied to random spatial centers. These two macroscopic states are fundamentally different and cannot be analytically continued into each other.

If the Fermi energy is situated within the localized states, the system is an insulator. It might be argued that in a real physical situation, the role of electron–electron interaction will become more and more important as the system approaches localization and the notion of Anderson localization will loose its validity. In fact, quite the opposite may sometimes be true. A rigorous, but a simple example of spinless fermions, was recently studied[3,4] where interactions lead to a broken symmetry in the pure system, generating a gap, hence an insulator. But it was shown that for arbitrary disorder this gap is washed out, and there are gapless localized excitations resembling an "Anderson insulator". In any case, Anderson insulator has proven to be a powerful paradigm for metal–insulator transition.

Because the Anderson transition is a quantum phase transition, it is natural to develop a theoretical framework that comes as close as possible to any other thermodynamical quantum phase transitions. Although there are other theoretical approaches, including powerful numerical simulations of an electron in a random potential,[5] interesting insights can be gained by contrasting and comparing with more conventional models of phase transitions. In order to study Anderson localization, I shall focus on the scaling properties of the von Neumann entropy (vNE), which is a fundamental concept in quantum mechanics and quantum information theory.

2. Statistical Field Theory of Localization

It is well known that the properties of a Brownian particle can be understood from a free Euclidean field theory. The free fields act as a generating function for the Brownian motion. The Green's function of interacting fields, on the other hand, reflect particles with suitable constraints.[6] A particularly pretty example is that of the self avoiding random walks that can be described in terms of the correlation functions of the $O(n)$ spin model in the limit $n \to 0$, even though the partition function is exactly unity in that limit.[7] The lesson is that the language of statistical field theory and its scaling behavior can provide important insights. Similarly, a replica field theory discussed elsewhere in this volume maps the Anderson problem of a single particle with disorder to a suitable non-linear σ-model, which depends on the relevant symmetries, with the proviso that the number of replicas N has to be set to zero at the end of all calculations. It is only in the limit $N \to 0$ that the effect of randomness appear; as long as $N \neq 0$, the model is translationally invariant. In spite of the subtleties of the replica limit, much has been learnt as far as the criticality of the Anderson transition is concerned by drawing analogies with the problem of critical phenomena in statistical mechanics.[8]

One can also reverse the chain of reasoning and learn about the statistical mechanics of critical phenomena from the Anderson problem. As an example, let us consider the universal conductance fluctuations in a mesoscopic system. It was shown that if we consider the disorder averaged conductance by $\langle G \rangle$ and its fluctuation by $\langle (\delta G)^2 \rangle$,[9] then the latter is independent of scale and is universal for dimension $D < 4$. A sample is considered to be mesoscopic if its linear dimension L is larger than the mean free path but smaller than the scale at which the phase coherence of the electrons is lost. The relative fluctuation $\langle (\delta G)^2 \rangle / \langle G \rangle^2$ is proportional to L^{4-2D} and is independent of scale at $D = 2$.[9] In fact, it was shown that this result along with many others can be obtained from a replica field theory of an extended non-linear sigma model defined on a Grassmannian manifold.[10] This raises the possibility that perhaps a similar result should also hold on a much simpler manifold, namely the coset space of $O(n)/O(n-1)$.[11] For $n = 3$, this is the familiar $O(3)$ σ-model of classical n-vector spins of unit length $\hat{\Omega}^2 = 1$, which is a faithful description of the long wavelength behavior of the classical Heisenberg model. What could possibly be the analog of the conductance for the Heisenberg model? It was argued that it is the spin stiffness constant, defined by the response of the system with respect to a twist in the boundary condition and measures the rigidity of the system. By a meoscopic sample we now mean L such that it is much larger than the microscopic cutoff of the order of lattice spacing and much smaller than the correlation length ξ of the Heisenberg model. From a one-loop calculation, it is easy to show that the absolute fluctuation of the spin stiffness constant ρ_s is independent of the scale and its relative fluctuation is given by

$$\frac{\overline{\delta \rho_s^2}}{\overline{\rho_s}^2} \propto L^{4-2D}, \tag{2.1}$$

where the overline now represents the average with respect to the thermal fluctuations. More explicitly in the interesting case of $D = 2$ we get, including the logarithmic correction,

$$\frac{\overline{\delta\rho_s^2}}{\overline{\rho_s}^2} = \frac{2\pi}{(n-2)[\ln(\xi/L)]^2}.$$ (2.2)

One can find many more interesting connections between these two disparate systems, which behooves us to take a closer look at the "thermodynamics" of the quantum phase transition in the Anderson model, leading us to a discussion of the vNE.

3. von Neumann Entropy

A set of brief remarks seem to be appropriate to place our subsequent discussion in a more general context. In a landmark paper on black hole entropy, Bekenstein[12] demonstrated the power of the notion of information entropy. The concept can also be applied to any quantum mechanical ground state. Given a unique ground state, the thermodynamic entropy is of course zero. To distinguish various ground states one usually studies the analyticity of the ground state energy as a function of a tuning parameter. In most cases, a quantum phase transition is characterized by the non-analyticity of the ground state energy. In some cases, for example in the Anderson transition and in the integer quantum Hall plateau transitions, the ground state energy is analytic through the transitions and does not provide any indication of their existence. Yet we know that the wave function encodes special correlations internal to its state. How can we quantify such correlations? In particular how do they change across these quantum phase transitions? We shall show that in these cases the non-analyticity of vNE can be used as a fingerprint of these quantum phase transitions.[13]

For a pure state $|\Psi\rangle$, the density matrix is $\rho = |\Psi\rangle\langle\Psi|$. Consider partitioning the system into A and B, where A denotes the subsystem of interest and B the environment whose details are of no interest. The reduced density matrix ρ_A is constructed by tracing over the degrees of freedom of B, similar to integrating out the microstates corresponding to a set of macroscopic thermodynamic variables. The vNE, $S = -\mathrm{Tr}(\rho_A \ln \rho_A)$, is a measure of the bipartite entanglement and therefore contains information about the quantum correlations present in the ground state. The interesting point is that the reduced density matrix is a mixture if the state $|\Psi\rangle$ is entangled, that is, it cannot be factored into $|\Psi\rangle_A \otimes |\Psi\rangle_B$. Of course, partitioning a mixed state will also lead to a mixed state; there is nothing new here. Since ρ_A is a mixture, we can perform a statistical analysis of it and obtain a nontrivial of entropy that can summarize the essential features of an entangled state. The result follows from the Schmidt decomposition theorem: for a bipartition of a pure state there exist sets of orthonormal states $\{|i_A\rangle\}$ of A and orthonormal states $\{|i_B\rangle\}$ of B such that

$$|\Psi\rangle = \sum_i \lambda_i |i_A\rangle \otimes |i_B\rangle, \qquad (3.1)$$

where λ_i are non-negative real numbers satisfying $\sum_i \lambda_i^2 = 1$. The result that a state can be fully known, yet its subsystem is in a mixed state is a remarkable consequence of entanglement. Unfortunately, there no such theorems if we partition the system into more than two parts, say A, B, and C. Multipartite entanglements are consequently less understood.

As we have argued above, the mapping of the Anderson localization to a problem of a statistical field theory has been quite successful. It leaves us with little doubt that the notion of criticality and scaling are correct. We might pursue this argument further and ask does this transition fit into the general framework of a quantum phase transition? If we define such a transition in terms of the non-analyticity of the ground state energy as a function of disorder, the answer to this question is no. Edwards and Thouless[14] have shown rigorously that the ground state energy, which depends on the average density of states, is smooth through the localization transition. We believe that the closest we can come is the non-analyticity of vNE,[13] which is of great current interest in regard to quantum phase transitions.[15] One expects that vNE must play a role in understanding the correlations that exist on all length scales at a quantum critical point. But a state can be entangled without being critical — consider, for instance, the singlet state of two spin-1/2 particles. It is the special critical scaling property of entanglement that we are interested here. Even more paradoxical, it may sound, is that Anderson localization is a single particle problem, and the conventional notion entanglement of particles does not apply. Clearly, the notion of entanglement will have to be extended, and this extension will be the theory of entanglement defined using the site occupation number basis in the second-quantized Fock space.[16]

As noted above, we shall consider two important models to illustrate our expectation. Our first example is Anderson localization in dimension greater than two, which has been extensively studied and is known to have a quantum critical point. At the critical point the wave function exhibits a fractal character.[17] The second example is the plateau-to-plateau transition in the integer quantum Hall effect in which the Anderson localization plays a crucial role in establishing the very existence of the plateaus.[18] We shall see that vNE is nonanalytic at these transitions and exhibits the correct scaling behavior when compared to other approaches. vNE and its scaling behavior characterize the entanglement associated with these quantum phase transitions. Because they are determined by single-particle properties in the presence of disorder, their vNE is different from that associated with disorder-free interacting systems.

Consider the single-particle probability $|\psi_E(r)|^2$ at energy E and position r for a noninteracting electronic system. In the neighborhood of a critical point governed by disorder, it fluctuates so strongly that it has a broad (non-Gaussian) distribution even in the thermodynamic limit.[19] This non-self-averaging nature of the wave function intensity can be seen in the scaling of its moments.[20] In particular, the moments, P_ℓ, defined as the generalized inverse participation ratios, obey the finite-size scaling *Ansatz*,

$$P_\ell(E) \equiv \sum_r \overline{|\psi_E(r)|^{2\ell}} \sim L^{-\tau_\ell} \mathcal{G}_\ell\big[(E - E_c)L^{1/\nu}\big], \qquad (3.2)$$

where L is the system size and ν is the exponent characterizing the divergence of the correlation length at the critical point E_c, $\xi_E \sim |E - E_c|^{-\nu}$. The quantity τ_ℓ is the multifractal spectrum, and the overline denotes the disorder average. $\mathcal{G}_\ell(x)$ is a scaling function with $\mathcal{G}_\ell(x \to 0) \to 1$ as $E \to E_c$. As E deviates from the critical point, the system either tends to an ideal metallic state with $P_\ell(E) \sim L^{-D(\ell-1)}$ (D is the dimensionality) or to a localized state with $P_\ell(E)$ that is independent of L. In the multifractal state, right at the Anderson transition, the intensity of the wave function has local exponents, defined by its sample-size dependence, which vary from point to point. A beautiful simulation of multifractality of the intensity of the wave function at the 3D Anderson transition is shown in Fig. 7.1 reproduced from Ref. 21. The parameters and the notations are defined in Sec. 5. In contrast, a single non-integer scaling exponent applicable to the entire volume corresponds to the fluctuations of a fractal. The multifractal spectrum uniquely characterizes the wildly complex spatial structure of the wave function. It is quite remarkable that the same multifractal spectrum determines the vNE.

4. von Neumann Entropy in Disordered Noninteracting Electronic Systems

We define entanglement[22] using the site occupation number basis in the second-quantized Fock space.[16] Let us partition a lattice of linear dimension L into two parts, A and B. A single particle eigenstate at energy E in the

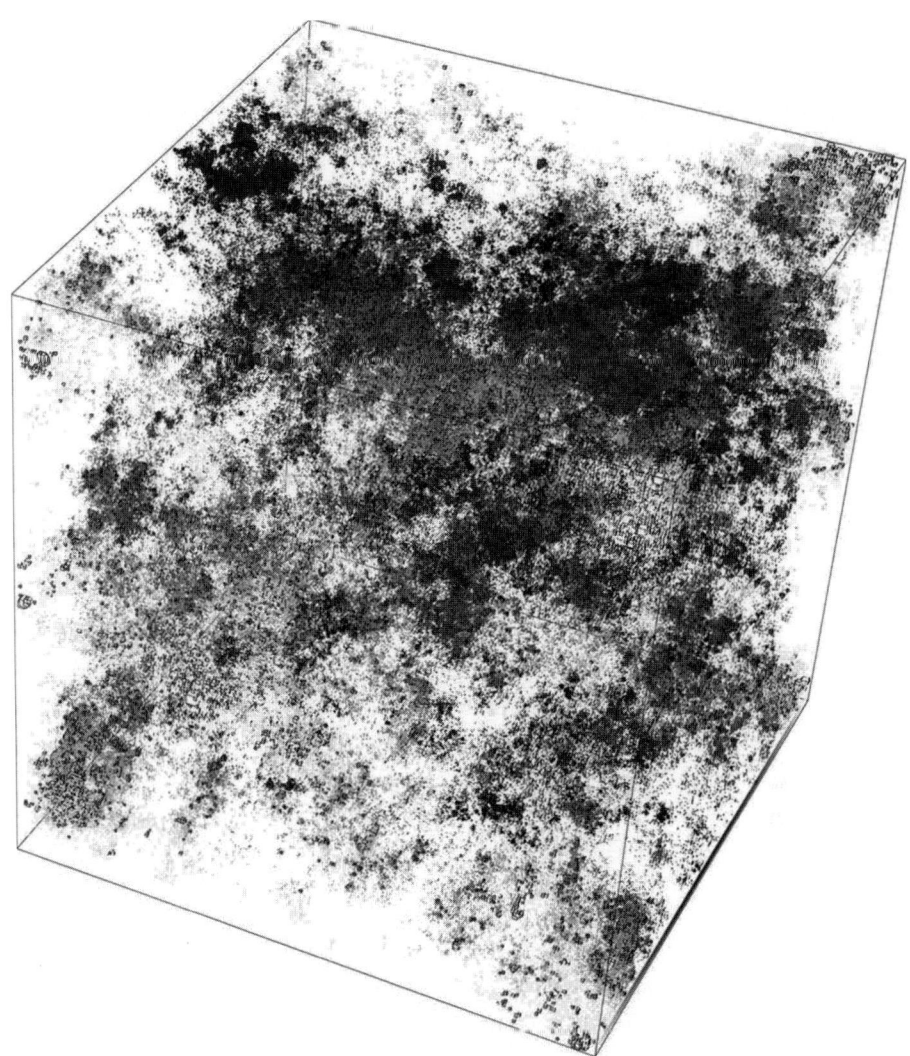

Fig. 7.1.   Multifractal eigenstate for the 3D Anderson model at $E = 0$ and $W_c =$ 16.5 for linear system size $L = 240$ with periodic boundary conditions. The 410 075 sites with probability Ψ_j^2 twice larger than the average$1/L^3$ are shown as boxes with volume $|\Psi_j|^2 L^3$. The 26 097 boxes with $|\Psi_j|^2 L^3 > 2\sqrt{1000}$ are plotted with black edges. The grayscale distinguishes between different slices of the system along the axis into the page. Reprinted (Fig. 1) with permission from L. J. Vasquez, A. Rodriguez and R. A. Römer *Phys. Rev. B* **78**, 195106 (2008). © American Physical Society.

site occupation number basis is

$$|\psi_E\rangle = \sum_{r \in A \cup B} \psi_E(r) |1\rangle_r \bigotimes_{r' \neq r} |0\rangle_{r'}. \tag{4.1}$$

Here $\psi_E(r)$ is the probability amplitude at the site r and $|n\rangle_r$ is the occupation number at site r, either 0 or 1. We rewrite the above sum over lattice sites r into mutually orthogonal parts,

$$|\psi_E\rangle = |1\rangle_A \otimes |0\rangle_B + |0\rangle_A \otimes |1\rangle_B \tag{4.2}$$

where

$$|1\rangle_A = \sum_{r \in A} \psi_E(r)|1\rangle_r \bigotimes_{r' \neq r} |0\rangle_{r'}, \quad |0\rangle_A = \bigotimes_{r \in A} |0\rangle_r \tag{4.3}$$

similarly for $|1\rangle_B$ and $|0\rangle_B$. Note that

$$\langle 0|0\rangle_A = \langle 0|0\rangle_B = 1, \quad \langle 1|1\rangle_A = p_A, \quad \langle 1|1\rangle_B = p_B, \tag{4.4}$$

where

$$p_A = \sum_{r \in A} |\psi_E(r)|^2, \tag{4.5}$$

and similarly for p_B with $p_A + p_B = 1$.

The reduced density matrix ρ_A is obtained from $\rho = |\psi_E\rangle\langle\psi_E|$, after tracing out the Hilbert space over B, is

$$\rho_A = |1\rangle_A\langle 1| + (1 - p_A)|0\rangle_A\langle 0|. \tag{4.6}$$

The corresponding vNE is given by

$$S_A = -p_A \ln p_A - (1 - p_A) \ln(1 - p_A). \tag{4.7}$$

Here, manifestly $S_A = S_B$, and either of them is bounded between 0 and $\ln 2$ for any eigenstate. Despite the use of a second-quantized language, we are considering a single particle state rather than a many body correlated state. The entanglement entropy can not grow arbitrarily large as the size of A increases, unlike the entanglement entropy in interacting quantum systems where it can be arbitrarily large close to the critical point.

If the system size becomes very large in comparison to the size of the subsystem A, we can restrict A to be a single lattice site and study scaling with respect to L. Thus, we consider the single site vNE[13]

$$S(E) = -\sum_{r \in L^d} \Big\{ |\psi_E(r)|^2 \ln |\psi_E(r)|^2$$

$$+ \big[1 - |\psi_E(r)|^2\big] \ln \big[1 - |\psi_E(r)|^2\big] \Big\}. \tag{4.8}$$

To study the leading critical behavior, the second term in the curly brackets in the right-hand side of Eq. (4.8) can be ignored since $|\psi_E(r)|^2 \ll 1$ for all r for states close to the critical energy. The disorder averaged entropy $\overline{S}$ can be expressed in terms of the multifractal scaling in Eq. (3.2), giving

$$\overline{S}(E) \approx -\left.\frac{dP_\ell}{d\ell}\right|_{\ell=1} \approx \left.\frac{d\tau_\ell}{d\ell}\right|_{\ell=1} \ln L - \left.\frac{\partial \mathcal{G}_\ell}{\partial \ell}\right|_{\ell=1}. \tag{4.9}$$

Although we do not know the analytical form of the scaling function $\mathcal{G}_\ell$, its approximate L dependence can be obtained in various limiting cases. Exactly at criticality, $\mathcal{G}_\ell \equiv 1$ for all values of ℓ and

$$\overline{S}(E) \sim \alpha_1 \ln L, \tag{4.10}$$

where the constant $\alpha_1 = d\tau_\ell/d\ell|_{\ell=1}$. The leading scaling behaviors of $\overline{S}(E)$ in both the metallic and the localized states can now be obtained, following the discussion below Eq. (3.2). The results are

$$\overline{S}_{\text{metal}}(E) \sim D \ln L, \quad \overline{S}_{\text{loc}}(E) \sim \alpha_1 \ln \xi_E. \tag{4.11}$$

We see that in general $\overline{S}(E)$ is of the form

$$\overline{S}(E) \sim \mathcal{Q}[(E - E_C)L^{1/\nu}] \ln L, \tag{4.12}$$

where the coefficient function $\mathcal{Q}(x)$ is D in the metallic state, decreases to α_1 at criticality and then goes to zero for the localized state. We now turn to numerical simulations to see the extent to which this scaling behavior is satisfied.

5. von Neumann Entropy in the Three-Dimensional Anderson Model

Let us consider the disordered Anderson model on a 3D cubic lattice.[22] The Hamiltonian is

$$H = \sum_i V_i c_i^\dagger c_i - t \sum_{\langle i,j \rangle} (c_i^\dagger c_j + H.c.), \tag{5.1}$$

where $c_i^\dagger (c_i)$ is the creation (annihilation) operator for an electron at site i and the $\langle i, j \rangle$ indicates that the second sum is over nearest neighbors. The V_i are random variables uniformly distributed in the range $[-W/2, W/2]$. In what follows, we set $t = 1$. Of course, the model has been extensively studied. Below a critical disorder strength W_c, there is a region of extended states at the band center.[5] The recent values of the critical disorder strength W_c and the localization length exponent are $W_c = 16.3$ and $\nu = 1.57 \pm 0.03$.[23]

To obtain the energy-averaged entropy, we average Eq. (4.8) over the entire band of energy eigenvalues. From this we construct the vNE,

$$\overline{S}(w, L) = \frac{1}{\mathcal{N}} \sum_E \overline{S}(E, w, L), \qquad (5.2)$$

where $\mathcal{N}$ counts the total number of states in the band. Here $w = |W - W_c|/W_c$. Near $w = 0$, we can show, using Eqs. (4.12) and (5.4), that

$$\overline{S}(w, L) \sim C + L^{-1/\nu} f_\pm(wL^{1/\nu}) \ln L, \qquad (5.3)$$

where C is a constant independent of L and $f_\pm(x)$ are two universal functions corresponding to the regimes $w > 0$ and $w < 0$. We numerically diagonalize Eq. (5.1) for systems of sizes $L \times L \times L$ with periodic boundary conditions. The maximum system size was $L = 13$, and the results were averaged over 20 disorder realizations. The scaling form of $\overline{S}(w, L)$ is given by Eq. (5.3). Figure 7.2 shows[22] the results of the data collapse with a choice of $\nu = 1.57$, and the nonuniversal constant $C = 12.96$ is determined by a powerful algorithm

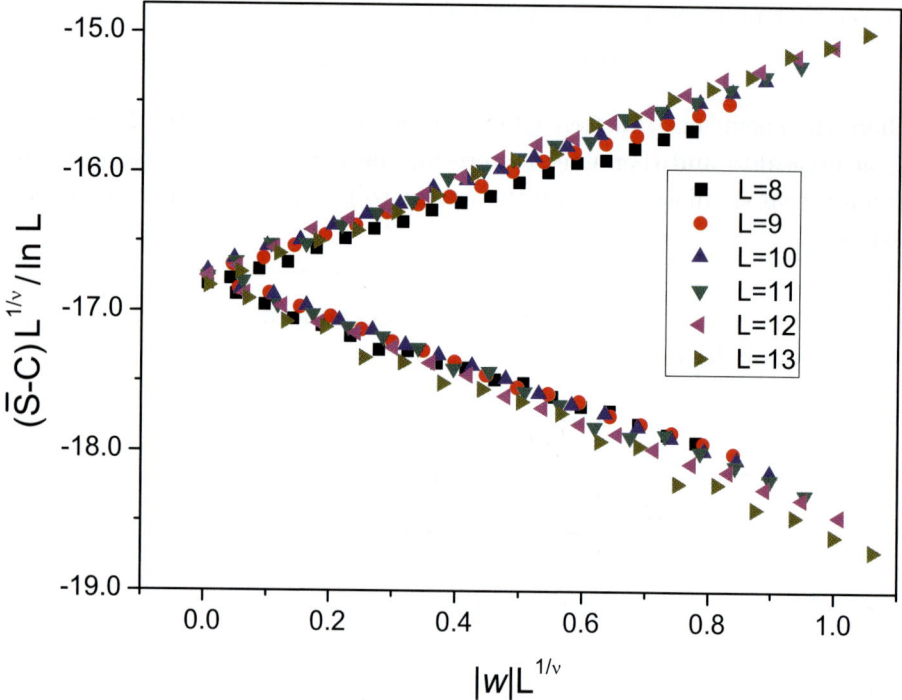

Fig. 7.2.   Scaling curve in the 3D Anderson model. With the choice of $\nu = 1.57$ and $C = 12.96$, all data collapse to a universal functions $f_\pm(x)$. The two branches correspond to $w < 0$ and $w > 0$.

described in the Appendix C of Ref. 24. The data collapse is reasonable and is consistent with the nonanalyticity of vNE and the multifractal analysis. Clearly, it would be useful to improve the numerics by increasing both the system sizes and the number of disorder realizations to attain a better data collapse.

We can also study vNE at the band center $E = 0$ by sweeping W across the critical value W_c. In this case, the states at $E = 0$ will evolve continuously from metallic to critical and then to localized states. The entanglement entropy will be given similarly by another scaling function

$$\overline{S}(E = 0, w, L) \sim \mathcal{C}(wL^{1/\nu}) \ln L, \qquad (5.4)$$

where $\mathcal{C}(x)$ is a scaling function, which as remarked earlier, $\to D$ as $w \to -1$ and $\to 0$ as $w \to \infty$, and $\mathcal{C} = \alpha_1$ when $w = 0$. For this purpose we use the transfer matrix method[25] to study the energy resolved $\overline{S}(E, w, L)$ by considering a quasi-one-dimensional (quasi-1D) system with a size of $(mL) \times L \times L$, $m \gg 1$; L up to 18, and $m = 2000$ were found to be reasonable. To compute vNE, we divide the quasi-1D system into m cubes labeled by $I = 1, 2, \ldots, m$, each containing L^3 sites. The wave function within each cube is normalized and the vNE, $\overline{S^I}(E, W, L)$ in the Ith cube is computed. Finally $\overline{S}(E, W, L)$ was obtained by averaging over all cubes. The validity of the scaling form in Eq. (5.4) is seen in Fig. 7.3.[22] In particular, the function $\mathcal{C}(x)$ shows the expected behavior, approaching $D = 3$ as $w \to -1$, and tending to 0 as $w \to \infty$.

6. von Neumann Entropy in the Integer Quantum Hall System

For the integer quantum Hall system, we use a basis defined by the states $|n, k\rangle$, where n is the Landau level index and k is the wave vector in the y-direction. The Hamiltonian can be expressed[26] in terms of the matrix elements in this basis as

$$H = \sum_{n,k} |n, k\rangle\langle n, k| \left(n + \frac{1}{2} \right) \hbar\omega_c + \sum_{n,k} \sum_{n',k'} |n, k\rangle\langle n, k|V|n', k'\rangle\langle n', k'| \quad (6.1)$$

where $\omega_c = eB/mc$ is the cyclotron frequency, and B is the magnetic field. $V(\mathbf{r})$ is the disorder potential. If we focus on the lowest Landau level, $n = 0$, and assume that the distribution of disorder is δ-correlated with zero mean, that is, $\overline{V(\mathbf{r})} = 0$ and $\overline{V(\mathbf{r})V(\mathbf{r}')} = V_0^2 \delta(\mathbf{r} - \mathbf{r}')$, the matrix elements,

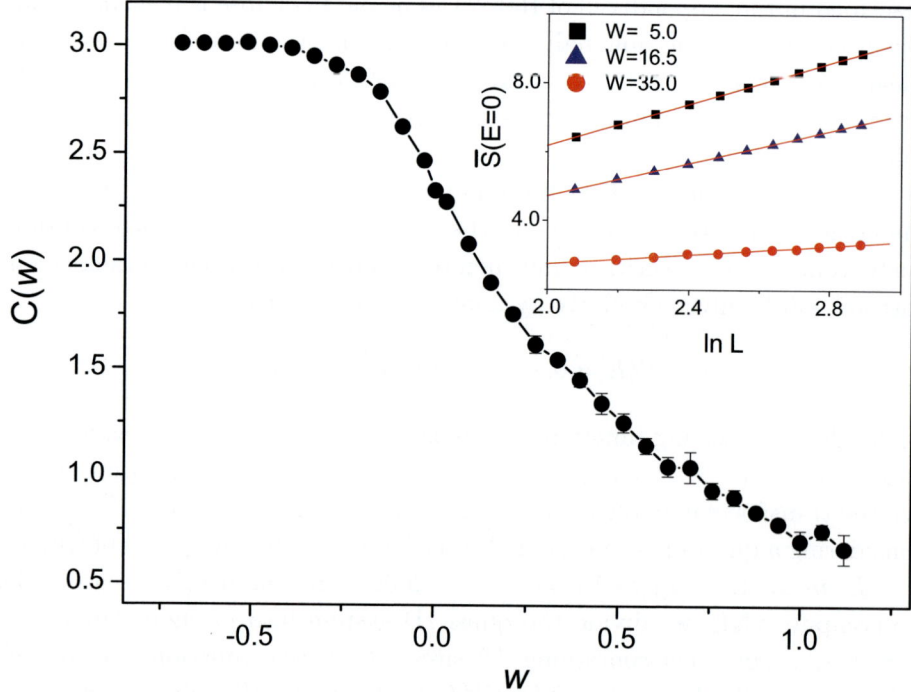

Fig. 7.3. The quantity $\mathcal{C}$ in Eq. (5.4). The system sizes are too small to observe the weak L dependence. Inset: $\overline{S}(E = 0, W, L)$ as a function of $\ln L$ for three different W.

$\langle 0, k | V | 0, k' \rangle$, are

$$
\langle 0, k | V | 0, k' \rangle = \frac{V_0}{\sqrt{\pi L_y}} \exp \left[-\frac{1}{4} l_B^2 (k - k')^2 \right]
$$
$$
\times \int \mathrm{d}\chi e^{-\chi^2} u_0 \left(l_B \chi + \frac{k + k'}{2} l_B^2, k' - k \right) \quad (6.2)
$$

where $l_B = (\hbar c / eB)^{1/2}$ is the magnetic length, and $u_0(x, k)$ is the Fourier transform of $V(x, y)$ along the y direction,

$$
u_0(x, k) = \frac{1}{\sqrt{L_y}} \int dy V(x, y) e^{iky}. \quad (6.3)
$$

We choose a two-dimensional square with a linear dimension $L = \sqrt{2\pi M} l_B$, where M is an integer. We impose periodic boundary conditions in both directions and discretize by a mesh of size $\sqrt{\pi} l_B / \sqrt{2M}$. The Hamiltonian matrix is diagonalized and the eigenstates $|\phi_a\rangle = \sum_k \alpha_{k,a} |0, k\rangle$, $a = 1 \ldots M^2$, are obtained along with the corresponding eigenvalues E_a. The zero of the energy is at the center of the lowest Landau band[27] and the unit of energy is $\Gamma = 2V_0/\sqrt{2\pi}l_B$. For each eigenstate the wave function in real space is constructed:

$$\phi_a(x, y) = \langle x, y | \phi_a \rangle = \sum_k \alpha_{k,a} \phi_{0,k}(x, y), \tag{6.4}$$

where $\phi_{0,k}(x, y)$ is the wave function with quantum number k in the lowest Landau level. The dimension of the Hamiltonian matrix increases as $N_k \sim M^2$, making it difficult to diagonalize fully. We circumvent this difficulty[22] by computing only those states $|\phi_a\rangle$ whose energies lie within a window Δ around a fixed value E thus: $E_a \in [E - \Delta/2, E + \Delta/2]$. We take Δ to be sufficiently small (0.01), but still large enough such that it spans large number of states in the interval Δ (at least 100 eigenstates).

We now uniformly break up the $L \times L$ square into smaller squares $\mathcal{A}_i$ of size $l \times l$, where $l = l_B\sqrt{\pi/2}$, independent of the system size L. The $\mathcal{A}_i$ do not overlap. For each of the states, we compute the coarse grained quantity $\int_{(x,y)\in\mathcal{A}_i} |\psi_a(x, y)|^2 dxdy$. The vNE for a given eigenstate is calculated by the same procedure described above for the Anderson localization. The vNE $\overline{S}(E, L)$ is obtained at energy E by averaging over states in the interval Δ. $\overline{S}(E, L)$ has a scaling form given by Eq. (4.12) with $E_c = 0$; it is $\overline{S}(E, L) = \mathcal{K}(|E|L^{1/\nu}) \ln L$. Figure 7.4 shows reasonably good agreement with the numerical simulations.[22] The exponent ν is consistent with that obtained by other approaches.[26] Thus, the criticality of the vNE at the center of the Landau band is demonstrated. There is only one branch of the scaling curve because all states are localized, except at the center of the band. Again, more extensive numerical calculations are necessary to obtain more definitive results.

7. A Brief Note on the Single-Site von Neumann Entropy

The scaling of single-site vNE can lead to some misunderstanding in regard to universality. This can be illustrated by considering Ising chain in a transverse field for which the Hamiltonian is

$$\mathcal{H} = -J\lambda \sum_i S_i^z S_{i+1}^z - J \sum_i S_i^x \tag{7.1}$$

where S_x, S_y and S_z are spin-1/2 matrices. The sum is over all sites $N \to \infty$. It is well known that the second derivative of the ground state energy[28]

$$\frac{E_0}{N} = -\frac{2J}{\pi}(1 + \lambda)\mathbb{E}\left(\frac{2\sqrt{\lambda}}{1 + \lambda}\right) \tag{7.2}$$

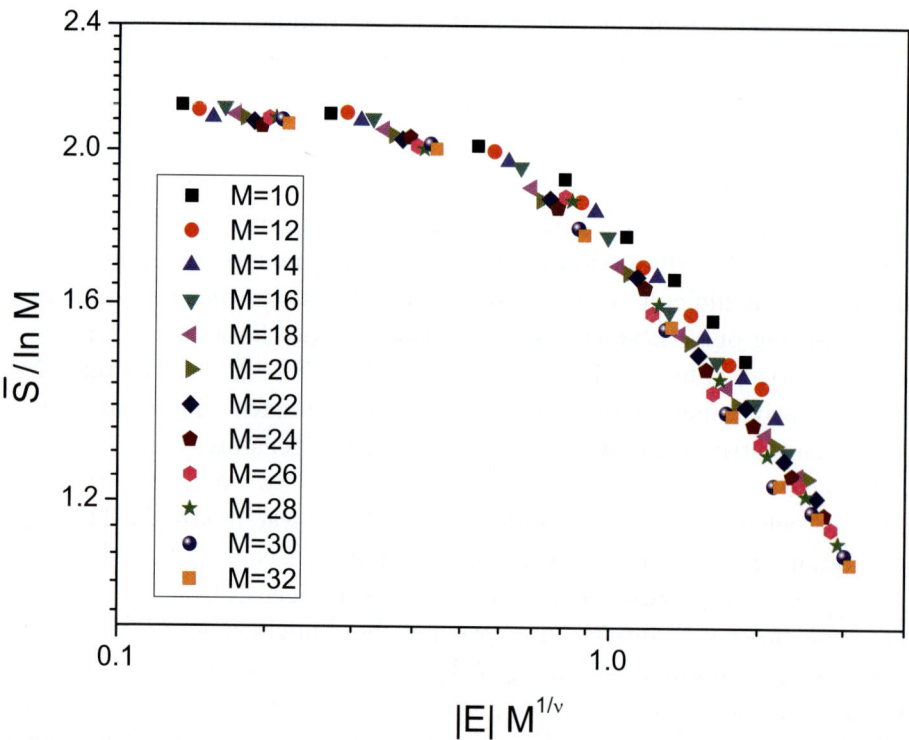

Fig. 7.4. Scaling of the von Neumann entropy $\overline{S}(E)$ for the integer quantum Hall effect. M instead of L is used in the data collapse with the accepted value of $\nu \approx 2.33$; see Ref. 26.

has a logarithmic singularity at $\lambda = 1$, signifying a quantum critical point in the conventional sense of the non-analyticity of the ground state energy. This non-analyticity is symmetric as $\lambda \to 1\pm$. Here, $\mathbb{E}$ is the complete Elliptic integral of the second kind. It is also simple to calculate the single site vNE. A given site constitutes part A of the system, and part B is the rest of the Ising chain of $N - 1$ sites. The reduced density matrix ρ_A is[29]

$$\rho_A = \frac{1}{2} \begin{pmatrix} 1 + \langle \sigma_i^z \rangle & \langle \sigma_i^x \rangle \\ \langle \sigma_i^x \rangle & 1 - \langle \sigma_i^z \rangle \end{pmatrix}, \tag{7.3}$$

where the exact known results are[28]

$$\langle \sigma_i^z \rangle = (1 - 1/\lambda^2)^{1/8}, \; \lambda > 1; 0, \; \text{otherwise}, \tag{7.4}$$

$$\langle \sigma_i^x \rangle = \frac{1-\lambda}{\pi} \mathbb{K}\left(\frac{2\sqrt{\lambda}}{1+\lambda}\right) + \frac{1+\lambda}{\pi} \mathbb{E}\left(\frac{2\sqrt{\lambda}}{1+\lambda}\right), \tag{7.5}$$

where $\mathbb{K}$ is the complete elliptic integral of the first kind. The vNE, S, can now be easily computed from the 2×2 reduced density matrix ρ_A. The singularities approaching the critical point are

$$\lim_{\lambda \to 1-} \frac{\partial S}{\partial \lambda} = -\frac{1}{2\pi} \ln \left(\frac{\pi + 2}{\pi - 2} \right) \ln |\lambda - 1|, \tag{7.6}$$

$$\lim_{\lambda \to 1+} \frac{\partial S}{\partial \lambda} = -\frac{\pi}{2^{19/4}} \ln \left(\frac{\pi + 2}{\pi - 2} \right) (\lambda - 1)^{-3/4}. \tag{7.7}$$

The exponents differ as to how we approach the critical point. Nonetheless, the exponents are pure numbers independent of the coupling constant, as are the amplitudes. This then is a perfectly legitimate case of universality. The reason for the asymmetry at the critical point is clear: the magnetization (a local order parameter) vanishes for $\lambda \leq 1$, while it is non-zero for $\lambda > 1$. For the case of Anderson localization and the integer quantum Hall effect, there are no such local order parameters that vanish at the transition. If we regard the average density of states as an order parameter, it is smooth through both the Anderson transition and the plateau-to-plateau transition for the integer quantum Hall effect. Thus, the single site vNE has a scaling function that is symmetric around the transition as deduced from the multifractal scaling. The moral is that the single site entropy is an important and useful quantity to compute.

8. Epilogue

Entropy measures uncertainty in a physical system. It is therefore not surprising that it is a central concept in quantum information theory. That it may be an essential concept at a quantum critical point can also be anticipated. At a critical point a system cannot decide in which phase it should be. At the Anderson transition the wave function is a highly complex multifractal, and it is not surprising that vNE exhibits non-analyticity in the infinite volume limit, even though the ground state energy in which the complexity of the wave function is averaged over is smooth through it. The non-analyticity of vNE is perfectly consistent with other measures of entanglement, for example the linear entropy,[30] which is $S_L = 1 - \text{Tr}\rho_A^2$. The inverse participation ratio $P^{(2)} = \frac{1}{\mathcal{N}} \sum_{r,E} |\psi_r(E)|^4 = 1 - S_L/2$, where $\mathcal{N}$ is the total number of states in the band. In the extreme localized case, only one site participates and $S_L = 0$. In the opposite limit $S_L = 2 - 2/\mathcal{N} \to 2$, when $\mathcal{N} \to \infty$. The participation ratio, hence S_L, exhibits scaling at the metal–insulator transition: $P^{(2)} = L^{-x} g_\pm (L^{1/\nu} w)$, where $g_\pm$ is another universal function. This scaling was also verified with $x \approx 1.4$, $\nu \approx 1.35$, and $W_c \approx 16.5$ for the $3D$ Anderson transition.[13]

A key question now is what happens when we have both disorder and interaction. In recent years there has been much progress in one-dimensional systems, especially from the perspective of vNE in the ground state.[31] The quantum criticalities of these disordered interacting systems belong to universality classes different from their counterparts with interaction but no disorder and are generally described by infinite disorder fixed points. Little is understood for similar higher dimensional systems. The principal difficulty in constructing a universal theory when both interactions and disorder are present is qualitatively clear. When interactions are strong and disorder is absent, the ground state can break many symmetries and organize itself into a variety of phases. Introducing disorder may affect the stability of these many body correlated phases in different ways.[32,33] Although the symmetry of the order parameters can guide us, the strongly correlated nature of these phases makes theories difficult to control. As mentioned above, in a simple case of spinless fermion, we were able to provide some rigorous answers: no matter how strong the interaction is there appears to be gapless excited states and the broken symmetry is broken. In the opposite limit, we have to examine how weak interaction affects the Anderson problem. Here there has been progress in recent years; see the contribution by Finkelstein in the present volume.

Acknowledgments

I would like to thank my collaborators A. Kopp, X. Jia, A. Subramanium, D. J. Schwab and I. Gruzberg. This work was supported by a grant from the National Science Foundation, DMR-0705092.

References

1. P. W. Anderson, Absence of diffusion in certain random lattices, *Phys. Rev.* **109**, 1492 (1958).
2. E. Abrahams, P. W. Anderson, D. C. Licciardello and T. V. Ramakrishnan, Scaling theory of localization: Absence of quantum diffusion in two dimensions, *Phys. Rev. Lett.* **42**, 673 (1979).
3. R. Shankar, Solvable model of a metal–insulator transition, *Int. J. Mod. Phys. B* **4**, 2371 (1990).
4. D. J. Schwab and S. Chakravarty, Glassy states in fermionic systems with strong disorder and interactions, *Phys. Rev. B* **79**, 125102 (2009).

5. A. MacKinnon and B. Kramer, One-parameter scaling of localization length and conductance in disordered systems, *Phys. Rev. Lett.* **47**, 1546 (1981).

6. C. Itzykson and J.-M. Drouffe, *Statistical Field Theory*, Vol. 1 (Cambridge University Press, Cambridge, 1989).

7. P. G. de Gennes, Exponents for the excluded volume problem as derived by the Wilson method, *Phys. Lett. A* **38**, 339 (1972).

8. F. Wegner, Mobility edge problem — continuous symmetry and a conjecture, *Z. Phys. B* **35**, 207 (1979).

9. P. A. Lee and A. D. Stone, Universal conductance fluctuations in metals, *Phys. Rev. Lett.* **55**, 1622 (1985).

10. B. L. Altshuler, V. E. Kravtsov and I. V. Lerner, Statistics of mesoscopic fluctuations and instability of one-parameter scaling, *Zh. Eksp. Teor. Fiz.* **91**, 2276 (1986); *Sov. Phys. JETP* **64**, 1352 (1986).

11. S. Chakravarty, Scale-independent fluctuations of spin stiffness in the Heisenberg model and its relationship to universal conductance fluctuations, *Phys. Rev. Lett.* **66**, 481 (1991).

12. J. D. Bekenstein, Black holes and entropy, *Phys. Rev. D* **7**, 2333 (1973).

13. A. Kopp, X. Jia and S. Chakravarty, Replacing energy by von Neumann entropy in quantum phase transitions, *Ann. Phys.* **322**, 1466 (2007).

14. J. T. Edwards and D. J. Thouless, Regularity of density of states in Andersons localized electron model, *J. Phys. C: Solid State Phys.* **4**, 453 (1971).

15. L. Amico, R. Fazio, A. Osterloh and V. Vedral, Entanglement in many-body systems, *Rev. Mod. Phys.* **80**, 517 (2008).

16. P. Zanardi, Quantum entanglement in fermionic lattices, *Phys. Rev. A* **65**, 042101 (2002).

17. F. Wegner, Inverse participation ratio in $(2 + \epsilon)$-dimensions, *Z. Phys. B* **36**, 209 (1980).

18. R. B. Laughlin, Nobel lecture: Fractional quantization, *Rev. Mod. Phys.* **71**, 863, (1999).

19. C. Castellani and L. Peliti, Multifractal wavefunction at the localisation threshold, *J. Phys. A* **19**, L429 (1986).

20. F. Evers and A. D. Mirlin, Anderson transitions, *Rev. Mod. Phys.* **80**, 1355 (2008).

21. L. J. Vasquez, A. Rodriguez and R. A. Römer, Multifractal analysis of the metal–insulator transition in the three-dimensional anderson model. I. Symmetry relation under typical averaging, *Phys. Rev. B* **78**, 195106 (2008).

22. X. Jia, A. R. Subramaniam, I. A. Gruzberg and S. Chakravarty, Entanglement entropy and multifractality at localization transitions, *Phys. Rev. B* **77**, 014208 (2008).

23. K. Slevin, P. Markos and T. Ohtsuki, Reconciling conductance fluctuations and the scaling theory of localization, *Phys. Rev. Lett.* **86**, 3594 (2001).

24. P. Goswami, X. Jia and S. Chakravarty, Quantum hall plateau transition in the lowest landau level of disordered graphene, *Phys. Rev. B* **76**, 205408 (2007).

25. B. Kramer and M. Schreiber, Transfer-matrix methods and finite-size scaling for disordered systems, in *Computational Physics*, eds. K. H. Hoffmann and M. Schreiber (Springer, Berlin, 1996), p. 166.

26. B. Huckestein, Scaling theory of the integer quantum hall effect, *Rev. Mod. Phys.* **67**, 357 (1995).
27. T. Ando and Y. Uemura, Theory of quantum transport in a two-dimensional electron system under magnetic fields. I. Characteristics of level broadening and transport under strong fields, *J. Phys. Soc. Jpn.* **36**, 959 (1974).
28. P. Pfeuty, The one-dimensional ising model with a transverse field, *Ann. Phys. (NY)* **57**, 79 (1970).
29. T. J. Osborne and M. A. Nielsen, Entanglement in a simple quantum phase transition, *Phys. Rev. A* **66**, 032110 (2002).
30. W. H. Zurek, S. Habib and J. P. Paz, Coherent states via decoherence, *Phys. Rev. Lett.* **70**, 1187 (1993).
31. G. Refael and J. E. Moore, Criticality and entanglement in random quantum systems, *J. Phys. A: Math. Theor.* **42**, 504010 (2009).
32. S. Chakravarty, S. Kivelson, C. Nayak and K. Voelker, Wigner glass, spin liquids and the metal-insulator transition, *Phil. Mag. B* **79**, 859 (1999).
33. K. Voelker and S. Chakravarty, Multiparticle ring exchange in the Wigner glass and its possible relevance to strongly interacting two-dimensional electron systems in the presence of disorder, *Phys. Rev. B* **64**, 235125 (2001).

Chapter 8

FROM ANDERSON LOCALIZATION
TO MESOSCOPIC PHYSICS

Markus Büttiker

University of Geneva, Department of Theoretical Physics
24 Quai E. Ansermet, 1211 Geneva, Switzerland
Markus.Buttiker@unige.ch

Michael Moskalets

Department of Metal and Semiconductor Physics
NTU "Kharkiv Polytechnic Institute"
21 Frunze Street, 61002 Kharkiv, Ukraine
moskalets@kpi.kharkov.ua

In the late seventies an increasing interest in the scaling theory of Anderson localization led to new efforts to understand the conductance of systems which scatter electrons elastically. The conductance and its relation to the scattering matrix emerged as an important subject. This, coupled with the desire to find explicit manifestations of single electron interference, led to the emergence of mesoscopic physics. We review electron transport phenomena which can be expressed elegantly in terms of the scattering matrix. Of particular interest are phenomena which depend not only on transmission probabilities but on both amplitude and phase of scattering matrix elements.

1. Introduction

Theories and experiments on Anderson localization[1] have been a vibrant topic of solid state physics for more than five decades. For a long time, it seems, it was the only problem in which disorder and phase coherence were brought together to generate macroscopic effects in electron transport through solids. In particular the scaling theory initially advanced by Thouless[2] and further developed towards the end of the seventies and early eighties brought the conductance of disordered systems into focus.[3,4] The key argument of the single parameter scaling theory of localization is that the conductance (not conductivity or anything else) is in fact the only parameter conductance (not conductivity or anything else) is in fact the only parameter that counts. Long after the early work of Landauer,[5,6] this emphasis on conductance eventually brought to the forefront the question[7-14] of a "correct formulation of conductance." It is on this background that the notion of quantum channels and the scattering matrix as a central object of electrical transport theory emerged. A multi-terminal formulation[15,16] of conductance brought much clarity and permitted to connect the scattering theory of electrical conductance to the Onsager theory of irreversible processes. This triggered questions about dynamic fluctuations and led to a very successful theory of current noise in mesoscopic structures.[17]

The quest to understand the quantum coherent conductance through a disordered region led to more than simply a relation between conductance and transmission. While in the discussions of the localization the calculation of conductance always implied an average over an ensemble of disorder configurations, eventually, the basic question was raised whether we can observe interference effects directly rather than indirectly by considering an ensemble averaged quantity. This brought the specific sample into focus. Early work proposed samples with the shape of a ring with the hole of the ring penetrated by an Aharonov–Bohm flux (see Figs. 8.1 and 8.2).

Of interest was the sample specific response to the Aharonov–Bohm flux. The oscillation of the quantity of interest with the period of the single particle flux quantum h/e would be the direct signature of single particle interference. These expectations led to the prediction[18] of persistent currents and Josephson (Bloch) oscillations in closed disordered rings and Aharonov–Bohm oscillations with a period of a single charge flux quantum in rings connected to leads.[19,20] We mention here only two recent experiments: one on persis-

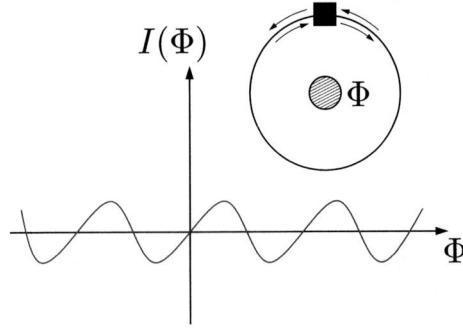

Fig. 8.1.   Persistent current $I(\Phi)$ as a function of an Aharonov–Bohm flux Φ of a closed ring with an elastic scatterer (square).

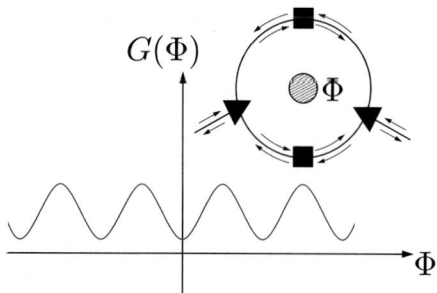

Fig. 8.2. Single charge Aharonov–Bohm conductance oscillations of a normal ring with leads. Triangles indicate electron wave beam splitters and elastic scatterers are symbolized by squares.

tent currents[21] and one on Aharonov–Bohm conductance oscillations.[22] For additional references and a broader account of this development, we refer to Imry's book.[23]

The effect of an Aharonov–Bohm flux on the properties of a disordered systems was not entirely novel in mesoscopic physics: it occurs already in discussions of the Anderson transition. The sensitivity to boundary conditions played a very important part in localization theories. Different transverse boundary conditions for wires are already investigated by Dorokhov[24] in the form of an AB-flux along the axes of a cylindrical wire. Nevertheless, these discussions were a purely technical means to arrive at an ensemble average conductance. However, there is an Aharonov–Bohm effect which shows up even in the ensemble averaged conductance. In the absence of a magnetic field, interference effects of self-intersecting trajectories are not averaged out since there exists a time reversed trajectory which traverses the loop in the opposite direction with exactly the same phase. This leads to enhanced backscattering and, if a magnetic flux is applied to the sample, to oscillations with decaying amplitude and a period in magnetic field determined by the size of the most likely self-intersecting loop. The enhanced backscattering is known as weak localization. The magnetic field oscillations were first discussed by Altshuler, Aronov and Spivak,[25] and were soon demonstrated in an ingenious experiment by Sharvin and Sharvin[26] who created a cylinder by coating a glass fiber with a metallic film. This lead to the first observations of an Aharonov–Bohm effect in a dissipative, metallic diffusive conductor.

The concern with the conductance of an elastic scatterer led to a novel approach in which transport quantities are expressed in terms of the scattering matrix. From initial efforts to discuss the conductance in the presence of stationary applied voltages, there exist now formulations of transport problems for a wide range of situations. We will discuss only a few topics — the response to oscillating voltages, which implies a concern with capacitance, quantum pumping, and non-linear dynamic transport leading to a description of single particle emission from localized states. There are many additional extensions which we will not further discuss here — extensions to superconducting-normal structures, conductors with spin orbit interactions, systems like graphene or topological insulators and superconductors. The message is simply to point to the wider application of scattering theory. Many of these novel developments concern the variation of the scattering matrix due to a variation of one or several parameters. The scattering matrix can be varied through a displacement of an impurity, a change in a gate voltage or due to a magnetic field. Changes of the scattering matrix lead to transfer of charge out of the scatterer into the contacts or from contacts into the sample.

2. Charge Transfer from the Scattering Matrix

Let S be the scattering matrix of a conductor. Its elements are $S_{\alpha\beta}$ where $\alpha = 1, 2, \ldots, N$ numbers the leads connected to the scattering region. Let there be a small variation in a parameter of the S-matrix that leads to a change in the scattering matrix which we can denote by dS. A variation of the scattering matrix typically implies a transfer of charge from the sample into the leads. An elegant expression for the charge transferred from the scattering region into a contact[27,28] α is

$$d\overline{Q}_\alpha = -\frac{e}{2\pi i}(dS S^\dagger)_{\alpha\alpha}. \tag{2.1}$$

This is the emittance of the sample into contact α. Similarly, the charge injected into the sample from a lead α is

$$d\underline{Q}_\alpha = -\frac{e}{2\pi i}(S^\dagger dS)_{\alpha\alpha}. \tag{2.2}$$

In the absence of a magnetic field, injected and emitted charge are the same. In the presence of a magnetic field B, the emittance is equal to the injectance in the reversed magnetic field, $d\overline{Q}_\alpha(B) = d\underline{Q}_\alpha(-B)$. These expressions are close relatives of the Wigner–Smith matrix,[29] a time-shift matrix,[28] scattering matrix expressions of the ac-response of a conductor to a slow potential variation[30] and they are now widely used to describe quantum pumping.[27,28,31–37] Equations (2.1) and (2.2) are for non-interacting electrons. To account for the Coulomb interaction, dS must be determined self-consistently. Below we discuss how this leads to expressions for the capacitance of mesoscopic conductors.[38]

For a single channel, two terminal conductor we can parameterize the scattering matrix as follows:

$$S = e^{i\gamma} \begin{pmatrix} e^{i\alpha} \cos\theta & i\,e^{-i\phi} \sin\theta \\ i\,e^{i\phi} \sin\theta & e^{-i\alpha} \cos\theta \end{pmatrix}. \tag{2.3}$$

A variation of γ, α and ϕ gives each rise to a separate response and allows for an elementary interpretation.[28] The resulting emittance into the left contact is:

$$d\overline{Q}_L = \frac{e}{2\pi}(-\cos^2\theta\,d\alpha + \sin^2\theta\,d\phi - d\gamma) \tag{2.4}$$

and into the right contact it is

$$d\overline{Q}_R = \frac{e}{2\pi}(\cos^2\theta\,d\alpha - \sin^2\theta\,d\phi - d\gamma). \tag{2.5}$$

The total charge emitted by the sample is $dQ = d\overline{Q}_R + d\overline{Q}_L = -e\,d\gamma/\pi$. This is nothing but the Friedel sum rule[39] which relates the variation of the phase γ to the charge of the scatterer (Krein formula in mathematics).

The phase α describes the asymmetry in reflection at the scatterer of carriers incident from the left as compared to those incident from the right. A displacement dx of the scatterer to the right changes α by $d\alpha = 2k_F dx$. It absorbs charge from the left lead and emits it into the right lead. Note that the transmission can be completely suppressed[28] and the scatterer than acts as a "snow plow".

The phase ϕ can be non-vanishing only if time-reversal symmetry is broken, that is, if a magnetic field acts on the conductor. Therefore we can suppose that ϕ is the result of a vector potential, $\phi = -2\pi \int dx A/\Phi_0$ where Φ_0 is the single charge flux quantum. Suppose that A is generated by an Aharonov–Bohm flux Φ which depends linearly on time. The electric field generated by the flux induces a voltage V between the reservoirs connected to the sample. The phase ϕ increases as $\phi = -Vt/\Phi_0$ with V the voltage induced across the sample. We then have[28,40] $d\overline{Q}_\alpha/dt = -e\sin^2\theta\,d\phi/dt = (e^2/h)TV$ which with the transmission probability $T = \sin^2\theta$ is the Landauer conductance $G = (e^2/h)T$.

The phase γ multiples all channels. It is a phase that depends on what we consider the scatterer and the outside regions. In one dimension, on the line, phases of the scattering matrix are defined from two points to the right and left of the scatterer. (More generally scattering phases depend on the choice of cross-sections). If we extend the region that we attribute to the scatterer by dx on both sides of the scatterer, the phase γ changes by

$d\gamma = 2k_F dx$. This operation removes charges from the leads and we have $d\overline{Q}_L = d\overline{Q}_R = -e\,k_F dx/\pi$.

A variation of the parameter θ which corresponds to a variation of the transmission probability has no effect on the emittance if all phases are kept fixed.

3. The Wigner–Smith Delay Time and Energy Shift Matrix

There are two close relatives of the expressions for charge transfer given by Eqs. (2.1) and (2.2). With both the emittance Eq. (2.1) and the injectance Eq. (2.2), we can associate a Wigner–Smith delay time matrix.[29] In the first case we focus on the time-delay in the outgoing channel regardless of the input channel and in the second case we specify the incoming channel but sum over all outgoing channels. Multiplying the emittance per unit energy with $\hbar/e$ gives the time delay matrix

$$\overline{\mathcal{T}} = \frac{\hbar}{i}\frac{\partial S}{\partial E}S^\dagger . \tag{3.1}$$

The Wigner phase delay times are the diagonal elements of this matrix. The off-diagonal elements also have physical significance and going back to charge we can related them to charge density fluctuations.[41,42] There is a dual to the Wigner–Smith matrix which Avron *et al.*[28] call the energy shift matrix. It is important once the scattering matrix is taken to be time-dependent. The energy shift matrix is

$$\overline{\mathcal{E}} = i\hbar\frac{\partial S}{\partial t}S^\dagger . \tag{3.2}$$

The two matrices do not commute, reflecting the time-energy uncertainty. Since they are matrices, we can define two commutators (Poisson brackets) depending on the order in which they are taken. These commutators play an important role whenever there is a time-variation of the scattering matrix. One of them,

$$\mathcal{P}\left\{S, S^\dagger\right\} = \frac{i}{\hbar}\left(\overline{\mathcal{T}}\,\overline{\mathcal{E}} - \overline{\mathcal{E}}\,\overline{\mathcal{T}}\right) = i\hbar\left(\frac{\partial S}{\partial t}\frac{\partial S^\dagger}{\partial E} - \frac{\partial S}{\partial E}\frac{\partial S^\dagger}{\partial t}\right), \tag{3.3}$$

defines the spectral densities of currents generated by a parametric variation of the scattering matrix.[28,43] The other order, $\mathcal{P}\left\{S^\dagger, S\right\}$, defines corrections to the scattering matrix proportional to the frequency ω with which parameters of the scattering matrix are varied.[43,44]

4. The Internal Response

The energy shift matrix multiplied by c/h defines a matrix of currents. The diagonal elements $\alpha\alpha$ are the currents generated in contact α in response to a temporal variation of the S-matrix,

$$dI_\alpha = \frac{e}{2\pi i} \left(\frac{\partial S}{\partial t} S^\dagger \right)_{\alpha\alpha}. \tag{4.1}$$

Assuming that the scattering matrix S is a function of the electrostatic potential $U(t)$ we can write

$$dI_\alpha = \frac{e}{2\pi i} \left(\frac{\partial S}{\partial U} S^\dagger \right)_{\alpha\alpha} \frac{dU}{dt}, \tag{4.2}$$

or if we take the Fourier transform

$$dI_\alpha(\omega) = -ie\omega \frac{1}{2\pi i} \left(\frac{\partial S}{\partial U} S^\dagger \right)_{\alpha\alpha} dU(\omega). \tag{4.3}$$

Equation (4.3) gives the current in contact α in response to a potential oscillating over the entire region of the scatterer. The potential U typically is a function not only of time but also of space r. This can be taken into account by testing the scattering matrix with respect to a small localized potential at every point r in the sample and integrating the result over the entire scattering region,

$$dI_\alpha(\omega) = -ie\omega \frac{1}{2\pi i} \int dr^3 \left(\frac{\delta S}{\delta U(r)} S^\dagger \right)_{\alpha\alpha} \delta U(r, \omega). \tag{4.4}$$

This is the result obtained by one of the authors in collaboration with H. Thomas and A. Prêtre.[30] Equation (4.4) is a zero temperature result. It depends on the scattering matrix only at the Fermi energy.

4.1. *Quantum pumping*

For a sinusoidal potential the time average current vanishes. However, if $U(r, t)$ cannot be written as a product of a time-dependent function times a spatial function, but effectively varies the potential of the conductor at several points and out of phase, the current integrated over a period $\mathcal{T}$ does not average to zero, and Eq. (4.1) describes a pump current[31]

$$I_{dc,\alpha} = \frac{e}{2\pi i \mathcal{T}} \oint_{\mathcal{L}} \left(dS\, S^\dagger \right)_{\alpha\alpha}, \tag{4.5}$$

where $\mathcal{L}$ is the "pump path". It is often stated that pumping needs at least two parameters which oscillate out of phase. These two parameters describe the pumping path $\mathcal{L}$ in the parameter space. A two parameter formulation of pumping based on the concept of emittance has been given by Brouwer.[31] In terms of two parameters $X_1(t) - X_1\cos(\omega t + \varphi_1)$ and $X_2(t) = X_2\cos(\omega t + \varphi_2)$ Brouwer obtained for weak pumping

$$I_{dc,\alpha} = \frac{e\omega\sin(\Delta\varphi)X_1 X_2}{2\pi}\sum_{\beta=1}^{N_r}\Im\left(\frac{\partial S_{\alpha\beta}^*}{\partial X_1}\frac{\partial S_{\alpha\beta}}{\partial X_2}\right)_{X_1=0,X_2=0}, \qquad (4.6)$$

where $\Delta\varphi \equiv \varphi_1 - \varphi_2 \neq 0$ is the phase lag. However, single parameter pumps are also possible. A typical example is an Archimedes screw where we turn only one handle to generate a pumped current.[45,46] If the rotation is with a constant speed, $\chi = \omega t$, and the scattering matrix is periodic in a rotation angle, $S(\chi) = S(\chi + 2\pi)$, then according to Eq. (4.5), the dc current is not zero, $I_{dc,\alpha} \neq 0$, if the Fourier expansion for a diagonal element $(\partial S/\partial\chi S^\dagger)_{\alpha\alpha}$ includes a constant term.

Equation (4.5) is valid for both weak and strong pumping but requires that the pump is driven slowly. At higher pumping frequency or pulsed excitation, the generated currents can not be described merely via a temporal variation of a scattering matrix dependent on a single energy. In contrast the scattering matrix dependent on two energies[47,48] or, equivalently, two times[49,50] has to be used.

4.2. *Pumping in insulators and metals*

Thouless investigated pumping in band insulators.[51] A specific example considered is a potential that is periodic in space with lattice constant a and moves with a constant velocity v and is (in one dimension) of the form $U(x,t) = U(x - vt)$. Depending on the number n of bands filled the charge moved by advancing the potential by a period a in space or by $\mathcal{T} = a/v$ in time is quantized and given by a Chern number ne. This formulation assumes that the Fermi energy lies in a spectral gap. On the other hand in gapless Anderson insulators, quantized pumping is also possible. It is due to tunneling resonances through the sample.[52]

Thouless's result seems hard to reconcile with a pump formula that is based explicitly only on the properties of the scattering matrix at the Fermi energy. However, Graf and Ortelli[53] have shown that by taking a periodic lattice of some finite length L and coupling it on either side to metallic contacts, the scattering approach converges, in the limit of a very long lattice, to the quantized pump current of Thouless. The point is of course, that as soon as the lattice is taken to be finite, there are exponentially decaying states which penetrate from the contacts into the sample. Considering the scattering matrix due to these states gives in the long length limit the same result as found from Thouless's approach. While the one-channel case permits to solve this problem by direct calculation, the many-channel case requires more thought.[54]

4.3. *Scattering formulation of ac response*

Consider a conductor connected to multiple contacts[30,55] and suppose that the voltages $V_\beta(\omega)$ at the contacts oscillate with frequency ω. In addition, the conductor is capacitively coupled to a gate at voltage V_g. The coupling is described in terms of a geometrical capacitance C. The currents $dI_\alpha(\omega)$ are related to the voltages by a dynamical conductance matrix $G_{\alpha\beta}(\omega)$. In linear response the total current has in general two contributions: a current that is solely the response to the external voltage applied to the contact defines a conductance $G_{\alpha\beta}^{ext}(\omega)$ and a current that arises because the application of external voltages charges the sample which leads through interaction to an internal potential $U(t)$,

$$dI_\alpha(\omega) = \sum_\beta G_{\alpha\beta}^{ext} \, dV_\beta + i\omega \, \Pi_\alpha \, dU . \qquad (4.7)$$

Both response functions $G_{\alpha\beta}^{ext}$ and Π_α remain to be determined. The current at the gate is

$$I_g(\omega) = -i\omega \, C \, (dV_g - dU) . \qquad (4.8)$$

The internal potential is found self-consistently by the requirement that potentials are defined only up to a constant (an overall shift in potentials does not change a physical quantity). This implies

$$i\omega \, \Pi_\alpha = - \sum_\beta G_{\alpha\beta}^{ext} . \qquad (4.9)$$

The external response can be expressed in terms of the scattering matrix and the Fermi function of the contacts[55]

$$G_{\alpha\beta}^{ext}(\omega) = \frac{e^2}{h} \int dE \, Tr[1_\alpha \delta_{\alpha\beta} - S_{\alpha\beta}^\dagger(E) S_{\alpha\beta}(E + \hbar\omega)] \frac{f_\beta(E) - f_\beta(E + \hbar\omega)}{\hbar\omega} . \qquad (4.10)$$

Here we now consider leads with several transverse channels. The trace Tr is a sum over transverse channels in the leads α and β. Using the above and requiring that current is conserved determines the internal potential. Eliminating the internal potential leads to a total conductance[55]

$$G_{\alpha\beta} = G_{\alpha\beta}^{ext} + \frac{\sum_\gamma G_{\alpha\gamma}^{ext} \sum_\delta G_{\delta\beta}^{ext}}{i\omega C - \sum_{\gamma\delta} G_{\gamma\delta}^{ext}} . \qquad (4.11)$$

The second term is the internal response due to the self-consistent potential. It depends on the capacitance C.

A voltage at contact β of the conductor does, in the absence of interactions, not induce a response at the gate. Thus if $\alpha = g$ in Eq. (4.11) the external response $G_{g\beta}^{ext}$ vanishes and $\sum_\delta G_{\delta\beta}^{ext}$ has to be replaced by $-i\omega C$. Similarly, an oscillating gate voltage does not generate an external response $G_{\alpha g}^{ext}$ but due to capacitive coupling generates a current at contact α due to second term of Eq. (4.11) in which $\sum_\delta G_{\delta g}^{ext}$ is replaced by $-i\omega C$. The rows and columns of the conductance matrix add up to zero. Next we now discuss the low frequency limit of these results and show the role played by the emittances and injectances.

4.4. *Capacitance, emittances, partial density of states*

An expansion of the external conductance to first order in ω leads to

$$G_{\alpha\beta}^{ext}(\omega) = G_{\alpha\beta} - i\omega e^2 dN_{\alpha\beta}/dE + ... \qquad (4.12)$$

where the first term is the dc conductance and

$$\frac{dN_{\alpha\beta}}{dE} = \frac{1}{4\pi i} Tr \left\{ \frac{dS_{\alpha\beta}}{dE} S_{\alpha\beta}^\dagger - S_{\alpha\beta} \frac{dS_{\alpha\beta}^\dagger}{dE} \right\} \qquad (4.13)$$

is a partial density of states. In contrast to the emittance for which the contact into which carriers are emitted are specified or the injectance for which the contact from which carriers are injected is specified, here both the contact β from which carriers are injected and the contact α into which they are emitted are specified. For the injectance we have a pre-selection, for the emittance a post-selection, but for the partial density of states we have both a pre- and post-selection of the contacts. The sum of partial density of states over the second index is the emittance $d\overline{N}_\alpha/dE = \sum_\beta dN_{\alpha\beta}/dE$. The sum over the first index of the partial density of states is the injectance $d\underline{N}_\alpha/dE = \sum_\beta dN_{\beta\alpha}/dE$. The sum over both indices is the total density of states $dN/dE = \sum_{\alpha\beta} dN_{\alpha\beta} dE$. Using this we find for the screened emittance[55]

$$E_{\alpha\beta} = e^2 \left[\frac{dN_{\alpha\beta}}{dE} - \frac{d\overline{N}_\alpha}{dE} \frac{e^2}{C + e^2 dN/dE} \frac{d\underline{N}_\beta}{dE} \right] . \qquad (4.14)$$

A positive diagonal term $E_{\alpha\alpha} > 0$ (a negative off-diagonal term, $E_{\alpha\beta} < 0$) signals a capacitive response. A negative diagonal term $E_{\alpha\alpha} < 0$ (a positive off-diagonal term, $E_{\alpha\beta} > 0$) describes an inductive response. We have not taken into account magnetic fields and hence the inductive response is of purely kinetic origin. The main point for our discussion is the physical relevance of the different density of states which is brought out by Eq. (4.14). The Coulomb term illustrates that an additional injected charge generates an electrical potential which in turn emits charge.

4.5. *Ac response of a localized state*

Modern sample fabrication permits to investigate a single localized state coupled only via one contact of arbitrary transmission to a metallic contact (see Figs. 8.3 and 8.4). Such a structure can be viewed as a mesoscopic version of a capacitor. A particular realization consists of a quantum point contact (QPC) which determines the coupling to the inside of a cavity. In the presence of a high magnetic field, there is a single edge state coupled to the contact (see Fig. 8.4). The dc conductance of such a localized state is zero since every carrier that enters the cavity is after some time reflected. However by coupling the cavity to a (macroscopic) gate, an ac voltage can be applied and the ac conductance of the localized state can be investigated. For a macroscopic capacitor, the contact of the cavity to the metallic reservoir would act as a resistor and the low frequency conductance would be

$$G(\omega) = -i\omega C + \omega^2 C^2 R + \dots , \qquad (4.15)$$

with C the geometrical capacitance and R the series resistance. A mesoscopic capacitor has a response of the same form but now with quantum corrections to the capacitance and the resistance,

$$G(\omega) = -i\omega C_\mu + \omega^2 C_\mu^2 R_q + \dots . \qquad (4.16)$$

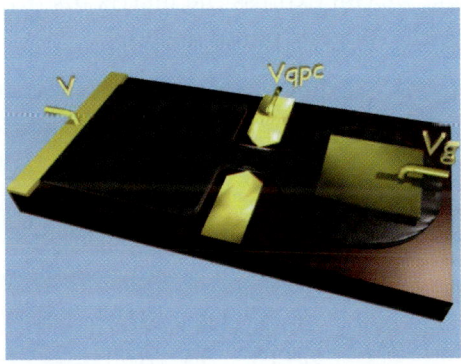

Fig. 8.3. Mesoscopic capacitor. A cavity separated via a quantum point contact at voltage V_{QPC} from a lead which is in turn coupled to a metallic contact at a voltage V. The cavity is capacitively coupled to a gate V_g.

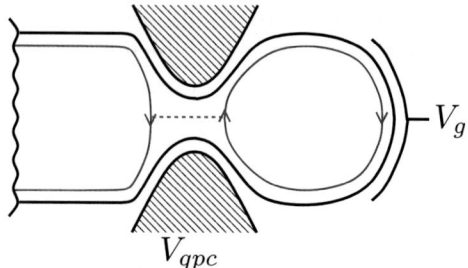

Fig. 8.4.   A localized state (edge state) coupled via a quantum point contact to a lead and reservoir.

Here C_μ is an electrochemical capacitance and R_q is the charge relaxation resistance. Specializing Eq. (4.7) to a single contact, yields[56]

$$C_\mu^{-1} = C^{-1} + \left[e^2 \nu(\mu)\right]^{-1} , \qquad (4.17)$$

where $\nu(E) \equiv dN/dE = (1/2\pi)\, Tr[S^\dagger dS/dE]$ is the density of states of the cavity. For weak interaction, C is large and C_μ^{-1} is entirely determined by the density of states. For strong interaction, C is small and the density of states provides a small correction to the geometrical capacitance.

The charge relaxation resistance is a particularly interesting transport coefficient. From Eq. (4.7) we find[55,56]

$$R_q = \frac{h}{2e^2} \frac{Tr[dS^\dagger/dE\, dS/dE]}{(|Tr[S^\dagger dS/dE]|)^2} . \qquad (4.18)$$

For the single contact considered here, the injectance and emittance are identical. In the basis in which the scattering matrix is diagonal, its elements are $e^{i\phi_n}$ $n = 1, 2, ..$ and we can express the matrix $d\overline{N}/dE$ in terms of energy derivatives of phases, the Wigner phase delays,

$$\left[\frac{dN}{dE}\right] = \frac{1}{2\pi i} Tr\left[\frac{\partial S}{\partial E} S^\dagger\right] = \frac{1}{2\pi} \sum_n \frac{\partial \phi_n}{\partial E} ,$$

$$\frac{1}{(2\pi)^2} Tr\left[\frac{\partial S^\dagger}{\partial E} \frac{\partial S}{\partial E}\right] = \frac{1}{(2\pi)^2} \sum_n \left(\frac{\partial \phi_n}{\partial E}\right)^2 ,$$

and thus

$$R_q = \frac{h}{2e^2} \frac{\sum_n (\partial \phi_n/\partial E)^2}{(\sum_n \partial \phi_n/\partial E)^2} . \qquad (4.19)$$

Interestingly this is now a resistance that is not expressed in terms of transmission probabilities but in terms of derivatives of phases. It is the ratio of the mean square time of carriers in the cavity divided by the mean delay time squared. For a single quantum channel, this two quantities are equal and

$$R_q = \frac{h}{2e^2} \tag{4.20}$$

is universal,[55,56] i.e. independent of the scattering properties of the quantum channel!! This derivation assumes that the quantum coherence can be maintained long compared to the time it takes a carrier to be reflected from the cavity. If the phase breaking time becomes comparable to the dwell time, theory predicts deviations from the above quantized resistance value.[57] Note the factor 2 in Eq. (4.20) is not a consequence of spin: the quantization at half of a resistance value is for a single spin channel.

An experiment with a setup as shown in Fig. 8.3 has been carried out by Gabelli *et al.*[58] and has provided good evidence of the resistance quantization of R_q. In the experiment[58] the quantum point contact is modeled by a scattering matrix with transmission amplitude t and reflection amplitude r (taken to be real) and a phase ϕ which is accumulated by an electron moving along the edge state in the cavity. The amplitude of the current incident from the metallic contact on the QPC a, the amplitude of the transmitted wave leaving the QPC b, which after completing a revolution is $\exp(i\phi)b$, and the amplitude of the current leaving the QPC sa, with s the scattering matrix element, are related by

$$\begin{pmatrix} sa \\ b \end{pmatrix} = \begin{pmatrix} r & -t \\ t & r \end{pmatrix} \begin{pmatrix} a \\ \exp(i\phi)b \end{pmatrix}. \tag{4.21}$$

This determines the scattering matrix element

$$s(\epsilon) = -e^{-i\phi} \frac{r - e^{i\phi}}{r - e^{-i\phi}} \tag{4.22}$$

and gives a density of states

$$\nu(E) = \frac{1}{2\pi i} s^\dagger \frac{\partial s}{\partial E} = \frac{1}{2\pi i} s^\dagger \frac{ds}{d\phi} \frac{\partial \phi}{\partial E} = \frac{1}{2\pi} \frac{\partial \phi}{\partial E} \frac{1 - r^2}{1 - 2r \cos(\phi) + r^2}. \tag{4.23}$$

Since the scattering matrix is needed only around the Fermi energy it is reasonable to assume that the phase is a linear function of energy

$$\phi = 2\pi E/\Delta, \tag{4.24}$$

where Δ is the level spacing. For small transmission, the density of states is sharply peaked at the energies for which ϕ is a multiple of 2π. In the experiment, the capacitance C_μ and R_q are investigated as functions of the voltage applied to the QPC. This dependence enters through the specification of the transmission probability of the QPC,

$$T = t^2 = 1/\left(1 + \exp(-(V_{QPC} - V_0)/\Delta V_0)\right) . \tag{4.25}$$

Here V_0 determines the voltage at which the QPC is half open and ΔV_0 determines the voltage scale over which the QPC opens.

Equations (4.16), (4.20)–(4.24) permit a very detailed description of the experimental data.[58] Indeed the experiment is in surprisingly good agreement with scattering theory. The main reason for this agreement is the weak interaction which in the experimental arrangement is generated by a top gate that screens the cavity. In the presence of side gates, interactions might become more important and it is then of interest to provide theories which go beyond the random phase approximation discussed above. We refer the reader to recent discussions.[59–65]

5. Nonlinear AC Response of a Localized State

An experiment by Fève *et al.*,[66] with the same setup as shown in Fig. 8.3, demonstrated *quantized particle emission* from the localized state when the capacitor was subjected to a large periodic potential modulation, $U(t) = U(t + T)$, comparable with Δ. Fève *et al.* used either sinusoidal or pulsed modulations. We consider the former type of modulation. The theory for the latter one can be found in Ref. 67.

Let the gate V_g, see Fig. 8.4, induce a sinusoidal potential on the capacitor, $U(t) = U_0 + U_1 \cos(\omega t + \varphi)$. If $U_1 \sim \Delta$ then we can not use a linear response theory and instead of conductance $G(\omega)$ we have to consider directly the current $I(t)$ that is generated. At low frequency, $\omega \to 0$, we can expand $I(t)$ in powers of ω in the same way as we do in Eq. (4.16). However, now the density of states depends on the potential $U(t)$ and thus on time, $\nu(t, E) = \nu[U(t), E]$, where $\nu[U(t), E]$ is given in Eq. (4.23) with $\phi(E)$ being replaced by $\phi[U(t), E] = 2\pi[E - eU(t)]/\Delta$. At zero temperature and assuming the geometrical capacitance $C \to \infty$ we obtain up to ω^2 terms,[67]

$$I(t) = C_\partial \frac{dU}{dt} - R_\partial C_\partial \frac{\partial}{\partial t}\left[C_\partial \frac{dU}{dt}\right] , \tag{5.1}$$

with a differential capacitance

$$C_\partial = e^2 \nu(t, \mu) , \tag{5.2}$$

and a differential resistance

$$R_\partial = \frac{h}{2e^2} \left\{ 1 + \frac{\frac{\partial \nu(t,\mu)}{\partial t} \frac{dU}{dt}}{\frac{\partial}{\partial t} \left[\nu(t,\mu) \frac{dU}{dt} \right]} \right\}. \tag{5.3}$$

We see that in the non-linear regime the dissipative current through the capacitor is defined by the resistance R_∂ which, even in a single-channel case, unlike the linear response resistance R_q, Eq. (4.20), is not universal and depends on the parameters of the localized state, on the strength of coupling to the extended state, and on the driving potential $U(t)$.

Remarkably, in both linear and non-linear regimes the charge relaxation resistance quantum $R_q = h/(2e^2)$ defines the Joule heat which is found in the leading order in ω to dissipate with a rate,[68]

$$I_E = R_q \langle I^2 \rangle = \frac{h}{2e^2} \int_0^{\mathcal{T}} \frac{dt}{\mathcal{T}} I^2(t). \tag{5.4}$$

Note I_E is calculated as a work done by the potential $U(t)$ under the current $I(t)$ during the period $\mathcal{T}$.

6. Quantized Charge Emission from a Localized State

If the potential of a capacitor is varied in time, then the energy of the localized state (LS) changes. With increasing potential energy $eU(t)$, the level of the LS can rises above the Fermi level and an electron occupying this level leaves the capacitor. When $eU(t)$ decreases, the empty LS can sink below the Fermi level and an electron enters the capacitor leaving a hole in the stream of electrons in the linear edge state. Therefore, under periodic variation of a potential the localized state can emit non-equilibrium electrons and holes propagating away from the LS within the edge state which acts here similar to a waveguide.

Consider the localized state weakly coupled to the extended state. In our model this means that the transparency of a QPC is small, $T \to 0$. Then the generated AC current consists of positive and negative peaks with width $\Gamma_\tau \sim T/(2\pi\omega)$ corresponding to emitted electrons and holes, respectively. Assume the amplitude U_1 of an oscillating potential to be chosen such that during the period only one quantum level E_n crosses the Fermi level. The time of crossing t_0 is defined by the condition $\phi(t_0) = 0 \mod 2\pi$. There are two times of crossing. At time $t_0^{(-)}$, when the level rises above the Fermi level, an electron is emitted, and at time $t_0^{(+)}$, when the level sinks below the Fermi level, a hole is emitted. If without the potential U the level E_n

aligns with the Fermi energy μ, then the times of crossing are defined by $U\left(t_0^{(\mp)}\right) = 0$. For $|eU_0| < \Delta/2$ and $|eU_0| < |eU_1| < \Delta - |eU_0|$ we find the emission times, $t_0^{(\mp)} = \mp t_0^{(0)} - \varphi/\omega$, where $\omega t_0^{(0)} = \arccos\left(-U_0/U_1\right)$. Here φ is the phase lag of the oscillating potential introduced above. With these definitions we find the scattering amplitude, Eq. (4.22), for electrons with the Fermi energy

$$
s(t,\mu) = e^{i\theta_r}
\begin{cases}
\dfrac{t - t_0^{(+)} - i\Gamma_\tau}{t - t_0^{(+)} + i\Gamma_\tau}, & \left|t - t_0^{(+)}\right| \lesssim \Gamma_\tau, \\[3mm]
\dfrac{t - t_0^{(-)} + i\Gamma_\tau}{t - t_0^{(-)} - i\Gamma_\tau}, & \left|t - t_0^{(-)}\right| \lesssim \Gamma_\tau, \\[3mm]
1, & \left|t - t_0^{(\mp)}\right| \gg \Gamma_\tau,
\end{cases}
\tag{6.1}
$$

where $\omega\Gamma_\tau = T\Delta/\left(4\pi|e|\sqrt{U_1^2 - U_0^2}\right)$ and $0 < t < \mathcal{T}$. The density of states,

$$
\nu(t,\mu) = \frac{4}{\Delta T}\left\{\frac{\Gamma_\tau^2}{\left(t - t_0^{(-)}\right)^2 + \Gamma_\tau^2} + \frac{\Gamma_\tau^2}{\left(t - t_0^{(+)}\right)^2 + \Gamma_\tau^2}\right\},
\tag{6.2}
$$

peaks at $t = t_0^{(\mp)}$ when the particles are emitted. Then from Eq. (5.1) we find to leading order in $\omega\Gamma_\tau \ll 1$,

$$
I(t) = \frac{e}{\pi}\left\{\frac{\Gamma_\tau}{\left(t - t_0^{(-)}\right)^2 + \Gamma_\tau^2} - \frac{\Gamma_\tau}{\left(t - t_0^{(+)}\right)^2 + \Gamma_\tau^2}\right\}.
\tag{6.3}
$$

This current consists of two pulses of Lorentzian shape with width Γ_τ corresponding to emission of an electron and a hole. Integrating over time, it is easy to check that the first pulse carries a charge e while the second pulse carries a charge $-e$.

The emitted particles carry energy from the dynamical localized state to the extended state and further to the reservoir which this extended state flows to. The energy carried by the particles emitted during the period defines a heat generation rate I_E. Substituting Eq. (6.3) into Eq. (5.4), we calculate

$$
\mathcal{T}I_E = \frac{\hbar}{\Gamma_\tau}.
\tag{6.4}
$$

Since there are two particles emitted during the period $\mathcal{T}$, the emitted particle (either an electron or a hole) carries an additional energy $\hbar/(2\Gamma_\tau)$ over the Fermi energy. Since $I_E \neq 0$ the emitted particles are non-equilibrium particles.

6.1. *Multi-particle emission from multiple localized states*

If several localized states are placed in series along the same extended state (see inset to Fig. 8.5), then such a combined structure can act as a multi-particle emitter. Let the corresponding gates V_{gL} and V_{gR} induce the potentials $U_j(t) = U_{j0} + U_{j1} \cos(\omega t + \varphi_j)$, $j = L, R$, on the respective capacitors. Then at times $t_{0j}^{(\zeta)}$ the cavity j emits an electron ($\zeta = $ '$-$') and a hole ($\zeta = $ '$+$'). Since the emission times $t_{0j}^{(\zeta)}$ are defined by φ_j, then depending on the phase difference $\Delta\varphi = \varphi_L - \varphi_R$ between the potentials $U_L(t)$ and $U_R(t)$, such a double-cavity capacitor can emit electron and hole pairs, or electron–hole pairs, or emit single particles, electrons and holes.

To recognize the emission regime it is convenient to analyze the mean square current generated by this structure,[69]

$$\langle I^2 \rangle = \lim_{\Delta t \to \infty} \frac{1}{\Delta t} \int_0^{\Delta t} dt\, I^2(t) \equiv \int_0^{\mathcal{T}} \frac{dt}{\mathcal{T}} I^2(t). \qquad (6.5)$$

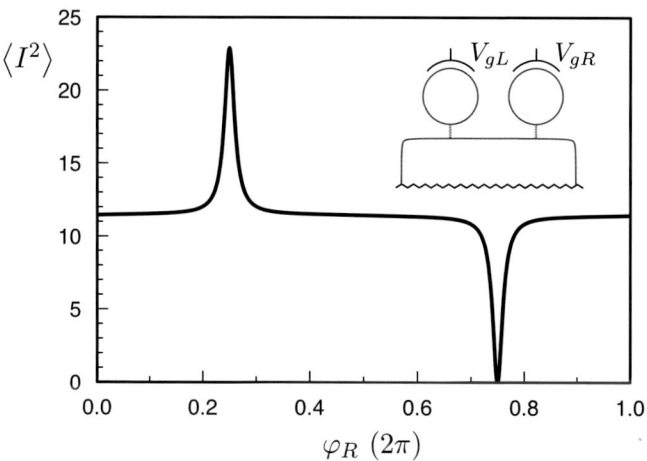

Fig. 8.5.   Inset: Two localized states (two edge states) coupled to the same extended state and reservoir. Main: The mean square current $\langle I^2 \rangle$ in units of $(e^2\omega U_{j1}/\Delta)^2$ as a function of the phase φ_R. The parameters are: $\omega\Gamma_j = 1/(20\pi)$, $eU_{j1} = 0.5\Delta$, $\Delta_j = \Delta$ $(j = L, R)$, $\varphi_L = \pi/2$.

In leading order in ω, the current $I(t)$ produced by the coupled localized states is a sum of currents generated by each localized state separately, $I(t) = I_L(t) + I_R(t)$, where $I_j(t)$ is defined by Eq. (6.3) with $t_0^{(\mp)}$ and Γ_τ being replaced by $t_{0j}^{(\mp)}$ and $\Gamma_{\tau j}$, correspondingly. The advantage of considering $\langle I^2 \rangle$ rather than $I(t)$ is the average over a long time instead of a more complicated time-resolved measurement.

Introducing the difference of times, $\Delta t_{L,R}^{(\zeta,\zeta')} = t_{0L}^{(\zeta)} - t_{0R}^{(\zeta')}$, we find to leading order in $\omega \Gamma_{\tau j} \ll 1$,

$$\langle I^2 \rangle = \frac{e^2}{\pi \mathcal{T}} \left(\frac{1}{\Gamma_{\tau L}} + \frac{1}{\Gamma_{\tau R}} \right)$$
$$\times \left\{ 1 - L\left(\Delta t_{L,R}^{(-,+)} \right) - L\left(\Delta t_{L,R}^{(+,-)} \right) + L\left(\Delta t_{L,R}^{(-,-)} \right) + L\left(\Delta t_{L,R}^{(+,+)} \right) \right\},$$
$$(6.6)$$

where $L(\Delta t) = 2\Gamma_{\tau L}\Gamma_{\tau R} \left\{ (\Delta t)^2 + (\Gamma_{\tau L} + \Gamma_{\tau R})^2 \right\}^{-1}$. If both cavities emit particles at different times, $\left| \Delta t_{L,R}^{(\zeta,\zeta')} \right| \gg \Gamma_{\tau j}$, then they contribute to the mean square current additively, $\langle I^2 \rangle_0 = e^2/(\pi \mathcal{T}) \left(\Gamma_{\tau L}^{-1} + \Gamma_{\tau R}^{-1} \right)$. Below we use this quantity as a reference point.

Changing the phase lag $\Delta\varphi$ one can enter the regime when one cavity emits an electron (a hole) at the time when the other cavity emits a hole (an electron), $\left| \Delta t_{L,R}^{(-,+)} \right| \lesssim \Gamma_{\tau j} \left(\left| \Delta t_{L,R}^{(+,-)} \right| \lesssim \Gamma_{\tau j} \right)$. We expect that the source comprising both cavities does not generate a current, since the particle emitted by the first (L) cavity is absorbed by the second (R) cavity. Indeed, in this regime $\langle I^2 \rangle = \langle I^2 \rangle_0 \left\{ 1 - L\left(\Delta t_{L,R}^{(-,+)} \right) - L\left(\Delta t_{L,R}^{(+,-)} \right) \right\}$ is reduced indicating a reabsorption regime. For identical cavities, $\Gamma_{\tau L} = \Gamma_{\tau R}$, emitting in synchronism, $\Delta t_{L,R}^{(-,+)} = \Delta t_{L,R}^{(+,-)} = 0$, the mean square current vanishes, $\langle I^2 \rangle = 0$. In Fig. 8.5, this synchronized regime shows up as a dip in the mean squared current. Therefore, in this case the second cavity re-absorbs all the particles emitted by the first cavity.

An additional regime when the two particles of the same kind are emitted near simultaneously, $\left| \Delta t_{L,R}^{(\zeta,\zeta)} \right| \lesssim \Gamma_{\tau j}$ is interesting. In this case, the mean square current is enhanced and for identical cavities emitting in synchronism it is, $\langle I^2 \rangle = 2\langle I^2 \rangle_0$, and shows up as a peak in Fig. 8.5. The enhancement of the mean square current can be explained from the energy perspective. If two electrons (holes) are emitted simultaneously, then due to the Pauli exclusion principle they should be in different states, and that for spinless particles means different energies. The second emitted particle should have an energy larger than the first one. To be more precise, the electron (hole) pair has a larger energy then the sum of energies of two separately emitted electrons (holes). Therefore, the heat flow I_E generated by the double-cavity capacitor should be enhanced, that, by virtue of Eq. (5.4), implies an enhanced mean square current.

7. Conclusion

The success of scattering theory of electrical transport for conductance and noise is well known. We have shown that interesting conductance problems, of a qualitatively different nature, like the movement of charge in and out of a conductor, the ac response to small amplitude voltage oscillations and even the large amplitude non-linear response find an elegant formulation in scattering theory. Novel experiments in a 2D electron gas in the integer quantum Hall effect regime demonstrate[58] quantization of the charge relaxation resistance of a mesoscopic cavity at half a von Klitzing resistance quantum. A successor experiment demonstrates[66] that a localized state can serve as a sub-nanosecond, single-electron source for coherent quantum electronics. Both of these phenomena seem to be well described by scattering theory. We only mention a few additional elements of a coherent quantum electronics using such sources in this paper. The shot noise produced by the particles emitted by a dynamical localized state due to scattering at a quantum point contact is quantized, i.e., it is proportional to the number of emitted particles during an oscillation period $\mathcal{T}$.[70,71] If two emitters are placed at different sides of a QPC then they contribute to shot noise additively, unless they are synchronized such that both cavities emit electrons (holes) at the same time. In this case, the shot noise vanishes.[70] This effect arises due to Fermi correlations between electrons (holes) colliding at the QPC and propagating to different contacts. This effect looks similar to the Hong, Ou, and Mandel[72] effect in optics. However for fermions (electrons) the two-particle probability peaks while for bosons (photons) it shows a dip. In collaboration with J. Splettstoesser, we demonstrated that synchronized sources of uncorrelated particles can produce orbitally entangled pairs of electrons (holes).[73] The amount of entanglement can be varied from zero to a maximum by a simple variation of the difference of phase $\Delta\varphi$ of the potentials driving the localized states.

Acknowledgments

We are grateful to Simon Nigg for discussion and technical assistance. We thank Janine Splettstoesser and Geneviève Fleury for a careful reading of

M. Büttiker & M. Moskalets

the manuscript. This work is supported by the Swiss NSF, MaNEP, and by the European Marie Curie ITN NanoCTM.

References

1. P. W. Anderson, *Phys. Rev.* **109**, 1492 (1958).
2. D. J. Thouless, *Phys. Rep.* **13**, 93 (1974).
3. F. J. Wegner, *Z. Phys. B* **25**, 327 (1976).
4. E. Abrahams, P. W. Anderson, D. C. Licciardello and T. V. Ramakrishnan, *Phys. Rev. Lett.* **42**, 673 (1979).
5. R. Landauer, *IBM J. Res. Dev.* **1**, 223 (1957).
6. R. Landauer, *Philos. Mag.* **21**, 863 (1970).
7. D. C. Langreth and E. Abrahams, *Phys. Rev. B* **24**, 2978 (1981).
8. P. W. Anderson, *Phys. Rev. B* **23**, 4828 (1981).
9. H.-L. Engquist and P. W. Anderson, *Phys. Rev. B* **24**, 1151 (1981).
10. M. Ya. Azbel, *J. Phys. C* **14**, L225 (1981).
11. E. N. Economou and C. M. Soukoulis, *Phys. Rev. Lett.* **46**, 618 (1981).
12. D. S. Fisher and P. A. Lee, *Phys. Rev. B* **23**, 6851 (1981).
13. R. Landauer, *Phys. Lett. A* **85**, 91 (1981).
14. M. Büttiker, Y. Imry, R. Landauer and S. Pinhas, *Phys. Rev. B* **31**, 6207 (1985).
15. M. Büttiker, *Phys. Rev. Lett.* **57**, 1761 (1986).
16. M. Büttiker, *IBM J. Res. Developm.* **32**, 317 (1988).
17. Ya. M. Blanter and M. Büttiker, *Phys. Rep.* **336**, 1 (2000).
18. M. Büttiker, Y. Imry and R. Landauer, *Phys. Lett. A* **96**, 365 (1983).
19. Y. Gefen, Y. Imry and M. Ya. Azbel, *Phys. Rev. Lett.* **52**, 129 (1984).
20. M. Büttiker, Y. Imry and M. Ya. Azbel, *Phys. Rev. A* **30**, 1982 (1984).
21. A. C. Bleszynski-Jayich, W. E. Shanks, B. Peaudecerf, E. Ginossar, F. von Oppen, L. Glazman and J. G. E. Harris, *Science* **326**, 272 (2009).
22. K.-T. Lin, Y. Lin, C. C. Chi, J. C. Chen, T. Ueda and S. Komiyama, *Phys. Rev. B* **81**, 035312 (2010).
23. Y. Imry, *Introduction to Mesoscopic Physics* (Oxford University Press, USA, 2002).
24. O. N. Dorokhov, *Sov. Phys. JETP* **58**, 606 (1983).
25. B. L. Al'tshuler, A. G. Aronov and B. Z. Spivak, *Pis'ma Zh. Eksp. Teor. Fiz.* **33**, 101 (1981) [*JETP Lett.* **33**, 94 (1981)].
26. A. G. Aronov and Yu. V. Sharvin, *Rev. Mod. Phys.* **59**, 755 (1987).
27. J. E. Avron, A. Elgart, G. M. Graf and L. Sadun, *Phys. Rev. B* **62**, R10618 (2000).
28. J. E. Avron, A. Elgart, G. M. Graf and L. Sadun, *J. Stat. Phys.* **116**, 425 (2004).
29. F. T. Smith, *Phys. Rev.* **118**, 349 (1960).
30. M. Büttiker, H. Thomas and A. Prêtre, *Z. Phys. B* **94**, 133 (1994).
31. P. W. Brouwer, *Phys. Rev. B* **58**, 10135 (1998); *Phys. Rev. B* **63**, 121303 (2001).
32. M. L. Polianski and P. W. Brouwer, *Phys. Rev. B* **64**, 075304 (2001).
33. Y. Makhlin and A. Mirlin, *Phys. Rev. Lett.* **87**, 276803 (2001).

34. M. Moskalets and M. Büttiker, *Phys. Rev. B* **66**, 035306 (2002).
35. H.-Q. Zhou, S. Y. Cho and R. H. McKenzie, *Phys. Rev. Lett.* **91**, 186803 (2003).
36. A. Agarwal and D. Sen, *J. Phys. Condens. Matter* **19**, 046205 (2007).
37. M. Jääskeläinen, F. Corvino, Ch. P. Search and V. Fessatidis, *Phys. Rev. B* **77**, 155319 (2008).
38. M. Büttiker, *J. Phys. Condens. Matter* **5**, 9361 (1993).
39. J. Friedel, *Nuovo Cimento Suppl.* **7**, 287 (1958).
40. D. Cohen, *Phys. Rev. B* **68**, 201303 (2003).
41. M. Büttiker, *J. Math. Phys.* **37**, 4793 (1996).
42. M. H. Pedersen, S. A. van Langen and M. Büttiker, *Phys. Rev. B* **57**, 1838 (1998).
43. M. Moskalets and M. Büttiker, *Phys. Rev. B* **69**, 205316 (2004).
44. M. Moskalets and M. Büttiker, *Phys. Rev. B* **72**, 035324 (2005).
45. X.-L. Qi and S.-C. Zhang, *Phys. Rev. B* **79**, 235442 (2009).
46. L. Oroszlány, V. Zólyomi and C. J. Lambert, arXiv:0902.0753 (unpublished).
47. M. Moskalets and M. Büttiker, *Phys. Rev. B* **66**, 205320 (2002); *Phys. Rev. B* **78**, 035301 (2008).
48. L. Arrachea and M. Moskalets, *Phys. Rev. B* **74**, 245322 (2006).
49. M. L. Polianski and P. W. Brouwer, *J. Phys. A: Math. Gen.* **36**, 3215 (2003).
50. M. G. Vavilov, *J. Phys. A: Math. Gen.* **38**, 10587 (2005).
51. D. J. Thouless, *Phys. Rev. B* **27**, 6083 (1983).
52. C.-H. Chern, S. Onoda, S. Murakami and N. Nagaosa, *Phys. Rev. B* **76**, 035334 (2007).
53. G. M. Graf and G. Ortelli, *Phys. Rev. B* **77**, 033304 (2008).
54. G. Bräunlich, G. M. Graf and G. Ortelli, *Commun. Math. Phys.* **295**, 243 (2010).
55. A. Prêtre, H. Thomas and M. Büttiker, *Phys. Rev. B* **54**, 8130 (1996).
56. M. Büttiker, H. Thomas and A. Prêtre, *Phys. Lett. A* **180**, 364 (1993).
57. S. E. Nigg and M. Büttker, *Phys. Rev. B* **77**, 085312 (2008).
58. J. Gabelli, G. Fève, J.-M. Berroir, B. Plaçais, A. Cavanna, B. Etienne, Y. Jin and D. C. Glattli, *Science* **313**, 499 (2006).
59. S. E. Nigg, R. Lopez and M. Büttiker, *Phys. Rev. Lett.* **97**, 206804 (2006).
60. M. Büttiker and S. E. Nigg, *Nanotechnology* **18**, 044029 (2007).
61. Z. Ringel, Y. Imry and O. Entin-Wohlman, *Phys. Rev. B* **78**, 165304 (2008).
62. Ya. I. Rodionov, I. S. Burmistrov and A. S. Ioselevich, *Phys. Rev. B* **80**, 035332 (2009).
63. C. Mora and K. Le Hur, arXiv:0911.4908 (unpublished).
64. Y. Hamamoto, T. Jonckheere, T. Kato and T. Martin, arXiv:0911.4101 (unpublished).
65. J. Splettstoesser, M. Governale, J. König and M. Büttiker, arXiv:1001.2664 (unpublished).
66. G. Fève, A. Mahé, J.-M. Berroir, T. Kontos, B. Plaçais, D. C. Glattli, A. Cavanna, B. Etienne and Y. Jin, *Science* **316**, 1169 (2007).
67. M. Moskalets, P. Samuelsson and M. Büttiker, *Phys. Rev. Lett.* **100**, 086601 (2008).
68. M. Moskalets and M. Büttiker, *Phys. Rev. B* **80**, 081302 (2009).

69. J. Splettstoesser, S. Ol'khovskaya, M. Moskalets and M. Büttiker, *Phys. Rev. B* **78**, 205110 (2008).
70. S. Ol'khovskaya, J. Splettstoesser, M. Moskalets and M. Büttiker, *Phys. Rev. Lett.* **101**, 166802 (2008).
71. J. Keeling, A. V. Shytov and L. S. Levitov, *Phys. Rev. Lett.* **101**, 196404 (2008).
72. C. K. Hong, Z. Y. Ou and L. Mandel, *Phys. Rev. Lett.* **59**, 2044 (1987).
73. J. Splettstoesser, M. Moskalets and M. Büttiker, *Phys. Rev. Lett.* **103**, 076804 (2009).

Chapter 9

THE LOCALIZATION TRANSITION AT FINITE TEMPERATURES: ELECTRIC AND THERMAL TRANSPORT

Yoseph Imry and Ariel Amir

Department of Condensed Matter Physics,
Weizmann Institute of Science, Rehovot 76100, Israel

The Anderson localization transition is considered at finite temperatures. This includes the electrical conductivity as well as the electronic thermal conductivity and the thermoelectric coefficients. An interesting critical behavior of the latter is found. A method for characterizing the conductivity critical exponent, an important signature of the transition, using the conductivity and thermopower measurements, is outlined.

1. Introduction

Anderson localization[1,2] is a remarkable, and very early, example of a quantum phase transition (QPT), where the nature of a system at $T = 0$ changes abruptly and nonanalytically at a point, as a function of a control parameter. Here it implies that the relevant quantum states of the system acquire an exponential decay at large distances (similar to, but much more complex than, the formation of a bound state). In the original paper this phenomenon was discovered via a change in the convergence properties of the "locator" expansion. In the disordered tight-binding model, with short-range hopping, this is the expansion about the bound atomic, or Wannier-type orbitals. As long as the expansion converges, the relevant eigenstates are localized; its divergence signifies the transition to "extended", delocalized states. This was reviewed, hopefully pedagogically, in Ref. 3. Later, Mott introduced the very useful picture of the "mobility edge" within the band of allowed energies, separating localized from extended states.[2] In the lower part of the band, the states below the "lower mobility edge" are localized, while those above it are extended. When the disorder and/or the position of the Fermi level, E_F, are changed, a point where they cross each other is where the states at E_F change nature from extended to localized, and this is a simple and instructive model for a metal–insulator transition at vanishing temperature T.

The analysis of the above localization transition centers on the behavior of $\sigma_0(E)$, the conductivity at energy E, which would be the $T = 0$ conductivity of the sample with $E_F = E$. $\sigma_0(E)$ vanishes for E below the lower mobility edge E_m and goes to zero when E approaches E_m from above[4]

$$\sigma_0(E) = A(E - E_m)^x. \tag{1.1}$$

The characteristic exponent for this, x, is an important parameter of the theory. It is expected to be universal for a large class of noninteracting models, but its value is not really known, in spite of the several analytical, numerical and experimental methods used to attempt its evaluation.

The electron–electron interaction is certainly relevant near this transition, which may bring it to a different universality class. This is a difficult problem. The benchmark treatment is the one by Finkel'stein.[5] Since it is likely that the situation in at least most experimental systems is within this class, it would appear impossible to determine the value of x for the pure Anderson transition (without interactions).[6]

In this paper, we still analyze the thermal and thermoelectric transport for a general model with $\sigma_0(E)$ behaving as in Eq. (1.1). This is certainly valid for noninteracting electrons and should be valid also including the interactions, as long as some kind of Fermi liquid quasiparticles exist. In that case, $\sigma_0(E)$ for quasiparticles is definable and thermal averaging with Fermi statistics holds. Even then, however, various parameter renormalizations and interaction corrections[7] should come in. The effect of the interactions on the thermopower for a small system in the Coulomb blockade picture was considered in Ref. 8, and correlations were included in Ref. 9. Even in the latter case, the simple Cutler–Mott[10] formula (Eq. (3.6), derived in Sec. 3) was found to work surprisingly well.

It should be mentioned that the sharp and asymmetric energy-dependence of $\sigma_0(E)$ near the mobility edge[10,11] should and does[10,12,13] lead to rather large values of the thermopower. Exceptions will be mentioned and briefly discussed later. Large thermopowers are important for energy conversion and refrigeration applications[14] and this clearly deserves further studies.

A serious limitation on the considerations presented here is that the temperature should be low enough so that all the inelastic scattering (electron–phonon, electron–electron, etc.) is negligible. For simplicity, we consider here *only* longitudinal transport (currents parallel to the driving fields). Thus, no Hall or Nernst–Ettinghausen effect! We also do not consider the thermoelectric transport in the hopping regime. It might be relevant even in the metallic regime (chemical potential $\mu > E_m$), once $k_B T \gtrsim \mu - E_m$. This may well place limitations on the high temperature analysis we make in the following.

In Sec. 2, we review the basic concepts behind the scaling theory[4] for the transition, and reiterate the critical behavior of the conductivity as in Eq. (1.1), obtaining also its temperature dependence. In Sec. 3, we derive all results for the thermal and thermoelectric transport and analyze the scaling critical behavior of the latter as function of temperature and distance from the transition. A brief comparison with experiment is done in Sec. 4 and concluding remarks are given in Sec. 5. In the appendix, we present a proof that the heat carried by a quasiparticle is equal to its energy measured from the chemical potential, μ, hoping that this also clarifies the physics of this result. This derivation is valid for Bose quasiparticles (phonons, photons, magnons, excitons, plasmons, etc.) as well.

2. The Zero and Finite Temperature Macroscopic Conductivity Around the Anderson Localization Transition

2.1. *The Thouless picture within the tunnel-junction model*

We start this section by briefly reviewing the tunnel-junction picture of conduction[15,16] at $T = 0$, which is a useful way to understand the important Thouless[17] picture for such transport. Consider first two pieces (later referred to as "blocks") of a conducting material with a linear size L, connected through a layer of insulator (usually an oxide) which is thin enough to allow for electron tunneling. The interfaces are assumed rough, so there is no conservation of the transverse momentum: each state on the left interacts with each state on the right with a matrix element t with a roughly uniform absolute value. The lifetime τ_L for an electron on one such block for a transition to the other one is given by the Fermi golden rule (at least when tunneling is a weak perturbation):

$$\tau_L^{-1} = \frac{2\pi}{\hbar} \overline{|t|^2} N_r(E_F), \tag{2.1}$$

where $\overline{|t|^2}$ is the average of the tunneling matrix element squared and $N_r(E_F)$ is the density of states on the final (right-hand) side. Taking the density of states (DOS) in the initial side to be $N_\ell(E_F)$, we find that when a voltage V is applied, $eV N_\ell(E_F)$ states are available, each decaying to the right with a time constant τ_L, so that the current is

$$I = e^2 N_\ell(E_F)\tau_L^{-1}V, \qquad (2.2)$$

and the conductance is

$$G = e^2 N_\ell(E_F)/\tau_L = \frac{2\pi e^2}{\hbar}\overline{t^2}N_\ell(E_F)N_r(E_F), \qquad (2.3)$$

which is an extremely useful result. This equality is well-known in the tunnel junction theory.[15,16] Clearly, G is symmetric upon exchanging l and r, as it should.[18] Note that Eqs. (2.1) and (2.3) are valid in any number of dimensions. An important remark is that Eq. (2.1) necessitates a continuum of final states, while the final (RHS) block is finite and has a discrete spectrum. One must make the assumption that the interaction of that system with the outside world leads to a level broadening larger than, or on the same order of, the level spacing.[19–21] This is the case in most mesoscopic systems. One then naively assumes that this condition converts the spectrum to an effectively continuous one (the situation may actually be more subtle[22]). Otherwise, when levels really become discrete, one gets into the *really microscopic (molecular) level.*

The result of Eq. (2.3) is very general. Let us use it for the following scaling picture[17]: Divide a large sample to (hyper) cubes or "blocks" of side L. We consider the case $L \gg \ell, a$; ℓ being the elastic mean free path and a the microscopic length. The typical level separation for a block at the relevant energy (say, the Fermi level), d_L, is given by the inverse of the density of states (per unit energy) for size L, $N_L(E_F)$. Defining an energy associated with the transfer of electrons between two such adjacent systems by $V_L \equiv \pi\hbar/\tau_L$ (τ_L is the lifetime of an electron on one side against transition to the other side) the dimensionless interblock conductance $g_L \equiv G_L/(e^2/\pi\hbar)$ is:

$$g_L = V_L/d_L \qquad (2.4)$$

i.e. g_L is the (dimensionless) ratio of the only two relevant energies in the problem. The way Thouless argued for this relation is by noting that the electron's diffusion on the scale L is a random walk with a step L and characteristic time τ_L, thus

$$D_L \sim L^2/\tau_L.$$

Note that as long as the classical diffusion picture holds, D_L is independent of L and $\tau_L = L^2/D$, which is the diffusion time across the block. It will turn out that the localization or quantum effects, when applicable, cause D_L to decrease with L. For metals, the conductivity, σ_L, on the scale of the block size L, is given by the Einstein relation $\sigma = D_L e^2 dn/d\mu$ (where μ is the chemical potential and $dn/d\mu$ is the density of states per unit volume), and the conductance in d dimensions is given by:

$$G_L = \sigma_L L^{d-2}. \tag{2.5}$$

Putting these relations together and remembering that $N_L(E_F) \sim L^d dn/d\mu$, yields Eq. (2.4). To get some physical feeling for the energy h/τ_L we note again that, at least for the weak coupling case, the Fermi golden rule yields Eq. (2.1) or:

$$V_L = 2\pi^2 \overline{|t|^2}/d_R. \tag{2.6}$$

Thus, V_L is defined in terms of the interblock matrix elements. Clearly, when the blocks are of the same size, Eq. (2.6) is also related to the order of magnitude of the second order perturbation theory shift of the levels in one block by the interaction with the other. For a given block this is similar to a surface effect — the shift in the block levels due to changes in the boundary conditions on the surface of the block. Indeed, Thouless has given appealing physical arguments for the equivalence of V_L with the sensitivity of the block levels to boundary conditions. This should be valid for L much larger than ℓ and all other microscopic lengths.

Since in this scaling picture the separations among the blocks are fictitious for a homogeneous system, it is clear that the interblock conductance is just the conductance of a piece whose size is of the order of L, i.e., this is the same order of magnitude as the conductance of the block itself.

The latter can also be calculated using the Kubo linear response expression. It has to be emphasized that the Kubo formulation also applies strictly only for an infinite system whose spectrum is continuous. For a finite system, it is argued again that a very small coupling of the electronic system to some large bath (e.g. the phonons, or to a large piece of conducting material) is needed to broaden the discrete levels into an effective continuum. Edwards and Thouless,[23] using the Kubo–Greenwood formulation, made the previously discussed relationship of V_L with the sensitivity to boundary conditions very precise.

The above picture is at the basis of the finite-size scaling[4,24] theory of localization. It can also can be used for numerical calculations of $g(L)$, which is a most relevant physical parameter of the problem, for non-interacting electrons, as we shall see. Alternatively, Eq. (2.6) as well as generalizations thereof can and have been used for numerical computations. Powerful numerical methods exist to this end.[25]

It is important to emphasize that $g_L \gg 1$ means that states in neighboring blocks are tightly coupled while $g_L \ll 1$ means that the states are essentially single-block ones. g_L is therefore a good general dimensionless measure of the strength of the coupling between two quantum systems. Thus, if $g_L \gg 1$ for small L and $g_L \to 0$ for $L \to \infty$, then the range of scales L where $g_L \sim 1$ gives the order of magnitude of the localization length, ξ.

Although the above analysis was done specifically for non-interacting electrons, it is of greater generality. g_L (with obvious factors) may play the role of a conductance also when a more general entity (e.g. an electron pair) is transferred between the two blocks. The real limitations for the validity of this picture seem to be the validity of the Fermi-liquid picture and that no inelastic effects (e.g with phonons or electron–hole pairs) occur.

The analysis by Thouless[17] of the consequences of Eq. (2.4) for a long thin wire has led to extremely important results. First, it showed that 1D localization should manifest itself not only in "mathematically 1D" systems but also in the conduction in realistic, finite cross-section, thin wires, demonstrating also the usefulness of the block-scaling point of view. Second, the understanding of the effects of finite temperatures (as well as other experimental parameters) on the relevant scale of the conduction, clarifies the relationships between $g(L)$ and experiment in any dimension. Third, defining and understanding the conductance $g(L)$ introduces the basis for the scaling theory of the Anderson localization transition.[4] Here, we use the results for the (macroscopic) $T = 0$ conductivity around the localization transition to get the finite temperature conductivity there.

2.2. *The critical behavior of the $T = 0$ conductivity*

Near, say, the lower mobility edge, E_m, the conductivity $\sigma_0(E)$ vanishes for $E < E_m$ and approaches zero for $E \to E_m$ from above, in the manner:

$$\sigma_0(E) = A(E - E_m)^x, \tag{2.7}$$

A being a constant and x the conductivity critical exponent for localization, which has so far eluded a precise determination either theoretically or experimentally. Within the scaling theory,[4] x is equal to the critical exponent of the characteristic length (ξ), because

$$\sigma \sim \frac{e^2}{\pi\hbar\xi}. \tag{2.8}$$

In that case, an appealing intuitive argument by Mott[27] and Harris[26] places a lower bound on x:

$$x \geq 2/d. \tag{2.9}$$

In fact, Eq. (2.8) may be expected to hold on dimensional grounds for any theory which does not generate another critical quantity with the dimension of length. This should be the case for models which effectively do not have electron–electron interactions. With electron–electron interactions, for example, we believe that the critical exponent for the characteristic length should satisfy an inequality such as Eq. (2.9).[28] However, this may no longer be true for the conductivity exponent.

2.3. The conductivity at finite temperatures

In Eqs. (2.2) and (2.3) we calculated, at $T = 0$, the current in an infinitesimal (linear response) energy strip of width eV around the Fermi energy. Generalizing this to an arbitrary energy at finite temperature, we find that the current due to a strip dE at energy E is

$$I(E)dE = eN_\ell(E_F)\tau_L^{-1}(E)[f_l(E) - f_r(E)]dE, \tag{2.10}$$

$f_l(E)$ ($f_r(E)$) being the Fermi function at energy E at the left (right). The total current is obtained by integrating Eq. (2.10) over energy. For linear response $f_l(E) - f_r(E) = eV[-\frac{\partial f}{\partial E}]$. This gives

$$\sigma(T) = \int_{-\infty}^{\infty} dE\sigma_0(E)\left[-\frac{\partial f}{\partial E}\right], \tag{2.11}$$

where $\sigma_0(E) \equiv (e^2/\pi\hbar)\frac{V_L(E)}{d_L(E)}L^{(2-d)}$ is the conductivity (using Eqs. 2.4 and 2.5) at energy E, which would be the $T = 0$ conductivity of the sample with $E_F = E$.

2.4. Analysis of $\sigma(T, \mu - E_m)$

From now on we assume Eq. (2.7) to hold. Measuring all energies from the chemical potential μ and scaling them with T, we rewrite Eq. (2.11) in the manner (we shall employ units in which the Boltzmann constant, k_B is unity, and insert it in the final results)

$$\sigma(T, \mu - E_m) = AT^x\Sigma([\mu - E_m]/T), \tag{2.12}$$

where the function $\Sigma(z)$ is given by:

$$\Sigma(z) \equiv \int_{-z}^{\infty} dy (y+z)^x \left[-\frac{\partial[1 + \exp(y)]^{-1}}{\partial y} \right]. \tag{2.13}$$

Figure 9.1 shows a numerical evaluation of this integral. Let us consider the low and high temperature limits of this expression, that will also be useful later for the analysis of the thermopower.

At low temperatures, we can use the Sommefeld expansion, to obtain:

$$\sigma_{\text{low}}(T)/A = (\mu - E_m)^x + \frac{\pi^2}{6} T^2 x(x-1)(\mu - E_m)^{x-2}. \tag{2.14}$$

Notice that $\frac{\partial \sigma}{\partial T}$ is negative for $x < 1$: this comes about since in this case the function $\sigma(E)$ is concave.

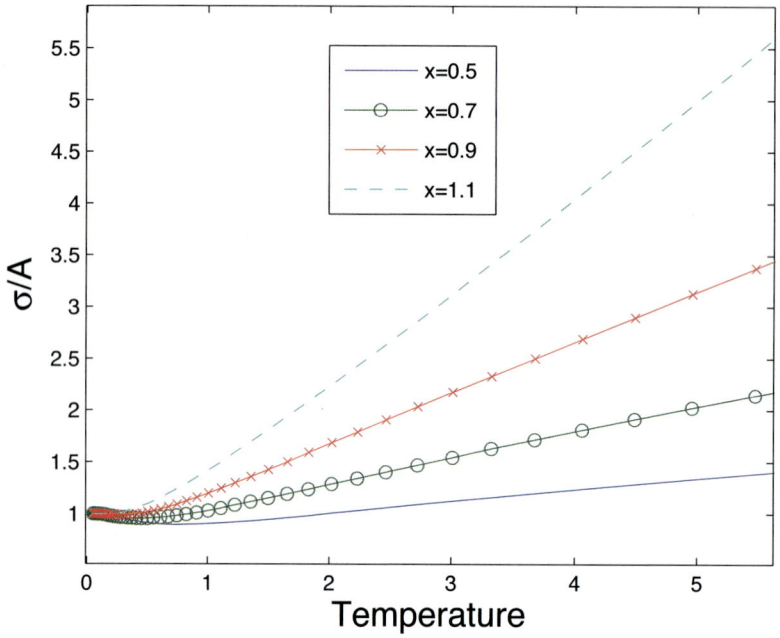

Fig. 9.1.   The integral of Eq. (2.13) was evaluated numerically, for different values of the exponent x, for $\mu - E_m = 1$. Shown is the dependence of the conductivity, measured in units of A (see Eq. (2.7)) as a function of temperature (the energy scale is set by $\mu - E_m$). At low temperatures, the conductivity saturates to a value given by Eq. (2.14), while at high temperatures it scales as T^x, see Eq. (2.15). The derivative at low temperatures can be positive or negative, depending on x (see Eq. (2.14)).

At high temperatures, one can set $E_m = 0$, since the contribution to the integrals comes from energies smaller or of the order of the temperature, and $T \gg \mu - E_m$. Therefore we have to evaluate:

$$\sigma_{\text{high}}(T)/A = \int_0^\infty E^x \left(-\frac{\partial f}{\partial E} \right) dE. \tag{2.15}$$

Thus, at high temperatures $\sigma_{\text{high}}/A \sim T^x$, with the coefficient given by $\int_0^\infty \frac{q^x}{2+\cosh(q)} dq$. In Sec. 3.3 we show that this integral can be connected with the Riemann Zeta function, and its value is given by Eq. (3.20).

This scaling could be used for a determination of the exponent x. However, a much closer determination of that exponent would follow from the scaling of both the conductivity and the thermal and thermoelectric transport coefficients, which will be studied in the coming sections.

3. Thermal and Thermoelectric Transport

3.1. *General relationships*

Consider now the case where both a voltage V and a temperature difference ΔT are applied between the two blocks. We choose for convenience $k_B \equiv 1$. Both are small enough for linear response to hold. Here we have to replace the Fermi function difference in Eq. (2.10) by

$$f_l(E) - f_r(E) = eV \left[-\frac{\partial f}{\partial E} \right] + \Delta T \left[-\frac{\partial f}{\partial T} \right].$$

Then generalizing Eq. (2.11) yields for the electrical current

$$I = \int dE \frac{G(E)}{e} \left\{ \left[-e \frac{\partial f}{\partial E} \right] V + \left[-\frac{\partial f}{\partial T} \right] \Delta T \right\}, \tag{3.1}$$

with G given in Eq. (2.5). The first term is the ordinary ohmic current and the second one is the thermoelectric charge current due to the temperature gradient.

Next, we derive the heat current. The heat carried by an electron with energy E (measured from the chemical potential, μ) is equal to E. This is shown, for example in Ref. 30 by noting that the heat is the difference between the energy and the free energy. Sivan and Imry[11] verified it in their Landauer-type model by calculating the flux of TS along the wire connecting the two reservoirs. In the appendix, we obtain the same result in our block model, from the time derivative of the entropy of each block. Thus, we obtain the heat current I_Q,

$$I_Q = \int E dE \frac{G(E)}{e^2} \left\{ \left[-e \frac{\partial f}{\partial E} \right] V + \left[-\frac{\partial f}{\partial T} \right] \Delta T \right\}. \tag{3.2}$$

Here the first term is the thermoelectric heat current due to the voltage, while the second one is the main contribution to the usual electronic thermal conductivity κ.

In this model the ratio of *thermal* to electrical conductivities is of the order of $(k_B/e)^2 T$. This is because a typical transport electron carries a charge e and an excitation energy of the order of $k_B T$ and the driving forces are the differences in eV and $k_B T$. This ratio is basically the Wiedemann–Franz law.

It is convenient to summarize Eqs. (3.1) and (3.2) in matrix notation[29]:

$$\begin{pmatrix} I \\ I_Q \end{pmatrix} = \begin{pmatrix} L_{11} & L_{12} \\ L_{21} & L_{22} \end{pmatrix} \begin{pmatrix} V \\ \Delta T \end{pmatrix}, \tag{3.3}$$

where the coefficients L_{ij} can be read off Eqs. (3.1) and (3.2). Since f is a function of E/T, we see that

$$-\frac{\partial f}{\partial T} = \frac{E}{T} \frac{\partial f}{\partial E}. \tag{3.4}$$

Therefore, the two "nondiagonal" thermoelectric coefficients: the one relating I to ΔT, L_{12}, and the one relating I_Q to V, L_{21}, are equal within a factor T.

$$L_{12} = L_{21}/T. \tag{3.5}$$

This is an Onsager[29-32] relationship, which holds very generally for systems obeying time-reversal symmetry (and particle conservation — unitarity). The case where time-reversal symmetry is broken, say by a magnetic field, is briefly discussed in the next subsection.

We conclude this subsection by defining and obtaining an expression for the absolute thermoelectric power (henceforth abbreviated as just "thermopower") of a material. Suppose we apply a temperature difference ΔT across a sample which is open circuited and therefore no current can flow parallel to ΔT. To achieve that, the sample will develop a (usually small) voltage V, so that the combined effect of both ΔT and V will be a vanishing current. From Eqs. (3.1), (3.3) and (3.4) we find that the ratio between V and ΔT, which is defined as the thermopower, S, is given by

$$S \equiv \frac{V}{\Delta T} = -\frac{L_{12}}{L_{11}} = \frac{\int dE \, E \sigma_0(E) \frac{\partial f}{\partial E}}{eT \int dE \sigma_0(E) \frac{\partial f}{\partial E}}. \tag{3.6}$$

3.2. *Onsager relations in a magnetic field*

From time-reversal symmetry at $H = 0$ and unitarity (particle conservation) follows the Onsager relation[29–32] for the $T = 0$ conductance

$$\sigma(E, H) = \sigma(E, -H). \tag{3.7}$$

This can be proven for our model from the basic symmetries of the interblock matrix elements. This symmetry obviously follows for the temperature-dependent electrical and thermal conductivities $\sigma(T)$ and $\kappa(T)$.

For the nondiagonal coefficients, the usual Onsager symmetry reads

$$L_{12}(H) = L_{21}(-H)/T. \tag{3.8}$$

In our case, since the nondiagonal coefficients are expressed as integrals over a symmetric function (Eq. (3.7)), they also obey

$$L_{ij}(H) = L_{ij}(-H). \tag{3.9}$$

i.e. the nondiagonal coefficients are symmetric in H as well.

3.3. *Analysis of the thermopower*

Equation (3.6) for the thermopower is identical to the one derived in two-terminal linear transport within the Landauer formulation in Ref. 11, which is equal in the appropriate limit to the Cutler–Mott[10,27] expression:

$$S = \frac{\int_{E_m}^{\infty} dE(E - \mu)\sigma_0(E)(-\frac{\partial f}{\partial E})}{e\sigma(T)T}, \tag{3.10}$$

where μ is the chemical potential, $\sigma_0(E)$ is the conductivity for carriers having energy E and σ is the total conductivity. The physics of this formula is clear for the (Onsager-dual) Peltier coefficient: a carrier at energy E carries an excitation energy (similar to heat, see the appendix) of $E - \mu$.

Clearly, electrons and holes contribute to S with opposite signs. S will tend to vanish with electron–hole symmetry and will be small, as happens in many metals, especially in ordered ones, when the variation in energy of $\sigma_0(E)$ around μ is weak.

Having a strong energy dependence of $\sigma_0(E)$, and being very different above and below μ will cause relatively large values of S. We believe that this is what happens in disordered narrow-gap semiconductors, which feature in many present-day good thermoelectrics. As noted in Refs. 10 and 11, the Anderson metal–insulator transition (or at least its vicinity) offers an almost ideal situation for large thermopowers. There, $\sigma_0(E)$ vanishes below the mobilty edge E_M (for electrons) and approaches zero, probably with an infinite slope, above it. Hopping processes in the localized phase are not considered here. A brief analysis in Ref. 11 demonstrated that S scales with $z \equiv (\mu - E_M)/T$:

$$S = Y\left(\frac{\mu - E_M}{T}\right), \qquad (3.11)$$

(Y being a universal scaling function) and assumes the two limits:

$$S \sim (\mu - E_m)^{-1}, \quad for \quad z \gg 1); \qquad (3.12)$$

and

$$S \sim \text{const} - z, \quad for \quad z \ll 1. \qquad (3.13)$$

Of course, there is no "real" divergence of S,[33] since when $(\mu - E_m) \to 0$ (and the slope of $S(T)$ diverges), it will eventually become smaller than T and the large-slope linear behavior will saturate as in Eq. (3.13).

Figure 9.2 shows a numerical evaluation of Eq. (3.10), demonstrating the linear low temperature regime and the saturation at high temperatures. Let

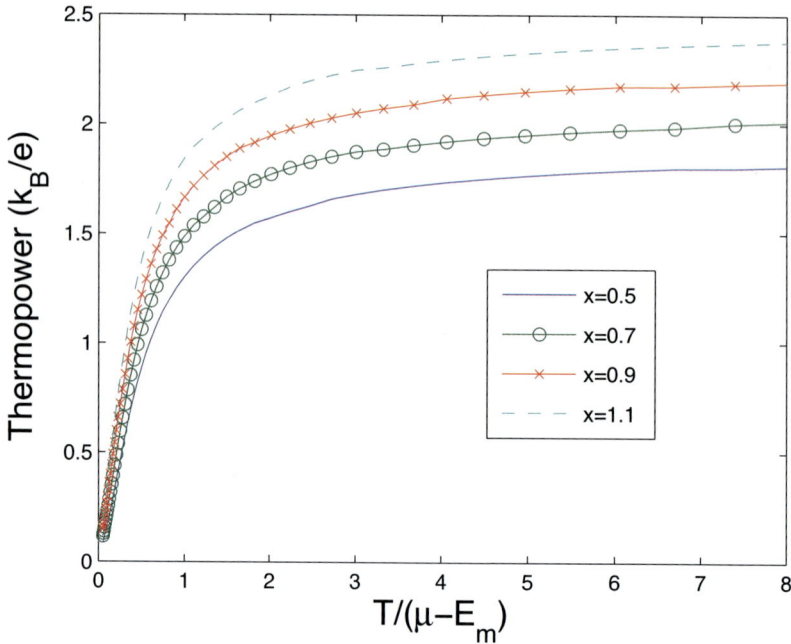

Fig. 9.2. The integrals of Eq. (3.10) were evaluated numerically, for different values of the exponent x. At low temperatures, the thermopower is linear with temperature, while at high temperatures it saturates, to a value which depends on x.

us now make a more thorough investigation of the low and high temperature regimes.

For low temperatures, we can use, as before, the Sommerfeld expansion for the nominator and denominator, to obtain:

$$S_{\text{low}} \approx \frac{\frac{\pi^2}{3}Tx(\mu - E_m)^{x-1} + O(T^3)}{e[(\mu - E_m)^x + \frac{\pi^2}{6}T^2x(x-1)(\mu - E_m)^{x-2} + O(T^4)]}. \tag{3.14}$$

A more complete expression is give in Eq. (3.24) below. Thus, at $T \ll \mu - E_m$, the thermopower is linear in temperature:

$$S_{\text{low}} \approx \frac{\pi^2 xT}{3e(\mu - E_m)} + O(T^3). \tag{3.15}$$

Figure 9.3 compares this expression with the numerically evaluated slope.

At high temperatures, one can set $E_m = 0$, as before. Therefore we have to evaluate:

$$S_{\text{high}} = \frac{\int_0^\infty E^{x+1}(-\frac{\partial f}{\partial E})dE}{eT \int_0^\infty E^x(-\frac{\partial f}{\partial E})dE}. \tag{3.16}$$

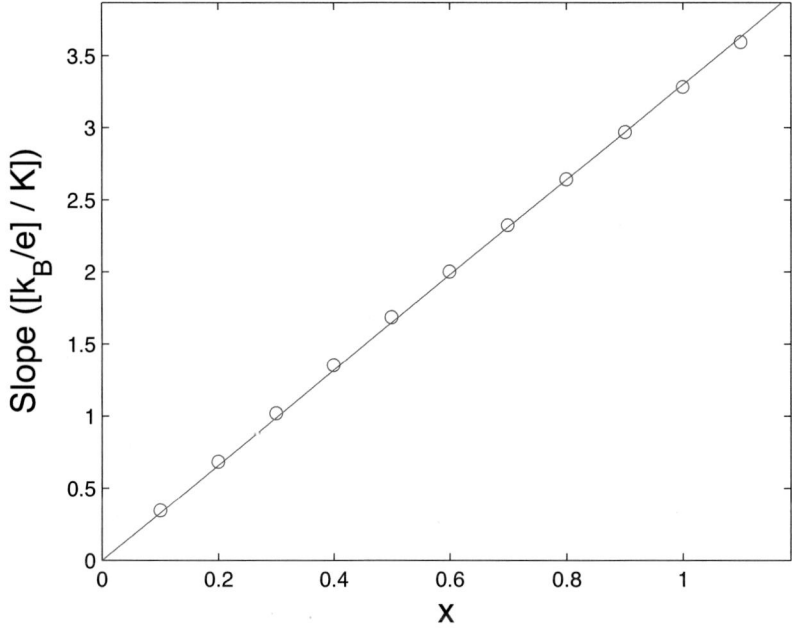

Fig. 9.3.   At low temperatures, the thermopower is described by Eq. (3.14). The plot shows a comparison between this expression and the evaluation of the slope extracted from the numerics demonstrated in Fig. 9.2.

We can write $\int_0^\infty E^\beta(-\frac{\partial f}{\partial E})dE = T^{\beta+2}G(\beta)$, with the dimensionless function $G(\beta)$ defined as:

$$G(\beta) = \int_0^\infty \frac{m^\beta}{2 + \cosh(m)}dm. \tag{3.17}$$

We then have $S_{\text{high}} = \frac{G(x+2)}{eG(x+1)}$.

In fact, the integral of Eq. (3.17) can be related to the Riemann Zeta function ζ:

$$\zeta(\beta) = \frac{1}{\Gamma(\beta)}\int_0^\infty \frac{m^{\beta-1}}{e^m - 1}dm. \tag{3.18}$$

Defining $C = \int_0^\infty \frac{m^{\beta-1}}{e^m-1}dm$, we find that:

$$C - G/\beta = \int_0^\infty \frac{2m^{\beta-1}}{e^{2m} - 1}dm = C/2^\beta, \tag{3.19}$$

therefore:

$$G(\beta) = \beta C(1 - 1/2^{\beta-1}) = \beta\zeta(\beta)\Gamma(\beta)(1 - 1/2^{\beta-1}). \tag{3.20}$$

This gives an exact formula for the thermopower at high temperatures:

$$S_{\text{high}} = (1 + x)\frac{\zeta(1+x)(2^x - 1)}{e\zeta(x)(2^x - 2)}. \tag{3.21}$$

At $x = 0$, one obtains $S = 2\log(2)$, while for $x \gg 1$, one obtains $S_{\text{high}} \approx 1+x$.

Actually, understanding the behavior for large x is simple: If we were to approximate the derivative of the Fermi function by $e^{-E/T}$, we would have $G(\beta) = \Gamma(\beta)$, where Γ stands for the Gamma function. Then, by its properties, we immediately have that $S_{\text{high}} \approx (1+x)/e$.

It turns out that a good approximation to $S_{\text{high}}(x)$ can be obtained by interpolating the exact $x = 0$ result and the large x result, by the form:

$$S_{\text{high}} \approx \frac{1}{e}[2\log(2) + x]. \tag{3.22}$$

Figure 9.4 compares the exact saturation values of Eq. (3.21) with this approximate form. We found that the difference for all values of x is less than 6 percent, and therefore Eq. (3.22) provides a practical working formula for the saturation value of the thermopower.

An interesting feature of the crossover from the low to high temperature regime is the possibility of an inflection point in the thermopower dependence. Similar to the behavior of the conductance, which grew for $x > 1$ but diminished for $x < 1$, here there will be an inflection point for $x < 1$. To see

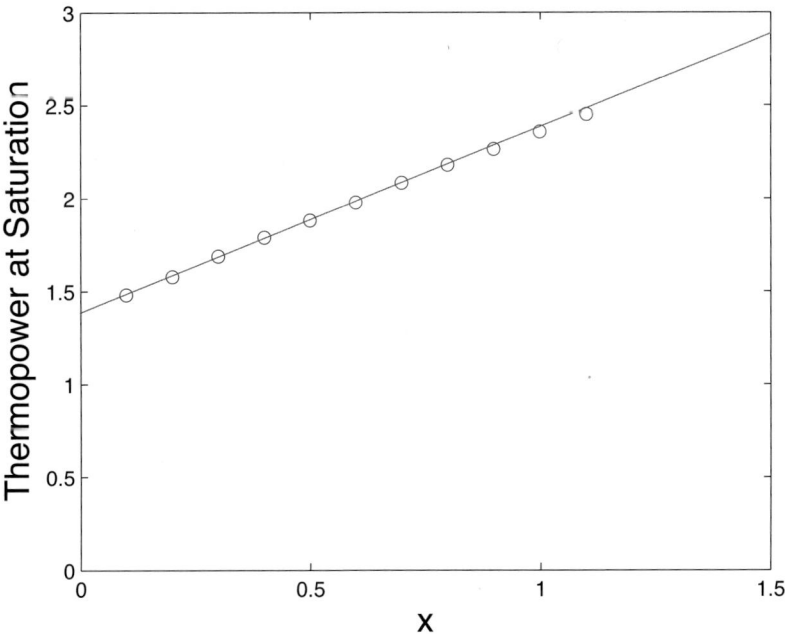

Fig. 9.4. At high temperatures, the thermopower saturates, at a value dependent on x. The plot shows the saturation values extracted by evaluating Eq. (3.10) numerically (see Fig. 9.2), and a linear dependence corresponding to Eq. (3.22).

this, we have to calculate the next order in the Sommerfeld expansion in the nominator Q of Eq. (3.14), to obtain, up to corrections of order $O(T^5)$:

$$Q = \frac{\pi^2}{3}Tx(\mu - E_m)^{x-1} + \frac{7\pi^4}{90}T^3x(x-1)(x-2)(\mu - E_m)^{x-3}. \quad (3.23)$$

This leads to the following low temperature correction of the thermopower:

$$S_{\text{low}} \approx \frac{\pi^2 xT}{3e(\mu - E_m)} + \frac{\pi^4}{45}x(x-1)(x \quad 7)\frac{T^3}{e(\mu - E_m)^3} + O(T^5), \quad (3.24)$$

implying an inflection point for $x < 1$.

4. Brief Discussion of Experiments

Large thermopowers that are linear in the temperature, at least in the metallic regime, were already found in the pioneering extensive work on Cerium sulfide compounds by Cutler and Leavy,[12] and analyzed by Cutler and Mott.[10] It is interesting to address specifically the behavior around the localization transition. An experiment performed on In_2O_{3-x} (both amorphous and crystalline), approaching the Anderson MIT, shortly after Ref. 11, confirmed qualitatively the main features of Eqs. (3.12) and (3.13).[13] Values of S exceeding $100\frac{\mu V}{K}$ were achieved. It should be kept in mind that for a good determination of the critical exponent x, one needs data at low temperatures and small $\mu - E_m$. Data too far from the QPT, which is at both $T = 0$ and $\mu - E_m = 0$, will not be in the critical region and may be sensitive to other effects, as will be discussed later.

It has been customary to use only the low temperature conductivity to determine the critical exponent x. Using similar In_2O_{3-x} samples, the conductivity was extrapolated in Ref. 34 to $T = 0$ and those values were plotted against a control parameter which should be proportional to $\mu - E_m$ when both are small. A value of $x = 0.75 - 0.8$ was found.

It would be much better to use both the above conductivity values and the slopes of $S(T)$ near $T = 0$, according to Eq. (3.15). An even better way to do that would be to eliminate the control parameter $\mu - E_n$ from Eqs. (2.7) and (3.12), getting

$$\frac{dS}{dT}_{T \to 0} \sim [\sigma(T = 0)]^{-1/x}, \tag{4.1}$$

not having to determine the additional parameter $\mu - E_m$ for each case. The data allowed us to effect this only approximately, see Fig. 9.5, giving $x \cong 1 \pm .2$. However getting near the QPT, this procedure appears to be the one of choice.

Obviously, using the two full functions $\sigma(\mu - E_m, T)$ and $S(\mu - E_m, T)$, in the critical region would place even more strict constraints on x. Below, we do this for the existing data, to demonstrate the method. Their scaling works well, but the value of x obtained is not likely to be the real critical value. This is due to a few caveats which will be mentioned.

Figure 9.6 compares the above predictions to the experimental data, taking the exponent x and the values of E_m as fitting parameters. Figure 9.7 shows the approximate data collapse obtained by rescaling the temperature axis of each of the measurements (corresponding to the appropriate value of E_M) as in Eq. (3.11), and the theoretical curve corresponding to $x = 0.1$.

The fit is certainly acceptable. However, the value of $x = 0.1$ is both in disagreement with the previously determined value and impossible for noninteracting electrons, since there $x > 2/3$ in three dimensions. Although, as explained at the end of Sec. 2.2, this constraint may not be valid with interactions, we do not take this last value of x seriously. Since the saturation value was shown be approximately given by a $x + C$, with $C \sim 1.39$ (see

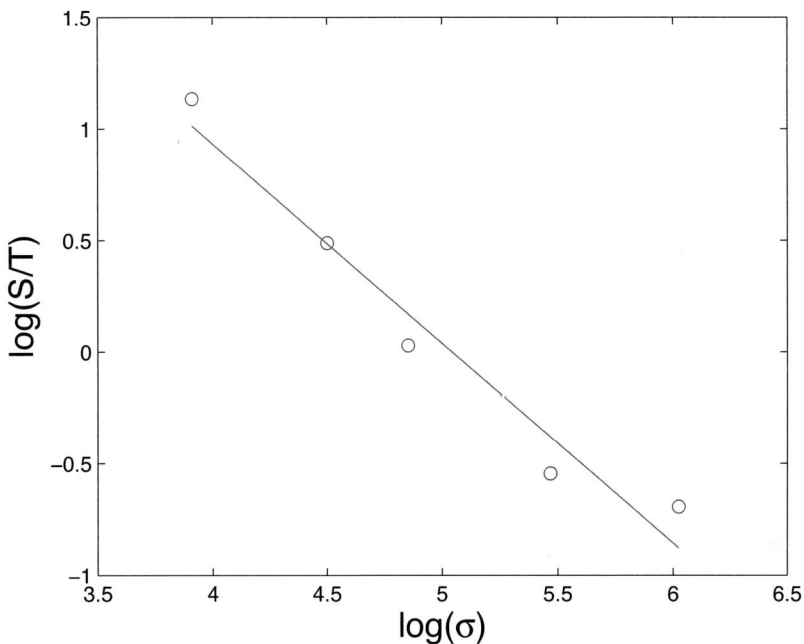

Fig. 9.5. A small set of data, which should be taken as a preliminary to more extensive studies, was used to find the critical exponent x. Equations (2.7) and (3.12) show that the linear (low-temperature) regime of the thermopower has a slope which has an inverse power-law dependence on the distance from the transition. The slope of the log-log plot gives $x \sim 1.1$, with a significant error.

Eq. (3.22)), a change of the thermopower by tens of percents will cause a large change in the deduced value of x. These last fits should be regarded only as demonstrating our recommendations for a possible extension of the analysis of future experimental studies.

At higher temperatures, the analysis will be influenced not only by data that are not in the critical region, but two further relevant physical processes may well come in. Obviously, inelastic scattering (by both phonons and other electrons) will be more important. Moreover, for $T \gtrsim \mu - E_m$, some of the transport will occur via hopping of holes below the mobility edge. Their thermopower might cancel some of the contribution of electrons above μ and thus reduce the thermopower below the values considered here. This clearly needs further treatment.

At any rate, the interactions appear to be strongly relevant and may give unexpected values for x. Measurements at lower temperatures and closer to the transition are clearly needed. Checking simultaneously the behavior of

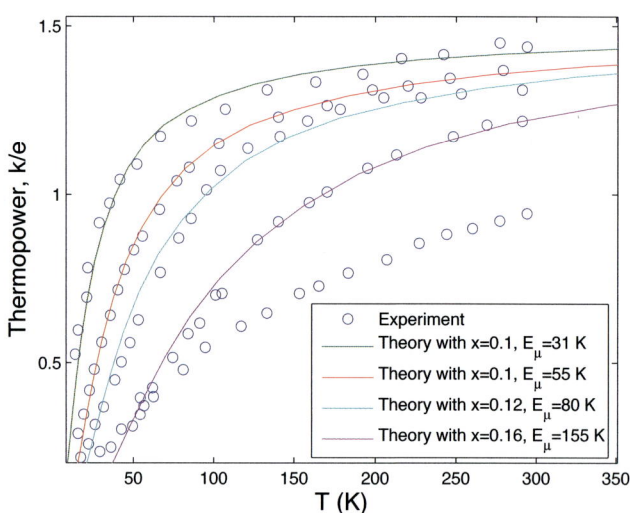

Fig. 9.6. A fit of the theoretical expression of Eq. (3.10) to the experimental data. The data was taken from Ref. 13. The thermopower is measured in units of $k_B/e \sim 86\mu V/K$.

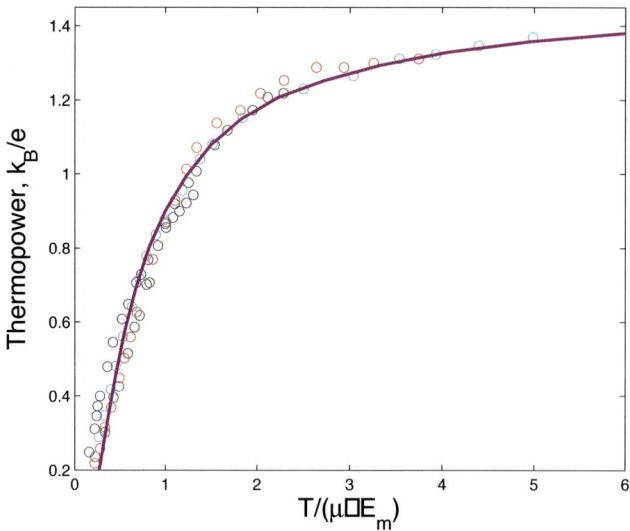

Fig. 9.7. The 4 data sets of Fig. 9.6 which are closest to the transition are shown, scaled to lie on a single universal curve, as function of T/E_m. x was taken as 0.1 and each curve was given a single value of E_m. Possible explanations for the unrealistic value of x deduced in this way, which we do not believe to be correct, are given in Sec. 4.

both the conductivity and the thermopower is suggested as the method of choice for this problem.

It must be mentioned that similar experiments on granular Al did not show the expected behavior. It should be kept in mind that the resistivities needed to approach the MIT for $\sim 100A$ grains are larger[21] than the ones for microscopic disorder. Another relevant issue, which we are going to examine in detail in future work, is that while E_F is in the eV range, all the energies (without electron–electron interactions) relevant for localization are smaller by several orders of magnitude than for microscopic disorder. Thus, the temperature range for the enhanced thermopower might well be in the sub K range, which was not addressed in the experiments. The Coulomb blockade may partially alleviate this, but only when it is operative (not in the metallic regime).

The later experiment of Ref. 35, on Si:P obtained very modest enhancement of S, but were stated to have been dominated by effective magnetic impurities, which are known to be strongly relevant for the Anderson transition (e.g. eliminating the weak localization contributions). All these issues have to be clarified.

5. Concluding Remarks

In this paper we have shown how to include the thermal current in the Thouless scaling picture of conduction in disordered systems. Expressions were given for the 2×2 matrix of longitudinal thermoelectric coefficients, in terms of $\sigma_0(E)$, the $T = 0$ conductivity of the system were its Fermi level fixed at the energy E. The Onsager relations were shown to hold within this formulation. For the usual critical behavior of $\sigma_0(E)$, given by Eq. (1.1), these behaviors were analyzed for an arbitrary ratio of T to the distance to the mobility edge. They were shown to satisfy scaling relationships which were confirmed numerically along with their limiting behaviors.

It was shown how the conductivity and thermopower data close to the Anderson QPT should be analyzed simultaneously to yield a better estimate of the critical exponent x than the determination based on $\sigma(T)$ alone. This was done for the low-temperature limits of existing data[13,34] on the transition in In_2O_{3-x}, giving already a good ballpark estimate of x, namely $x \sim 1.1$. The data going to higher (probably too high) temperatures do scale and collapse according to Eq. (3.11), but the resulting values of x appear to be too small. One may speculate that this is due to interaction effects, but we prefer to postpone this to after having done this analysis with lower temperature data closer to the transition.

Similar experiments on granular Al do not produce a large and interesting thermopower as above. This is certainly a matter for concern. The explanation might well be due to the smaller microscopic conductivity scales ($\frac{e^2}{\hbar R}$, R being the grain size), or to the different energy scales relevant for these systems.[21] Alternatively, the inelastic scattering, not treated in this paper, may be relevant as well.

The sharp and asymmetric behavior of $\sigma_0(E)$ near the transition is ideal for getting large thermopowers. The predicted values approach $\sim 200 \frac{\mu V}{K}$. While the experimental results[13] on In_2O_{3-x} are smaller by $\sim 40\%$, this is still encouraging. Were it possible to increase these values say by phonon drag effects, this might even become applicable. Clearly, a treatment of the effects of inelastic scattering on the thermopower is called for, especially including the hopping conductivity regime.

Appendix — The Heat Carried by a Transport Quasiparticle

To make this analysis useful also for heat transport by phonons, etc., we display the equations for both fermions and bosons. The entropy associated with a state of a given equilibrium system at energy E, having a population f is

$$S_E = -k_B[f ln f + (1 \pm f) ln (1 \pm f)], \qquad (A.1)$$

where the upper (lower) sign is for bosons (fermions). When the population f changes with time, the change of S_E with time is

$$\dot{S}_E = -k_B \dot{f} ln \frac{f}{1 \pm f} = -\frac{E}{T} \dot{f}, \qquad (A.2)$$

where to get the last equality we used the equilibrium $f = (e^{\frac{E}{T}} \mp 1)^{-1}$. The outgoing heat current $T\dot{S}$ is the time derivative of the population times the excitation energy. Thus, each particle leaving the system carries "on its back" an amount of heat E which is its energy (measured from μ). Summing $\dot{S}_E$ over all energies shows that the outgoing heat current is given by the outgoing particle current where the contribution of each energy is multiplied by $E - \mu$.

It should be noted that the equality of the amounts of $E - \mu$ and $-TS$ carried by the excitation implies that the relevant free energy does not change when the (quasi)particle moves to another system which is in equilibrium with the first one. This is true for equilibrium fluctuations and also for linear response transport ($V \rightarrow 0$ and $\Delta T \rightarrow 0$), between the two systems.

As remarked, the result that the heat carried by an electron is given by its energy measured from the chemical potential, μ, is valid also for bosons. As a small application, one can easily calculate the net thermal current carried by a single-mode phonon/photon waveguide fed by thermal baths at $T \pm \Delta T/2$. The result is a thermal conductance of $k_B^2 T \pi/(6\hbar)$ (per mode), with no reflections. This agrees with the result of Ref. 36. The sound/light velocity cancels between the excitation velocity and its (1D) DOS, exactly as in the electronic case. This is why this result and the one based on the Wiedemann–Franz law for electrons are of the same order of magnitude. That their numerical factors are equal is just by chance. With reflections due to disorder, once the waveguide's length is comparable to or larger than the localization length (mean free path for a single mode), its thermal conductance drops markedly.

Acknowledgments

We thank Uri Sivan, Ora Entin–Wohlman, Amnon Aharony and the late C. Herring for discussions. Special thanks are due to Zvi Ovadyahu for instructive discussions and for making his data available to us. This work was supported by the German Federal Ministry of Education and Research (BMBF) within the framework of the German–Israeli project cooperation (DIP), by the Humboldt Foundation, by the US-Israel Binational Science Foundation (BSF), by the Israel Science Foundation (ISF) and by its Converging Technologies Program. YI is grateful to the Pacific Institute of Theoertical Physics (PITP) for its hospitality when some of this work was done.

References

1. P. W. Anderson, *Phys. Rev.* **109**, 1492 (1958).
2. N. F. Mott, *Adv. Phys.* **16**, 49 (1967); *Phil. Mag.* **17**, 1259 (1968).
3. A. Shalgi and Y. Imry, in *Mesoscopic Quantum Physics*, Les Houches Session LXI, eds. E. Akkermans, G. Montambaux and J.-L. Pichard (North-Holland, Amsterdam, 1995), p. 329.
4. E. Abrahams, P. W. Anderson, D. C. Licciardello and T. V. Ramakrishnan, *Phys. Rev. Lett.* **42**, 673 (1979).
5. A. M. Finkelstein, *Disordered Electron Liquid with Interactions*, this volume; *Sov. Phys. J. Exp. Theor. Phys.* **57**, 97 (1983); A. M. Finkelstein, *Springer Proc. Phys.* **28**, 230 (1989); A. Punnoose and A. M. Finkelstein, *Science* **310**, 289 (2005); K. Michaeli and A. M. Finkelstein, *Europhys Lett.* **86**, 27007 (2009).

6. Were one lucky to have very weak interaction in some system, the crossover region to the interaction-dominated behavior would be small and approximate information on the noninteracting critical behavior could be obtained from experiments on this system.

7. B. L. Altshuler and A. G. Aronov, *Sov. Phys.* **50**, 968 (1980).

8. C. W. J. Beenakker and A. A. M. Staring, *Phys. Rev. B* **46**, 9667 (1992).

9. A. M. Lunde, K. J. Flensberg and L. I. Glazman, *Phys. Rev. Lett.* **97**, 256802 (2005).

10. M. Cutler and N. F. Mott, *Phys Rev.* **181**, 1336 (1969).

11. U. Sivan and Y. Imry, *Phys. Rev. B* **33**, 551 (1986).

12. M. Cutler and J. F. Leavy, *Phys. Rev.* **133**, A1153 (1964).

13. Z. Ovadyahu, *J. Phys. C: Solid State Phys.* **19**, 5187 (1986).

14. T. C. Harman and J. M. Honig, *Thermoelectric and Thermomagnetic Effects and Applications* (McGraw-Hill, New York, 1967).

15. J. Bardeen, *Phys. Rev. Lett.* **6**, 57 (1961).

16. W. A. Harrison, *Solid State Theory* (McGraw Hill, New York, 1970).

17. D. J. Thouless, *Phys. Rev. Lett.* **39**, 1167 (1977).

18. This symmetry is valid also when there is no symmetry between the blocks (say, when $N_l(E) \neq N_r(E)$). We shall however mainly consider here the case where the blocks are of the same size ($N_l(E) = N_r(E)$), but they usually do have different defect configurations.

19. G. Czycholl and B. Kramer, *Solid State Commun.* **32**, 945 (1979).

20. D. J. Thouless and S. Kirkpatrick, *J. Phys. C: Solid State Phys.* **14**, 235 (1981).

21. Y. Imry, *Introduction to Mesoscopic Physics*, 2nd edn. (Oxford, 2002).

22. A. Amir, Y. Oreg and Y. Imry, *Phys. Rev. A* **77**, 050101 (2008).

23. J. T. Edwards, and D. J. Thouless, *J. Phys. C: Solid State Phys.* **5**, 807 (1972).

24. B. Kramer, A. MacKinnon, T. Ohtsuki and K. Slevin, *Finite-Size Scaling Analysis of the Anderson Transition*, this volume.

25. D. S. Fisher and P. A. Lee, *Phys. Rev. B.* **23**, 6851 (1981).

26. A. B. Harris, *J. Phys. C: Solid State Phys.* **7**, 1671 (1974).

27. N. F. Mott, *Phil Mag.* **13**, 989 (1966); N. F. Mott, *Phil. Mag.* **19**, 835 (1969); N. F. Mott, *Phil. Mag.* **22**, 7 (1970).

28. This is because this inequality assures that the characterisic length, ξ, grows quickly enough near the transition to average out the disorder fluctuations between different volumes of ξ^d. This implies that no intrinsic inhomogeneity is generated when the transition is approached.

29. H. B. Callen, *An Introduction to Thermodynamics and Thermo-Statistics* (Wiley, New York, 1985).

30. J. M. Ziman *Principles of the Theory of Solids* (Cambridge, 1969).

31. L. Onsager, *Phys. Rev.* **38**, 2265 (1931).

32. L. D. Landau and E. M. Lifshitz, *Statistical Physics*, Part 1 (Pergamon, Oxford, 1980).

33. J. E. Enderby and A. C. Barnes, *Phys Rev. B* **49**, 5062 (1994).

34. E. Tousson and Z. Ovadyahu, *Phys. Rev. B* **38**, 12290 (1988).

35. M. Lakner and H. V. Löhneysen, *Phys. Rev. Lett.* **70**, 3475 (1993).

36. L. G. C. Rego and G. Kirczenow, *Phys. Rev. Lett.* **81**, 232 (1998).

Chapter 10

LOCALIZATION AND THE METAL–INSULATOR TRANSITION — EXPERIMENTAL OBSERVATIONS

R. C. Dynes

Department of Physics, University of California San Diego,
La Jolla Ca. 92093, USA
rdynes@ucsd.edu

This article describes a series of experiments, mostly performed by the author and his collaborators over a period of 15 years, attempting to understand with the increase of disorder, the evolution from a Fermi liquid, to a state of localized electrons. During this period we relied heavily on theoretical advances, and the experiments were designed to test those theoretical models and challenge them in the regime of a highly correlated system. Experimentally, we were able to continuously tune through this regime into the insulating state and this allows measurements all the way from weak localization in two dimensions to strong localization and the insulating state. My efforts were focused mainly on electron transport and tunneling and will be described in this article. For this author, the work of my colleagues (and myself) during this period changed the way I think about electron transport.

1. Introduction

Landau liquid theory, or even Drude theory has been remarkably successful in describing electron transport in conventional metals and semiconductors. With increased disorder it has been common to describe electronic transport as a free electron liquid using straightforward band theory with weak interactions and in the relaxation-time approximation. Increasing disorder results in electronic mean free path reductions all the way to as much disorder as there is in an amorphous material (when mean free path λ is of order a_0, the interatomic spacing). Mott and his colleagues[1] recognized that in this highly disordered limit, these simple descriptions are not likely to be valid and suggested that near this point it did not make any sense to think of a metal as a liquid of free electrons. This reasoning led him to suggest the concept of a minimum metallic conductivity.[2] At some value of conductivity where the electron mean free path became comparable to the equivalent electronic wave length, the extended wave description of an electron no longer made physical sense, and the electron would localize. In two dimensions the idea was particularly appealing as the 2d conductivity in a simple theory is given by

$$\sigma = 2ne^2\tau/m_e = (e^2/h)k_F\lambda, \tag{1.1}$$

where $n = \pi k_F^2/(2\pi)^3$ is the number of electrons of a given spin per unit area, τ is the relaxation time, λ is the mean free path, and k_F the electronic wave number. Mott then invoked the Ioffe–Regel condition[3] that it did not make any physical sense to have the product $k_F\lambda$ assume a value $<\sim 1$ and he obtained a value in two dimensions for the minimum metallic conductivity

$$\sigma = e^2/h. \tag{1.2}$$

In three dimensions, there is an extra factor of k_F which comes from an additional dimension in the expression for n and the conductivity has a material dependent length. Hence the expression for the minimum metallic conductivity has a material dependent length in it. It is this simple physically appealing picture which launched the experiments that are described in this article. The minimum metallic conductivity in two dimensions is 3.86×10^{-5} mhos or a resistance per square of 26000 ohms/square. The appealing aspect of this idea in two dimensions is that it is material independent. A simple ratio of fundamental constants allows simple tests of the concept with a variety of materials parameters. In the course of the early experiments described in this paper,[4] the landmark work of Abrahams, Anderson, Licciardello and Ramakrishnan[5] was published. This paper changed the way we think about electrical transport in the dirty limit. I will describe how that work influenced the author's subsequent investigations in two and three dimensions and the insight that resulted.

2. Two Dimensions

Following the earlier work of Strongin and collaborators,[6] we chose to study transport in ultra-thin low temperature quench-condensed metallic films. We chose to study both normal metals and superconductors as the question of the existence of superconductivity and whether it existed in a two-dimensional metallic film with a conductivity below the Mott limit was (and still is!) an interesting question. Some of the results of our original investigation are shown in Fig. 10.1. The sheet resistance of these quench-condensed films drops exponentially with increasing average film thickness.

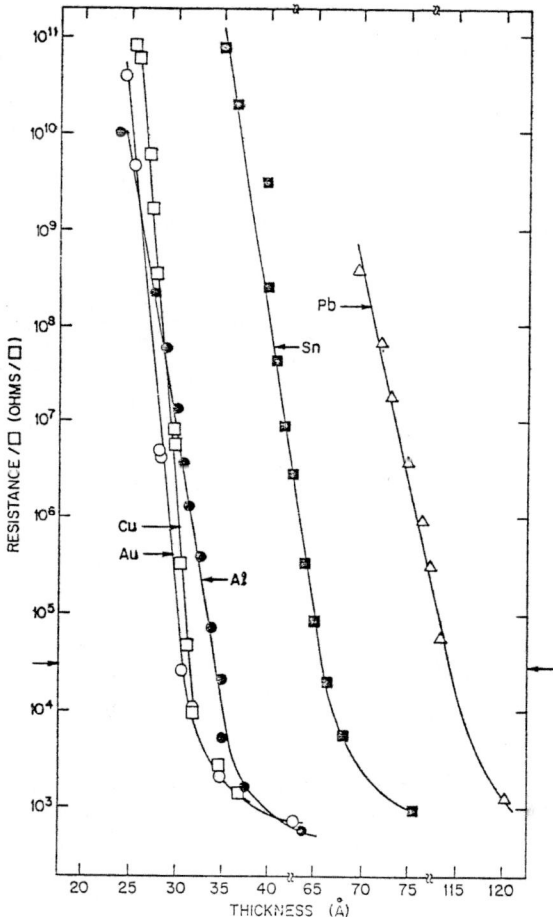

Fig. 10.1. Sheet resistance of low temperature quench condensed metal films as a function of average thickness. Reprinted (Fig. 1) with permission from *Phys. Rev. Lett.* **40**, 479 (1978). © American Physical Society.

The films produced in this experiment are disordered but granular in nature with the resistivity dominated by tunneling or hopping between grains. The grain size varied from material to material and eventually the grains joined together. The variation of grain size explains the different average thicknesses where the crossover from exponential behavior to (1/thickness) occurs in Fig. 10.1. We later learned how to grow them statistically smooth and these films showed similar transport results. The arrows on each side of the graph correspond to the value $R = h/e^2$. Interestingly, the resistance crosses over from exponential behavior to (1/thickness) behavior in this regime, independent of the material. The temperature dependence of

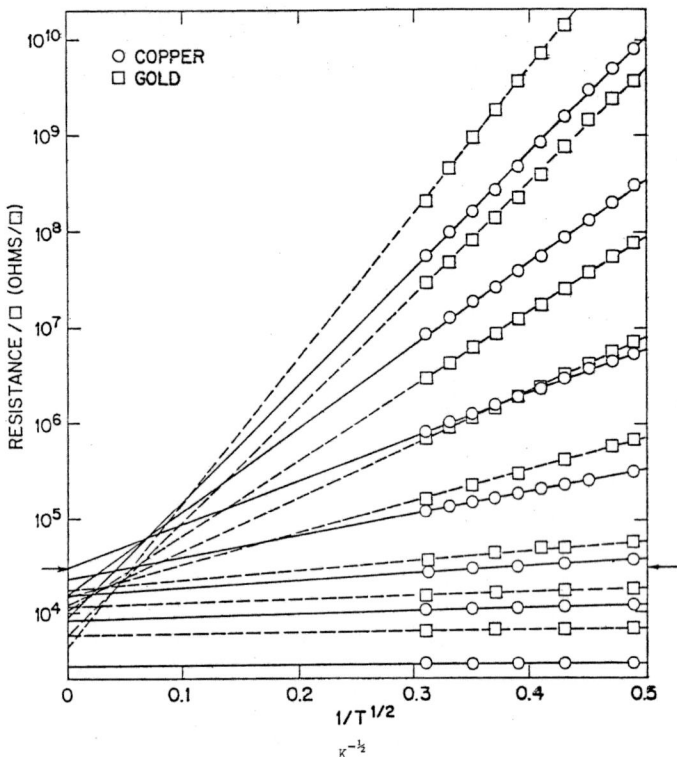

Fig. 10.2. Sheet resistance vs $1/T^{1/2}$. Note that the extrapolation of this exponential behavior all converges at infinite T to the Mott resistance. Reprinted (Fig. 2) with permission from *Phys. Rev. Lett.* **40**, 479 (1978). © American Physical Society.

the resistance/square also indicates a crossover in the vicinity of this Mott resistance. This behavior is shown in Fig. 10.2. Here the temperature dependence follows a $\exp(-1/T^{1/2})$ behavior, clearly localized and indicative of variable range hopping.[7] For resistances below the Mott number (again indicated by arrows on the side of the figures), the dependence appears flat and no longer extrapolates to the Mott number at high T and insulating at low T.

These data suggested to us that there was a clear transition from localized (insulating) to metallic behavior *independent* of the material. Furthermore, for the materials studied in Fig. 10.1 that were superconductors, we discovered that superconductivity appeared for resistances below the Mott number and the materials were insulating for $R >$ Mott resistance. This is illustrated in Fig. 10.3.

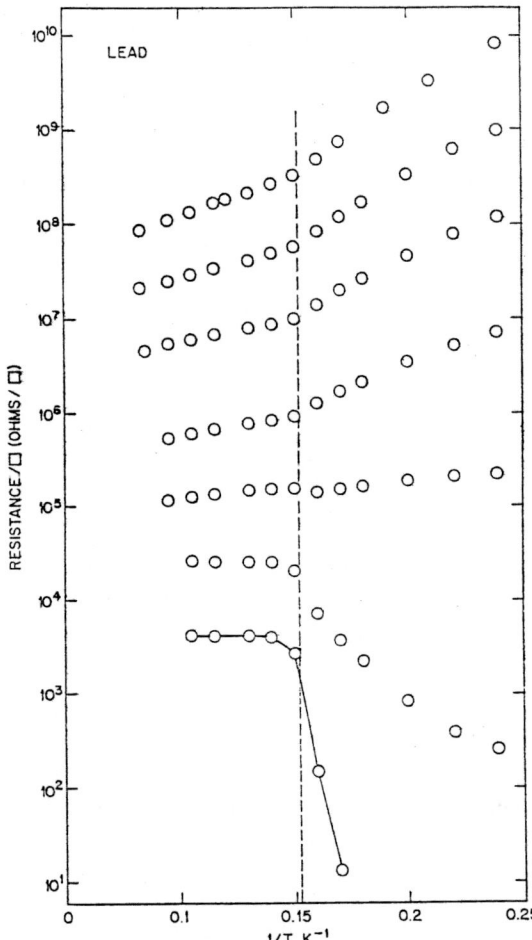

Fig. 10.3. Resistance of quench condensed lead (Pb) films as a function of T. Note that for films with $R <$ Mott number, they become superconducting at the T_c for Pb. For $R >$ Mott number, they are insulating. Reprinted (Fig. 4) with permission from *Phys. Rev. Lett.* **40**, 479 (1978). © American Physical Society.

These results and a variety of others consistent with those shown here led us to conclude that electrons were localized for those two-dimensional metals that had a sheet resistance at finite T above h/e^2 and that there was a transition at this resistance from metallic (and even superconducting) to insulating. The even more remarkable observation was that this result was independent of the choice of material of study.

Before the ink was dry on this paper, Abrahams, Anderson, Licciardello and Ramakrishnan published their remarkable scaling argument[5] that

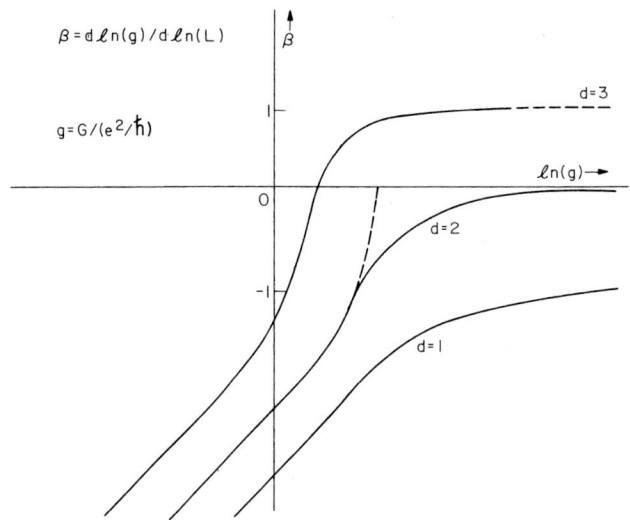

Fig. 10.4.   The central result from the work of Abrahams, Anderson, Licciardello and Ramakrishnan. The parameter β scales the conductance with length for different dimensionality d. For two dimensions the trajectory never crosses (or meets) the x-axis and that means that a two-dimensional electronic system is never metallic. The crossover at the Mott number is from exponential to logarithmic (weak) localization. For dimension greater than 2 ($d > 2$) there is a continuous metal–insulator transition. Reprinted (Fig. 1) with permission from E. Abrahams, P. W. Anderson, D. C. Licciardello and T. V. Ramakrishnan, *Phys. Rev. Lett.* **42**, 673 (1979). © American Physical Society.

indicated that the physics in two dimensions was much more subtle and that in the vicinity of this Mott minimum metallic conductivity, the crossover was not from insulator to metal but from exponentially localized to very weakly logarithmically localized. Their now-famous figure that summarized their conclusions in one, two and three dimensions is reproduced in Fig. 10.4. In the very interesting case of two dimensions, Mott imagined the trajectory for $d = 2$ would meet the horizontal axis at the minimum metallic conductivity. This scaling result in Fig. 10.4 concluded that the $d = 2$ trajectory approaches the axis logarithmically but never intersects the g-axis.

While the results of their work was not inconsistent with the data of Figs. 10.1–10.3, it led us to look more carefully at the resistive behavior in the region $R < h/e^2$ where we earlier had concluded that two dimensional metallic behavior was observed. A more careful look at Fig. 10.2 on a linear (not exponential) scale showed that indeed there was a weak dependence on temperature which was not metallic but increased logarthmically

n-channel MOSFET

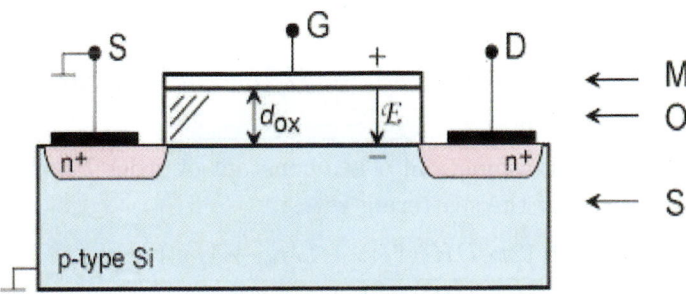

Fig. 10.5. A silicon n-channel MOSFET. A gate (G) voltage induces a charge under the oxide electrically connecting the source (S) to the drain (D). The carrier density can be continuously tuned by the gate voltage thus allowing a wide range of sheet resistances in a single device. This allows the range of resistances illustrated in Fig. 10.1 with the tuning of the gate voltage.

with reducing temperature. We later showed from magnetoresistance measurements in these two dimensional metallic films, that they were, indeed weakly localized.

This observation led us to an extensive study of a different type of two dimensional electronic conductor, the inversion layer of a 2d silicon MOSFET[8,9] schematically illustrated in Fig. 10.5. Several other research groups were engaged in parallel (and competitive) efforts at the same time[10,11] and there was eventually very little disagreement in the results. As illustrated in Fig. 10.5, the range of two-dimensional resistances allowed by this device permitted studies like those in Fig. 10.1 on a very different type of two dimensional conductor and the results were remarkably similar to those "metal" films. Data like that shown in Fig. 10.1 were generated in a system with a very different electron density and mobility and yet the results were similar. A transition from exponential localization to what appeared to be metallic (we know now it was weak logarithmic localization) occurred again at the Mott number $R_{\text{Mott}} \sim h/e^2$.

The weak localization implied a long localization length and this implied an orbital magnetoresistance on magnetic field scales that were much smaller than any other effects (Zeeman splitting, for example). Several theoretical calculations implied that the localization length scales of interest were ~ 0.1 to 1 microns which corresponds to fields of 1–100 gauss. Lee and Ramakrishnan,[12] and Hikami, Larkin and Nagaoka[13] calculated the magnetoresistance using this localization theory and in the weak localization limit ($k_F \lambda \gg 1$)

the change in conductance with perpendicular H field and temperature is given by

$$\delta\sigma(H,T) = -2(\alpha e^2/\pi h)\{[\psi(a_1 + 1/2) - \psi(a_2 + 1/2)]$$
$$+ 1/2\,[\psi(a_3 + 1/2) - \psi(a_4 + 1/2)]\} + (\alpha e^2/\pi h)\ln(a_1 a_3^{1/2}/a_2 a_4^{1/2}). \tag{2.1}$$

Here, ψ is the digamma function, α is a constant of order 1 and the a_n's are linear combinations of the scattering rates:

$$a_1 = hc/(12\pi eDH)(1/\tau_e + 1/\tau_{so} + 1/\tau_s),$$
$$a_2 = hc/(12\pi eDH)[4/3(1/\tau_{so}) + 2/3(1/\tau_s) + 1/\tau_i],$$
$$a_3 = hc/(12\pi eDH)(1/\tau_i + 2/\tau_s),$$
$$a_4 = a_2.$$

Here, D is the electron diffusivity, $1/\tau_e$ is the elastic scattering rate, $1/\tau_s$ is the spin flip scattering rate, $1/\tau_i$ is the inelastic scattering rate, and $1/\tau_{so}$ is the spin orbit scattering rate. Interestingly enough, if $1/\tau_e$ dominates the scattering, logarithmic negative magnetoresistance results signaling logarithmic localization. Inelastic and spin-flip scattering will cause decoherence and thus delocalization when these processes begin to dominate. Spin-orbit scattering reverses the sign of the magnetoresistance and causes antilocalization.[14]

As a result, careful magnetoresistance measurements as a function of field H and temperature T can measure these scattering rates. A sampling of the

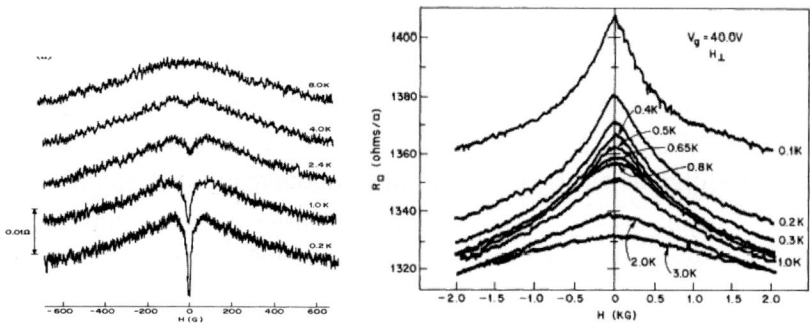

Fig. 10.6. Perpendicular magnetoresistance in a thin Mg film (22.3 ohms/square) and a 2d electron layer of a MOSFET. Several scattering rates can be determined from data like this. It is remarkable that such dissimilar systems show logarithmic localization and the amplitude scales with R/square. Reprinted (Figs. 1a and 2) with permission from *Phys. Rev. B* **29**, 3694 (1984); *Phys. Rev. B* **26**, 773 (1982). © American Physical Society.

results acquired in measurements on two very different systems is shown in Fig. 10.6. Here we show the magnetoresistance[15] of a thin Mg film (22.3 ohms/sq) and a 2d electron gas on a MOSFET.[9] While they are two very different systems, they show similar magnetoresistance behavior. The effects of the various scattering processes appear in these data. The elastic scattering determines the overall resistance. The temperature dependent inelastic processes cause an overall rounding of the magnetoresistance around $H = 0$. Raising the temperature causes more frequent inelastic scattering processes and delocalizes the electrons. The distance an electron diffuses between inelastic scattering processes becomes shorter with increasing T and this is observed by a less sensitive magnetoresistance at low fields. Lower fields probe longer distances and with increasing T, the inelastic diffusion length becomes shorter and the magnetoresistance disappears. Fitting expression (2.1) to these curves allows a determination of the inelastic scattering rates and hence the length scale of the localization effects. Furthermore, in the Mg magnetoresistance shown in Fig. 10.6, we see at the lowest fields and temperatures, the inverting effect of spin-orbit scattering. The H field dependence of the magnetoresistance changes sign causing a narrow dip around $H = 0$. This occurs when the temperature is lowered sufficiently that the spin-orbit scattering dominates over the inelastic scattering rate. The length scale over which this dominates is then measured directly by the H field.

As a result of measurements of this nature and fits to Eq. (2.1), the inelastic scattering rates, spin-orbit scattering rates and spin-flip scattering rates have been measured in a variety of systems and we have learned much about the length scales over which diffusing electrons have maintained their coherence before inelastic or spin-flip scattering occurs. An example of results of such fits is shown in Fig. 10.7, where we show inelastic scattering times for a high mobility Si MOSFET as a function of temperature for various gate voltages (electron densities).[9] From these studies it was determined that the dominant inelastic scattering process was electron–electron scattering and the theoretical predictions of the scattering rate were confirmed. While the elastic scattering rate was the dominant scattering, these careful magnetoresistance studies allowed a quantitative determination of the inelastic rate. Similar studies on a variety of two dimensional systems allowed a determination of inelastic, spin orbit and spin-flip scattering rates. For example, in Fig. 10.7 for a thin Mg film, the crossover from inelastic scattering to spin-orbit scattering can be seen with lowering temperature and thus allows a determination of both $1/\tau_i$ and $1/\tau_{so}$. Some very elegant experiments by Bergmann[14] where he "tuned" the spin-orbit scattering by depositing a

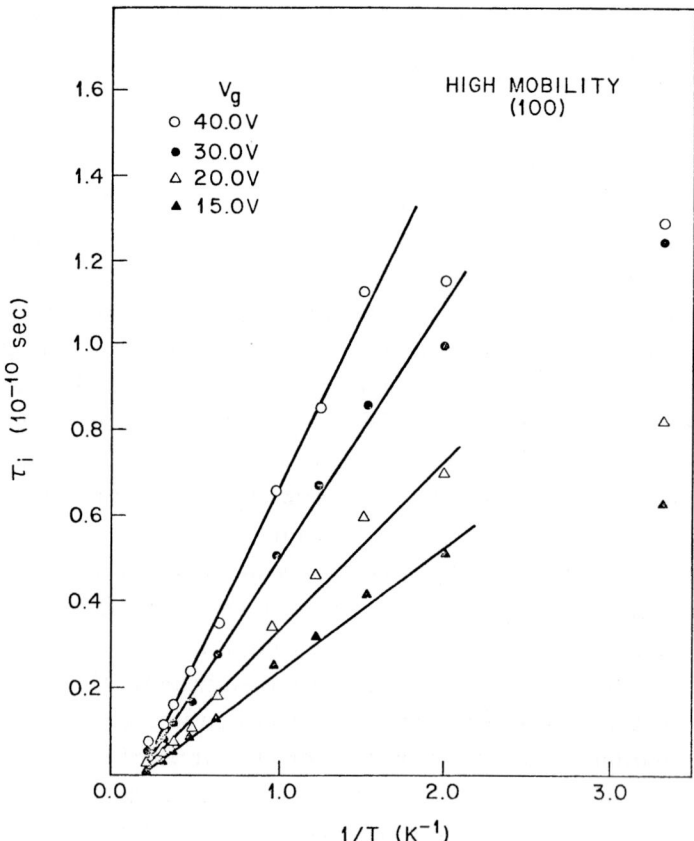

Fig. 10.7.   Inelastic scattering times determined by a fit to Eq. (2.1) from data from a Si MOSFET like that illustrated in Fig. 10.6. Reprinted (Fig. 6) with permission from *Phys. Rev. B* **26**, 773 (1982). © American Physical Society.

low concentration of high Z material on the surface of such films made the case compelling. In addition, by depositing a magnetic, spin-flip material on the surface,[18] he showed that the long range coherence can be destroyed.

This has given us a much broader insight into the nature of conduction in disordered materials and a quantitative spectroscopy in the determination of scattering rates. Confirmation that these are all orbital effects required simply the equivalent magnetoresistance measurements in a parallel magnetic field where these effects are not observed.[9]

A second interesting effect is observed in the data shown in Fig. 10.3. These data illustrate a superconducting-insulator transition in the same vicinity of the Mott number. The work briefly summarized above shows that in the two dimensional case, the electrons are localized, albeit weakly for resistances less than h/e^2 . The pairing interaction for superconductivity has a length scale of the superconducting coherence length which can be shorter than the localization length thus allowing the long range coherence of the superconductor to dominate. Hence, while a two dimensional metallic film may have weakly localized electrons, the superconducting pairing dominates in the weak localization regime and superconductivity results.[4]

3. Three Dimensions

Mott's original thoughts about the existence of a minimum metallic conductivity led to a variety of experiments to study the critical behavior in systems where a metal–insulator transition was understood to occur. For decades it has been well known that impurity doping of semiconductors at low doping concentration results in bound electrons (or holes). Depending upon details of the host material and the donor, the binding energy can be large (comparable to the band gap) or quite small a few meV). It is also well known that at some donor concentration, the bound states overlap and a transition occurs to a metallic state (at $T = 0$ K, the conductivity is non-zero). Mott's notion of a minimum metallic conductivity suggested that with increasing doping, the material would transition from being insulating ($\sigma = 0$ at $T = 0$ K) to metallic ($\sigma = \sigma_{\min}$ at $T = 0$ K). Referring to Fig. 10.4, the metal–insulator transition is expected to occur for dimensionality $d > 2$ and in the trajectory for $d > 2$, there is a metal–insulator transition (at g_c in the figure) but no suggestion of a minimum metallic conductivity below which the system abruptly transitions to the insulating state.

I will describe experiments designed to search for this minimum metallic conductivity. None found an abrupt drop in conductivity at that point but instead found that the conductivity $\sigma(T = 0)$ went continuously to 0 as a function of a critical variable (concentration of donors in two cases and disorder in the third) and then transitioned continuously to the insulating state. The three examples, while demonstrating detailed differences on approach to $\sigma(0) = 0$, showed a continuous transition. The systems are very different but their similarities made a convincing argument for the lack of a conductance jump to zero at $\sigma_{\min}$. The three systems are P doped crystalline Si,[19] the amorphous glass $Nb_{1-x}Si_x$[20] and magnetic field tuning of $Gd_{1-x}Si_x$.[21] In three dimensions, simple Drude considerations expect

$$\sigma = 2ne^2\tau/m_e = (\pi^2/3)(k_F^2 e^2 \lambda). \tag{3.1}$$

Unlike in the two dimensional case, Eq. (1.2), when you invoke the Ioffe–Regel condition ($k_F\lambda \sim 1$), Mott's minimum metallic conductivity has a

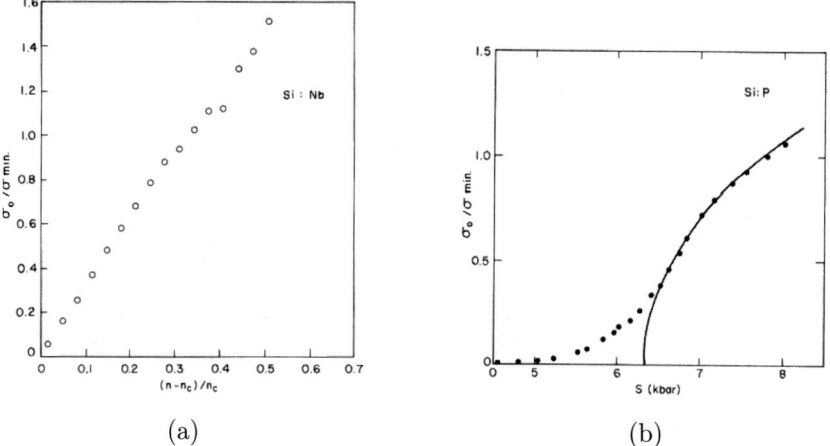

(a)   (b)

Fig. 10.8. $T = 0$ K conductivity $\sigma_0/\sigma_{\min}$ in (a) amorphous Nb:Si and (b) crystalline doped Si:P. In the Nb:Si case, the variable is the excursion from the critical concentration. In the Si:P case it is uniaxial pressure. Both cases show $\sigma(T = 0)$ continuously approaching zero conductivity well below the Mott minimum metallic conductivity $\sigma_{\min}$. Reprinted (Fig. 2) with permission from M. A. Paalanen, T. F. Rosenbaum, G. A. Thomas and R. N. Bhatt, *Phys. Rev. Lett.* **48**, 1284 (1982). © American Physical Society.

material dependent quantity (k_F) remaining in the expression for the Mott limit $\sigma_{\min}$. In the three experiments discussed here (Si:P, Nb$_{1-x}$Si$_x$ and Gd$_{1-x}$Si$_x$), the quantity k_F is expected to be vastly different. The metal–insulator transition occurs in Si:P at a P concentration of $\sim 3.4 \times 10^{18}$/cc while in the amorphous Nb:Si case a niobium concentration of $\sim 11\%$ is the critical concentration and Gd:Si is similar. This results in 1–2 orders of magnitude difference in the k_F for the expected Mott minimum metallic conductivity in the three systems.. The results of systematic studies in two of these cases are shown in Fig. 10.8. Here we show the $T = 0$ (extrapolated) values for $\sigma(0)$ from low temperature measurements. While the behavior of the conductivity $\sigma(0)$ as a function of concentration is different, both data sets show compelling evidence for a continuous transition to the insulating state. The functional form of the critical behavior near the transition has been a subject of research and debate for several years but the significant result in Fig. 10.8 clearly shows measured conductivities well below the Mott limit ($\sigma_{\min}$) all the way to the metal–insulator transition as anticipated in Fig. 10.4.

In the studies of the Nb:Si system, in addition to the low temperature transport studies, we were able to investigate the electronic density

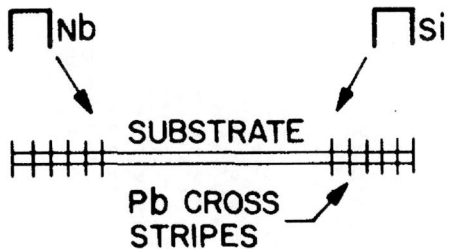

Fig. 10.9.   Geometry of fabrication of phase-spread Nb:Si with tunnel junction configurations.

of states[20] in this critical regime. The Nb:Si alloys were fabricated by a technique where the films of Nb:Si were low temperature deposited by two sputter guns (one for Nb, one for Si) spatially separated in such a way that across a 10 cm substrate we could vary the relative concentrations substantially. With practice, we could position the concentrations such that the spatial variation spanned the metal–insulator transition and we could then fabricate tunnel junctions on the very samples that we performed transport studies. A schematic of the sample and tunnel junction configuration is shown in Fig. 10.9. The tunnel junctions (the cross Pb stripes) served as voltage probes for the transport measurements as well as tunnel probes at each point on the phase spread.

The first suggestion that in this highly disordered regime the tunneling density of states could yield insight into the nature of strongly interacting electrons came from the paper of Altshuler and Aronov.[22] In this work, the authors considered the properties of a disordered electronic system in the Mott regime. They showed that the electron–electron interaction in this low diffusivity regime resulted in an anomaly in the tunneling density of states, symmetric about the Fermi energy, with a $(eV)^{1/2}$ dependence. This work also implied that commensurate with this $(eV)^{1/2}$ dependence there would be a $T^{1/2}$ dependence to the conductance in the critical regime. In temperature dependence studies, both the Si:P and amorphous Nb:Si systems showed a conductivity

$$\sigma(T) = \sigma_0 + \sigma_1 T^{1/2}. \tag{3.2}$$

The consistency of this observation allowed careful fits of the temperature dependence and a reliable determination of σ_0.

The results of tunneling measurements along the sample shown schematically in Fig. 10.9 are shown in Fig. 10.10. In this figure, it can be clearly observed that the tunneling density of states follows a $N(E) = N(0) + N_1 E^{1/2}$

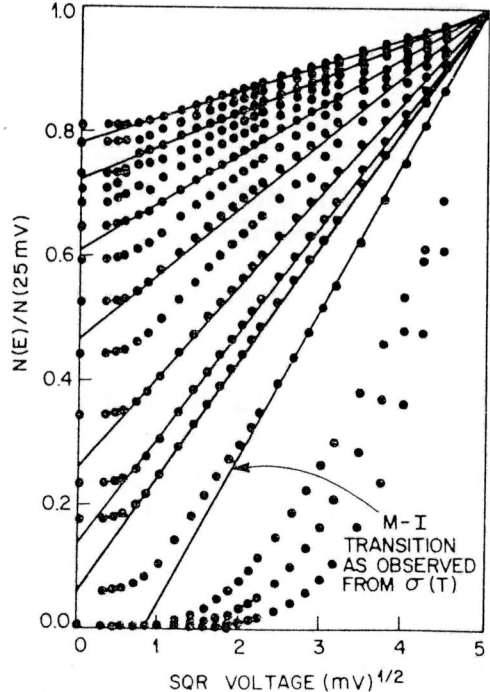

Fig. 10.10.   Tunneling density of states for tunnel junctions along the sample shown in Fig. 10.9. The straight lines represent a $(V)^{1/2}$ dependence. The extrapolated zero bias density of states reduces systematically to zero at the metal–insulator transition. Reprinted (Fig. 2) with permission from *Phys. Rev. Lett.* **50**, 743 (1983). © American Physical Society.

behavior. Approaching the metal–insulator transition, the Coulomb effects become more dominant and the $E^{1/2}$ behavior becomes stronger. At the point where the transport measurements (i.e. σ_0) signal the metal–insulator transition, $N(0)$ goes to zero. Apparently the highly correlated system gains total energy by moving electronic states away from the Fermi energy until none remain, resulting in an insulator. At non-zero temperature, measurements can be extended some distance into the insulating state as long as the tunnel junction resistance is much greater than the resistance of the Nb:Si in the spatial region of the measurement. On the insulating side of the metal–insulator transition, a soft Coulomb gap $[N(E) \sim E^2]$ is visible and increases in size as we move further away on the insulating side of the transition. The coincidence of the collapse of the availability of extended states at the Fermi level and the metal–insulator transition is illustrated in Fig. 10.11, where within the experimental resolution, the conductivity $\sigma(0)$

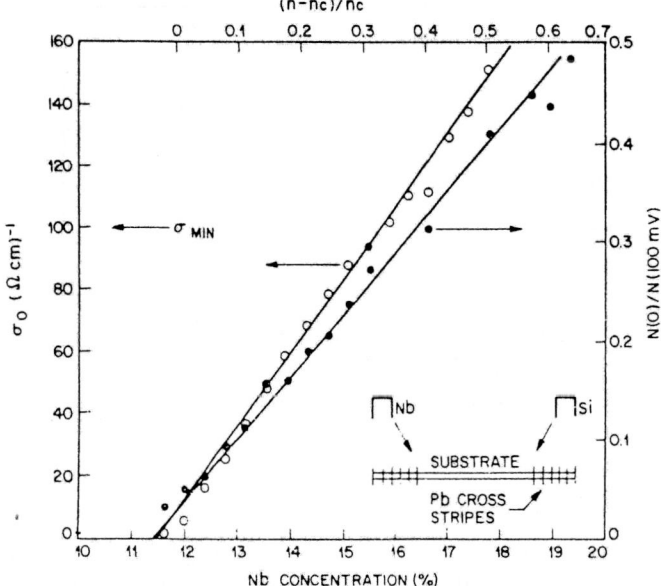

Fig. 10.11.   The $T = 0$ conductance σ_0 and the $V = 0$ tunneling density of states $N(0)$ as a function of Nb concentration in amorphous Si, $N(0)$ and $s0$ both collapse to zero at the same Nb concentration signalling the metal–insulator transition. Both indicate strong electron–electron interactions driving the transition. Reprinted (Fig. 1) with permission from *Phys. Rev. Lett.* **50**, 743 (1983). © American Physical Society.

at $T = 0$, and the density of electronic states $N(0)$ both go to zero at the same critical density of Nb.

A more extensive set of measurements on a system that could be magnetically tuned through the metal–insulator transition were performed on the amorphous Gd:Si alloy system. This system is particularly well suited because, besides the structural disorder, there is an additional degree of dis order due to the random orientation of the Gd magnetic moments, which can be continuously aligned with magnetic field. By aligning the Gd moments, the disorder is reduced and with careful choice of the Gd concentration in amorphous Si, the system can be magnetically tuned through the transition. This has been demonstrated[21] and the magnetoconductance of various different samples at $T = 100$ mK is shown in Fig. 10.11. All these samples are at conductances below the Ioffe–Regel limit and are below Mott's $\sigma_{\min}$. For samples near the metal–insulator transition, the transition can be seen where the conductivity is a stronger than linear function of H to where it becomes sublinear on the metallic side. Also on the metallic side the conductivity

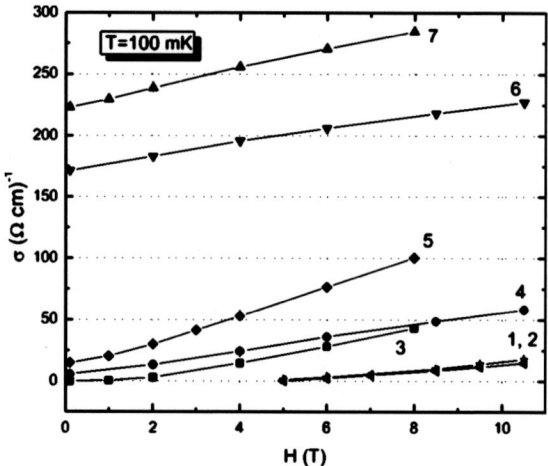

Fig. 10.12. Low temperature magnetoconductivity of seven different Gd:Si samples. At $H = 0.1$ T, samples 1 and 2 are insulating, 3 is at the transition and 4, 5, 6 and 7 are metallic. Reprinted (Fig. 1) with permission from *Phys. Rev. B* **69**, 235111 (2004). © American Physical Society.

follows a $\sigma = \sigma_0 + \sigma_1 T^{1/2}$ where σ_0 is magnetic field dependent. The tunneling conductance for the same samples is shown in Fig. 10.12 where the crossover from insulating to conducting behavior is signaled by the disappearance of a soft coulomb gap and the existence of states at the Fermi level. The samples that are clearly metallic as demonstrated in the conductance curves of Fig. 10.13 again show a tunneling conductance that behaves like $N(V) = N(0) + N_1 V^{1/2}$ where $N(0)$ is magnetic field dependent. These data are interpreted by a model where the Gd moment is ordered by the application of magnetic field. The details right at the transition are not so clear as $N(0)$ goes to zero at a slightly different point than σ_0 does. Nevertheless, the overview that this highly correlated system goes continuously through the transition from extended to localized states is clear.

4. Summary

This chapter attempts to describe the evolution and my own thinking as we explored the early insights of Mott on electrical transport in highly disordered metals. It is clear that the Mott criterion for a minimum metallic conductivity signaled a point, not where the system transitioned to an insulator, but to a system where we could no longer think in terms of independent particles. This highly correlated system brought new challenges to our

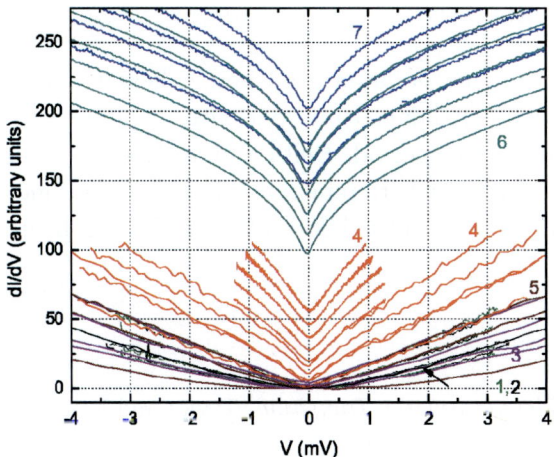

Fig. 10.13. Tunneling conductance vs voltage for the seven samples in Fig. 10.12. The different tunneling conductances for the same sample correspond to different applied fields. The transition from insulating to conducting can be clearly seen. Reprinted (Fig. 6) with permission from *Phys. Rev. B* **69**, 235111 (2004). © American Physical Society.

community; some concepts we now understand, some we don't. One of my more vivid memories during this period was a message left on my desk one day at Bell Laboratories which is shown in Fig. 10.14. It was left there by Phil Anderson.

Fig. 10.14. A note appearing on my desk one afternoon at Bell Laboratories. The text of the note reads: He plotted T_c vs. Resistance Ratio to hide his results, instead of resistivity. — Phil.

It was an admonition I will not forget.

Acknowledgments

I am grateful to all my collaborators and colleagues, students and post-doctoral fellows who challenged me, patiently explained to me, and shared the hard work during this period. Bell Laboratories was an exciting and ever-stimulating environment where much of this work was done. The University of California, San Diego and the Office of Naval Research supported the later work.

References

1. N. F. Mott, *Metal–Insulator Transition* (Taylor and Francis, London, England, 1974).
2. D. C. Licciardello and D. J. Thouless, *Phys. Rev. Lett.* **35**, 1475 (1975).
3. A. F. Ioffe and A. R. Regel, *Prog. Semicond.* **4**, 237 (1960).
4. R. C. Dynes, J. M. Rowell and J. P. Garno, *Phys. Rev. Lett.* **40**, 479 (1978).
5. E. Abrahams, P. W. Anderson, D. C. Licciardello and T. V. Ramakrishnan, *Phys. Rev. Lett.* **42**, 673 (1979).
6. M. Strongin, R. S. Thompson, O. F. Kammerer and J. E. Crow, *Phys. Rev. B* **1**, 1078 (1970).
7. N. F. Mott, *Phil. Mag.* **19**, 835 (1969).
8. D. J. Bishop, D. C. Tsui and R. C. Dynes, *Phys. Rev. Lett.* **44**, 1153 (1980).
9. D. J. Bishop, R. C. Dynes and D. C. Tsui, *Phys. Rev. B* **26**, 773 (1982).
10. R. Wheeler, *Phys. Rev. B* **24**, 4645 (1981).
11. R. A. Davies, M. J. Uren and M. Pepper, *J. Phys. C* **14**, L531 (1981).
12. P. A. Lee and T. V. Ramakrishnan, *Phys. Rev. B* **26**, 4009 (1982).
13. S. Hikami, A. I. Larkin and Y. Nagaoka, *Prog. Theor. Phys.* **63**, 707 (1980).
14. G. Bergmann, *Phys. Rev. B* **25**, 2937 (1982).
15. A. E. White, R. C. Dynes and J. P. Garno, *Phys. Rev. B* **29**, 3694 (1984).
16. A. Schmidt, *Z. Phys.* **271**, 251 (1974).
17. E. Abrahams, P. W. Anderson, P. A. Lee and T. V. Ramakrishnan, *Phys. Rev. B* **24**, 6383 (1981).
18. G. Bergmann, *Phys. Rev. Lett.* **49**, 162 (1982).
19. T. F. Rosenbaum, R. F. Milligan, M. A. Paalanen, G. A. Thomas and R. N. Bhatt, *Phys. Rev. B* **27**, 7509 (1983).
20. G. Hertel, D. J. Bishop, E. G. Spencer, J. M. Rowell and R. C. Dynes, *Phys. Rev. Lett.* **50**, 743 (1983).
21. L. Bokacheva, W.Teizer, F. Hellman and R. C. Dynes, *Phys. Rev. B* **69**, 235111 (2004).
22. B. L. Altshuler and A. G. Aronov, *Solid State Commun.* **39**, 115 (1979).

Chapter 11

WEAK LOCALIZATION AND ITS APPLICATIONS AS AN EXPERIMENTAL TOOL

Gerd Bergmann

Department of Physics, University of Southern California,
Los Angeles, California 90089-0484, USA
bergmann@usc.edu

The resistance of two-dimensional electron systems such as thin disordered films shows deviations from Boltzmann theory, which are caused by quantum corrections and are called weak localization. The theoretical origin of weak localization is the Langer–Neal graph in Kubo formalism. It represents an interference experiment with conduction electrons split into pairs of waves interfering in the back-scattering direction. The intensity of the interference (integrated over the time) can easily be measured by the resistance of the film. The application of a magnetic field B destroys the phase coherence after a time which is proportional to $1/B$. For a field of 1 T this time is of the order of 1 ps. Therefore with a dc experiment, one can measure characteristic times of the electron system in the range of picoseconds. Weak localization has been applied to measure dephasing, spin-orbit scattering, tunneling times, etc. One important field of application is the investigation of magnetic systems and magnetic impurities by measuring the magnetic dephasing time and its temperature dependence. Here the Kondo maximum of spin-flip scattering, spin-fluctuations, Fermi liquid behavior and magnetic d-resonances have been investigated. Another field is the detection of magnetic moments for very dilute alloys and surface impurities. This article given a brief survey of different applications of weak localization with a focus on magnetic impurities.

1. Introduction

In 1961, Anderson[1] introduced electron localization in three dimensions. In the meantime this concept has found many applications in different fields of physics. In 1977, Thouless[2] showed that a thin wire with a finite mean free path yields electron localization. While the three-dimensional case required a large disorder to obtain localization, in one dimension an arbitrarily small disorder already yields (at zero temperature) a complete confinement of the electrons. The critical case was the two-dimensional metal. This case was investigated by Wegner[3] and Abrahams *et al.*[4] The latter analysed a special Kubo diagram, the so-called fan diagram, and using a scaling approach concluded that a two-dimensional electron system with weak disorder becomes insulating for large (infinite) sample size.

The fan diagram had been studied more than a decade earlier by Langer and Neal.[5] Now it was analyzed in detail, and it developed a life of its own. Its physics is now known as weak localization and weak anti-localization. There are a number of excellent review articles in the field of weak localization.[6-18] Figure 11.1 shows an experimental result of weak localization in a thin Mg film. At low temperatures one observes an increased resistance and a negative magnetoresistance. If one covers the Mg film with 1/100 of a mono-layer of Au, then one introduces a finite spin-orbit scattering, and the magnetoresistence shows an interesting structure with a positive magnetoresistence at small fields.

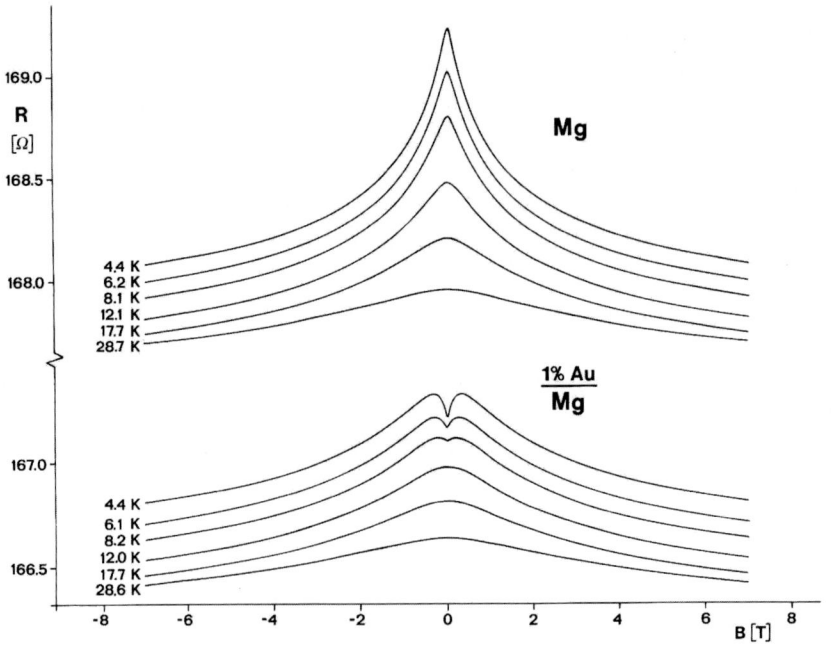

Fig. 11.1. The magnetoresistance of a thin pure Mg film (upper part) and the same film with a cover of 1/100 atomic layers of Au. Reprinted (Fig. 6) with permission from *Phys. Rev. B* **28**, 2914 (1983). © American Physical Society.

2. The Physics of Weak Localization

In a disordered metal, the conduction electrons are scattered by the impurities. If we consider the conduction electrons as plane waves then the scattered waves propagate in all directions. The usual Boltzmann theory neglects interference between the scattered partial waves and assumes that the momentum of the electron wave disappears exponentially after the elastic scattering time τ_0. The neglect of interference is, however, not quite correct. There is a coherent superposition of the scattered electron wave that results in back-scattering of the electron wave and lasts as long as its coherence is not destroyed. This causes a correction to the conductance which is generally calculated in the Kubo formalism by evaluating Kubo graphs. The most important correction was already discussed by Langer and Neal[5] in 1966 and is shown in the top of Fig. 11.2. Anderson *et al.*[19] and Gorkov *et al.*[20] showed that at low but finite temperature the Langer–Neal diagram (fan diagram) yields a quantum correction to the conductance

$$\Delta L = -\frac{\Delta R}{R^2} = L_{00} \ln\left(\frac{\tau_i}{\tau_0}\right), L_{00} = \frac{e^2}{2\pi^2\hbar} \tag{2.1}$$

where τ_i is the inelastic dephasing time. This correction is temperature dependent because the dephasing time depends on the temperature (for example $1/\tau_i \propto T^2$).

2.1. *The echo of a scattered electron wave*

The author gave the Langer–Neal diagram a physical interpretation which represents a phascinating interference experiment with the conduction electrons.[10,11] Let us consider at the time $t = 0$ an electron of momentum $\mathbf{k}$ which has the wave function $\exp(i\mathbf{kr})$. The electron in state $\mathbf{k}$ is scattered after the time τ_0 into a state $\mathbf{k}'_1$, after a time $2\tau_0$ into the state $\mathbf{k}'_2$, etc. There is a finite probability that the electron will be scattered into the vicinity of the state $-\mathbf{k}$; for example after n scattering events. This scattering sequence (with the final state $-\mathbf{k}$) $\mathbf{k} \to \mathbf{k}'_1 \to \mathbf{k}'_2 ... \to \mathbf{k}'_n = -\mathbf{k}$ is drawn in Fig. 11.2 in $\mathbf{k}$-space. The momentum transfers are $\mathbf{g}_1, \mathbf{g}_2, .. \mathbf{g}_n$. There is an equal probability for the electron $\mathbf{k}$ to be scattered in n steps from the state $\mathbf{k}$ into $-\mathbf{k}$ via the sequence $\mathbf{k} \to \mathbf{k}''_1 \to \mathbf{k}''_2 ... \to \mathbf{k}''_n = -\mathbf{k}$ where the momentum transfers are $\mathbf{g}_n, \mathbf{g}_{n-1}, .. \mathbf{g}_1$. This complementary scattering series has the same changes of momenta in opposite sequence. If the final state is $-\mathbf{k}$, then the intermediate states for both scattering processes lie symmetric to the origin. The important point is that the amplitude in the final state

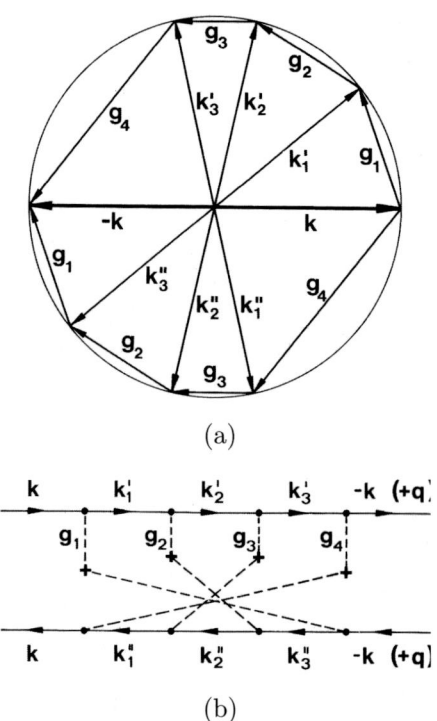

(a)

(b)

Fig. 11.2.   The fan diagram (a), introduced by Langer and Neal, which allows calculations of quantum corrections to the resistance within the Kubo formalism. (b) The physical interpretation of the fan diagram in (a). The electron in the eigenstate of momentum $\mathbf{k}$ is scattered via two complementary series of intermediate scattering states into the state $-\mathbf{k}$. The two scattering series are $\mathbf{k} \rightarrow \mathbf{k}'_1 \rightarrow .\mathbf{k}'_{n-1} \rightarrow \mathbf{k}'_n = -\mathbf{k}$ and $\mathbf{k} \rightarrow \mathbf{k}''_1 \rightarrow \mathbf{k}''_2 \rightarrow .\mathbf{k}''_{n-1} \rightarrow \mathbf{k}''_n = -\mathbf{k}$. The momentum changes in the two series in opposite sequence, i.e. by $\mathbf{g}_1, \mathbf{g}_2, \mathbf{g}_3, .., \mathbf{g}_n$ for the first series and by $\mathbf{g}_n, .., \mathbf{g}_3, \mathbf{g}_2, \mathbf{g}_1$ for the second. The amplitudes in the final state $-\mathbf{k}$ are identical, $A' = A'' = A$ and interfere constructively, yielding an echo in back-scattering direction which decays as $1/t$ in two dimensions. Only for times longer than the inelastic lifetime τ_i the coherence is lost and the echo disappears. Reprinted (Fig. 2.2) with permission from *Phys. Rep.* **107**, 1 (1984). © Elsevier.

$-\mathbf{k}$ is the same for both scattering sequences. Since the final amplitudes A' and A'' are phase coherent and equal, $A' = A'' = A$, the total intensity is $I = |A' + A''|^2 = |2A|^2 = 4|A|^2$. If the two amplitudes were not coherent then the total scattering intensity of the two complementary sequences would only be $2|A|^2$. This means that the scattering intensity into the state $-\mathbf{k}$ is larger by factor 2 than in the case of incoherent scattering. In Ref. 10, a semi-quantitative calculation of the back-scattering intensity is performed in simple physical terms. This additional scattering intensity exists only in the back-scattering direction.

At finite temperature the scattering processes are partially inelastic. As a consequence the amplitudes A' and A'' lose their phase coherence (after the time τ_i) and the coherent backscattering disappears after τ_i. The integrated momentum of the electron decreases with increasing τ_i. The coherent back-scattering is not restricted to the exact state $-\mathbf{k}$, one has a small spot around the state $-\mathbf{k}$ which contributes. Its radius is inversely proportional of the diffusion length in real space $\sqrt{Dt}$ (where $D = v_F\tau_0/2$ is the diffusion constant in two dimensions). The spot of coherent final states, i.e. its radius in $\mathbf{k}$-space, shrinks with increasing time as $1/\sqrt{Dt}$. Therefore in two dimension the portion of coherent back-scattering is proportional to $1/t$. In the presence of an electric field, the coherent back-scattering reduces the contribution of the electron $\mathbf{k}$ to the current, and the conductance is decreased. The important consequence of the above consideration is that the conduction electrons perform a typical interference experiment. The (incoming) wave $\mathbf{k}$ is split into two complementary waves $\mathbf{k}'_1$ and $\mathbf{k}''_1$. The two waves propagate individually, experience changes in phase, spin orientation, etc. and are finally unified in the state $-\mathbf{k}$ where they interfere. The intensity of the interference is simply measured by the resistance. In the situation which has been discussed above the interference is constructive in the time interval from τ_0 to τ_i.

2.2. *Time of flight experiment in a magnetic field*

One of the interesting possibilities for an interference experiment is to shift the relative phase of the two interfering waves. For charged particles this can be easily done by an external magnetic field. In a magnetic field the phase coherence of the two partial waves is weakened or destroyed. In real space the two partial waves propagate on a closed loop in opposite directions.[21] See Fig. 11.3.

When the two partial waves surround the area A containing the magnetic flux Φ, then the relative change of the two phases is $(2e/\hbar)\,\Phi$. The factor of 2 arises because the two partial waves surround the area twice. Altshuler *et al.*[22] suggested performing such an "interference experiment" with an cylindrical film in a magnetic field parallel to the cylinder axis. Then the magnetic phase shift between the complementary waves is always a multiple of $2e\Phi/\hbar$ (Φ=flux in the area of the cylinder). Sharvin and Sharvin[23] showed in a beautiful experiment that the resistance of a hollow Hg cylinder oscillates with a flux period of $\Phi_0 = h/(2e)$. This is shown in Fig. 11.4.

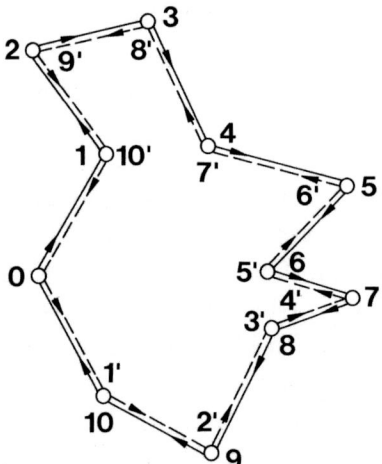

Fig. 11.3. Diffusion path of the conduction electron in the disordered system. The electron propagates in both directions (full and dashed lines). In the case of quantum diffusion, the probability to return to the origin is twice as great as in classical diffusion since the amplitudes interfer coherently. Reprinted (Fig. 2.5) with permission from *Phys. Rep.* **107**, 1 (1984). © Elsevier.

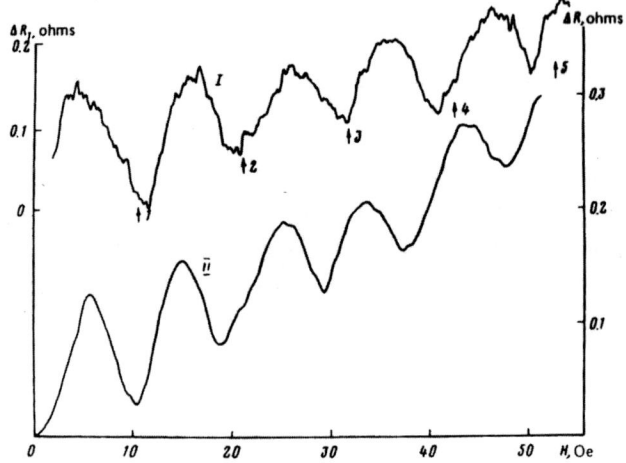

Fig. 11.4. The Sharvin and Sharvin experiment. Reprinted (Fig. 7) with permission from A. G. Aronov and Yu V. Sharvin, *Rev. Mod. Phys.* **59**, 755 (1987). © American Physical Society.

For a thin film in a perpendicular magnetic field, the pairs of partial waves enclose areas between $-2Dt$ and $2Dt$. When the largest phase shift exceeds the value of 1, the interference is constructive and destructive as well and the average cancels. This happens roughly after the time $t_s = \hbar/(4eDH)$. This means essentially that the conductance correction in the field H i.e. $L(H)$ yields the coherent back-scattering intensity integrated from τ_0 to t_H

$$\Delta L\left(H\right) \approx \int_{\tau_0}^{t_H} I_{coh} dt \approx -L_{00} \ln\left(t_H/\tau_0\right).$$

This means that the magnetic field allows a time of flight experiment. If a magnetic field H is applied the contribution of coherent back scattering is integrated in the time interval between τ_0 and $t_H = \hbar/(4eDH)$. If one reduces the field from the value H' to the value H'' and measures the change of resistance this yields the contribution of the coherent back-scattering in the time interval t'_H and t''_H. Since the magnetic field introduces a time t_H into the electron system all characteristic times τ_n of the electrons can be expressed in terms of magnetic fields H_n where $\tau_n H_n = \hbar/(4eD)$. In a thin film this product is given by $\hbar e\rho N/4$ which is of the order of 10^{-12} to $10^{-13} Ts$ (ρ = resistivity of the film and N = density of electron states for both spin directions). Formally one can express the contribution of the fan-diagram as the sum of two-electron pair amplitudes, or "cooperons", a cooperon in a singlet state with the weight one and a cooperon in the triplet state with the weight three. Over time the pair amplitudes decay with the singlet and the triplet dephasing rates: $1/\tau_S$ and $1/\tau_T$. The magnetoresistance in two dimensions (thin films or 2D-electron gases) is given by

$$\frac{\Delta L\left(H\right)}{L_{00}} = \frac{3}{2} f_2\left(H/H_T\right) - \frac{1}{2} f_2\left(H/H_S\right) \qquad (2.2)$$

where the function $f_2\left(x\right)$ is given by $f_2\left(x\right) = \ln\left(x\right) + \Psi\left(1/2 + 1/x\right)$ and $\Psi\left(z\right)$ is the digamma function. The characteristic fields H_T and H_S are the triplet and singlet fields

$$H_S = H_i + 2H_s = H_i^* \qquad (2.3)$$

$$H_T = \frac{4}{3} H_{so} + \frac{2}{3} H_s + H_i = H_i^* + \frac{4}{3} H_{so}^*.$$

Here H_i, H_{so} and H_s are the characteristic fields for the inelastic dephasing rate $1/\tau_i$, the spin-orbit scattering rate $1/\tau_{so}$ and the magnetic scattering rate $1/\tau_s$. Again the products of $H_i \tau_i = H_{so}\tau_{so} = H_s \tau_s$ have the constant value $\hbar/\left(4eD\right)$ so that the rates can be expressed by the corresponding fields. Since the characteristic fields are directly obtained from the evaluation

of the magnetoresistance curves it is often more convenient to plot these
characteristic fields.

A magnetoresistance measurement can only determine the two fields H_S
and H_T which yield H_i^* and H_{so}^*. If one wants to determine the scattering
rate of magnetic impurities, H_s, then one needs an independent measurement
for H_i.

Magnetoresistance measurements on thin films have been performed by
many groups.[10,22–98]   In order to avoid the influence of spin-orbit coupling, the magnetoresistance experiment must be performed with a very
light metal, because spin-orbit scattering causes interesting complications.
In Fig. 11.5, the magneto-resistance of a Cu film is plotted for different
temperatures.[51] The full points are measurements and the full curves are
theoretical curves, fitted with Eq. (2.2). One temperature-independent value
for H_{so} and a fitted value of H_i for each temperature are used. As the 20.1 K

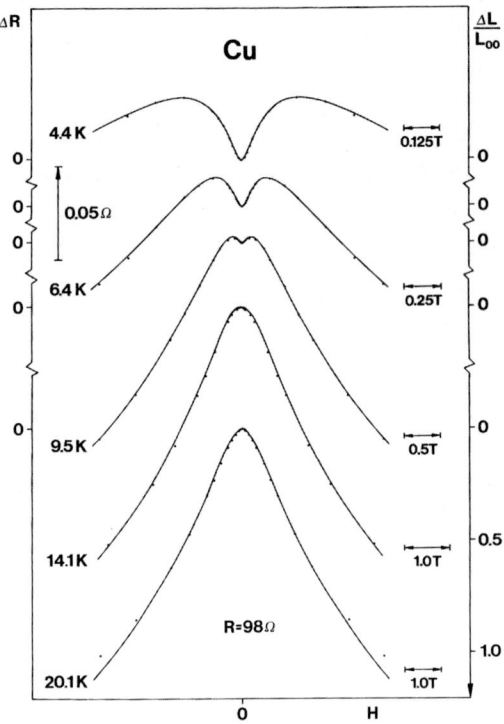

Fig. 11.5.   The magneto-resistance of a quench-condensed Cu film at different temperatures. The full points are measured and the full curves are theoretical curves,
fitted with Eq. (2.2). Reprinted (Fig. 8) with permission from *Z. Phys. B* **48**, 5
(1982). © American Physical Society.

curve shows in Fig. 11.5, one cannot extract the value of H_{so} from the experimental curve when H_{so} considerably smaller than H_i.

The Cu is quench condensed at helium temperature, because the quenched condensation yields homogeneous films with high resistances. The agreement between the experimental points and the theory is very good. The experimental result proves the destructive influence of a magnetic field on weak localization. It measures the area in which the coherent electronic state exists as a function of temperature and allows the quantitative determination of the dephasing time τ_i. The temperature dependence follows a $T^{-1.7}$ law for Cu.

3. Spin-Orbit and Inelastic Scattering

One of the most interesting questions in weak localization is the influence of spin-orbit coupling. While originally it was thought that spin-orbit coupling has only a minor influence on weak localization, Hikami *et al.*[25] and Maekawa and Fukuyama[43] calculated within perturbation theory that spin-orbit coupling should reverse the sign of the correction to the resistivity and reduce its magnitude by a factor of 1/2. This decrease of the resistance with decreasing temperature appeared to contradict the original picture of weak localization as a precursor of localization. The relevance of this effect of spin-orbit coupling and its physical origin was one of the striking questions at LT XVI. Experimentally the author[34] showed, by magnetoresistance measurements, that indeed the spin-orbit coupling reverses the sign of the quantum correction to the resistance and agrees well with the theoretical prediction (see Fig. 11.6).

The physical origin of the change in sign is due to the fact that the spin-orbit coupling rotates the spin of the two complementary waves in opposite direction. If the relative rotation of the spins is 2π then their spin states have opposite signs since spin 1/2 particles have a rotational periodicy of 4π. This basic quantum theoretical law has been experimentally proved by a neutron experiment. Since weak localization is due to the interference of scattered conduction electrons, it represents a rather compact interference experiment. The reversal of the magnetoresistance in the presence of spin-orbit scattering is an experimental proof of the sign reversal of an electron wave whose spin is rotated by 2π.

The strength of the spin-orbit scattering plays an important role in many areas of solid state physics because it determines whether the electron spin is a good quantum number. From the Knight shift in disordered

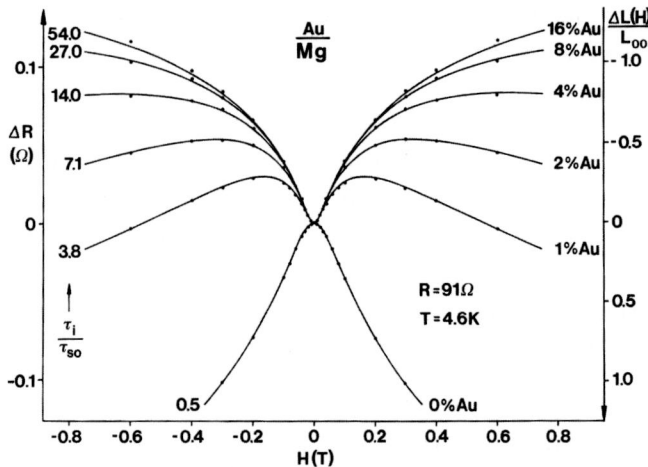

Fig. 11.6. The magnetoresistance of a thin Mg-film at 4.5 K for different coverages with Au. The Au thickness is given in % of an atomic layer on the right side of the curves. The Au increases the spin-orbit scattering. The points are measured. The full curves are obtained with the theory by Hikami *et al.* The ratio τ_i/τ_{so} on the left side gives the strength of the adjusted spin-orbit scattering. It is essentially proportional to the Au thickness. Reprinted (Fig. 2.10) with permission from *Phys. Rep.* **107**, 1 (1984). © Elsevier.

superconductors to the calculation of the upper critical field B_{c2}, the destruction of the Clogston limit and formation of spin-polarized excitations in high magnetic fields, the whole field of superconductivity is strongly influenced by spin-orbit scattering. But there are other areas in solid state physics such as the Hall effect of heavy liquid metals like Tl, Pb and Bi which show a deviation from the free electron Hall constant (for a survey, see for example, Refs. 99 and 100). Furthermore the Hall effect of liquid transition metals and the anomalous Hall effect are, according to our present understanding, determined by the spin-orbit scattering processes.

Because of the sensitivity to spin-orbit scattering, weak localization is probably the most sensitive method to measure the strength of spin-orbit scattering. Previously very little was known about spin-orbit scattering in metals. Abrikosov and Gorkov[101] suggested several decades ago that the spin-orbit scattering strength in metals with disorder is proportional to the fourth power of the atomic number. Prior to the use of weak localization, the only systematic investigation had been performed by nuclear magnetic resonance (see for example Ref. 102) and for pure metals with lattice defects by Meservey *et al.*[103] using polarized electrons in superconducting tunneling junctions. The Z^4 law as suggested by Abrikosov and Gorkov was only intended as a rule of thumb. It would be very surprising if there is no dependence of the spin-orbit scattering cross section on the valence of the impurity.

Our group used weak localization for a systematic investigation of Mg with different impurities.[104,105] In Fig. 11.7, the spin-orbit scattering cross section $\sigma_{so}k_F^2/4\pi$ (right scale) is shown as a function of the nuclear charge Z of different (s,p) impurities in Mg in a log–log plot. One recognizes that each impurity row (the $4sp$, $5sp$ and $6sp$ row) shows a strong increase of the spin-orbit scattering cross section with increasing valence. If one compares impurities with the same valence as a function of Z one finds a strong increase with increasing Z. For the noble metals the power law is roughly $\sigma_{so} \propto Z^5$.

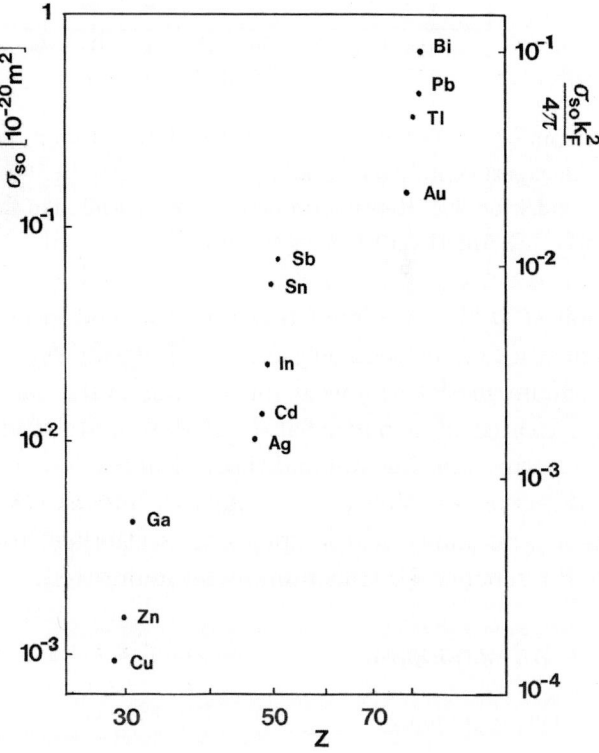

Fig. 11.7. The spin-orbit scattering cross section $\sigma_{so}k_F^2/4\pi$ (right scale) as a function of the nuclear charge Z of different (s,p) impurities in Mg on a log–log plot. Reprinted (Fig. 2) with permission from *Phys. Rev. Lett.* **68**, 2520 (1992). © American Physical Society.

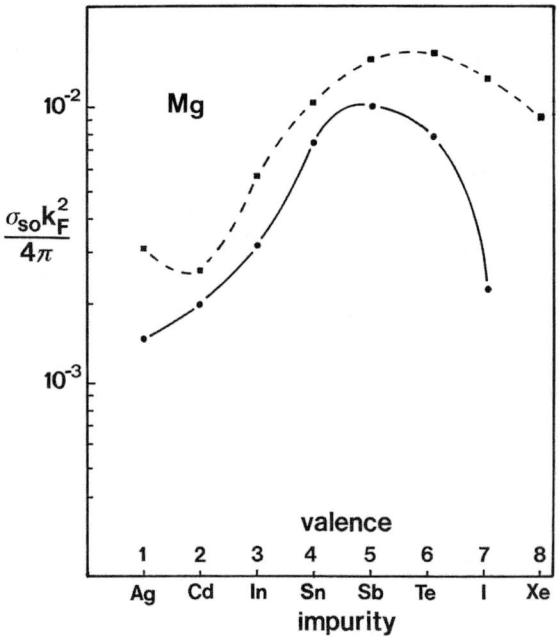

Fig. 11.8.   The spin-orbit scattering cross section of $5(s,p)$ impurities in Mg. the full points are the experimental results, and the squares give the theoretical results using a jellium model for Mg. Reprinted (Fig. 1) with permission from *Phys. Rev. B* **49**, 8377 (1994). © American Physical Society.

Papanikolaou *et al.*[106] calculated the spin-orbit scattering cross section within the framework of the self-consistent local density function (SCLDF) theory. The jellium model was used for the Mg host and the spin-orbit interaction was treated as a perturbation of the non-relativistic Hamiltonian. In Fig. 11.8, the experimental and theoretical results for the spin-orbit scattering cross section are shown. For Mg with different (s,p) impurities one recognizes a p-resonance in the spin-orbit scattering cross section (in analogy to the d-resonance for transition metal impurities).

3.1. *The inelastic dephasing*

The physics of the dephasing of weak localization is a large field on its own. Among others, the dephasing rate has been investigated experimentally in Refs. 17,26,30,34,35,40,48,49,51–54,61,65,67,69,70,75,78,81,83,86,87, 90,91,94–96,98,107–116. I refer the reader to the experimental reviews of the dephasing in thin films by Gershenson[17] and Lin.[18,115] There is also a large number of theoretical investigations of the effect of the electron–phonon interaction[21,117–123] and the electron–electron interaction[118,122,124–133] on the dephasing rate. Recent years have seen a large body of experimental investigations in nanowires and nanotubes (see for example Ref. 134).

4. Magnetic Scattering

According to the theory by Hikami *et al.*[25] in first Born approximation, a magnetic impurity causes a constant dephasing of the weak localization, i.e. the cooperon dephasing rate is $2/\tau_s$ where τ_s is the magnetic scattering time. In this repect, a weak localization experiment can detect whether an impurity is magnetic. An example is surface impurities. While a number of $3d$ atoms show a magnetic moment in (s,p)-hosts the $4d$ and $5d$ impurities are generally not magnetic in an (s,p)-host. However, there were theoretical predictions that a number of $4d$ and $5d$ impurities on the surface of Ag and Au should show a magnetic moment.

Another application of weak localization is the detection of very small magnetic dephasing. Let us make the hypothetical assumption that bulk gold is superconducting with a transition temperature of $T_c = 1$ mK. The superconducting state with such a low T_c can be easily destroyed by a very small concentration of magnetic impurities when the magnetic scattering time τ_s is about $\hbar/(k_B T_c) \approx 10^{-8}$ s. This corresponds to an extremely small concentration of magnetic impurities which is hard to detect. If the Au sample has such a small concentration of magnetic impurities, then one could cool down the Au sample to arbitrarily low temperature and would never observe the superconductivity. However, this experiment would not yield any certainty that Au is not a superconductor. Here weak localization would help. Already at 20 mK, a magnetoresistance measurement would detect a magnetic dephasing with $\tau_s \approx 10^{-8}$ s because the regular (non-magnetic) dephasing rate $1/\tau_i$ is already much smaller. Therefore a weak localization experiment tells the experimentalist whether his sample is clean enough to observe the hypothetical superconductivity.

The properties of a magnetic impurity desolved in a metal can be very complicated and represent a lively field of research for many decades. The impurity can loose its magnetic moment at low temperatures due to the Kondo effect. Its average moment might disappear but still show thermal spin fluctuations. Weak localization is a tool to investigate these phenomena.

4.1. *Magnetism of 3d, 4d and 5d surface impurities*

It is well known that $3d$ impurities such as V, Mn, Cr, Fe and Co possess a magnetic moment in most (s,p)-hosts as long as the temperature is kept above the Kondo temperature. The impurities of $4d$ and $5d$ transition metals are generally not magnetic in simple (s, p)-hosts. Lang *et al.*[135] investigated the properties of these impurities on the surface of the host. They predicted that not only the majority of $3d$ atoms but many of the $4d$ and $5d$ transition metal atoms should possess a magnetic moment on the surface of Cu or Ag.

Our group used a cryostat which allowed us to evaporate thin films in situ. This opened the door to investigate this new class of magnetic impurities. At first a thin film of the host is quench condensed at He-temperatures. The properities of the host film are determined with a set of magnetoresistance measurements yielding the characteristic times, τ_i and τ_{so} of the host. This has the important advantage that the properties of the actual host are measured (and not of a similar host film in a different experiment). In the next step, the (potentially magnetic) impurity is condensed in a very small concentration on the surface of the film (between 0.01 and 0.001 atomic layers). It is now a very reasonable and well founded assumption that the τ_i and τ_{so} of the host have hardly changed. A new set of magnetoresistance measurements yield now an effective dephasing rate $(1/\tau_i)^*$ and an effective spin-orbit scattering rate $(1/\tau_{so})^*$. The effective dephasing rate $(1/\tau_i)^*$ is composed of the inelastic dephasing of the host and the magnetic scattering due to the impurities, $(1/\tau_i)^* = 1/\tau_i + 2/\tau_s$.

In Fig. 11.9, the magnetoresistance curves are shown for the system Au/Mo.[136] The top curve is for the pure Au film. It has a narrow magnetoresistance curve corresponding to a small dephasing rate. The middle curve shows the result for the same Au film covered with $1/100$ of a monolayer of Mo. One recognizes the broadening indicating a strong additional dephasing. The dephasing is roughly the same as for $1/100$ mono-layers of Fe on Au. The bottom curve demonstrates that bulk Mo impurities in Au are non-magnetic. Here five additional mono-layers of Au are condensed on top of the Au/Mo double layer making the Mo a bulk impurity. The dephasing in the bottom curve is essentially the same as for the initial Au film.

For the Mo on the surface of the Au film, the dephasing rate goes through a maximimum around 0.1 atomic mono-layers of Mo on Au as shown in Fig. 11.10.

It turned out that Mo was the only $4d$ and $5d$ surface impurity in our experiments that showed a magnetic moment at the surface of Cu, Ag or Au. But we observed a broad variety of different behaviors, and the magnetic dephasing strength spanned a range over three orders of magnitude. Figure 11.11 shows a summery of the different experimental results. The dephasing cross sections of the $3d$, $4d$, and $5d$ surface impurities on noble metals (generally Au) are plotted.

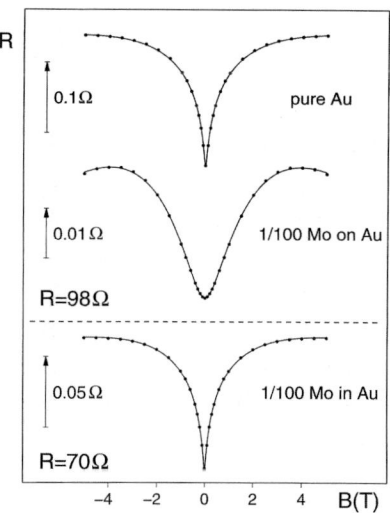

Fig. 11.9. Three magnetoresistance curves for the system of Au/Mo/Au measured at 4.5 K. The top curve is for pure Au. For the second curve, the Au is covered with 1/100 atomic layers of Mo. In the third curve, the Au/Mo is covered with 5 atomic layers of Au. The points are the experimental values. The curves are fitted with the theory of weak localization. Reprinted (Fig. 1) with permission from *Europhys. Lett.* **33**, 563 (1996). © EDP Sciences.

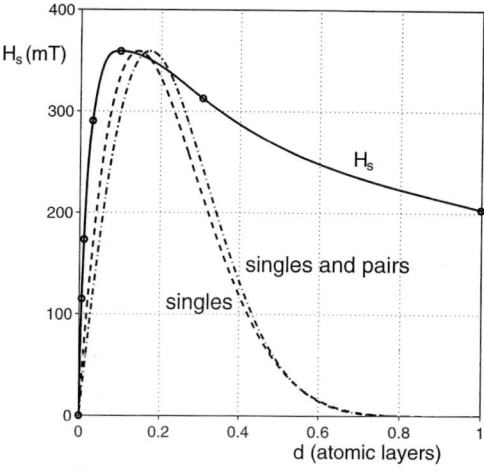

Fig. 11.10. The strength of the dephasing by magnetic Mo surface atoms as a function of Mo coverage. The strength is given in terms of the magnetic dephasing field H_s which is directly proportional to $1/\tau_s$. The dashed and dash-dotted curves are discussed in the original paper. Reprinted (Fig. 3) with permission from *Europhys. Lett.* **33**, 563 (1996). © EDP Sciences.

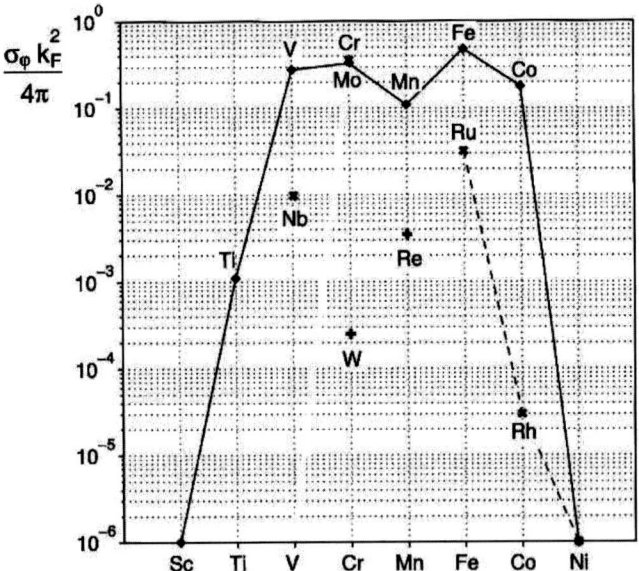

Fig. 11.11. The dephasing cross sections of $3d$, $4d$, and $5d$ surface impurities on noble metals (generally Au). ($3d$ full line, $4d$ dashed line, and $5d$ crosses). Reprinted (Fig. 7) from *J. Low Temp. Phys.* **110**, 1173 (1998). © Springer.

In the following I am going to briefly summarize the results for the transition impurity rows.

4.1.1. *3d surface impurities Ti, V, Cr, Mn, Fe, Co, Ni*

In Fig. 11.12, the magnetic scattering cross section is plotted for the $3d$ surface impurities Ti, V, Cr, Mn, Fe, Co, Ni.[137] The magnetic dephasing due to the $3d$ surface impurities has two maxima, one for Fe and a smaller one for Cr impurities. A clear minimum lies at Mn. This behavior can be qualitatively described within the Friedel–Anderson model for magnetic impurities. However, the experimental values for the magnetic dephasing cross section are smaller roughly by a factor of five than the theoretical estimates within this model.

The magnetic dephasing of Ti and Ni surface impurities is almost two orders of magnitude smaller.[138] In Fig. 11.13, the temperature dependence of the magnetic scattering cross section is shown for these surface impurities. One observes a linear temperature dependence, indicating spin fluctuating impurities.

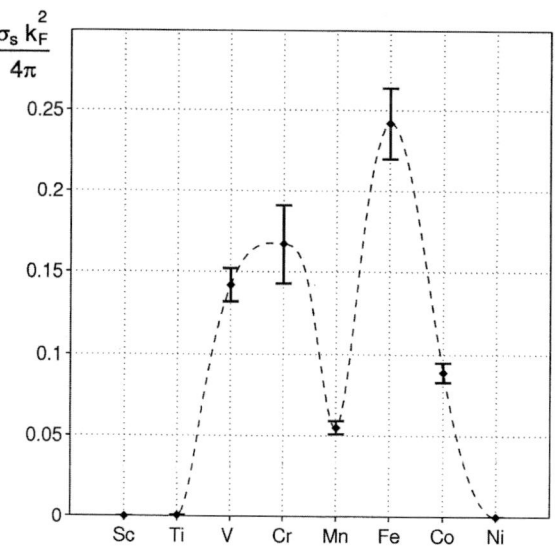

Fig. 11.12.   The magnetic scattering cross section σ_s in units of $4\pi/k_F^2$ for different $3d$ atoms on the surface of Au.[137,138] Reprinted (Fig. 2) with permission from *Phys. Rev. B* **54**, 368 (1996).   © American Physical Society.

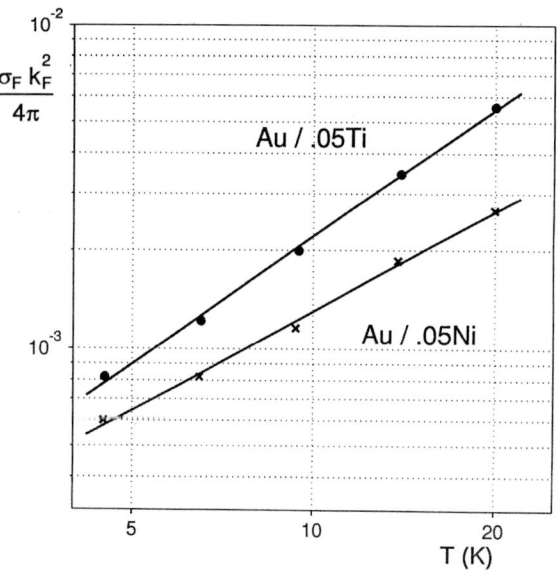

Fig. 11.13.   The temperature dependence of the dephasing cross section (in units of $4\pi/k_F^2$) of Ti and Ni surface impurities on the surface of Au. Reprinted (Fig. 1) with permission from *Phys. Rev. B* **52**, 15687 (1995).   © American Physical Society.

4.1.2. *4d surface impurities Nb, Mo, Ru, Rh, Pd*

The magnetism of Mo surface impurities has already been shown in Fig. 11.9. Although the other 4d surface impurities (Nb, Ru, and Rh) have also been predicted to be magnetic, experimentally they showed a magnetic dephasing which was at least a factor of 10 smaller than that of Mo surface impurities. This means that the other 4d surface impurities do not possess a full magnetic moment. For Nb[139] the results are particularly interesting. In Fig. 11.14, the broadening of the magnetoresistance curve of Ag films due to different

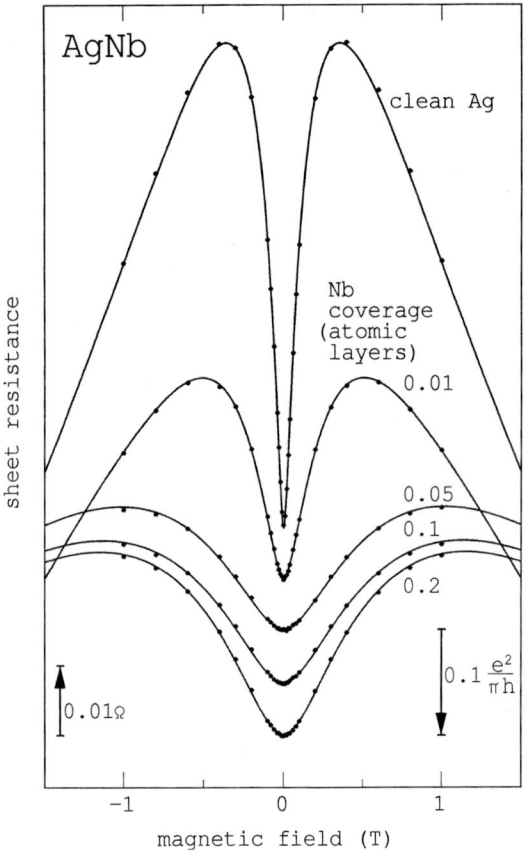

Fig. 11.14.   Several magnetoresistance curves for a Ag film with Nb surface impurities. The top film is for pure Ag. In the following curves the Ag is covered with of 0.01, 0.05, 0.1 and 0.2 atomic layers of Nb. The curves are measured at 4.5 K. The points are the experimental values. The curves are fitted with the theory of weak localization. Reprinted (Fig. 1) with permission from *Solid State. Commun.* **98**, 45 (1996). © Elsevier.

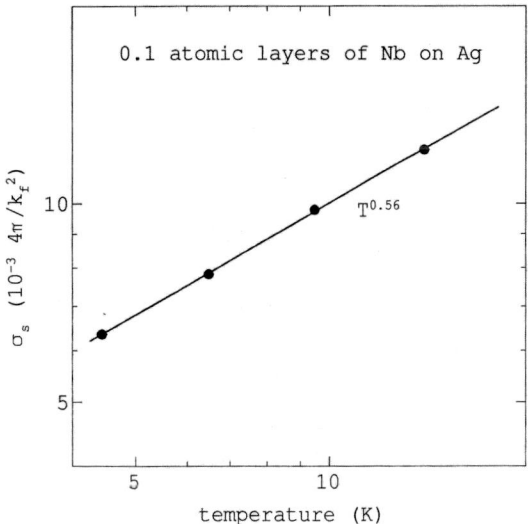

Fig. 11.15. The temperature dependence of the magnetic scattering cross section of Nb on the surface of Ag (in units of $4\pi/k_F^2$) for a Nb coverage of 0.05 atomic layers in a log–log plot. Reprinted (Fig. 2) with permission from *Solid State Commun.* **98**, 45 (1996). © Elsevier.

coverages with Nb are shown. An additional dephasing due to the Nb is clearly visible. The temperature dependence of the additional dephasing follows a $T^{1/2}$ law and is shown in Fig. 11.15. The data suggest that Nb surface impurities show a Kondo behavior with a Kondo temperature clearly above 20 K.

Rh and Ru surface impurities show both no magnetic moment. In Figs. 11.16 and 11.17, the normalized dephasing cross sections are plotted for these surface impurities. The difference between Rh and Ru surface impurities is that the cross section of Rh approaches zero for zero coverage whereas Ru has a finite dephasing cross section in the limit of small coverage. However, its values is much smaller than for magnetic surface impurities such as Fe, Mo, etc. For a coverage of about 1/10 of an atomic layer, both impurities show a maximum in the dephasing cross section.[140] It appears that only pairs (and possible larger clusters) of Rh show a magnetic dephasing.

4.1.3. *5d surface impurities W and Re*

The 5d surface impurities are very hard to condense onto the surface of a noble metal film. In Ref. 141, the two 5d impurities W and Re are investigated. Fig. 11.18 shows the dephasing cross section of Re surface impurities

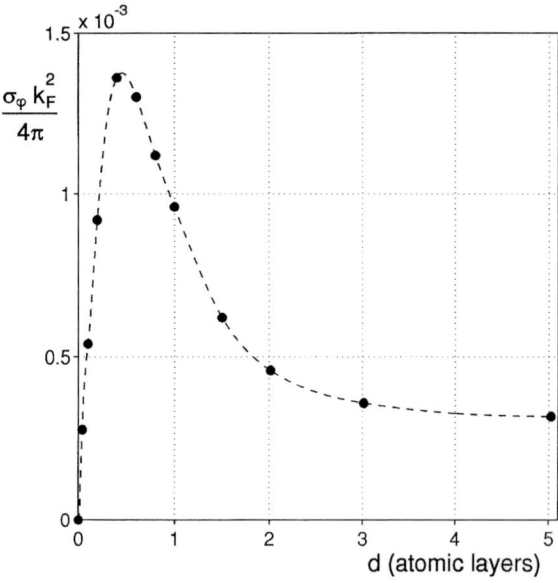

Fig. 11.16.   The dephasing cross section per Rh atom on top of a Au film as a function of the Rh coverage. The data are taken at 4.5 K. Reprinted (Fig. 3) with permission from *Phys. Rev. B* **55**, 14350 (1997). © American Physical Society.

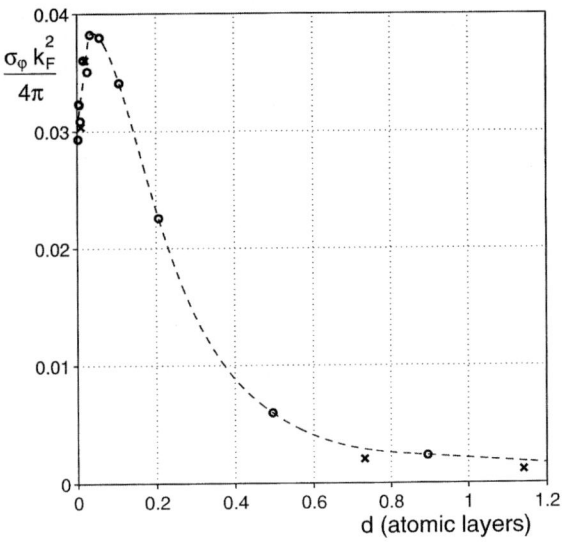

Fig. 11.17.   The dephasing cross section per Ru atom on top of a Au surface (open circles) and a Ag surface (crosses) as a function of the Ru coverage. The data are taken at 4.5 K. Reprinted (Fig. 9) with permission from *Phys. Rev. B* **55**, 14350 (1997). © American Physical Society.

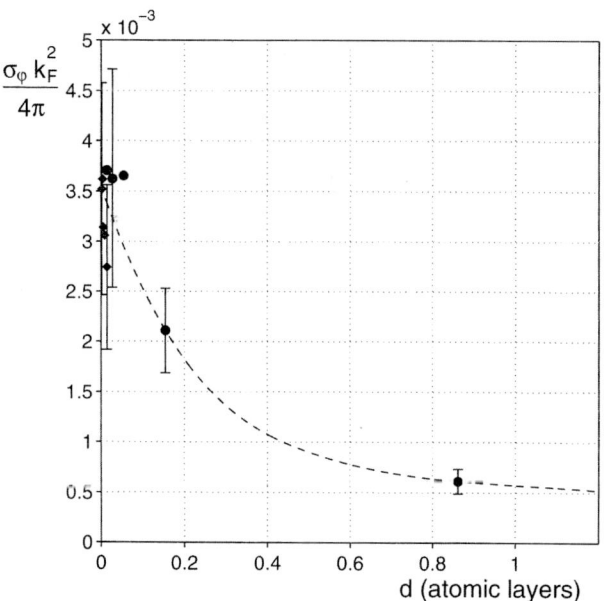

Fig. 11.18. The dephasing cross section of Re on Ag as a function of the Re coverage. Reprinted (Fig. 4) with permission from *J. Low. Temp. Phys.* **110**, 1173 (1998). © Springer.

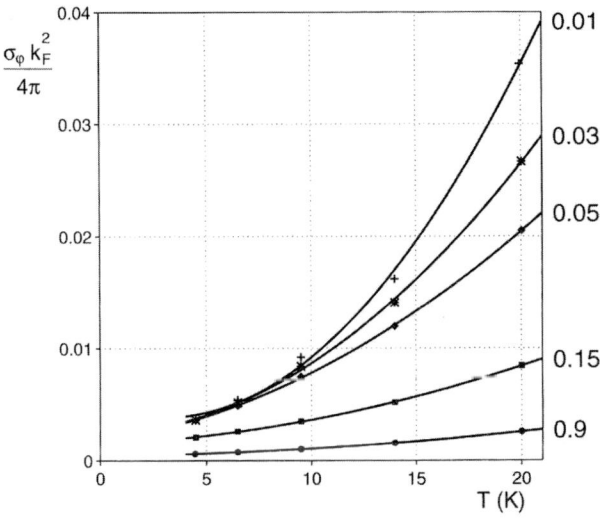

Fig. 11.19. The temperature dependence of the dephasing cross section of Re on Ag for different coverages. The coverages are given at the right side of the curves. Reprinted (Fig. 5) with permission from *J. Low Temp. Phys.* **110**, 1173 (1998). © Springer.

on Ag as a function of the coverage. The temperature dependence is shown in Fig. 11.19.

The experimental result that only one of the $4d$ and $5d$ surface impurities showed a stable magnetic moment lead the author to the conclusion that the mean-field component of the theory[135] was the origin for the incorrect prediction. To resolve this question, the author to developed the FAIR theory for the formation of magnetic moments in impurities and the Kondo effect[142,143] (FAIR stands for Friedel artificially inserted resonance). In this approach one obtains a many electron state with considerably lower energy then in mean field, and the formation of a magnetic state is much less likely than in mean-field theory.

4.2. *Kondo impurities*

The scattering of conduction electrons by Kondo impurities can be separated into two contributions, the non-spin-flip and the spin-flip parts. The latter causes a strong dephasing in superconductivity and in weak localization. This has been intensively studied in superconducting alloys where it is possible to obtain this rate from the T_c depression at different concentrations[144] and the dependence of H_{c2} on temperature.[145] These measurements show an increase of the spin-flip rate with decreasing temperature[146] in the temperature range above the Kondo temperature T_K. The predicted maximum of the rate at T_K and its decrease for $T < T_K$ could, however, not be verified by this method. Therefore the theoretical prediction of a maximum in the spin-flip scattering waited almost 20 years for an experimental proof by weak localization measurements.[147,148] In Fig. 11.20, the magnetic dephasing field H_s is plotted for 0.0003 atomic layers of Fe on the surface of Au (top curve). The lower curve in Fig. 11.20 shows the same system covered with 5 atomic layers of Au so that the Fe atoms become bulk impurities. A small maximum of the magnetic dephasing at the Kondo temperature of about 0.5 K can be observed. The dashed line gives the theoretical curve according to the Nagaoka and Suhl theory

$$1/\tau_s \propto \frac{1}{ln^2(T_K/T) + \pi^2 S(S+1)}.$$

The extraction of H_s above the Kondo temperature is very difficult because in the experiment one measures only the sum of $H_i^* = H_i + 2H_s$. Since H_i increases dramatically with temperature, and since one has to use a very small concentration of Kondo impurities to avoid interaction, the evaluation requires the difference between two very similar fields, H_i^*, the dephasing

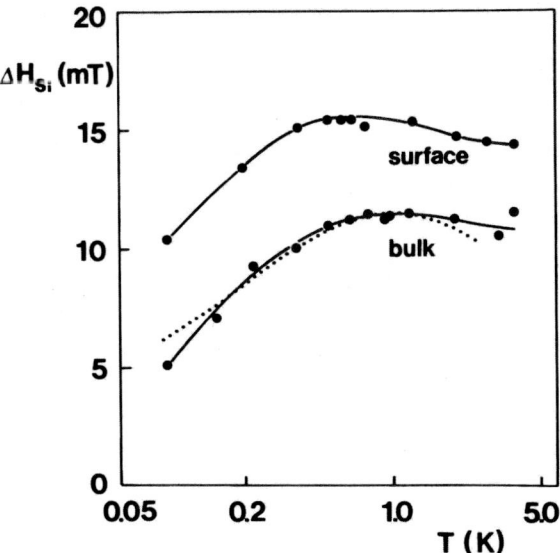

Fig. 11.20. The increase of the singlet field $H_S = H_i^*$ due to the 0.0003 atomic layers of Fe at the surface of Au (film "A2") and in the bulk (film "A3") as a function of temperature. ΔH_{Si} is proportional to the singlet spin-flip scattering rate. The absolute values of the two curves differ because the additional Au layers dilute the Fe. Reprinted (Fig. 3) with permission from *Phys. Rev. Lett.* **58**, 1964 (1987). © American Physical Society.

field with the impurities and H_i, the dephasing field without impurities. If one uses *in situ* evaporation (as in Fig. 11.20) then one is measuring the two fields in the same physical sample and statistical random variations between different samples can be excluded. In the majority of experiments the measurements of H_i^* and H_i are performed on different samples. This introduces a considerable uncertainty in the evaluation.

At that time our group discovered another Kondo system, Co impurities on the surface of Cu,[149] which became rather famous in later STM experiments.[150] In Fig. 11.21, the magnetic dephasing rate for a Cu film covered with Co impurities is shown. The investigation of the electron dephasing in the Kondo effect is still a very active field (see for example, Refs. 151 and 152).

4.2.1. *Interacting Kondo impurities*

The investigation of a single Kondo impurity with weak localization is difficult because different Kondo impurities interact over the range of the Kondo

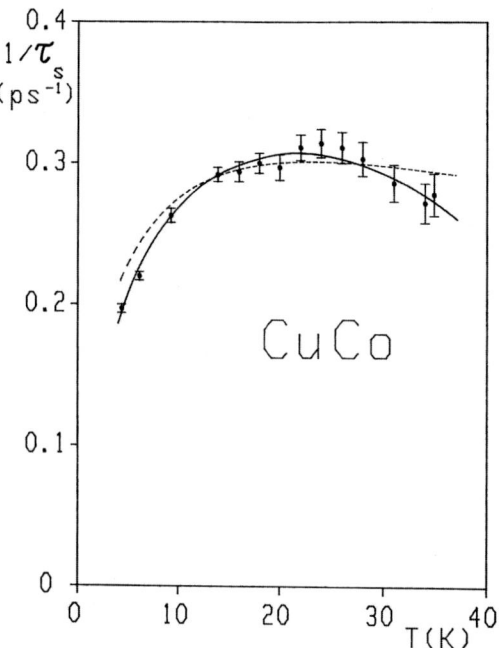

Fig. 11.21. The inverse magnetic scattering time as a function of temperature for 0.01 atomic layers of Co on top of a Cu film. The magnetic scattering rate shows a Kondo maximum at $T_K = 23$ K. The dashed line gives the theoretical curve according to the the Nagaoka and Suhl theory. Reprinted (Fig. 3) with permission from *Phys. Rev. B* **37**, 5990 (1988). © American Physical Society.

length r_K where

$$r_K = \frac{\hbar v_F}{k_B T_K}.$$

Here $\hbar/(k_B T_K) = \tau_K$ is the characteristic time corresponding to the Kondo energy and $v_F \tau_K$ is the distance an electron travels with Fermi velocity v_F during this time. (In disordered metals, the range is equal to the diffusion distance during the time τ_K which yields $\sqrt{v_F l \tau_K/3} = \sqrt{D\tau_K}$). For a Kondo temperature of 10 K, the Kondo length is already $2\,\mu$m. A Kondo alloy in which the impurities do not interact requires an extremely small impurity concentration which makes their dephasing effect completely invisible. However, Kondo alloys with interacting impurities represent experimentally as well as theoretically a very interesting system. A system which has been intensively studied is Fe impurities on the surface of Mg.[153–155]

If one condenses Fe atoms onto the surface of Mg, the Fe introduces an additional spin-orbit scattering and a magnetic scattering. While the

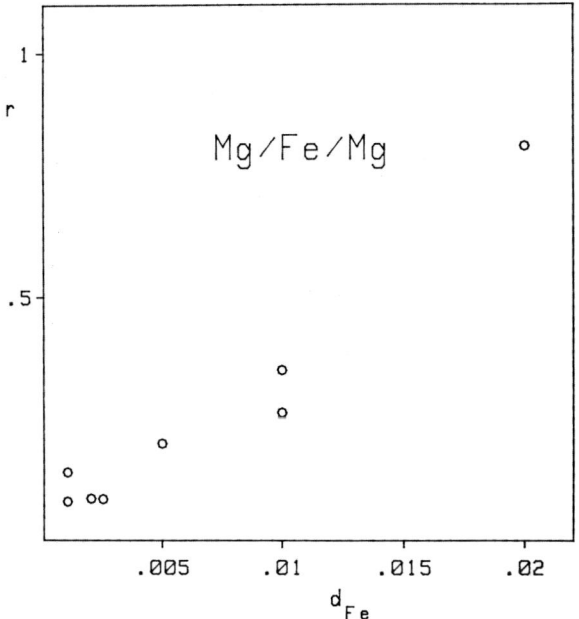

Fig. 11.22. The ratio of the magnetic scattering to the spin-orbit scattering r= $(1/\tau_s)/(1/\tau_{so}) = H_s/H_{so}$ for the sandwich Mg/Fe/Mg is plotted versus the Fe-coverage in atomic layers. This plot shows the ratio of the two cross sections of a Fe atom. Since the spin-orbit scattering rate is proportional to the Fe coverage, the magnetic rate is proportional to the square of the Fe concentration. Reprinted (Fig. 2) with permission from *Phys. Rev. Lett.* **57**, 1460 (1986). © American Physical Society.

addtional spin-orbit scattering rate is proportional to the Fe coverage the magnetic scattering rate is not. In Fig. 11.22, the ratio of the two (additional) rates $(1/\tau_s)/(1/\tau_{so})$ is plotted as a function of the Fe coverage (in units of atomic layers) at 4.5 K. This ratio is essentially proportional to d_{Fe}. This means that the magnetic scattering strength is essentially proportional to the square of the Fe coverage. This means that the magnetic scattering is partially suppressed at small Fe concentrations. In addition the magnetic scattering strength is strongly temperature dependent as Fig. 11.23, shows where $1/\tau_s$ is plotted for two different Fe coverages, 0.005 and 0.01 atomic layers of Fe.

Both effects, the suppression of the magnetic scattering scattering and its temperature dependence indicate that the magnetic moment of the Fe is partially screened and this screening is reduced for larger Fe coverage and for higher temperatures. The Mg/Fe film represents a disordered Kondo alloy.

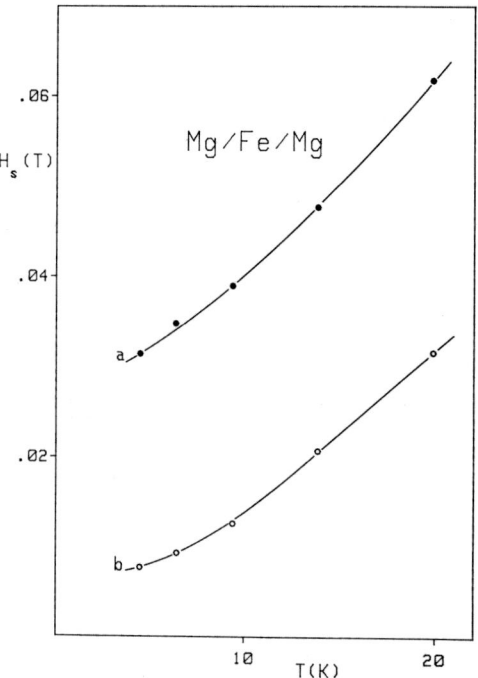

Fig. 11.23. The magnetic scattering strength (in units of a magnetic field) as a function of temperature. The upper curve corresponds to 0.01 atomic layers of Fe and the lower to 0.005 atomic layers of Fe. The factors relating H_s and $1/\tau_s$ are 0.42 and 0.44 psT respectively. Reprinted (Fig. 2) with permission from *Phys. Rev. Lett.* **58**, 1236 (1987). © American Physical Society.

These experimental results of the low concentration interacting Kondo alloys raise two essential questions: (i) is the dephasing scattering magnetic or purely inelastic and (ii) is the dephasing in Fig. 11.23 the residual dephasing that is essentially constant down to zero temperature?

Based on Wilson's[156] renormalization approach, Nozieres[157] pointed out that a Kondo system shows Fermi liquid behavior at low temperature. This means that the impurity loses its magnetic character. It then yields a change in the density of states and an interaction between the electrons which introduces an additional inelastic scattering. (This is restricted to low energies. Phenomena which include virtual high energy excitations such as superconductivity may still feel the magnetic impurity.)

This Fermi liquid theory means that there is no magnetic scattering at sufficiently low temperatures but only inelastic scattering. It is quite interesting to check whether an interacting Kondo system at low temperature shows magnetic or only inelastic scattering. The weak localization method is in principle capable of distinguishing between the "magnetic" and the "Fermi liquid" model. The inelastic scattering and the magnetic scattering affect the two relevant fields H_S and H_T differently. According to Eq. (2.3), the difference between H_T and H_S is

$$(3/4)[H_T - H_S] = H_{so} - H_s.$$

The two models have the following properties: (i) in the case of the Fermi liquid model, H_s is zero at sufficiently low temperature and $(3/4)[H_T - H_S]$ is equal to H_{so} and should be temperature independent. (ii) for the magnetic model the term $(3/4)[H_T - H_S]$ is equal to $H_{so} - H_s$ and should show the same temperature dependence as H_s since the spin-orbit scattering is (in first approximation) temperature independent. In this model, one has $H_s = \Delta H_S/2$ where ΔH_S is the change in the singlet field due to the magnetic impurities.

Plotting $(3/4)[H_T - H_S]$ as a function of $\Delta H_S/2$ should yield a straight line with either the slope zero for the Fermi liquid model or the slope minus one in the case of the magnetic model (provided that ΔH_S is temperature dependent).

A multilayer of Mg/Fe/Mg/Au is well suited for this check. The Fe impurities between the Mg layers yield an additional temperature dependent dephasing and an additional spin-orbit scattering. The additional coverage with Au is an experimental trick to obtain the singlet field H_S with high accuracy because it brings the system into the large spin-orbit scattering limit. This yields the total H_S in the presence of the magnetic impurities. H_S does not change with the Au coverage and therefore has the same value in the Mg/Fe/Mg sandwich. Then one can determine H_T for the Mg/Fe/Mg sandwich. Although the magnetoresistance curves show little structure, particularly at higher temperatures, the value of H_T can be determined without ambiguity. In Fig. 11.24, the expression $(3/4)[H_T - H_S]$ is plotted as a function of $\Delta H_S/2$. The experimental points lie roughly on a straight line with the slope -0.7. The uncertainty of the measurement becomes greater with increasing singlet field H_S. The data do not agree with a Fermi liquid model in the temperature range between 4.5 and 20 K. (The Fe impurities appear to be 70% magnetic).

The second question whether the dephasing in an interacting Kondo alloy is essentially constant down to zero temperature has been studied for Fe impurities on the surface of Mg. In Fig. 11.23 this dephasing strength is plotted down to 4.5 K. If the temperature in a He^4 cryostate is reduced by pumping, the dephasing hardly changes and appears to be a residual dephasing.

G. Bergmann

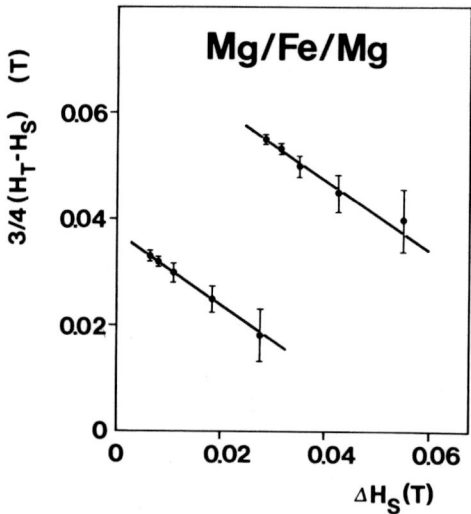

Fig. 11.24. $(3/4)[H_T - H_S]$ as a function of $\Delta H_S/2$. For the Fermi liquid model of the Kondo effect the straight line should be horizontal, for the magnetic model its slope should be -1. Reprinted (Fig. 3) with permission from *Phys. Rev. Lett.* **58**, 1236 (1987). © American Physical Society.

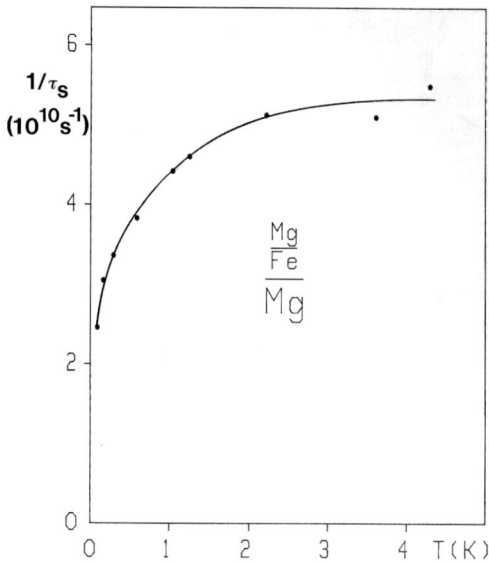

Fig. 11.25. Temperature dependence of the magnetic scattering rate in the MgFeMg sandwich. The points show the measurements; the line is a guide for the eye. Reprinted (Fig. 2) with permission from *Phys. Rev. Lett.* **60**, 1093 (1988). © American Physical Society.

However, if one performs the experiment (including the quenched conden-sation) in a He^3–He^4 cryostat,[155] then one observes a dramatic reduction of the dephasing as shown in Fig. 11.25 The low temperature behavior of the dephasing follows a $T^{0.5}$ law. The same law was observed for the in-teracting Kondo alloy of Co on the surface of Cu. It is well known that for a single Kondo impurity and for a periodic Kondo lattice the magnetic character of the Kondo atoms disappears at low temperatures. The above experiment shows that a similar behavior emerges for disordered interacting Kondo alloys.

5. Tunneling Effect

Two metals separated by a thin insulating layer represent a tunnel junction. Generally one investigates the tunneling probability by applying a voltage between the metals and measuring the current through the barrier. Weak localization provides a new method in which the properties of the tunneling junction are investigated by a current parallel to the insulating layer.[158–160] This is achieved in a sandwich of two thin Mg films separated by an insulating Sb layer. The tunneling through the Sb junction couples the dephasing rates of the electrons in the Mg layers. This rate is determined by means of weak localization. One observes a splitting of the dephasing rates in analogy to the energy splitting of coupled quantum systems. Tunnel junctions with a resistance of less than $1\mu\Omega$ per mm^2 can be investigated by this new method. In Fig. 11.26, the tunneling time is plotted as a function of the thickness of the Sb barrier.

5.1. *Proximity effect*

If there is no barrier between the two films then one observes the proximity effect of weak localization between the two films. This method can be used to determine the electron dephasing in systems that in the isolated form can not be investigated with weak localization. Amorphous Bi films are such an example. They become superconducting at 6 K and their superconducting Aslamazov–Larkin fluctuations totally overshadow the magnetoresistance of weak localization. In Fig. 11.27 the total dephasing of a MgBi double layer is shown as a function of the the the Bi layer.[161]

Another example is the magnetism of a Ni layer on top of a polyvalent metal film such as Pb. In a recent investigation[162] our group investigated this question by means of weak localization. Using thin multilayers of Ag/Pb/Ni, the superconductivity of Pb is suppressed by the proximity effect with the

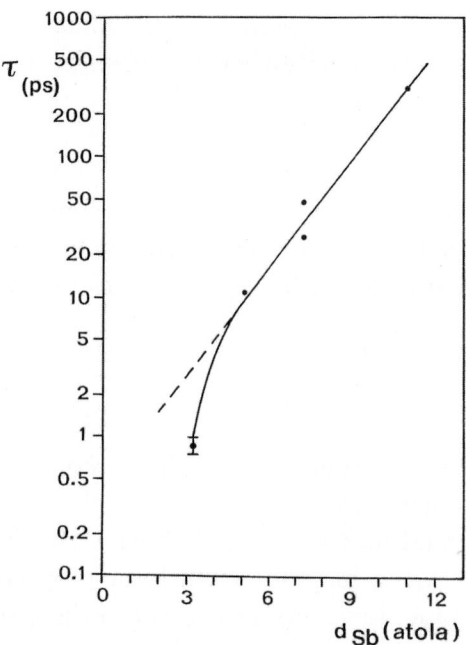

Fig. 11.26. The logarithm of the tunneling time for a Mg/Sb/Mg sandwich as a function of the thickness of the insulating Sb layer. The Mg films have a thickness of about 13 atomic layers. The tunneling time is determined with the theory of weak localization. Reprinted (Fig. 1) with permission from *Solid State Commun.* **71**, 1011 (1989). © Elsevier.

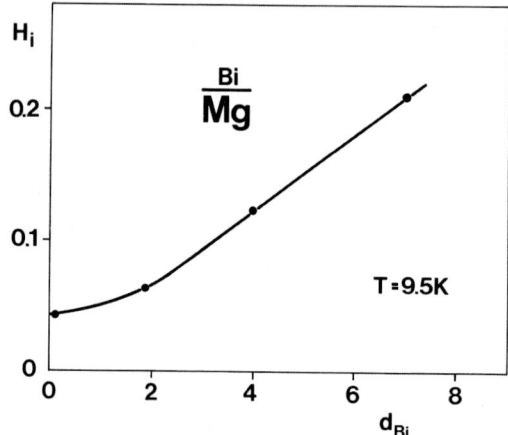

Fig. 11.27. The inelastic field H_i of the Mg/Bi sandwich at 9.5 K as a function of the Bi-thickness (in units of atomic layers). Reprinted (Fig. 2) with permission from *Phys. Rev. Lett.* **53**, 1100 (1984). © American Physical Society.

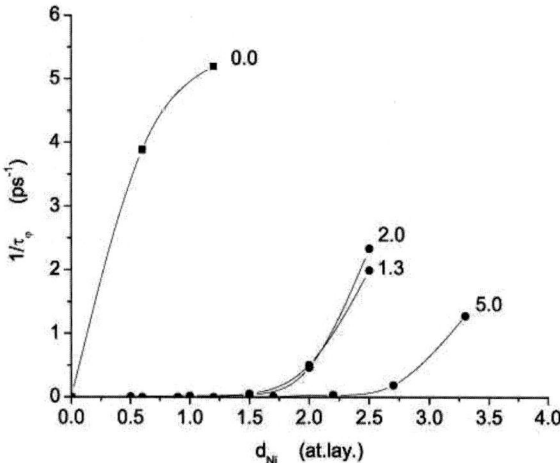

Fig. 11.28. The additional dephasing rate $1/\tau_\varphi$ of AgPbNi multilayers as a function of d_{Ni} for different Pb thicknesses d_{Pb} (full circles). The full squares give $1/\tau_\varphi$ for AgNi layers. Reprinted (Fig. 4) with permission from *Eur. Phys. J. B* **73**, 155 (2010). © Springer.

Ag film. The thickness of the Ag film is about 9 nm and the thickness of the Pb films varies between 1.3, 2.0 and 5.0 atomic layers. The additional magnetic dephasing due to the Ni is plotted in Fig. 11.28 as a function of the Ni coverage. Up to about 1.5–2.0 atomic layers of Ni there is no detectable magnetic dephasing for the Ni on top of a Pb film. However, for the Ni on top of the Ag (curve denoted with 0.0) one observes right away an additional dephasing that is due to magnetic Ni on top of the monovalent Ag.

6. Conclusion

In the investigation of strong Anderson localization in three dimensions and one-dimensional localization with arbitrary small disorder, the Langer–Neal diagram was originally just an auxilary mechanism to extrapolate the physics of the two-dimensional localization. However, this quantum interference effect in the conduction process, commonly called weak localization, has provided us with a new tool to measure important physical properties in solid state physics. New materials are constantly developed in particular in the nanoscale range. Here weak localization continues to give important insight into their properties. It is now about 50 years since Anderson's paper on localization in three dimensions, 40 years since Langer and Neal's paper on the fan diagram and about 30 years since Abrahams *et al.*[4] published the scaling paper and introduced weak localization. Anderson localization and weak localization are alive. Alone in the year 2009, Google yields more than 500 references for each of them. Actually the number has been slowly increasing over the past couple of years.

Finally I like to conclude with a historical comment. After my PhD, I worked as a post-doc (assistant) at the Institut of Physik in Goettingen which was at that time the center of quench-condensed thin film physics. With the support and expertise of my "Dokktor-Vater", Prof. Hilsch, we combined an existing cryostat for quenched condensation with an external iron-core magnet. This combined system permitted us to measure the critical magnetic field of quench-condensed superconductors. At that time I was considering using the system for measuring the magnetoresistance of quench-condensed Cu films. But then I recalled Kohler's rule about the magnetoresistence. (Dr. Kohler was at that time in Goettingen). According to Kohler's rule the magnetoresistence would be of the order of $(\omega\tau)^2$. With $\omega = eB/m \approx 2 \times 10^{11}$ for our magnetic field and $\tau \approx 0.3 \times 10^{-15}$, this would yield a relative magnetoresistence of less than 1×10^{-8}, much too small to be experimentally visible. I still remember that I was kind of proud that by pure theoretical considerations I avoided wasting any helium for such a useless experiment. Actually now I think that the time would not have been ready for an experimental result as shown in Fig. 11.5.

Quite a few years later I was looking for the predicted surface magnetism in Pd using quench-condensed Pd films. To my original delight, I observed a magnetic reaction, a strong and temperature dependent magnetoresistence as shown in Fig. 11.29.[24]

However, one effect was very puzzling; applying a magnetic field increased the temperature dependence of the resistance. I expected that a large magnetic field would either saturate or suppress the underlying physical effect and remove the temperature dependence.

Of course, looking backwards we know that I encountered two very different phenomena at the same time, the magnetoresistance and temperature effect of weak localization and the temperature effect of the electron–electron interaction. As a free bonus, the weak localization was in the strong spin-orbit limit which transformed it into weak anti-localization. This is a very good example that nature does not lie but it surely loves to tease its servants. Life is not made easier by the fact that our journals and referees generally object strongly to publishing an experimental observation that cannot be explained at the same time.

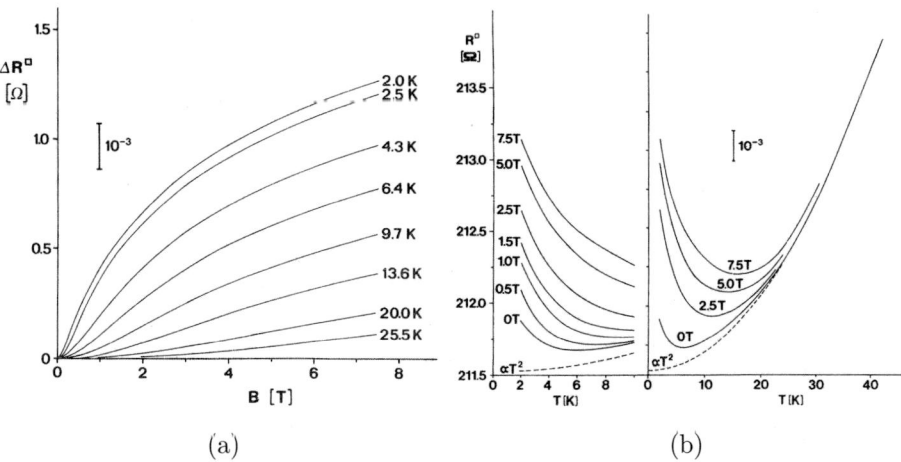

Fig. 11.29. (a) The magnetoresistance of a thin quench condensed Pd film (4 nm thick) at different temperatures. (b) The temperature dependent resistance of a thin quench-condensed Pd at different magnetic fields. Reprinted (Fig. 2) with permission from *Phys. Rev. Lett.* **43**, 1357 (1979). © American Physical Society.

Fortunately, there were independently two different theories for two-dimensional anomalies developed at the same time (which initially were competing to explain the experimentally observed temperature dependence of the resistance of thin disordered films). I encountered these theories delayed at the LT conference in Los Angeles (which enlightened my heart for California) where I had also a short conversation with Phil Anderson about a measurement of the spin-orbit effect. Inspired by this experience, I enjoyed very much to study the Kubo formalism to understand the Langer–Neal diagram and then interpret the physics as an interference experiment as described in Sec. 2. The greatest challenge was the spin-orbit effect in the Langer–Neal diagram, which appeared to puzzle everybody because it transformed the precursor of localization into anti-localization. Half a year later during a private work-surf shop at the beach of Senegal came the insight that it is the spin 1/2 rotation with its reversal of the sign of the spin-function that causes weak anti-localization.

References

1. P. W. Anderson, *Phys. Rev.* **124**, 41 (1961).
2. D. J. Thouless, *Phys. Rev. Lett.* **39**, 1167 (1977).
3. F. J. Wegner, *Z. Phys.* **35**, 207 (1979).

4. E. Abrahams, P. W. Anderson, D. C. Licciardello and T. V. Ramakrishnan, *Phys. Rev. Lett.* **42**, 673 (1979).
5. J. S. Langer and T. Neal, *Phys. Rev. Lett.* **16**, 984 (1966).
6. H. Fukuyama, *Surf. Sci.* **113**, 489 (1982).
7. B. L. Altshuler, A. G. Aronov, D. E. Khmelnitskii and A. I. Larkin, in *Quantum Theory of Solids*, ed. I. M. Lifshits (MIR Publishers, Moscow, 1982), p. 130.
8. R. C. Dynes, *Physica B and C* **109**, 1857 (1982).
9. Y. Nagaoka and H. Fukuyama (eds.), *Proc. 4th Taniguchi Int. Symp.*, Sandashi, Japan, 1981 (Springer-Verlag, Heidelberg, 1982), p. 89.
10. G. Bergmann, *Phys. Rev. B* **28**, 2914 (1983).
11. G. Bergmann, *Phys. Rep.* **107**, 1 (1984).
12. K. B. Efetov, *Adv. Phys.* **32**, 53 (1983).
13. H. Fukuyama, in *Modern Problems in Condensed Sciences*, eds. A. L. Efros and M. Pollak (North-Holland, Amsterdam, 1985), p. 155.
14. P. A. Lee and T. V. Ramakrishnan, *Rev. Mod. Phys.* **57**, 287 (1985).
15. B. L. Altshuler, A. G. Aronov, M. E. Gershenson and Y. E. Sharvin, *Sov. Sci. Rev. A* **39**, 223 (1987).
16. A. G. Aronov and Yu. V. Sharvin, *Rev. Mod. Phys.* **59**, 755 (1987).
17. M. E. Gershenson, *Annalen der Physik* **8**, 559 (1999).
18. J. J. Lin and J. P. Bird, *J. Phys. Condens. Matter* **14**, R501 (2002).
19. P. W. Anderson, E. Abrahams and T. V. Ramakrishnan, *Phys. Rev. Lett.* **43**, 718 (1979).
20. L. P. Gorkov, A. I. Larkin and D. E. Khmelnitzkii, *Pis'ma Zh. Eksp. Teor. Fiz.* **30**, 248 (1979); *JETP Lett.* **30**, 228 (1979).
21. B. L. Altshuler, A. G. Aronov and D. Khmelnitskii, *Solid State Commun.* **39**, 619 (1981).
22. B. L. Altshuler, A. G. Aronov and B. Z. Zpivak, *JETP Lett.* **33**, 94 (1981); *Pisma Zh. Eksp. Teor. Fiz.* **33**, 101 (1981).
23. D. Y. Sharvin and Y. V. Sharvin, *JETP Lett.* **34**, 272 (1981); *Pis'ma Zh. Eksp. Teor. Fiz.* **34**, 285 (1981).
24. G. Bergmann, *Phys. Rev. Lett.* **43**, 1357 (1979).
25. S. Hikami, A. I. Larkin and Y. Nagaoka, *Prog. Theor. Phys.* **63**, 707 (1980).
26. Y. Kawaguchi and S. Kawaji, *J. Phys. Soc. Jpn.* **48**, 699 (1980)
27. B. L. Altshuler, D. Khmelnitskii, A. I. Larkin and P. A. Lee, *Phys. Rev. B* **22**, 5142 (1980).
28. B. L. Altshuler and A. G. Aronov, *Solid State Commun.* **38**, 11 (1981).
29. B. L. Altshuler, A. G. Aronov, A. I. Larkin and D. Khmelnitskii, *Sov. Phys. JETP* **54**, 411 (1981); *Zh. Eksp. Teor. Fiz.* **81**, 768 (1981).
30. Y. Kawaguchi and S. Kawaji, in *Proc. 15th Int. Conf. Phys. Semicond.*, Kyoto, 1980; *J. Phys. Soc. Jpn.* **49** (Suppl. A), 983 (1980).
31. A. I. Larkin, *Pisma Zh. Eksp. Teor. Fiz.* **31**, 239 (1980); *JETP Lett.* **31**, 219 (1980).
32. M. J. Uren, R. A. Davies and M. Pepper, *J. Phys. C: Solid State Phys.* **13**, L985 (1980).

33. B. L. Altshuler, A. G. Aronov and B. Z. Zpivak, *JETP Lett.* **33**, 499 (1981), *Pisma Zh. Eksp. Teor. Fiz.* **33**, 515 (1981).
34. G. Bergmann, *Phys. Rev. Lett.* **48**, 1046 (1982).
35. G. Bergmann, *Phys. Rev. B* **25**, 2937 (1982).
36. G. Bergmann, *Phys. Rev. Lett.* **49**, 162 (1982).
37. H. Fukuyama, *J. Phys. Soc. Jpn.* **50**, 3407 (1981).
38. C. Van Haesendonck, L. Van den dries and Y. Bruynseraede, *Phys. Rev. B* **25**, 5090 (1982).
39. C. Van Haesendonck, L. Van den Dries, Y. Bruynseraede and G. Deutscher, *Physica B* **107**, 7 (1981).
40. Y. Kawaguchi and S. Kawaji, *Surf. Sci.* **113**, 505 (1982).
41. T. Kawaguti and Y. Fujimori, *J. Phys. Soc. Jpn.* **51**, 703 (1982).
42. P. A. Lee and T. V. Ramakrishnan, *Phys. Rev. B* **26**, 4009 (1982).
43. S. Maekawa and H. Fukuyama, *J. Phys. Soc. Jpn.* **50**, 2516 (1981).
44. W. C. McGinnis, M. J. Burns, R. W. Simon, G. Deutscher and P. M. Chaikin, in *Proc. 16th Int. Conf. Low Temp. Phys.*, Los Angeles, 1981, ed. W. G. Clark (North-Holland Amsterdam, Los Angelos, 1981), p. 5.
45. Y. Ono, D. Yoashioka and H. Fukuyama, *J. Phys. Soc. Jpn.* **50**, 2143 (1981).
46. Z. Ovadyahu, *Phys. Rev. B* **24**, 7439 (1981).
47. Z. Ovadyahu, S. Moehlecke and Y. Imry, *Surf. Sci.* **113**, 544 (1982).
48. M. J. Uren, R. A. Davies, M. Kaveh and M. Pepper, *J. Phys. C: Solid State Phys.* **14**, L395 (1980).
49. D. Abraham and R. Rosenbaum, *Phys. Rev. B* **27**, 1413 (1983).
50. B. L. Altshuler, A. G. Aronov, B. Z. Spivak, D. Y. Sharvin and Y. V. Sharvin, *JETP Lett.* **35**, 588 (1982); *Pisma Zh. Eksp. Teor. Fiz.* **35**, 476 (1982).
51. G. Bergmann, *Z. Phys. B* **48**, 5 (1982).
52. G. Bergmann, *Z. Phys. B* **49**, 133 (1982).
53. D. J. Bishop, R. C. Dynes and D. C. Tsui, *Phys. Rev. B* **26**, 773 (1982).
54. Y. Bruynseraede, M. Gijs, C. Van Haesendonck and G. Deutscher, *Phys. Rev. Lett.* **50**, 277 (1983).
55. R. A. Davies and M. Pepper, *J. Phys. C: Solid State Phys.* **16**, L361 (1983).
56. M. E. Gershenson, V. N. Gubankov and J. E. Juravlev, *Pis'ma Zh. Eksp. Teor. Fiz.* **35**, 467 (1982); *JETP Lett.* **35**, 467 (1982).
57. M. E. Gershenson and V. N. Gubankov, *Solid State Commun.* **41**, 33 (1982).
58. M. E. Gershenson, B. N. Gubankov and Y. E. Zhuravlev, *Sov. Phys. JETP* **56**, 1362 (1982); *Zh. Eksp. Teor. Fiz.* **83**, 2348 (1982).
59. H. Hoffman, F. Hofmann and W. Schoepe, *Phys. Rev. B* **25**, 5563 (1982).
60. Y. Isawa, K. Hoshino and H. Fukuyama, *J. Phys. Soc. Jpn.* **51**, 3262 (1982).
61. T. Kawaguti and Y. Fujimori, *J. Phys. Soc. Jpn.* **52**, 722 (1983).
62. S. Kobayashi S. Okuma and F. Komori, *J. Phys. Soc. Jpn.* **52**, 20 (1983).
63. Y. F. Komnik, E. I. Bukhshtab, A. V. Butenko and V. V. Andrievsky, *Solid State Commun.* **44**, 865 (1982).
64. F. Komori, S. Kobayashi and W. Sasaki, *J. Phys. Soc. Jpn.* **52**, 368 (1983).
65. D. A. Poole, M. Pepper and A. Hughes, *J. Phys. C: Solid State Phys.* **15**, L1137 (1982).

66. A. K. Savchenko, V. N. Lutskii and V. I. Sergeev, *Pis'ma Zh. Eksp. Teor. Fiz.* **36**, 150 (1982); *JETP Lett.* **36**, 185 (1982).
67. P. H. Woerlee, G. C. Verkade and A. G. M. Jansen, *J. Phys. C: Solid State Phys.* **16**, 3011 (1983).
68. G. Bergmann, *Solid State Commun.* **46**, 347 (1983).
69. G. Bergmann, *Phys. Rev. B* **28**, 515 (1983).
70. J. B. Bieri, A. Fert, G. Creuzet and J. C. Ousset, *Solid State Commun.* **49**, 849 (1984).
71. H. Ebisawa and H. Fukuyama, *J. Phys. Soc. Jpn.* **52**, 3304 (1983).
72. H. Ebisawa and H. Fukuyama, *J. Phys. Soc. Jpn.* **53**, 34 (1984).
73. S. V. Morozov, K. S. Novoselov, M. I. Katsnelson, F. Schedin, L. A. Ponomarenko, D. Jiang and A. K. Geim, *Phys. Rev. Lett.* **97**, 016801 (2006).
74. V. K. Dugaev, P. Bruno and J. Barnaś, *Phys. Rev. B* **64**, 144423 (2001).
75. J. M. Gordon, C. J. Lobb and M. Tinkham, *Phys. Rev. B* **28**, 4046 (1983).
76. Y. Isawa and H. Fukuyama, *J. Phys. Soc. Jpn.* **53**, 1415 (1984).
77. Y. Koike, M. Okamura, T. Nakamomyo and T. Fukase, *J. Phys. Soc. Jpn.* **52**, 597 (1983).
78. F. Komori, S. Kobayashi and W. Sasaki, *J. Phys. Soc. Jpn.* **52**, 4306 (1983).
79. F. Komori, S. Kobayashi and W. Sasaki, *J. Magn. Magn. Mater.* **35**, 74 (1983).
80. R. S. Markiewicz and C. J. Rollins, *Phys. Rev. B* **29**, 735 (1984).
81. D. S. McLachlan, *Phys. Rev. B* **28**, 6821 (1983).
82. S. Moehlecke and Z. Ovadyahu, *Phys. Rev. B* **29**, 6203 (1984).
83. Z. Ovadyahu, *J. Phys. C: Solid State Phys.* **16**, L845 (1983).
84. H. R. Raffy, R. B. Laibowitz, P. Chaudhari and S. Maekawa, *Phys. Rev. B* **28**, 6607 (1983).
85. A. C. Sacharoff, R. M. Westervelt and J. Bevk, *Phys. Rev. B* **26**, 5976 (1982).
86. A. E. White, R. C. Dynes and J. P. Garno, *Phys. Rev. B* **29**, 3694 (1984).
87. G. Bergmann, *Phys. Rev. B* **29**, 6114 (1984).
88. G. Bergmann and C. Horriar-Esser, *Phys. Rev. B* **31**, 1161 (1985).
89. L. Dumoulin, H. Raffy, P. Nedellec, D. S. MacLachlan and J. P. Burger, *Solid State Commun.* **51**, 85 (1984).
90. M. Gijs, C. Van Haesendonck and Y. Bruynseraede, *Phys. Rev. Lett.* **52**, 2069 (1984).
91. J. M. Gordon, C. J. Lobb and M. Tinkham, *Phys. Rev. B* **29**, 5232 (1984).
92. F. Komori, S. Okuma and S. Kobayashi, *J. Phys. Soc. Jpn.* **53**, 1606 (1984).
93. J. M. B. Lopes dos Santos and E. Abrahams, *Phys. Rev. B* **31**, 172 (1985).
94. B. Pannetier, J. Chaussy, R. Rammal and P. Gandit, *Phys. Rev. Lett.* **53**, 718 (1984).
95. P. Santhanam and D. E. Prober, *Phys. Rev. B* **29**, 3733 (1984).
96. J. C. Licini, G. J. Dolan and D. J. Bishop, *Phys. Rev. Lett.* **54**, 1585 (1985).
97. G. J. Dolan, J. C. Licini and D. J. Bishop, *Phys. Rev. Lett.* **56**, 1493 (1985).
98. R. P. Peters and G. Bergmann, *J. Phys. Soc. Jpn.* **54**, 3478 (1985).
99. L. E. Ballentine and M. Huberman, *J. Phys. C: Solid State Phys.* **10**, 4991 (1977).
100. M. Huberman and L. E. Ballentine, *Can. J. Phys.* **56**, 704 (1978).
101. A. A. Abrikosov and L. P. Gorkov, *Sov. Phys. JETP* **42**, 1088 (1962).

102. J. R. Asik, M. A. Ball and C. P. Slichter, *Phys. Rev.* **181**, 645 (1969).
103. R. Meservey and P. M. Tedrow, *Phys. Lett. A* **58**, 131 (1976).
104. S. Geier and G. Bergmann, *Phys. Rev. Lett.* **68**, 2520 (1992).
105. G. Bergmann, *Phys. Rev. B* **49**, 8377 (1994).
106. N. Papanikolaou, N. Stefanou, P. H. Dederichs, S. Geier and G. Bergmann, *Phys. Rev. Lett.* **69**, 2110 (1992).
107. M. E. Gershenson, V. N. Gubankov and Y. E. Zhuravlev, *Solid State Commun.* **45**, 87 (1983).
108. R. A. Davies and M. Pepper, *J. Phys. C: Solid State Phys.* **16**, L353 (1983).
109. G. Deutscher, M. Gijs, C. Van Haesendonck and Y. Bruynseraede, *Phys. Rev. Lett.* **52**, 2069 (1984).
110. W. C. McGinnis and P. M. Chaikin, *Phys. Rev. B* **32**, 6319 (1985).
111. J. B. Bieri, A. Fert, G. Creuzet and A. Schuhl, *J. Phys. F* **16**, 2099 (1986).
112. R. M. Mueller, R. Stasch and G. Bergmann, *Solid State Commun.* **91**, 255 (1994).
113. P. Mohanty, E,M,Q. Jariwala and R. A. Webb, *Phys. Rev. Lett.* **78**, 3366 (1997).
114. P. Mohanty and R. A. Webb, *Phys. Rev. B* **55**, R13452 (1997).
115. J. J. Lin, T. J. Li and Y. L. Zhong, *J. Phys. Soc. Jpn.* **72** (Suppl. A), 7 (2003).
116. P. Mohanty and R. A. Webb, *Phys. Rev. Lett.* **91**, 066604 (2003).
117. H. Takayama, *Z. Phys.* **263**, 329 (1973).
118. A. Schmid, *Z. Phys.* **271**, 251 (1973).
119. M. J. Kelly, *J. Phys. C: Solid State Phys.* **16**, L517 (19883).
120. S. Rammer and A. Schmid, *Phys. Rev. B* **34**, 13525 (1986).
121. P. B. Allen, *Phys. Rev. Lett.* **59**, 1460 (1987).
122. D. Belitz and S. D. Sarma, *Phys. Rev. B* **36**, 7701 (1987).
123. A. Sergeev and V. Mitin, *Phys. Rev. B* **61**, 6041 (2000).
124. B. L. Altshuler and A. G. Aronov, *Pis'ma Zh. Eksp. Teor. Fiz.* **30**, 514 (1979); *JETP Lett.* **30**, 482 (1979).
125. E. Abrahams, P. W. Anderson, P. A. Lee and T. V. Ramakrishnan, *Phys. Rev. B* **24**, 6783 (1981).
126. B. L. Altshuler, A. G. Aronov and D. E. Khmelnitskii, *J. Phys. C: Solid State Phys.* **15**, 7367 (1982).
127. H. Ebisawa, S. Maekawa and H. Fukuyama, *Solid State Commun.* **45**, 75 (1983).
128. H. Fukuyama and E. Abrahams, *Phys. Rev. B* **27**, 5976 (1083).
129. G. F. Giuliani and J. J. Quinn, *Phys. Rev. B* **26**, 4421 (1982).
130. W. Eiler, *J. Low Temp. Phys.* **56**, 481 (1984).
131. W. Eiler, *Solid State Commun.* **56**, 917 (1985).
132. M. Y. Reizer, *Phys. Rev. B* **39**, 1602 (1989).
133. I. L. Aleiner, B. L. Altshuler and M. E. Gershenson, *Wave Random Media* **9**, 201 (1999).
134. S. Roche, J. Jiang, F. Triozon and R. Saito, *Phys. Rev. Lett.* **95**, 076803 (2005).
135. P. Lang, V. S. Stepanyuk, K. Wildberger, R. Zeller and P. H. Dederichs, *Solid State Commun.* **92**, 755 (1994).

136. H. Beckmann, R. Schaefer, W. Li and G. Bergmann, *Europhys. Lett.* **33**, 563 (1996).
137. H. Beckmann and G. Bergmann, *Phys. Rev. B* **54**, 368 (1996).
138. G. Bergmann and H. Beckmann, *Phys. Rev. B* **52**, 15687 (1995).
139. R. Schaefer and G. Bergmann, *Solid State Commun.* **98**, 45 (1996).
140. H. Beckmann and G. Bergmann, *Phys. Rev. B* **55**, 14350 (1997).
141. H. Beckmann and G. Bergmann, *J. Low Temp. Phys.* **110**, 1173 (1998).
142. G. Bergmann, *Phys. Rev. B* **74**, 144420 (2006).
143. G. Bergmann, *Phys. Rev. B* **73**, 092418 (2006).
144. B. Maple, *Appl. Phys.* **9**, 179 (1976).
145. P. M. Chaikin and T. W. Mihalisin, *Solid State Commun.* **10**, 465 (1972).
146. M. B. Maple, W. A. Fertig, A. C. Mota, D. E. DeLong, D. Wohlleben and R. Fitzgerald, *Solid State Commun.* **11**, 829 (1972).
147. R. P. Peters, G. Bergmann and R. M. Mueller, *Phys. Rev. Lett.* **58**, 1964 (1987).
148. C. V. Haesendonck, J. Vranken and Y. Bruynseraede, *Phys. Rev. Lett.* **58**, 1968 (1986).
149. W. Wei and G. Bergmann, *Phys. Rev. B* **37**, 5990 (1988).
150. H. C. Manoharan, C. P. Lutz and D. M. Eigler, *Nature (London)* **403**, 512 (2000).
151. G. M. Alzoubi and N. O. Birge, *Phys. Rev. Lett.* **97**, 226803 (2006).
152. F. Mallet, J. Ericsson, D. Mailly, S. Ünlübayir, D. Reuter, A. Melnikov, A. D. Wieck, T. Micklitz, A. Rosch, T. A. Costi, L. Saminadayar and C. Bäuerle, *Phys. Rev. Lett.* **97**, 226804 (2006).
153. G. Bergmann, *Phys. Rev. Lett.* **57**, 1460 (1986).
154. G. Bergmann, *Phys. Rev. Lett.* **58**, 1236 (1987).
155. R. P. Peters, G. Bergmann and R. M. Mueller, *Phys. Rev. Lett.* **60**, 1093 (1988).
156. K. G. Wilson, *Rev. Mod. Phys.* **47**, 773 (1975).
157. P. Nozieres, *J. Low Temp. Phys.* **17**, 31 (1974).
158. G. Bergmann, *Phys. Rev. B* **39**, 11280 (1989).
159. W. Wei and G. Bergmann, *Phys. Rev. B* **40**, 3364 (1989).
160. G. Bergmann and W. Wei, *Solid State Commun.* **71**, 1011 (1989).
161. G. Bergmann, *Phys. Rev. Lett.* **53**, 1100 (1984).
162. G. Tateishi and G. Bergmann, *Eur. Phys. J. B* **73**, 155 (2010).

Chapter 12

WEAK LOCALIZATION AND ELECTRON–ELECTRON INTERACTION EFFECTS IN THIN METAL WIRES AND FILMS

N. Giordano

Department of Physics, Purdue University,
West Lafayette, IN 47907, USA
giordano@purdue.edu

A brief and selective review of experimental studies of electrical conduction in thin metal wires and films at low temperatures is given. This review will illustrate the importance of various length scales and of dimensionality in determining the properties disordered metals. A few intriguing and still unresolved experimental findings are also mentioned.

1. Introduction

The basic notion of what is now termed "localization" was first described more than 50 years ago, when Anderson[1] showed how disorder can lead to spatially localized electronic states. That profound discovery changed the way we think about electron wave functions and conduction in disordered systems, and led directly to an enormous body of work, much of which is described and reviewed in this volume. While Anderson's work immediately attracted a great deal of interest, the vast majority of the experiments in the decade following Anderson's paper did not involve what might be termed "conventional" metals, such as Au or Cu, or their alloys, since it was thought that the levels of disorder needed to observe localization in such systems would be difficult to attain or perhaps prohibitively large.

The importance of dimensionality in localization was recognized and emphasized by Mott, Twose and Ishii,[2,3] who argued that all states are localized in one dimension, for any amount of disorder (as long as it is nonzero). However, this result was viewed as being mainly of theoretical importance (although its relevance to conduction in polymers was discussed) until the seminal work of Thouless.[4] Thouless considered localization in thin (i.e., small diameter) metal wires, and argued that the states in a wire would all be localized, provided that the wire is sufficiently long so that its residual resistance exceeds about 10 kΩ. Furthermore, Thouless argued that if the temperature is very low, so that inelastic processes are very weak, localization would cause the resistance of such a wire to be greater than the value caused by ordinary elastic scattering. This extra resistance was expected to increase as the inelastic scattering rate becomes smaller, that is, as the temperature is reduced.

The work of Thouless stimulated a new wave theoretical and experimental work, much of which is discussed in this volume. On the theoretical front, the effect of dimensionality on the nature of the electron states was examined in detail, and a scaling theory of localization for noninteracting electrons was formulated by Abrahams *et al.*[5] The scaling theory, which has since been developed further by many authors (see, for example, Refs. 6 and 7), predicts that all states are localized in both one and two dimensions, with a transition from localized to extended states occurring only in dimensions above two. The extreme sensitivity to even small amounts of disorder in one and two dimensions led to the term "weak localization" to describe the behavior of these systems. Hence, thin metal wires and films should both exhibit weak localization, that is, an increasing resistance at low temperatures. There were many other theoretical discoveries, some of which will be mentioned below. We have termed these "theoretical discoveries" since in many cases the theory recognized the importance and implications of various effects before their experimental observation.

On the experimental front, a number of attempts were made to observe the effects predicted by Thouless. Using the best estimates for the inelastic electron scattering rates, he predicted that the increase in the resistance produced by localization should be easily observable with wires of diameter 500 Å at 1 K.[4,8] Wires with such small diameters were near the limits of fabrication methods available at the time. Some creative methods for making such wires were devised (see, e.g., Ref. 9), but the early results were negative. As we will explain in the next section, the initial estimates for the magnitude of the resistance change at low temperatures were much too large, due to an incomplete understanding of electron phase coherence. It turns out that there were a host of hitherto unknown or unappreciated electron scattering processes, in addition to the inelastic processes considered by Thouless. These processes can play a dominant role in localization, and at the same time, localization has turned out to be a unique tool with which to study such processes.

We have so far mentioned only localization, which involves noninteracting electrons in a disordered potential. We will see that systems for which localization is important also exhibit effects due to the interactions between electrons. These electron–electron interaction effects were not anticipated until the work of Altshuler *et al.*[10,11] and will have an important role in our story.

This paper describes, from the author's perspective, the evolution of our understanding of electron localization and electron–electron interaction effects in thin metal wires and (in a few cases) thin films. We begin with the first observations of localization and electron–electron interaction effects in thin wires, and progress through studies of electron phase coherence lengths. These experiments led naturally to studies of systems that are now termed *mesoscopic*. These are systems for which the electron phase coherence lengths are all larger than all dimensions of the system. We close with a discussion of several effects in thin films that remain puzzling and unexplained. In the spirit of this volume, this paper is a selective review that focuses mainly on the author's own work.

2. Resistance of Thin Wires: Dependence on Temperature, Diameter, and Length

The earliest studies of localization effects in thin wires were aimed at simply observing an increase in the resistance at low temperatures. Thouless' initial estimates suggested that for a wire composed of a strongly disordered metal, there would be an easily measurable resistance increase for a 500 Å diameter wire at 1 K, but initial experiments with such samples gave null results. The first observation of a low temperature resistance increase in thin wires[12] came in experiments with wires composed of a AuPd alloy. This has turned out to be a very useful material for such studies, as it is easily deposited as a very fine-grained film and resists oxidation. Some results for AuPd wires with a range of diameters are shown in Fig. 12.1.

The low temperature resistance increase in Fig. 12.1 is smaller than the estimates of Thouless,[4,8] and the temperature dependence is much weaker. Thouless' predictions were based on the inelastic scattering processes that were expected to dominate in metals. The results in Fig. 12.1 were thus an early indication that our understanding of inelastic processes in these systems was far from complete. Despite the differences between the experimental results and the theoretical expectations, the fact that the resistance increase in Fig. 12.1 becomes larger as the wire diameter is made smaller rules out "bulk" mechanisms such as the Kondo effect, and confirms the importance of dimensionality.

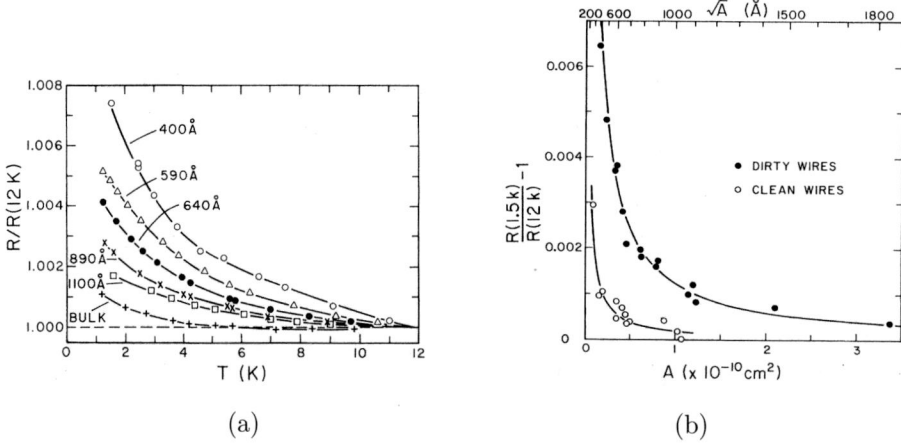

Fig. 12.1.   (a) Low temperature resistance of AuPd wires with various diameters.
Also shown for comparison is the behavior of a AuPd film, labeled here as "bulk".
The increase in resistance at low temperatures for the film is caused by localization and electron–electron interactions in two dimensions. (b) Resistance increase
$\Delta R/R_0$ at 1.5 K as a function of A, where A is the cross-sectional area, for AuPd
wires with a short elastic mean free path (the "dirty" wires) and long mean free
path ("clean" wires). From Ref. 12.

The dependence of the low temperature resistance increase on the cross-
sectional area A of the wire is also shown in Fig. 12.1. The fractional resistance increase $\Delta R/R_0$ at 1.5 K (relative to the resistance at the "high"
temperature of 12 K) was found to vary as $1/A$.

Figure 12.1(b) shows results for wires composed of AuPd alloys prepared
in two different ways, giving different levels of disorder. A convenient measure of the disorder in such alloys is the residual resistivity; this can also be
used to infer the elastic mean free path ℓ_e, which was approximately 5 Å in
the "dirty" AuPd wires and about 25 Å in the "clean" wires. The results in
Fig. 12.1(b) show that $\Delta R/R_0$ increases approximately as $1/\ell_e$, as expected
for localization in one dimension.

We have already mentioned that the temperature dependence in Fig. 12.1
was much weaker than expected based on the inelastic electron processes
known at the time. In fact, over the limited temperature range available
in the first experiments, the resistance increase could not be distinguished
from a logarithmic dependence on temperature, which contrasted with the
expected power law (e.g., T^{-1} or stronger). Experiments with AuPd wires
over a wider temperature range (Fig. 12.2; Ref. 13) made it possible to rule
out a logarithmic dependence, and were adequately described by the very

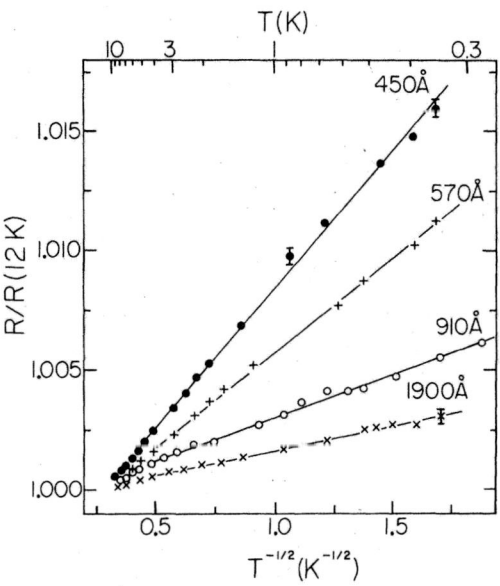

Fig. 12.2.   Temperature dependence of the resistance of several "dirty" (short mean free path) AuPd wires. The resistance increase at low temperatures is consistent with a $T^{-1/2}$ temperature dependence. From Ref. 13.

weak power law $\Delta R/R_0 \sim T^{-1/2}$, although other exponents close to 1/2 were also consistent with the measurements.

At about the same time as the $T^{-1/2}$ temperature dependence in Fig. 12.2 was observed, new theoretical work demonstrated the importance of electron–electron interaction effects in disordered systems.[7,10,11] Interaction effects were predicted to affect the resistance of thin wires and films in two ways. One was through electron–electron scattering processes which are very weakly inelastic. These are processes in which relatively little energy is exchanged, but the electron phase is randomized, just as in a typical inelastic event. The electron–electron phase coherence time from this process depends on temperature as T^{-1} in two dimensions (thin films), and $T^{-2/3}$ in one dimension (thin wires). The latter would then lead to a localization effect $\Delta R_{\mathrm{loc}}/R_0 \sim T^{-1/3}$ which was consistent with the temperature dependence found in Fig. 12.2.

A second effect predicted for electron–electron interactions in the presence of disorder was a low temperature resistance increase ΔR_{ee} separate from that due to localization.[10,11] It was predicted that $\Delta R_{ee} \sim T^{-1/2}/A$, which is precisely the dependence on both temperature and wire diameter as found in AuPd (Figs. 12.1 and 12.2). Moreover, the predicted magnitude of ΔR_{ee} was quite close to that found in the experiments.

The conclusion after our early experiments with AuPd was that there is indeed a low temperature resistance increase that depends on dimensionality, but the cause of this increase could be either localization or electron–electron interaction effects. The latter provided a slightly better account of the experiments, but it seemed likely that both effects should be present. The question of how to determine the relative importance of weak localization and electron–electron interaction effects was addressed by Bergmann[14] who showed how magnetoresistance measurements can be used to distinguish the two mechanisms. Bergmann's work involved thin films; we will describe the results of similar experiments with thin wires in Sec. 4.

3. Early Observation of Mesoscopic Effects

In a disordered metal, the electron motion is described by diffusion with a diffusion constant D determined by the elastic mean free path ℓ_e. If an electron maintains phase coherence for a time τ_ϕ, one can then define a diffusive length $L_\phi = \sqrt{D\tau_\phi}$. Various types of scattering processes can affect or limit the phase coherence time, including electron–phonon scattering, electron–electron scattering, and scattering from localized spins. The length L_ϕ plays a central role in localization, since it determines the effective dimensionality of a system. So, for example, a thin wire will behave one-dimensionally as far as localization is concerned if the diameter is less than L_ϕ. It is also turns out that $\Delta R_{\text{loc}} \sim L_\phi$ (where ΔR_{loc} is the extra resistance due to localization).

A similar diffusive length scale plays a similar role in electron–electron interaction effects. In that case the diffusive thermal length $L_T = \sqrt{\hbar D/k_B T}$ determines the effective dimensionality of a system and $\Delta R_{ee} \sim L_T$ (where ΔR_{ee} is the extra resistance due to electron–electron interactions).

The length scales L_ϕ and L_T determine the effective dimensionality of a sample. Hence, a wire will behave one dimensionally if its diameter is smaller than L_ϕ and/or L_T, while a film will behave two-dimensionally if its thickness is smaller less than L_ϕ and/or L_T. One can also consider the behavior if the *length* of a wire is comparable to or smaller than L_ϕ and L_T. In this case, the "wire" would be effectively zero dimensional, i.e., mesoscopic. The first studies in this regime appear to be those of Masden.[15] He started with wires similar to those considered in Figs. 12.1 and 12.2 (although Masden studied Pt wires instead), and then placed very closely spaced contacts along the wire. Some of his results are shown in Fig. 12.3, which shows

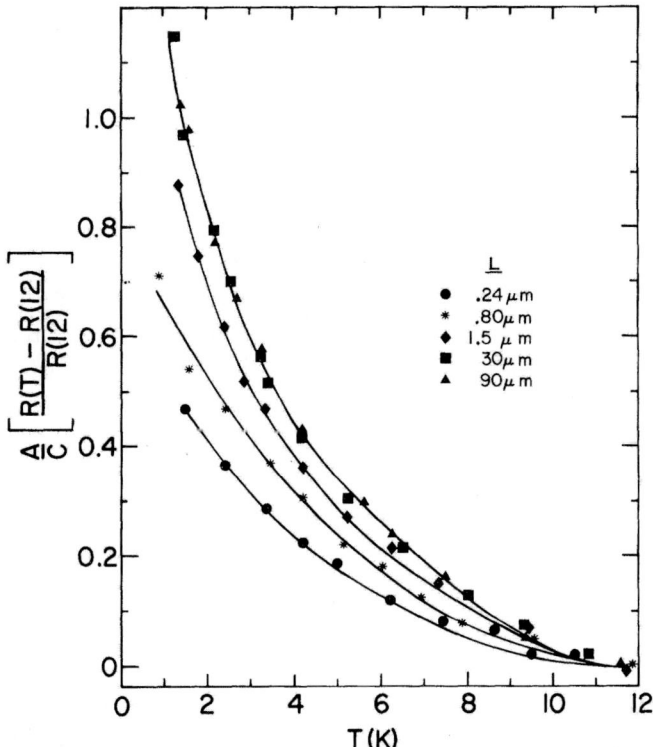

Fig. 12.3. Low temperature resistance increase $\Delta R/R_0$ for a series of Pt wires of different lengths L. These wires had slightly different cross-sectional areas, all near 500 Å × 500 Å. To remove the dependence of ΔR on A, ΔR has been multiplied by A so as to remove the dependence on area, leaving only the dependence on sample length. After Ref. 15.

the behavior for wires of different lengths. The resistance increase is seen to become smaller as the wire is made shorter than about 1.5 μm, and the resistance increase is reduced by about a factor of two when the wire length was reduced to 0.24 μm = 240 nm. This provides a direct measurement of the length scale that governs the behavior in these wires, and the result is in good agreement with that inferred from the temperature dependence of the resistance. Masden's experiment was the first to observe such mesoscopic effects. However, such length dependent behavior is expected for both localization and electron–electron interactions, so these results do not tell us which mechanism is causing the low temperature resistance changes in these wires.

4. Magnetoresistance and Electron Phase Coherence

The results in Figs. 12.1–12.3 show a variety of size dependent effects in
the resistance of thin metal wires. However, these experiments do not allow
us to unambiguously distinguish between localization and electron–electron
interaction effects. It turns out that measurements of the magnetoresistance
do allow one to distinguish between the two effects. Such studies were carried out by Bergmann in an elegant set of experiments with thin films[14]
that confirmed nicely the theoretical predictions.[10,11] Bergmann also gave
a intuitive explanation of why the magnetoresistance is an important tool
for exploring localization and electron phase breaking in disordered systems.
He emphasized how weak localization can be understood in terms of the
interference of time-reversed diffusive electron trajectories. Of particular
importance are trajectories that return to the "origin" (i.e., their starting
point). These trajectories form irregular "loops" that can be traversed in either of two directions. Electrons that follow such a loop in one direction can
be thought of as a partial wave that interferes with its time-reversed partner
that traverses the same loop in the opposite sense, when both return to the
origin.

This interference has two important consequences. First, if the phase
breaking due to inelastic and other processes is very weak, the electron
partial waves will maintain their coherence as they traverse their loop. In the
simplest case, the two electron partial waves interfere constructively when
they return to the origin, since they undergo the same series of phase shifts
(but in opposite order) as they scatter elastically on their way around the
loop. This constructive interference gives an "extra" probability to return to
the origin (as compared to the purely classical diffusive probability), which
means there is a lower probability to diffuse away from the origin and hence
a lower conductance. This is just the resistance increase caused by weak
localization. This simple case turns out to be relatively difficult (but not
impossible) to observe in metal wires and films. In most metals, spin-orbit
scattering connected with the normal elastic scattering process serves (in
a classical sense) to rotate the spin of an electron as it diffuses around its
loop. The electron partial wave that navigates the loop in the opposite
direction also undergoes spin rotation, but due to the way successive spin
rotations combine, the spin rotations of the two electron partial waves lead,
on average, to *destructive* interference of the partial waves when they return
to the origin.[14] This leads to the surprising result that in the presence of
spin-orbit scattering, the resistance is *lower* than expected classically, an
effect known as antilocalization. Spin-orbit scattering is strongest in high Z

metals, and in practice all but the lightest (low Z) metals are dominated by antilocalization.

An important consequence of this picture of interfering partial waves is that these waves lead to Aharonov–Bohm-like experiments. Application of a magnetic field perpendicular to the plane of the loop produces electron phase shifts that are of opposite sign for the two counterpropagating partial waves. The total conductance depends on the interference of many such loops, with different areas, so the application of a field will, in sum, quench the interference and suppress the resistance change associated with weak localization. The characteristic field for this suppression is the field that gives one flux quantum through an area of order L_ϕ^2. Hence, by measuring how ΔR_{loc} varies with field, one can directly measure the size of the largest loops that can give interference, which just the phase breaking length L_ϕ. (The dependence of ΔR_{loc} on magnetic field, while somewhat complex, has been calculated analytically; see Refs. 11 and 14.)

Measurements of the magnetoresistance of thin wires was carried out by several groups (see, e.g., Ref. 16). In our group, Lin studied AuPd wires, and showed that: (1) the resistance changes observed previously in AuPd and Pt as a function of temperature (Figs. 12.1–12.3) was dominated by electron–electron interaction effects; and (2) AuPd exhibits antilocalization, as expected since it is a high Z material. Some of Lin's results for the phase breaking length in a AuPd wire[17] are shown in Fig. 12.4. Here we show the phase breaking time τ_ϕ, inferred from the relation $L_\phi = \sqrt{D\tau_\phi}$.

Lin's results showed that τ_ϕ saturates at low temperatures, which can be explained as follows. The total phase breaking in AuPd in this temperature range is caused by two mechanisms, with two different phase breaking times. One mechanism gives a temperature dependent τ_ϕ, and may be due to spin-spin scattering. This is a process in which the spin of the electron interacts with the spin of local moments in the AuPd. At temperatures above the Kondo temperature of the local moment, such scattering should give a temperature independent contribution to the phase breaking time. When two different scattering mechanisms are at play, the total phase breaking time will be $1/\tau_{\phi,\mathrm{total}} = 1/\tau_{\phi,1} + 1/\tau_{\phi,2}$. In this case $\tau_{\phi,1}$ is due to spin-spin scattering (and is thus constant) while $\tau_{\phi,2}$ is believed to be due to electron–electron scattering. Fitting the total phase breaking time to this functional form (with a constant $\tau_{\phi,1}$) gives the phase breaking times plotted as open circles in Fig. 12.4. These follow the $T^{-2/3}$ dependence predicted for electron–electron scattering[18] with a magnitude that also agrees with the theory.

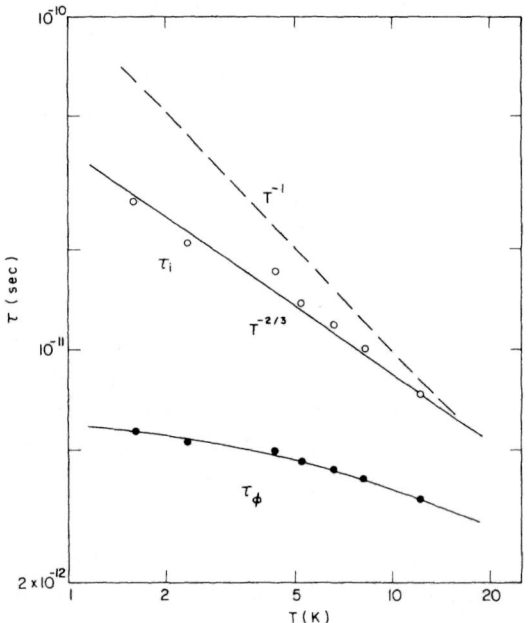

Fig. 12.4.   Phase breaking time τ_ϕ for a AuPd wire (solid circles). The solid circles show the measured electron phase coherence time. Fitting this to a combination of spin-spin scattering (which is temperature independent) and a temperature dependent scattering time gives a $T^{-2/3}$ dependence for the temperature dependent time, which is shown by the open circles. The open circles are in good agreement with the theory of weakly inelastic electron–electron scattering by Altshuler *et al.*,[18] Results from Ref. 17.

The phenomenon of antilocalization is counterintuitive, since it implies that the addition of disorder causes a *reduction* in the resistance. In the vast majority of systems, including the AuPd and Pt wires we have considered above, the simultaneous presence of electron–electron interaction effects, which increase the resistance, causes the total resistance to increase in the presence of disorder. However, there is no fundamental reason why interaction effects must always dominate the contribution of antilocalization. A case in which antilocalization dominates was discovered by Beutler in his studies of Bi wires.[19] Bulk pure Bi is normally a semimetal, but when Bi is deposited as an evaporated thin film it behaves as a metal with a very low carrier concentration. For this reason, the screening effects that play a central role in electron–electron interaction effects are (relatively) weaker than in metals such as Au, AuPd, or Pt.

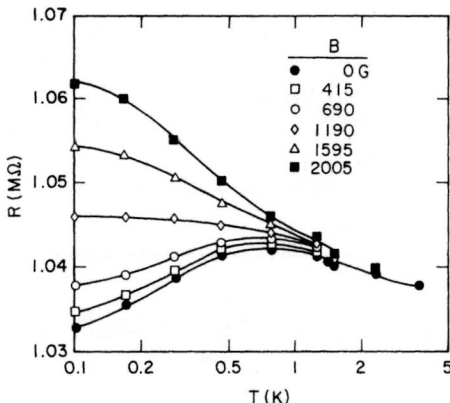

Fig. 12.5.   Resistance as a function of temperature for a Bi wire at several different values of the magnetic field, with the field applied perpendicular to the wire. From Ref. 19.

Figure 12.5 shows results for the resistance of a thin Bi wire as a function of temperature, for various values of the applied magnetic field. When a field greater than about 1 kOe is applied (and at temperatures below about 1 K) antilocalization is suppressed, leaving only an increase in the resistance (ΔR_{ee}) due to electron–electron interactions. However, in zero field, the resistance *decreases* at low temperatures, as antilocalization dominates. This experiment thus demonstrates the existence of antilocalization in a very striking way, with disorder *decreasing* the resistance.

Experiments like those of Lin and Beutler, and those of other groups (e.g., Ref. 16), provided detailed results for localization and antilocalization, and for various phase breaking times and mechanisms in thin wires. Magnetoresistance has thus proven to be a powerful tool for the study of electron scattering mechanisms.

5. Universal Conductance Fluctuations

The diffusive nature of electron motion in disorder metals is central to weak localization and antilocalization. It also leads to interesting fluctuations in the conductance. These fluctuations are caused by various types of "changes" in a system, which can be produced by changes in an applied magnetic field (which affects the relative phases of different electron trajectories), or by changes in the random potential (Refs. 20–23). When measured in terms of the conductance these fluctuations can assume a "universal"

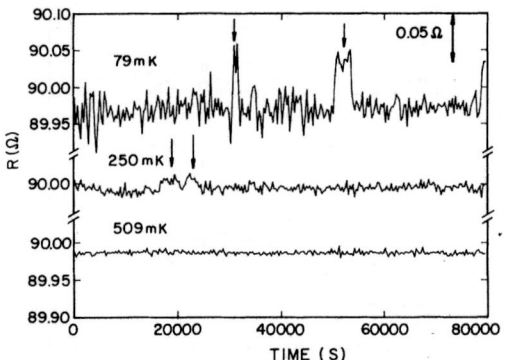

Fig. 12.6. Time dependent resistance fluctuations in a mesoscopic Ag sample. From Ref. 25.

value of order $< (\delta G)^2 >^{1/2} \sim e^2/\hbar$, and are termed universal conductance fluctuations.

Experimentally it is simplest to measure the fluctuations as a function of magnetic field (see Ref. 23). It is more complicated to measure the fluctuations produced by changes in the random potential, but we were fortunate to encounter samples that were very cooperative. Our group[24] was the first to observe such fluctuations through the time dependent changes in the conductance, in studies of thin Bi wires. Later studies of fluctuations with a similar origin in Ag samples[25,26] are shown in Fig. 12.6. The samples used here were thin (100 Å) Ag films whose width and length were both less than 1 μm, making this another case of a mesoscopic sample.

Figure 12.6 shows the resistance as a function of time at several temperatures. The resistance is "noisy" at all three temperatures, but this noise is not due to the measurement electronics, but is instead a property of the sample. Roughly speaking, there are two types of fluctuations in Fig. 12.6; slow fluctuations (indicated by the arrows) in which the resistance abruptly changes to a new (average) value and then changes back to its original value, and fast fluctuations which appear more like random noise. It is quite striking that the magnitude of the fast fluctuations *increases* as the temperature is reduced. The is also true for the slow fluctuations in Fig. 12.6.

The origin of these resistance fluctuations is the motion of individual scattering centers (impurities, etc.) with time.[20] This motion is expected to be take place over a wide range of time scales and produce fluctuations with a range of magnitudes. Indeed, the theory[20,22] suggests that the power spectrum should be approximately $1/f$, where f is the frequency, and spectral analysis of data like that in Fig. 12.6 confirms this prediction.[25,26] The arrows in Fig. 12.6 are thus believed to be particularly large fluctuations in which an impurity moves from one location to another, and then back to its original position.

In this description of the time dependent fluctuations we have given only a very brief explanation. It is quite fascinating that the motion of a single impurity in a sample containing $\sim 10^9$ atoms can produce a fluctuation nearly as large as that produced by a complete change in the random potential.[20–22] The basic reason is that with diffusive classical paths in low dimensions, each path has a significant probability of visiting every scattering center in a sample. As a result, the motion of a single scattering center can affect a large number of classical paths, giving changes that are large and having (under certain conditions) a universal value.

6. Mesoscopic Photovoltaic Effect

One of the important lessons learned from the study of universal conductance fluctuations is that even small changes in the distribution of impurities/scattering centers can have a large effect. This applies not just to the conductance, but to other properties as well. One such property is the photovoltaic effect, which was first studied in mesoscopic systems by Fal'ko and Khmel'nitskii.[27] The photovoltaic effect involves the response of a system to an ac field. Typically a sample is exposed to a very high frequency electric field (e.g., microwave or optical) and a dc voltage is produced in response. In ordinary samples, a nonzero dc voltage is possible only in systems that lack inversion symmetry, so many systems do not exhibit a photovoltaic effect. However, it would be very rare for an actual sample to have perfect inversion symmetry. Even if the crystal structure itself has inversion symmetry, the presence of even one impurity or a surface, etc., will break this symmetry.

In practice, the effect of this breaking of inversion symmetry by impurities is usually small. The surprise is that in mesoscopic systems, the photovoltaic effect can be quite substantial. This is because of its connection to universal conductance fluctuations. In the limit of a weak ac field, each absorbed photon will excite an electron, which will then diffuse through the sample. Because of the asymmetry induced by disorder, different numbers of electrons will diffuse to the two contacts attached to a sample, producing a dc voltage. Fal'ko and Khmel'nitskii[27] have shown that this asymmetry in the numbers of electrons that reach the contacts is proportional to the magnitude of the

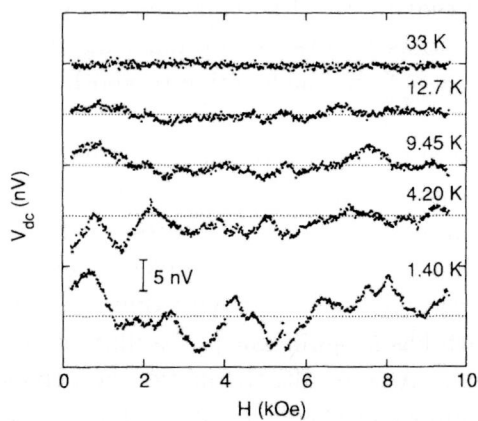

Fig. 12.7.   Photovoltaic effect for a Ag sample with dimensions 100 Å×1 μm ×1 μm. The dotted lines are the horizontal axes ($V_{\rm dc} = 0$) for the different temperatures. The magnetic field was applied perpendicular to the plane of the sample and the ac frequency was 8.4 GHz. After Ref. 28.

universal conductance fluctuations, e^2/h. Other features of the photovoltage can also be derived from the properties of universal conductance fluctuations discussed in Sec. 5.

Our group studied the photovoltaic effect in small samples of various metals, including Au, Ag, and Au with a small concentration of magnetic impurities such as Fe.[28] These samples were typically 100 Å thick and 1 μm on a side. At the lowest temperatures studied, the phase breaking length was also about 1 μm, so these samples were mesoscopic. Results for a Ag sample are shown in Fig. 12.7, which shows the photovoltage (the dc voltage measured across the sample) as a function of magnetic field at several temperatures.

Drawing on the analogy with universal conductance fluctuations, we expect that different samples, i.e., samples with different impurity distributions, will exhibit different values of $V_{\rm dc}$ (different in magnitude and sign). The application of a magnetic field also effectively changes the disorder, since the field affects the phases accumulated along the interfering electron paths. Hence $V_{\rm dc}$ fluctuates as a function of field, similar to the resistance fluctuations in Fig. 12.6.

Most striking in Fig. 12.7 is that the fluctuation magnitude increases as the temperature is reduced. This is because of the temperature dependence of the phase coherence length L_ϕ. At high temperatures, where L_ϕ is smaller than the sample size, the sample can be viewed as many fluctuating subunits, each of order $L_\phi \times L_\phi$, and each having a photovoltaic voltage with a certain average size. These voltages are themselves fluctuations, and the voltage of different subunits have random signs, giving a total photovoltaic voltage that is less than that due to a single subunit.[27] As the temperature is reduced, the number of subunits decreases, and the total photovoltaic voltage increases. The temperature dependence of V_{dc} is linked to the temperature dependence of L_ϕ, and the experimental results are in reasonable agreement with separate measurements of L_ϕ.[28]

Overall, the results for the photovoltaic effect agree well with the theory of Fal'ko and Khmel'nitskii, and are another example of the universal nature of many effects in mesoscopic systems.

7. Behavior of Parallel Metal Layers: Some Puzzles

The interaction between electrons plays a central role in localization, since electron–electron scattering often determines the electron phase coherence length, L_ϕ. This interaction is, of course, also at the heart of electron–electron interaction effects. Qualitatively, interaction effects are different in disordered systems than in ordered ones because of the diffusive nature of the electron motion. Roughly speaking, diffusion increases the interaction time, enhancing the electron–electron scattering rate. A related and important aspect of electrons in a disordered system is that the screening behavior is different than in a pure system, again because of the diffusive nature of electron motion. There are various ways to probe this change in the screening behavior in disordered metals. One approach was taken by Missert and Beasley[29] in their study of the critical temperature of disordered superconducting films placed very near a high conductance ground plane. No affect on T_c was found, contrary to expectations. Another intriguing way to probe these issues is to study electrons confined in nearby but "separated" layers. Studies of this kind have been carried out by Bergmann and Wei[30] in experiments with sandwich samples in which Mg films were separated by insulating layers, and in semiconductor heterostructures (e.g., Ref. 31) in studies of electron drag effects. All of this work prompted our group to explore electron–electron interactions in metal/insulator/metal sandwich samples.

In our first experiments, we studied samples composed of Sb/SiO/Sb trilayer films.[32] (The evaporated Sb films in these trilayers behaved as metals with a very low carrier concentration.) The goal was to see how the electron–electron scattering rate in one Sb film is affected by the presence of a second,

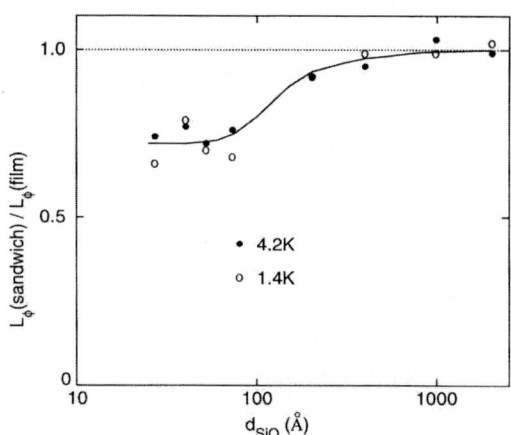

Fig. 12.8. Electron phase coherence length in Sb/SiO/Sb sandwich samples, normalized by the phase coherence length for a single Sb layer, as a function of the thickness of the SiO layer. That is, L_ϕ of the sandwich samples is normalized by the phase coherence length for an identical Sb film that is not part of a sandwich sample. In all of these samples, L_ϕ is dominated by electron–electron scattering, and these results show that this scattering rate is increased (and hence L_ϕ is made shorter) when the Sb films in a sandwich are within about 100 Å. From Ref. 32.

nearby Sb layer. Here the two layers were separated by an insulating layer of SiO. The SiO layers in these samples contained a small number of pinholes, so it was only possible to measure the resistance of the two conducting layers (the Sb) in parallel, but it was still possible to use measurements of the magnetoresistance to determine the electron phase coherence length. Some results for L_ϕ are given in Fig. 12.8, which shows the phase coherence length in a sandwich sample relative to L_ϕ in a single, isolated Sb film. To within the uncertainties, L_ϕ(sandwich) is equal to L_ϕ(film) when the SiO layer is thicker than about 200 Å, but for thinner SiO layers, L_ϕ(sandwich) falls to about 70% of L_ϕ(film).

The results in Fig. 12.8 show that electron–electron scattering becomes stronger when the Sb layers are brought close together (the scattering rate increases), suggesting that electrons in one layer are able to scatter from electrons in the other layer. To probe this further, we developed a way to make trilayers in which the middle (insulating) layer was free of pinholes, so that the two outer (conducting) layers were electrically isolated. We then arranged to contact the two conducting layers separately, so that their resistances could be measured independently.[33]

In two dimensions, electron–electron interactions make a contribution to the conductance of the form[10]

$$\Delta G_\square = \frac{e^2}{2\pi^2 \hbar} A_{ee} \ln T \,, \tag{7.1}$$

where $G_\square$ is the conductance per square. The factor A_{ee} depends weakly on the details of the screening, and is typically around 0.9–1.0 in a metal film.

Monnier studied Al/SiO/Sb sandwich samples, and performed measurements of the the conductance of the Sb layers in magnetic fields large enough to suppress the contribution from weak localization. The Al layers served as highly conducting ground planes, and the goal was to study how this ground plane affects the screening and electron–electron interactions in a nearby Sb layer. (Here the temperatures and magnetic fields were such that the Al film was not superconducting.) The temperature dependence of the Sb conductance was described well by Eq. (7.1), and results for A_{ee} are shown in Fig. 12.9.

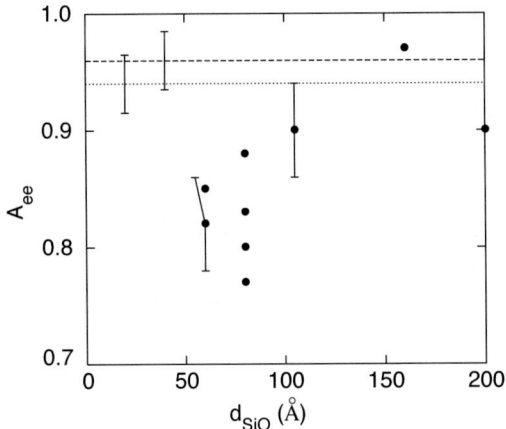

Fig. 12.9.   Results for A_{ee} as defined in Eq. (7.1) in several Sb films. The dotted line shows results reported by various workers for Sb films, while the dashed line shows results obtained for Sb films by Giordano and Monnier.[33] The filled circles show values of A_{ee} measured for Sb films that were part of Al/SiO/Sb trilayer samples, as a function of d_{SiO}, the thickness of the SiO layer, and representative error bars are given. The point shown at $d_{SiO} = 200$ Å actually corresponds to $d_{SiO} = 1000$ Å, but has been shifted to avoid a greatly compressed horizontal scale. In these samples, the Al films were deposited first, at the "bottom" of the trilayer, and there was an Al oxide layer estimated as 40 Å between the Al and SiO layers. From Ref. 33. Reprinted (Fig. 1) with permission from *Europhys. Lett.* **24**, 127 (1993). © EDP Sciences.

The dashed and dotted lines show results for single, isolated Sb films from our group and from other workers, while the filled circles show results for our Al/SiO/Sb sandwich samples as a function of the thickness of the SiO layer, d_{SiO}. When d_{SiO} was greater than about 100 Å, we found $A_{ee} \approx 0.95$ which was in agreement with previous studies of films of Sb (and other metals), and with the theory. With smaller values of d_{SiO} the temperature dependence of $\Delta G_\square$ was still logarithmic, but the value of A_{ee} fell to about 0.83 ± 0.05 for the smallest values of d_{SiO} that we were able to study. This change in A_{ee} indicates that the electron–electron interactions were weaker in the sandwich samples, due to the enhanced screening from the Al layer. While such an effect on the screening was broadly expected, it runs opposite to the *higher* electron–electron scattering rates found in the Sb/SiO/Sb trilayers in Fig. 12.8. It is still not clear how to reconcile these two results.

8. Conclusions

This paper has highlighted the work of our group on various aspects of weak localization and electron–electron interaction effects in one and two dimensional disordered metals. One of the remarkable aspects of this field has been the very productive and rapid interplay of theory and experiment. The vast majority of the experiments are now accounted for extremely well by the theory, which makes the results with multilayer samples in Figs. 12.8 and 12.9 all the more interesting. These experiments provide powerful ways to probe electron–electron scattering and screening, and could probably be expanded upon in various ways.

Acknowledgments

It is a pleasure to thank W. Gilson, J. T. Masden, W. D. Williams, J. J. Lin, D. E. Beutler, T. L. Meisenheimer, J. Liu, M. Pennington, N. R. Dilley and J. D. Monnier, for their essential contributions to the work discussed in this paper. Many colleagues, too numerous to mention, have provided important discussions and comments on all aspects of this work. This work was supported by the National Science Foundation, the Research Corporation, and the Alfred P. Sloan Foundation. I also thank the editors for the invitation to contribute to this volume, and to recognize again some of the work stimulated by the seminal insights of P. W. Anderson more than five decades ago.

References

1. P. W. Anderson, *Phys. Rev.* **109**, 1492 (1958).
2. N. F. Mott and W. D. Twose, *Adv. Phys.* **10**, 107 (1961).
3. K. Ishii, *Prog. Theor. Phys.*, *Suppl.* **53**, 77 (1973).
4. D. J. Thouless, *Phys. Rev. Lett.* **39**, 1167 (1977).
5. E. Abrahams, P. W. Anderson, D. C. Licciardello and T. V. Ramakrishnan, *Phys. Rev. Lett.* **42**, 673 (1979).
6. P. W. Anderson, D. J. Thouless, E. Abrahams and D. S. Fisher, *Phys. Rev. B* **22**, 3519 (1980).
7. P. A. Lee and T. V. Ramakrishnan, *Rev. Mod. Phys.* **57**, 287 (1985).
8. D. J. Thouless, *Solid State Commun.* **34**, 683 (1980).
9. J. C. Garland, W. J. Gully and D. B. Tanner, *Phys. Rev. B* **22**, 507 (1980).
10. B. L. Altshuler, A. G. Aronov and P. A. Lee, *Phys. Rev. Lett.* **44**, 1288 (1980).
11. B. L. Altshuler, D. Khmel'nitskii, A. I. Larkin and P. A. Lee, *Phys. Rev. B* **22**, 5142 (1980).
12. N. Giordano, W. Gilson and D. E. Prober, *Phys. Rev. Lett.* **43**, 725 (1979).
13. N. Giordano, *Phys. Rev. B* **22**, 5635 (1980).
14. G. Bergmann, *Phys. Rep.* **107**, 1 (1984).
15. J. T. Masden and N. Giordano, *Phys. Rev. Lett.* **49**, 819 (1982).
16. S. Wind, M. J. Rooks, V. Chandrasekhar and D. E. Prober, *Phys. Rev. Lett.* **57**, 633 (1986).
17. J. J. Lin and N. Giordano, *Phys. Rev. B* **33**, 1519 (1986).
18. B. L. Altshuler, A. G. Aronov and D. E. Khmel'nitskii, *J. Phys. C: Solid State Phys.* **15**, 7367 (1982).
19. D. E. Beutler and N. Giordano, *Phys. Rev. B* **36**, 7705 (1987).
20. S. Feng, P. A. Lee and A. D. Stone, *Phys. Rev. Lett.* **56**, 1960 (1986).
21. P. A. Lee, A. D. Stone and H. Fukuyama, *Phys. Rev. B* **35**, 1039 (1987).
22. B. L. Altshuler, P. A. Lee and R. A. Webb (eds.), *Mesoscopic Phenomena* (Elsevier, 1991).
23. S. Washburn and R. A. Webb, *Rep. Prog. Phys.* **55**, 1311 (1992).
24. D. E. Beutler, T. L. Meisenheimer and N. Giordano, *Phys. Rev. Lett.* **58**, 1240 (1987).
25. T. L. Meisenheimer and N. Giordano, *Phys. Rev. B* **39**, 9929 (1989).
26. N. Giordano, in *Mesoscopic Phenomena*, eds. B. L. Altshuler, P. A. Lee and R. A. Webb (North-Holland, Amsterdam, 1991), p. 131.
27. V. I. Fal'ko and D. E. Khmel'nitskii, *Zh. Eksp. Teor. Fiz.* **95**, 328 (1989); *Sov. Phys. JETP* **68**, 186.
28. J. Liu, M. Pennington and N. Giordano, *Phys. Rev. B* **45**, 1267 (1992).
29. N. Missert and M. R. Beasley, *Phys. Rev. Lett.* **63**, 672 (1989).
30. G. Bergmann and W. Wei, *Solid State Commun.* **71**, 1011 (1989).
31. T. J. Gramila, J. P. Eisenstein, A. H. MacDonald, L. N. Pfeiffer and K. W. West, *Phys. Rev. Lett.* **66**, 1216 (1991).
32. N. Giordano and N. R. Dilley, *Phys. Rev. B* **48**, 4675 (1993).
33. N. Giordano and J. D. Monnier, *Europhys. Lett.* **24**, 127 (1993).

Chapter 13

INHOMOGENEOUS FIXED POINT
ENSEMBLES REVISITED

Franz J. Wegner

Institute for Theoretical Physics, Ruprecht-Karls-University,
Philosophenweg 19, D-69120 Heidelberg, Germany

The density of states of disordered systems in the Wigner–Dyson classes approaches some finite non-zero value at the mobility edge, whereas the density of states in systems of the chiral and Bogolubov-de Gennes classes shows a divergent or vanishing behavior in the band centre. Such types of behavior were classified as homogeneous and inhomogeneous fixed point ensembles within a real-space renormalization group approach. For the latter ensembles, the scaling law $\mu = d\nu - 1$ was derived for the power laws of the density of states $\rho \propto |E|^\mu$ and of the localization length $\xi \propto |E|^{-\nu}$. This prediction from 1976 is checked against explicit results obtained meanwhile.

1. Introduction

Some time ago I used real-space renormalization group arguments in analogy to the cell model of Kadanoff[1] in order to investigate the critical behavior[2] close to the mobility edge of the Anderson model.[3] Two types of ensembles were considered, a homogeneous and an inhomogeneous one.

Homogeneous fixed point ensemble (HFPE). This ensemble is homogeneous in energy ϵ. It is invariant under the transformation $\epsilon \to \epsilon + $ constans. Since the density of states ρ stays constant during the renormalization group (RG) procedure the scale change

$$r \to r/b \quad \text{implies} \quad \epsilon \to \epsilon b^d \qquad (1.1)$$

with dimension d of the system. We assume one relevant perturbation to this system which grows like

$$\tau \to \tau b^y. \qquad (1.2)$$

Depending on the sign of τ the perturbation produces localized and extended states, resp. This perturbation is added to the HFPF in a strength increasing

289

linearly in energy E

$$\tau = cE, \tag{1.3}$$

where the mobility edge is taken at $E = 0$, and extended states at $\tau > 0$ and localized ones for $\tau < 0$. c transforms under RG.

Inhomogeneous fixed point ensemble (IHFPE). In this ensemble the scale factors for length and energies are independent from each other. The ensemble is inhomogeneous in the energy,

$$r \to r/b, \quad \epsilon \to \epsilon b^y. \tag{1.4}$$

It is assumed that there is no relevant perturbation to such an ensemble.

Both ensembles yield power and homogeneity laws. The density of states obeys

$$\rho_{\text{hom}} = \text{const.}, \quad \rho_{\text{inh}} \propto |E|^\mu, \quad \mu = d/y - 1. \tag{1.5}$$

The localization length yields in both cases

$$\xi \propto |E - E_c|^{-\nu}, \quad \nu = 1/y. \tag{1.6}$$

The low-temperature a.c. conductivity obeys the homogeneity relation

$$\sigma(\omega, \tau) = \begin{cases} b^{2-d}\sigma(\omega b^d, \tau b^y) & \text{HFPE} \\ b^{2-d}\sigma(\omega b^y, \tau b^y) & \text{IHFPE} \end{cases}. \tag{1.7}$$

One deduces the d.c. conductivity in the region of extended states

$$\sigma(0, \tau) \sim \tau^s, \quad s = (d-2)/y = (d-2)\nu. \tag{1.8}$$

What comes out correctly on the basis of these ideas? Not only the scaling and homogeneity laws shown above can be deduced, but also such laws for averaged correlations, including the inverse participation ratio and long-range correlations between states energetically close to each other including those in the vicinity of the mobility edge. What has to be added are averages of matrix elements and of their powers for the transformation step by the linear scale factor b of the cell model.

A short historical digression may be allowed. The oldest paper on the mobility edge i.e. the separation of localized and extended states of a disordered system was given by Phil Anderson[3] (1958) (well aware of possible complications by the Coulomb interaction he considered the transition from spin diffusion to localized spin excitations). It is a nice accident that its page number 1492 coincides with the year of another important discovery. Earlier papers on disordered systems, which became important for the development of this field was Wigner's[4-6] Gaussian matrix ensemble (1951) for nuclei. Probably the oldest paper on chiral systems is Dyson's paper[7] on disordered chains (1953). Other early contributions on disordered chains were by Schmidt[8] and arguments that states in one dimension are localized.[9,10] In 1962, Dyson gave the threefold classification of ensembles of orthogonal, unitary and symplectic symmetry depending on the behaviour under time-reversal invariance.[11,12]

Since these early developments a lot of progress has been made. There are numerous calculations for the behaviour around the mobility edge both analytic and numerical. I refer to the review by Evers and Mirlin.[13] 1979 marked important break-throughs: the scaling theories of localization by Abrahams *et al.*[14] and by Oppermann and Wegner[15] appeared. The mapping onto a non-linear sigma-model was conjectured,[16] brought into its bosonic-replica,[17] its fermionic-replica[18] and finally in its supersymmetric[19] form. A self-consistent approximation for the Anderson transition was put forward by Götze,[20] Vollhardt and Wölfle.[21,22] A numerical renormalization scheme was devised by MacKinnon and Kramer.[23]

Since then many more results and techniques were developped. Here I mention only a few: the complete classification of ten symmetry classes of random matrix theories, σ-models, and Cartan's symmetric spaces was given by Zirnbauer and Altland[24,25] and by Schnyder, Ryu, Furusaki, and Ludwig[26] after several occurences of chiral and Bogolubov-de Gennes classes.[27–34] These classes are listed in Table 13.1 since I will refer later to this nomenclature.

Transfer matrix approaches originally used for linear chains were developped for the non-linear σ-model[19] as well as for the for the distribution function of the transfer matrix of chains with many channels (DMPK-equation[35,36]). These techniques allowed the determination of correlations, wave-function statistics and transport properties. Such chains can have broad distributions of conductivities and even cases of perfect transmissions were found.[37,38]

In two dimensions, some of these symmetry classes allow the inclusion of a topological θ-term. As observed by Pruisken *et al.*,[39,40] the Wigner–Dyson unitary class with this term describes the integer quantum Hall effect. Another term which may be added is a Wess–Zumino term. Such terms are of importance in the study of disordered Dirac fermions, which appear in dirty d-wave superconductors[41–43] and in disordered graphene.[44–49]

Network models originally introduced by Shapiro[50] are very useful for the description of quantum Hall systems as first shown for the integer quantum Hall effect in the Chalker–Coddington model.[51]

Table 13.1. Symmetry classes of single particle Hamiltonians defined in terms of presence or absence of time-reversal symmetry (TRS) and particle-hole symmetry (PHS). Absence is denoted by 0, presence by the symmetry square ±1. SLS indicates absence (0) and presence (1) of sublattice or chiral symmetry. After Ref. 26.

System	Cartan Nomenclature	Symmetry	TRS	PHS	SLS
Standard	A	unitary	0	0	0
(Wigner–Dyson)	AI	orthogonal	+1	0	0
	AII	symplectic	−1	0	0
Chiral	AIII	unitary	0	0	1
(sublattice)	BDI	orthogonal	+1	+1	1
	CII	symplectic	−1	−1	1
Bogolubov-	D		0	+1	0
de Gennes	C		0	−1	0
	DIII		−1	+1	1
	CI		+1	−1	1

Obviously the HFPE applies to the Wigner–Dyson classes, whereas the IHFPE applies to the chiral and the Bogoliubov-de Gennes classes.

The main object of this paper is the comparison of the scaling law for the IHFPE

$$\mu = d\nu - 1 \tag{1.9}$$

derived from Eqs. (1.5) and (1.6) with various results meanwhile obtained. I will shortly comment on the scaling law (1.8) for the conductivity in Sec. 4.1.

2. One-Dimensional Chains

2.1. *Thouless relation*

Thouless[52] following Herbert and Jones[53] considered a one-dimensional chain governed by the Hamiltonian

$$H = \sum_{i=1}^{N} \epsilon_i |i\rangle\langle i| - \sum_{i=1}^{N-1} (V_{i,i+1}|i\rangle\langle i+1| + V_{i+1,i}|i+1\rangle\langle i|) \tag{2.1}$$

and found in the limit $N \to \infty$ that the function

$$K(z) = \int \mathrm{d}x \rho(x) \ln(x - z) - \overline{\ln|V|}, \quad -\pi < \arg \ln(x - z) < \pi \quad (2.2)$$

connects both the integrated density of states $I(E)$ and the exponential decrease of eigenfunctions $\lambda(E)$ (inverse correlation length ξ)

$$K(E \pm \mathrm{i}0) = \lambda(E) \mp \mathrm{i}\pi I(E). \quad (2.3)$$

The density of states is symmetric in chiral and Bogoliubov-de Gennes classes $\rho(-E) = \rho(E)$. Then besides $K(z^*) = K^*(z)$ also $K(-z) = K(z) + \mathrm{i}\pi s(z)$ with $s(z) = \text{sign}\Im(z)$ holds. If $K(z) + \mathrm{i}\pi s/2 \propto z^\gamma$ for small z, then

$$K(z) + \mathrm{i}\frac{\pi s}{2} = cr^\gamma \mathrm{e}^{\mathrm{i}\gamma(\phi - s\pi/2)}, \quad z = r\mathrm{e}^{\mathrm{i}\phi} \quad (2.4)$$

with real c. Then

$$K(E + \mathrm{i}0) = c|E|^\gamma \left(\cos(\frac{\pi}{2}\gamma) - \mathrm{i}\,\text{sign}(E)\sin(\frac{\pi}{2}\gamma) \right) - \frac{\pi}{2}\mathrm{i}, \quad (2.5)$$

from which $\lambda \propto |E|^\gamma$, $\rho(E) \propto |E|^{\gamma-1}$ follows in agreement with Eq. (1.9). One observes that

$$\frac{\mathrm{d}K(z)}{\mathrm{d}z} = \int \mathrm{d}x \frac{\rho(x)}{z - x}. \quad (2.6)$$

Thus the sign of the imaginary part of Eq. (2.6) is opposite to the sign of $\Im z$. This implies that $\gamma \leq 1$. If a contribution with $\gamma > 1$ appears, then there is also a contribution with $\gamma = 1$, which according to Eq. (2.5) does not contribute to λ, but to a finite density of states in the center of the spectrum. Such a system is described by the homogeneous fixed point ensemble.

For $\gamma \leq 0$, the integrated density of states would diverge. Thus these arguments can only be applied for $0 < \gamma < 1$.

In a number of cases the asymptotic behaviour is given by a power multiplied by some power of the logarithm. Then

$$K(z) + \mathrm{i}\frac{\pi s}{2} = cr^\gamma \mathrm{e}^{\mathrm{i}\gamma(\phi - s\pi/2)} (\ln r + \mathrm{i}(\phi - s\pi/2))^g. \quad (2.7)$$

This yields for $\gamma = 0$ and $\gamma = 1$

γ	$K(E + \mathrm{i}0) + \mathrm{i}\pi/2$
0	$c(\ln|E|)^g - \mathrm{i}cg\frac{\pi}{2}\text{sign}(E)(\ln|E|)^{g-1}$
1	$-\mathrm{i}cE(\ln|E|)^g - cg\frac{\pi}{2}|E|(\ln|E|)^{g-1}$

(2.8)

and thus

γ	$\lambda \sim$	$\rho \sim$
0	$(\ln|E|)^g$	$\frac{(\ln|E|)^{g-2}}{|E|}$
1	$|E|(\ln|E|)^{g-1}$	$(\ln|E|)^g$

(2.9)

Dyson[7] calculated the averaged density of states for the chain (2.1) with $\epsilon_i = 0$ and random independently distributed matrix elements V, for which he assumed a certain distribution and obtained

$$\rho(E) \sim \frac{1}{|E(\ln|E|)^3|}, \tag{2.10}$$

which corresponds to the case in Eq. (2.9) with $\gamma = 0$ and $g = -1$. Indeed Theodorou and Cohen[54] and Eggarter and Ridinger[55] found the averaged localization length diverging

$$\xi \sim |\ln|E|| \tag{2.11}$$

in agreement with Eq. (2.9).

2.2. *Ziman's model*

Ziman[56] (compare also Alexander *et al.*[57]) considered a one-dimensional tight-binding model (his case II) Eq. (2.1) requiring the diagonal matrix elements to vanish, $\epsilon_i = 0$, and the hopping matrix-elements to agree pairwise $V_{2m,2m+1} = V_{2m+1,2m+2}$. Apart from this restriction he assumed the Vs to be independently distributed with probability distribution

$$p(V) = (1 - \alpha)V^{-\alpha}, \quad 0 < V < 1, \quad -\infty < \alpha < 1. \tag{2.12}$$

He obtained for these distributions

	ν	μ
$-1 < \alpha < 1$	$\frac{2(1-\alpha)}{3-\alpha}$	$\frac{-1-\alpha}{3-\alpha}$
$-3 < \alpha < -1$	$\frac{1-\alpha}{2}$	0
$\alpha < -3$	2	0

$$\tag{2.13}$$

Obviously the first row describes models in accordance with the IHFPE, the second and third row with the HFPE.

2.3. *Further one-dimensional results*

Titov *et al.*[58] have summarized and completed results for the density of states of all classes of chains with N channels as shown in Table 13.2.

All chiral classes are equivalent by a gauge transformation for $N = 1$ and yield the Dyson result (2.10) and $\xi \propto |\ln|E||$ for this case in agreement with Eq. (2.9). Due to Gruzberg *et al.*[59] also the BdG classes BD and DIII fall into the same universality class. The localization length does not diverge for the chiral classes, if N is even. The same holds for (general N) for the BdG classes C and CI.

Table 13.2. Density of states close to $E = 0$ for various universality classes (after Titov *et al.*[58])

Class		$\rho(E)$ $x = E\tau$
Chiral	All classes, odd N	$|x^{-1}\ln^{-3}(x)|$
	AIII, even N	$|x\ln x|$
	CII, even N	$|x^3 \ln x|$
	BDI, even N	$|\ln x|$
BdG	CI	$|x|$
	C	x^2
	D,DIII, $N \neq 2$	$|x^{-1}\ln^{-3}(x)|$
	D,DIII, $N = 2$	Two mean free paths

3. Bosons From One To Two Dimensions

3.1. *One-dimensional chain*

Whereas the Hamiltonian (2.1) yields the equation for eigenstates $|\psi\rangle = \sum_i \psi_i |i\rangle$

$$E\psi_i = \epsilon_i \psi_i - V_{i,i-1}\psi_{i-1} - V_{i,i+1}\psi_{i+1} \qquad (3.1)$$

one obtains a similar equation for harmonic phonons governed by the Hamiltonian

$$H = \sum_i \frac{p_i^2}{2m} + \sum_i \frac{W_i}{2}(x_{i+1} - x_i)^2, \qquad (3.2)$$

which reads

$$m\omega^2 x_i = W_i(x_i - x_{i+1}) + W_{i-1}(x_i - x_{i-1}). \qquad (3.3)$$

Thus Thouless' arguments can be applied again with $x = \omega^2$. Since there are no states for $\omega^2 < 0$, one has

$$K(z) = cr^\gamma e^{i\gamma(\phi - s\pi)}, \quad z = re^{i\phi}, \qquad (3.4)$$

which yields

$$K(\omega^2 + i0) = c\omega^{2\gamma} e^{-i\pi\gamma} \qquad (3.5)$$

and thus

$$\lambda(\omega) = c\omega^{2\gamma}\cos(\pi\gamma), \quad I(\omega) = c\omega^{2\gamma}\sin(\pi\gamma), \quad 0 < \gamma < 1. \qquad (3.6)$$

Alexander *et al.*[57] (cases a, b, and c) and Ziman[56] (case II) determined the density of states $\rho(\omega) \propto \omega^\mu$ for harmonically coupled phonons with independently distributed spring constants

$$p(W) = (1 - \alpha)W^{-\alpha}. \tag{3.7}$$

Ziman moreover determined the localization length $\xi \propto \omega^{-\nu}$ and obtained

$$
\begin{array}{c|cc}
 & \nu & \mu \\
\hline
0 < \quad \alpha < 1 & \frac{2(1-\alpha)}{2-\alpha} & -\frac{\alpha}{2-\alpha} \\
-1 < \quad \alpha < 0 & 1 - \alpha & 0 \\
\alpha < -1 & 2 & 0
\end{array} \tag{3.8}
$$

Again the first line is in agreement with the IHFPE, whereas the two other cases correspond to the HFPE.

3.2. *Bosonic excitations discussed by Gurarie and Chalker*

Gurarie and Chalker[60] point out that bosonic systems with and without Goldstone modes show different localization behavior.

John *et al.*[61] investigated localization in an elastic medium with randomly varying masses. For $d > 2$ they found extended states for small frequencies ω. The phonon states are localized beyond some critical ω_c. This transition is described by the orthogonal ensemble. For $d < 2$ all states are localized and obey[61] $\nu = 2/(2-d)$. The density of states for phonons shows the same power law $\rho(\omega) \propto \omega^{d-1}$ as in the ordered case. In this system with Goldstone modes the critical density below which the density of states would differ from that of the ordered system is $d_c = 0$.

In a disordered antiferromagnet one obtains below the critical dimension $d_c = 2$

$$\rho(\omega) \sim \omega^\mu, \quad \mu = \frac{3d-4}{4-d}, \quad \xi(\omega) \sim \omega^{-\nu}, \quad \nu = \frac{2}{4-d}, \tag{3.9}$$

where the result of[62] and the argument of[60] have been generalized from $d = 1$ to general $d < d_c$. This is in agreement with the IHFPE. These results rest on the assumption that there is a single length scale $\xi \propto 1/k$.

4. Electronic Systems In Two Dimensions

4.1. *Conductivity in two dimensions*

From the homogeneity law $s = (d-2)\nu$, (1.8), which works well for $d > 2$, I concluded[2] $s = 0$ for dimensionality $d = 2$ and thus a jump to a minimum metallic conductivity. At that time I did not expect that ν may diverge as d approaches 2. This was found three years later by means of explicit renormalization group calculations.[14,15] Thus the idea of a minimum metallic conductivity was in error for the orthogonal and unitary Wigner–Dyson classes, where all states are localized in $d = 2$. The critical conductivity in the symplectic class shows some distribution[63] and is of order e^2/h.

In many two-dimensional models of chiral and Bogolubov-de Gennes classes including the classes applying to d-wave superconductors and graphene, one obtains a finite non-zero conductivity of order e^2/h at criticality. This is to a large extend due to edge currents, as first observed by Pruisken *et al.*[39,40] for the integer quantum Hall effect. Thus although the prediction[2] turns out to be correct, the true mechanism is more complex.

4.2. *Chiral and Bogolubov-de Gennes models in* $d = 2$ *dimensions*

The unitary case of chiral models (Gade and Wegner, Gade[31,32]) yields at intermediate energies effective exponents

$$\nu = \frac{1}{B}, \quad \mu = -1 + \frac{2}{B} \tag{4.1}$$

in accordance with Eq. (1.9). At asymptotically low energies $\rho \propto E^{-1}\xi^2(E)$ corresponds to the limit $B \to \infty$. These results as well as many similar results for various disordered Dirac hamiltonians are obtained under the assumption that the localization length is given by the cross-over length from chiral to Wigner–Dyson behaviour without taking further renormalization into account[32]; see also the argument after Eq. (6.60) of the review by Evers and Mirlin.[13] Alternatively the integrated density of states from the band center up to energy E is set to ξ^{-2} as in Motrunich *et al.*,[64] which yields Eq. (1.9) per definition. It is important that only one coupling yields a relevant perturbation. The conductivity itself stays constant for the chiral models in $d = 2$. The exponent which drives the renormalization of the energy is usually called dynamical exponent z, which is identical to the exponent y of Eq. (1.9). A more rigorous investigation of the localization of such systems taking into account any dependence of the initial couplings on the energy and of the cross-over would be of interest.

The spin quantum Hall effect yields at the percolation transition point[65–67]

$$\nu = 4/7, \quad \mu = 1/7 \tag{4.2}$$

in agreement with Eq. (1.9). The same behavior is obtained for the Bogolubov-de Gennes class C if two of the four nodes of a dirty d-wave superconductor are coupled.[41,43]

4.3. *Power law for density of states, finite localization length*

The two fixed point ensembles describe the situation, in which the localization length diverges and the density of states either approaches some finite non-zero value (HFPE) or diverges or goes to zero by a power law, which may be augmented by a logarithmic term. As mentioned above this holds for chains with an even number of channels in the chiral classes and for the Bogolubov-de Gennes classes C and CI. Certain single-channel models of class D and DIII show a divergence of the density of states $\rho \propto |E|^{-1+\delta}$ without divergence of the localization length.[68]

Gurarie and Chalker[69] found that bosonic excitations in random media, which are not Goldstone modes, obey $\rho \propto \omega^4$ with finite localization length at low frequencies. Apparently this type of behavior is not covered by HFPE and IHFPE.

5. Conclusion

The scaling prediction (1.9) of the IHFPE relating the exponent of the density of states and of the localization length yields correct results in the cases, in which I found both exponents. The author appreciates the wealth of systems, which has been found and investigated over the years.

Acknowledgments

The author enjoyed part of the summer program *Mathematics and Physics of Anderson Localization: 50 years after* at the Newton Institute of Mathematical Sciences in Cambridge. He gratefully acknowledges a Microsoft Fellowship. He thanks for many useful discussions in particular with John Chalker, Alexander Mirlin, Tom Spencer, and Martin Zirnbauer.

References

1. L. P. Kadanoff, *Physics* **2**, 263 (1966).
2. F. J. Wegner, *Z. Phys.* B **25**, 327 (1976).
3. P. W. Anderson, *Phys. Rev.* **109**, 1492 (1958).
4. E. P. Wigner, *Ann. Math.* **53**, 36 (1951).

5. E. P. Wigner, *Ann. Math.* **62**, 548 (1955).
6. E. P. Wigner, *Ann. Math.* **67**, 325 (1958).
7. F. J. Dyson, *Phys. Rev.* **92**, 1331 (1953).
8. H. Schmidt, *Phys. Rev.* **105**, 425 (1957).
9. R. F. Borland, *Proc. R. Soc. London, Scr. A* **274**, 529 (1963).
10. H. Furstenberg, *Trans. Amer. Math. Soc.* **108**, 377 (1963).
11. F. J. Dyson, *J. Math. Phys.* **3**, 140 (1962).
12. F. J. Dyson, *J. Math. Phys.* **3**, 1199 (1962).
13. F. Evers and A. D. Mirlin, *Rev. Mod. Phys.* **80**, 1355 (2008).
14. E. Abrahams, P. W. Anderson, D. C. Licciardello and T. V. Ramakrishnan, *Phys. Rev. Lett.* **42**, 673 (1979).
15. R. Oppermann and F. Wegner, *Z. Phys. B* **34**, 327 (1979).
16. F. Wegner, *Z. Phys. B* **35**, 207 (1979).
17. L. Schäfer and F. Wegner, *Z. Phys. B* **38**, 113 (1980).
18. K. B. Efetov, A. I. Larkin and D. E. Khmel'nitskii, *Zh. Eksp. Teor. Fiz.* **79**, 1120 (1980); *Sov. Phys. JETP* **52**, 568 (1980).
19. K. B. Efetov, *Adv. Phys.* **32**, 53 (1983).
20. W. Götze, *J. Phys. C.: Solid State Phys.* **12**, 1279 (1979).
21. D. Vollhardt and P. Wölfle, *Phys. Rev. Lett.* **45**, 842 (1980).
22. D. Vollhardt and P. Wölfle, *Phys. Rev. Lett.* **48**, 699 (1982).
23. A. MacKinnon and B. Kramer, *Phys. Rev. Lett.* **47**, 1546 (1981).
24. M. R. Zirnbauer, *J. Math. Phys.* **37**, 4986 (1996).
25. A. Altland and M. R. Zirnbauer, *Phys. Rev. B* **55**, 1142 (1997).
26. A. P. Schnyder, S. Ryu, A. Furusaki and A. W. W. Ludwig, *Phys. Rev. B* **78**, 195125 (2008).
27. S. Hikami, *Nucl. Phys. B* **215**, 555 (1983).
28. R. Oppermann, *Nucl. Phys. B* **280**, 753 (1987).
29. R. Oppermann, *Physica A* **167**, 301 (1990).
30. F. Wegner, *Nucl. Phys. B* **316**, 663 (1989).
31. R. Gade and F. Wegner, *Nucl. Phys. B* **360**, 213 (1991).
32. R. Gade, *Nucl. Phys. B* **398**, 499 (1993).
33. K. Slevin and T. Nagao, *Phys. Rev. Lett.* **70**, 635 (1993).
34. J. J. M. Verbaarschot and I. Zahed, *Phys. Rev. Lett.* **70**, 3852 (1993).
35. O. N. Dorokhov, *JETP Lett.* **36**, 318 (1982).
36. P. A. Mello, P. Pereyra and N. Kumar, *Ann. Phys. (N.Y.)* **181**, 290 (1988).
37. M. R. Zirnbauer, *Phys. Rev. Lett.* **69**, 1584 (1992).
38. A. D. Mirlin and Y. V. Fyodorov, *Phys. Rev. Lett.* **72**, 526 (1994).
39. H. Levine, S. B. Libby and A. M. M. Pruisken, *Phys. Rev. Lett.* **51**, 1915 (1983).
40. A. M. M. Pruisken, *Nucl. Phys. B* **235**, 277 (1984).
41. A. A. Nersesyan, A. M. Tsvelik and F. Wenger, *Nucl. Phys. B* **438**, 561 (1995).
42. M. Bocquet, D. Serban and M. R. Zirnbauer, *Nucl. Phys. B* **578**, 628 (2000).
43. A. Altland, *Phys. Rev. B* **65**, 104525 (2002).
44. I. L. Aleiner and K. B. Efetov, *Phys. Rev. Lett.* **97**, 236801 (2006).
45. D. V. Khveshchenko, *Phys. Rev. Lett.* **97**, 036802 (2006).
46. E. McCann, K. Kechedzhi, V. I. Fal'ko, H. Suzuura, T. Ando and B. L. Altshuler, *Phys. Rev. Lett.* **97**, 146805 (2006).

47. P. M. Ostrovsky, I. V. Gornyi and A. D. Mirlin, *Phys. Rev. B* **74**, 235443 (2006).
48. P. M. Ostrovsky, I. V. Gornyi and A. D. Mirlin, *Phys. Rev. Lett.* **98**, 256801 (2007).
49. P. M. Ostrovsky, I. V. Gornyi and A. D. Mirlin, *Eur. Phys. J. Spec. Top.* **148**, 63 (2007).
50. B. Shapiro, *Phys. Rev. Lett.* **48**, 823 (1982).
51. J. T. Chalker and P. D. Coddington, *J. Phys. C.: Solid State Phys.* **21**, 2665 (1988).
52. D. J. Thouless, *J. Phys. C.: Solid State Phys.* **5**, 77 (1972).
53. D. C. Herbert and R. Jones, *J. Phys. C.: Solid State Phys.* **4**, 1145 (1971).
54. G. Theodorou and M. H. Cohen, *Phys. Rev. B* **13**, 4597 (1976).
55. T. P. Eggarter and R. Riedinger, *Phys. Rev. B* **18**, 569 (1978).
56. T. L. A. Ziman, *Phys. Rev. Lett.* **49**, 337 (1982).
57. S. Alexander, J. Bernasconi, W. R. Schneider and R. Orbach, *Rev. Mod. Phys.* **53**, 175 (1981).
58. M. Titov, P. W. Brouwer, A. Furusaki and C. Mudry, *Phys. Rev. B* **63**, 235318 (2003).
59. I. A. Gruzberg, N. Read and S. Vishveshwara, *Phys. Rev. B* **71**, 245124 (2005).
60. V. Gurarie and J. T. Chalker, *Phys. B* **68**, 134207 (2003).
61. S. John, H. Sompolinsky and M. J. Stephen, *Phys. Rev. B* **27**, 5592 (1983).
62. R. B. Stinchcombe and I. R. Pimentel, *Phys. Rev. B* **38**, 4980 (1988).
63. B. Shapiro, *Phil. Mag. B* **56**, 1031 (1987).
64. O. Motrunich, K. Damle and D. A. Huse, *Phys. Rev. B* **65**, 064206 (2002).
65. H. Saleur and B. Duplantier, *Phys. Rev. Lett.* **58**, 2325 (1987).
66. I. A. Gruzberg, N. Read and A. W. W. Ludwig, *Phys. Rev. Lett.* **82**, 4524 (1999).
67. E. J. Beamond, J. Cardy and J. T. Chalker, *Phys. Rev. B* **65**, 214301 (2002).
68. O. Motrunich, K. Damle and D. A. Huse, *Phys. Rev. B* **63**, 224204 (2001).
69. V. Gurarie and J. T. Chalker, *Phys. Rev. Lett.* **89**, 136801 (2002).

Chapter 14

QUANTUM NETWORK MODELS AND CLASSICAL LOCALIZATION PROBLEMS

John Cardy*

Rudolph Peierls Centre for Theoretical Physics,
1 Keble Road, Oxford OX1 3NP, United Kingdom
and All Souls College, Oxford
j.cardy1@physics.ox.ac.uk

A review is given of quantum network models in class C which, on a suitable 2d lattice, describe the spin quantum Hall plateau transition. On a general class of graphs, however, many observables of such models can be mapped to those of a classical walk in a random environment, thus relating questions of quantum and classical localization. In many cases it is possible to make rigorous statements about the latter through the relation to associated percolation problems, in both two and three dimensions.

1. Introduction

Lattice models of spatially extended systems have a long record of usefulness in condensed matter physics. Even when the microscopic physics is not necessarily related to a crystalline lattice, it can be very useful to concentrate the essential degrees of freedom onto a regular lattice whose length scale is larger than the microscopic one yet much smaller than that the expected scale of the physical phenomena the model is designed to address. In many cases, the phenomenon of universality ensures that this idealization can nevertheless reproduce certain aspects exactly. The classic example is that of a lattice gas, where a coarse-grained lattice on the scale of the particle interaction radius is introduced and used to make predictions for continuum systems, in cases in which the correlation length is large, for example close to the liquid–gas critical point.

The lattice models discussed in this article — in this context called network models — were first introduced by Chalker and Coddington[1] as a theoretical model for non-interacting electrons in two dimensions in a strong

---

*Address for correspondence.

transverse magnetic field and in the presence of disorder: the physical setting for the integer quantum Hall effect. The starting point is to consider
non-interacting electrons moving in two dimensions in a disordered potential $V(r)$ and a strong perpendicular magnetic field B. We assume that the
length scale of variation of $V(r)$ is much larger than the magnetic length. In
this limit the electronic motion has two components with widely separated
time scales[2]: the cyclotron motion and the motion of the guiding center,
along contours of $V(r)$. The total energy of the electron in this approxima-
tion is $E = (n + \frac{1}{2})\hbar\omega_c + V(r)$, where $\omega_c = eB/m$ and n labels the Landau
levels, and we therefore expect to find extended states at energies E_c corresponding to those values of V at which the contours of $V(r)$ percolate.
In 2d this is expected to occur at one particular value of V, which can be
taken to be $V = 0$. Otherwise, away from the percolation threshold the
guiding centers are confined to the neighborhoods of the closed contours,
corresponding to bulk insulating phases which conduct only along the edges
of the sample. This immediately provides a simple explanation of the experimental result that extended states occur only at the transition and not in
the Hall plateaux. If this were literally correct the plateau transition would
be in the same universality class as classical percolation.

However, this picture is modified for energies close to the transition, since
quantum tunnelling is expected to be important[3] where closed contours approach each other, see Fig. 14.1. The network model idealizes this picture
by distorting the percolating contours of $V(r)$ into a regular square lattice,
known as the L-lattice, shown in Fig. 14.2. In this approximation the potential V takes a checkerboard form, being > 0 on (say) even squares and
< 0 on odd squares. In the limit of large magnetic field the spin degree of
freedom of the electrons can be ignored and there is a one-dimensional vector

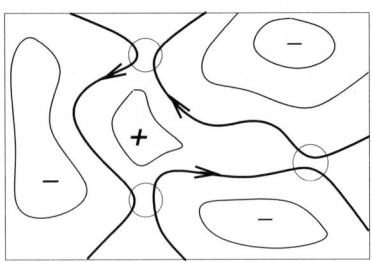

Fig. 14.1.   A typical configuration of contours of the random potential $V(r)$. In
the guiding center approximation the particle follows these. Only those with $V \approx 0$
(thicker lines) are important for the plateau transition. Quantum tunnelling can
occur close to the saddle points of $V(r)$ (circled). Figure adapted from Ref. 1.

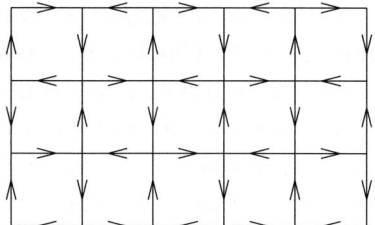

Fig. 14.2. The idealization of Fig. 14.1 on the L-lattice. The edges correspond to the contours with $V \approx 0$ and tunnelling occurs at the nodes.

space associated with each edge. The quantum tunnelling at each node is taken into account by an S-matrix which connects the spaces on the adjacent incoming and outgoing edges. This depends on E in such a way that for $E > E_c$ tunnelling between edges bordering regions with $V > 0$ is enhanced, and *vica versa* for $E < E_c$. Apart from this, the most important quantum feature of the problem is the phase which the electron wave function picks up on traversing a closed contour, which is proportional to the magnetic flux through the loop, and therefore its area. On idealizing the loops to a regular lattice, this is represented by a quenched uncorrelated random flux through each plaquette, or, equivalently, a quenched random U(1) phase as the particle traverses a given edge. While in principle the connectivity of the lattice, the S-matrix elements at the nodes, and the U(1) phases on the edges are all quenched random variables, in fact only the latter appear to be the most relevant in describing the universal properties of the transition.

The Chalker–Coddington model was initially analyzed numerically[1] using transfer matrix methods. Its predictions appear to agree remarkably well with experimental results, perhaps embarrassingly so since it ignores electron–electron interactions which may become important near the transition. However it has so far resisted all attempts at an analytic solution (as have other more sophisticated field theoretic formulations of the integer quantum Hall plateau transition[4]), which by now perhaps elevates this to being one of the outstanding unsolved problems of mathematical physics.

Later, following interest in various forms of exotic superconductivity, it was suggested that certain disordered spin-singlet superconductors, in which time-reversal symmetry is broken for orbital motion but Zeeman splitting is negligible, should exhibit a quantum spin Hall effect, in which the role of the electric current is replaced by that of a spin current.[5] The single-particle hamiltonians for such a system then turn out to possess an Sp(2) (or equivalently SU(2)) symmetry. In the classification scheme of localization universality classes due to Altland and Zirnbauer[6] they are labelled as class C. The corresponding variant of the Chalker–Coddington model is straightforward to write down, and was studied, once again numerically.[5,7] Then, in a remarkable paper, Gruzberg, Read and Ludwig[8] argued that several important ensemble-averaged properties of this model (including the conductance) are simply related to those of critical *classical* 2d percolation. This is a powerful result because many of the universal properties of percolation are known rigorously.[9] It therefore gives exact information about a non-trivial quantum localization transition.

The arguments of Gruzberg *et al.*[8] were based on a transfer matrix formulation of the problem and therefore restricted to the particular oriented lattice (the L-lattice) used by the Chalker–Coddington model, which is appropriate to the quantum Hall problem in 2d. They also used supersymmetry to perform the quenched average. Subsequently, Beamond, Cardy and Chalker[10] gave an elementary, albeit long-winded, proof of their main result which holds for any lattice of coordination number four, and any orientation of this lattice as long each node has two incoming and two outgoing edges. Later, this was shown in a slightly more elegant fashion using supersymmetry.[11] In each case certain quenched averages of the quantum problem are related to observables of a certain kind of *classical* random walk on the same lattice. If the quantum states are localized, the corresponding classical walks close after a finite number of steps. If the quantum states are extended, in the classical problem the walks can escape to infinity.

Since this correspondence holds on a very general set of graphs and lattices, it can be used more generally to improve our understanding of quantum localization problems. In particular it can test the generally accepted notion that in two dimensions all states are localized, except in certain cases with special symmetries (such as at the Hall plateau transition). It can help understand why in higher dimensions there should be in general a transition between localized and extended states, and possibly illumine the nature of that transition. More mathematically, it may shed light on the search for a rigorous proof of the existence of extended states. Apart from the L-lattice considered by Gruzberg *et al.*, it is possible in several cases to use known information about percolation to place bounds on the behavior of the classical walks and hence the quantum network model. Up until recently, these arguments have been restricted to two dimensions, but now suitable three-dimensional lattices have been identified in which the correspondence to classical percolation is explicit.

The layout of this paper is as follows. In Sec. 2, we describe general network models and observables which are related to experimentally measurable quantities. In Sec. 3, we summarize the supersymmetric proof[11] of the main theorem which relates suitable quenched averages of these observables in the Sp(2) network model on a general graph to averages in a classical random walk problem on the same graph. The next section, Sec. 4 describes how these classical models on certain lattices (the L-lattice, relevant to the quantum Hall effect, and the Manhattan lattice) relate to 2d classical percolation which can then be used to bound their behavior. In Sec. 5, we extend this to some special 3d lattices, describing relatively recent work, some of it so far unpublished. Finally in Sec. 6, we discuss some outstanding problems.

2. General Network Models

In this section, we define a general network model on a graph $\mathcal{G}$ and discuss the kind of observables we would like to calculate. The graph $\mathcal{G}$ consists of nodes n and oriented edges. Initially suppose that $\mathcal{G}$ is closed, that is every edge connects two nodes, and that each node has exactly two incoming and two outgoing edges. In fact the general theorem to be proved in Sec. 3 holds for more general graphs, but it can be shown[11] that the corresponding classical problem has non-negative weights (and so admits a probabilistic interpretation) if and only if each node in $\mathcal{G}$ and its correspond transition amplitudes can be decomposed into a skeleton graph with only $2 \to 2$ nodes. See Fig. 14.3.

On each edge e of $\mathcal{G}$ is an N-dimensional Hilbert space $\mathcal{H}_e$. We assume these are all isomorphic. The Hilbert space of the whole system is then $\otimes_{e \in \mathcal{G}} \mathcal{H}_e$. We consider a single particle whose wave function at at time t is a superposition of the basis states in this space. The dynamics is discrete: if the particle is at the center of edge e at time t, at the next time $t + 1$ it must move in the direction of the orientated edges through a node to the

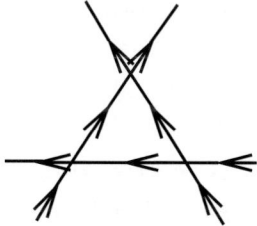

Fig. 14.3.　A $3 \to 3$ node which can be decomposed into $2 \to 2$ nodes.

center of a neighboring edge e'. Because of the discrete dynamics we should consider the unitary time-evolution operator $\mathcal{U}$. This has an off-diagonal block elements $\mathcal{U}_{e'e}$ which have the form

$$\mathcal{U}_{e'e} = U_{e'}^{1/2} S_{e'e} U_e^{1/2}.$$

Here U_e is a unitary $N \times N$ matrix which maps $\mathcal{H}_e \to \mathcal{H}_e$ and describes the rotation of the wave function in the internal space as the particle moves along the edge e, and $S_{e'e}$ maps $\mathcal{H}_e \to \mathcal{H}_{e'}$ and describes the transmission through a node. The evolution matrix after time t therefore has non-zero block elements

$$\mathcal{U}_{e_f,e_i}^t = \sum_{\gamma(e_t,e_0)} \delta_{e_t,e_f} \delta_{e_0,e_i} U_{e_t}^{1/2} S_{e_t,e_{t-1}} U_{e_{t-1}} \cdots U_{e_1} S_{e_1,e_0} U_{e_0}^{1/2},$$

where the sum is over all Feynman paths $\gamma(e_t, e_0)$ on $\mathcal{G}$ of length t, starting at e_i and ending at e_f. Note that in this sum a given edge can be traversed an arbitrary number of times.

Both the matrices U_e and $S_{e'e}$ are quenched random variables, assumed chosen from the invariant measure on some subgroup of $U(N)$. This is chosen according to the symmetry class under consideration. For spinless, or spin-polarized, electrons, where electric charge is conserved, we can take $N = 1$ and the $U_e \in U(1)$. For models with class C symmetry,[6] corresponding for example to the spin quantum Hall effect, the single-particle Hilbert space is even-dimensional, and there is an action of σ_y such that the single-particle hamiltonian $\mathcal{H}$ satisfies $\mathcal{H}^* = -\sigma_y \mathcal{H} \sigma_y$. This implies a symmetry between states with energies $\pm E$, and that the time-evolution operator $\mathcal{U}^t = e^{-i\mathcal{H}t}$ satisfies

$$\mathcal{U}^* = \sigma_y \mathcal{U} \sigma_y,$$

implying that the matrices U_e should be symplectic, in $Sp(N)$, which for $N = 2$ is isomorphic to $SU(2)$.

For a given node n with incoming edges (e_1, e_2) and outgoing edges (e_1', e_2'), the S-matrix has the block form

$$\begin{pmatrix} S_{e_1'e_1} & S_{e_2'e_1} \\ S_{e_1'e_2} & S_{e_2'e_2} \end{pmatrix}, \tag{2.1}$$

where each element is an $N \times N$ matrix. However, since all the matrices are chosen at random, there is the gauge freedom of redefining $S_{e'e} \to V_{e'}^{-1} S_{e'e} V_e$, $U_e \to V_e^{-1} U_e V_e$, which allows us to choose each $S_{e'e}$ to be proportional to the unit $N \times N$ matrix $\mathbf{1}_N$. The remaining 2×2 matrix can further be chosen to be real and therefore orthogonal. Thus in fact (2.1) can be replaced by

$$1_N \otimes \begin{pmatrix} \cos\theta_n & \sin\theta_n \\ -\sin\theta_n & \cos\theta_n \end{pmatrix}. \tag{2.2}$$

We are left with the gauge-transformed U_e and the θ_n as quenched random variables. However, we shall treat them differently, first keeping the θ_n fixed while performing quenched averages over the U_e. In fact we shall see that in most cases is suffices to take all the θ_n as fixed and equal on each sublattice.

Because we consider discrete rather than continuous time evolution, the usual Green function is replaced by the resolvent $(1 - z\mathcal{U})^{-1}$ of the unitary evolution operator, whose matrix elements we shall however continue to refer to as the Green function:

$$G(e_f, e_i; z) \equiv \langle e'|(1 - z\mathcal{U})^{-1}|e\rangle. \tag{2.3}$$

Here $|e\rangle \in \mathcal{H}$ has non-zero components only in $\mathcal{H}_e$. Note that $G(e_f, e_i; z)$ is an $N \times N$ matrix mapping $\mathcal{H}_{e_i} \to \mathcal{H}_{e_f}$. The parameter z is the analog of the energy (roughly $z \sim e^{iE}$). For $|z| \ll 1$, the expansion of (2.3) in powers of z gives G as a sum over Feynman paths from e_i to e_f. Each path γ is weighted by $z^{|\gamma|}$ times an ordered product of the U_e with $e \in \gamma$ and the factors of $\cos\theta_n$ or $\pm\sin\theta_n$ for $n \in \gamma$. For a finite closed graph $\mathcal{G}$, this expansion is convergent for $|z| < 1$ and therefore defines G as an analytic function in this region. In general, for a finite $\mathcal{G}$, there are poles on the circle $|z| = 1$ corresponding to the eigenvalues of $\mathcal{U}$.

However, for $|z| > 1$, G admits an alternative expansion in powers of z^{-1} by writing it as

$$G(e_f, e_i; z) \equiv -\langle e'|z^{-1}\mathcal{U}^\dagger(1 - z^{-1}\mathcal{U}^\dagger)^{-1}|e\rangle. \tag{2.4}$$

This is given by a sum over paths γ with length ≥ 1 of a product of $z^{-|\gamma|}$ with ordered factors $U_e^\dagger$ along the path, and $\cos\theta_n$ or $\pm\sin\theta_n$ as before.

The eigenvalues of $\mathcal{U}$ have the form $e^{i\epsilon_j}$, where the $-\pi < \epsilon_j \leq \pi$ with $j - 1, \ldots, \mathcal{N}$ are discrete for a finite graph $\mathcal{G}$. We define the *density of states* by

$$\rho(\epsilon) \equiv \frac{1}{\mathcal{N}} \sum_j \delta(\epsilon - \epsilon_j).$$

In the standard way, the density of states is given by the discontinuity in the trace of the Green function, this time across $|z| = 1$ rather than $\text{Im } E = 0$:

$$\rho(\epsilon) = \frac{1}{2\pi N|\mathcal{G}|} \sum_{e \in \mathcal{G}} \lim_{\eta \to 0+} \left(\text{Tr } G(e, e; z = e^{i\epsilon - \eta}) - \text{Tr } G(e, e; z = e^{i\epsilon + \eta}) \right),$$

$$\tag{2.5}$$

where the trace is in the N-dimensional space $\mathcal{H}_e$. In the case where $\mathcal{G}$ is a regular lattice, in the thermodynamic limit we expect the eigenvalues to be continuously distributed around the unit circle.

We note that in the U(1) case, when the U_e are pure phases $e^{i\phi_e}$, in each term in the Feynman path expansion of Eqs. (2.3) and (2.4), a given edge e occurs with a weight $e^{in_e^\gamma \phi_e}$, where n_e^γ is the number of times the path γ traverses this edge. On averaging a given path γ will contribute to the mean density of states only if $n_e^\gamma = 0$, that is it has length zero. Thus, in the U(1) network models, $\overline{G(e,e,z)} = 1$ for $|z| < 1$, and zero for $|z| > 1$, and the mean density of states is constant, and completely independent of the θ_n. This is consistent with the general result that at the plateau transition in the charge quantum Hall effect, the density of states is non-singular.

We now turn to the conductance. In order to define this, we must consider an open graph, which can be obtained from a given closed graph $\mathcal{G}$ by breaking open a subset $\{e\}$ of the edges, relabelling each broken edge e as e_{in} and e_{out}. External contacts are subsets C_{in} and C_{out} of these. The transmission matrix T is a rectangular matrix with elements

$$T = \langle e_{\text{out}}|(1-\mathcal{U})^{-1}|e_{\text{in}}\rangle \,,$$

where $e_{\text{in}} \in C_{\text{in}}$, $e_{\text{out}} \in C_{\text{out}}$. Note that for an open graph the resolvent $(1-z\mathcal{U})^{-1}$ generally has poles inside the unit circle $|z| = 1$. In the thermodynamic limit, however, as long as the fraction of broken edges is zero (for example if we have contacts only along part of the edge of the sample), it is believed that the limit as $|z| \to 1$ can be taken.

The multi-channel Landauer formula then gives the conductance between the contacts as

$$g = (Q^2/h)\operatorname{Tr} T^\dagger T \,,$$

where Q is the quantum of charge carried by the particle. For the integer quantum Hall effect, $Q = e$, and for the spin Hall effect $Q = \frac{1}{2}\hbar$.

3. The Main Theorems

In this section, we focus on the Sp(2) (=SU(2)) case and summarize the method of proof of the main theorem relating the quenched averages of the density of states and conductances to observables of a classical random walk model on the same graph $\mathcal{G}$. We restrict attention to graphs with exactly two incoming and two outgoing edges. The general case is considered in Ref. 11 and is considerably more verbose.

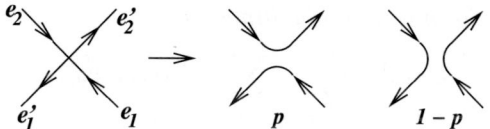

Fig. 14.4.   The two ways to decompose a node of $\mathcal{G}$. Note that in general this does not have to be planar, although for the L-lattice (Fig. 14.2) it is.

Let us first define the corresponding classical problem. Starting with a given closed graph $\mathcal{G}$, to each node with incoming edges (e_1, e_2) and outgoing edges (e'_1, e'_2), associate its two possible *decompositions* $((e'_1, e_1), (e'_2, e_2))$ and $((e'_2, e_1), (e'_1, e_2))$, corresponding to the two distinct ways of passing twice through the node without using a given edge more than once, irrespective of the order in which the edges are used. This is illustrated in Fig. 14.4. To each decomposition assign a probability $p_n = \cos^2 \theta_n$ or $1 - p_n = \sin^2 \theta_n$, corresponding to the S-matrix in (2.2). Doing this for each node in $\mathcal{G}$ gives a decomposition of $\mathcal{G}$ into a union of closed loops. There are $2^{|\mathcal{G}|}$ such decompositions, where $|\mathcal{G}|$ is the total number of nodes, and the assigned probabilities give a product measure on the set of decompositions.

We now state the two main theorems. We use the symbol $\overline{A}$ to denote a quenched average in the original quantum network model. Let $P(L, e)$ be the probability that the edge e lies on a closed loop of length L in the classical problem.

Theorem 14.1. $\overline{G(e_f, e_i, z)}$ *vanishes unless* $e_f = e_i$, *when it is given by*

$$\text{Tr}\,\overline{G(e, e, z)} = \begin{cases} 2 - \sum_{L>0} P(L, e) z^{2L} : |z| < 1 \\ \sum_{L>0} P(L, e) z^{-2L} \quad : |z| > 1 \end{cases} .$$

If we apply this to the density of states in (2.5) we find simply

$$\rho(\epsilon) = (1/2\pi)\big(1 - \sum_{L>0} P(L) \cos(2L\epsilon)\big), \tag{3.1}$$

where $P(L) = (|\mathcal{G}|)^{-1} \sum_e P(L, e)$, the probability that a edge chosen at random lies on a loop of length L. Note that if this distribution decays sufficiently fast for large L, $\rho(\epsilon)$ is expected to be analytic, while if it decays as a power law, $\rho(\epsilon)$ will have a power law singularity at $\epsilon = 0$.

For an open graph with external contacts C_{in} and C_{out}, the decomposition of $\mathcal{G}$ can also lead to open paths connecting edges in C_{in} to those in C_{out}. In this case, for the conductance, we have

Theorem 14.2. *The mean conductance is*

$$\bar{g} = 2 \sum_{e_{\text{in}} \in C_{\text{in}}} \sum_{e_{\text{out}} \in C_{\text{out}}} P(e_{\text{out}}, e_{\text{in}}),$$

where $P(e_{\text{out}}, e_{\text{in}})$ *is the probability that an open path from* e_{in} *to* e_{out} *exists. That is,* $\bar{g}$ *is twice the mean number of open paths connecting* C_{in} *to* C_{out}.

We describe the proof of these two Theorems using the supersymmetric path integral method of Ref. 11: the combinatorial method in Ref. 10 is perhaps more illustrative of why the result holds, but it involves heavier algebra.

In the standard way, the Green function G, being the inverse of a matrix, may be written as a gaussian integral over commuting (bosonic) variables. The notation is a little cumbersome but the basic idea is simple. Label each end of a given edge e by e_R and e_L, in the direction of propagation $e_R \to e_L$. Introduce complex integration variables $b_R(e)$ and $b_L(e)$, each of which is a 2-component column vector in the SU(2) space. Then

$$G(e_f, e_i; z) = \langle b_L(e_f) b_L(e_i)^\dagger \rangle = \frac{\int \prod_e [db_L(e)][db_R(e)] b_L(e_f) b_L(e_i)^\dagger \, e^{W[b]}}{\int \prod_e [db_L(e)][db_R(e)] e^{W[b]}} \tag{3.2}$$

where $W[b] = W_{\text{edge}} + W_{\text{node}}$ and

$$W_{\text{edge}} = z \sum_e b_L^\dagger(e) U_e b_R(e), \qquad W_{\text{node}} = \sum_n \sum_{ji} b_R^\dagger(e_j') S_{ji} b_L(e_i).$$

We use the notation $\langle \cdots \rangle$ to denote averages with respect to this gaussian measure. The measure for each integration is

$$\int [db] = (1/\pi^2) \int e^{-b^\dagger b} \, d\mathrm{Re}\, b \, d\mathrm{Im}\, b.$$

For a finite graph there are a finite number of integrations and the integral is convergent for $|z| < 1$.

The next step is to average over the quenched random matrices U_e. As usual this is difficult because these appear in both the numerator and denominator of (3.2). This can be addressed using replicas, or, much more effectively in this case, by adding an anticommuting (fermionic) copy $(f, \bar{f})$ of each pair bosonic variables $(b, b^\dagger)$. Note that each f is also a 2-component column vector in SU(2) space, and each $\bar{f}$ a row vector. The Grassman integration over these is defined by

$$\int [df] = \int e^{-\bar{f} f} \, df \, d\bar{f},$$

where

$$\int [df] f = \int [df] \bar{f} = 0 \quad \text{and} \quad \int [df] 1 = \int [df] f \bar{f} = 1 \,.$$

Integrating over these cancels the denominator in (3.2) so that

$$G(e_f, e_i; z) = \int \prod_e [db_L(e)][db_R(e)][df_L(e)][df_R(e)] f_L(e_f) \bar{f}_L(e_i) \, e^{W[b] + W[f]} \,.$$

$$(3.3)$$

The action $W[b] + W[f]$ is supersymmetric under rotating bosons into fermions and *vice versa*, so we can replace the bosonic fields at e_i and e_f by fermionic ones and consider $\langle f_L(e_f) \bar{f}_L(e_i) \rangle$.

Now consider the average over U on a given edge. This has the form

$$\int dU \exp(z b_L^\dagger U b_R + z f_L U f_R) \,.$$

$$(3.4)$$

Because the anticommuting fields square to zero, the expansion of the exponential in powers of the second term terminates at second order. The gauge symmetry discussed earlier shows that the purely bosonic zeroth order term is in fact independent of $b_L^\dagger$ and b_R and is in fact unity. The second order term is also straightforward: carefully using the anticommuting property we see that is actually proportional to $\det U = 1$, times the determinant of the SU(2) matrix of fermion bilinears with elements $\bar{f}_{Li} f_{Rj}$, where i and j take the values 1 or 2. The first order term then simply converts this into something supersymmetric. The conclusion is that the integral (3.4) equals $1 + \frac{1}{2} z^2 \det \mathbf{M}$ where the 2×2 matrix $\mathbf{M}$ has components $M_{ij} = b_{Li}^\dagger b_{Rj} + \bar{f}_{Li} f_{Rj}$. This can be rearranged in the form

$$1 + z^2 \left[(1/\sqrt{2})(b_{L1}^\dagger \bar{f}_{L2} - b_{L2}^\dagger \bar{f}_{L1}) \right] \left[(1/\sqrt{2})(b_{R1} f_{R2} - b_{R2} f_{R1}) \right]$$
$$+ z^2 \left[\bar{f}_{L1} \bar{f}_{L2} \right] \left[f_{R2} f_{R1} \right] \,.$$

Each expression in square brackets is an antisymmetric SU(2) singlet. The three terms above describe the propagation of either nothing, a fermion boson (fb) singlet, or a fermion-fermion (ff) singlet along the edge e. This is a remarkable simplification: before the quenched average, each edge may be traversed many times, corresponding to the propagation of multi-particle states. It is this which gives one of the principle simplifications of the SU(2) case, that does not happen for U(1), one of the reasons this is much more difficult.

This result shows that single fermions or bosons cannot propagate alone, so $\overline{G(e_f, e_i)} = 0$ if $e_f \neq e_i$. A non-zero correlation function is however

$$\langle f_1(e_f) f_2(e_f) \bar{f}_2(e_i) \bar{f}_1(e_i) \rangle = \overline{G_{11} G_{12} - G_{21} G_{22}} = \overline{\det G(e_f, e_i, z)} \quad (3.5)$$

Now G, being a real linear combination of products of SU(2) matrices, can in fact always be written in the form $\lambda \widetilde{G}$ where λ is real and $\widetilde{G} \in$ SU(2).[a] Hence $\det G = \lambda^2$ and $G^\dagger G = \lambda^2 \mathbf{1}$, so $\text{Tr}\, G^\dagger G = 2 \det G$. When $z = 1$ this gives the point conductance between e_i and e_f.

When $e_i = e_f = e$, however, we can always insert a pair of fermion fields $f_2(e)\bar{f}_2(e)$ into the correlator $\langle f_1(e)\bar{f}_1(e)\rangle$ at no cost, so that in fact

$$\overline{G(e,e;z)_{11}} = \overline{G(e,e;z)_{22}} = \overline{\det G(e,e;z)}\,.$$

Thus both the mean density of states and the mean conductance are proportional to $\overline{\det G}$ and therefore are given by the correlation function (3.5).

The next step is to consider propagation through the nodes. Note that we can now drop the distinction between b_R and b_L, etc. The contribution from a given node takes the form

$$\prod_i A_{\alpha_i'}(e_i')\, \mathcal{S} \prod_j A_{\alpha_j}^\dagger(e_j)\,, \tag{3.6}$$

where $A_1 = 1$, $A_2 = (1/\sqrt{2})(b_1 f_2 - b_2 f_1)$, $A_3 = f_1 f_2$, and

$$\mathcal{S} = \exp\left(\sum_{ij} b_i^\dagger S_{ij} b_j + \bar{f}_i S_{ij} f_j\right),$$

where S_{ij} is the matrix in (2.2). Since this expression conserves fermion and boson number, it follows that the total numbers of (fb) and (ff) singlets are also conserved. Also, only terms second order in the S_{ij} survive. In fact, after a little algebra (3.6) reduces to[b]

$$\delta_{\alpha_1'\alpha_1}\delta_{\alpha_2'\alpha_2}S_{11}S_{22} - \delta_{\alpha_2'\alpha_1}\delta_{\alpha_1'\alpha_2}S_{21}S_{12} = \delta_{\alpha_1'\alpha_1}\delta_{\alpha_2'\alpha_2}\cos^2\theta_n + \delta_{\alpha_2'\alpha_1}\delta_{\alpha_1'\alpha_2}\sin^2\theta_n\,.$$

These two terms correspond to the decomposition of the node n described earlier. It shows that, on performing the quenched average, the quantum network model is equivalent to a classical one in which $\mathcal{G}$ is decomposed into disjoint loops, and along each loop propagates either an (fb) singlet, an (ff) singlet, or nothing. Theorems 14.1 and 14.2 now follow straightforwardly. We argued above that the mean diagonal Green function $\overline{G(e,e;z)}$ is given by the (ff) correlation function $\langle [f_1(e)f_2(e)][\bar{f}_1(e)\bar{f}_2(e)]\rangle$. So, in each decomposition of $\mathcal{G}$, there must be an (ff) pair running around the unique loop containing e. This gets weighted by a factor z^{2L}. Around all the other loops we can have either an (ff) pair, a (bf) pair, or just 1. The (ff) pair,

[a]This follows from the representation $U = \cos\alpha + i(\sigma \cdot \mathbf{n})\sin\alpha$.

[b]In Ref. 11, this was carried out for a general node of arbitrary coordination, with excruciating algebra.

being itself bosonic, gives z^{2L} for a loop of length L, while the (bf) pair, being fermionic, gives $-z^{2L}$. These two cancel (by supersymmetry), leaving a factor 1 for every loop other than the one containing e. The argument for Theorem 14.2 for the conductance works in the same way.

Note that these methods may be extended to the quenched averages of other observables in the quantum model, although the density of states and the conductance are most important. However not all quantities of interest can be treated in this fashion. For example, the fluctuations in the conductance involve

$$\overline{\left(G(e_{\text{out}}, e_{\text{in}})^\dagger G(e_{\text{out}}, e_{\text{in}})\right)^2},$$

and, in order to treat this, we would need to double the number of degrees of freedom in the integral representation. Many of the formulas which are special to SU(2) integrations then no longer hold. In this context, it is important to note that the conductance fluctuations in the quantum model are *not* given by the fluctuations in the number of paths connecting the two contacts in the classical model. (If this were the case, the quantum system would be behaving completely classically!)

An amusing application[10] of the general theorems is to consider single edge e, closed on itself, but take $U_e \in \text{Sp}(N)$ with N even and > 2 in general. A general $\text{Sp}(N)$ matrix may be built up in terms of successive $\text{Sp}(2)$ rotations in overlapping two-dimensional subspaces. For example for $\text{Sp}(4)$ we may write a general matrix in the form

$$\frac{1}{\sqrt{2}} \begin{pmatrix} U_1 & 0 \\ 0 & U_2 \end{pmatrix} \begin{pmatrix} 1 & 1 \\ -1 & 1 \end{pmatrix} \tag{3.7}$$

where U_1 and U_2 are independent $\text{Sp}(2)$ matrices. If these are drawn from the invariant measure on $\text{Sp}(2)$, then the product of a large number of independent such matrices will converge to the invariant measure on $\text{Sp}(4)$. Thus, in a particular $\text{Sp}(2)$ basis, $\mathcal{G}$ has the form shown in Fig. 14.5. After applying Theorem 14.1, each decomposition corresponds to a permutation of the different channels corresponding to the basis chosen in (3.7). This generalizes to arbitrary N. If we now connect opposite ends of Fig. 14.5 to

Fig. 14.5. Graph corresponding to a single link in the Sp(4) model, and its topologically distinct decompositions. Each node corresponds to the S-matrix which is the second factor in (3.7), so each term in the decomposition is equally weighted.

make a closed graph, we find, after the decomposition, all possible lengths L of loops from 1 to $\frac{1}{2}N$, with equal probabilities. Thus $P(L) = 2/N$ for $1 \leq L \leq \frac{1}{2}N$, and zero otherwise. Using (3.1), this gives for the density of eigenvalues of a random $Sp(N)$ matrix

$$\rho(\epsilon) = \frac{N+1}{2\pi N} \left(1 - \frac{\sin(N+1)\epsilon}{(N+1)\sin\epsilon} \right),$$

in agreement with Ref. 12.

4. Two-Dimensional Models

In this section we discuss the consequences of the main theorems for specific 2d lattices relevant to physical problems.

4.1. *The L-lattice*

This is the lattice, illustrated in Fig. 14.2, used by the original Chalker–Coddington model for the quantum Hall plateau transition. The reasons for choosing this lattice were discussed in Sec. 1. In the class C version of this, the same lattice is used, the only difference being that the quenched random U(1) phases on each edge are replaced with SU(2) matrices. We also recall that, because of the checkerboard nature of the potential, where even plaquettes correspond to $V > 0$ and odd ones to $V < 0$, the angles θ_n, which represent the degree of anisotropy of the tunnelling at the nodes, are in fact staggered: $\theta_n = \theta$ on the even sublattice, and $(\pi/2) - \theta$ on the odd sublattice. Thus for $\theta = 0$ all the loops in the decomposition of $\mathcal{G}$ will be the minimum size allowed, surrounding the even plaquettes, and for $\theta = \pi/2$ they will surround the odd plaquettes. Away from these extreme values, the loops will be larger. If there is a single transition it must occur at $\theta = \pi/4$.

The mapping to square lattice bond percolation for this model is exact. Consider independent bond percolation on the square lattice $\mathcal{L}'$, rotated by $45°$ with respect to the original, whose sites lie at the centers of the even plaquettes of the original lattice. See Fig. 14.6. Each edge of $\mathcal{L}'$ intersects a node of the original one. We declare it to be open, with probability $p = \cos^2\theta$, or closed, with probability $1 - p = \sin^2\theta$, according to the way the node is decomposed in the classical loop model on $\mathcal{G}$. There is thus a 1-1 correspondence between decompositions of $\mathcal{G}$ and bond percolation configurations on $\mathcal{L}'$. We can also consider the dual lattice $\mathcal{L}''$ whose vertices are at the centers of the odd plaquettes of the original lattice. Each edge of this lattice crosses a unique edge of $\mathcal{L}'$, and we declare it to be open, with probability $1 - p$, if

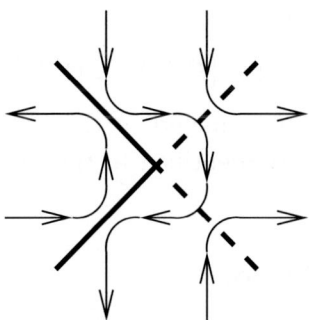

Fig. 14.6.   The mapping between decompositions of the L-lattice and bond percolation on the lattice $\mathcal{L}'$. Open bonds are shown as thick lines, closed as dashed lines.

the corresponding edge of $\mathcal{L}'$ is closed, and *vice versa*. For each percolation configuration on $\mathcal{L}'$, there is a corresponding dual configuration on $\mathcal{L}''$. The clusters and dual clusters fill the plane without intersecting each other. (For this purpose it is necessary to regard single isolated sites as clusters.) Close to $p_c = \frac{1}{2}$ many clusters nest inside dual clusters and *vice versa*.

For a given decomposition of $\mathcal{G}$, the loops give the *hulls* of percolation clusters on $\mathcal{L}'$ and dual clusters on $\mathcal{L}''$. For a closed simply connected lattice $\mathcal{G}$, these are closed curves which simultaneously circumscribe a cluster and inscribe a dual cluster, or *vice versa*. For open boundary conditions, some of these curves may begin and end on the boundary.

From this mapping to 2d percolation, and known exact and conjectured results about the latter, many results about the SU(2) network model on the L-lattice may be deduced.[8] At the transition, the conductance between two bulk points a distance r apart is given by the probability that they are on the same loop, which is known to decay as $|r|^{-2x_1}$ where $x_1 = \frac{1}{4}$.[13] The conductance of a rectangular sample with contacts along opposite edges is given by the mean number of hulls which cross the sample. At the critical point this depends only its aspect ratio L_1/L_2, in a complicated but calculable way.[14] For $L_1/L_2 \gg 1$ (where L_1 is the length of the contacts) it goes like $\bar{g}(L_1/L_2)$ where $\bar{g}$ is the universal critical conductance which, from conformal field theory results applied to percolation, takes the value $\sqrt{3}/2$.[14] In Sec. 3, we showed that the mean density of states is given by the probability $P(L)$ that a given edge is on a loop of length L. At criticality, this probability scales like L^{-x_1} where $d_f = 2 - x_1 = \frac{7}{4}$ is the fractal dimension of percolation hulls.[13] Also $|z - 1| \propto |E|$, which is conjugate to L, has RG eigenvalue $y_1 = 2 - x_1$. This means that the singular part of the mean density of states behaves like[8]

$$\overline{\rho(E)} \sim |E|^{x_1/y_1} = |E|^{1/7} \,.$$

A number of other exponents, including the usual percolation correlation length exponent $\nu = \frac{4}{3}$, were identified in the physics of the spin quantum Hall transition in Ref. 8.

4.2. *The Manhattan Lattice*

Although the L-lattice is the natural candidate for studying the spin quantum Hall transition, the mapping discussed in Sec. 3 is of course valid for any orientation of the edges of a square lattice, and one can legitimately ask whether other possibilities lead to interesting physics. The expectation, based on the continuum classification of localization universality classes, is that unless there is some special symmetry, such as occurs for the L-lattice with its sublattice symmetry, all states in 2d will be exponentially localized and therefore almost all loops in the classical model will have finite length. This was studied for the Manhattan lattice in Ref. 15. On this lattice all edges in the same row or column are oriented in the same direction, and these alternate, see Fig. 14.7. This resembles the one-way system of streets and avenues in Manhattan. At each corner, the driver can go straight on, or turn either left or right according to the parity of the intersection.

Consider an SU(2) network model on this lattice. The decomposition of the lattice corresponds to replacing each node either by a crossing, with probability $1 - p$, say, or an avoidance, with probability p (see Fig. 14.8). Note that in this case the loops on the decomposed lattice are in general nonplanar. Nevertheless it is possible to make rigorous progress using a mapping to percolation, owing to the sublattice structure. Consider once again bond percolation on the 45°-rotated lattice $\mathcal{L}'$ (see Fig. 14.8). An edge is declared open if the corresponding node is decomposed in such a way that the paths turn by 90°, as shown. The open edges once again form connected clusters

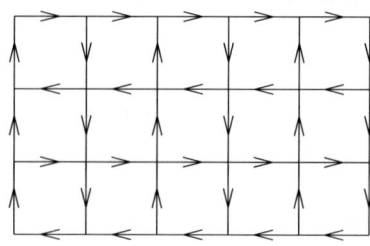

Fig. 14.7.   The Manhattan lattice.

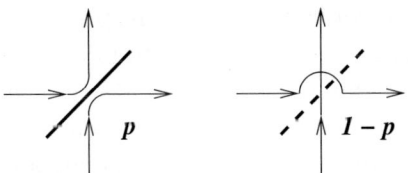

Fig. 14.8. The decomposition of each node of the Manhattan lattice and its relation to percolation on $\mathcal{L}'$. Open bonds shown as solid thick lines, closed as dashed.

which enclose, and are enclosed by, percolation hulls, and similarly for the dual clusters. A little reflection shows that each loop of the decomposed Manhattan lattice is constrained to lie on or between neighboring hulls which enclose a dual cluster. For $p > p_c - \frac{1}{2}$ the dual clusters are almost surely finite, and therefore so are their surrounding hulls, and therefore also the loops of the decomposed $\mathcal{G}$. Therefore for $p > \frac{1}{2}$ the SU(2) network model on the Manhattan lattice is in the localized phase. An unproven conjecture, consistent with our expectations for Anderson localization in 2d, is that this happens for all $p > 0$. Simulations[15] of the classical loop model indicate that this is the case for $p > 0.2$. The field theory arguments, discussed in Refs. 15 and 24, indicate that there should always be a finite localization length, diverging as $\xi \sim \exp\left(\text{const.}/p^3\right)$ as $p \to 0$.

4.3. *Other 2d lattices*

We may consider other orientations of the square lattice. In general a given orientation corresponds to a configuration of the six-vertex model, which satisfies the "ice rule" that there are two incoming and two outgoing arrows at each vertex, or node. The number of such allowed configurations grows exponentially with the size of the lattice, but we could, for example, consider a randomly oriented lattice in which the weights for different types of node are given by the six-vertex model. In that case the L-lattice and Manhattan lattice are just particular extreme points of the parameter space. Once again we can associate an edge of the lattice $\mathcal{L}'$ or $\mathcal{L}''$ with the decomposition of a node where the path has to turn. These edges can be thought of as two-sided mirrors, reflecting all the paths which impinge on either side. The study of these "mirror models" as models for classical localization has been extensive (see, for example Ref. 16, and references quoted therein), although on the whole only simulational results are available. However it is important to realize that arbitrary mirror models do not in general lead to a unique orientation for the edges: a path may traverse a given edge in both directions.

Thus the set of mirror configurations corresponding to quantum network models is a subset, and it would be wrong to infer general conclusions about these from the study of the wider problem. Nevertheless, one expects that, for a sufficiently high density of randomly oriented mirrors, the paths are finite and so the states are localized. An interesting and unresolved question, however, is what happens at low mirror density, when the mean free path is large. Expectations from quantum localization would then suggest that the paths in such models are still localized on large enough scales, unless there is some special symmetry like that of the L-lattice.

5. Three-Dimensional Models

We now discuss some results for class C network models on 3d lattices. It is possible, of course, to consider layered 2d lattices which might be used as models for quantum Hall physics in multilayered systems. For example, in a bilayer system consisting of two coupled L-lattices, depending on the strength of the coupling between the layers, one expects to see either two separate transitions between states of Hall conductance 0, 1 and 2. These can be simply understood in terms of the classical model and an equivalent percolation problem.[17] However, such models do not capture some of the important properties of real bilayer systems which depend on electron–electron interactions.

One motivation for studying truly 3d class C network models is to shed light on the physics of the localization transition in 3d. So far, these have been carried out either numerically or by mapping to 3d percolation. However the restriction to four-fold coordination in order that the equivalent classical model has non-negative weights[11] means that it is necessary to use loose-packed lattices.

5.1. *Diamond lattice*

The most extensive numerical simulations have been carried out on the diamond lattice.[18] Although this lattice has cubic symmetry, assigning the orientations of the edges breaks this down to tetragonal, inducing an anisotropy. However, numerical tests show that this is not very great, giving, for example, a ratio of about 1.1 between the conductances in the two distinct directions in the conducting phase. As expected, the model exhibits a sharp transition at $p = p_c$ between an insulating phase and a conducting phase. This is shown, for example, in data for the conductance $G(p, L)$ of a cubic sample of linear size L, which, according to the Theorem 14.2, is given by the mean number of open paths between two opposite faces. In the insulating phase $p < p_c$ this should approach zero as $L \to \infty$, while for $p > p_c$ we expect ohmic behavior with $G(p, L) \sim \sigma(p)L$. Close to the critical points we expect finite-size scaling of the form

$$G(p, L) = f\left(L/\xi(p)\right),$$

where the localization length $\xi(p) \sim |p - p_c|^{-\nu}$. Thus the data should show collapse when plotted as a function of $(p - p_c)L^{1/\nu}$, and this is clearly exhibited in Fig. 14.9. The best fitted value for ν, taking into account corrections to scaling, is[18] $\nu = 0.9985 \pm 0.0015$. The closeness of this value to unity, the value predicted by a first-order result $\nu^{-1} = \epsilon + O(\epsilon^2)$ of the $2 + \epsilon$-expansion (see Refs. 15 and 24) is remarkable, but perhaps a coincidence.

At $p = p_c$, the weighted number of return paths of length L, behaves as $P(L) \sim L^{-x_1/y_1}$, as discussed in Sec. 4.1. In Ref. 18, this exponent is denoted by $2 - \tau$, where numerically $\tau = 2.184 \pm 0.003$. This exponent is related to the fractal dimension $d_f = y_1 = 3 - x_1$ of the paths at p_c, giving $d_f = 2.534 \pm 0.009$. By the same arguments as in Sec. 4.1, this gives, for example, the singular part of the density of states $\rho(E) \sim |E|^{\tau-2}$. Note that, because of the mapping to the classical problem for which far larger

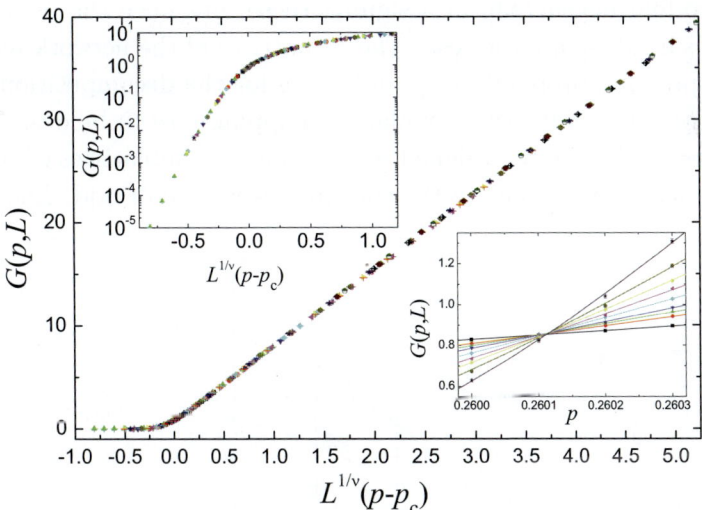

Fig. 14.9. Conductance as a function of $(p - p_c)L^{1/\nu}$, illustrating scaling collapse. Upper inset: same data on a logarithmic scale. Lower inset: conductance as a function of p for several values of L. Lines are a scaling fit described in detail in Ref. 18. Reprinted (Fig. 1) with permission from M. Ortuño, A. M. Somoza and J. T. Chalker, *Phys. Rev. Lett.* **102**, 070603 (2009). © American Physical Society.

systems can be studied, the error bars on these exponents are much smaller than those quoted for the conventional 3d Anderson transition.

5.2. *3d L-lattice and Manhattan lattice*

It is possible to construct 3d oriented lattices with cubic symmetry which are direct analogues of the 2d L-lattice and Manhattan lattices discussed in Sec. 4 and for which the arguments relating the classical models to percolation can be generalized.

Consider two interpenetrating cubic lattices $C_1 \equiv \mathbf{Z}^3$ and $C_2 \equiv (\mathbf{Z} + \frac{1}{2})^3$. Each face of C_1 intersects an edge of C_2 at its midpoint. The four faces of C_2 which meet along this edge intersect the given face of C_1 along two mutually perpendicular lines, also perpendicular to the edge (see Fig. 14.10). These lines form part of the lattice $\mathcal{G}$. The same is true, interchanging the roles of C_1 and C_2. The full graph $\mathcal{G}$ is the lattice formed by the intersection of the faces of C_1 with those of C_2. The nodes of $\mathcal{G}$ lie on the midpoints of the edges of C_1 (the centers of the faces of C_2) and *vice versa*, and have coordination number 4. Clearly $\mathcal{G}$ has cubic symmetry.

For the L-lattice on $\mathcal{G}$, the orientation of the edges is chosen so that each node looks like the nodes of the 2d L-lattice, as in Fig. 14.10. There is an overall two-fold degeneracy in assigning these, but once the orientation at one node is fixed, so are the rest. The S-matrices of the network model, and the corresponding probabilities p and $1 - p$ for the decomposition of $\mathcal{G}$ are assigned consistent with the percolation mapping now to be described.

The sites of C_1 may be assigned to even and odd sublattices C_1' and C_1'' according to whether the sum of the coordinates is even or odd. Each of these

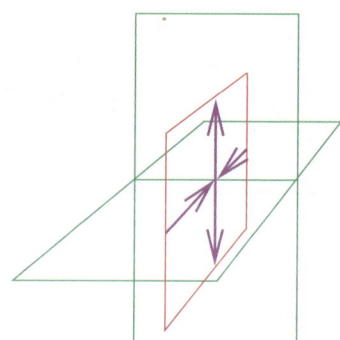

Fig. 14.10.   A node of $\mathcal{G}$ (purple) is formed by the intersection of four faces of C_2 (green) and one face of C_1 (red), and *vice versa*. The orientation shown corresponds to the 3d L-lattice.

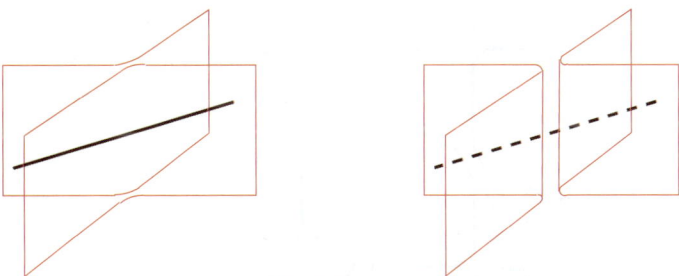

Fig. 14.11.   Decomposition of the four faces (in red) at an edge of C_1 corresponding to bond percolation (bonds shown in brown) on an fcc sublattice of C_2.

lattices is in fact a face-centered cubic (fcc) lattice. Now consider nearest neighbor bond percolation on C'_1. Each nearest neighbor edge of C'_1 intersects the midpoint of an edge of C_2, along which 4 faces of C_2 intersect. A decomposition of this edge consists in connecting up these faces in neighboring pairs, as in Fig. 14.11.

For each edge there are two possible decompositions, and we can do this in such a way that if the corresponding edge of the percolation problem on C'_1 is open, it passes between the connecting pairs (see Fig. 14.11), just as in Fig. 14.6 in 2d. If on the other hand the edge is closed, then it intersects both pairs. Equivalently, we can consider percolation on the "dual" fcc lattice C''_1. Each edge of this lattice intersects one edge of C'_1 at the midpoint of an edge of C_2. We declare the edge of C''_1 to be open if the intersecting edge of C'_1 is closed, and *vice versa*.

For a finite lattice, each decomposition of the edges of C_2 divides the faces of C_2 into a union of dense, non-intersecting, closed surfaces, in the same way that a decomposition of the corresponding nodes of the 2d square lattice divides the edges into non-intersecting closed loops. These closed surfaces form the hulls of the bond percolation clusters on the fcc lattice C'_1 and its dual C''_1. That is, each closed surface either touches a unique cluster externally and a unique dual cluster internally, or *vice versa*.

However, this is only half the description. The edges of $\mathcal{G}$ are formed by the intersection of the faces of C_1 with those of C_2. Therefore we need to also decompose the faces of C_1. This is carried consistent with another, independent, percolation problem on an fcc sublattice C'_2 of C_2, and its dual C''_2. To each double decomposition of the faces of C_1 into closed non-intersecting surfaces, and similarly of the faces of C_2, corresponds a unique decomposition of $\mathcal{G}$ into closed loops.

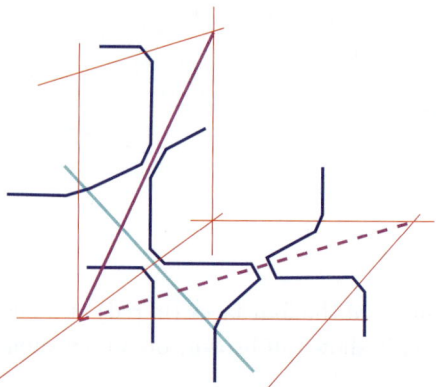

Fig. 14.12. A decomposition of $\mathcal{G}$ on the 3d L-lattice, showing how the loops (blue) reflect off the open bonds of percolation on the fcc lattices $\mathcal{C}_1'$ (solid magenta lines) and $\mathcal{C}_2'$ (solid turquoise lines).

This is the classical loop model which we seek. Unlike the 2d case, it corresponds to *two* independent percolation problems, one on an fcc lattice of $\mathcal{C}_1$, the other on an fcc sublattice of $\mathcal{C}_2$. Each closed loop of $\mathcal{G}$ is formed by the intersection of a closed surface made up of faces of $\mathcal{C}_1$ and a closed surface made up of faces of $\mathcal{C}_2$. (It is of course possible that such pairs of closed surfaces intersect in more than one loop of $\mathcal{G}$, or not at all.) It alternately "reflects" off an open edge of percolation on $\mathcal{C}_1'$ (or an open dual edge of $\mathcal{C}_1''$), then an open edge of $\mathcal{C}_2'$ or of $\mathcal{C}_2''$, and so on (see Fig. 14.12).

We now discuss the physics of this model and its implications for the class C network model. In principle we can assign different probabilities, p_1 and p_2, to the two independent percolation problems on $\mathcal{C}_1$ and $\mathcal{C}_2$. The phase diagram is symmetric under the duality symmetry $p_j \to 1 - p_j$, so we can restrict attention to the quadrant $p_1 \leq \frac{1}{2}$, $p_2 \leq \frac{1}{2}$. For either p_1 or p_2 less than the bond percolation threshold $p_c^{\text{fcc}} \approx 0.12$,[19] since the clusters are finite so are the hulls on either $\mathcal{C}_1$ or $\mathcal{C}_2$ (or both), and, since the loops on $\mathcal{G}$ are formed by their intersection, these must be finite also, corresponding to a localized phase. Preliminary simulations of the model[20] for values of p_j close to $\frac{1}{2}$ appear to show that the loops are no longer finite, but, unlike the case of the diamond lattice,[18] neither are they simple random walks on large scales with fractal dimension 2, as would be expected of a sample exhibiting ohmic behavior. In fact, their fractal dimension appears to be close to 3, indicating that they are space filling. This may be a pathology of this model. It is known that the fractal dimension of cluster boundaries for $p > p_c$ is 3 — that is, a finite fraction of the infinite cluster, which has dimension 3 because it contains a finite fraction of the sites, is on its boundary. The loops for $p > p_c$ are formed by the intersection of random such $d_f = 3$ objects, so it is perhaps not surprising that they should also have $d_f = 3$.

For this reason it may be more useful to consider a 3d version of the Manhattan lattice. This is based on the same graph $\mathcal{G}$, but the edges are oriented so that along each line through the lattice they point in the same direction, and the lines alternate direction transversally. As in the 2d case, the probability of turning at a given node is p and of going straight on $1 - p$. (In principle we could again take different values of p at the two types of nodes corresponding to $\mathcal{C}_1$ and $\mathcal{C}_2$.) Analogously with the 2d case (see Sec. 4.2), we can introduce an associated percolation problem on fcc sublattices of $\mathcal{C}_2$ and $\mathcal{C}_1$, so that if, in the decomposition of $\mathcal{G}$, the paths turn at a given node, they reflect off an open edge. This means that, as in 2d, they are constrained to lie on regions occupied by the dual clusters of each percolation problem. Thus, for $p > 1 - p_c^{\text{fcc}} \approx 0.88$, the loops of $\mathcal{G}$ are almost surely finite in length, corresponding to the existence of a localized phase of the corresponding class C network model.

On the other hand, for small p we expect a finite fraction of the paths to escape to infinity on a infinite lattice, corresponding to extended states. In this case the particle following a path will, almost, all the time, go straight ahead, with only a small probability $p \ll 1$ of turning. In this case, the whether the particle has traversed an even or odd number of edges should be unimportant, leading to an effective simple random walk with diffusion constant $O(p^{-1})$ on intermediate distance scales. On larger scales, the walk may revisit regions it has in the past, but, unlike the case of 2d, this is unlikely because the 3d random walk is not recurrent. However, as appealing as this argument may be, it is not rigorous, and indeed its proof appears to be of the same order of difficulty as showing that the "true" self-avoiding walk in 3d is asymptotically Gaussian.[21,22] Thus, at this stage, a proof of the existence of extended states in this class of network models remains elusive.

6. Summary and Further Remarks

We have shown how the quantum $\rightarrow$ classical mapping for class C network models helps to gain insight into the nature of Anderson localization in both two and three dimensions. These models have direct physical relevance in systems where time-reversal symmetry is broken but spin rotational symmetry is preserved. The classical models correspond to deterministic motion in a random medium with two-sided mirrors, or, equivalently, certain kinds of history-dependent random walks. On some lattices, including the important examples of the L and Manhattan lattices, the paths are associated with the hulls of percolation clusters, and rigorous information can be inferred on whether the corresponding quantum model is in a localized or extended phase.

However, there are a number of unresolved questions. Although the diamond lattice and 3d Manhattan lattice discussed in Sec. 5 are expected to exhibit an Anderson transition, there is as yet no proof of the existence of an extended phase in which the motion is asymptotically diffusive, although this is strongly indicated on both numerical and other grounds. The relation of these models to other types of history-dependent random walks, such as the "true" self-avoiding walk, in which the walk avoids regions it has visited in the past, is also unclear. Like Anderson localization, $d = 2$ is a critical dimension for the true self-avoiding walk. However in this case the RG flows[22,23] are to free random walks for $d \geq 2$, and to a non-trivial stable fixed point for $d < 2$, while for Anderson localization we expect to find a non-trivial unstable fixed point for $d > 2$. This suggests that the two problems are related by a change of sign of the interaction. However the analysis of Peliti and Obukhov[23] shows that for history-dependent random walks there are in fact three coupling constants which are potentially important near $d = 2$. An attempt to fit the walks on the 2d Manhattan lattice into this picture was made in Ref. 15. However this was not systematic and further work needs to be done in this direction. It should be noted, however, that a sigma-model analysis of the original class C quantum model does give the expected unstable fixed point for $d > 2$.[24] A related question is that of the *upper* critical dimension for the transition in this model. The relation to interacting random walks suggests that this might be $d = 4$, as for ordinary polymers, but in this case the interactions are not simply repulsive, so this conclusion may not hold.

Acknowledgments

I am especially grateful to John Chalker for many informative discussions of this subject over the years, as well as Ilya Gruzberg, Yacine Ikhlef, Andreas Ludwig, Adam Nahum, Aleks Owczarek, Nick Read, Tom Spencer, Bob Ziff and Martin Zirnbauer. This work was supported in part by EPSRC Grant EP/D050952/1.

References

1. J. T. Chalker and P. D. Coddington, Percolation, tunnelling and the integer quantum Hall effect, *J. Phys. C: Solid State Phys.* **21**, 2665–2679 (1988).

2. R. E. Prange and R. Joynt, Conduction in a strong field in two dimensions: The quantum Hall effect, *Phys. Rev. B* **25**, 2943–2946 (1982).
3. S. A. Trugman, Localization, percolation, and the quantum Hall effect, *Phys. Rev. B* **27**, 7539–7546 (1983).
4. A. M. M. Pruisken, On localization in the theory of the quantized hall effect: A two-dimensional realization of the θ-vacuum, *Nucl. Phys. B* **235**, 277–298 (1984); I. Affleck, Critical behaviour of SU(n) quantum chains and topological non-linear σ-models, *Nucl. Phys. B* **305**, 582–596 (1988); A. W. W. Ludwig, M. P. A. Fisher, R. Shankar and G. Grinstein, Integer quantum Hall transition: An alternative approach and exact results, *Phys. Rev. B* **50**, 7526–7552 (1994); M. R. Zirnbauer, Conformal field theory of the integer quantum Hall plateau transition, arXiv:hep-th/9905054 (unpublished).
5. T. Senthil, J. B. Marston and M. P. A. Fisher, Spin quantum Hall effect in unconventional superconductors, *Phys. Rev. B* **60**, 4245–4254 (1999).
6. A. Altland and M. R. Zirnbauer, Nonstandard symmetry classes in mesoscopic normal-superconducting hybrid structures, *Phys. Rev. B* **55**, 1142–1161 (1997); M. R. Zirnbauer, Riemannian symmetric superspaces and their origin in random-matrix theory, *J. Math. Phys.* **37**, 4986–5018 (1996).
7. V. Kagalovsky, B. Horovitz and Y. Avishai, Landau-level mixing and spin degeneracy in the quantum Hall effect, *Phys. Rev. B* **55**, 7761–7770 (1997); V. Kagalovsky, B. Horovitz, Y. Avishai and J. T. Chalker, Quantum Hall plateau transitions in disordered superconductors, *Phys. Rev. Lett.* **82**, 3516–3519 (1999).
8. I. A. Gruzberg, A. W. W. Ludwig and N. Read, Exact exponents for the spin quantum Hall transition, *Phys. Rev. Lett.* **82**, 4524–4527 (1999).
9. S. Smirnov and W. Werner, Critical exponents for two-dimensional percolation, *Math. Res. Lett.* **8**, 729–744 (2001).
10. E. J. Beamond, J. Cardy and J. T. Chalker, Quantum and classical localization, the spin quantum Hall effect and generalizations, *Phys. Rev. B* **65**, 214301–214310 (2002).
11. J. Cardy, Network models in class C on arbitrary graphs, *Comm. Math. Phys.* **258**, 87–102 (2005).
12. M. R. Zirnbauer, Supersymmetry for systems with unitary disorder: circular ensembles, *J. Phys. A* **29**, 7113–7136 (1996).
13. B. Nienhuis, Critical behavior of two-dimensional spin models and charge asymmetry in the Coulomb gas, *J. Stat. Phys.* **34**, 731–761 (1984).
14. J. Cardy, Linking numbers for self-avoiding loops and percolation: application to the spin quantum Hall transition, *Phys. Rev. Lett.* **84**, 3507–3510 (2000).
15. E. J. Beamond, J. Cardy and A. L. Owczarek, Quantum and classical localization and the manhattan lattice, *J. Phys. A* **36**, 10251–10267 (2003).
16. M. S. Cao and E. G. D. Cohen, Scaling of particle trajectories on a lattice, *J. Stat. Phys.* **87**, 147–178 (1997).
17. E. A. Beamond, Ph.D. thesis, unpublished (2003).
18. M. Ortuño, A. M. Somoza and J. T. Chalker, Random walks and Anderson localization in a three-dimensional class C network model, *Phys. Rev. Lett.* **102**, 070603–070606 (2009).

19. C. D. Lorenz and R. M. Ziff, Precise determination of the bond percolation thresholds and finite-size scaling corrections for the sc, fcc, and bcc lattices, *Phys. Rev. E* **57**, 230–236 (1998).
20. Y. Ikhlef, private communication (2009).
21. R. T. Durrett and L. C. Rogers, Asymptotic behaviour of Brownian polymers, *Probab. Theory Related Fields* **92**, 337–349 (1992.)
22. D. J. Amit, G. Parisi and L. Peliti, The asymptotic behaviour of the "true" self-avoiding walk, *Phys. Rev. B* **27**, 1635–1645 (1983).
23. S. P. Obukhov and L. Peliti, Renormalisation of the "true" self-avoiding walk, *J. Phys. A* **16**, L147–L151 (1983).
24. T. Senthil, M. P. A. Fisher, L. Balents and C. Nayak, Quasiparticle transport and localization in high-T_c superconductors, *Phys. Rev. Lett.* **81**, 4704–4707 (1998); T. Senthil and M. P. A. Fisher, Quasiparticle density of states in dirty high-T_c superconductors, *Phys. Rev. B* **60**, 6893–6900 (1999).

Chapter 15

MATHEMATICAL ASPECTS OF ANDERSON LOCALIZATION

Thomas Spencer

Institute for Advanced Study

Princeton, NJ 08540, USA

This article discusses mathematical results and conjectures motivated by Anderson localization and by related problems for deterministic and nonlinear systems. Finite volume criteria for Anderson localization are explained for random potentials. Recent results on a phase transition for a hyperbolic, supersymmetric sigma model on a 3D lattice are also presented. This transition is analogous to the Anderson transition.

1. Introduction

Philip Anderson's landmark 1958 paper[1] has inspired thousands of articles in theoretical and experimental physics. It has also motivated a substantial body of mathematical research. The aim of this article is to give a brief mathematical view of Anderson localization and related problems. We begin by giving a history of mathematical developments in the theory of localization in Sec. 2. Section 3 presents two finite volume criteria for localization. The first is based on a multi-scale analysis and the second uses the fractional moment technique. Problems and results concerning localization for "deterministic" potentials and the effects of nonlinearity are discussed in Secs. 2 and 4.

In 1983, K. Efetov introduced supersymmetric (SUSY) lattice field models which provide a dual representation for many disordered quantum systems.[2] Section 5 of this article is devoted to the description of a hyperbolic SUSY statistical mechanics model. This model was introduced by Zirnbauer[3] and may be thought of as a simplified version of one of Efetov's models. It will be referred to as the $\mathbb{H}^{2|2}$ sigma model. $\mathbb{H}^{2|2}$ is the target space — a hyperbolic space in two bosonic and two fermionic fields. In collaboration with M. Disertori and M. Zirnbauer,[4,5] we showed that in 3D this model exhibits the analog of the Anderson localization–delocalization transition. Moreover, it admits a probabilistic interpretation. The dynamics of a quantum particle in a random environment at energy E roughly corresponds to a random walk in a highly correlated random environment. The environment is determined by a classical but nonlocal statistical mechanics model at inverse temperature β where β is approximately the local conductance across an edge. The proof of this transition is based on Ward identities and is briefly explained in Sec. 6.

The final section will briefly comment on a *history dependent* random walk called linearly edge reinforced random walk, ERRW. This is a walk on $\mathbb{Z}^d$ which favors edges it has visited more frequently in the past. Diaconis showed that it has the remarkable feature that it can also be expressed as a random walk in a correlated random environment.[6–8] This process has a parameter β which is inversely proportional to the strength of the reinforcement. It is known to localize for all values of β in 1 dimension.[8] It has a phase transition on the Cayley tree[9] from transient, for large β, to recurrent, for small β. In three dimensions it may also have an Anderson-like transition.

2. One Dimension: History, Results and Conjectures

To fix notation, let H be the tight binding Hamiltonian with a random potential $v(j)$, $j \in \mathbb{Z}^d$ which are independent identically distributed random variables of mean 0 and variance 1. The Hamiltonian is given by

$$Hf(j) = -\Delta f(j) + \lambda\, v(j) f(j), \tag{2.1}$$

and $\Delta f(j) = \sum_{i \sim j}[f(i) - f(j)]$ is the finite difference Laplacian on $\mathbb{Z}^d$. The relation $i \sim j$ will be used to denote nearest neighbor vertices on the lattice. The parameter λ measures the strength of the disorder.

Following Anderson's paper, Mott and Twose[10] argued that all eigenstates of a one dimensional tight binding model are localized for any nonzero disorder. In 1963, Furstenberg[11] proved a fundamental theorem stating that products of independent $Sl(n, R)$ matrices have have a positive Lyapunov exponent. This implies that with probability one, solutions of the initial value problem of a 1D tight binding model grow exponentially fast. This is a key ingredient for proving localization in 1 dimension. However, a positive Lyapunov exponent does not necessarily imply that eigenstates are localized. For example, for any irrational α, the quasi-periodic potential $v(j) = \cos(\pi \alpha j + \omega)$ produces a positive Lyapunov exponent when $\lambda > 2$.[12] But if α is very well approximated by rationals, there are no localized eigenstates. It took some time to understand that such pathologies cannot occur for random potentials with a regular density.

In 1977, Goldsheid, Molchanov and Pastur[13] were the first to prove that the Schrödinger equation on the line with a stochastic Markovian potential has only localized eigenstates for any nonzero disorder. In 1980, Kunz and Souillard[14] proved localization for the tight binding Hamiltonian in one dimension assuming that the potential $v(j)$ are independent random variables. Their proof drew on ideas of Borland.[15]

In one dimension, there is a considerable literature about Lyapunov exponents for general ergodic potentials, i.e., let v in (2.1) be $v(n, \omega) = V(\tau^n \omega)$ where τ is an ergodic measure preserving transformation of the measure space $(\Omega, d\mu(\omega))$ and V is a real valued function on Ω. For the case of a quasi-periodic potential, Ω is the circle with the uniform measure and τ is rotation by an irrational α. For independent random variables, $(\Omega, d\mu)$ is the product measure over Z, τ is a shift and $v(n, \omega) = \omega_n$. Kotani's main result[16,17] states that if an ergodic potential is non-deterministic, i.e., the process is not determined by its past, then the Lyapunov exponent is positive for almost all energies. Under additional assumptions one can prove this implies localization.

There are many challenging open problems concerning 1D localization and positive Lyapunov exponents for deterministic systems. For example, if Ω is the two torus and τ is the skew shift $\tau(\omega_1, \omega_2) = (\omega_1 + \pi\alpha, \omega_1 + \omega_2)$ and define $V(\omega_1, \omega_2) = \cos(\omega_2)$ then we have $v(n, \omega) = \cos(\frac{\pi}{2}n(n-1)\alpha + n\omega_1 + \omega_2)$. It is expected[18,19] that this potential has a positive Lyapunov exponent for all $\lambda > 0$. It is clearly deterministic in the sense described above. For large λ, Bourgain established localization at all energies and has partial results for small λ. See the book of Bourgain[20] and references therein for results on quasi-periodic and deterministic potentials.

Another class of problems concerns Lyapunov exponents for nonlinear systems such as the Chirikov or Standard map which may be written as a discrete time pendulum:

$$\Delta x_n = x_{n+1} + x_{n-1} - 2x_n = K \sin(x_n). \tag{2.2}$$

It may be thought of as an area preserving map of the two torus or cylinder to itself. The linearized equation for $\psi_n \equiv \partial x_n / \partial x_0$ is given by $H\psi = 0$. Here H is the tight binding Hamiltonian in 1D with $v_n = \cos(x_n)$, with $K = \lambda$, but the dynamics τ on the 2 torus is given by Eq. (2.2) and x_n depends on the initial condition (x_0, x_1). For small K, it is known from KAM theory that most orbits are quasi-periodic corresponding to a 0 Lyapunov exponent. Note that τ defined via Eq. (2.2) is area preserving but not necessarily ergodic.

It is a fundamental open problem in dynamical systems to prove that for a set of initial conditions (x_0, x_1) of *positive* measure, the Lyapunov exponent at $E = 0$ is positive. This is the positive metric entropy conjecture. It would imply that the standard map has robust chaotic behavior. This problem is unsolved even for large values of K for which chaotic behavior certainly appears to be abundant numerically. At a mathematical level it is straightforward to show that if one assumes some Hölder continuity of the density of states then the metric entropy is positive[21] for large K. Similarly, if we introduce an energy E into the linearized equation, then the set of energies at which the Lyapunov exponent (integrated over (x_0, x_1)) is 0 has measure $\approx e^{-K}$.[22] Both of these approaches rely on the Herbert–Jones–Thouless formula[23] but do not resolve the basic problem of chaotic behavior.

In order to establish localization for *quasi* 1D models, such as the tight binding model on a strip of width W, one needs to show that all 2W Lyapunov exponents are nonzero. This was established for products of random matrices of under certain irreducibility conditions see Refs. 24–26. These results imply that for independent random potentials all states are localized. When $W = 1$, there are good asymptotic expansions for the localization length. However, for large W, good lower bounds on the smallest positive Lyapunov exponent (inverse localization length) are lacking. The best rigorous results are for non-resonant energies and very weak disorder $\lambda \ll 1/W$, see Ref. 27. At a more qualitative level, it is known that if the dynamics of a quasi 1D tight binding model is sub-ballistic, i.e., if $\langle X^{2p}(t) \rangle^{1/2p} \leq C_p t^{1-\delta}$ for large p and $\delta > 0$, then it follows that all states are localized and the localization length can be estimated in terms of C_p and $\delta > 0$.[28,29] Similarly if the motion in 2D is strictly subdiffusive, localization follows. This result is closely related to the Thouless scaling theory which implicitly assumes that dynamics are diffusive or sub-diffusive. Although such an assumption is of course physically natural, it has not been established mathematically.

3. Finite Volume Criteria for Localization on $\mathbb{Z}^d$

The first mathematical proof of the absence of diffusion on $\mathbb{Z}^d$ for strong disorder $(\lambda \gg 1)$ appeared in a 1983 paper of Fröhlich and Spencer.[30] The potential v is assumed to consist of independent random variables with a common regular distribution density $g(v_j)$. The absence of diffusion was also established for $\lambda > 0$ at energies where the density of states is small, such as in the band tails. This paper developed a multi-scale technique to establish exponential decay of the Green's function at long distances. It assumes that for some finite box, the Green's function does not "feel" the boundary: Let $B \subset \mathbb{Z}^d$ with side L, the Green's function of H restricted to B is small:

$$|G_B(E; x, y)| = |(E - H_B)^{-1}(x, y)| \ll 1 \qquad \text{for } |x - y| \geq L/2 \qquad (3.1)$$

holds with high probability. Once a precise version of this initial criterion is established, exponential decay of the Green's function can be established with probability rapidly approaching one at larger scales. The initial hypothesis is easily checked for strong disorder or at energies where the density of states is sufficiently small.

Exponential localization of the eigenstates near E was shown to follow from the decay of the Green's function G(E). See Refs. 31–33. Numerous improvements and simplifications of the multi-scale methods have appeared. See Refs. 34–38. Related multi-scale arguments also appear in the analysis of deterministic potentials see Ref. 20. Such multi-scale techniques are more complicated for deterministic potentials because of a lack of independence and control of the density of states.

Mathematical proofs of localization rely on two basic facts: Wegner's estimate of the density of states and the resolvent identity. As above, assume $v(j)$ are independent random variables with a common bounded density $g(v)$. Let $\mathbb{E}$ denote the expectation with respect to this product measure. Wegner's[39] estimate uses the fact that for a fixed vertex x, v_x is a rank one perturbation of H, hence for $E_\epsilon = E - i\epsilon$, $\epsilon > 0$

$$\text{Im} \int G_B(E_\epsilon; x, x)\, g(v_x) dv_x = \text{Im} \int \frac{1}{(\sigma_x - \lambda v_x)}\, g(v_x) dv_x \qquad (3.2)$$

where, $\text{Im}\, \sigma_x \leq -\epsilon$ is independent of v_x. It is easy to see that the right side of Eq. (3.4) is less than $\pi \lambda^{-1} \max g(v_x)$ for all such values of σ_x. Thus the average density of states in a box B is

$$\mathbb{E}\, \rho_B(E) = \frac{1}{\pi |B|} \text{Im}\, tr\, \mathbb{E}\, G_B(E_\epsilon) \leq \frac{1}{\lambda} \max g \equiv C_\lambda. \qquad (3.3)$$

This implies the probability that operator norm of G is large is bounded by

$$\text{Prob} \{||G_B(E)|| \geq r^{-1}\} \leq r|B|C_\lambda. \qquad (3.4)$$

The resolvent identity allows us to take information at scale L and to get information a longer scales. If B denotes a box with $x \in B$ and $y \notin B$

$$G(E_\epsilon, x, y) = \sum_{j,j' \in \partial B} G_B(E_\epsilon; x, j) G(E_\epsilon, j', y). \qquad (3.5)$$

Here, G denotes the Green's function of some much larger domain containing B. The sum ranges over the nearest neighbor boundary pairs j, j' separating B and its complement. If $|x - y| \gg L$ then we can apply the identity again with a new box B' of side L centered at j'. This will give us a block walk from x to y with steps of size L.

To state a finite volume criterion for localization more precisely, let B be a box of side $2\,L$ centered at 0. For a given potential v, we say that $G_B(E)$ is *regular* if

$$\sum_{y,\, |x-y| \geq L/2} |G_B(E; x, y)| \leq (L^2 + 3)^{-1}. \tag{3.6}$$

We assume that Eq. (3.6) holds for x, $y \in B$ and y ranges over the inside boundary of B.

Theorem 1. *Suppose that for some L:*

$$P(E, L) \equiv \mathrm{Prob} \left\{ \sum_{y,\, |x-y| \geq L/2} |G_B(E; x, y)| \geq (L^2 + 3)^{-1} \right\} \leq \delta. \tag{3.7}$$

Then for δ sufficiently small (independent of L), (3.7) implies localization:

$$|G(E + i\epsilon; 0, x)| \leq R\, e^{-m|x|}. \tag{3.8}$$

Here, $m > 0$ and R is a positive random variable with finite moments. All estimates are uniform in $\epsilon > 0$.

We shall give a brief sketch of Eq. (3.8) and refer to Refs. 30, 34–36 for details. Let b be an integer with $b > 10d$. We first show that for $\delta \ll 1$, $P(E, b^n L) \to 0$. This fact follows from the inequality

$$P(E, bL) \leq C P(E, L)^2 b^{2d} + 2\lambda^{-1} (2bL)^d L^{-b/2}. \tag{3.9}$$

For a suitable choice of δ depending on b, P^2 drives the right hand side to zero very quickly. The two terms on the right side of Eq. (3.9) can be understood as follows: Divide the cube bB of side bL into b^d subcubes of side L. If all subcubes are regular then it is easy to see that $G_{bB}(E)$ is regular by iterating Eq. (3.5). If there are two disjoint subcubes which are not regular then this gives rise to the first term on the right side of Eq. (3.9). Finally, if there is just one subcube which is singular (not regular) then it can shown that $G_{bB}(E)$ is regular unless this singular block has a very large norm. In this case, Wegner's estimate (3.4) proves that this probability is small and accounts for the second term on the right of Eq. (3.9). To obtain exponential localization, the idea is similar except that at some stage the length scales must grow more rapidly, and the power law in Eq. (3.7) is replaced by an exponential. See Refs. 35 and 36.

In 2001, Aizenman, Friedrich, Hundertmark and Schenker[40] proved an elegant finite volume criterion for localization using the what is now called the *fractional moment method*. It also relies on Wegner's estimate and the resolvent identity. However, it has the advantage that multi-scale analysis can be replaced by a closure scheme. This idea builds on earlier work of Aizenman and Molchanov[41] and uses the fact that that $\mathbb{E}[|G_\Lambda(E_\epsilon; x, x)|^\alpha]$ is bounded for $0 \leq \alpha < 1$. This is easily seen from Eq. (3.2). The fractional moment criterion may stated as follows:

Theorem 2. *Suppose that for some box B of side $2L$, centered at 0, and $\alpha = 1/2$ we have*

$$\mathbb{E} \sum_{j \in \partial B} |G_B(E_\epsilon; 0, j)|^{1/2} \leq C_2 L^{-2(d-1)}, \tag{3.10}$$

where $C_2 > 0$ is an explicit constant depending on λ. Then

$$\mathbb{E}|G(E_\epsilon; 0, x)|^{1/2} \leq \text{Const.} \, e^{-m|x|} \tag{3.11}$$

for some constants Const. *and $m > 0$. This estimate implies that all eigenstates with energy near E are exponentially localized.*

We conclude this section by discussing finite volume criteria for exponential decay for Ising and rotator models with random ferromagnetic exchange couplings. Although the randomness does not have the dramatic effects it has for the tight binding model, some features are similar to localization. Let

$$H_\Lambda(s) = -\sum_{j,j' \in \Lambda} J_{jj'} s_j \cdot s_{j'} \tag{3.12}$$

denote the Hamiltonian for an Ising or rotator with independent but positive nearest neighbor interaction, $J_{jj'}$. A finite volume criterion for exponential decay is given by

$$\mathbb{E} \sum_{j \in \partial B} \langle s_0 s_j \rangle_B \leq C_1 L^{-(d-1)}. \tag{3.13}$$

Here $\mathbb{E}$ denotes the expectation over the J_{ij}. This criterion is sharp[42–44] and enables one to prove that the correlation length exponent $\nu \geq 2/d$. If one defines the finite size localization length by the smallest box for which say Eq. (3.10) holds, then we also have $\nu \geq 2/d$.[42]

4. The Nonlinear Schrödinger Equation with a Random Potential

Let us now consider the discrete nonlinear Schrödinger equation on $\mathbb{Z}^d$ given by

$$i\,\partial\psi(j,t)/\partial t = -\Delta\,\psi(j,t) + \lambda\,v(j)\,\psi(j,t) + \beta|\psi|^2\,\psi(j,t). \qquad (4.1)$$

Such models arise in the study of nonlinear optics and mean field models of Bose–Einstein condensates in a disordered background. The challenge is to understand whether localization can survive the effects of nonlinearity. The nonlinearity may appear to be irrelevant, but it formally introduces a time dependent effective potential $\lambda v_j + \beta|\psi|^2(j,t)$ which can change the character of the evolution even if $|\psi(j,t)|^2$ becomes small. The existence of nonlinear time periodic localized eigenstates for which $|\psi|^2(j,t)$ is independent of t has been established by Albanese and Fröhlich.[45] For strong disorder (large λ), there are theorems of Bourgain and Wang[46] which state that with high probability there are multi-parameter families of small amplitude solutions which are quasi-periodic in t and localized about some prescribed finite subset of $\mathbb{Z}^d$. These solutions may be thought of as a nonlinear superposition of localized eigenstates. Although there is good control of such solutions, their initial data may be rather special and typical initial conditions may not give rise to quasi-periodic solutions. In general, one expects that an N dimensional Hamiltonian dynamical system will have KAM tori. However, the measure of these tori may go to zero as $N \to \infty$.

Suppose that at time $t = 0$ our initial state $|\psi(j,0)|^2 = \delta(j,0)$ is localized at 0. Will the solution remain exponentially localized? It is known[47] that for large β the max over j of $|\psi(j,t)|^2$ does not go to zero. This fact relies the conservation of the norm $\sum_j |\psi(j,t)|^2$ and the conservation of energy. However, the growth of the mean square displacement

$$\sum_j |\psi(j,t)|^2 |j|^2 \equiv R^2(t) \qquad (4.2)$$

as a function of β is not well understood. If the parameter β in Eq. (4.1) is j dependent and $\beta_j \leq |j|^{-\tau}$ with $\tau > 0$, then Bourgain and Wang[48] proved that for large disorder, $R^2(t)$ grows at most like $|t|^p$ with p small. Recent results of Wang and Zhang[49] show that in one dimension for large λ and small β of the size of the essential support of $\psi(j,t)$, grows very slowly for very long time scales with high probability. This does not rule out the possibility that the wave packet propagates at a faster rate at longer time scales for fixed β. See also earlier related work of Benettin, Fröhlich and Giorgili[50] which obtained Nekhoroshev type estimates for infinite dimensional Hamiltonian systems.

Theoretical and numerical work by Shepelyansky and others[51,52] suggests that there is subdiffusive propagation of Eq. (4.2) for intermediate values of β and λ, whereas for small β, localization was observed numerically for very long time scales. Note that the existence of localized (periodic) states of Ref. 45 established for a wide range of parameters does not contradict Shepelyansky's assertion since the initial condition of Eq. (4.2) not tuned to the random potential. We refer to Refs. 53 and 54 for recent reviews of the perturbative and numerical analysis of this system.

5. A Simple SUSY Model of the Anderson Transition in 3D

In theoretical physics, the Anderson transition is frequently analyzed in terms of statistical mechanics models with an internal hyperbolic super-symmetry, such as $SU(1,1|2)$. Wegner[55] and Efetov[2] showed how to recast expectations of Green's functions in terms of correlations of SUSY matrix field models on the lattice. These models provide valuable insight to the physics because many features of disordered quantum systems can be understood by analyzing fluctuations about a saddle manifold. We refer to the review of Evers and Mirlin[56] for recent developments in the analysis of the Anderson transition.

The Efetov SUSY models are quite difficult to analyze with mathematical rigor. Below we shall study a simpler version of these models due to Zirnbauer.[3] We shall refer to this model as the $\mathbb{H}^{2|2}$ model where $\mathbb{H}$ is a hyperboloid. This model is expected to qualitatively reflect the phenomenology of Anderson's tight binding model. The great advantage of this model is that the fermion or Grassmann degrees of freedom can be explicitly integrated out to produce a real effective action in bosonic variables. Thus probabilistic methods can be applied. In 3D we shall prove that this model has the analog of the Anderson transition.

In order to define the $\mathbb{H}^{2|2}$ sigma model, let u_j be a vector at each lattice point $j \in \Lambda \subset \mathbb{Z}^d$ with three bosonic components and two fermionic components

$$u_j = (z_j, x_j, y_j, \xi_j, \eta_j) \ , \tag{5.1}$$

where ξ, η are odd elements and z, x, y are even elements of a real Grassmann algebra (see Ref. 4 for more details). The scalar product is defined by

$$(u, u') = -zz' + xx' + yy' + \xi\eta' - \eta\xi' \ , \qquad (u, u) = -z^2 + x^2 + y^2 + 2\xi\eta \tag{5.2}$$

and the action is obtained by summing over nearest neighbors j, j'

$$S[u] = \frac{1}{2} \sum_{(j,j')\in\Lambda} \beta(u_j - u_{j'}, u_j - u_{j'}) + \sum_{j\in\Lambda} \varepsilon_j(z_j - 1) . \tag{5.3}$$

The sigma model constraint, $(u_j, u_j) = -1$, is imposed so that the field lies on a SUSY hyperboloid, $\mathbb{H}^{2|2}$.

We choose the branch of the hyperboloid so that $z_j \geq 1$ for each j. It is very useful to parametrize this manifold in horospherical coordinates:

$$x = \sinh t - e^t \left(\tfrac{1}{2}s^2 + \bar\psi\psi\right) , \quad y = se^t, \quad \xi = \bar\psi e^t, \quad \eta = \psi e^t, \tag{5.4}$$

and

$$z = \cosh t + e^t \left(\tfrac{1}{2}s^2 + \bar\psi\psi\right), \tag{5.5}$$

where t and s are even elements and $\bar\psi$, ψ are odd elements of a real Grassmann algebra.

In these coordinates, the sigma model action is given by

$$S[t, s, \psi, \bar\psi] = \sum_{(ij)\in\Lambda} \beta(\cosh(t_i - t_j) - 1) + \tfrac{1}{2}[s; D_{\beta,\varepsilon}s]$$

$$+ [\bar\psi D_{\beta,\varepsilon}\psi] + \sum_{j\in\Lambda} \varepsilon_j(\cosh t_j - 1) . \tag{5.6}$$

Note that the action is quadratic in the Grassmann and s variables. Here $D_{\beta,\varepsilon} = D_{\beta,\varepsilon}(t)$ is the generator of a random walk in random environment, given by the quadratic form

$$[v; D_{\beta,\varepsilon}(t) v]_\Lambda \equiv \beta \sum_{(jj')} e^{t_j + t_{j'}} (v_j - v_{j'})^2 + \sum_{k\in\Lambda} \varepsilon_k e^{t_k} v_k^2 . \tag{5.7}$$

The weights, $e^{t_j + t_{j'}}$, are the local conductances across an nearest neighbor edge j, j'. The $\varepsilon_j e^{t_j}$ term is a killing rate for the walk at j.

After integrating over the Grassmann variables ψ, $\bar\psi$ and the variables $s_j \in \mathbb{R}$, we get the effective bosonic field theory with action $S_{\beta,\varepsilon}(t)$ and partition function

$$Z_\Lambda(\beta, \epsilon) = \int e^{-S_{\beta,\epsilon}(t)} \prod e^{-t_j} dt_j = \int e^{-\beta\mathcal{L}(t)} \cdot [\det D_{\beta,\varepsilon}(t)]^{1/2} \prod_j e^{-t_j} \frac{dt_j}{\sqrt{2\pi}}, \tag{5.8}$$

where

$$\mathcal{L}(t) = \sum_{j\sim j'} [\cosh(t_j - t_{j'}) - 1] + \sum_j \frac{\varepsilon_j}{\beta}[(\cosh(t_j - 1)) - 1].$$

Note that the determinant is a positive but nonlocal function of t_j, hence the effective action S is also nonlocal. The additional factor of e^{-t_j} in Eq. (5.8) arises from a Jacobian. Because of the internal supersymmetry, we know that for all values of β, ε the partition function $Z(\beta, \varepsilon) \equiv 1$. This fact holds even if β is edge dependent.

The analog of the Green's function $\langle |G(E_\epsilon; 0, x)|^2 \rangle$ of the Anderson model is the average of the Green's function of $D_{\beta,\varepsilon}$,

$$\langle s_0 e^{t_0} s_x e^{t_x} \rangle (\beta, \varepsilon) = \langle e^{(t_0 + t_x)} D_{\beta,\varepsilon}(t)^{-1}(0, x) \rangle (\beta, \varepsilon) \equiv \mathcal{G}_{\beta,\varepsilon}(0, x) \qquad (5.9)$$

where the expectation is with respect to the SUSY statistical mechanics weight defined above. The parameter $\beta = \beta(E)$ is roughly the bare conductance across an edge and we shall usually set $\varepsilon = \varepsilon_j$ for all j. In addition to the identity $Z(\beta, \varepsilon) \equiv 1$, there are additional Ward identities

$$\langle e^{t_j} \rangle \equiv 1, \qquad \varepsilon \sum_x \mathcal{G}_{\beta,\varepsilon}(0, x) = 1, \qquad (5.10)$$

which hold for all values of β and ε.

Note that if the $|t_j| \leq \textit{Const}$, then the conductances are uniformly bounded from above and below and

$$D_{\beta,\varepsilon}(t)^{-1}(0, x) \approx (-\beta \Delta + \varepsilon)^{-1}(0, x)$$

is the diffusion propagator. Thus the Anderson transition can only occur due to the large deviations of the t field.

An alternative Schrödinger like representation of Eq. (5.9) is given by

$$\mathcal{G}_{\beta,\varepsilon}(0, x) = \langle \tilde{D}_{\beta,\varepsilon}^{-1}(t)(0, x) \rangle \qquad (5.11)$$

where

$$e^{-t} D_{\beta,\varepsilon}(t) e^{-t} \equiv \tilde{D}_{\beta,\varepsilon}(t) = -\beta \Delta + \beta V(t) + \varepsilon e^{-t}, \qquad (5.12)$$

and $V(t)$ is a diagonal matrix (or 'potential') given by

$$V_{jj}(t) = \sum_{|i-j|=1} (e^{t_i - t_j} - 1).$$

In this representation, the potential is highly correlated and $\tilde{D} \geq 0$ as a quadratic form.

Some insight into the transition for the $\mathbb{H}^{2|2}$ model can be obtained by finding the configuration $t_j = t^*$ which minimizes that action $\mathcal{S}_{\beta,\varepsilon}(t)$ appearing in Eq. (5.8). It is shown in Ref. 4 that this configuration is unique and does not depend on j. For large β

$$\text{1D:} \quad \varepsilon e^{-t^*} \simeq \beta^{-1}, \qquad \text{2D:} \quad \varepsilon e^{-t^*} \simeq e^{-\beta} \qquad \text{3D:} \quad t^* \simeq 0. \qquad (5.13)$$

Note that in one and two dimensions, t^* depends sensitively on ϵ and that negative values of t_j are favored as $\varepsilon \to 0$. This means that at t^*, a mass

εe^{-t^*} in Eq. (5.12) appears even as $\varepsilon \to 0$. Another interpretation is that the classical conductance $e^{t_j+t_{j'}}$ should be small in *some sense*. This is a somewhat subtle point. Due to large deviations of the t field in 1D and 2D, $\langle e^{t_j+t_{j'}} \rangle$ is expected to diverge, whereas $\langle e^{t_j/2} \rangle$ should become small as $\varepsilon \to 0$.

When β is small, $\varepsilon e^{-t^*} \simeq 1$ in any dimension. Thus the saddle point t^* suggests localization occurs in both 1D and 2D for all β and in 3D for small β. In 2D, this agrees with the predictions of localization by Abrahams, Anderson, Licciardello and Ramakrishnan[57] at any nonzero disorder. Although the saddle point analysis has some appeal, it does not account for the large deviations away from t^* and seems incompatible with the sum rule $\langle e^{t_j} \rangle = 1$. In 3D, large deviations away from $t^* = 0$ are controlled for large β. See the discussion below.

For later discussion, it is interesting to consider the case in which $\varepsilon_0 = 1$ but $\varepsilon_j = 0$ otherwise. This corresponds to a random walk starting at O with no killing. In this case the saddle point is not translation invariant. In one and two dimensions, we have $e^{t_j^*}$ goes to 0 exponentially fast for large $|j|$. Thus the conductance becomes small as we move away from 0. We expect that this implies $\langle e^{t_j/2} \rangle \to 0$ exponentially fast in 1D and 2D producing localization.

The main theorem established in Ref. 4 states that in 3D fluctuations around $t^* = 0$ are rare. See Eq. (5.14) below. Let $G_0 = (-\beta\Delta + \epsilon)^{-1}$ be the Green's function for the Laplacian.

Theorem 3. *If $d \geq 3$, and the volume $\Lambda \to \mathbb{Z}^d$, there is a $\bar{\beta} \geq 0$ such that for $\beta \geq \bar{\beta}$ then for all j*

$$\langle \cosh^8(t_j) \rangle \leq \text{Const}, \tag{5.14}$$

where the constant is uniform in ϵ. This implies quasi-diffusion: Let $\mathcal{G}$ be given by Eq. (5.9). There is a constant K so that we have the quadratic form bound

$$\frac{1}{K}[\tilde{f}; G_0\tilde{f}] \leq \sum_{x,y} \mathcal{G}_{\beta,\varepsilon}(x,y)\, f(x)f(y) \leq K[f; G_0 f], \tag{5.15}$$

where $f(x)$ is nonnegative function and $\tilde{f}(x) = (1+|x|)^{-\alpha} f(x)$. The constant $\alpha > 0$ is small for large β.

Remarks. The power 8 can be increased by making β larger. The lower bound is not sharp (α should be 0) and one expects point wise diffusive bounds on $\mathcal{G}_{\beta,\varepsilon}(x,y)$ to hold. However, in order to prove this one needs to show that the set $(j : |t_j| \geq M \gg 0)$ does not percolate. This is expected to be true but has not yet been mathematically established partly because of the high degree of correlation in the t field.

The next theorem establishes localization for small β in any dimension. See Ref. 5.

Theorem 4. *Let $\varepsilon_x > 0$, $\varepsilon_y > 0$ and $\sum_{j \in \Lambda} \varepsilon_j \leq 1$. Then for all $0 < \beta < \beta_c$ (β_c defined below) the correlation function $\mathcal{G}_{\beta,\varepsilon}(x,y)$, (5.11), decays exponentially with the distance $|x - y|$. More precisely:*

$$\mathcal{G}_{\beta,\varepsilon}(x,y) = \langle \tilde{D}_{\beta,\varepsilon}^{-1}(t)(x,y) \rangle \leq C_0 \left(\varepsilon_x^{-1} + \varepsilon_y^{-1} \right) \left[I_\beta\, e^{\beta(c_d - 1)}\, c_d \right]^{|x-y|} , \quad (5.16)$$

where $c_d = 2d - 1$, C_0 is a constant and

$$I_\beta = \sqrt{\beta} \int_{-\infty}^{\infty} \frac{dt}{\sqrt{2\pi}} e^{-\beta(\cosh t - 1)} . \quad (5.17)$$

Finally β_c is defined so that:

$$\left[I_\beta\, e^{\beta(c_d - 1)}\, c_d \right] < \left[I_{\beta_c}\, e^{\beta_c(c_d - 1)}\, c_d \right] = 1 \quad \forall \beta < \beta_c. \quad (5.18)$$

These estimates hold uniformly in the volume.

Remarks. The first proof of localization for the $\mathbb{H}^{2|2}$ model in 1D was given by Zirnbauer in Ref. 3. Note that in 1D, $c_d - 1 = 0$ and inequality holds for any $\beta_c \geq 0$. The above estimate is sharp in 1D. Thus the decay for small β is proportional to $|\sqrt{\beta}\, ln\beta|^{|x-y|}$ rather than $\beta^{|x-y|}$ which is typical for lattice sigma models with compact targets. The divergence of ε^{-1} is compatible with the sum rule in Eq. (5.10) and is a signal of localization.

The proof of the above theorem relies heavily on the supersymmetric nature of the action. It is known that a purely hyperbolic sigma model of the kind studied in Ref. 58 cannot have a phase transition. The action for the purely hyperbolic case looks like that of the $\mathbb{H}^{2|2}$ model except that $[DetD_{\beta,\varepsilon}(t)]^{1/2}$ is replaced by $[DetD_{\beta,\varepsilon}(t)]^{-1/2}$. D. Brydges has pointed out that since the logarithm of $DetD_{\beta,\varepsilon}(t)$ is convex as a functional of t, the action for the hyperbolic sigma model is always convex and therefore no transition can occur. See Ref. 4 for details. In Wegner's hyperbolic model, the replica number must be 0 in order to see localization.

6. Role of Ward Identities in the Proof

The proof of Theorems 3 and 4 above rely heavily on Ward identities. For Theorem 3 we use Ward identities to bound fluctuations of the t field by getting bounds in 3D on $\langle \cosh^m(t_i - t_j) \rangle$. This is done by induction on the distance $|i - j|$. For Theorem 4, we use the fact that for any region Λ, the partition function $Z_\Lambda = 1$.

If a function S of the variables x, y, z, ψ, $\bar{\psi}$ is supersymmetric, i.e., it is invariant under transformations preserving

$$x_i x_j + y_i y_j + \bar{\psi}_i \psi_j - \psi_i \bar{\psi}_j ,$$

then $\int S = S(0)$. In horospherical coordinates the function S_{ij} given by

$$S_{ij} = B_{ij} + e^{t_i + t_j}(\bar{\psi}_i - \bar{\psi}_j)(\psi_i - \psi_j), \qquad B_{ij} = \cosh(t_i - t_j) + \frac{1}{2}e^{t_i + t_j}(s_i - s_j)^2$$
(6.1)

is supersymmetric. If i and j are nearest neighbors, $S_{ij} - 1$ is a term in the action and it follows that the partition function $Z_\Lambda(\beta, \epsilon) \equiv 1$. More generally for each m we have

$$\langle S_{ij}^m \rangle = \langle B_{ij}^m [1 - m B_{ij}^{-1} e^{t_i + t_j}(\bar{\psi}_i - \bar{\psi}_j)(\psi_i - \psi_j)] \rangle \equiv 1. \qquad (6.2)$$

The integration over the Grassmann variables above is explicitly given by

$$G_{ij} = \frac{e^{t_i + t_j}}{B_{xy}}\left[(\delta_i - \delta_j); D_{\beta,\varepsilon}(t)^{-1}(\delta_i - \delta_j)\right]_\Lambda \qquad (6.3)$$

since the action is quadratic in $\bar{\psi}$, ψ. Thus we have the identity

$$\langle B_{ij}^m(1 - m G_{ij}) \rangle \equiv 1. \qquad (6.4)$$

Note that $0 \leq \cosh^m(t_i - t_j) \leq B_{ij}^m$. From the definition of $D_{\beta,\varepsilon}$ given in Eq. (5.7), we see that for large β, G is typically proportional to $1/\beta$ in 3D. However, there are rare configurations where $t_k \approx -\infty$ for k on a closed surface $\subset \mathbb{Z}^3$ separating i and j for which G_{ij} can diverge as $\epsilon \to 0$. If this surface is of finite volume enclosing i, then there is a finite volume 0 mode producing a divergence in $D_{\beta,\varepsilon}(t)^{-1}(i, i)$. If i, j are nearest neighbors then it is easy to show that G_{ij} is less than β^{-1} for all t configurations. Thus if $m/\beta \leq 1/2$ then Eq. (6.4) implies that $0 \leq \cosh^m(t_i - t_j) \leq 2$. In general, there is no uniform bound on G_{ij} and we must use induction on $|i - j|$ to prove that configurations for which $1/2 \leq m G_{ij}$ are rare for large β in 3D. In this way fluctuations of the t field can be controlled and quasi-diffusion is established, see Ref. 4 for details.

The proof of the localized phase is technically simpler than the proof of Theorem 3. Nevertheless, it is of some interest because it shows that $\mathbb{H}^{2|2}$ sigma model reflects the localized as well as the extended states phase in 3D. The main idea relies on the following lemma. Let M be an invertible matrix indexed by sites of Λ and let γ denote a self avoiding path starting at i and ending at j. Let M_{ij}^{-1} be matrix elements of the inverse and let M_{γ^c} be the matrix obtained from M by striking out all rows and columns indexed by the vertices covered by γ.

Lemma. *Let M and M_{γ^c} be as above, then*

$$\frac{\partial}{\partial M_{ji}} \det M = [M_{ij}^{-1} \det M] = \sum_{\gamma_{ij}} [(-M_{ij_1})(-M_{j_1j_2}) \cdots (-M_{j_mj})] \det M_{\gamma^c},$$

(6.5)

where the sum ranges over all self-avoiding paths γ connecting i and j, $\gamma_{ij} = (i, j_1, \ldots j_m, j)$, with $m \geq 0$.

Apply this lemma to

$$M = e^{-t} D_{\beta,\varepsilon}(t) e^{-t} \equiv \tilde{D}_{\beta,\varepsilon}(t) = -\beta\Delta + \beta V(t) + \varepsilon \, e^{-t} \tag{6.6}$$

and notice that with this choice of M, for all nonzero contributions, γ are nearest neighbor self-avoiding paths and that each step contributes a factor of β. The proof of Eq. (6.5) comes from the fact the determinant of M can be expressed as a gas of non overlapping cycles covering Λ. The derivative with respect to M_{ji} selects the cycle containing j and i and produces the path γ_{ij}. The other loops contribute to $\det M_{\gamma^c}$. By Eqs. (5.11) and (6.6) we have

$$\mathcal{G}_{\beta,\varepsilon}(x,y) = < M_{xy}^{-1} > = \int e^{-\beta\mathcal{L}(t)} M_{xy}^{-1} [\det M]^{1/2} \prod_j \frac{dt_j}{\sqrt{2\pi}}. \tag{6.7}$$

Note the factors of e^{-t_j} appearing in Eq. (5.8) have been absorbed into the determinant. Now write

$$M_{xy}^{-1}[\det M]^{1/2} = \sqrt{M_{xy}^{-1}} \sqrt{M_{xy}^{-1} \det M}.$$

The first factor on the right hand side is bounded by $\epsilon_x^{-1/2} e^{t_x/2} + \epsilon_y^{-1/2} e^{t_y/2}$. For the second factor, we use the lemma. Let $\mathcal{L} = \mathcal{L}_\gamma + \mathcal{L}_{\gamma^c} + \mathcal{L}_{\gamma,\gamma^c}$ where $\mathcal{L}_\gamma$ denotes the restriction of $\mathcal{L}$ to γ. Then using the fact that

$$\int e^{-\beta\mathcal{L}_{\gamma^c}} [\det M_{\gamma^c}]^{1/2} \prod_j \frac{dt_j}{\sqrt{2\pi}} \equiv 1,$$

we can bound

$$0 \leq \mathcal{G}_{\beta,\varepsilon}(x,y)$$

$$\leq \sum_{\gamma_{xy}} \sqrt{\beta}^{|\gamma_{xy}|} \int e^{-\beta\mathcal{L}_\gamma + \beta\mathcal{L}_{\gamma,\gamma^c}} [\epsilon_x^{-1/2} e^{t_x/2} + \epsilon_y^{-1/2} e^{t_y/2}] \prod_j \frac{dt_j}{\sqrt{2\pi}},$$

where $|\gamma_{xy}|$ is the length of the self-avoiding path from x to y. The proof of Theorem 4 follows from the fact that the integral along γ is one dimensional and can be estimated as a product. See Ref. 5 for further details.

7. Edge Reinforced Random Walk and Localization

Linearly edge reinforced random walk (ERRW) is a history-dependent walk which prefers to visit edges it has visited more frequently in the past. Consider a discrete time walk on $\mathbb{Z}^d$ starting at the origin and let $n(e,t)$ denote the number of times the walk has visited the edge e up to time t. Then the probability $P(v,v',t+1)$ that the walk at vertex v will visit a neighboring edge $e = (v,v')$ at time $t+1$ is given by

$$P(v,v',t+1) = (\beta + n(e,t))/S_\beta(v,t),$$

where S is the sum of $\beta + n(e',t)$ over all the edges e' touching v. The parameter β is analogous to β in the $\mathbb{H}^{2|2}$ model. Note that if β is large, the reinforcement is weak. This process was defined by Diaconis and is partially exchangeable which means that any two paths with the same stating point and same values of $n(e,t)$ have the same probability. Thus the order in which the edges were visited is irrelevant. Such processes can be expressed as a superposition of Markov processes.[59] In fact Coppersmith and Diaconis proved that this ERRW can be expressed as a random walk in a random environment. There is an explicit formula for the Gibbs weight of the local conductances across each edge, see Refs. 6, 7 and 60 which is quite close to that for $\mathbb{H}^{2|2}$ model with $\varepsilon_j = 0$ except at 0 where $\varepsilon_0 = 1$. It is nonlocal and also expressed in terms of a square root of a determinant. Moreover, the partition function can be explicitly computed and there are identities similar to Ward identities (5.10). These presumably reflect conservation of probability.

In 1D and 1D strips, ERRW is *localized* for any value of $\beta > 0$. This means that the probability of finding an ERRW, $W(t)$, at a distance r from the origin at fixed time t is exponentially small in r, thus

$$\text{Prob}\left[|W(t)| \geq r\right] \leq Ce^{-mr}.$$

Merkl and Rolles[8] established this result by proving that the conductance across an edge goes to zero exponentially fast with the distance of the edge to the origin. More precisely they show that the conductance c satisfies

$$\langle c_{jj'}^{1/4}\rangle \leq Ce^{-m|j|}.$$

The local conductance $c_{jj'}$ roughly corresponds to $e^{t_j + t_{j'}}$ hence the decay of $\langle c_{jj'}^{1/4} \rangle$ should be closely related to that of $\langle e^{t_j/2} \rangle$. See the discussion just before Theorem 3. Note that the factor $1/2$ is important, otherwise we have $\langle e^{t_j} \rangle \equiv 1$. Their argument is based on a Mermin–Wagner like deformation of the Gibbs measure. It also shows that in 2D, $\langle c_{jj'}^{1/4} \rangle \to 0$. In 3D, there are no rigorous theorems for ERRW. However, by analogy with Theorem 2, localization is expected to occur for strong reinforcement, i.e., for β small. It is natural to conjecture that in 2D ERRW is always exponentially localized for all values of reinforcement. On the Bethe lattice, Pemantle[9] proved that ERRW has a phase transition. For $\beta \gg 1$ the walk is weakly reinforced and transient whereas for $0 < \beta \ll 1$ the walk is recurrent. It is an open question whether ERRW has the analog of the Anderson transition in 3D. See Refs. 61 and 60 for reviews of this subject.

To conclude, we mention another interesting classical walk defined on the oriented Manhattan lattice. In this model disorder occurs by placing obstructions at each vertex, independently with probability $0 < p < 1$.[62] This model is closely related to a disordered quantum network model (class C). The renormalization group analysis of Beamond, Owczarek and Cardy,[62] indicates that for all $p > 0$, every path of this walk is closed with probability one and has a finite expected diameter.

Acknowledgments

I would like to thank my colleagues, Jean Bourgain, Margherita Disertori, Jürg Fröhlich, and Martin Zirnbauer for sharing their insights into the many facets of Anderson localization. I would also like to dedicate this article to Philip Anderson.

References

1. P. W. Anderson, *Phys. Rev.* **109**, 1492 (1958).
2. K. B. Efetov, *Adv. Phys.* **32**, 53 (1983).
3. M. R. Zirnbauer, *Commun. Math. Phys.* **141**, 503 (1991).
4. M. Disertori, T. Spencer and M. R. Zirnbauer, arXiv:0901.1652 (2009).
5. M. Disertori and T. Spencer, arXiv:0910.3325 (2009).
6. P. Diaconis, in *Bayesian Statistics* (Oxford University Press, New York, 1988), p. 111.
7. M. S. Keane and S. W. W. Rolles, in *Infinite Dimensional Stochastic Analysis*, eds. P. Clement, F. den Hollander, J. van Neerven and B. de Pagter (Koninklijke Nederlandse Akademie van Wetenschappen, 2000), p. 217.
8. F. Merkl and S. W. W. Rolles, *Probab. Theory Relat. Fields* **145**, 323 (2009).

9. R. Pemantle, *Ann. Probab.* **16**, 1229 (1988).

10. N. F. Mott and W. D. Twose, *Adv. Phys.* **10**, 107 (1961)

11. H. Furstenburg, *Trans. Amer. Math. Soc.* **108**, 377 (1963).

12. M. Herman, *Comment. Math. Helv.* **58**, 453 (1983).

13. I. Goldsheid, S. Molchanov and L. Pastur, *Funct. Anal. Appl.* **11**, 10 (1977).

14. H. Kunz and B. Souillard, *Commun. Math. Phys.* **78**, 201 (1980).

15. R. E. Borland, *Proc. R. Soc. London, Ser. A* **274**, 529 (1963).

16. S. Kotani, in *Stochastic Analysis*, ed. K. Ito (North Holland, Amsterdam, 1984), p. 225.

17. B. Simon, *Commun. Math. Phys.* **89**, 277 (1983).

18. M. Grinasty and S. Fishman, *Phys. Rev. Lett.* **60**, 1334 (1988)

19. D. J. Thouless, *Phys. Rev. Lett.* **61**, 2141 (1988).

20. J. Bourgain, *Annals Mathematics Studies*, Vol. 158 (Princeton University Press, 2004).

21. J. Avron, W. Craig and B. Simon, *J. Phys. A: Math. Gen.* **16**, L209 (1983).

22. T. Spencer, in *Analysis et Cetera*, Vol. 623, eds. E. Zehnder and P. Rabinowitz (Academic Press, 1990).

23. D. J. Thouless, *J. Phys. C: Solid State Phys.* **5**, 77 (1972).

24. P. Bougerol and J. Lacroix, *Progress in Probability and Statistics*, Vol. 8 (Birkhaüser, Boston-Basel-Stuttgart, 1985).

25. Y. Guivarch, *Lect. Notes Math.* **1064**, 161 (1984).

26. I. Goldsheid and G. Margulis, *Russ. Math. Surv.* **44**, 11 (1989).

27. H. Schulz-Baldes, *GAFA* **14**, 1089 (2004).

28. W.-M. Wang, Ph.D. thesis, Princeton University (1992).

29. F. Germinet and A. Klein, *Duke Math. J.* **124**, 309 (2004).

30. J. Fröhlich and T. Spencer, *Commun. Math. Phys.* **88**, 151 (1983).

31. J. Fröhlich, F. Martinelli, E. Scoppola and T. Spencer, *Commun. Math. Phys.* **101**, 21 (1985).

32. B. Simon and T. Wolff, *Commun. Pure Appl. Math.* **39**, 75 (1986).

33. F. Delyon, Y. Levy and B. Souillard, *Commun. Math. Phys.* **100**, 463 (1985).

34. H. von Dreifus, Ph.D. thesis, New York University (1987).

35. T. Spencer, *J. Stat. Phys.* **51**, 1009 (1988).

36. H. von Dreifus and A. Klein, *Commun. Math. Phys.* **124**, 285 (1989).

37. F. Germinet and S. de Bievre, *Commun. Math. Phys.* **194**, 323 (1998)

38. D. Damanik and P. Stollmann, *GAFA* **11**, 11 (2001).

39. F. Wegner, *Z. Phys. B* **44**, 9 (1981).

40. M. Aizenman, J. H. Schenker, R. H. Friedrich and D. Hundertmark, *Commun. Math. Phys.* **224**, 219 (2001).

41. M. Aizenman and S. Molchanov, *Commun. Math. Phys.* **157**, 245 (1993).

42. J. Chayes, L. Chayes, D. Fisher and T. Spencer, *Phys. Rev. Lett.* **57**, 2999 (1986).

43. J. Chayes, L. Chayes, D. Fisher and T. Spencer, *Commun. Math. Phys.* **120**, 501 (1989)

44. H. von Dreifus, *Ann. Inst. Henri Poincare* **55**, 657 (1991).

45. C. Albanese and J. Fröhlich, *Commun. Math. Phys.* **138**, 193 (1991).

46. J. Bourgain and W.-M. Wang, *J. Eur. Math. Soc.* **10**, 1 (2008).

47. G. Kopidakis, S. Komineas, S. Flach and S. Aubry, *Phys. Rev. Lett.* **100**, 4103 (2008).
48. J. Bourgain and W.-M. Wang, *Annals Mathematics Studies*, Vol. 163 (Princeton University Press, 2007), p. 21.
49. W.-M. Wang and Z. Zhang, *J. Stat. Phys.* **134**, 953 (2009).
50. G. Benettin, J. Fröhlich and A. Giorgili, *Commun. Math. Phys.* **119**, 95 (1988).
51. D. L. Shepelyansky, *Phys. Rev. Lett.* **70**, 1787 (1993).
52. I. Garcia-Mata and D. L. Shepelyansky, *Phys. Rev. E* **79**, 6205 (2009).
53. S. Fishman, Y. Krivolapov and A. Soffer, arXiv:0901.4951v2 (2009).
54. C. Skokos, D. O. Krimer, S. Komineas and S. Flach, arXiv:0901.4418 (2009).
55. F. Wegner, *Z. Phys. B* **35**, 207 (1979).
56. F. Evers and A. Mirlin, *Rev. Mod. Phys.* **80**, 1355 (2008), arXiv:0707.4378.
57. E. Abrahams, P. W. Anderson, D. C. Licciardello and T. V. Ramakrishnan, *Phys. Rev. Lett.* **42**, 673 (1979).
58. T. Spencer and M. R. Zirnbauer, *Comm. Math. Phys.* **252**, 167 (2004).
59. P. Diaconis and D. Freedman, *IMS* **8**, 115 (1980).
60. F. Merkl and S. W. W. Rolles, arXiv:0608220 (2006).
61. R. Pemantle, *Probab. Surv.* **4**, 1 (2007).
62. E. J. Beamond, A. L. Owczarek and J. Cardy, *J. Phys. A: Math. Gen.* **36**, 10251 (2003).

Chapter 16

FINITE SIZE SCALING ANALYSIS OF THE ANDERSON TRANSITION

B. Kramer[*], A. MacKinnon[†], T. Ohtsuki[‡] and K. Slevin[§]

[*]*School of Engineering and Sciences, Jacobs University Bremen,*
Campus Ring 1, 28759 Bremen, Germany

[†]*Blackett Laboratory, Imperial College London,*
South Kensington Campus, London SW7 2AZ, UK

[‡]*Physics Department, Sophia University,*
Kioi-cho 7-1, Chiyoda-ku, Tokyo, Japan

[§]*Department of Physics, Graduate School of Science, Osaka University,*
1-1 Machikaneyama, Toyonaka, Osaka 560-0043, Japan
[*]*b.kramer@jacobs-university.de*
[§]*slevin@phys.sci.osaka-u.ac.jp*

This chapter describes the progress made during the past three decades in the finite size scaling analysis of the critical phenomena of the Anderson transition. The scaling theory of localization and the Anderson model of localization are briefly sketched. The finite size scaling method is described. Recent results for the critical exponents of the different symmetry classes are summarised. The importance of corrections to scaling are emphasised. A comparison with experiment is made, and a direction for future work is suggested.

1. Introduction

Originally, the phenomenon of localization is a property of quantum mechanical wave functions bound in potential wells of finite range. At infinity, where the potential vanishes, the wave functions decay exponentially for negative energies indicating that the probability of finding the particle far from the potential well vanishes. This is called "potential localization". It had already been suggested in the 1950s that potentials with infinite range can also support the existence of localized wave functions at positive energies provided that the spatial variation of the potential is random. This localization

---

[*]Corresponding author. Permanent affiliation: Institut für Theoretische Physik, Universität Hamburg, Jungiusstraße 9, 20355 Hamburg, Germany.

phenomenon is due to destructive interference of randomly scattered partial waves and is now referred to as "Anderson localization". The most important physical consequence of Anderson localization is the suppression of diffusion at zero temperature, which was conjectured by P.W. Anderson in his seminal paper.[1] Perhaps, the most striking example of Anderson localization is in one dimensional random potentials where all the states are localized, irrespective of their energy. The study of one dimensional localization was pioneered by Mott and Twose[2] and by Gertsenshtein and Vasilev.[3] It can be treated exactly and has been the subject of several reviews.[4-6]

In higher dimensions the problem is more subtle, with the possibility of energy regions corresponding to localized states only, and to extended states only, separated by critical energies, called "mobility edges". The zero temperature and zero frequency electrical conductivity σ_0 of the system vanishes if the Fermi energy is located in a region of localized states. In the region of extended states, $\sigma_0 \neq 0$. In the absence of interactions, the system is an electrical insulator in the former case while in the latter case metallic conductivity is expected. It was conjectured in a seminal work[7] that this metal–insulator transition exists only in three dimensions, while in dimensions $d \leq 2$ systems are always insulating. This conjecture was based on the hypothesis of one parameter scaling of the conductance $g(L)$ of a system of size L, i.e. that the dependence of the conductance on system size can be described by a beta-function,

$$\beta(g) = \frac{\mathrm{d}\ln g(L)}{\mathrm{d}\ln L}, \tag{1.1}$$

that depends only on the conductance. The behaviour of $\beta(g)$ with g was conjectured based on perturbation theory in the limits of weak and strong disorder (large and small conductance), and assuming continuity and monotonicity in between. Moreover, according to the scaling theory, at the mobility edge a continuous quantum phase transition between an insulator and a metal occurs accompanied by the power law behaviour of physical quantities, described by critical exponents, that is typical of critical phenomena at continuous phase transitions. The critical exponents of the conductivity

$$\sigma_0 \sim (E - E_0)^s, \tag{1.2}$$

and the localization length

$$\xi \sim (E_0 - E)^{-\nu}, \tag{1.3}$$

were predicted to obey Wegner's previously conjectured scaling law[8]

$$s = (d - 2)\nu. \tag{1.4}$$

While this work was a great leap forward in our understanding of Anderson localization, it remained to establish the validity of the central assumption of the theory, namely the one parameter scaling hypothesis.

This question was addressed numerically by simulating the Anderson model[1] of disordered quantum systems which consists of a delocalizing kinetic energy modelled by a hopping term V and a localizing random potential energy ϵ_j, commonly assuming a white noise distribution of width W, on a discrete square lattice $\{j\}$,

$$H = V \sum_{j,\delta} | j \rangle\langle j + \delta | + \sum_{j} \epsilon_j | j \rangle\langle j |, \tag{1.5}$$

where δ denotes the nearest neighbours of the lattice site j. Such simulations allowed the one parameter scaling hypothesis to be verified with a reasonable numerical precision,[9–12] in the center of the band, at energy $E = 0$, and also to confirm the prediction $s = \nu$.[9] The critical disorder in three dimensions was initially found to be $W_c(E = 0) = 16 \pm 0.5$ while $s = \nu = 1.2 \pm 0.3$. Although this latter value seemed to be consistent with $\nu = 1$, there were subsequently substantial doubts about whether or not this was indeed the case. It was found necessary to improve the precision of the estimate of the critical exponent and to study in detail and with high precision the conditions for the validity of the one parameter scaling hypothesis. Later, it was found that the exponent was in fact *not* unity and this intriguing discrepancy was the reason for numerous further numerical as well as analytical efforts, especially since the experimental situation was also far from clear.[13]

In the following sections, we briefly review the development of the finite size scaling analysis of the Anderson transition, paying particular attention to the role of symmetry and the estimation of the critical exponents. We stress the importance of the taking proper account of corrections to scaling, which has been found to be essential in order to estimate the critical exponents precisely. Finally, we tabulate the "state of the art" estimates for the critical exponents of the different universality classes. Some of the early results have been described in previous review articles.[14–17]

2. The Anderson Model of Disordered Systems

In this section we briefly explain the Anderson model of localization. We generalize Eq. (1.5) in order to describe more general systems with different symmetries. The most general form of Eq. (1.5) is

$$H = \sum_{j\mu,j'\mu'} V_{j\mu,j'\mu'} | j\mu \rangle\langle j'\mu' | + \sum_{j\mu} \epsilon_{j\mu} | j\mu \rangle\langle j\mu |. \tag{2.1}$$

The states $\mid j\mu\rangle$ that are associated with the sites of a regular lattice j — usually for simplicity a square or a cubic is assumed — are assumed to form a complete set such that $\langle j\mu \mid j'\mu'\rangle = \delta_{j,j'}\delta_{\mu,\mu'}$. Indices μ denote additional degrees of freedom associated with the lattice sites which lead to several states per site. If there are n states the above Hamiltonian describes Wegner's n-orbital model.[8,18] In general, the potential energies $\epsilon_{j\mu}$ and the hopping integrals $V_{j\mu,j'\mu'}$ are random variables described by some statistical distributions.

If the energy bands emerging due to the broadening by the kinetic terms are not strongly overlapping, we may use the single band approximation Eq. (1.5). In addition, if we assume that sufficiently close to the Anderson critical point the critical phenomena are *universal*, i.e. independent of the microscopic details of the system, Eq. (1.5) is the simplest model that can describe the critical behaviour at the Anderson transition. In principle, these assumptions have to be verified *a posteriori*, and to some extent this has indeed been done during the past decades.

If the Anderson transition is a genuine phase transition, the critical behaviour can be expected to depend only on symmetry and dimensionality. For a disordered system, spatial symmetry is absent and only two important symmetries remain: invariance with respect to time reversal, and invariance with respect to spin rotations. Three symmetry classes are distinguished[a]: the orthogonal class which is invariant with respect to both time reversal and spin rotations, the symplectic class which is invariant with respect to time reversal but where spin rotation symmetry is broken, and the unitary class where time reversal symmetry is broken. Note that, if time reversal symmetry is broken, the system is classified as unitary irrespective of its invariance, or otherwise, under spin-rotations.

When the kinetic energy parameter V is a real number, the Anderson model Eq. (1.5) is time reversal invariant and belongs to the *orthogonal class*. In this case, universality has been verified by showing that a Gaussian, Cauchy and a box distribution of the disorder potential give the same critical exponents.[20]

When the kinetic terms $V_{j,j'}$ become complex, the system is no longer time reversal invariant and thus belongs to the *unitary class*. This can be physically realised by applying a magnetic field. Then, the hopping term has to be replaced by the Peierls substitution

---

[a]In fact, the classification is more complicated.[16,17,19] However, for the present purposes, the following classification is sufficient.

$$V_{jj'} = V \exp \left[\mathrm{i} \frac{e}{\hbar} \int_j^{j'} \mathbf{A} \cdot \mathrm{d}\mathbf{x} \right], \tag{2.2}$$

where the vector potential $\mathbf{A}$ describes the magnetic field, $\mathbf{B} = \nabla \times \mathbf{A}$. Two different unitary models can be constructed using the Peierls Hamiltonian, namely a *random phase model* which is characterised by

$$V_{jj'} = V \exp \left(\mathrm{i} \varphi_{jj'} \right), \tag{2.3}$$

with the uncorrelated phases $\varphi_{jj'}$ as random variables, and a model of a uniform magnetic field that leads to a similar expression for the kinetic term but with *correlated phases*. Whether or not these two unitary models have the same critical behaviour has been the subject of numerous studies.

In the presence of spin-orbit interaction, spin rotation symmetry is broken. The simplest Hamiltionian for such a *symplectic* case is[21-23]

$$H = \sum_{j,\sigma} \epsilon_j \mid j, \sigma \rangle \langle j, \sigma \mid + V \sum_{jj',\sigma\sigma'} U_{j,j'} \mid j, \sigma \rangle \langle j', \sigma' \mid, \tag{2.4}$$

where $U_{j,j'}$ is an SU(2) matrix and σ and σ' are the spin indices. This model describes a two dimensional electron system in the presence of Rashba[24] and Dresselhaus[25] spin-orbit couplings.

If the Anderson transition is a genuine quantum phase transition, we expect that the critical behaviour is universal and that the critical exponents depend only on the symmetry class and the dimensionality.

3. Finite Size Scaling Analysis of the Anderson Transition

In principle, phase transitions occur only in the thermodynamic limit, i.e. in an infinite system. In practice, computer simulations are limited to systems of small size. This necessitates an extrapolation to the thermodynamic limit. This extrapolation is far from trivial. It requires a numerically stable procedure which, at least in principle, allows control of the errors involved. This is especially the case when the goal is precise estimates of the critical exponents. Finite size scaling is such a procedure.

3.1. *Finite size scaling*

The raw data for the finite size scaling procedure is some appropriate physical quantity in a system of finite size. For some physical quantities it may be necessary to take a statistical average. An example is the two terminal conductance where an average over a large number of realisations of the random potential is required. For self-averaging quantities an average may not be required. An example is the quasi-one dimensional localization length of the electrons on a very long bar where simulation of a single realisation is sufficient.

This physical quantity Γ to be analysed depends on the system size L and a set of parameters $\{w_i\}$

$$\Gamma = \Gamma(\{w_i\}), L).\tag{3.1}$$

These latter parameters characterise the distribution function of the potential energies and also other system parameters such as the energy E, applied magnetic field $\mathbf{B}$, spin-orbit couplings, etc. The extrapolation to the thermodynamic limit is performed by assuming that that Γ obeys a scaling law

$$\Gamma = F(\chi L^{1/\nu}, \phi_1 L^{y_1}, \phi_2 L^{y_2}, \dots).\tag{3.2}$$

Here, for the sake of simplicity, we assume that Γ is dimensionless. The hope is that, in the thermodynamic limit, only one of the many scaling variables $(\chi, \phi_1, \phi_2, \dots)$ turns out to be relevant, say χ, and the others $\{\phi_i\}$ irrelevant. Here, the words relevant and irrelevant are used in the technical sense that the exponent of the relevant scaling variable is positive $\nu > 0$ and the exponents of the irrelevant scaling variables are negative $y_i < 0$. This *ad hoc* assumption has, of course, to be verified during the numerical analysis.

For very large systems the contribution of the irrelevant scaling variables can be neglected and we obtain a one parameter scaling law

$$\Gamma = f(L/\xi),\tag{3.3}$$

with a correlation length,

$$\xi \sim |\chi|^{-\nu},\tag{3.4}$$

that depends on the parameters $\{w_i\}$. This limit is rarely reached in numerical simulations and we are forced to deal with the corrections to this one parameter scaling behaviour due to the irrelevant scaling variables. (Below we shall refer rather loosely to "corrections to scaling"; strictly speaking we mean corrections to one parameter scaling.)

In practice, we need to simulate not too small systems such that consideration of at most one irrelevant scaling variable is sufficient. In this case, the scaling form Eq. (3.2) reduces to

$$\Gamma = F(\chi L^{1/\nu}, \phi L^y).\tag{3.5}$$

We then fit numerical data for the region close to the phase transition by Taylor expanding the scaling function and the scaling variables, and performing a non-linear least squares fit. It is important to control the errors in this fitting procedure carefully and to specify the precision of all numerical estimates, if the results are to be scientifically meaningful. For details we refer the reader to the article by Slevin and Ohtsuki.[20]

Such finite size scaling analyses have been used successfully to analyse the Anderson transition in three dimensional systems in various symmetry classes,[20,26–28] the Anderson transition in two dimensional systems with spin-orbit coupling[21,22] and the plateau transition in the integer quantum Hall effect.[29]

3.2. *Quasi-one dimensional localization length*

The next question is which physical quantity to use in the finite size scaling analysis. It must be sensitive to the nature, localized or extended, of the eigenstates. (This rules out the average of the density of states, for example.) It should also be easily determined numerically with a high precision. There are several possibilities. One is the localization length of electrons on a very long bar. Another possibility is the level spacing distribution.[30,31] Yet another possibility is the Landauer conductance of a hypercube.[26,32] In this section we discuss the first of these possibilities in detail.

Consider a very long d-dimensional bar with linear cross-section L. This is a quasi-one dimensional system in which all states, irrespective of the values of the parameters w_i are known to be exponentially localized with a quasi-one dimensional localization length $\lambda(L; w_1, w_2, \ldots)$. Using this quasi-one dimensional localization length we define a dimensionless quantity, sometimes called the MacKinnon–Kramer parameter,

$$\Lambda(L; w_1, w_2, \ldots) = \frac{\lambda(L; w_1, w_2, \ldots)}{L}. \tag{3.6}$$

In practice, the error analysis of the simulation is simplified by working directly with the inverse of the MacKinnon–Kramer parameter

$$\Gamma = \Lambda^{-1}. \tag{3.7}$$

In the localized phase, Γ increases with L for large enough L, while in the extended phase, it decreases. Exactly at the critical point we have scale invariance for sufficiently large L

$$\lim_{L \to \infty} \Gamma(L) = \text{const} = \Gamma_c. \tag{3.8}$$

3.3. *The transfer matrix method*

The transfer matrix method is the most efficient way of calculating the quasi-one dimensional localization length.[9,14] The Schrödinger equation for the

Anderson Hamiltonian on a d-dimensional bar is rewritten as

$$\mathbf{V}_{n,n+1}\mathbf{a}_{n+1} = (E - \mathbf{H}_n)\mathbf{a}_n - \mathbf{V}_{n,n-1}\mathbf{a}_{n-1}\,. \tag{3.9}$$

Here, $\mathbf{a}_n$ is the vector consisting of the L^{d-1} amplitudes on the lattice sites of the cross sectional plane of the bar at n, $\mathbf{V}_{n,n+1}$ is the $M \times M$ ($M = L^{d-1}$) dimensional matrix of inter-layer couplings between sites on the cross sections at n and $n+1$, and $\mathbf{H}_n$ is the matrix of intra-layer couplings between sites on the cross section at n. Equation (3.9) couples the amplitudes of a state at energy E on the cross section $n + 1$ to those at the cross sections n and $n - 1$. We rewrite (3.9) to define the $2M \times 2M$ transfer matrix,

$$\mathbf{T}_n = \begin{pmatrix} \mathbf{V}_{n,n+1}^{-1}(E\mathbf{1} - \mathbf{H}_n)\,, & -\mathbf{V}_{n,n+1}^{-1}\mathbf{V}_{n,n-1} \\ \mathbf{1} & , & \mathbf{0} \end{pmatrix}, \tag{3.10}$$

and the transfer matrix product for the whole bar of length N

$$\mathbf{Q}_N = \prod_{n=1}^{N} \mathbf{T}_n\,. \tag{3.11}$$

With this, we write

$$\begin{pmatrix} \mathbf{a}_{N+1} \\ \mathbf{a}_N \end{pmatrix} = \mathbf{Q}_N \begin{pmatrix} \mathbf{a}_1 \\ \mathbf{a}_0 \end{pmatrix}. \tag{3.12}$$

As a consequence of Oseledec's theorem,[33–36] the eigenvalues λ_i of the matrix,

$$\Omega = \ln\left(\mathbf{Q}_N \mathbf{Q}_N^{\dagger}\right), \tag{3.13}$$

obey the following limit

$$\gamma_i = \lim_{N\to\infty} \frac{\lambda_i}{2N}\,. \tag{3.14}$$

Here, i indexes the $2M$ eigenvalues of Ω. The values on the left hand side are called Lyapunov exponents. They occur in pairs of opposite sign. The smallest positive Lyapunov exponent is the inverse of the quasi-one dimensional localization length, i.e.

$$\gamma_M = \frac{1}{\lambda}\,, \tag{3.15}$$

where we have assumed that the exponents are labelled in decreasing order.

Some typical high precision numerical data for the Anderson model in three dimensions obtained using the transfer matrix method are shown in Fig. 16.1. For weak disorder Γ decreases, which indicates that in the three dimensional limit the system is in the metallic phase. For strong disorder

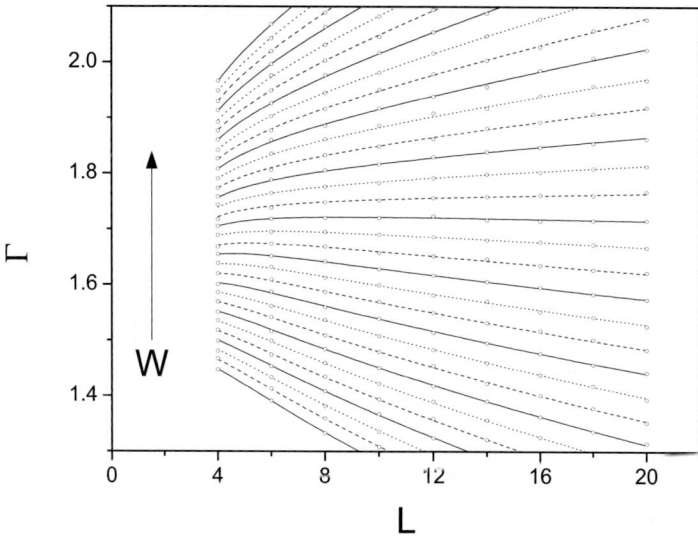

Fig. 16.1. Numerical data for the three dimensional Anderson model with box distributed random potential, width $W = 15 - 18$ in steps of 0.1. The precision of the data is 0.1%. The lines are a finite size scaling fit that includes corrections to scaling.

Γ increases, which indicates that in the three dimensional limit the system is now in the localized phase. At the critical disorder, we see that Γ is independent of system size. Note that a transient behaviour for small system sizes is clearly resolved, which must be taken into account by including corrections to scaling when fitting the numerical data.

3.4. *The correlation length*

In addition to the critical exponent ν and the scaling functions, one of the principal results of the finite size scaling analysis is the correlation length ξ. We find in the localized regime that

$$\lim_{L \to \infty} \lambda(L) = \xi. \tag{3.16}$$

Thus, provided the system is in the localized phase, we can identify ξ with the localization length in the infinite d-dimensional system. Note that it is important to distinguish the quasi-one dimensional localization λ on a long bar, which is always finite, from the localization length ξ in the the infinite d-dimensional system, which diverges at the Anderson transition. Equation (3.16) applies only in the localized phase.

Physically the localization length ξ describes the exponential decay of the transmission probability $t(E; \mathbf{x}, \mathbf{x}')$ of a quantum particle between two sites $\mathbf{x}$ and $\mathbf{x}'$ in an infinite d-dimensional system that is in the localized phase

$$\frac{2}{\xi} = - \lim_{|\mathbf{x}-\mathbf{x}'|\to\infty} \frac{\langle \ln t(E; \mathbf{x}, \mathbf{x}')\rangle}{|\mathbf{x} - \mathbf{x}'|}. \tag{3.17}$$

Thus, the transmission probability, and hence the diffusion constant, vanish in the thermodynamic limit and the system is an insulator.[14]

In the metallic phase, the correlation length is again finite and can be related to the resistivity.

4. The Critical Exponents

4.1. *Numerical results*

Most strikingly, although corrections to scaling had not been considered extensively at that time, already the first works dealing with the orthogonal symmetry class showed that the finite size scaling method was able to confirm the most important result of the scaling theory of localization: whereas in three dimensions clear evidence for the existence of a critical point was found, none was found in two dimensions.[9,10] During subsequent years, the universality of the critical behaviour for the orthogonal class was explicitly demonstrated by analysing orthogonal models with different disorder distributions.[20] It was also demonstrated that the high precision estimates of the critical exponents could also be obtained by analysing the finite size scaling of various statistics of the conductance distribution.[26,32]

In addition, the critical behaviours of the other universality classes have been extensively studied. As can be seen by reference to Table 16.1, in a given dimension, the values of the exponents in the different symmetry classes differ only by several percent. Success in clearly distinguishing the critical exponents for the different universality classes is a triumph of the finite size scaling method. This is in sharp contrast to other methods of estimating the exponents, in particular, the ϵ expansion, which have singularly failed to yield precise estimates of the exponents and even in some cases predicted values that violate the well established inequality[37,38]

$$\nu \geq \frac{2}{d}. \tag{4.1}$$

Table 16.1. List of critical exponents for different universality classes and in different dimensions. The error is a 95% confidence interval.

$\nu = 1.57 \pm 0.02$	3D orthogonal symmetry[20]
$\nu = 1.43 \pm 0.04$	3D unitary symmetry[27]
$\nu = 1.375 \pm 0.016$	3D symplectic symmetry[28]
$\nu = 2.73 \pm 0.02$	2D symplectic symmetry[21]
$\nu = 2.593 \pm 0.006$	Integer quantum Hall effect[29]

4.2. *Remarks concerning experiments*

Measurement of the conductivity at finite temperature on the metallic side of the transition and extrapolation to zero temperature permits an estimate of s. Measurement of the temperature dependence of the conductivity on the insulating side of the transition and fitting to the theory of variable range hopping[39,40] permits an estimate of ν.

An alternative approach, called finite temperature scaling,[41] is to fit finite temperature conductivity data on both sides of the transition to

$$\sigma(T) = T^{s/z\nu} f(\chi/T^{1/z\nu}). \qquad (4.2)$$

This permits estimates of s and the product $z\nu$. Here, χ is the relevant scaling variable, which is a function of the parameter used to drive the transition. For example, for a transition driven by varying the carrier concentration, we can approximate

$$\chi \approx \frac{(n - n_{\rm c})}{n_{\rm c}}, \qquad (4.3)$$

for doping concentrations n sufficiently close to the critical concentration $n_{\rm c}$. The exponent z, which is called the dynamical exponent, describes the divergence of the phase coherence length as the temperature tends to zero

$$L_\varphi \sim T^{-1/z}. \qquad (4.4)$$

Fitting the temperature dependence of the conductivity precisely at the critical point, and assuming the validity of Wegner's scaling law Eq. (1.4), permits an estimate of z. In quantum Hall effect experiments, z has been estimated by exploiting the fact that a crossover in the temperature dependence can be observed in very small systems when the phase coherence length becomes comparable to the systems size.

The most recent experiments on doped semiconductors[13] have yielded values of s and ν in the range between 1 and 1.2 that are consistent with

Wegner's scaling law Eq. (1.4). However, there is a clear deviation of the values of ν from those in Table 16.1. The most recent experimental estimate of the critical exponent for the plateau transition in the integer quantum Hall effect is $\nu = 2.38 \pm 0.06$.[42] Again this differs from the numerical estimate given in Table 16.1.

The limitations of models of non-interacting electrons as a description of the critical behaviour of the Anderson transition in electronic systems is clearly seen in the disagreement between the predicted and measured values of the dynamical exponent z. Whereas models of non-interacting electrons predict $z = d$,[43] where d is the dimensionality, the experimentally observed value is often smaller. Itoh *et al.*[13] found $z \approx 3$ in vanishing magnetic field, which agrees with non-interacting theory, but $z \approx 2$ in applied magnetic field, which does not. For the plateau transition Li *et al.*[42] found $z \approx 1$, which again disagrees with non-interacting theory.

The advent of experiments with cold atomic gases,[44] and also with ultrasound in random elastic media,[45] have allowed Anderson localization and the Anderson transition to be measured in systems that can be reasonably described as non-interacting. In particular, Chabe *et al.*[44] recently measured the critical behaviour of the Anderson transition in a quasi-periodic kicked rotor that was realised in a cold gas of cesium atoms. For this system, which is in the three dimensional orthogonal universality class, Chabe *et al.* found $\nu = 1.4 \pm 0.3$; a result that is consistent with the numerical estimate in Table 16.1.

5. Conclusions

The finite size scaling method combined with high precision numerical simulations has permitted the successful verification of the fundamental assumptions underlying the scaling theory of localization and provided high precision estimates of the critical exponents. The advent of cold atomic gasses has permitted the experimental observation of the Anderson transition in a system that can be reasonably described as non-interacting. Describing the critical behaviour observed at the Anderson transition in electronic systems remains a challenge and would seem to require the development of numerically tractable models that include the long range Coulomb interaction between the electrons.

References

1. P. W. Anderson, *Phys. Rev.* **109**, 1492 (1958).
2. N. F. Mott and W. D. Twose, *Adv. Phys.* **10**, 107 (1961).

3. M. E. Gertsenshtein and V. B. Vasilev, *Theor. Probab. Appl.* **4**, 391 (1959).
4. K. Ishi, *Prog. Theor. Phys.* **53**, 77 (1973).
5. A. A. Abrikosov and I. A. Ryshkin, *Adv. Phys.* **27**, 147 (1978).
6. P. Erdös and R. C. Herndon, *Adv. Phys.* **31**, 65 (1982).
7. E. Abrahams, P. W. Anderson, D. C. Licciardello and T. V. Ramakrishnan, *Phys. Rev. Lett.* **42**, 673 (1979).
8. F. Wegner, *Z. Phys. B* **35**, 207 (1979).
9. A. MacKinnon and B. Kramer, *Phys. Rev. Lett.* **47**, 1546 (1981).
10. A. MacKinnon and B. Kramer, *Z. Phys. B* **53**, 1 (1983).
11. J. L. Pichard and G. Sarma, *J. Phys. C: Solid State Phys.* **14**, L127 (1981).
12. J. L. Pichard and G. Sarma, *J. Phys. C: Solid State Phys.* **14**, L617 (1981).
13. See for example, N. Itoh *et al.*, *J. Phys. Soc. Jpn.* **73**, 173 (2004), and references therein.
14. B. Kramer and A. MacKinnon, *Rep. Progr. Phys.* **56**, 1496 (1993).
15. B. Huckestein, *Rev. Mod. Phys.* **67**, 357 (1995).
16. B. Kramer, T. Ohtsuki and S. Kettemann, *Phys. Rep.* **417**, 211 (2005).
17. F. Evers and A. D. Mirlin, *Rev. Mod. Phys.* **80**, 1355 (2008).
18. F. Wegner, *Phys. Rev. B* **19**, 783 (1979).
19. A. Altland and M. R. Zirnbauer, *Phys. Rev. B* **55**, 1142 (1997).
20. K. Slevin and T. Ohtsuki, *Phys. Rev. Lett.* **82**, 382 (1999).
21. Y. Asada, K. Slevin and T. Ohtsuki, *Phys. Rev. Lett.* **89**, 256601 (2002).
22. Y. Asada, K. Slevin and T. Ohtsuki, *Phys. Rev. B* **70**, 035115 (2004).
23. J. Ohe, M. Yamamoto and T. Ohtsuki, *Phys. Rev. B* **68**, 165344 (2003).
24. Y. A. Bychkov and E. I. Rashba, *J. Phys. C* **17**, 6039 (1984).
25. G. Dresselhaus, *Phys. Rev.* **100**, 580 (1955).
26. K. Slevin, P. Markos and T. Ohtsuki, *Phys. Rev. Lett.* **86**, 3594 (2001).
27. K. Slevin and T. Ohtsuki, *Phys. Rev. Lett.* **78**, 4083 (1997).
28. Y. Asada, K. Slevin and T. Ohtsuki, *J. Phys. Soc. Jpn.* **74** (Supp.), 258 (2005).
29. K. Slevin and T. Ohtsuki, *Phys. Rev. B* **80**, 041304 (2009).
30. B. I. Shklovskii, B. Shapiro, B. R. Sears, P. Lambrianides and H. B. Shore, *Phys. Rev. B* **47**, 11487 (1993).
31. I. Kh. Zharekeshev and B. Kramer, *Phys. Rev. Lett.* **79**, 717 (1997).
32. K. Slevin, P. Markos and T. Ohtsuki, *Phys. Rev. B* **67**, 155106 (2003).
33. V. I. Oseledec, *Trans. Moscow Math. Soc.* **19**, 197 (1968).
34. D. Ruelle, *Ann. Math.* **155**, 243 (1982).
35. U. Krengel, *Ergodic Theorems* (de gryter, Berlin, 1985).
36. R. Carmona and J. Lacroix, *Spectral Theory of Random Schrödinger Equations* (Birkhauser, Boston, 1990).
37. J. T. Chayes, L. Chayes, D. S. Fisher and T. Spencer, *Phys. Rev. Lett.* **57**, 2999 (1986).
38. B. Kramer, *Phys. Rev. B* **47**, 9888 (1993).
39. N. F. Mott, *Metal–Insulator Transitions*, 2nd edn. (Taylor and Francis, London, 1990).
40. B. I. Shklovskii and A. L. Efros, *Electronic Properties of Doped Semiconductors* (Springer-Verlag, Berlin, 1984).

41. S. Bogdanovich, M. P. Sarachik and R. N. Bhatt, *Phys. Rev. Lett.* **82**, 137 (1999).
42. W. Li *et al.*, *Phys. Rev. Lett.* **102**, 216801 (2009).
43. F. Wegner, *Z. Phys. B* **25**, 327 (1976).
44. J. Chabe *et al.*, *Phys. Rev. Lett.* **101**, 255702 (2008).
45. S. Faez, A. Strybulevych, J. H. Page, A. Lagendijk and B. A. van Tiggelen, *Phys. Rev. Lett.* **103**, 155703 (2009).

Chapter 17

A METAL–INSULATOR TRANSITION IN 2D: ESTABLISHED FACTS AND OPEN QUESTIONS

S. V. Kravchenko

Physics Department, Northeastern University
Boston, MA 02115, USA

s.kravchenko@neu.edu

M. P. Sarachik

Physics Department, City College of the City University of New York
New York, NY 10031, USA

sarachik@sci.ccny.cuny.edu

The discovery of a metallic state and a metal–insulator transition (MIT) in two-dimensional (2D) electron systems challenges one of the most influential paradigms of modern mesoscopic physics, namely, that "there is no true metallic behavior in two dimensions". However, this conclusion was drawn for systems of noninteracting or weakly interacting carriers, while in all 2D systems exhibiting the metal–insulator transition, the interaction energy greatly exceeds all other energy scales. We review the main experimental findings and show that, although significant progress has been achieved in our understanding of the MIT in 2D, many open questions remain.

1. Introduction

In two-dimensional (2D) electron systems, the electrons move in a plane in the presence of a weak random potential. According to the scaling theory of localization of Abrahams *et al.*,[1] these systems lie on the boundary between high and low dimensions insofar as the metal–insulator transition is concerned. The carriers are always strongly localized in one dimension, while in three dimensions, the electronic states can be either localized or extended. In the case of two dimensions the electrons may conduct well at room temperature, but a weak logarithmic increase of the resistance is expected as the temperature is reduced. This is due to the fact that, when scattered from impurities back to their starting point, electron waves interfere constructively with their time reversed paths. Quantum interference becomes increasingly important as the temperature is reduced and leads to localization of the electrons, albeit on a large length scale; this is generally referred to as "weak localization". Indeed, thin metallic films and many of the 2D electron systems fabricated on semiconductor surfaces display the predicted logarithmic increase of resistivity.

The scaling theory[1] does not consider the effects of the Coulomb interaction between electrons. The strength of the interactions is usually characterized by the dimensionless Wigner–Seitz radius,

$$r_s = \frac{1}{(\pi n_s)^{1/2} a_B},$$

(here n_s is the electron density and a_B is the Bohr radius in a semiconductor). As the density of electrons is reduced, the Wigner–Seitz radius grows and the interactions provide the dominant energy of the system. In the early 1980's, Finkelstein[2,3] and Castellani et al.[4] found that for weak disorder and sufficiently strong interactions, a 2D system scales towards a *conducting* state as the temperature is lowered. However, the scaling procedure leads to an increase in the effective strength of the spin-related interactions and to a divergent spin susceptibility, so that the perturbative approach breaks down as the temperature is reduced toward zero. Therefore, the possibility of a 2D metallic ground state stabilized by strong electron–electron interactions was not seriously considered at that time, particularly as there were no experimental observations to support the presence of a metallic phase.

Progress in semiconductor technology has enabled the fabrication of high quality 2D samples with very low randomness in which measurements can be made at very low carrier densities. The strongly-interacting regime ($r_s \gg 1$) has thus become experimentally accessible. The first observation of a metal–insulator transition in strongly-interacting, low-disordered 2D systems on a silicon surface was reported in 1987 by Zavaritskaya and Zavaritskaya.[5] Although identified by the authors as a metal–insulator transition, the discovery went by unnoticed. Subsequent experiments on even higher mobility silicon samples[6–10] confirmed the existence of a metal–insulator transition in 2D and demonstrated that there were surprising and dramatic differences between the behavior of strongly interacting systems with $r_s > 10$ as compared with weakly-interacting systems. These results were met with great skepticism and were largely overlooked until they were confirmed in other strongly-interacting 2D systems in 1997.[11–16] Moreover, it was found[17–20] that in the strongly-interacting regime, an external in-plane magnetic field strong enough to polarize the spins of the electrons or holes induces a giant positive magnetoresistance and completely suppresses the metallic behavior, implying that the spin state is central to the high conductance of the metallic state. Experiments[21–28] have shown that there is a sharp enhancement of the spin susceptibility as the metal–insulator transition is approached. Interestingly, this enhancement is due to a strong increase of the effective mass, while the g-factor remains essentially constant.[25,27,29,30] Therefore, the effect is not related to the Stoner instability.[31]

In this article, we summarize the main experimental findings. Of the many theories that have been proposed to explain the observations, we provide a detailed discussion of the theory of Punnoose and Finkelstein,[32] as it provides numerical predictions with which experimental results can be compared directly. We end with a brief discussion of some of the unsolved problems.

2. Experimental Results in Zero Magnetic Field

The first experiments that demonstrated the unusual temperature dependence of the resistivity[5–9] were performed on low-disordered MOSFETs with maximum electron mobilities reaching more than 4×10^4 cm^2/Vs; these mobilities were considerably higher than in samples used in earlier investigations. The very high quality of the samples allowed access to the physics at electron densities below 10^{11} cm^{-2}. At these low densities, the Coulomb energy, E_C, is the dominant parameter. Estimates for Si MOSFETs at $n_s = 10^{11}$ cm^{-2} yield $E_C \approx 10$ meV, while the Fermi energy, E_F, is about 0.6 meV (a valley degeneracy of two is taken into account when calculating the Fermi energy, and the effective mass is assumed to be equal to the band mass.) The ratio between the Coulomb and Fermi energies, $r^* \equiv E_C/E_F$, thus assumes values above 10 in these samples.

The earliest data that clearly show the MIT in 2D are shown in Fig. 17.1(a). Depending on the initial ("high-temperature") value of conductivity, σ_0, the temperature dependence of conductivity $\sigma(T)$ in a low-disordered Si MOSFET exhibits two different types of behavior: for $\sigma_0 < e^2/h$, the conductivity decreases with decreasing temperature following Mott's hopping law in 2D, $\sigma \propto \exp(T^{-1/3})$; on the other hand, for $\sigma_0 > e^2/h$, the conductivity increases with decreasing T by as much as a factor of 7 before finally saturating at sub-kelvin temperatures. Fig. 17.1(b) shows the temperature dependence of the resistivity (the inverse of the conductivity) measured in units of h/e^2 of a high-mobility MOSFET for 30 different electron densities n_s varying from 7.12×10^{10} to 13.7×10^{10} cm^{-2}.

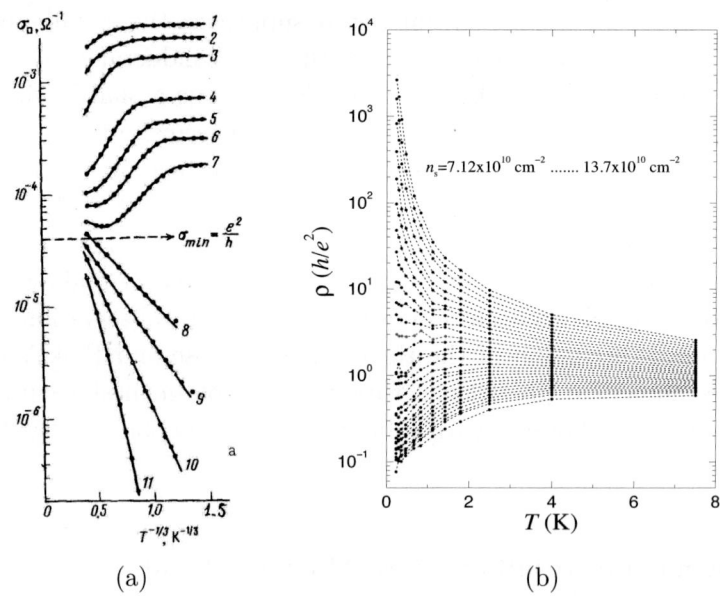

(a)   (b)

Fig. 17.1.   (a) Conductivity vs. the inverse cube root of temperature in silicon inversion channel for electron densities, n_s ranging from 10^{12} cm^{-2} (the upper curve) to less than 10^{11} cm^{-2} (the lowest curve); adapted from Ref. 5. (b) Temperature dependence of the $B = 0$ resistivity in a dilute low-disordered Si MOSFET for 30 different electron densities ranging from 7.12×10^{10} cm^{-2} to 13.7×10^{10} cm^{-2}; adapted from Ref. 7. Figure 17.1(a) reprinted (Fig. 1a) with permission from T. N. Zavaritskaya and E. I. Zavaritskaya, *JETP Lett.* **45**, 609 (1987). © Springer. Figure 17.1(b) reprinted (Fig. 3) with permission from *Phys. Rev. B* **51**, 7038 (1995). © American Physical Society.

If the resistivity at high temperatures exceeds the quantum resistance h/e^2, $\rho(T)$ increases monotonically as the temperature decreases, behavior that is characteristic of an insulator. However, for n_s above a certain "critical" value, n_c (the curves below the "critical" curve that extrapolates to $3h/e^2$), the temperature dependence of $\rho(T)$ is non-monotonic: with decreasing temperature, the resistivity first increases (at $T > 2$ K) and then decreases as the temperature is further reduced. At yet higher density n_s, the resistivity is almost constant at $T > 4$ K but drops by an order of magnitude at lower temperatures, showing strongly metallic behavior as $T \to 0$.

A metal–insulator transition similar to that seen in clean Si MOSFETs has also been observed in other low-disordered, strongly-interacting 2D systems: p-type SiGe heterostructures,[11,33] p-GaAs/AlGaAs heterostructures,[12,34,35] n-GaAs/AlGaAs heterostructures,[13,36] AlAs heterostructures,[14] and n-SiGe heterostructures.[15,16] The values of the resistivity are

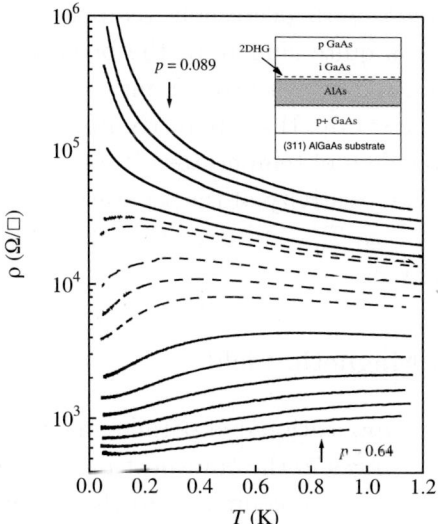

$\rho~(\Omega/\square)$

T (K)

Fig. 17.2.   For low-disordered 2D hole systems in p-GaAs/AlGaAs, the resistivity per square is shown as a function of temperature for $B = 0$ at various fixed hole densities, p.   Data for an ISIS (inverted semiconductor-insulator-semiconductor) structure with hole densities (from top to bottom) $p = 0.89$, 0.94, 0.99, 1.09, 1.19, 1.25, 1.30, 1.50, 1.70, 1.90, 2.50, 3.20, 3.80, 4.50, 5.10, 5.70, and $6.40{\cdot}10^{10}$ cm^{-2}. The inset shows a schematic diagram of the ISIS structure: the carriers are accumulated in an undoped GaAs layer situated on top of an undoped AlAs barrier, grown over a p^{+} conducting layer which serves as a back-gate; the hole density, p, is varied by applying a voltage to the back gate.   From Ref. 12.   Reprinted (Fig. 1) with permission from Y. Hanein *et al.*, *Phys. Rev. Lett.* **80**, 1288 (1998). © American Physical Society.

quite similar in all systems. In Fig. 17.2, the resistivity is shown as a function of temperature for a p-type GaAs/AlGaAs sample; here the interaction parameter, r_s, changes between approximately 12 and 32.[a] The main features are very similar to those found in Si MOSFETs: when the resistivity at "high" temperatures exceeds the quantum resistance, h/e^2 (i.e., at hole densities below some critical value, p_c), the $\rho(T)$ curves are insulating-like in the entire temperature range; for densities just above p_c, the resistivity shows insulating-like behavior at higher temperatures and then drops by a factor of 2 to 3 at temperatures below a few hundred mK; and at yet higher hole densities, the resistivity is metallic in the entire temperature range. Note that the curves that separate metallic and insulating behavior have

[a]These r_s values were calculated assuming that the effective mass is independent of density and equal to $0.37\,m_e$, where m_e is the free-electron mass.

resistivities that increase with decreasing temperature at the higher temperatures shown; this is quite similar to the behavior of the separatrix in Si MOSFETs when viewed over a broad temperature range (see Fig. 17.1). Below approximately 150 mK, the separatrix in p-type GaAs/AlGaAs heterostructures is independent of temperature,[12] as it is in Si MOSFETs below approximately 2 K. The resistivity of the separatrix in both systems extrapolates to ≈ 2 or $3\,h/e^2$ as $T \to 0$, even though the corresponding carrier densities are very different.

3. The Effect of a Magnetic Field

In ordinary metals, the application of a parallel magnetic field ($B_{\parallel}$) does not lead to any dramatic changes in the transport properties: if the thickness of the 2D electron system is small compared to the magnetic length, the parallel field couples largely to the electrons' spins while the orbital effects are suppressed. Only weak corrections to the conductivity are expected due to electron–electron interactions.[38] It therefore came as a surprise when Dolgopolov *et al.*[17] observed a dramatic suppression of the conductivity in dilute Si MOSFETs by a parallel in-plane magnetic field $B_{\parallel}$. The magnetoresistance

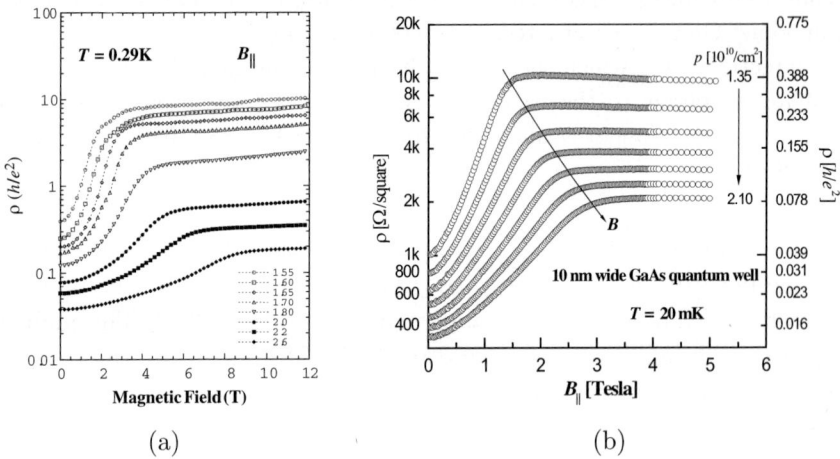

(a)   (b)

Fig. 17.3.   (a) Resistivity vs. parallel magnetic field measured at $T = 0.29$ K in a Si MOSFET. Different symbols correspond to densities from 1.01 to $2.17 \cdot 10^{11}$ cm^{-2}; adapted from Ref. 19 (b) Resistivity as a function of $B_{\parallel}$ of a 10 nm wide p-GaAs quantum well at 50 mK; adapted from Ref. 37. Figure 17.3(a) reprinted (Fig. 1) with permission from V. M. Pudalov *et al.*, *JETP Lett.* **65**, 932 (1997). © Springer. Figure 17.3(b) reprinted (Fig. 3) with permission from X. P. A. Gao *et al.*, *Phys. Rev. B* **73**, 241315(R) (2006). © American Physical Society.

in a parallel field was studied in detail by Simonian *et al.*[18] and Pudalov *et al.*,[19] also in Si MOSFETs. In the left hand part of Fig. 17.3, the resistivity is shown as a function of parallel magnetic field at a fixed temperature of 0.3 K for several electron densities. The resistivity increases sharply as the magnetic field is raised, changing by a factor of about 4 at the highest density shown and by more than an order of magnitude at the lowest density, and then saturates and remains approximately constant up to the highest measuring field, $B_\parallel = 12$ tesla. The magnetic field where the saturation occurs, B_{sat}, depends on n_s, varying from about 2 tesla at the lowest measured density to about 9 tesla at the highest. The metallic conductivity is suppressed in a similar way by magnetic fields applied at any angle relative to the 2D plane[39] independently of the relative directions of the measuring current and magnetic field.[18,40] All these observations suggest that the giant magnetoresistance is due to coupling of the magnetic field to the electrons' spins. Indeed, from an analysis of the positions of Shubnikov-de Haas oscillations in tilted magnetic fields[21,41,42] it was concluded that in MOSFETs, the magnetic field B_{sat} is equal to that required to fully polarize the electrons' spins.

In p-type GaAs/AlGaAs heterostructures, the effect of a parallel magnetic field is similar, as shown in the right hand part of Fig. 17.3. As in the case of Si MOSFETs, there is a distinct knee above which the resistivity remains constant. For high hole densities, Shubnikov-de Haas measurements[43] have shown that this knee is associated with full polarization of the spins by the in-plane magnetic field. However, unlike Si MOSFETs, the magnetoresistance in p-GaAs/AlGaAs heterostructures has been found to depend on the relative directions of the measuring current, magnetic field, and crystal orientation[44]; one should note that the crystal anisotropy of this material introduces added complications. In p-SiGe heterostructures, the parallel field was found to induce negligible magnetoresistance[33] because in this system the parallel field cannot couple to the spins due to very strong spin-orbit interactions.

Over and above the very large magnetoresistance induced by an in-plane magnetic field, an even more important effect of a parallel field is that it causes the zero-field 2D metal to become an insulator.[18,45–48] The extreme sensitivity to parallel field is illustrated in Fig. 17.4. The top two panels compare the temperature dependence of the resistivity in the absence and in the presence of a parallel magnetic field. For $B_\parallel = 0$, the resistivity displays the familiar, nearly symmetric (at temperatures above 0.2 K) critical behavior about the separatrix (the dashed line). However, in a

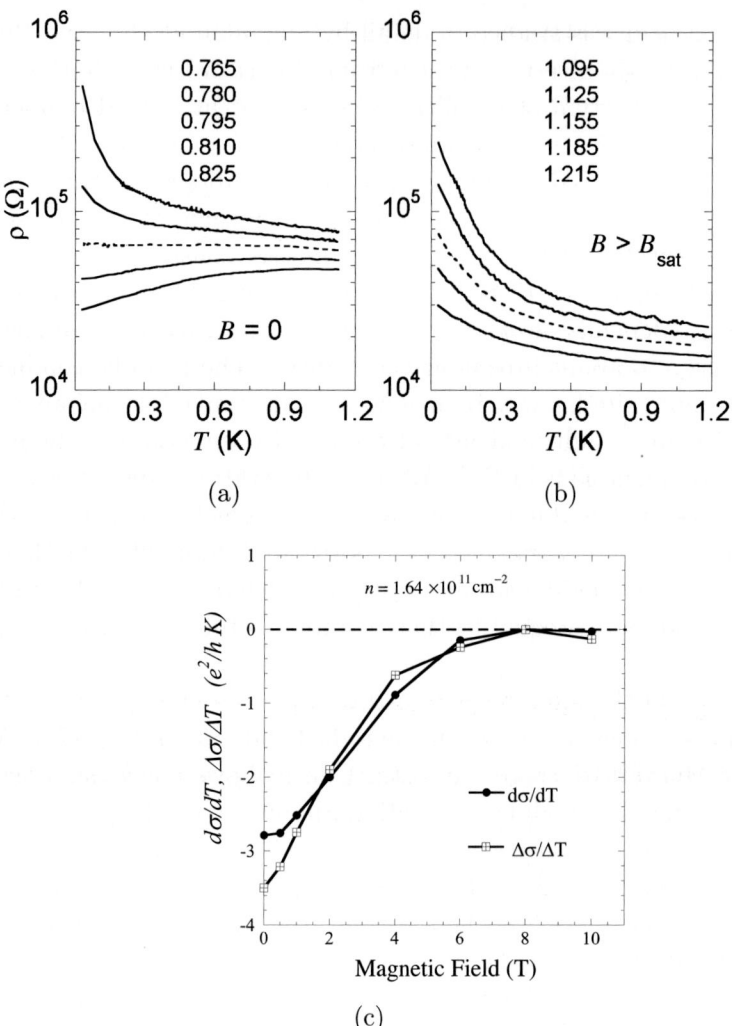

Fig. 17.4. (a) Temperature dependence of the resistivity of a Si MOSFET at different electron densities near the MIT in zero magnetic field and (b) in a parallel magnetic field of 4 tesla. The electron densities are indicated in units of 10^{11} cm^{-2}. Dashed curves correspond to $n_s = n_{c1} = 0.795 \times 10^{11}$ cm^{-2} in zero field and 1.155×10^{11} cm^{-2} in $B_\parallel = 4$ tesla; taken from Ref. 46. (c) Data taken as a function of parallel magnetic field for a Si MOSFET of density 1.64×10^{11} cm^{-2}; closed symbols denote the slope $d\sigma/dT$ and open symbols denote $\Delta\sigma/\Delta T$ calculated for the temperature interval 0.27 to 1.35 K; from Ref. 48. Figures 17.4(a) and 17.4(b) reprinted (Fig. 1) with permission from *Phys. Rev. Lett.* **87**, 266402 (2001). Figure 17.4(c) reprinted (Fig. 3) with permission from *Phys. Rev. Lett.* **71**, 13308 (2005). © American Physical Society.

parallel magnetic field of $B_\parallel = 4$ tesla, which is high enough to cause full spin polarization at this electron density, all the $\rho(T)$ curves display "insulating-like" behavior, including those which start below h/e^2 at high temperatures. There is no temperature-independent separatrix at any electron density in a spin-polarized electron system.[18,46] The effect of a parallel magnetic field is further demonstrated in the bottom panel of Fig. 17.4, where the slope of the resistivity calculated for the temperature interval 0.27 K to 1.35 K is plotted as a function of magnetic field at a fixed density; these data show explicitly and quantitatively that a magnetic field applied parallel to the plane of the electrons reduces the temperature dependence of the conductivity to near zero. Moreover, this was shown to be true over a broad range of electron densities extending deep into the metallic regime where the high-field conductivity is on the order of $10(e^2/h)$. The clear difference in behavior with and without in-plane magnetic field convincingly demonstrates that the spin-polarized and unpolarized states behave very differently and rules out explanations that predict similar behavior of the resistance regardless of the degree of spin polarization.

4. Spin Susceptibility Near the Metal–Insulator Transition

4.1. *Experimental measurements of the spin susceptibility*

In Fermi-liquid theory, the electron effective mass and the g-factor (and, therefore, the spin susceptibility, $\chi \propto g^* m^*$) are renormalized due to electron–electron interactions.[49] Earlier experiments,[50,51] performed at relatively small values of $r_s \sim 2$ to 5, confirmed the expected increase of the spin susceptibility. More recently, Okamoto *et al.*[21] observed a renormalization of χ by a factor of ~ 2.5 at r_s up to about 6. At yet lower electron densities, in the vicinity of the metal–insulator transition, Kravchenko *et al.*[22] have observed a disappearance of the energy gaps at "cyclotron" filling factors which they interpreted as evidence for an increase of the spin susceptibility by a factor of at least 5.

It was noted many years ago by Stoner that strong interactions can drive an electron system toward a ferromagnetic instability.[31] Within some theories of strongly interacting 2D systems,[2–4,32,52] a tendency toward ferromagnetism is expected to accompany metallic behavior. The easiest way to estimate the spin magnetization of 2D electrons (or holes) is to measure the magnetic field above which the magnetoresistance saturates (and thus full spin polarization is reached) as a function of electron density. For non-interacting electrons, the saturation field is proportional to the electron density:

$$B^* = \frac{\pi \hbar^2 n_s}{g m \mu_B}.$$

Here g is the Landé g-factor, m is the effective mass, and μ_B is the Bohr magneton. Experiments[23,24] have shown, however, that in strongly correlated 2D systems in Si MOSFETs, the parallel field required for full spin polarization extrapolates to zero at a non-zero electron density, n_χ. The left-hand panel of Fig. 17.5 shows that the field B^* for full polarization obtained by Shashkin et al.[23] extrapolates to zero at a finite electron density; the dashed line indicates the calculated $B^*(n_s)$ for comparison; the fact that the measured B^* lies significantly lower than the calculated value indicates that either g or m (or both) are larger than their band values. Using a different method of analysis, Vitkalov et al.[24] obtained a characteristic energy $k_B \Delta$ associated with the magnetic field dependence of the conductivity plotted as a function of electron density, as shown in the right-hand panel of Fig. 17.5; the parameter Δ decreases with decreasing density, and extrapolates to zero at a critical density labeled n_o. That B^* and $k_B \Delta$, both measures of the field required

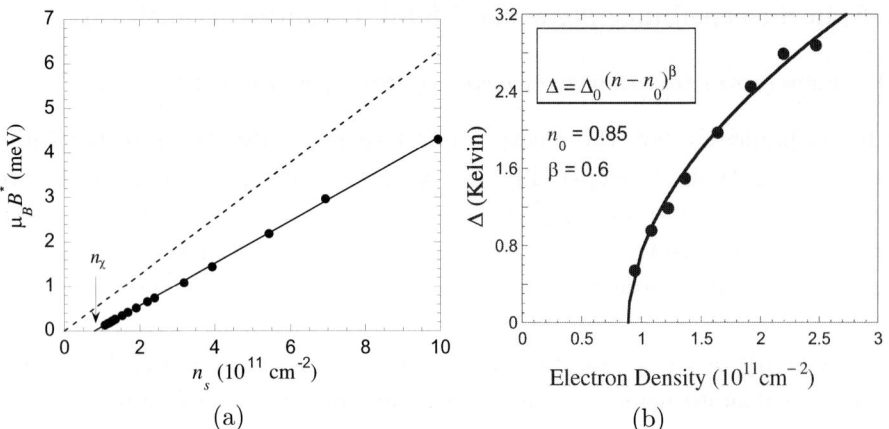

(a)   (b)

Fig. 17.5.   (a) Magnetic field for the onset of complete spin polarization vs. electron density; the dashed line depicts B^* calculated assuming that g and m are not renormalized; adapted from Ref. 23. (b) Characteristic energy $k_B \Delta$ associated with the magnetic field dependence of the conductivity plotted as a function of electron density; the parameter Δ decreases with decreasing density, and extrapolates to 0 at a critical density n_o; adapted from Ref. 24. Figure 17.5(a) reprinted (Fig. 3) with permission from *Phys. Rev. Lett.* **87**, 086801 (2001). Figure 17.5(b) reprinted (Fig. 3b) with permission from *Phys. Rev. Lett.* **87**, 086401 (2001). © American Physical Society.

to obtain complete spin polarization, extrapolate to zero at a finite density implies there is a spontaneous spin polarization at $n_s = n_\chi = n_0$. Many Si MOSFET samples of different quality have been tested, and the results indicate that $n_\chi \approx 8 \times 10^{10}$ cm^2 is independent of disorder. In the highest quality samples, n_χ was found to be within a few percent of the critical density for the metal–insulator transition, n_c (but consistently below).

It is easy to calculate the renormalized spin susceptibility using the data for $B^*(n_s)$:

$$\frac{\chi}{\chi_0} = \frac{n_s}{n_s - n_\chi},$$

where χ_0 is the "non-interacting" value of the spin susceptibility. Such critical behavior of a thermodynamic parameter usually indicates that a system is approaching a phase transition, possibly of magnetic origin. However, direct evidence of a phase transition can only be obtained from measurements of thermodynamic properties. Given the tiny number of electrons in a dilute 2D layer, magnetic measurements are very hard to perform. A clever technique was designed and implemented by Prus *et al.*[53] and Shashkin *et al.*[54] These authors modulated the parallel magnetic field with a small ac field, B_{mod}, and measured the tiny induced current between the gate and the 2D electron system. The imaginary (out-of-phase) component of the current is proportional to $d\mu/dB$, where μ is the chemical potential of the 2D gas. By applying the Maxwell relation $dM/dn_s = -d\mu/dB$, one can obtain the magnetization M from the measured current. Full spin polarization corresponds to $dM/dn_s = 0$. Yet another way of finding the density for complete spin polarization is related to measurements of the thermodynamic density of states of the 2D system obtained from measurements[54] of the capacitance of a MOSFET: the thermodynamic density of states was found to change abruptly with the onset of complete spin polarization of the electrons' spins.

The results obtained for the spin susceptibility are shown in Fig. 17.6. One can see that upon approaching to the critical density of the metal–insulator transition, the spin susceptibility increases by almost an order of magnitude relative to its "non-interacting" value. This implies the occurrence of a spontaneous spin polarization (either Wigner crystal or ferromagnetic liquid) at low n_s, although in currently available samples, the formation of a band tail of localized electrons at $n_s \lesssim n_c$ conceals the origin of the low-density phase. In other words, so far, one can only reach an incipient transition to a new phase.

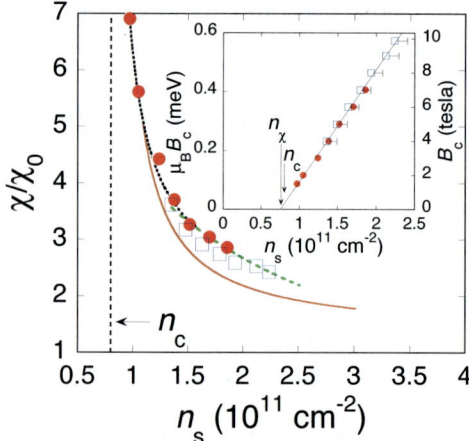

Fig. 17.6.   The Pauli spin susceptibility as a function of electron density obtained by thermodynamic methods: direct measurements of the spin magnetization (dashed line), $d\mu/dB = 0$ (circles), and density of states (squares). The dotted line is a guide to the eye. Also shown by a solid line is the transport data of Ref. 57. Inset: Field for full spin polarization as a function of the electron density determined from measurements of the magnetization (circles) and magnetocapacitance (squares). The data for B_c are consistent with a linear fit which extrapolates to a density n_χ close to the critical density n_c for the $B = 0$ MIT. Adapted from Ref. 54. Reprinted (Fig. 4) with permission from *Phys. Rev. Lett.* **96**, 036403 (2006). © American Physical Society.

A strong dependence of the magnetization on n has also been seen[28,55,56] in other types of devices for n_s near the critical density for the metal–insulator transition.

4.2. *Effective mass or g-factor?*

In principle, the strong increase of the Pauli spin susceptibility at low electron densities can be due to either the increase of the effective mass or the Lande g-factor (or both). The effective mass was measured by several groups employing different methods[25,29,30] which gave quantitatively similar results. The values g/g_0 and m/m_b as a function of the electron density are shown in Fig. 17.7 (here $g_0 = 2$ is the g factor in bulk silicon, m_b is the band mass equal to $0.19m_e$, and m_e is the free electron mass). In the high n_s region (relatively weak interactions), the enhancement of both g and m is relatively small, with both values increasing slightly with decreasing electron density, in agreement with earlier data.[58] Also, the renormalization of the g factor is

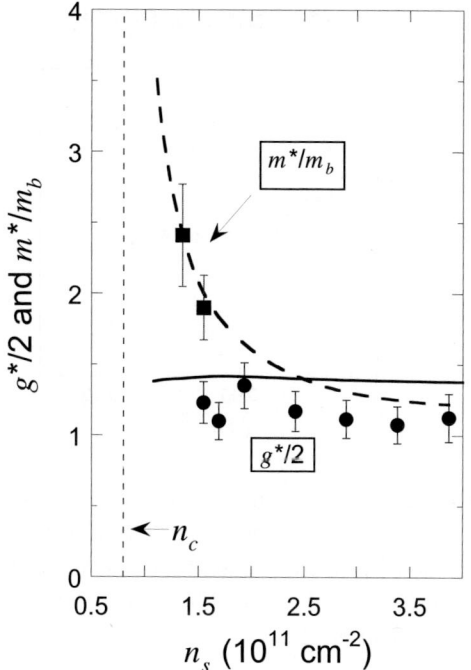

Fig. 17.7. The Landé g-factor and the effective mass as a function of electron density obtained from transport measurements[25] (solid and dashed lines, respectively). Also shown are the effective g-factor (circles) and the cyclotron mass (squares) obtained by measurements of thermodynamic magnetization.[30] The critical density n_c for the metal–insulator transition is indicated by the arrow. Reprinted (Fig. 4) with permission from *Phys. Rev. Lett.* **96**, 046409 (2006). © American Physical Society.

dominant compared to that of the effective mass, consistent with theoretical studies.[59–61]

In contrast, the renormalization at low n_s (near the critical region), where $r_s \gg 1$, is striking. As the electron density is decreased, the effective mass increases dramatically while the g factor remains essentially constant and relatively small, $g \approx g_0$. Hence, it is the effective mass, rather than the g factor, that is responsible for the drastically enhanced spin susceptibility near the metal–insulator transition.

4.3. *Effective mass as a function of r_s*

The effective mass has also been measured in a dilute 2D electron system in (111)-silicon. This system is interesting because the band electron mass

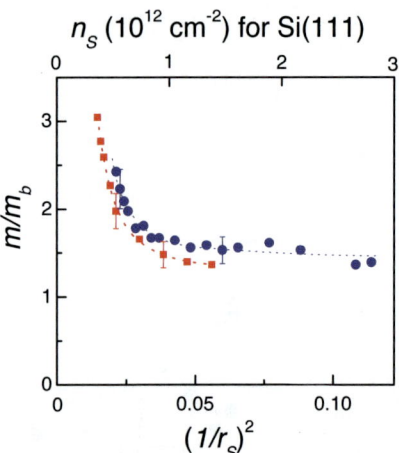

Fig. 17.8. The effective mass (dots) in units of m_b as a function of $(1/r_s)^2 \propto n_s$ for a (111)-Si MOSFET. Also shown by squares is the data obtained in (100)-Si MOSFETs.[29] The dashed lines are guides to the eye; from Ref. 62. Reprinted (Fig. 3) with permission from A. A. Shashkin *et al.*, *Phys. Rev. B* **76**, 241302(R) (2007). © American Physical Society.

$m_b = 0.358 m_e$ is approximately a factor of two larger than it is in (100)-silicon. In addition, the (111)-silicon samples used in these experiments have a much higher level of disorder. Remarkably, the relative enhancement of the effective mass, i.e., m^*/m_b, was found to be essentially the same function of the interaction parameter, r_s, as in (100) samples of Si MOSFETs. Shown in Fig. 17.8, the effective mass plotted in units m_b as a function of $(1/r_s)^2 \propto n_s$ is essentially the same for the two systems within the experimental uncertainty, despite the fact that the band mass differs by about a factor of two and the level of disorder differs by almost one order of magnitude. This implies that the relative mass enhancement is determined solely by the strength of the electron–electron interactions.

5. Comparison with Theory and Open Questions

Many theories have been proposed to account for the experimental observations summarized above. These include exotic superconductivity,[63] the formation of a disordered Wigner solid,[64] microemulsion phases,[65–67] percolation,[68] and a non-Fermi liquid state.[69] In what follows, we restrict our discussion to theories that provide numerical predictions with which experimental can be compared directly.

5.1. Ballistic regime ($k_B T \gg \hbar/\tau$)

When the resistivity of a sample is much smaller than h/e^2, the electrons are in a ballistic regime for temperatures $T > h/k_B\tau$; this encompasses most of the experimentally accessible range except in samples with electron densities that are very close to the critical density. Theories[70–76] that invoke the effect of electron screening attempt to explain the transport results in the ballistic regime by extrapolating classical formulas for the resistivity, valid for $r_s < 1$, to the regime where $r_s \gg 1$. Indeed, quantitatively successful comparisons with experiment have been reported.[72,74–76] However, at large r_s, the screening length λ_{sc} obtained using a random-phase approximation becomes parametrically smaller than the spacing between electrons: $\lambda_{sc}\sqrt{\pi n} = (1/4)r_s^{-1} \ll 1$.[67] Screening lengths smaller than the distance between electrons are clearly unphysical.

This approach was corrected by Zala *et al.*,[77] who considered the contribution due to the scattering from the Friedel oscillations induced by impurities. At small r_s, the "insulating-like" sign of $d\rho/dT$ is obtained, in agreement with experiments on samples with $r_s \sim 1$. However, when extrapolated to large enough r_s, $d\rho/dT$ changes sign, and $\rho(T)$ becomes a linearly increasing function of T. This theory[77] predicts the complete suppression of the metallic behavior in parallel magnetic fields sufficiently strong to completely polarize spins, again in agreement with the experiments. However, one should keep in mind that this theory considers *corrections* to the conductivity that are small compared to the Drude conductivity. By contrast, changes in resistivity by an order of magnitude are often observed experimentally.

5.2. Scaling theory of the metal–insulator transition in 2D: diffusive regime ($k_B T \ll \hbar/\tau$)

The two-parameter scaling theory[32,78] of quantum diffusion in an interacting disordered system is based on the scaling hypothesis that both the resistivity and the electron-electron scattering amplitudes, γ_2, become scale (temperature) dependent. Essentially, this is the theory of Anderson localization in the presence of electron–electron interactions. The renormalization-group (RG) equations describing the evolution of the resistance and the scattering amplitude in 2D have the form[78]

$$\frac{d\ln\rho^*}{d\xi} = \rho^*\left[n_v + 1 - (4n_v^2 - 1)\left(\frac{1+\gamma_2}{\gamma_2}\ln(1+\gamma_2) - 1\right)\right], \qquad (5.1)$$

$$\frac{d\gamma_2}{d\xi} = \rho^* \frac{(1+\gamma_2)^2}{2}, \tag{5.2}$$

where $\xi = -\ln(T\tau/\hbar)$, τ is the elastic scattering time, $\rho^* = (e^2/\pi h)\rho$, and n_v is the number of degenerate valleys in the spectrum. The resistance, $\rho(T)$, is a nonmonotonic function of temperature, reaching a maximum value ρ_{max} at some temperature T_{max} with a metallic temperature dependence $(d\rho/dT > 0)$ for $T < T_{\text{max}}$. Furthermore, $\rho(T)$ can be written as

$$\rho = \rho_{\text{max}} F\left(\rho_{\text{max}} \ln(T_{\text{max}}/T)\right), \tag{5.3}$$

where F is a *universal* function shown by the solid curve in Fig. 17.9(a). The strength of spin-dependent interactions, γ_2, is also a universal function of $\ln(T_{\text{max}}/T)$, shown by the solid line in Fig. 17.9(b).

This theory can account for the large changes in resistivity observed experimentally, and provides quantitative functions that can be directly compared with the experimental data. Such a comparison was made by Anissimova et al.,[79] who deduced the interaction amplitude from the magnetoresistance; the results are presented in Figs. 17.9(a) and 17.9(b). The agreement between theory and experiment is especially striking given that the

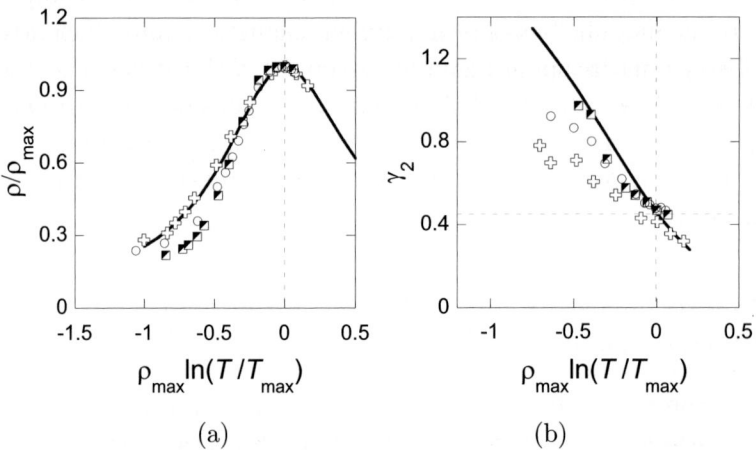

Fig. 17.9. Comparison between theory (lines) and experiment (symbols). (a) ρ/ρ_{max} as a function of $\rho_{\text{max}} \ln(T/T_{\text{max}})$. (b) γ_2 as a function of $\rho_{\text{max}} \ln(T/T_{\text{max}})$. Vertical dashed lines correspond to $T = T_{\text{max}}$, the temperature at which $\rho(T)$ reaches maximum. Note that at this temperature, the interaction amplitude $\gamma_2 \approx 0.45$ (indicated by the horizontal dashed line in (b)), in excellent agreement with theory. Electron densities are 9.87 (squares), 9.58 (circles), and 9.14×10^{10} cm^{-2} (crosses). Adapted from Ref. 79. Reprinted (Fig. 4) with permission from *Nature Phys.* **3**, 707 (2007). © Macmillan Publishers Ltd.

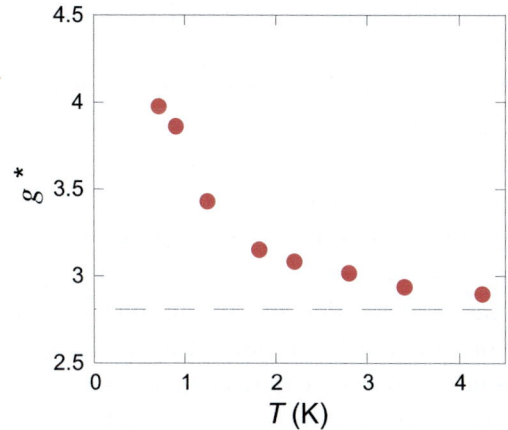

Fig. 17.10. *g*-factor vs. temperature in the diffusive regime. The dashed line shows the value of the *g*-factor obtained in the ballistic regime. Adapted from Ref. 80.

theory has no adjustable parameters. Systematic deviations from the universal curves occur at lower densities as higher order corrections in ρ become important. Furthermore, the resistivity reaches a maximum at $\gamma_2 \approx 0.45$, in excellent agreement with theory[78] for $n_v = 2$.

The data for γ_2 enables one to calculate the renormalized Landé *g*-factor $g^* = 2(1 + \gamma_2)$. As shown in Fig. 17.10, at $n_s = 9.87 \times 10^{10}$ cm^{-2}, the Landé *g*-factor increases from $g^* \approx 2.9$ at the highest temperature to $g^* \approx 4$ at the lowest. Therefore, the *g*-factor becomes temperature-dependent in the diffusive regime and increases with decreasing temperature, in agreement with the predictions of Punnoose and Finkelstein.[32,78] Note that at higher temperatures it is close to the temperature-independent "ballistic" value of about 2.8.

However, it should be noted that a theory based on electron interference effects can predict substantial changes of resistivity only in the near vicinity of the critical point, where $\rho \sim h/e^2$. At much lower resistivities $\rho \ll h/e^2$, only small logarithmic corrections are possible, while the experiments demonstrate very large changes in resistivity even deep in the metallic region. Therefore, although the theory of Punnoose and Finkelstein[32,78] quantitatively describes experimental data in the close vicinity of the transition, it cannot explain the large effects observed far from the transition. It should be further noted that most of the experiments in GaAs and n-SiGe devices are performed in the ballistic regime where this theory is irrelevant. Given the conspicuous similarity of the results obtained on different 2D systems, whether diffusive or ballistic, it seems clear that a unified theory applicable in both regimes is needed to account for the transport data.

Such a theory, based on the Pomeranchuk effect, was proposed by Spivak and Kivelson.[65–67] In essence, this theory proposes that when the interaction energy is much higher than the Fermi energy, the short-range interactions are of the Wigner-crystalline type. As the temperature or magnetic field is increased, the volume fraction occupied by the insulating Wigner crystallites grows relative to that of the conducting electron liquid, explaining the metallic temperature dependence of the resistance and the giant positive magnetoresistance observed at low temperatures. Moreover, this theory predicts that the metallic temperature dependence of the resistance is quenched in magnetic fields strong enough to completely polarize the electrons' spins. The Pomeranchuk effect provides a qualitative explanation of all the major experimental observations; however, a quantitative theory is not currently available.

5.3. *Spin susceptibility and the effective mass enhancement*

There are several mechanism that could lead to the strong enhancement of the effective mass at low carrier densities (high r_s). Within Fermi liquid theory, the enhancement of g and m is due to spin exchange effects. Extension of the Fermi liquid model to relatively large r_s[59–61] is problematic, the main outcome being that the renormalization of g is large compared to that of m. In the limiting case of high r_s, one may expect a divergence of the g factor that corresponds to the Stoner instability. These predictions are in obvious contradiction to the experimental data. The divergence of the effective mass and spin susceptibility follow also from the Gutzwiller variational approach[81] (see also Ref. 82). Recent theoretical developments include the following. Using a renormalization group analysis for multi-valley 2D systems, it has been found that the spin susceptibility increases dramatically as the density is decreased toward the critical density for the metal–insulator transition, while the g factor remains nearly intact.[32] However, this prediction is made for the diffusive regime, $T\tau/\hbar \ll 1$, while the spin susceptibility enhancement has been observed well into the ballistic regime, $T\tau/\hbar \gg 1$. In the Fermi-liquid-based model of Khodel *et al.*,[83] a flattening at the Fermi energy in the spectrum that leads to a diverging effective mass has been predicted, in qualitative agreement with experiment, but a detailed microscopic theory is needed before conclusions can be drawn. The strong increase of the effective mass has also been obtained (in the absence of disorder) by solving an extended Hubbard model using dynamical mean-field theory.[84,85] This is consistent with experiment, especially taking into account that the relative mass enhancement has been experimentally found to be independent of the level of the disorder.[62]

6. Summary

Although the behavior in the close vicinity of the transition is quantitatively described by the renormalization-group theory of Punnoose and Finkelstein[32] without any fitting parameters, there is currently no microscopic theory that can explain the whole range of observed phenomena. The origin of the large changes in the resistance deep in the metallic phase remains unclear, with suggested explanations ranging from temperature-dependent screening[72–76] to an analog of the Pomeranchuk effect.[65–67] From an empirical perspective, numerous experiments on various strongly correlated 2D electron and hole systems strongly suggest the existence of a metal–insulator transition and a metallic phase in two-dimensions, despite the persistent view on the part of many that such a transition is impossible in 2D.

Acknowledgments

The authors acknowledge the support of the Department of Energy under grant DE-FG02-84ER45153.

References

1. E. Abrahams, P. W. Anderson, D. C. Licciardello and T. V. Ramakrishnan, Scaling theory of localization: Absence of quantum diffusion in two dimensions, *Phys. Rev. Lett.* **42**, 673–676 (1979).
2. A. M. Finkel'shtein, Influence of Coulomb interaction on the properties of disordered metals, *Sov. Phys. JETP* **57**, 97–108 (1983).
3. A. M. Finkel'shtein, Weak localization and Coulomb interaction in disordered-systems, *Z. Phys. B* **56**, 189–196 (1984).
4. C. Castellani, C. Di Castro, P. A. Lee and M. Ma, Interaction-driven metal–insulator transitions in disordered fermion systems, *Phys. Rev. B* **30**, 527–543 (1984).
5. T. N. Zavaritskaya and E. I. Zavaritskaya, Metal–insulator transition in inversion channels of silicon MOS structures, *JETP Lett.* **45**, 609–613 (1987).
6. S. V. Kravchenko, G. V. Kravchenko, J. E. Furneaux, V. M. Pudalov and M. D'Iorio, Possible metal–insulator-transition at $B = 0$ in 2 dimensions, *Phys. Rev. B* **50**, 8039–8042 (1994).

7. S. V. Kravchenko, W. E. Mason, G. E. Bowker, J. E. Furneaux, V. M. Pudalov and M. D'Iorio, Scaling of an anomalous metal–insulator-transition in a 2-dimensional system in silicon at $B = 0$, Phys. Rev. B **51**, 7038–7045 (1995).

8. S. V. Kravchenko, D. Simonian, M. P. Sarachik, W. Mason and J. E. Furneaux, Electric field scaling at a $B = 0$ metal–insulator transition in two dimensions, *Phys. Rev. Lett.* **77**, 4938–4941 (1996).

9. D. Popović, A. B. Fowler and S. Washburn, Metal–insulator transition in two dimensions: Effects of disorder and magnetic field, *Phys. Rev. Lett.* **79**, 1543–1546 (1997).

10. R. N. McFarland, T. M. Kott, L. Sun, K. Eng and B. E. Kane, Temperature-dependent transport in a sixfold degenerate two-dimensional electron system on a H-Si(111) surface, *Phys. Rev. B* **80**, 161310(R) (2009).

11. P. T. Coleridge, R. L. Williams, Y. Feng and P. Zawadzki, Metal–insulator transition at $B = 0$ in p-type SiGe, *Phys. Rev. B* **56**, R12764–R12767 (1997).

12. Y. Hanein, U. Meirav, D. Shahar, C. C. Li, D. C. Tsui and H. Shtrikman, The metalliclike conductivity of a two-dimensional hole system, *Phys. Rev. Lett.* **80**, 1288–1291 (1998).

13. Y. Hanein, D. Shahar, J. Yoon, C. C. Li, D. C. Tsui and H. Shtrikman, Observation of the metal–insulator transition in two-dimensional n-type GaAs, *Phys. Rev. B* **58**, R13338–R13340 (1998).

14. S. J. Papadakis and M. Shayegan, Apparent metallic behavior at $B = 0$ of a two-dimensional electron system in AlAs, *Phys. Rev. B* **57**, R15068–R15071 (1998).

15. T. Okamoto, M. Ooya, K. Hosoya and S. Kawaji, Spin polarization and metallic behavior in a silicon two-dimensional electron system, *Phys. Rev. B* **69**, 041202(R) (2004).

16. K. Lai, W. Pan, D. C. Tsui, S. A. Lyon, M. Muhlberger and F. Schäffler, Two-dimensional metal–insulator transition and in-plane magnetoresistance in a high-mobility strained Si quantum well, *Phys. Rev. B* **72**, 081313(R) (2005).

17. V. T. Dolgopolov, G. V. Kravchenko, A. A. Shashkin and S. V. Kravchenko, Properties of electron insulating phase in Si inversion-layers at low-temperatures, *JETP Lett.* **55**, 733–737 (1992).

18. D. Simonian, S. V. Kravchenko, M. P. Sarachik and V. M. Pudalov, Magnetic field suppression of the conducting phase in two dimensions, *Phys. Rev. Lett.* **79**, 2304–2307 (1997).

19. V. M. Pudalov, G. Brunthaler, A. Prinz and G. Bauer, Instability of the two-dimensional metallic phase to a parallel magnetic field, *JETP Lett.* **65**, 932–937 (1997).

20. M. Y. Simmons, A. R. Hamilton, M. Pepper, E. H. Linfield, P. D. Rose, D. A. Ritchie, A. K. Savchenko and T. G. Griffiths, Metal–insulator transition $B = 0$ in a dilute two dimensional GaAs-AlGaAs hole gas, *Phys. Rev. Lett.* **80**, 1292–1295 (1998).

21. T. Okamoto, K. Hosoya, S. Kawaji and A. Yagi, Spin degree of freedom in a two-dimensional electron liquid, *Phys. Rev. Lett.* **82**, 3875–3878 (1999).

22. S. V. Kravchenko, A. A. Shashkin, D. A. Bloore and T. M. Klapwijk, Shubnikov-de Haas oscillations near the metal–insulator transition in a two-dimensional electron system in silicon, *Solid State Commun.* **116**, 495–499 (2000).

23. A. A. Shashkin, S. V. Kravchenko, V. T. Dolgopolov and T. M. Klapwijk, Indication of the ferromagnetic instability in a dilute two-dimensional electron system, *Phys. Rev. Lett.* **87**, 086801 (2001).

24. S. A. Vitkalov, H. Zheng, K. M. Mertes, M. P. Sarachik and T. M. Klapwijk, Scaling of the magnetoconductivity of silicon MOSFETs: Evidence for a quantum phase transition in two dimensions, *Phys. Rev. Lett.* **87**, 086401 (2001).

25. A. A. Shashkin, S. V. Kravchenko, V. T. Dolgopolov and T. M. Klapwijk, Sharp increase of the effective mass near the critical density in a metallic two-dimensional electron system, *Phys. Rev. B* **66**, 073303 (2002).

26. S. A. Vitkalov, M. P. Sarachik and T. M. Klapwijk, Spin polarization of strongly interacting two-dimensional electrons: The role of disorder, *Phys. Rev. B* **65**, 201106(R) (2002).

27. V. M. Pudalov, M. E. Gershenson, H. Kojima, N. Butch, E. M. Dizhur, G. Brunthaler, A. Prinz and G. Bauer, Low-density spin susceptibility and effective mass of mobile electrons in Si inversion layers, *Phys. Rev. Lett.* **88**, 196404 (2002).

28. J. Zhu, H. L. Stormer, L. N. Pfeiffer, K. W. Baldwin and K. W. West, Spin susceptibility of an ultra-low-density two-dimensional electron system, *Phys. Rev. Lett.* **90**, 056805 (2003).

29. A. A. Shashkin, M. Rahimi, S. Anissimova, S. V. Kravchenko, V. T. Dolgopolov and T. M. Klapwijk, Spin-independent origin of the strongly enhanced effective mass in a dilute 2D electron system, *Phys. Rev. Lett.* **91**, 046403 (2003).

30. S. Anissimova, A. Venkatesan, A. A. Shashkin, M. R. Sakr, S. V. Kravchenko and T. M. Klapwijk, Magnetization of a strongly interacting two-dimensional electron system in perpendicular magnetic fields, *Phys. Rev. Lett.* **96**, 046409 (2006).

31. E. C. Stoner, Ferromagnetism, *Rep. Prog. Phys.* **11**, 43–112 (1946).

32. A. Punnoose and A. M. Finkel'stein, Metal–insulator transition in disordered two-dimensional electron systems, *Science* **310**, 289–291 (2005).

33. V. Senz, U. Dötsch, U. Gennser, T. Ihn, T. Heinzel, K. Ensslin, R. Hartmann and D. Grützmacher, Metal–insulator transition in a 2-dimensional system with an easy spin axis, *Ann. Phys. (Leipzig)* **8**, 237–240 (1999).

34. J. Yoon, C. C. Li, D. Shahar, D. C. Tsui and M. Shayegan, Wigner crystallization and metal–insulator transition of two-dimensional holes in GaAs at $B = 0$, *Phys. Rev. Lett.* **82**, 1744–1747 (1999).

35. A. P. Mills, Jr., A. P. Ramirez, L. N. Pfeiffer and K. W. West, Nonmonotonic temperature-dependent resistance in low density 2D hole gases, *Phys. Rev. Lett.* **83**, 2805–2808 (1999).

36. M. P. Lilly, J. L. Reno, J. A. Simmons, I. B. Spielman, J. P. Eisenstein, L. N. Pfeiffer, K. W. West, E. H. Hwang and S. Das Sarma, Resistivity of dilute 2D electrons in an undoped GaAs heterostructure, *Phys. Rev. Lett.* **90**, 056806 (2003).

37. X. P. A. Gao, G. S. Boebinger, A. P. Mills, Jr., A. P. Ramirez, L. N. Pfeiffer and K. W. West, Spin-polarization-induced tenfold magnetoresistivity of highly metallic two-dimensional holes in a narrow GaAs quantum well, *Phys. Rev. B* **73**, 241315(R) (2006).

38. P. A. Lee and T. V. Ramakrishnan, Disordered electronic systems, *Rev. Mod. Phys.* **57**, 287–337 (1985).

39. S. V. Kravchenko, D. Simonian, M. P. Sarachik, A. D. Kent and V. M. Pudalov, Effect of tilted magnetic field on the anomalous $H = 0$ conducting phase in high-mobility Si MOSFETs, *Phys. Rev. B* **58**, 3553–3556 (1998).

40. V. M. Pudalov, G. Brunthaler, A. Prinz and G. Bauer, Weak anisotropy and disorder dependence of the in-plane magnetoresistance in high-mobility (100) Si-inversion layers, *Phys. Rev. Lett.* **88**, 076401 (2002).

41. S. A. Vitkalov, H. Zheng, K. M. Mertes, M. P. Sarachik and T. M. Klapwijk, Small-angle Shubnikov-de Haas measurements in a 2D electron system: The effect of a strong in-plane magnetic field, *Phys. Rev. Lett.* **85**, 2164–2167 (2000).

42. S. A. Vitkalov, M. P. Sarachik and T. M. Klapwijk, Spin polarization of two-dimensional electrons determined from Shubnikov-de Haas oscillations as a function of angle, *Phys. Rev. B* **64**, 073101 (2001).

43. E. Tutuc, E. P. De Poortere, S. J. Papadakis and M. Shayegan, In-plane magnetic field-induced spin polarization and transition to insulating behavior in two-dimensional hole systems, *Phys. Rev. Lett.* **86**, 2858–2861 (2001).

44. S. J. Papadakis, E. P. De Poortere, M. Shayegan and R. Winkler, Anisotropic magnetoresistance of two-dimensional holes in GaAs, *Phys. Rev. Lett.* **84**, 5592–5595 (2000).

45. K. M. Mertes, H. Zheng, S. A. Vitkalov, M. P. Sarachik and T. M. Klapwijk, Temperature dependence of the resistivity of a dilute two-dimensional electron system in high parallel magnetic field, *Phys. Rev. B* **63**, 041101(R) (2001).

46. A. A. Shashkin, S. V. Kravchenko and T. M. Klapwijk, Metal–insulator transition in 2D: equivalence of two approaches for determining the critical point, *Phys. Rev. Lett.* **87**, 266402 (2001).

47. X. P. A. Gao, A. P. Mills, Jr., A. P. Ramirez, L. N. Pfeiffer and K. W. West, Weak-localization-like temperature-dependent conductivity of a dilute two-dimensional hole gas in a parallel magnetic field, *Phys. Rev. Lett.* **88**, 166803 (2002).

48. Y. Tsui, S. A. Vitkalov, M. P. Sarachik and T. M. Klapwijk, Conductivity of silicon inversion layers: Comparison with and without an in-plane magnetic field, *Phys. Rev. B* **71**, 13308 (2005).

49. L. D. Landau, The theory of a Fermi liquid, *Sov. Phys. - JETP* **3**, 920–925 (1957).

50. F. F. Fang and P. J. Stiles, Effects of a tilted magnetic field on a two-dimensional electron gas, *Phys. Rev.* **174**, 823–828 (1968).

51. J. L. Smith and P. J. Stiles, Electron–electron interactions continuously variable in the range $2.1 > r_s > 0.9$, *Phys. Rev. Lett.* **29**, 102–104 (1972).

52. A. M. Finkel'shtein, Spin fluctuations in disordered systems near the metal–insulator transition, *JETP Lett.* **40**, 796–799 (1984).

53. O. Prus, Y. Yaish, M. Reznikov, U. Sivan and V. M. Pudalov, Thermodynamic spin magnetization of strongly correlated two-dimensional electrons in a silicon inversion layer, *Phys. Rev. B* **67**, 205407 (2003).

54. A. A. Shashkin, S. Anissimova, M. R. Sakr, S. V. Kravchenko, V. T. Dolgo-polov and T. M. Klapwijk, Pauli spin susceptibility of a strongly correlated 2D electron liquid, *Phys. Rev. Lett.* **96**, 036403 (2006).

55. K. Vakili, Y. P. Shkolnikov, E. Tutuc, E. P. De Poortere and M. Shayegan, Spin susceptibility of two-dimensional electrons in narrow AlAs quantum wells, *Phys. Rev. Lett.* **92**, 226401 (2004).

56. T. M. Lu, L. Sun, D. C. Tsui, S. Lyon, W. Pan, M. Muhlberger, F. Schäffler, J. Liu and Y. H. Xie, In-plane field magnetoresistivity of Si two-dimensional electron gas in Si/SiGe quantum wells at 20 mK, *Phys. Rev. B* **78**, 233309 (2008).

57. A. A. Shashkin, S. V. Kravchenko and T. M. Klapwijk, Metal–insulator transition in 2D: equivalence of two approaches for determining the critical point, *Phys. Rev. Lett.* **87**, 266402 (2001).

58. T. Ando, A. B. Fowler and F. Stern, Electronic-properties of two-dimensional systems, *Rev. Mod. Phys.* **54**, 437–672 (1982).

59. N. Iwamoto, Static local-field corrections of two-dimensional electron liquids, *Phys. Rev. B* **43**, 2174–2182 (1991).

60. Y. Kwon, D. M. Ceperley and R. M. Martin, Quantum Monte Carlo calculation of the Fermi-liquid parameters in the two-dimensional electron gas, *Phys. Rev. B* **50**, 1684–1694 (1994).

61. G.-H. Chen and M. E. Raikh, Exchange-induced enhancement of spin-orbit coupling in two-dimensional electronic systems, *Phys. Rev. B* **60**, 4826–4833 (1999).

62. A. A. Shashkin, A. A. Kapustin, E. V. Deviatov, V. T. Dolgopolov and Z. D. Kvon, Strongly enhanced effective mass in dilute two-dimensional electron systems: System-independent origin, *Phys. Rev. B* **76**, 241302(R) (2007).

63. P. Phillips, Y. Wan, I. Martin, S. Knysh and D. Dalidovich, Superconductivity in a two-dimensional electron gas, *Nature (London)* **395**, 253–257 (1998).

64. S. Chakravarty, S. Kivelson, C. Nayak and K. Voelker, Wigner glass, spin-liquids, and the metal–insulator transition, *Phil. Mag. B* **79**, 859–868 (1999).

65. B. Spivak, Phase separation in the two-dimensional electron liquid in MOS-FETs, *Phys. Rev. B* **67**, 125205 (2003).

66. B. Spivak and S. A. Kivelson, Intermediate phases of the two-dimensional electron fluid between the Fermi liquid and the Wigner crystal, *Phys. Rev. B* **70**, 155114 (2004).

67. B. Spivak and S. A. Kivelson, Transport in two-dimensional electronic micro-emulsions, *Ann. Phys.* **321**, 2071–2115 (2006).

68. Y. Meir, Percolation-type description of the metal–insulator transition in two dimensions, *Phys. Rev. Lett.* **83**, 3506–3509 (1999).

69. S. Chakravarty, L. Yin and E. Abrahams, Interactions and scaling in a disordered two-dimensional metal, *Phys. Rev. B* **58**, R559–R562 (1998).

70. F. Stern, Calculated temperature dependence of mobility in silicon inversion layers, *Phys. Rev. Lett.* **44**, 1469–1472 (1980).

71. A. Gold and V. T. Dolgopolov, Temperature dependence of the conductivity for the two-dimensional electron gas: Analytical results for low temperatures, *Phys. Rev. B* **33**, 1076–1084 (1986).

72. S. Das Sarma and E. H. Hwang, Charged impurity scattering limited low temperature resistivity of low density silicon inversion layers, *Phys. Rev. Lett.* **83**, 164–167 (1999).
73. V. T. Dolgopolov and A. Gold, Magnetoresistance of a two-dimensional electron gas in a parallel magnetic field, *JETP Lett.* **71**, 27–30 (2000).
74. S. Das Sarma and E. H. Hwang, Parallel magnetic field induced giant magnetoresistance in low density quasi-two-dimensional layers, *Phys. Rev. B* **61**, R7838–R7841 (2000).
75. S. Das Sarma and E. H. Hwang, Low density finite temperature apparent insulating phase in 2D semiconductor systems, *Phys. Rev. B* **68**, 195315 (2003).
76. S. Das Sarma and E. H. Hwang, Metallicity and its low temperature behavior in dilute 2D carrier systems, *Phys. Rev. B* **69**, 195305 (2004).
77. G. Zala, B. N. Narozhny and I. L. Aleiner, Interaction corrections at intermediate temperatures: Longitudinal conductivity and kinetic equation, *Phys. Rev. B* **64**, 214204 (2001).
78. A. Punnoose and A. M. Finkelstein, Dilute electron gas near the metal–insulator transition: Role of valleys in silicon inversion layers, *Phys. Rev. Lett.* **88**, 016802 (2001).
79. S. Anissimova, S. V. Kravchenko, A. Punnoose, A. M. Finkel'stein and T. M. Klapwijk, Flow diagram of the metal–insulator transition in two dimensions, *Nature Phys.* **3**, 707–710 (2007).
80. A. Mokashi and S. V. Kravchenko, unpublished (2009).
81. W. F. Brinkman and T. M. Rice, Application of Gutzwiller's variational method to the metal–insulator transition, *Phys. Rev. B* **2**, 4302–4304 (1970).
82. V. T. Dolgopolov, On effective electron mass of silicon field structures at low electron densities, *JETP Lett.* **76**, 377–379 (2002).
83. V. A. Khodel, J. W. Clark and M. V. Zverev, Thermodynamic properties of Fermi systems with flat single-particle spectra, *Europhys. Lett.* **72**, 256–262 (2005).
84. S. Pankov and V. Dobrosavljević, Self-doping instability of the Wigner–Mott insulator, *Phys. Rev. B* **77**, 085104 (2008).
85. A. Camjayi, K. Haule, V. Dobrosavljević and G. Kotliar, Coulomb correlations and the Wigner-Mott transition, *Nature Phys.* **4**, 932–935 (2008).

Chapter 18

DISORDERED ELECTRON LIQUID
WITH INTERACTIONS

Alexander M. Finkel'stein

Department of Condensed Matter Physics, Weizmann Institute of Science,
Rehovot 76100, Israel
Department of Physics and Astronomy, Texas A&M University,
College Station, TX 77843-4242, USA

The metal–insulator transition (MIT) observed in a two-dimensional dilute electron liquid raises the question about the applicability of the scaling theory of disordered electrons, the approach pioneered by Phil Anderson and his collaborators,[8] for the description of this transition. In this context, we review here the scaling theory of disordered electrons with electron–electron interactions. We start with the disordered Fermi liquid, and show how to adjust the microscopic Fermi-liquid theory to the presence of disorder. Then we describe the non-linear sigma model (NLSM) with interactions. This model has a direct relation with the disordered Fermi liquid, but can be more generally applicable, since it is a minimal model for disordered interacting electrons. The discussion is mostly about the general structure of the theory emphasizing the connection of the scaling parameters entering the NLSM with conservation laws. Next, we show that the MIT, as described by the NLSM with interactions, is a quantum phase transition and identify the parameters needed for the description of the kinetics and thermodynamics of the interacting liquid in the critical region of the transition. Finally, we discuss the MIT observed in Si-MOSFETs. We consider it as an example of the Anderson transition in the presence of the electron interactions. We demonstrate that the two parameter RG equations, which treat disorder in the one-loop approximation but incorporate the full dependence on the interaction amplitudes, describe accurately the experimental data in Si-MOSFETs including the observed non-monotonic behavior of the resistance and its strong drop at low temperatures. The fact that this drop can be reproduced theoretically, together with the argument that Anderson localization should occur at strong disorder, justified the existence of the MIT within the scaling theory.

1. Disordered Fermi-Liquid: Dk^2, ω, $T < 1/\tau_{el}$

The original Fermi-liquid theory has been formulated in terms of quasiparticles labeled with momenta $\mathbf{p}$. The most distinctive feature of the Fermi-liquid is the jump in the occupation number $n(\mathbf{p})$ at the Fermi-surface. Since in the presence of disorder the Fermi-surface is smeared, for some people this means the end of applicability of Fermi-liquid theory. This is, however, not completely correct. Indeed, the description in terms of plane waves is not working well for low-lying excitations with energies less than the rate of collisions with static impurities. Still, some elements of Fermi-liquid theory hold as far as rescattering of electron–hole pairs is considered. The Fermi-liquid description stops working only when the production of multiple electron–hole pairs becomes important [N.1].[a] Elastic impurity scattering by itself does not generate electron–hole pairs and therefore, some elements of the Fermi-liquid description should be preserved even in the presence of disorder.

In this section, we show how to adjust the Fermi-liquid description to disordered electron systems.[1–3] While conventional Fermi-liquid theory has been constructed starting from single-particle excitations, in the case of a disordered Fermi-liquid, the focus shifts towards diffusing electron–hole pairs. In Landau's original microscopic theory of the clean Fermi-liquid the term $v_F\mathbf{nk}/(\omega - v_F\mathbf{nk})$ is used as the propagator of an electron–hole quasiparticle pair, see Chapter 2, Sec. 17 in Ref. 4. This expression describes propagation of the pair along the direction $\mathbf{n}$, when the momentum difference of the two quasiparticles is $\mathbf{k}$, and the frequency difference is ω. The combination $v_F\mathbf{nk}$ originates from the energy difference of the constituents of the pair, $\delta\epsilon_k(\mathbf{p}) = \epsilon(\mathbf{p}+\mathbf{k}) - \epsilon(\mathbf{p}) \approx v_F\mathbf{nk}$. The two quasiparticle poles sitting close-by make the discussed term singular. This in turn makes the two-particle vertex function $\Gamma(\omega, \mathbf{k})$ singular since it describes, among other processes, multiple rescattering of electron–hole pairs. The propagator $v_F\mathbf{nk}/(\omega - v_F\mathbf{nk})$ may be rewritten as the sum of a static and a dynamic part: $[-1 + \omega/(\omega - v_F\mathbf{nk})]$. In fact, it is more convenient to keep explicitly only the *dynamic* part of this propagator, $\omega/(\omega - v_F\mathbf{nk})$, and to delegate the static part (i.e., -1) to the amplitude of the electron–electron (e–e) interaction. This amplitude is denoted as Γ^k. Index k in Γ^k means that in the singular amplitude $\Gamma(\mathbf{k}, \omega)$ one first takes the limit $\omega = 0$ and only afterwards the limit $k \to 0$, i.e., $\Gamma^k = \Gamma(\mathbf{k} \to 0, \omega = 0)$. The choice to work with the *static* amplitude Γ^k is motivated by the following reasoning. Generally speaking, $\Gamma(\mathbf{k}, \omega)$ includes:

---

[a]See the list of Notes which follows the main text.

(i) a part irreducible with respect to particle–hole pair propagators (the diagram for such an amplitude cannot be separated into disconnected blocks by cutting two single-particle Green's functions only), along with contributions from incoherent background, and (ii) the contributions containing rescattering of quasi-particle pairs that has been already mentioned. The contributions from the irreducible part and from incoherent background are determined by short scales. Therefore, they are robust and, apart from small corrections, not sensitive to modifications of the electron spectrum near the Fermi-energy ϵ_F. On the contrary, the terms describing rescattering of quasi-particle pairs are fragile, and they require certain care. Remarkably, the amplitude Γ^k is also insensitive to a modification of the low-energy part of the energy spectrum unless the density of states changes significantly. Indeed, in the considered order of limits the combination $\delta\epsilon/(\omega - \delta c)$ is equal to -1 for any energy spectrum of electrons. These arguments led us to conclude[1,3] that the amplitude Γ^k is not influenced by not too strong disorder, $1/\tau_{el} \lesssim \epsilon_F$. The robustness of the static amplitude Γ^k makes it particularly convenient for the purpose of a microscopic analysis in the presence of disorder.

It is almost evident from the discussion above that disorder reveals itself most clearly in dynamics. Diagrammatically, the dynamic part of the particle–hole propagator can be obtained from a product of two Green's functions where one is retarded (R), while the other one is advanced (A). We will refer to such a product as RA-section. After integration over the energy variable $\xi = p^2/2m^* - \mu$, and summation over the fermionic frequency ϵ_n, the RA-section generates just the dynamic part of the electron–hole propagator, $\omega_n/(\omega_n + iv_F\mathbf{nk})$. [From now on, we prefer to use Matsubara frequencies for which $\omega/(\omega - v_F\mathbf{nk}) \Longrightarrow \omega_n/(\omega_n + iv_F\mathbf{nk})$]. In the presence of disorder, the dynamic part of the propagator given by an RA-section changes its functional form. When the effective collisions of electrons with impurities are frequent enough (in the sense of inequalities given in the section head) multiple impurity scattering leads to the diffusive propagation of the quasi-particles for times $t \gg \tau_{el}$; here τ_{el} is the elastic mean free time for scattering from static impurities. Under these circumstances, the propagator of an electron–hole pair changes in such a way that its denominator acquires the diffusive form:

$$\frac{\omega_n}{\omega_n + iv_F\mathbf{nk}} \Longrightarrow \frac{\omega_n}{\omega_n + Dk^2}. \tag{1.1}$$

Here, $D = v_F^2\tau_{el}/d$ is the diffusion coefficient for the spatial dimension d [N.2]. With this result at hand, let us consider the two-particle amplitude $\Gamma(\mathbf{k},\omega)$. The amplitude $\Gamma(\mathbf{k},\omega)$ can be represented as a series in which Γ^k

and RA-sections alternate with each other [N.3]. In a sense, in disordered systems, the process of multiple rescattering is even simpler than in clean ones. The point is that in the clean Fermi-liquid, because of the angular dependence contained in $\mathbf{nk}$, the angular harmonics of the interaction amplitudes, denoted as Γ_l^k, come into play. On the other hand, for the slow propagation of an electron–hole pair in the presence of disorder only the zeroth harmonic, $l = 0$, remains singular. Consequently, only the zeroth harmonic of the interaction amplitude, $\Gamma_{l=0}^k$, is relevant for the processes of rescattering of diffusing electron–hole pairs. As a result, the calculation of $\Gamma(\mathbf{k}, \omega)$ reduces to a geometric series.

The two-particle amplitude $\Gamma_{l=0}^k$ can be split into parts which can be classified according their spin structure:

$$\nu a^2 \Gamma_{l=0}^k \, {}_{\alpha_3\,\alpha_4}^{\alpha_1\alpha_2} = \widetilde{\Gamma}_1 \delta_{\alpha_1,\alpha_3}\delta_{\alpha_2,\alpha_4} - \Gamma_2 \delta_{\alpha_1,\alpha_2}\delta_{\alpha_3,\alpha_4}$$

$$= \frac{1}{2}[(2\widetilde{\Gamma}_1 - \Gamma_2)\delta_{\alpha_1,\alpha_3}\delta_{\alpha_2,\alpha_4} - \Gamma_2 \vec{\sigma}_{\alpha_1,\alpha_3} \vec{\sigma}_{\alpha_2,\alpha_4}]. \quad (1.2)$$

Here, ν is the single-particle density of states per one spin component at energy ϵ_F, and the factor a describes the weight (residue) of the quasi-particle part in Green's function $\mathcal{G}(i\epsilon, \mathbf{p})$ [N.4]; $\Gamma_{1,2}$ are dimensionless. The minus sign in the amplitude Γ_2 is due to the anti-commutation of the fermionic operators. The two-particle propagators can be classified in terms of the total spin of the particle–hole pairs. The combination $\widetilde{\Gamma}_\rho = 2\widetilde{\Gamma}_1 - \Gamma_2$ operates inside the singlet channel, $S = 0$, and controls propagation of the particle-number density $\rho(\mathbf{k}, \omega)$, while $\Gamma_\sigma = -\Gamma_2$ controls the spin density, i.e., the triplet channel, $S = 1$ [N.5]. To obtain the amplitude $\Gamma(\mathbf{k}, \omega)$, one has to sum, depending on the spin structure, a ladder of either $\widetilde{\Gamma}_\rho$ or Γ_σ with RA-sections in between; see Fig. 18.1. The resulting amplitudes $\widetilde{\Gamma}_\rho(\mathbf{k}, \omega)$ and $\Gamma_\sigma(\mathbf{k}, \omega)$ acquire the form:

$$\Gamma_\alpha(\mathbf{k}, \omega) = \Gamma_\alpha \frac{Dk^2 + \omega_n}{Dk^2 + (1 - \Gamma_\alpha)\omega_n}, \qquad \alpha = \rho, \sigma. \quad (1.3)$$

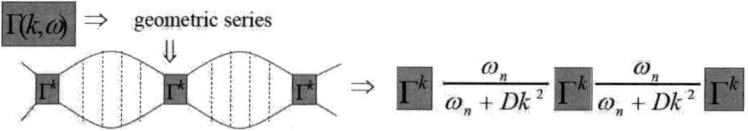

Fig. 18.1.   Disordered Fermi-liquid: geometric series leading to Eq. (1.3). Dashed lines describe impurity scattering.

Note the shift of the position of the diffusion poles in $\Gamma_{\rho,\sigma}(k,\omega)$ as a result of summation of the geometric series. Eventually, this shift is the origin of the renormalization of the diffusion coefficients in the disordered electron liquid.

Now that we know how disorder affects the amplitudes of the e–e interactions, we may study correlation functions. As we shall see, the conservation of particle-number (i.e., charge) and spin constrains the possible form of the corresponding correlation functions. We first consider the polarization operator $\Pi(k,\omega_n)$ in the presence of disorder. We discuss here a true electron liquid, i.e., a quantum liquid with *charged* current carriers. This is the reason why we are interested in $\Pi(k,\omega_n)$, the density–density correlation function irreducible with respect to the Coulomb interaction. To obtain this irreducible part, one has to exclude from $\widetilde{\Gamma}_1$ all terms that can be disconnected by cutting a single line of the Coulomb interaction. As a result of this operation, $\widetilde{\Gamma}_1$ transforms into Γ_1 and correspondingly, $\widetilde{\Gamma}_\rho$ transforms into $\Gamma_\rho = 2\Gamma_1 - \Gamma_2$. (The part of $\widetilde{\Gamma}_\rho$ that can be disconnected by cutting a single line will be denoted as Γ_ρ^0, while the amplitude Γ_ρ incorporates the remaining irreducible part, so that $\widetilde{\Gamma}_\rho = \Gamma_\rho^0 + \Gamma_\rho$.) As we shall see, the dimensionless parameters, Γ_ρ and Γ_σ, determine the Fermi-liquid renormalizations of the disordered charged liquid. The separation of the polarization operator $\Pi(k,\omega_n)$ into static and dynamic parts is performed in the same way as for $\Gamma(k,\omega_n)$: the static part does not contain RA-section, while all the rest goes to the dynamic part. Consequently, the dynamic part of the polarization operator contains two "triangle" vertices γ^ρ separated by a ladder of the RA-sections, see Fig. 18.2. In other words, both the left and right vertices γ^ρ are irreducible with respect to RA-sections (i.e., each of them extends from an external vertex to the first RA-section) [N.6]. Collecting the static and dynamic parts of the polarization operator, one gets

$$\Pi(k,\omega_n) = \Pi_{st} - 2\nu\left(\gamma^\rho\right)^2 \left[\frac{\omega_n}{Dk^2 + (1 - \Gamma_\rho)\omega_n}\right] \qquad (1.4a)$$

$$\longrightarrow \Pi_{st}\frac{Dk^2}{Dk^2 + (1 - \Gamma_\rho)\omega_n}. \qquad (1.4b)$$

The transition between the two lines will be commented upon below.

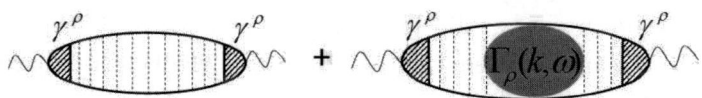

Fig. 18.2. Disordered Fermi-liquid: dynamic part of the polarization operator $\Pi(k,\omega_n)$.

In Eq. (1.4b), we have arrived at the canonical form for a correlation function of the density of any conserved quantity. This form recovers not only the static limit when $\omega_n = 0$ but, most remarkably, it vanishes in the limit $k \to 0$ when $\omega_n \neq 0$. Let us explain why such a vanishing necessarily occurs for any *conserved* quantity. The general form of a retarded correlation function is

$$\chi(\mathbf{k}, \omega) = i \int_0^\infty dt \ e^{i\omega t} \langle [x(t), x(0)] \rangle_\mathbf{k} \ . \tag{1.5}$$

In the limit $k \to 0$ the densities $x(t)$ and $x(0)$ transform into the quantities, $X(t)$ and $X(0)$, that are integrated over the space. In the case when $X(t)$ is conserved in time, it obviously commutes with $X(0)$ at any moment. Consequently, $\chi(k \to 0, \omega)$ should vanish at any frequency.

It remains to show how the expression (1.4b) follows from Eq. (1.4a). A specific cancelation between the static and dynamic parts of $\Pi(k, \omega_n)$ leading to the desired form of Eq. (1.4b) takes place only if the following relation is fulfilled:

$$\Pi_{st} = 2\nu \left(\gamma^\rho\right)^2 \frac{1}{(1 - \Gamma_\rho)} \ . \tag{1.6}$$

As is well known, the static part of the polarization operator, $\Pi_{st} = \Pi(k \to 0, \omega_n = 0)$, reduces to the thermodynamic quantity $\partial n / \partial \mu$, which is related to the compressibility and is also responsible for linear screening in the electron gas. Under the approximation of a constant density of states [N.7], $\partial n / \partial \mu$ is not sensitive to the disorder, if $1/\tau_{el} \ll \epsilon_F$. The point is that, generally speaking, μ can be measured with respect to an arbitrary energy, i.e., it can be shifted by an arbitrary value. For some quantity to be sensitive to a variation of μ, the chemical potential should be tied to a certain physical energy level which can serve as a reference point. In the discussed problem, the only special energy-level is the bottom of the conduction band. It is clear, however, that the information about moderate disorder cannot extend from ϵ_F up to the bottom of the band. Therefore, $\partial n / \partial \mu$ is not changed by disorder, unless it is very strong. Next, since the vertex γ^ρ is also connected with the derivative of the Green function with respect to the chemical potential, $\partial \mathcal{G} / \partial \mu$ [N.8], the arguments concerning insensitivity of $\partial n / \partial \mu$ to disorder remain valid for this quantity as well. Thus, we may use for $\Pi_{st} = \partial n / \partial \mu$ and γ^ρ their values known from the Fermi-liquid theory in the clean limit.

The Fermi-liquid theory connects $\partial n / \partial \mu$ with the Fermi-liquid parameter F_0^ρ as follows: $\partial n / \partial \mu = 2\nu / (1 + F_0^\rho)$, see Chapter 2, Sec. 2 in Ref. 4. Then, the relation connecting F_0^ρ with Γ_ρ yields $\Pi_{st} = 2\nu / (1 + F_0^\rho) = 2\nu (1 -$

Γ_ρ). Furthermore, it is known from identities for derivatives of the Green functions (see Chapter 2, Sec. 19 in Ref. 4 and Refs. 5 and 6) that $\gamma^\rho = (1 - \Gamma_\rho)$. Thus, the necessary relation holds with, one may say, excessive strength: $\Pi_{st}/2\nu = \gamma^\rho = (1 - \Gamma_\rho)$.

One may rewrite the expression given in Eq. (1.4b) in the more conventional form corresponding to diffusion (see also [N.2]):

$$\Pi(k, \omega_n) = \Pi_{st} \frac{D_\rho k^2}{D_\rho k^2 + \omega_n} \qquad D_\rho = \frac{D}{1 - \Gamma_\rho}. \tag{1.7}$$

Here, D_ρ is the diffusion coefficient of the particle-number density ρ. The next step is to relate $\Pi(k, \omega_n)$ through the continuity equation, $\partial\rho/\partial t + div\mathbf{j} = 0$, to the current–current correlation function. Then, with the help of the Kubo formula, one can obtain the Fermi-liquid expression for the electric conductivity (e is electron charge):

$$\frac{\sigma_{\text{charge}}}{e^2} = \lim_{k \to 0} \frac{\omega_n}{k^2} \Pi(k, \omega_n) =$$

$$= \frac{\partial n}{\partial \mu} D_\rho = 2\nu D. \tag{1.8}$$

The above equation is nothing else but the Einstein relation for the electric conductivity σ; $\sigma \equiv \sigma_{\text{charge}}$. It is worth emphasizing that the product $(\partial n/\partial \mu)D_\rho$ is equal to σ/e^2 rather than $(\partial n/\partial \mu)D$. This point is very important in view of contemporary experiment in heterostructures hosting two-dimensional (2d) electron gas. In these systems the electron gas is often studied under conditions when $\partial n/\partial \mu$ becomes negative, i.e., $1/(1 + F_0^\rho) = (1 - \Gamma_\rho) < 0$. However, as we have observed, in σ the two negative renormalizations exactly cancel each other, so that conductivity is unquestionably positive [N.9].

It is worth mentioning that the arguments presented above about the insensitivity of $\partial n/\partial \mu$ as well as γ^ρ to disorder are not restricted to the Fermi-liquid. Under the approximation of a constant density of states, $\partial n/\partial \mu$ and γ^ρ are not changed even if one goes beyond the framework of the disordered Fermi-liquid theory; we will come back to this point later.

The scheme outlined above can be straightforwardly applied for the analysis of the spin-density correlation function.[2,7] We will now rely on the arguments that lead us to the conclusion that the static amplitude Γ^k is not affected by moderately strong disorder. Actually, these arguments carry over to any static Fermi-liquid parameter. In the discussed case, the external vertices contain a spin operator $\sigma^x/2$ that corresponds to a probing magnetic field directed along x-axis. These vertices are renormalized by the e–e

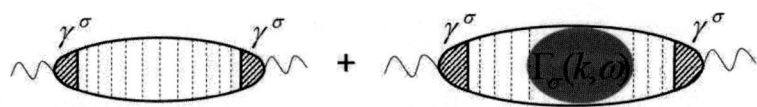

Fig. 18.3. Disordered Fermi-liquid: dynamic part of the spin-density correlation function $\chi_s(k, \omega_n)$.

interactions, and the corresponding renormalization factor is denoted below as γ^σ, see Fig. 18.3. In spite of this modification, all formulas are similar to those obtained in the case of $\Pi(k, \omega_n)$. The only needed change is to substitute in the above expressions Γ_ρ by Γ_σ. The spin susceptibility χ_s determines the static limit of the spin-density correlation function $\chi_s^{xx}(k, \omega_n)$, just like $\partial n/\partial \mu$ determines the static limit of the polarization operator. The spin susceptibility χ_s is modified by the Stoner factor equal to $(1 - \Gamma_\sigma)$. As a result, $\chi_s = \chi_s^{xx}(k \to 0, \omega_n = 0) = \chi_s^0(1 - \Gamma_\sigma) = (g\mu_B/2)^2 2\nu(1 - \Gamma_\sigma)$. The vertex γ^σ is equal to the same renormalization factor, $\gamma^\sigma = (1 - \Gamma_\sigma)$. As a result, the sum of the static and dynamic parts acquires the structure already familiar from the calculation of $\Pi(k, \omega_n)$:

$$\chi_s^{xx}(k, \omega) = \chi_s^0(1 - \Gamma_\sigma) \frac{Dk^2}{Dk^2 + (1 - \Gamma_\sigma)\,\omega_n}$$

$$= \chi_s \frac{D_\sigma k^2}{\omega_n + D_\sigma k^2}\,, \tag{1.9}$$

where $D_\sigma = D/(1 - \Gamma_\sigma)$. Note that Γ_σ is connected with the standard Fermi-liquid parameter F_0^σ as follows: $\Gamma_\sigma = F_0^\sigma/(1 + F_0^\sigma)$. Usually F_0^σ is negative. Then, $(1 - \Gamma_\sigma)$ describes the Stoner enhancement of the spin susceptibility due to the e–e interaction, as well as the suppression of the spin-diffusion coefficient $D_\sigma = D/(1 - \Gamma_\sigma)$.

Since we discuss the case when spin is conserved, we may now derive the Einstein relation for the spin-density current by following the route outlined previously for the electric conductivity, see Eq. (1.8):

$$\frac{\sigma_{\text{spin}}}{(\mu_B/2)^2} = \frac{1}{(g\mu_B/2)^2} \lim_{k \to 0} \frac{\omega_n}{k^2} \chi_s^{xx}(k, \omega_n)$$

$$= 2\nu(1 - \Gamma_\sigma)D_\sigma = 2\nu D. \tag{1.10}$$

Taken together, Eqs. (1.8) and (1.10) reflect the fact that both the charge and the spin are carried by the same particles.

Conclusion: The theory of the disordered Fermi-liquid focuses on diffusing electron–hole pairs. In the diffusion regime, i.e., for temperatures (frequencies) less than the elastic scattering rate, $T \lesssim 1/\tau_{el}$, diffusion modes and not quasi-particles are the low lying propagating modes. The conservation of particle-number (i.e., charge) and the conservation of spin constrains the possible form of the corresponding correlation functions. Besides ν and D, the theory contains two dimensionless parameters, Γ_ρ and Γ_σ, which describe Fermi-liquid renormalizations in the charge- and spin-density channels, respectively.

2. Beyond Fermi-Liquid Theory: Non-Linear Sigma Model and Renormalized Fermi-Liquid Theory

Let us explain why the theory of the disordered Fermi-liquid discussed above is incomplete. Obviously, the expression for the diffusion coefficient, $D = v_F^2 \tau_{el}/d$, has to be modified by the interference (weak localization) corrections, which in $d = 2$ are logarithmic.[8,9] This, by itself, does not affect the described above structure of the Fermi-liquid, and could easily be repaired. However, there is a number of other effects, which demand certain care. *Up to now*, averaging over disorder both in the polarization operator and the spin-density correlation function has been performed in a very particular fashion. Namely, in the ladders given in Figs. 18.2 and 18.3, the interaction amplitudes and disorder-averaged propagators appear in separate blocks. In fact, matrix elements determining amplitudes of the *e–e* interaction are seriously modified by disorder, especially for states that are close in energy. Two examples showing how it happens *after averaging* over disorder are presented in Fig. 18.4. One can see from these examples that ladder-diagram propagators describing diffusion of electron–hole pairs play a special role in modifying (renormalizing) the interaction amplitudes. Such propagators, see Fig. 18.5, contain a diffusion pole and are, therefore, called diffusion modes or just "diffusons". Technically, diffusion modes participating in the processes similar to those shown in Fig. 18.4 have to be integrated over their momentum **q** within the interval determined by $1/\tau_{el} > Dq^2 \gtrsim T$.

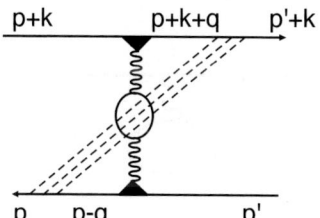
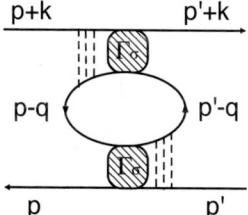

Fig. 18.4. Examples of the *e–e* interaction amplitudes modified by disorder.

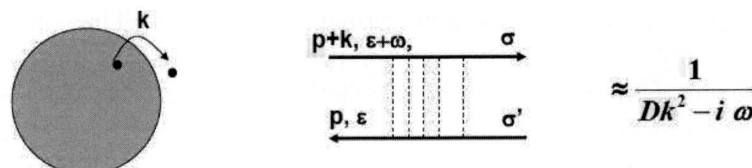

Fig. 18.5. Diffuson: disorder-averaged propagator of an electron-hole pair. These propagators capture the diffusive evolution of the quasiparticles at large times and length scales.

The scattering rate $1/\tau_{el}$ acts as a high-energy cutoff because only states with energy/frequency less than $1/\tau_{el}$ are relevant in the diffusive regime. On the other hand, temperature always enters as a low-energy cutoff in the effects related to the e–e interactions, because it determines smearing of the *energy* distribution of electrons [N.10]. (We emphasize energy, because momentum-smearing of single-particle states is already irrelevant when we use description in terms of diffusons.) As a result of the outlined integrations over the momenta [N.11], amplitudes of the e–e interaction acquire corrections that are non-analytic in temperature.[1,2,7]

As is well known,[10,11] the electric conductivity and, correspondingly, the diffusion constant D also acquire corrections (that are non-analytic in temperature) due to the combined action of the e–e interaction and disorder in the diffusive regime [N.12]. Two diagrams illustrating the origin of the effect are shown in Fig. 18.6.

In addition, there are corrections to conductivity due to the interference processes determined by "cooperon" modes. Diagrammatically, cooperons are described by a disorder-averaged particle–particle propagator with small total momentum of the scattering particles, see Fig. 18.7. These propagators also contain a diffusion pole [N.13]. In $d = 2$, *all* corrections both to the electric conductivity and interaction amplitudes are logarithmically divergent in

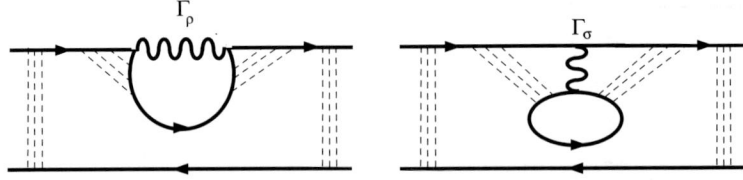

Fig. 18.6. Diagrams illustrating the origin of corrections to the diffusion constant D due to combined action of the e–e interactions and disorder.

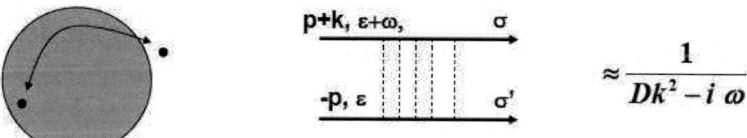

Fig. 18.7. Cooperon: disorder-averaged propagator of a particle–particle pair with small total momentum. The cooperons capture the effects of quantum interference which lead to the weak-localization corrections.

temperature, i.e., $\propto \ln 1/T\tau_{el}$ [N.14]. In higher dimensions, the problem becomes logarithmic near the metal–insulator transition, and it still contains several running parameters. The corrections divergent in temperature signal the breakdown of perturbation theory, and the need for a resummation of the divergent terms.

We reached an important point: Diffusing electrons dwell long in each other's vicinity, becoming more correlated at low enough energies. As a result, the e–e scattering amplitudes Γ_ρ and Γ_σ characterizing the interaction of diffusion modes acquire corrections, which are the more significant the stronger disorder is. Conversely, resistivity — which is a measure of the effective strength of disorder — in its turn also gets corrections which depend on the value of the interaction amplitudes. We see that one needs a scheme that can account for both effects in a self-consistent fashion. Such a scheme is the renormalization group (RG) analysis of the problem. The RG-theory applied to the dirty Fermi-liquid system is able to capture scale dependences originating from the interplay of disorder and interactions to all orders in the interaction amplitudes, making it a highly effective analytical tool for understanding the physics of the metal–insulator transition in disordered electron systems. Pedagogical reviews of the RG-theory can be found in Refs. 3, 12–14. For more recent advances, see Refs. 15 and 16.

The RG-analysis of the disordered electron liquid is best described by the matrix non-linear sigma model (NLSM).[17–20] In matrix terms, the disorder-averaged N-replica partition function of the interacting electrons reads as follows[1–3]:

$$\langle Z_N \rangle = \int dQ \; e^{-S[Q]} , \tag{2.1}$$

$$S[Q] = \frac{\pi}{8} \int d^d r \; \nu \mathrm{Tr} \left[D(\nabla \hat{Q})^2 - 4z(\hat{\epsilon}\hat{Q}) \right.$$

$$- \frac{\pi}{16} \int d^d r \; \nu \{ \hat{Q}(\hat{\Gamma}_\rho^0 + \hat{\Gamma}_\rho)\hat{Q} + \hat{Q}\hat{\Gamma}_\sigma\hat{Q} + \hat{Q}\hat{\Gamma}_c\hat{Q} \}. \tag{2.2}$$

Here, the functional integration has to be performed over an auxiliary matrix field $\hat{Q}$ within the manifold limited by the constraints: $\hat{Q}^2 = 1$, $\hat{Q} = \hat{Q}^\dagger$, and Tr $\hat{Q} = 0$. These constraints make the problem non-linear as well as very non-trivial. The components of $\hat{Q}$ are defined as $Q_{n_1,n_2}^{ij,\alpha\beta}$, where n_1, n_2 are the Matsubara fermionic energy indices with $\epsilon_n = (2n+1)\pi T$; i, j are the replica indices, and α, β include the spin and quaternion indices. The quaternion indices are needed to incorporate both the diffuson and cooperon modes. The trace is taken over all these variables. Eventually, the replica limit, $N \to 0$, should be performed [N.15]. The frequency matrix $\hat{\epsilon} = \epsilon_n \delta_{nm} \delta_{ij} \delta_{\alpha\beta}$. All the interaction terms are restricted by the energy and momentum conservation laws. The interaction terms are written symbolically omitting such important details as the Pauli matrices acting in the spaces of spin and quaternion degrees of freedoms; the description of the matrix structure can be found in Refs. 2 and 3. Note that the ρ-term is split into two pieces: The part that can be disconnected by cutting a single line of the Coulomb interaction is denoted as $\hat{\Gamma}_\rho^0$, while the term $\hat{\Gamma}_\rho$ incorporates the irreducible part; see previous discussion in connection with the polarization operator. The additional term $\hat{\Gamma}_c$ describes the interaction in the Cooper channel. The interaction amplitudes $\Gamma_{\alpha=\rho,\sigma,c}$ are dimensionless, but elements of the forms $(\hat{Q}\hat{\Gamma}_a\hat{Q})$ contain a factor of $2\pi T$ which appears as a result of the discrete Fourier transform from the Matsubara time to frequency [N.16]. Last point to be commented is the parameter z introduced in front of the frequency matrix in the action $S[Q]$. For free electrons, $z = 1$; also in the course of the Fermi-liquid analysis of a disordered electron liquid we have not met it so far. As we shall see soon, this parameter is absolutely needed to make the RG-procedure compatible with the charge- and spin-conservation laws. Moreover, since z determines the relative scaling of the frequency with respect to the length scale,[21] it plays a central role for both kinetic and thermodynamic quantities in the critical region of the metal–insulator transition.

The equilibrium (i.e., saddle-point) value of the matrix $\hat{Q}$, usually denoted as $\hat{\Lambda}$, is fixed by the frequency term $(\hat{\epsilon}\hat{Q})$ in the above action; $\Lambda_{n,m}^{ij,\alpha\beta} = \text{sign}n \, \delta_{nm} \delta_{ij} \delta_{\alpha\beta}$. It is clear that for small ϵ_n the strength of the fixation of $\hat{Q}$ along the equilibrium position is weak and, correspondingly, fluctuations are strong [N.17]. The fluctuations of the Q-field are nothing else but diffusons and cooperons. Their propagators, $\mathcal{D}(k, \omega_n) = 1/(Dk^2 + z\omega_n)$, can be obtained by expansion of the first two terms in $S[Q]$ up to quadratic order in $\delta\hat{Q} = (\hat{Q} - \hat{\Lambda})$. These two terms yield a diffusion-like singularity in the propagators $\mathcal{D}(k, \omega_n)$ [N.18]. Furthermore, with the use of quadratic expansion of the $\hat{\Gamma}$-terms in $\delta\hat{Q}$, one can reproduce the scattering amplitudes $\Gamma_{\rho,\sigma}(k,\omega)$ given by Eq. (1.3) and illustrated in Fig. 18.1; see [N.19].

The functional $S[Q]$ describes disordered interacting electrons with energies less than $1/\tau_{el}$. The coefficients in the action $S[Q]$ incorporate the Fermi-liquid renormalizations of the clean liquid as the input parameters. One may look on this from the RG point of view: "integrating out" the high-energy states till the energy interval $\sim 1/\tau_{el}$ around the Fermi-level leads to the Fermi-liquid renormalizations. The next step is integrating out the interval starting from $1/\tau_{el}$ down to temperature, which should result in the "true" RG-descriptions.

One may conclude that the Fermi-liquid description of the disordered electron liquid is given by the quadratic expansion of the action $S[Q]$ in deviations of $\hat{Q}$ from its equilibrium value. As to the renormalization corrections, they are determined by non-quadratic (i.e., anharmonic) terms in the action. The disorder-averaged interaction amplitudes, the diffusion coefficient D as well as the parameter z, are all scale-dependent at low energies $\lesssim 1/\tau_{el}$. Note that splitting into independent channels occurs only on the level of the quadratic form of the action. During the course of the RG-procedure different channels mix [N.20].

The parameter z gives the frequency renormalization in the propagators of the diffusion and cooperon modes. To some extent, z is similar to $(1 - \partial\Sigma/\partial\epsilon)$ in the single-particle Green function $\mathcal{G}(i\epsilon,\mathbf{p})$. There is an important difference, however. According to the Migdal theorem, the combination $(1 - \partial\Sigma/\partial\epsilon)^{-1} = a$ is proportional to the jump of the occupation numbers $n(\mathbf{p})$ at the Fermi-surface; see Chapter 2, Sec. 10 in Ref. 4. This fact constrains $(1 - \partial\Sigma/\partial\epsilon)$ to be larger than 1. On the contrary, the frequency renormalization factor in the two-particle propagators is not constrained, and z may be both smaller and larger than 1. It is known that $z < 1$ in systems with magnetic impurities,[21] in a spin-polarized system,[15] or in the presence of the spin-orbit scattering.[22] Only in the generic case of a purely potential impurity scattering when the spin degrees of freedom are not constrained, $z > 1$.

With the frequency renormalization parameter z being included, the action $S[Q]$ preserves its form in the course of RG-transformations. We, thereby, may come back to the discussion of a density-correlation function of a conserved quantity. The analysis includes a few steps: one has to find (i) the RG-modified static part of the correlation function, (ii) the renormalized triangle vertex, and then (iii) with the use of the quadratic expansion of already renormalized action $S[Q]$ to find the dynamic part of the correlation function. Performing all these steps, one will get the expressions similar to those given in Eqs. (1.4b) and (1.9):

$$\chi_a(k,\omega) = \chi_a^{\text{static}} \frac{Dk^2}{Dk^2 + (z - \Gamma_a)\,\omega_n}$$

$$= \chi_a^{\text{static}} \frac{D_a k^2}{\omega_n + D_a k^2}\,, \qquad \alpha = \rho, \sigma. \qquad (2.3)$$

Here, the coefficients of diffusion $D_a = D/(z - \Gamma_a)$. For $\chi_a(k,\omega)$ to acquire this form, a relation similar to the discussed above in Eq. (1.6) has to be fulfilled for the renormalized values of $\chi_a^{\text{static}}, \gamma^a$ and $(z - \Gamma_a)$:

$$\frac{\chi_a^{\text{static}}}{\chi_a^0} = \frac{(\gamma^a)^2}{(z - \Gamma_a)}\,. \qquad (2.4)$$

Let us consider how it works for the polarization operator and spin-density correlation function. As we have already explained, the static limit of the polarization operator as well as γ^ρ are not changed by disorder. There-fore, the amplitude Γ_ρ and the parameter z are renormalized in such a way that

$$z - \Gamma_\rho = \frac{1}{1 + F_0}\,. \qquad (2.5)$$

This, by the way, implies that the relation $\sigma/e^2 = 2\nu D$, see Eq. (1.8), still holds even in the course of the renormalizations. In the case of the spin-density correlation function, the RG-calculation yields[2,7]:

$$\frac{\chi_\sigma^{\text{static}}}{\chi_\sigma^0} = \gamma^\sigma = z - \Gamma_\sigma. \qquad (2.6)$$

One observes that the condition of Eq. (2.4) is indeed fulfilled, although these relations carry more information than would be needed for one relation. One may notice, however, that together these relations make the charge and spin conductivities equal to each other:

$$\frac{\sigma_{\text{spin}}}{(\mu_B/2)^2} = 2\nu(z - \Gamma_\sigma)D_\sigma = 2\nu D\,. \qquad (2.7)$$

As we have already mentioned, this is a manifestation of the fact that charge and spin are transported by the same carriers.

Finally, let us turn back to the amplitude Γ_ρ^0 which carries information about the screened Coulomb interaction. In view of the singular behavior of the Fourier component $V_C(k)$ at small momenta, this part of the interaction is equal to

$$\Gamma_\rho^0 = 2\nu \frac{(\gamma^\rho)^2}{\Pi_{st}} = \frac{1}{1 + F_0}. \qquad (2.8)$$

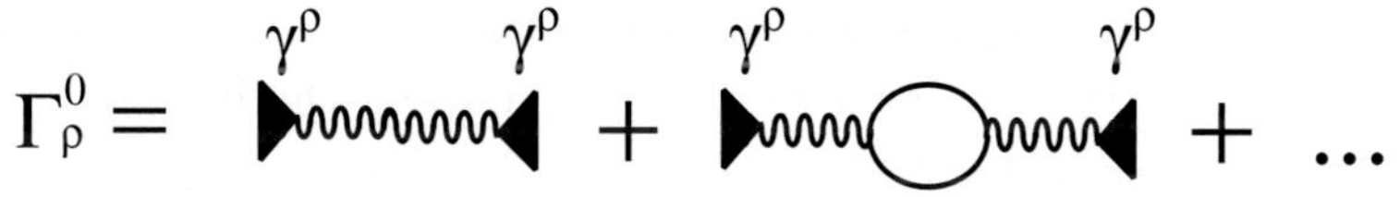

$$\Gamma^0_\rho = \quad \text{(diagram)} \quad + \quad \text{(diagram)} \quad + \quad \cdots$$

Fig. 18.8. The screened Coulomb interaction. The triangular vertices γ^ρ are attached to the ending points of the interaction line.

Here $(\gamma^\rho)^2$ originates from attaching triangular vertices γ^ρ to the ending points of the screened Coulomb interaction, see Fig. 18.8. Now, Eq. (2.5) can be rewritten as

$$z = \Gamma^0_\rho + \Gamma_\rho. \tag{2.9}$$

Thus, in $S[Q]$ the interaction amplitude in the density-channel can be substituted by z.[1,21] This implies that for unitary class systems where only fluctuations in the density channel are important (e.g., when magnetic scattering is present or in the case of spin-polarized electrons) the theory, apart from D, contains only one scaling parameter [N.21]. In other words, the theory of the electron gas interacting via the Coulomb interaction displays a high degree of universality. Pruisken and his coauthors connected this universality to a global symmetry of the problem which they called $\mathcal{F}$-invariance[23] and which is intimately related to gauge invariance.

Now that the structure of the theory has been established, it is useful to regroup its parameters by combining the frequency renormalization parameter z together with ν.[2] Then, z acquires the physical meaning of a parameter renormalizing the density of states of the diffusion modes, while $D_Q = D/z$ can be interpreted as the diffusion coefficient of the diffusion-mode "quasiparticles":

$$\nu \Longrightarrow z\nu, \qquad D \Longrightarrow D_Q = D/z. \tag{2.10}$$

It is natural to link $z\nu$ to the coefficient determining the specific heat c_V.[24,25] Furthermore, the Einstein relation, the renormalized susceptibilities, as well as the diffusion coefficients describing the evolution of the charge- and spin-densities at large scales, all acquire the form of the Fermi-liquid theory albeit with the renormalized coefficients equal to $(1 - \Gamma_a/z)$:

$$\sigma/e^2 = 2(\nu z)D_Q; \tag{2.11}$$

$$D_a = \frac{D_Q}{(1 - \Gamma_a/z)}, \qquad \alpha = \rho, \sigma; \tag{2.12}$$

$$\chi_a^{\text{static}} = z\nu(1 - \Gamma_a/z)(\chi_a^0/\nu), \qquad C_V/T = z\nu. \tag{2.13}$$

Here, χ_a^0/ν are factors that does not depend on the e–e interaction and ν [N.22]. Finally, notice that as a result of regrouping the interaction amplitudes appear always as Γ_a/z.[2]

Conclusion: The effective model that adequately describes the problem of electrons diffusing in the field of impurities is the NLSM with interactions. This model provides a compact but comprehensive description for disordered interacting electrons that is fully compatible with the constraints imposed by conservation laws. Parameters characterizing various properties of the disordered electron liquid preserve the structure of the Fermi-liquid theory although with renormalized coefficients determined by the RG-procedure.

3. Scaling Theory of the Metal–Insulator Transition in d = 2 + ε; Role of the Parameter z

In this section, we show how the scaling parameters D, Γ_σ, and z, together describe the transport and thermodynamic properties of the disordered electron liquid near the metal–insulator transition (MIT).

Let us start with the key points of the RG-analysis in $d = 2 + \epsilon$. As we have already mentioned, diffusion modes participating in the renormalization procedure have to be integrated over momenta, see Figs. 18.4 and 18.6 as examples. Each momentum integration involving diffusion propagators generates a factor $1/D_Q$ which eventually gives rise to the dimensionless parameter

$$\rho = \frac{r_d(\kappa)}{2\pi^2\hbar/e^2} \propto \frac{e^2}{\hbar\sigma}\kappa^{d-2}. \tag{3.1}$$

Here, ρ is equal to the resistance r_d of a d-dimensional cube of side length $\sim 2\pi/\kappa$ measured in units of $2\pi^2\hbar/e^2$ [N.23]; κ is the momentum cutoff which decreases during the renormalization [N.24].

It follows from the structure of the action $S[Q]$, when written with the help of Eqs. (2.9) and (2.10), that the RG-procedure can be performed in terms of the dimensionless resistance ρ and the reduced interaction amplitudes

$$\gamma_2 = -\Gamma_\sigma/z = \Gamma_2/z, \tag{3.2}$$

$$\gamma_c = \Gamma_c/z. \tag{3.3}$$

With these variables, the set of the RG equations takes the general form[2,3]:

$$d\ln\rho/dy = -\frac{\epsilon}{2} + \rho\beta_\rho(\rho; \gamma_2, \gamma_c; \epsilon), \tag{3.4a}$$

$$d\gamma_2/dy = \rho\beta_{\gamma_2}(\rho;\gamma_2,\gamma_c;\epsilon), \tag{3.4b}$$

and also

$$d\gamma_c/dy = -\gamma_c^2 + \rho\beta_{\gamma_c}(\rho;\gamma_2,\gamma_c;\epsilon). \tag{3.5}$$

The parameter z is described by an additional equation:

$$d\ln z/dy = \rho\beta_z(\rho;\gamma_2,\gamma_c;\epsilon). \tag{3.6}$$

Observe that β_z, as well as β_ρ and $\beta_{\gamma_2,\gamma_c}$ are all independent of z. In the above set of RG equations, the logarithmic variable $y = \ln 1/[\max(D\kappa^2/z,\omega_n)\tau_{el}]$ has been used. This choice of the logarithmic variable is convenient because it allows us to take T as a natural lower cutoff; the upper cutoff is $1/\tau_{el}$. The explicit factor ϵ in the equation determining ρ originates from κ^{d-2} entering the definition of the RG-charge ρ.

The β-functions in the above equations are multiplied by a factor ρ to emphasize that the sought-after corrections appear as a result of disorder [N.25]. The complete form of the β-functions is unknown. The general approach, however, is to expand the functions in a power series in ρ as $\beta(\rho;\gamma_{2,c}) = \beta_1(\gamma_2,\gamma_c) + \rho\beta_2(\gamma_2,\gamma_c) + ...$, such that for each power of ρ the full dependence on γ_2 and γ_c is retained. This is possible, in principle, because the maximal number of allowed interaction amplitudes (extended by ladders) is limited by the number of momentum integrations involving the diffusive propagators. Since each integration gives a factor of $1/D \sim \rho$, for a given order in ρ the number of (extended) interaction vertices is finite [N.26].

For a repulsive interaction in the Cooper channel, the amplitude γ_c scales rapidly to a ρ-dependent fixed point, which is determined by the competition of two terms in Eq. (3.5). In the following, we replace γ_c in the β-functions describing ρ, γ_2 and z by its fixed-point value $\gamma_c(\rho)$. As a result, the RG-evolution near the MIT can be described by only Eqs. (3.4a) and (3.4b) together with Eq. (3.6) for z.

To illustrate the scheme of finding the temperature or frequency behavior of the conductivity in the critical region of the MIT,[21] let us discuss an electron system in the presence of magnetic impurities. Then, Cooperons and fluctuations of the electron spin density are not effective because of a strong spin scattering. In this case, the parameter ρ representing the resistance of a d-dimensional cubic sample is described by a separate equation:

$$d\ln\rho/dy = -\frac{\epsilon}{2} + \rho\beta_\rho(\rho;\epsilon). \tag{3.7}$$

In the discussed case, corrections appearing as a result of the interplay of the e–e interaction and disorder[10,11] lead to an increase of the resistance as temperature decreases ($\beta_\rho > 0$, thus favoring localization). Therefore, the geometric factor ϵ competes with these corrections for $d > 2$. As a result, there is an unstable fixed point, $\rho = \rho_c$, which determines the critical behavior of the conductivity in the critical region of the MIT.

Then, as it follows from Eq. (3.1), in the vicinity of the transition,

$$\sigma(\kappa)/e^2 \propto \kappa^{d-2}. \tag{3.8}$$

In the $3d$ case, for example, on the metallic side of the transition the critical behavior develops when $\kappa \gg \sigma(T = 0)/e^2$. At non-zero temperature, in the critical regime of the MIT the process of renormalization ceases at a scale when

$$D(\kappa)\kappa^2/z(\kappa) \sim T \underset{\text{using Eq. (2.11)}}{\Longrightarrow} \kappa^d/\nu \sim zT. \tag{3.9}$$

For the electric conductivity measured at external frequency $\omega \gg T$, the renormalization is cut off by ω rather than T. The above relations are a consequence of (i) the form of the diffusion propagator $\mathcal{D}(k, \omega_n) = 1/(Dk^2 + z\omega_n)$, and (ii) the definition of the RG-parameter g which exhibits a fixed point. In addition, these relations take into account the result discussed in the previous section that (iii) all the renormalizations in between σ and D are canceled out: $\sigma/e^2 = 2\nu D$.

Thus, in order to find the temperature or frequency behavior of σ at the MIT, one has to connect the momentum and energy scales in the critical region, $\kappa \sim (z \max[\omega, T])^{1/d}$. However, z itself is a scaling parameter, see Eq. (3.6). Therefore, one needs to know the critical behavior of the parameter z at the transition, which is determined by the value of $\rho\beta_z$ at the critical point:

$$\sigma(\omega, T) \sim (z \max[\omega, T])^{\frac{d-2}{d}} \sim (\max[\omega, T])^{\frac{d-2}{d}(1+\tilde\zeta)}, \tag{3.10}$$

$$\tilde\zeta = -(\rho\beta_z)_{\text{critical point}}. \tag{3.11}$$

For free electrons z is not renormalized, and at zero temperature $\sigma(\omega) \sim \omega^{1/3}$ for $d = 3$.[26] The e–e interaction modifies this critical behavior of the conductivity through the critical exponent $\tilde\zeta$. If $\omega \lesssim T$, the renormalization procedure is cut off by the temperature

$$\sigma(T) \sim (zT)^{\frac{d-2}{d}} \sim T^{\frac{d-2}{d}(1+\tilde\zeta)}. \tag{3.12}$$

The scaling behavior described above suggests that the interplay between frequency and temperature can be described by a single function

$$\sigma(T, \omega)_{\text{critical}} = T^a f(\hbar\omega/k_b T), \tag{3.13}$$

where $f(x) \to$ const when $x \to 0$, and $f(x) \propto x^a$ when $x \to \infty$, with $a = \frac{d-2}{d}(1 + \widetilde{\zeta})$. This is a typical behavior near a quantum phase transition for which the MIT is, perhaps, a primary example.

To get an idea about the value of the critical exponent $\widetilde{\zeta}$, let us find it in the lowest order in ϵ. The equation describing resistance in the case of magnetic impurities in the lowest orders in ρ and ϵ is

$$d\ln t/dy = -\frac{\epsilon}{2} + \rho, \qquad \rho_c = \frac{\epsilon}{2}. \qquad (3.14)$$

Furthermore,

$$d\ln z/dy = -\frac{1}{2}\rho. \qquad (3.15)$$

At $d = 3$, this estimate yields $\widetilde{\zeta} = 1/4$ for the MIT in the presence of spin scattering and correspondingly, for $\epsilon = 1$, one gets $a = \frac{1}{3}(1 + \widetilde{\zeta}) \approx 0.4$.

Although we used for the purpose of illustration the case of magnetic scattering (a system where only fluctuations in the density channel are important), the conclusion of the above discussion is quite general: the frequency or temperature behavior of the conductivity in the critical region is determined by the right-hand side of Eq. (3.6) at the fixed point of the transition [N27]. Notice that, although the ϵ-expansion has been applied to estimate the value of $\widetilde{\zeta}$, the form of the combination $a = \frac{d-2}{d}(1 + \widetilde{\zeta})$ is determined by the general structure of the theory only and does not rely on the ϵ-expansion [N28].

Experimentally, the dependence of σ on the temperature in the critical region can be determined with a limited accuracy only [N29]. In Ref. 27, it was shown that in a persistent photoconductor where the carrier concentration can be controlled very neatly, $\sigma(T) \sim T^{1/2}$ at the transition, i.e., $a = 1/2$; this corresponds to $\widetilde{\zeta} = 1/2$. The direct measurements of $\sigma(\omega)$, are unfortunately, very rare. In Refs. 28 and 29, the temperature and frequency dependences were studied simultaneously in amorphous niobium-silicon alloys (Nb:Si) with compositions near the MIT. The measurements observed a one-to-one correspondence between the T- and ω-dependent conductivity thus confirming the above picture of the MIT as a quantum phase transition. The critical exponent a has also been found to be equal to $1/2$ for this system, i.e., $\sigma(T, \omega)_{\text{critical}} = T^{1/2} f(\hbar\omega/k_b T)$, see Eq. (3.13). The same scaling behavior should hold for the whole universality class which the discussed system represents. In measurements on the magnetic-field-induced MIT in GaAs and InSb semiconductors (representing a different universality class compared to the discussed measurements on Nb:Si) the critical behavior $\sigma(T) \sim T^{1/3}$ has been observed.[30] It may be worth mentioning that for this universality class $\widetilde{\zeta}$ is indeed equal to zero in the lowest order in the ϵ-expansion.[31]

Finally, let us note another important consequence of the fact that at the critical point of the MIT (determined by the fixed point of the set of Eqs. (3.4) and (3.5)) the only scaling parameter which continues to evolve is z. Since this parameter is directly related to the renormalization of the effective density of states of the diffusion modes, it follows immediately from Eq. (2.13) that the critical temperature dependence of thermodynamic quantities at the MIT is also described by the same critical exponent $\widetilde{\zeta}$; see Eq. (3.11) for the definition of $\widetilde{\zeta}$.

The content of this section may look like a simple dimensional analysis. In fact, it heavily relies on the structure of the theory based on the NLSM with the interaction terms, which was established in the previous section. As it was pointed out there, this low-energy field theory adequately describes the interacting electrons in the diffusive regime. In this context, the parameter z plays a special role. Since this parameter is responsible for the frequency renormalization, it is of particular importance in connection with the conservation laws of the particle-number and spin. Furthermore the law of number conservation allows to obtain the Einstein relation for the electron liquid in the appropriate form. Only with the information about the structure of the theory at hand, the critical behavior near the MIT can be found by a straightforward dimensional analysis.

Conclusion: The metal–insulator transition in a system of diffusing electrons is an example of a quantum phase transition[32] with a temperature-frequency scaling controlled by the parameter z. Precisely the same parameter describes the scaling behavior of both the conductivity and the thermodynamics in the critical region of the transition. The structure of the theory is very general and not related to the ϵ-expansion which can be used for the calculation of $\widetilde{\zeta}$.

4. Tunneling Density of States

The tunneling density of states (TDOS) or, as it is also called, the single-particle density of states, $\nu(\varepsilon)$, exhibits a rather pronounced critical behavior at the MIT.[33] This quantity can be obtained by measuring the differential conductance $G_j(V)$ of a tunneling junction at a finite voltage bias V: $G_j(V) \propto \nu(\varepsilon = V)$. In the early semi-phenomenological scaling theory of the MIT by McMillan,[34] the TDOS has been treated as a parameter which enters into the relation connecting the length and energy (or frequency) scales and gives rise to a critical exponent which has been replace of $\widetilde{\zeta}$ in the full microscopic theory. As we have already discussed, the parameter which connects these scales is z rather than the TDOS, see Eq. (3.9). Moreover, the TDOS stands actually outside the RG-scheme. Let us explain why. The TDOS is defined as

$$\nu(\varepsilon) = -\frac{1}{\pi} Im \int \mathcal{G}^R(\varepsilon, \mathbf{p}) \frac{d\mathbf{p}}{(2\pi)^d}. \tag{4.1}$$

As such, this quantity is not gauge invariant. It can be changed by a time-dependent gauge transformation. Thereby, it cannot enter the RG-scheme which operates only with truly gauge-invariant quantities. In the case of the TDOS, it is the external electrode with respect to which the measurement of the tunneling current is performed, that makes the TDOS a physically meaningful quantity [N30].

The combined effect of Coulomb interaction and disorder leads to a strong suppression of the TDOS. This observation allowed to explain the so-called zero-bias anomaly in the tunneling spectra of disordered systems.[10,35] Compared to other effects related to the interplay of the e–e interaction and disorder, corrections to the TDOS are the strongest. In particular, in two-dimensions the correction obtained in the lowest order in ρ appears to be log-squared rather than just logarithmic[11,36]:

$$\nu(\varepsilon) = \nu[1 - \frac{\rho}{4} \ln(1/|\varepsilon|\tau_{el}) \ln(\tau_{el}\omega_0^2/|\varepsilon|)]. \tag{4.2}$$

Here $\omega_0 = D\kappa_{scr}^2$, while κ_{scr} is the inverse of Thomas–Fermi screening radius.

To go beyond the perturbative correction, it is useful to apply the $\hat{Q}$-matrix technique. With this technique, $\nu(\varepsilon)$ can be expressed as an averaged product of two matrices:

$$\nu(\varepsilon) = \nu \left\langle \hat{\Lambda}\hat{Q} \right\rangle_{\varepsilon\varepsilon}. \tag{4.3}$$

Thus, by measuring the TDOS, one may study how an ε-component of the matrix $\hat{Q}$ fluctuates around its equilibrium position. Now compare with the physics of phonons: the quantity which measures the fluctuations of ions with respect to their equilibrium position is the Debye–Waller factor. It has been noted already in the early studies,[1,31] that the calculation of $\nu(\varepsilon)$ is indeed very similar to the calculation of the Debye–Waller factor, and can be reduced to a Gaussian integration. [By means of the $\hat{Q}$-technique, the right-hand-side of Eq. (4.3) can be expressed as $\nu(\varepsilon) = \nu \langle \exp W \rangle_{\varepsilon\varepsilon}$, where W is a matrix field that describes diffusion modes in the presence of the e–e interactions.] In fact, the formal similarity with the Debye–Waller factor reflects the physical essence of the TDOS. Measurement of a tunneling current is a kind of "*Mössbauer-type*" experiment which determines the effect of the zero-point fluctuations of the electromagnetic field on the probability of tunneling.

As it was first pointed out in Refs. 1 and 31, it follows from the structure of the discussed quantity that the perturbative correction to the TDOS should be exponentiated [N31]:

$$\nu(\varepsilon) = \nu \exp[-\frac{\rho}{4}\ln(1/|\varepsilon|\tau_{el})\ln(\tau_{el}\omega_0^2/|\varepsilon|)]. \qquad (4.4)$$

Examples of calculations of the critical exponent of the TDOS at the MIT using the ϵ-expansion are given in Ref. 31. The presence of the double-logarithmic corrections to $\nu(\varepsilon)$ when $d = 2$ leads to the fact that the ϵ-expansion of the critical exponent of the TDOS starts from a constant. The reason is that at $d = 2+\epsilon$ in the exponent of Eq. (4.4), the factor $1/\epsilon$ replaces one of the two logs and cancels a factor ϵ coming the charge $\rho_c \propto \epsilon$. It worth noting that the situation discussed above is specific for the long-range nature of the Coulomb interaction. In a model description when the dynamically screened Coulomb interaction $V_C(k, \omega_n)$ is replaced by a constant, double-logarithmic corrections do not arise.

Interestingly, the log-squared corrections cancel out when calculating any other physical quantities, except the TDOS. This occurs for the following reason. The discussed corrections accumulate from the momentum integration over the region of momenta that are much smaller than those typical for diffusion, $k \ll (\omega_n/D)^{1/2}$. Therefore, this integration does not involve the diffusion propagators but only the Coulomb interaction $V_C(k, \omega_n)$ [N32]. As a result of such an integration, the dynamically screened Coulomb interaction starts to depend effectively only on the frequency. However, as it was pointed out in Refs. 37, 3 and 38, any interaction of this kind, i.e., a purely time-dependent e–e interaction, can be completely eliminated by means of a time-dependent gauge transformation which can be performed exactly. [By a standard procedure, the four-fermion term $\psi^\dagger(\tau)\psi(\tau)V_{ee}(\tau - \tau')\psi^\dagger(\tau')\psi(\tau')$ can be decoupled by a time-dependent potential acting on the fermions, $\varphi(\tau)\psi^\dagger(\tau)\psi(\tau)$, which subsequently can be integrated out.] This is the reason why the corrections originating from the unscreened singularity of the Coulomb interaction at very small momenta cannot manifest themselves in transport or thermodynamic quantities: They cannot appear in gauge-invariant quantities. In fact, the physics of this observation is very close to the arguments presented in Sec. 1 about the insensitivity of the corrections induced by disorder and the e–e interactions to the variation of the chemical potential because of the absence of a reference point. The only difference is that now the variation of the potential is time-dependent.

Conclusion: The Coulomb interaction, $V_C(k, \omega_n)$, with a momentum transfer much smaller than those typical for diffusion, can contribute only to a quantity for which the condition of measurement makes it possible to detect the effect of the long-range time-dependent fluctuations of the electric potential. An example of such a quantity is the TDOS. By fabricating a counter-electrode of the tunneling junction, one creates a reference point which allows to study the effects of time-dependent long-range fluctuations of the electric potential which do not contribute to other physical quantities [N33].

5. The Anderson Transition in the Presence of Interactions in a Two-Dimensional System

Here we will apply the two-parameter scaling theory, in which ρ and γ_2 are the flowing parameters, for the discussion of the 2d-MIT. We use the data obtained in Si-MOSFETs for comparison with the theory. The MIT in a 2d electron gas, which does not occur for free electrons,[8] has been observed experimentally in dilute electron systems.[39,40] Obviously, this fact indicates that the e–e interactions are of crucial importance. The unexpected discovery of the 2d-MIT generated renewed interest in disordered electron systems with interactions (see the review articles[41–43] and references therein).

In Fig. 18.9, the data of Pudalov *et al.*[44] is presented which demonstrates clearly the existence of the MIT in 2d; different curves here correspond to different electron densities. In the metallic phase the resistance $\rho(T)$ drops noticeably as the temperature is lowered. [This drop is suppressed when a relatively weak in-plane magnetic field is applied.[45] The sensitivity to an in-plane magnetic field highlights the importance of the spin degrees of freedom for the MIT. Therefore, the spin related modes should be one of the ingredients of the theory of the transition.] A highly non-trivial feature revealed by the data shown in Fig. 18.9, is the non-monotonicity of $\rho(T)$ on the metallic side of the transition. This non-monotonic behavior is of the principle importance, because it points towards a competition between different mechanisms determining resistance [N34]. In the theory of the MIT developed by the author together with Alex Punnoose,[16,46] there is a competition between the charge-density diffusion modes and cooperons, on the one hand, and the fluctuations of the spin- (and valley-) degrees of freedom, on the other hand. The former favor

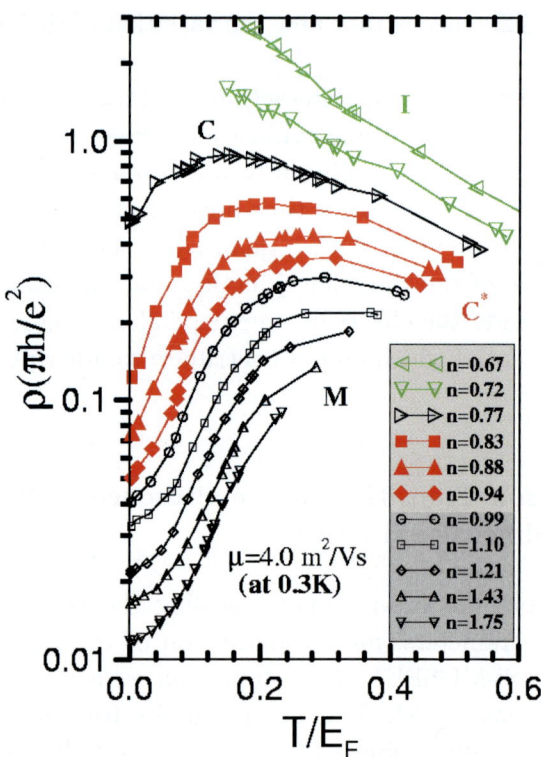

Fig. 18.9.   Resistivity of a high mobility Si-MOSFET sample for various densities as a function of temperature (adapted from Ref. 44). The data clearly demonstrates the existence of the metal-insulator transition. The electron densities are defined in units of 10^{11} cm^{-2}. Note that resistance is measured in units of $2\pi^2\hbar/e^2$. The three non-monotonous curves below the transition (shown in red color and denoted as C*) are fitted in Fig. 18.10. Reprinted (Fig. 1) with permission from *Phys. Rev. Lett.* **88**, 016802 (2002). © American Physical Society.

localization, while the latter act against it, thus stabilizing the metallic state.

The critical resistance at which the transition occurs in high-mobility Si-MOSFETs has been shown experimentally to be universal,[42] suggesting the applicability of the RG-description of the MIT of the kind discussed above in Sec. 3 [N35]. But can we use the disordered Fermi-liquid (or at least the NLSM with interactions) as a starting platform in a system with r_s of the order of 10? Measurements of the Shubnikov-de Haas oscillations and the Hall coefficient in Si-MOSFETS observe no anomalies in the properties characterizing the electron liquid on the metallic side of the MIT (at least, when applied magnetic fields are not too high). The Fermi-liquid renormalizations extracted as a result of these measurements were significant but not giant [N36]. We, therefore, have all reasons to apply the approach based on the RG-analysis of the NLSM discussed above for the analysis of the MIT in this material. Generally speaking, we believe that the NLSM description can be applicable even without the prerequisites of the Fermi-liquid. This is because the diffusion modes are more robust than the single-particle excitations and, hence, the NLSM with interactions, as a *minimal model*, can be valid even in the absence of the Fermi-liquid background.

For the discussion of Si-MOSFETs, the only necessary modification is due to the fact that Si-MOSFET is a multi-valley system. The conduction band of an n-(001) silicon inversion layer has two almost degenerate valleys. In the following we consider the number of equivalent valleys to be equal to n_v. Because inter-valley scattering requires a large change of the momentum, we assume that the interactions couple electrons in different valleys but do not mix them. This implies that inter-valley scattering processes, including those due to the disorder, are neglected. This assumption is appropriate for samples with high-mobility [N37]. In this limit, the RG equations describing the evolution of the resistance and the scattering amplitude γ_2 in $2d$ have the form[46]:

$$\frac{d\ln\rho}{dy} = \rho\left[n_v + 1 - (4n_v^2 - 1)\Phi(\gamma_2)\right], \tag{5.1a}$$

$$\frac{d\gamma_2}{dy} = \rho\frac{(1+\gamma_2)^2}{2}. \tag{5.1b}$$

The equations above are obtained in the lowest order in ρ (the one-loop order), but they incorporate the full dependence on the e–e amplitudes. Here, the amplitude γ_2 acts inside spin-valley "triplet channels." (The definition of γ_2 is given in Eq. (3.2); note that for repulsive interactions $\gamma_2 > 0$.) In the first equation, $\Phi(\gamma_2) = \frac{1+\gamma_2}{\gamma_2}\ln(1+\gamma_2) - 1$; the factor $(4n_v^2 - 1)$ corresponds to the number of spin-valley "triplet" channels, while the factor n_v corresponds to the weak-localization (cooperon) corrections. The factor of one entering the square brackets in Eq. (5.1a) is the contribution of the long-ranged Coulomb singlet-amplitude (after dynamic screening) and is, therefore, universal. Furthermore, it should be emphasized that the factor of one appearing in Eq. (5.1b) for γ_2 also originates from the Coulomb singlet-amplitude combined with scattering induced by disorder, see [N20]. Consequently, setting the initial value of γ_2 to zero does not imply the absence of interactions.

The following salient features should be noted: While the amplitude γ_2 increases monotonically as the temperature is reduced, the resistance, as a result, has a characteristic non-monotonic form changing from insulating behavior ($d\rho/dT < 0$) at high temperatures to metallic behavior ($d\rho/dT > 0$) at low temperatures. The change in slope occurs at a maximum value $\rho_{\max}$ at a temperature $T = T_{\max}$, neither of them is universal. The corresponding value of the amplitude γ_2 is, however, universal at the one-loop order, depending only on n_v; for $n_v = 1$, it is 2.08, whereas for $n_v = 2$, it has the considerably lower value 0.45. Next, it follows from the general form of Eqs. (5.1a) and (5.1b), that $\rho(T)/\rho_{\max}$ and $\gamma_2(T)$ can be presented as the universal functions $R(\eta_T)$ and $\tilde{\gamma}_2(\eta_T)$ when the argument η_T is introduced[1-3,46]:

$$R(\eta_T) \equiv \rho(T)/\rho_{\max} \qquad \tilde{\gamma}_2(\eta_T) \equiv \gamma_2(T),$$

$$\eta_T = \rho_{\max} \ln(T/T_{\max}).$$

(5.2)

The non-monotonic function $R(\eta)$ together with the fit of the resistance curves obtained for two samples of different origin are presented in Fig. 18.10. After re-scaling, the data at various densities is described by a *single* curve. The drop of $\rho(T)$ by a factor of five and the subsequent flattening of the curve at low T are captured in the correct temperature interval. The full

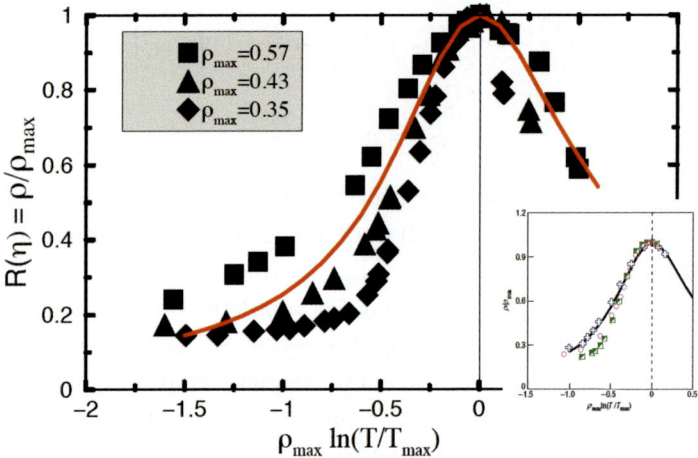

Fig. 18.10. RG-fitting of the resistivity for two different Si-MOSFETs. Main panel[46]: the data corresponding to $n = (0.83, 0.88, 0.94) \times 10^{11}$ cm^{-2} in Fig. 18.9 are scaled according to Eq. (5.2). The solid line (in red) is the solution of the RG equations (5.1a) and (5.1b) with $n_v = 2$; no adjustable parameters have been used in the fit of the data. Inset[49]: the same for a sample from a different wafer. Main figure reprinted (Fig. 2) with permission from *Phys. Rev. Lett.* **88**, 016802 (2002). © American Physical Society. Inset reprinted (Fig. 4a) with permission from *Nature Phys.* **707** (2007). © Macmillan Publishers Ltd.

temperature dependence of the resistance is completely controlled by its value $\rho_{\max}$ at the maximum; there are no other free (or fitting) parameters.

We can draw an important conclusion: We proved theoretically the existence of the MIT in $2d$. Since (i) the one-loop approximation gives a *drop* of the resistance at low temperatures for a moderate strength of disorder (i.e., in the region of the applicability of this approximation), and (ii) the Anderson localization at strong disorder is indisputable; it is therefore *logically unavoidable that the MIT should exist* in-between. Owing to the drop in the resistance, the reliability of the obtained RG equations is improved. Therefore, the conclusion about the existence of the MIT can be justified even within the one-loop approximation. Thus, the anti-localization effect of the e–e interactions fundamentally alters the common point of view that electrons in $2d$ are "eventually" (i.e., at $T = 0$) localized.

Let us turn now to the interaction amplitude $\gamma_2(T)$. Since this amplitude is related to the spin degrees of freedom, the information about the dependence of this amplitude on the temperature can be extracted from the in-plane magnetoconductance. This is because the fluctuations of the spin-density lead — with participation of γ_2 — to finite temperature corrections to the resistivity.[47] The spin-splitting induced by the in-plane magnetic field reduces spin-density fluctuations and leads, in this way, to a temperature dependent magnetoconductance. Hence, the magnetoconductance contains information about the value of the amplitude γ_2 and its evolution with temperature. In order to extract the value of γ_2, it is important, however, to perform measurements in weak magnetic fields, such that $g\mu_B(1 + \gamma_2)B/k_BT \ll 1$. Weak magnetic field is needed in order not to drive the electron liquid, which at large r_s is very "fragile", into some other state. As long as electrons are in the diffusive regime, $k_BT < h/\tau_{el}$, i.e., the temperature is less than the scattering rate on the static impurities, the expression for magnetoconductivity in the limit $b = g\mu_BB/k_BT \ll 1$ is given[47–49] as:

$$\Delta\sigma = -(e^2/\pi h)\, K_v C_{ee}(\gamma_2, \rho)\, b^2, \tag{5.3}$$

where in a system with n_v degenerate valleys, $K_v = n_v^2$. In the case when the resistance ρ is not too high, the coefficient determining the magnetoconductance, C_{ee}, is explicitly related to the amplitude γ_2 as follows: $C_{ee} = 0.091\gamma_2\,(\gamma_2 + 1)$. The experimental details and the results of the comparison with theory can be found in Refs. 49 and 50 where, for the first time, the scaling of the interaction amplitude was established. Not too close to the MIT, the extracted values of γ_2 are close to those predicted by the theory. Remarkably, the parameter γ_2 at $T = T_{\max}$ was found to correspond to 0.45 as predicted by theory for $n_v = 2$.

In spite of this success, Eqs. (5.1a) and (5.1b) have a limited applicability. Obviously, the *single-curve* solution $R(\eta)$ cannot provide the description of the MIT. To approach the critical region of the MIT, the disorder has to be treated beyond the lowest order in ρ, while adequately retaining the effects of the interaction.

An internally consistent theory of the MIT [38] which goes beyond the one-loop calculations was developed in Ref. 16 using the number of identical valleys as a large parameter, $n_v \to \infty$. The valley degrees of freedom are akin to flavors in standard field-theoretic models. Generally, closed loops play a special role in the diagrammatic RG-analysis in the limit when the number of flavors N is taken to be very large.[51] This is because each closed loop involves a sum over all the flavors, generating a large factor N per loop. It is then typical to send a coupling constant λ to zero in the limit $N \to \infty$ keeping λN finite. For interacting spin-1/2 electrons in the presence of n_v valleys ($N = 2n_v$), the screening makes the bare values of the interaction amplitude γ_2 to scale as $1/(2n_v)$. Furthermore, the increase in the number of conducting channels results in the resistance ρ to scale as $1/n_v$. It is, therefore, natural to introduce the amplitudes $\Theta = 2n_v\gamma_2$ together with the resistance parameter $t = n_v\rho$; the parameter t is thus the resistance per valley, $t = 1/[(2\pi)^2\nu D]$. Both quantities Θ and t remain finite in the large-n_v limit.

Following the large-n_v approximation scheme outlined above, the RG equations at order t^2 (i.e., in the two-loop approximation) have been derived. The obtained equations describe the competition between the e–e interactions and disorder in $2d$. The resulting resistance-interaction (t–Θ) flow diagram is plotted in Fig. 18.11. The arrows indicate the direction of the flow as the temperature is lowered. The quantum critical point, which corresponds to the fixed point of the equations describing the evolution of t and Θ, is marked by the circle. This quantum critical point separates the metallic phase, which is stabilized by electronic interactions, from the insulating phase where disorder prevails over the electronic interactions. The attractive ("horizontal") separatrix separate the metallic phase from the insulating phase. Crossing the separatrix by changing the initial values of t and Θ (e.g., by changing the carrier density) leads to the MIT.

In Ref. 49, the two-parameter scaling theory has been verified experimentally. In Fig. 18.12, the experimentally obtained flow diagram is presented. In this plot, the coefficient C_{ee} effectively represents the interaction amplitude in the spin-density channel. The authors used the fact that the coefficient C_{ee} reflects the strength of spin-related interactions of the diffusion

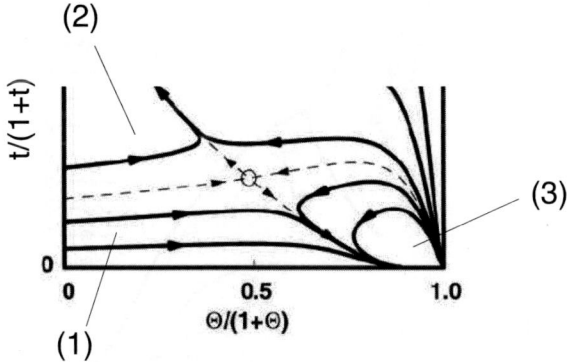

Fig. 18.11. The resistance-interaction (t-Θ) flow diagram obtained in the two-loop calculations.[16] The arrows indicate the direction of the flow as the temperature is lowered. The quantum critical point is marked by the circle. Area (1) is the metallic phase, which is stabilized by the interaction. Area (2) is the insulating phase where disorder prevails. Area (3) is the region of strong spin correlations. The attractive separatrices separate the metallic phase from the insulating phase.

modes at *any* value of the resistance. Therefore, one may get much broader insight into the MIT by studying the temperature dependence of the coefficient C_{ee} even without knowing exact relation connecting C_{ee} with γ_2. This procedure has been applied for the first time in Ref. 49, where the coefficient C_{ee} has been determined by fitting the $\Delta\sigma(B,T)$ traces to Eq. (5.3). Because the traces are taken at different temperatures, one obtains the RG-evolution of C_{ee} as a function of temperature.

We see that the flow diagram presented in Fig. 18.12 confirms all the qualitative features of the theoretical predictions, including the quantum critical point and the non-monotonic behavior of the resistance as a function of T on the metallic side of the transition. At not too high resistance, but still within the diffusive region, the data presented in this flow diagram can be accurately described by the RG theory without any fitting parameters, see the inset in Fig. 18.10. Most important, however, is that the possibility of presenting the data as a flow diagram gives a very strong argument in favor of the applicability of the two-parameter scaling theory in Si-MOSFETs.

So far, we described scaling in terms of two parameters, leaving aside the parameter z. Being related to the frequency renormalization of the diffusion modes, this parameter determines the transport and thermodynamic quantities in the critical region of the MIT. In the limit $N \to \infty$, the equation for z reads as follows: $d\ln z/dy = \beta_z(t, \Theta) = t\Theta$. Consequently, in the case discussed in this section, and unlike in the case of magnetic impurities

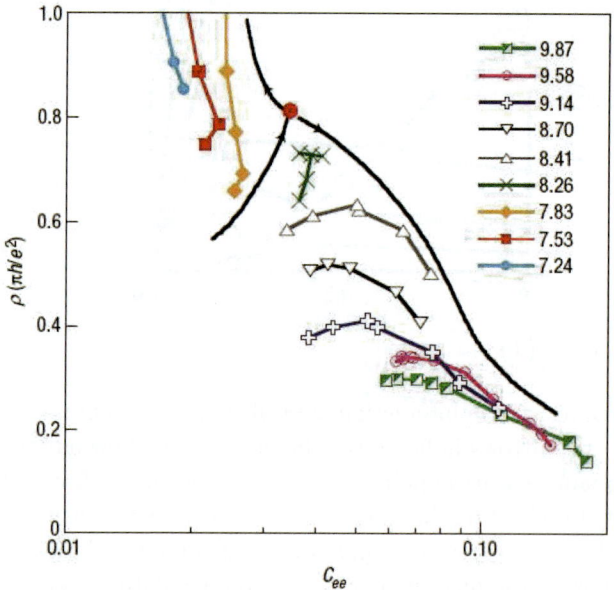

Fig. 18.12. The disorder–interaction flow diagram of the 2d electrons in Si-MOSFET.[49] The circle indicates the location of the quantum critical point from which the three separatrices emanate. The arrows shown on the separatrices indicate the direction of the flow as the temperature is lowered. The electron densities are indicated in units of 10^{10} cm^{-2}. The interaction amplitude in the spin-density channel is represented by the coefficient C_{ee} which was extracted from the magnetoresistance. Reprinted (Fig. 3) with permission from *Nature Phys.* **3**, 707 (2007). © Macmillan Publishers Ltd.

discussed in Sec. 3, z diverges in the vicinity of the MIT: $z \sim T^{\tilde{\zeta}}$ where $\tilde{\zeta} = -(t\Theta)_{\text{critical point}} < 0$. As we already mentioned, z can be interpreted as the parameter renormalizing the density of states of the diffusion modes and as such it controls the thermodynamic quantities. Hence, in the critical regime of the 2d-MIT, the specific heat C_V/T diverges (owing to the softening of the diffusion modes induced by z). Furthermore, a similar divergence is also expected in the Pauli spin susceptibility[16]:

$$C_V/T = \nu z(T) \propto T^\varsigma, \qquad \chi_{spin}/\chi_a^0 = z(T)(1 + \gamma_2) \propto T^\varsigma. \qquad (5.4)$$

Since the interaction parameter Θ is finite at the critical point, the divergence in the Pauli spin susceptibility is not related to any Stoner-like magnetic instability.

Before we conclude this section, let us touch upon a delicate point: How general is the discussed theory of the 2d-MIT? It is applicable only within the diffusive regime at low enough temperatures when $k_B T \ll \hbar/\tau_{el} \lesssim \epsilon_F$. Under these conditions, the charge and spin perturbations of the degenerate electron gas propagate diffusively. The boundary of the diffusive regime can be determined from the measurements of the magnetoconductivity described above. [The point is that the relation $\Delta\sigma \propto b^2$, where $b = g\mu_B B/k_B T \ll 1$, holds only inside the diffusive regime, $k_B T \ll \hbar/\tau_{el}$, while in the ballistic regime $\Delta\sigma \propto (T\tau_{el})b^2$.] In Si-MOSFETs, the diffusive regime extends up to a few Kelvin in a density range around the critical density of the MIT. The corresponding Fermi-temperature is of the order of 10 K and, therefore, at temperatures convenient for measurements electrons are already degenerate. In addition, the $2d$ electron gas in Si-MOSFETs (which in fact is a *moderately* high-mobility system) is *unique* in the sense that the scattering is mostly short-range in character, so that the *MIT occurs in the diffusive regime.* On the contrary, in *true* high-mobility systems like GaAs/AlGaAs or n-SiGe heterostructures the single particle scattering rate typically differs by a factor of ten compared to the transport scattering rate. Therefore, it is difficult to access the diffusive regime, because the smoothness of the disorder drives the system directly from the ballistic to the insulating phase. Furthermore, the presence of two equivalent valleys strongly enhances the anti-localization effect of the e–e interaction and disorder making MOSFETs ideal systems to study the MIT in $2d$ [N39].

We may thereby conclude that Si-MOSFETs provide an ideal playground to study the properties of a $2d$ disordered electron liquid and, in particular, the Anderson localization in the presence of the e–e interactions. Within the region of its applicability, the RG-theory gives not only a qualitative but also a quantitative description of the experimental data in these systems, see Figs. 18.10–18.12.

Still, a question arises — How can the one-loop theory work so well up to very high resistances [N40], in spite of the fact that the system is placed under such extreme conditions that r_s is as large as 10? In our opinion, it is maybe *not in spite but because* of these extreme conditions. The point is that at large r_s an electron is mostly trapped inside a temporary potential minimum created by other electrons as a result of Wigner-crystal like short range order. Naturally, the kinetic energy of the electrons temporarily trapped by the strong Coulomb interaction is larger than that of free electrons at the same density. Therefore disorder is actually small compared to both the kinetic and the interaction energies, even though the resistance is not small.

Summary: We demonstrated that the two-parameter RG-theory of the disordered electron liquid reviewed here captures both quantitatively (for moderate disorder) and qualitatively (for larger disorder) the physics of the disordered liquid in the diffusive regime. The possibility of presenting the data as a flow diagram is a strong argument in favor of the applicability of the two-parameter scaling theory in Si-MOSFETs. Finally, we showed that the existence of the MIT in $2d$ in Si-MOSFETs can be justified theoretically by combining the RG-analysis in the one-loop approximation with the fact of the existence of Anderson localization at strong disorder: The one-loop approximation gives a noticeable drop of the resistance at low temperatures for a moderate strength of disorder. Since, on the other hand, at very strong disorder the Anderson localization is unavoidable, it follows that in between the MIT should exist.

6. Notes

N.1   For example, when the rate of inelastic e–e collisions exceeds characteristic excitation energies, which are of the order of the temperature T or frequency ϵ.

N.2   In the diffusion coefficient D, both v_F and τ_{el} incorporate the Fermi-liquid renormalizations.

N.3   While the static part of the amplitude $\Gamma(\mathbf{k}, \omega)$ is equal to Γ^k, the dynamic part contains at least one RA-section. The amplitude Γ^k can be formally defined as the part of the two-particle amplitude $\Gamma(\mathbf{k}, \omega)$ which does not contain any RA-sections.

N.4   With the use of the effective mass m^* in the quasiparticle spectrum ϵ_p and redefining the interaction amplitudes, the explicit dependence on the residue a drops out from Fermi-liquid theory. This is the reason for attaching a^2 to the matrix $\Gamma^k_{l=0}$ in Eq. (1.2).

N.5   In the textbook notations, $2\widetilde{\Gamma}_1 - \Gamma_2 = B_{l=0}$ and $\Gamma_2 = -C_{l=0}$; see Eqs. (18.7) and (18.9) of Chapter 2, Sec. 18 in Ref. 4.

N.6   Also, for obvious reasons, γ^ρ does not contain any terms that can be disconnected by cutting a line of the Coulomb interaction.

N.7   Constant density of states is a usual approximation for Fermi-liquid theory, and it is in particular valid for a two-dimensional electron gas.

N.8   This is a standard Ward identity; see Chapter 2, Sec. 19 in Ref. 4.

N.9   Since we have touched a rather confusing question about negativity of $\partial n / \partial \mu$, it is worth mentioning that the stability of a liquid with charged carriers is determined by the combination $[V_C(k) + \partial \mu / \partial n] > 0$, rather than $\partial \mu / \partial n$ alone; $V_C(k)$, which is the Fourier component of the Coulomb interaction, stabilizes such a liquid.

N.10 In the Matsubara technique, temperature enters as the low-energy cutoff because of the discreteness of the fermionic frequencies, $\epsilon_n = (2n+1)\pi T$.

N.11 On the contrary, in the process of rescattering discussed in Sec. 1, see Figs. 18.1–18.3, there are no integrations over the momenta of the diffusion propagators.

N.12 In the ballistic region, $T > 1/\tau_{el}$, non-analytic temperature corrections due to the interplay of interaction and disorder also exist; they are linear in T. For studies of electric conductivity in the ballistic regime see Refs. 52–54. In our opinion, the effects in the ballistic and diffusive regions have little in common.

N.13 The term "cooperon" reflects relevance of these modes to the same channel in which the Cooper instability develops.

N.14 For our purposes, the difference between the temperature and the rate of de-coherence, $1/\tau_\varphi$, can be ignored.

N.15 An alternative to the replica description of the effects of the e–e interaction of disordered electrons exists; namely, the Keldysh formalism; see Refs. 55 and 56.

N.16 The factor $2\pi T$ which appears in $\hat{\Gamma}$ as a result of the Fourier transform from the Matsubara time to frequency has the same origin as 1/Length appearing in the transitions from spatial coordinates to the wave vectors.

N.17 To get some intuition, one may look on the NLSM as a sort of the Heisenberg functional used in the theory of magnetism. Then, the frequency term in $S[Q]$ is equivalent to interaction with the external magnetic field which determines the direction of the spontaneous magnetization. Furthermore, the fluctuations of the magnetization, i.e., magnons, are the counterparts of the diffusons and cooperons in the discussed problems.

N.18 Magnetic impurities, external magnetic field, or spin-orbit scattering induce additional terms in the action $S[Q]$, see e.g., Ref. 20. These terms make some of the diffusion modes gapped. Then a possible strategy is to preserve the general form of $S[Q]$ as given by Eq. (2.2), but to reduce the auxiliary matrix field $\hat{Q}$ to such a manifold that only singular diffusion modes remain, while all gapped modes will be excluded. Systems with different sets of singular fluctuation propagators (i.e., when the $\hat{Q}$-fields are elements of different manifolds) belong to different universality classes. When

only the fluctuations of density are important, such a system belongs
to the so-called unitary class. The richest case is the orthogonal class
in which fluctuations of the charge- and spin- densities as well as different kind of cooperons are relevant. A system where spin-orbit interaction reduces singular fluctuations to the charge-density mode and
the singlet cooperon belongs to the symplectic class.

N.19 In Fig. 18.1, which illustrates Eq. (1.3), the intermediate sections are
equal to $\frac{\omega_n}{\omega_n+Dk^2}$. Here, the denominator is determined by the diffusion
propagator $\mathcal{D}(k,\omega_n)$, while ω_n in the numerator appears as a result of
summation over a fermionic frequency within the interval available for
this propagator.

N.20 For example, in the diagram presented on the left side of Fig. 18.4, the
amplitude Γ_ρ^0 is converted into Γ_σ as a result of scattering induced by
disorder.

N.21 Naturally, the number of the e–e interaction terms involved in the
action $S[Q]$ is different for different classes. The most general form of
the e–e interaction presented in Eq. (2.2) is needed for the orthogonal
class systems, while for the unitary class only the ρ-term remains.

N.22 For the polarization operator, $\Pi_{st}^0/\nu = 2$. Similarly, in the case of the
spin susceptibility, $\chi_\sigma^0/\nu = 2(g_L\mu_B/2)^2$.

N.23 Compared to the standard definition of the quantum resistance, this
unit contains an additional factor π.

N.24 Decreasing κ corresponds to enlarging blocks in the real-space renormalization procedure.

N.25 The right hand side of Eq. (3.5) starts from the γ_c^2-term which describes the rescattering in the Cooper channel. This is the only term
in the RG equations that does not contain ρ. In the case of attraction,
$\gamma_c < 0$, this term is responsible for the superconducting instability at
low temperatures. Then, there is a competition between the two terms,
and β_{γ_c} describes the *suppression* of the temperature of the superconducting transition by disorder. In amorphous films superconductivity
can be totally suppressed by a moderate amount of disorder.[57]

N.26 The statement about the maximal number of the (extended) interaction vertices at a given number of momentum integrations demands
a certain clarification. As it has been explained above, the rescattering of the electron–hole pairs described by the ladder diagrams (see
Figs. 18.1–18.3) is not accompanied by integrations over momenta of
the diffusion propagators. Therefore, extending vertices by ladders, as

described in Eq. (1.3), does not generate any additional factors ρ. Owing to this fact, in the given order of ρ, the full dependences on the interaction amplitudes can be obtained by means of ladder extensions.

N.27 Connection of the frequency dependent conductivity in the critical region of the MIT with the dielectric constant on the insulating side of the transition was discussed in Refs. 31 and 3.

N.28 In our previous works, the combination $a = \frac{d-2}{d}(1+\widetilde{\zeta})$ was written as $\frac{d-2}{d-\varsigma}$.

N.29 For the analysis of the critical behavior, the data should be taken outside the region of perturbation corrections, $\sigma(T) - \sigma(T = 0) \gtrsim \sigma(T = 0)$ but, on the other hand, one should remain within the quantum transport region, $\sigma(T) < \sigma_{\min}$. (The Mott minimal conductivity $\upsilon_{\min}$ is a conditional boundary separating regions where transport is dominated by classical or quantum mechanisms.) In practice, these inequalities leave a limited window of $\sigma(T)$ appropriate for the analysis.

N.30 In short, the tunneling conductance is determined by the Fourier transform of a *product of two* Green functions of electrons located on the opposite sides of the tunneling junction. Each of them is not gauge invariant by itself, while the product is. Therefore, it is the presence of the counter-electrode, with respect to which the measurement is performed, that makes the TDOS a physically meaningful quantity.

N.31 This result was re-derived by many authors and in a different ways, see e.g., Refs. 58, 59 and 56.

N.32 In this region of momenta, the dynamically screened Coulomb interaction $V_C(k,\omega_n)$ is proportional to $(Dk^2 + \omega_n)/Dk^2$ i.e., it is singular despite of screening. This singularity is the origin of the log-squared corrections to the TDOS.

N.33 The other quantity that is sensitive to this kind of fluctuations is the thermal conductivity. The coordinate-dependent temperature invalidates the arguments about the absence of the energy reference level discussed in the main text.[60]

N.34 Therefore, any "universal" theory of the MIT in dilute electron systems that emphasizes only *one* aspect of the discussed systems — most often it is a very large r_s — cannot describe the observed nonmonotonic $\rho(T)$. An electron liquid characterized by a very strong Coulomb interaction *alone* is, in a sense, as featureless (and universal) as the free electron gas. Such a featureless description cannot provide a non-monotonic $\rho(T)$.

N.35 The universality has been confirmed by comparing the data obtained in samples from different wafers, see Fig. 3 in Ref. 42. Although the critical density at the MIT is sample dependent, the critical resistance has been found to be the same.

N.36 The g_L-factor is about 1.5 times larger than for free-electrons, i.e., $g_L/g_L^0 = \frac{1}{1+F_0^\sigma} = 1 - \Gamma_\sigma \approx 1.5$. The effective mass is about 3 times larger than the band mass, $m^*/m_b \approx 3$.

N.37 This is, actually, a crucial point. That is where the high mobility becomes important in the case of the Si-MOSFET. It was shown in Ref. 61 that in this device the ratio τ_v/τ_{el} monotonically increases as the electron density decreases; here τ_v is the time of the inter-valley scattering. High mobility allows to reach low densities such that for the temperature interval we are interested in the inter-valley scattering is negligible (i.e., the two distinct valleys are well defined).

N.38 The problematic feature of the scaling given by Eqs. (5.1a) and (5.1b) is that the amplitude γ_2 diverges at a finite temperature T^* and thereafter the RG-theory becomes uncontrolled.[2,7] Fortunately, the scale T^* decreases very rapidly with n_v; it was found in Ref. 46 that $\ln\ln(1/\tau_{el}T^*) \sim (2n_v)^2$. This observation makes the problem of the divergence of γ_2 irrelevant for all practical purposes, even for $n_v = 2$ which corresponds to Si-MOSFETs. At $n_v \to \infty$, the theory becomes internally consistent: $T^* \to 0$. Still, a delicate issue is the nature of the ground state of a system with finite n_v. For discussions of this question, see e.g., Refs. 62, 63, 3, 25, 13, and 64.

N.39 The measurements in Ref. 61 confirm our original idea[46] that the difference between high- and low-mobility MOSFET samples is in the strength of the inter-valley scattering rather than in r_s, which anyway is not too large even in the best Si-MOSFET samples.

N.40 For $n_v = 2$, $R(\eta)$ describes quantitatively the temperature dependence of the resistance of high-mobility Si-MOSFETs in the region of ρ up to $\rho \sim 0.5$,[46] which is not so far from the critical region.

Acknowledgments

I thank K. Michaeli and G. Schwiete for the critical reading the manuscript. Extended discussions with B. Spivak are gratefully acknowledged. Finally, I would like to thank Alex Punnoose for the fruitful collaboration. The author acknowledges support from the US-Israel BSF.

Personal Note

I am pleased to contribute this article to the volume celebrating 50 years of Anderson localization. Anderson's contributions to Science influenced my scientific work, especially in the beginning of my career as a many-body physicist. I would like to mention in particular his papers on the Kondo problem. These papers gave a very impressive example of mapping one problem onto another, an approach that in a general sense has also been applied here.

References

1. A. M. Finkel'stein, The influence of Coulomb interaction on the properties of disordered metals, *Zh. Eksp. Teor. Fiz.* **84**, 168 (1983) [*Sov. Phys. JETP* **57**, 97 (1983)].
2. A. M. Finkel'stein, Weak-localization and Coulomb interactions in disordered systems, *Z. Phys. B: Condens. Matter* **56**, 189 (1984).
3. A. M. Finkel'stein, Electron Liquid in Disordered Conductors, *Sov. Sci. Rev./Section A, Phys. Rev.* **14**, 1 (1990).
4. L. P. Pitaevskii and E. M. Lifshitz, Statistical Physics, Part 2, *Course of Theoretical Physics*, Vol. 9, eds. E. M. Lifshitz and L. D. Landau (Pergamon, Oxford, 1980).
5. P. Nozieres and J. M. Luttinger, Derivation of the Landau theory of Fermi-liquids. I and II, *Phys. Rev.* **127**, 1423 and 1431 (1962).
6. N. D. Mermin, Existence of zero sound in a Fermi liquid, *Phys. Rev.* **159**, 161 (1967).
7. C. Castellani, C. DiCastro, P. A. Lee, M. Ma, S. Sorella and E. Tabet, Spin fluctuations in disordered interacting electrons, *Phys. Rev. B* **30**, R1596 (1984).
8. E. Abrahams, P. W. Anderson, D. C. Licciardello and T. V. Ramakrishnan, Scaling theory of localization: Absence of quantum diffusion in two dimensions, *Phys. Rev. Lett.* **42**, 673 (1979).
9. L. P. Gor'kov, A. I. Larkin and D. E. Khmel'nitskii, Particle conductivity in a two-dimensional random potential, *Pis'ma ZhETF* **30**, 248 (1979) [*Sov. Phys. JETP Lett.* **30**, 228 (1979)].
10. B. L. Altshuler and A. G. Aronov, Electron–electron interactions in disordered systems, in *Modern Problems in Condensed Matter Physics*, eds. A. L. Efros and M. Pollak (Elsevier, North Holland, 1985), p. 1.
11. B. L. Altshuler, A. G. Aronov and P. A. Lee, Interaction effects in disordered Fermi systems in two dimensions, *Phys. Rev. Lett.* **44**, 1288 (1980).
12. C. Castellani, C. Di Castro, P. A. Lee and M. Ma, Interaction-driven metal–insulator transitions in disordered fermion systems, *Phys. Rev. B* **30**, 527 (1984).
13. D. Belitz and T. R. Kirkpatrick, The Anderson–Mott transition, *Rev. Mod. Phys.* **66**, 261 (1994).

14. C. Di Castro and R. Raimondi, Disordered electron systems, in *Proc. Int. School of Phys. "Enrico Fermi"*, eds. G. F. Giuliani and G. Vignale, Varenna, Italy, 29 July–8 August 2003 (IOS Press, Amsterdam) arXiv:cond-mat/0402203.

15. M. A. Baranov, I. S. Burmistrov and A. M. M. Pruisken, Non-Fermi-liquid theory for disordered metals near two dimensions, *Phys. Rev. B* **66**, 075317 (2002).

16. A. Punnoose and A. M. Finkel'stein, Metal–insulator transition in disordered two-dimensional electron systems, *Science* **310**, 289 (2005).

17. F. J. Wegner, The mobility edge: continuous symmetry and a conjecture, *Z. Phys. B: Condens. Matter* **35**, 207 (1979).

18. L. Schäfer and F. J. Wegner, Disordered system with n orbitals per site: Lagrange formulation, hyperbolic symmetry, and Goldstone modes, *Z. Phys. B: Condens. Matter* **38**, 113 (1980).

19. A. Houghton, A. Jevicki, R. D. Kenway and A. M. M. Pruisken, Noncompact σ models and the existence of a mobility edge in disordered electronic systems near two dimensions, *Phys. Rev. Lett.* **45**, 394 (1980).

20. K. B. Efetov, A. I. Larkin and D. E. Khmelnitskii, Diffusion mode interaction in the theory of localization, *Zh. Eksp. Teor. Fiz.* **79**, 1120 (1980) [*Sov. Phys. JETP* **52**, 568 (1980)].

21. A. M. Finkel'stein, On the frequency and temperature dependence of the conductivity near a metal–insulator transition, *Pis'ma ZhETF* **37**, 436 (1983) [*Sov. Phys. JETP Lett.* **37**, 517 (1983)].

22. C. Castellani, C. DiCastro, G. Forgacs and S. Sorella, Spin orbit coupling in disordered interacting electron gas, *Solid State Commun.* **52**, 261 (1984).

23. A. M. M. Pruisken, M. A. Baranov and B. Skoric', (Mis-)handling gauge invariance in the theory of the quantum Hall effect. I. Unifying action and the $\nu = 1/2$ state, *Phys. Rev. B* **60**, 16807 (1999).

24. C. Castellani and C. Di Castro, Effective Landau theory for disordered interacting electron systems: Specific-heat behavior, *Phys. Rev. B* **34**, 5935 (1986).

25. C. Castellani, G. Kotliar and P. A. Lee, Fermi-liquid theory of interacting disordered systems and the scaling theory of the metal–insulator transition, *Phys. Rev. Lett.* **59**, 323 (1987).

26. B. Shapiro and E. Abrahams, Scaling for frequence-dependent conductivity in disordered electronic system, *Phys. Rev. B* **24**, 4889–4891 (1981).

27. S. Katsumoto, F. Komori, N. Sano and S. Kobayashi, Fine tuning of metal–insulator transition in $Al_{0.3}Ga_{0.7}As$ using persistent photoconductivity, *J. Phys. Soc. Jap.* **56**, 2259–2262 (1987).

28. H.-L. Lee, J. P. Carini, D. V. Baxter and G. Grüner, Temperature-frequency scaling in amorphous niobium-silicon near the metal–insulator transition, *Phys. Rev. Lett.* **80**, 4261 (1998).

29. H.-L. Lee, J. P. Carini, D. V. Baxter, W. Henderson and G. Grüner, Quantum-critical conductivity scaling for a metal–insulator transition, *Science* **287**, 633 (2000).

30. D. J. Newson and M. Pepper, Critical conductivity at the magnetic field induced metal–insulator transition in n-GaAs and n-InSb, *J. Phys. C* **19**, 3983–3990 (1986).

31. A. M. Finkel'stein, Metal–insulator transition in a disordered system, *Zh. Eksp. Teor. Fiz.* **86**, 367 (1984) [*Sov. Phys. JETP* **59**, 212 (1984)].
32. D. Belitz, T. R. Kirkpatrick and T. Vojta, How generic scale invariance influences quantum and classical phase transitions, *Rev. Mod. Phys.* **77**, 579 (2005).
33. G. Hertel, D. J. Bishop, E. G. Spencer, J. M. Rowell and R. C. Dynes, Tunnelling and transport measurements at the metal–insulator transition of amorphous Nb:Si, *Phys. Rev. Lett.* **50**, 742–746 (1983).
34. W. L. McMillan, Scaling theory of the metal–insulator transition in amorphous materials, *Phys. Rev. B* **24**, 2739–2743 (1981).
35. B. L. Altshuler and A. G. Aronov, *Zh. Eksp. Teor. Fiz.* **77**, 2028 (1979) [*Sov. Phys. JETP* **50**, 968 (1979)].
36. B. L. Altshuler, A. G. Aronov and A. Yu. Zyuzin, *Zh. Eksp. Teor. Fiz.* **86**, 709 (1984) [*Sov. Phys. JETP* **59**, 4151 (1984)].
37. A. M. Finkel'stein, in *Proc. Int. Symp. on Anderson Localization*, Springer Proc. in Physics, Vol. 28, eds. T. Ando and H. Fukuyama (Springer-Verlag, Berlin, 1988), p. 230.
38. A. M. Finkel'stein, Suppression of superconductivity in homogeneously disordered systems, *Physica B* **197**, 636 (1994).
39. S. V. Kravchenko, G. V. Kravchenko, J. E. Furneaux, V. M. Pudalov and M. D'Iorio, Possible metal–insulator transition at $B = 0$ in two dimensions, *Phys. Rev. B* **50**, 8039 (1994).
40. S. V. Kravchenko, J. E. Furneaux, W. E. Mason, G. E. Bowker, J. E. Furneaux, V. M. Pudalov and M. D'Iorio, Scaling of an anomalous metal–insulator transition in a two-dimensional system in silicon at $B = 0$, *Phys. Rev. B* **51**, 7038 (1995).
41. E. Abrahams, S. V. Kravchenko and M. P. Sarachik, Metallic behavior and related phenomena in two dimensions, *Rev. Mod. Phys.* **73**, 251, (2001).
42. S. V. Kravchenko and M. P. Sarachik, Metal–insulator transition in two-dimensional electron systems, *Rep. Prog. Phys.* **67**, 1 (2004).
43. B. Spivak, S. V. Kravchenko, S. A. Kivelson and X. P. A. Gao, Transport in strongly correlated two-dimensional electron fluids, *Rev. Mod. Phys.*, to be published (2010); arXiv:0905.0414.
44. V. M. Pudalov, G. Brunthaler, A. Prinz and G. Bauer, Metal–insulator transition in two dimensions, *Physica (Amsterdam)* **3E**, 79 (1998).
45. D. Simonian, S. V. Kravchenko, M. P. Sarachik and V. M. Pudalov, Magnetic field suppression of the conducting phase in two dimensions, *Phys. Rev. Lett.* **79**, 2304 (1997).
46. A. Punnoose and A. M. Finkel'stein, Dilute electron gas near the metal–insulator transition: Role of valleys in silicon inversion layers, *Phys. Rev. Lett.* **88**, 016802 (2002).
47. P. A. Lee and T. V. Ramakrishnan, Magnetoresistance of weakly disordered electrons, *Phys. Rev. B* **26**, 4009 (1982).
48. C. Castellani, C. Di Castro and P. A. Lee, Metallic phase and metal–insulator transition in two-dimensional electronic systems, *Phys. Rev. B* **57**, 9381–9384 (1998).

49. S. Anissimova, S. V. Kravchenko, A. Punnoose, A. M. Finkel'stein and T. M. Klapwijk, Flow diagram of the metal–insulator transition in two dimensions, *Nat. Phys.* **3**, 707 (2007).
50. D. A. Knyazev, O. E. Omelyanovskii, V. M. Pudalov and I. S. Burmistrov, Critical behavior of transport and magnetotransport in 2D electron system in Si in the vicinity of the metal–insulator transition, *Pis'ma ZhETF* **84**, 780 (2006); [*JETP Lett.* **84**, 662 (2006)].
51. K. G. Wilson, Quantum field-theory models in less than 4 dimensions, *Phys. Rev. D* **7**, 2911 (1973).
52. F. Stern and S. Das Sarma, *Solid-State Electron.* **28**, 158 (1985); S. Das Sarma, Theory of finite-temperature screening in a disordered two-dimensional electron gas, *Phys. Rev. B* **33**, 5401 (1986).
53. A. Gold and V. T. Dolgopolov, Temperature dependence of the conductivity for the two-dimensional electron gas: Analytical results for low temperatures, *Phys. Rev. B* **33**, 1076 (1986).
54. G. Zala, B. N. Narozhny and I. L. Aleiner, Interaction corrections at intermediate temperatures: Longitudinal conductivity and kinetic equation, *Phys. Rev. B* **64**, 214204 (2001).
55. C. Chamon, A. W. Ludwig and C. Nayak, Schwinger-Keldysh approach to disordered and interacting electron systems: Derivation of Finkel'steins renormalization-group equations, *Phys. Rev. B* **60**, 2239 (1999).
56. A. Kamenev and A. Andreev, Electron-electron interactions in disordered metals: Keldysh formalism, *Phys. Rev. B* **60**, 2218 (1999).
57. A. M. Finkel'stein, Superconducting transition temperature in amorphous films, *Pis'ma ZhETF* **45**, 37 (1987) [*Sov. Phys. JETP Lett.* **45**, 63 (1987)].
58. Yu. V. Nazarov, Anomalous current-voltage characteristics of tunnel junctions, *Zh. Eksp. Teor. Fiz.* **96**, 975 (1989) [*Sov. Phys. JETP* **68**, 561 (1989)].
59. L. S. Levitov and A. V. Shytov, Semiclassical theory of the Coulomb anomaly, *Pis'ma ZhETF* **66**, 200 (1997) [*JETP Lett.* **66**, 214 (1997)].
60. K. Michaeli and A. M. Finkel'stein, Quantum kinetic approach for studying thermal transport in the presence of electron–electron interactions and disorder, *Phys. Rev. B* **80**, 115111 (2009).
61. A. Yu. Kuntsevich, N. N. Klimov, S. A. Tarasenko, N. S. Averkiev, V. M. Pudalov, H. Kojima and M. E. Gershenson, Intervalley scattering and weak localization in Si-based two-dimensional structures, *Phys. Rev. B* **75**, 195330 (2007).
62. A. M. Finkel'stein, Spin fluctuations in disordered systems near the metal–insulator transition, *Pis'ma ZhETF* **40**, 63 (1984) [*Sov. Phys. JETP Lett.* **40**, 796 (1984)].
63. A. M. Finkel'stein, ESR near a metal–insulator transition, *Pis'ma ZhETF* **46**, 407 (1987) [*Sov. Phys. JETP Lett.* **46**, 513 (1987)].
64. B. N. Narozhny, I. L. Aleiner and A. I. Larkin, Magnetic fluctuations in two-dimensional metals close to the Stoner instability, *Phys. Rev. B* **62**, 14898 (2000).

Chapter 19

TYPICAL-MEDIUM THEORY OF
MOTT–ANDERSON LOCALIZATION

V. Dobrosavljević

Department of Physics and National High Magnetic Field Laboratory,
Florida State University, Tallahassee, Florida 32310, USA

The Mott and the Anderson routes to localization have long been recognized as the two basic processes that can drive the metal–insulator transition (MIT). Theories separately describing each of these mechanisms were discussed long ago, but an accepted approach that can include both has remained elusive. The lack of any obvious static symmetry distinguishing the metal from the insulator poses another fundamental problem, since an appropriate *static* order parameter cannot be easily found. More recent work, however, has revisited the original arguments of Anderson and Mott, which stressed that the key diference between the metal end the insulator lies in the dynamics of the electron. This physical picture has suggested that the "typical" (geometrically averaged) escape rate $\tau_{\mathrm{typ}}^{-1} = \exp\langle \ln \tau_{\mathrm{esc}}^{-1} \rangle$ from a given lattice site should be regarded as the proper *dynamical order parameter* for the MIT, one that can naturally describe both the Anderson and the Mott mechanism for localization. This article provides an overview of the recent results obtained from the corresponding *Typical-Medium Theory*, which provided new insight into the the two-fluid character of the Mott–Anderson transition.

1. From Metal to Insulator: A New Perspective

Metal or insulator — and why? To answer this simple question has been the goal and the driving force for much of the physical science as we know it today. Going back to Newton's not-so-successful exercises in Alchemy, the scientist had tried to understand what controls the flow of electricity in metals and what prevents it in insulators.[1] To understand it and to control it — achieving this could prove more useful and lucrative than converting lead into gold. Indeed, the last few decades have witnessed some most amazing and unexpected advances in material science and technology.

And this ability — its intellectual underpinning — is what was indispensable in designing and fabricating the iPhone, the X-Box, and the MRI diagnostic tool. Today's kids have grown up in a different world than had their parents — all because we have learned a few basic ideas and principles of electron dynamics.

In almost every instance, these advances are based on materials that find themselves somewhere between metals and insulators. Material properties are easy to tune in this regime, where several possible ground states compete.[5] Here most physical quantities display unusual behavior,[6] and prove difficult to interpret using conventional ideas and approaches. Over the last few decades, scores of theoretical scenarios and physical pictures have been proposed, most of which will undoubtedly end up in back drawers of history. Last couple of years, however, have seen a veritable avalanche of new experimental results (Fig. 19.1), which provide compelling clues as to what the theorists should not overlook: the significant effects of spatial inhomogeneities in the midst of strongly correlated phases.

To understand many, if not most exotic new materials, one has to tackle the difficult problem of understanding the metal–insulator quantum phase transition, as driven by the combined effects of strong electronic correlations and disorder. Traditional approaches to the problem, which emerged in the

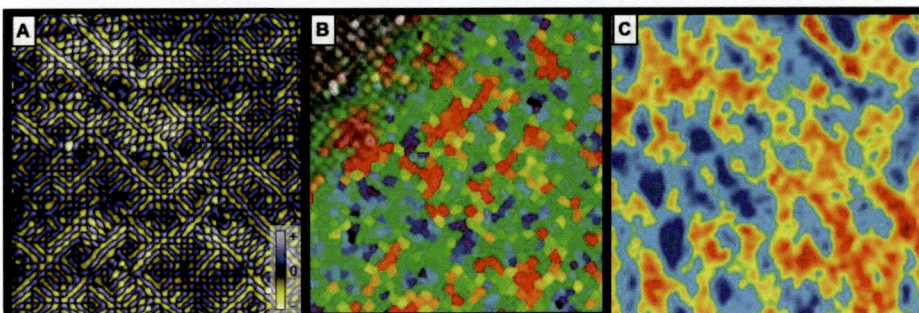

Fig. 19.1. Spectacular advances in scanning tunneling microscopy (STM) have revealed that many "bad metals" or barely-doped insulators are surprisingly inhomogeneous on the nano-scale. Understanding and controlling these materials will not be possible without coming to grips with the origin, the stability, and the statistics of such mesoscopic granularity. (A) "Tunneling asymmetry" imaging[2] provides evidence for the emergence of a low temperature "electronic cluster glass" within the superconducting phase of $Ca_{1.88}Na_{0.12}CuO_2Cl_2$; (B) Fourier-transform STM[3] reveals nano-scale Fermi surface variations in $Bi_2Sr_2CuO_{6-x}$; (C) Differential conductance maps[4] showing spatial variations of the local pseudogaps in the normal phase $(T \gg T_c)$ of $Bi_2Sr_2CaCu_2O_{8+\delta}$.

early 1980s, have focused on examining the perturbative effects of disorder within the Fermi liquid framework. Despite their mathematical elegance, these theories, unfortunately, prove ill-suited to describe several key physical processes, such as tendency to local magnetic moment formation and the approach to the Mott insulating state. In addition, such weak-coupling theories cannot easily describe strongly inhomogeneous phases, with behavior often dominated by broad distributions and rare disorder configurations.

This new insight, which is largely driven by experimental advances, seems to suggest that an alternative theoretical picture may provide a better starting point. In this article we describe recent advances based on a new theoretical method, which offers a complementary perspective to the conventional weak-coupling theories. By revisiting the original ides of Anderson and Mott, it examines the *typical escape rate* from a given site as the fundamental *dynamical order parameter* to distinguish between a metal and an insulator. This article describes the corresponding *Typical-Medium Theory (TMT)* and discusses some of its recent results and potential applications. We fist discuss discuss, in some detail, several experimental and theoretical clues suggesting that a new theoretical paradigm is needed. The formulation of TMT for Anderson localization of noninteracting electrons is then discussed, with emphasis on available analytical results. Finally, we review recent progress in applying TMT to the Mott–Anderson transition for disordered Hubbard models, and discuss resulting the two-fluid behavior at the critical point.

2. Theoretical Challenges: Beyond Cinderella's Slipper?

The existence of a sharp metal–insulator transition at $T = 0$ has been appreciated for many years.[1] Experiments on many systems indeed have demonstrated that a well defined critical carrier concentration can easily be identified. On the theoretical side, ambiguities on how to describe or even think of the metal–insulator transition have made it difficult to directly address the nature of the critical region. In practice, one often employs the theoretical tools that are available, even if possibly inappropriate. Even worse, one often focuses on those systems and phenomena that fit an available theoretical mold, ignoring and brushing aside precisely those features that seem difficult to understand. This "Cinderella's slipper" approach is exactly what one should not do; unfortunately it happens all too often. A cure is, of course, given by soberly confronting the experimental reality: what seems paradoxical at first sight often proves to be the first clue to the solution.

2.1. *Traditional approaches to disordered interacting electrons*

Most studies carried out over the last thirty years have focused on the limit of weak disorder,[7] where considerable progress has been achieved. Here, for non-interacting electrons the conductance was found to acquire singular (diverging) corrections in one and two dimensions, an effect known as "weak localization".[7,8] According to these predictions, for $d \leq 2$ the conductivity would monotonically decrease as the temperature is lowered, and would ultimately lead to an insulating state at $T = 0$. Interestingly, similar behavior was known in Heisenberg magnets,[9–11] where it resulted from $d = 2$ being the *lower critical dimension* for the problem. This analogy with conventional critical phenomena was first emphasized by the "gang of four",[8] as well as Wegner,[9,10] who proposed an approach to the metal–insulator transition based on expanding around two dimensions. For this purpose, an effective low energy description was constructed,[9,10,12] which selects those processes that give the leading corrections at weak disorder in and near two dimensions. This "non-linear sigma model" formulation[9,10,12] was subsequently generalized to interacting electrons by Finkelshtein,[13] and studied using renormalization group methods in $2 + \varepsilon$ dimensions.[13–15] In recent years, the non-linear sigma model of disordered interacting electrons has been extensively studied by several authors.[16,17]

While the sigma model approach presented considerable formal complexity, its physical content proved — in fact — to be remarkably simple. As emphasized by Castellani, Kotliar and Lee,[18] one can think of the sigma model of disordered interacting electrons as a low energy *Fermi liquid* description of the system. Here, the low energy excitations are viewed as a gas of diluted quasi-particles that, at least for weak disorder, can be described by a small number of Fermi liquid parameters such as the diffusion constant, the effective mass, and the interaction amplitudes. In this approach, one investigates the evolution of these Fermi liquid parameters as weak disorder is introduced. The metal–insulator transition is then identified by the *instability* of this Fermi liquid description, which in $d = 2 + \varepsilon$ dimensions happens at weak disorder, where controlled *perturbative* calculations can be carried out.

Remarkably, by focusing on such a stability analysis of the metallic state, one can develop a theory for the transition which does not require an *order parameter* description, in contrast to the standard approaches to critical phenomena.[11] This is a crucial advantage of the sigma model approach, precisely because of the ambiguities in defining an appropriate order parameter. We should stress, however, that by construction, the sigma model focuses on those physical processes that dominate the perturbative, weak disorder regime. In real systems, the metal–insulator transition is found at strong disorder, where a completely different set of processes could be at play.

2.2. *Anderson's legacy: strong disorder fluctuations*

From a more general point of view, one may wonder how pronounced are the effects of disorder on quantum phase transitions. Impurities and defects are present in every sample, but their full impact has long remained ill-understood. In early work, Griffiths discovered[19,20] that rare events due to certain types of disorder can produce nonanalytic corrections in thermodynamic response. Still, for classical models and thermal phase transitions he considered, these effects are so weak to remain unobservably small.[21] The critical behavior then remains essentially unmodified.

More recent efforts turned to quantum $(T = 0)$ phase transitions,[22] where the rare disorder configurations prove much more important. In some systems they give rise to "Quantum Griffiths Phases" (QGP)[6,23] (Fig. 19.2), associated with the "Infinite Randomness Fixed Point" (IRFP) phenomenology.[24] Here, disorder effects produce singular thermodynamic response not only at the critical point, but over *an entire region* in its vicinity. In other cases, related disorder effects are predicted[25,26] to result in "rounding" of the critical point, or to produce intermediate "cluster glass" phases[27,28] masking the critical point. Physically, QGP-IRFP behavior means[6,23] that very close to the critical point, the system looks increasingly inhomogeneous even in static response.

But how robust and generic may such pronounced sensitivity to disorder be in real systems? Does it apply only to (magnetic and/or charge) ordering transitions, or is it relevant also for the metal–insulator transitions (MITs)? A conclusive answer to these questions begs the ability to locally visualise the system on the nano-scale. Remarkably, very recent STM images provide striking evidence of dramatic spatial inhomogeneities in surprisingly many systems. While much more careful experimental and theoretical work is called for, these new insights makes it abundantly clear that strong disorder effects — as first emphasized by early seminal work of Anderson[32] — simply cannot be disregarded.

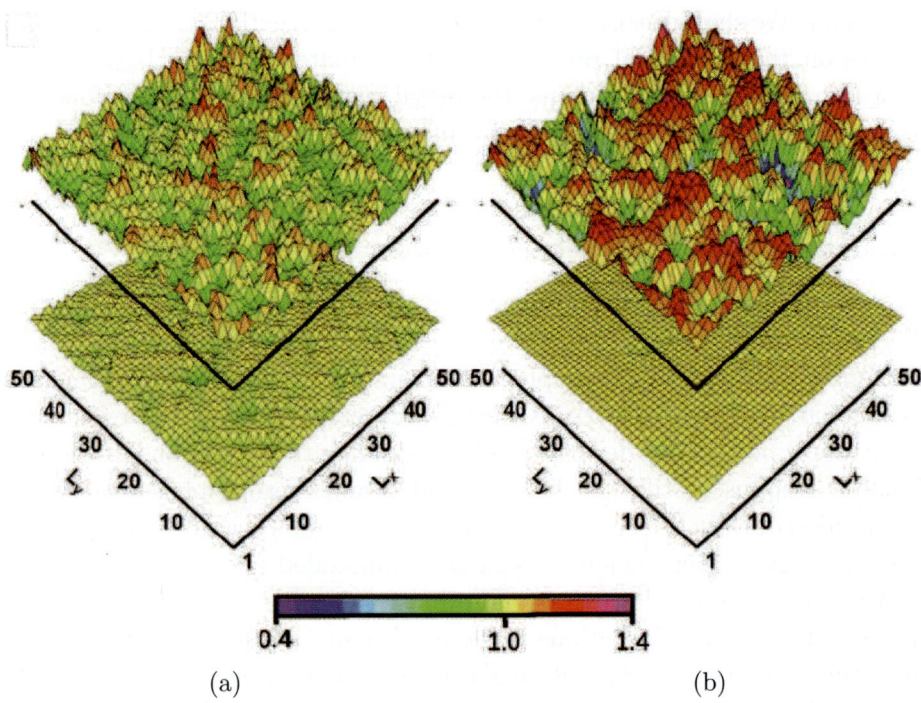

Fig. 19.2.   Theory predictions[29] for an "Electronic Griffiths Phase"[30] in a moderately disordered normal metal near a Mott meal–insulator transition. Local density
of states (LDOS) spectra look dramatically "smoother" near the Fermi energy (bottom) than away from it (top). This contrast is more pronounced close to the Mott
transition (a), than outside the critical region (b). Very similar behavior was recently observed by STM imaging of the superconducting phase of doped cuprates,[2]
but our results strongly suggest that such energy-resolved "disorder healing"[29–31]
is a much more general property of Mott systems.

2.3. The curse of Mottness: the not-so-Fermi liquids

One more issue poses a major theoretical challenge. According to Landau's
Fermi liquid theory, any low temperature metal behaves in a way very similar to a gas of weakly interacting fermions. In strongly correlated systems,
closer to the Mott insulating state, this behavior is typically observed only
below a modest crossover temperature $T^* \ll T_F$. Adding disorder typically
reduces T^* even further, and much of the experimentally relevant temperature range simply does not conform to Landau's predictions. Theoretically,
this situation poses a serious problem, since the excitations in this regime
no longer assume the character of diluted quasiparticles. Here perturbative

corrections to Fermi liquid theory simply do not work,[6] and a conceptually new approach is needed.

A new theoretical paradigm, which works best precisely in the incoherent metallic regime (see Fig. 19.3), has been provided by the recently developed *Dynamical Mean-Field Theory* (DMFT) methods.[36] Unfortunately, in its original formulation, which is strictly exact in the limit of infinite dimensions, DMFT is not able to capture Anderson localization effects. Over the last twelve years, this nonperturbative approach has been further extended[6,37–44] to incorporate the interplay between the two fundamental mechanisms for electron localization: the Mott (interaction-driven)[1] and the Anderson (disorder-driven)[32,45] route to arrest the electronic motion. In addition, the DMFT formulation can be very naturally extended to also describe strongly inhomogeneous and glassy phases of electrons,[46–55] and even capture some aspects of the QGP physics found at strong disorder.[6,27,28,39,40,44,56–63] In the following, we first discuss the DMFT method as a general order-parameter theory for the metal–insulator transition, and the explain how it needs to be modified to capture Anderson localization effects.

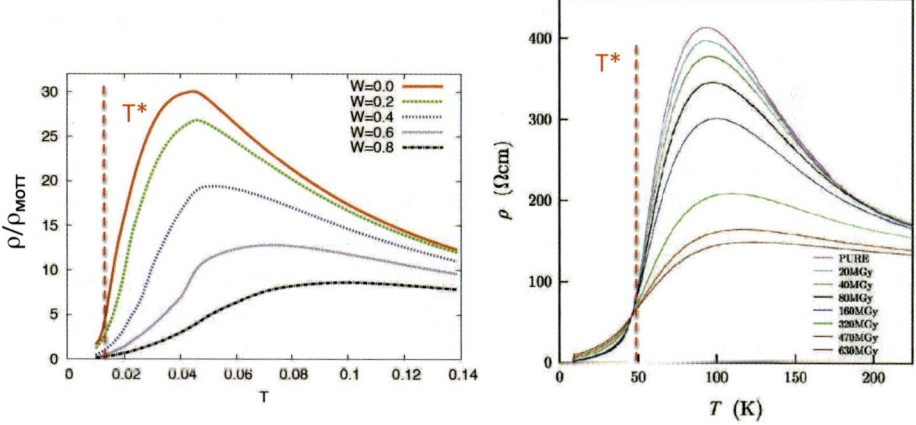

Fig. 19.3. Finite temperature metal–insulator crossover in transport close to a disordered Mott transition. Very high values of resistivity, strongly exceeding the "Mott limit"[1,33] are observed above the crossover temperature T^*. Remarkably, increasing disorder W reduces the resistivity maximum, rendering the system effectively more metallic. This behavior, which is clearly seen in our DMFT modelling[34] (left panel), has very recently been also observed in experiments[35] on organic charge-transfer salts (right panel), where disorder is systematically introduced by X-ray irradiation.

3. Order-Parameter Approach to Interaction-Localisation

3.1. *Need for an order-parameter theory: experimental clues*

In conventional critical phenomena, simple mean-field approaches such as
the Bragg–Williams theory of magnetism, or the van der Waals theory for
liquids and gases work remarkably well — everywhere except in a very arrow critical region. Here, effects of long wavelength fluctuations emerge
that modify the critical behavior, and its description requires more sophisticated theoretical tools, based on renormalization group (RG) methods. A
basic question then emerges when looking at experiments: is a given phenomenon a manifestation of some underlying mean-field (local) physics, or
is it dominated by long-distance correlations, thus requiring an RG description? For conventional criticality the answer is well know, but how about
metal–insulator transitions? Here the experimental evidence is much more
limited, but we would like to emphasize a few well-documented examples
which stand out.

3.1.1. *Doped semiconductors*

Doped semiconductors such as Si:P[64] are the most carefully studied examples
of the MIT critical behavior. Here the density-dependent conductivity extrapolated to $T = 0$ shows sharp critical behavior[65] of the form $\sigma \sim (n - n_c)^\mu$,
where the critical exponent $\mu \approx 1/2$ for uncompensated samples (half-filled
impurity band), while dramatically different $\mu \approx 1$ is found for heavily compensated samples of Si:P, B, or in presence of strong magnetic fields. Most
remarkably, the dramatically differences between these cases is seen over an
extremely broad concentration range, roughly up to several times the critical density. Such robust behavior, together with simple apparent values for
the critical exponents, seems reminiscent of standard mean-field behavior in
ordinary criticality.

3.1.2. *2D-MIT*

Signatures of a remarkably sharp metal–insulator transition has also been
observed[66–68] in several examples of two-dimensional electron gases (2DEG)
such as silicon MOSFETs. While some controversy regarding the nature or
even the driving force for this transition remains a subject of intense debate, several experimental features seem robust properties common to most
studied samples and materials. In particular, various experimental groups
have demonstrated[66,67] striking scaling of the resistivity curves (Fig. 19.5)

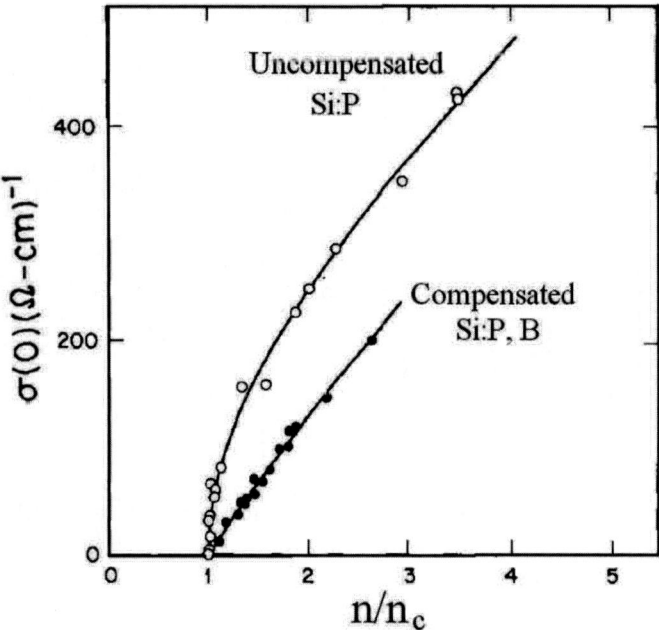

Fig. 19.4. Critical behavior of the conductivity for uncompensated Si:P and compensated Si:P, B.[65] The conductivity exponent $\mu \approx 1/2$ in absence of compensation, while $\mu \approx 1$ in its presence. Clearly distinct behavior is observed in a surprisingly broad range of densities, suggesting mean-field scaling. Since compensation essentially corresponds to carrier doping away from a half-filled impurity band,[64] it has been suggested[7] that the difference between the two cases may reflect the role of strong correlations.

in the critical region, which seems to display[69] remarkable mirror symmetry ("duality")[70] over a surprisingly broad interval of parameters. In addition, the characteristic behavior extends to remarkably high temperatures, which are typically comparable the Fermi temperature.[68] One generally does not expect a Fermi liquid picture of diluted quasiparticles to apply at such "high energies", or any correlation length associated with quantum criticality to remain long.

These experiments taken together provide strong hints that an appropriate mean-field description is what is needed. It should provide the equivalent of the van der Waals equation of state, for disordered interacting electrons. Such a theory has long been elusive, primarily due to a lack of a simple order-parameter formulation for this problem. Very recently, an alternative approach to the problem of disordered interacting electrons has been formulated, based on dynamical mean-field (DMFT) methods.[36] This formulation

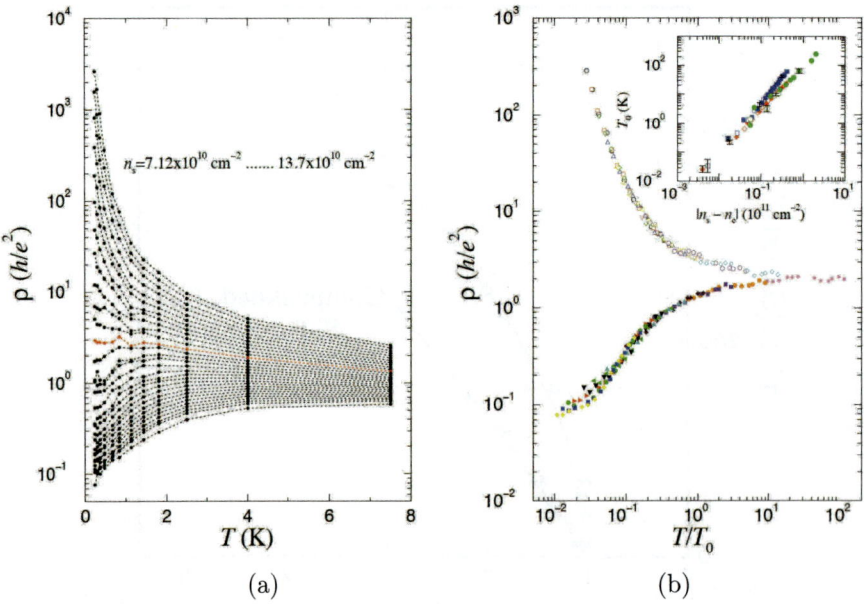

(a)　　　　　　　　(b)

Fig. 19.5.　The resistivity curves (left panel) for a two-dimensional electron system in silicon[66] show a dramatic metal–insulator crossover as the density is reduced below $n_c \sim 10^{11}$ cm^{-2}. Note that the system has "made up its mind" whether to be a metal or an insulator even at surprisingly high temperatures $T \sim T_F \approx 10$ K. The right panel displays the scaling behavior which seems to hold over a comparable temperature range. The remarkable "mirror symmetry"[69] of the scaling curves seems to hold over more then an order of magnitude for the resistivity ratio. This surprising behavior has been interpreted[70] as evidence that the transition region is dominated by strong coupling effects characterizing the insulating phase.

is largely complementary to the scaling approach, and has already resulting in several striking predictions. In the following, we briefly describe this method, and summarize the main results that have been obtained so far.

3.2. *The DMFT physical picture*

The main idea of the DMFT approach is in principle very close to the original Bragg–Williams (BW) mean-field theories of magnetism.[11] It focuses on a single lattice site, but replaces[36] its environment by a self-consistently determined "effective medium", as shown in Fig. 19.6.

In contrast to the BW theory, the environment cannot be represented by a static external field, but instead must contain the information about the dynamics of an electron moving in or out of the given site. Such a description can be made precise by formally integrating out[36] all the degrees

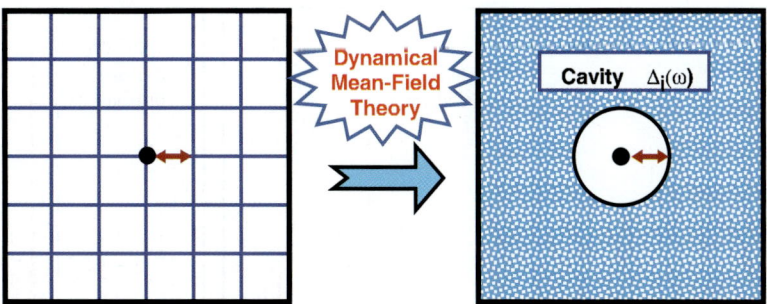

Fig. 19.6.   In dynamical mean-field theory, the environment of a given site is represented by an effective medium, represented by its "cavity spectral function" $\Delta_i(\omega)$. In a *disordered* system, $\Delta_i(\omega)$ for different sites can be very different, reflecting Anderson localization effects.

of freedom on other lattice sites. In presence of electron–electron interactions, the resulting local effective action has an arbitrarily complicated form. Within DMFT, the situation simplifies, and all the information about the environment is contained in the local single particle spectral function $\Delta_i(\omega)$. The calculation then reduces to solving an appropriate quantum impurity problem supplemented by an additional self-consistency condition that determines this "cavity function" $\Delta_i(\omega)$.

The precise form of the DMFT equations depends on the particular model of interacting electrons and/or the form of disorder, but most applications[36] to this date have focused on Hubbard and Anderson lattice models. The approach has been very successful in examining the vicinity of the Mott transition in clean systems, in which it has met spectacular successes in elucidating various properties of several transition metal oxides,[40] heavy fermion systems, and even Kondo insulators.[71]

3.3. *DMFT as an order-parameter theory for the MIT*

The central quantity in the DMFT approach is the local "cavity" spectral function $\Delta_i(\omega)$. From the physical point of view, this object essentially represents the *available electronic states* to which an electron can "jump" on its way out of a given lattice site. As such, it provides a natural order parameter description for the MIT. Of course, its form can be substantially modified by either the electron–electron interactions or disorder, reflecting the corresponding modifications of the electron dynamics. According to Fermi's golden rule, the transition rate to a neighboring site is proportional to the density of final states — leading to insulating behavior whenever

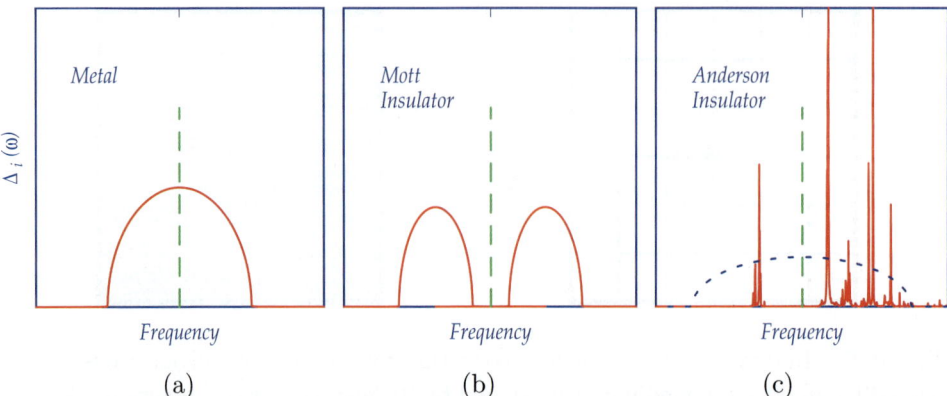

Fig. 19.7.   The local cavity spectral function $\Delta_i(\omega)$ as the order parameter for the MIT. In a metal (a) there are available electronic states near the Fermi level (dashed line) to which an electron from a given site can delocalize. Both for a Mott insulator (b) and the Anderson insulator (c) the Fermi level is in the gap, and the electron cannot leave the site. Note that the *averaged* spectral function (dotted line in (c)) has no gap for the Anderson insulator, and thus cannot serve as an order parameter.

$\Delta_i(\omega)$ has a gap at the Fermi energy. In the case of a Mott transition in the absence of disorder, such a gap is a direct consequence of the strong on-site Coulomb repulsion, and is the same for every lattice site.

The situation is more subtle in the case of disorder-induced localization, as first noted in the pioneering work of Anderson.[32] Here, the *average* value of $\Delta_i(\omega)$ has no gap and thus cannot serve as an order parameter. However, as Anderson noted a long time ago, "...no real atom is an average atom...".[45] Indeed, in an Anderson insulator, the environment "seen" by an electron on a given site can be very different from its average value. In this case, the *typical* "cavity" spectral function $\Delta_i(\omega)$ consists of several delta-function (sharp) peaks, reflecting the existence of localized (bound) electronic states, as shown in Fig. 19.7(c). Thus a *typical* site is embedded in an environment that has a *gap* at the Fermi energy — resulting in insulating behavior. We emphasize that the location and width of these gaps strongly vary from site to site. These strong fluctuations of the local spectral functions persist on the metallic side of the transition, where the typical spectral density $\Delta_{\text{typ}} = \exp\langle \ln(\Delta_i) \rangle$ can be much smaller than its average value. Clearly, a full *distribution function* is needed to characterize the system. The situation is similar as in other disordered systems, such as spin glasses.[72] Instead of simple averages, here the entire distribution function plays a role of an order parameter, and undergoes a qualitative change at the phase transition.

The DMFT formulation thus naturally introduces self-consistently defined order parameters that can be utilized to characterize the qualitative differences between various phases. In contrast to clean systems, these order parameters have a character of distribution functions, which change their qualitative form as we go from the normal metal to the non-Fermi liquid metal, to the insulator.

4. Typical Medium Theory for Anderson localization

In the following, we demonstrate how an appropriate local order parameter can be defined and self-consistently calculated, producing a mean-field like description of Anderson localization. This formulation is *not restricted* to either low temperatures or to Fermi liquid regimes, and in addition can be straightforwardly combined with well-known dynamical mean-field theories (DMFT)[36,38,39,73–75] of strong correlation. In this way, our approach which we call the *Typical Medium Theory* (TMT), opens an avenue for addressing questions difficult to tackle by any alternative formulation, but which are of crucial importance for many physical systems of current interest.

Our starting point is motivated by the original formulation of Anderson,[32] which adopts a *local* point of view, and investigates the possibility for an electron to *delocalize* from a given site at large disorder. This is most easily accomplished by concentrating on the (unaveraged) local density of electronic states (LDOS)

$$\rho_i(\omega) = \sum_n \delta(\omega - \omega_n)|\psi_n(i)|^2. \qquad (4.1)$$

In contrast to the global (averaged) density of states (ADOS) which is not critical at the Anderson transition, the LDOS undergoes a qualitative change upon localization, as first noted by Anderson.[32] This follows from the fact that LDOS directly measures the local amplitude of the electronic wavefunction. As the electrons localize, the local spectrum turns from a continuous to an essentially discrete one,[32] but the *typical* value of the LDOS vanishes. Just on the metallic side, but very close to the transition, these delta-function peaks turn into long-lived resonance states and thus acquire a finite *escape rate* from a given site. According to to Fermi's golden rule, this escape rate can be estimated[32] as $\tau_{esc}^{-1} \sim t^2 \rho$, where t is the inter-site hopping element, and ρ is the density of local states of the immediate neighborhood of a given site.

The *typical* escape rate is thus determined by the typical local density of states (TDOS), so that the TDOS directly determines the conductivity of the electrons. This simple argument strongly suggests that the TDOS should be recognized as an appropriate order parameter at the Anderson transition. Because the relevant distribution function for the LDOS becomes increasingly broad as the transition is approached, the desired typical value is well represented by the *geometric average* $\rho_{\text{TYP}} = \exp\{\langle \ln \rho \rangle\}$. Interestingly, recent scaling analyses[76,77] of the multi-fractal behavior of electronic wavefunctions near the Anderson transition has independently arrived at the same conclusion, identifying the TDOS as defined by the geometric average as the fundamental order parameter.

4.1. *Self-consistency conditions*

To formulate a self-consistent theory for our order parameter, we follow the "cavity method," a general strategy that we borrow from the DMFT.[36] In this approach, a given site is viewed as being embedded in an effective medium characterized by a local self energy function $\Sigma(\omega)$. For simplicity, we concentrate on a single band tight binding model of noninteracting electrons with random site energies ε_i with a given distribution $P(\varepsilon_i)$. The corresponding local Green's function then takes the form

$$G(\omega, \varepsilon_i) = [\omega - \varepsilon_i - \Delta(\omega)]^{-1}. \tag{4.2}$$

Here, the "cavity function" is given by

$$\Delta(\omega) = \Delta_o(\omega - \Sigma(\omega)) \equiv \Delta' + i\Delta'', \tag{4.3}$$

and

$$\Delta_o(\omega) = \omega - 1/G_o(\omega), \tag{4.4}$$

where the lattice Green's function

$$G_o(\omega) = \int_{-\infty}^{+\infty} d\omega' \, \frac{\rho_0(\omega')}{\omega - \omega'} \tag{4.5}$$

is the Hilbert transform of the bare density of states $\rho_0(\omega)$ which specifies the band structure.

Given the effective medium specified by a self-energy $\Sigma(\omega)$, we are now in the position to evaluate the order parameter, which we choose to be the TDOS as given by

$$\rho_{\text{typ}}(\omega) = \exp\left\{ \int d\varepsilon_i \, P(\varepsilon_i) \, \ln \rho(\omega, \varepsilon_i) \right\}, \tag{4.6}$$

where the LDOS $\rho(\omega, \varepsilon_i) = -\frac{1}{\pi}\mathrm{Im}G(\omega, \varepsilon_i)$, as given by Eqs. (4.2)–(4.5). To obey causality, the Green's function corresponding to $\rho_{typ}(\omega)$ must be specified by analytical continuation, which is performed by the Hilbert transform

$$G_{\text{typ}}(\omega) = \int_{-\infty}^{+\infty} d\omega' \, \frac{\rho_{\text{typ}}(\omega')}{\omega - \omega'}. \tag{4.7}$$

Finally, we close the self-consistency loop by setting the Green's functions of the effective medium be equal to that corresponding to the local order parameter, so that

$$G_{\text{em}}(\omega) = G_o(\omega - \Sigma(\omega)) = G_{\text{typ}}(\omega). \tag{4.8}$$

It is important to emphasize that our procedure defined by Eqs. (4.2)–(4.8) is not specific to the problem at hand. The same strategy can be used in any theory characterized by a local self-energy. The only requirement specific to our problem is the definition of the TDOS as a local order parameter given by Eq. (4.6). If we choose the *algebraic* instead of the geometric average of the LDOS, our theory would reduce to the well-known coherent potential approximation (CPA),[78] which produces excellent results for the ADOS for any value of disorder, but finds no Anderson transition. Thus TMT is a theory having a character very similar to CPA, with a small but crucial difference — the choice of the correct order parameter for Anderson localization.

In our formulation, as in DMFT, all the information about the electronic band structure is contained in the choice of the bare DOS $\rho_0(\omega)$. It is not difficult to solve Eqs. (4.2)–(4.8) numerically, which can be efficiently done using fast Fourier transform methods.[36] We have done so for several model of bare densities of states, and find that most of our qualitative conclusions do not depend on the specific choice of band structure. We illustrate these findings using a simple "semicircular" model for the bare DOS given by $\rho_0(\omega) = \frac{4}{\pi}\sqrt{1 - (2\omega)^2}$, for which $\Delta_o(\omega) = G_o(\omega)/16$.[36] Here and in the rest of this paper all the energies are expressed in units of the bandwidth, and the random site energies ε_i are uniformly distributed over the interval $[-W/2, W/2]$. The evolution of the TDOS as a function of W is shown in Fig. 19.10. The TDOS is found to decrease and eventually vanish even at the band center at $W \approx 1.36$. For $W < W_c$, the part of the spectrum where TDOS remains finite corresponds to the region of extended states (mobile electrons), and is found to shrink with disorder, indicating that the band tails begin to localize. The resulting phase diagram is presented in Fig. 19.8, showing the trajectories of the mobility edge (as given by the frequency

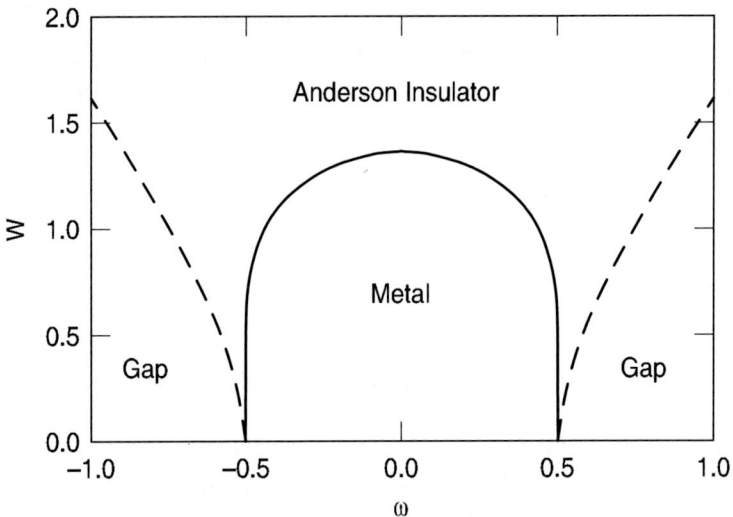

Fig. 19.8. Phase diagram for the "semicircular" model. The trajectories of the mobility edge (full line) and the CPA band edge (dashed line) are shown as a function the disorder strength W.

where the TDOS vanishes for a given W) and the band edge where the ADOS as calculated by CPA vanishes.

4.2. *Critical behavior*

Further insight in the critical behavior is obtained by noting that near $W = W_c$, it proves possible to analytically solve Eqs. (4.2)–(4.8). Here we discuss the the critical exponent of the Anderson metal–insulator transition within the TMT model. We will demonstrate that the critical exponent β with which the order parameter Δ'' vanishes at the transition is, in contradiction to the general expectations,[11] non-universal in this model.

4.2.1. *Critical behavior in the middle of the band $\omega = 0$*

To start with, let us concentrate at the band center ($\omega = 0$), and expand Eqs. (4.2)–(4.8) in powers of the order parameter Δ''. In the limit of $\omega = 0$, self-consistency equations quantities Δ, $\overline{G}$ and Σ become purely imaginary, and near the critical disorder typical Green's function can be expanded in powers of the parameter Δ'':

$$\overline{G(\omega, \varepsilon_i)} = i\Delta'' = \left\langle \frac{\Delta''}{(\omega - \varepsilon_i - \Delta')^2 + \Delta''^2} \right\rangle_{\text{typ}}$$

$$= i\Delta'' \exp\left[-\int d\varepsilon P(\varepsilon_i) \log[\varepsilon_i^2 + \Delta''^2]\right] = i\Delta'' f(\Delta'') \approx i\Delta''(a \quad b\Delta'')$$

(4.9)

where

$$a = f(0) = \exp\left[-2\int d\varepsilon P(\varepsilon) \log|\varepsilon|\right] \tag{4.10}$$

$$b = \left.\frac{\partial f}{\partial \Delta}\right|_{\Delta=0} = a \cdot \exp\left[-2\int d\varepsilon P(\varepsilon) \frac{-2\Delta''}{\varepsilon^2 + \Delta''^2}\right]$$

$$= -a \int d\varepsilon P(\varepsilon) 2\pi \delta(c) = -2\pi a P(0), \tag{4.11}$$

and after trivial algebraic operations our self-consistency Eqs. (4.2)–(4.8) reduce to a single equation for the order parameter Δ''

$$\Delta'' = \frac{\Delta''}{t^2}(a - b\Delta'') \int_{-2t}^{2t} \rho_0(\varepsilon)\varepsilon^2 d\varepsilon. \tag{4.12}$$

Equation (4.12) shows that near the transition along $\phi = 0$ direction, our order parameter Δ'' vanishes linearly (critical exponent $\beta = 1$) independently of the choice of bare lattice DOS ρ_0. In specific case of semicircular bare DOS, where

$$\Delta'' = a\Delta'' - b\Delta''^2, \tag{4.13}$$

the transition where Δ'' vanishes is found at $a = 1$, giving $W = W_c = e/2 = 1.3591$, consistent with our numerical solution. Near the transition, to leading order

$$\rho_{\text{typ}}(W) = -\frac{\Delta''}{\pi} = \left(\frac{4}{\pi}\right)^2 (W_c - W). \tag{4.14}$$

4.2.2. *Critical behavior near the band edge* $\omega = \omega_c$

In order to analytically examine scaling of the critical behavior at finite ω, we focus on a semi-circular bare DOS (for simplicity), where self-consistency Eqs. (4.2)–(4.8) are greatly simplified

$$\overline{G(\omega, \varepsilon_i)} = \Delta' + i\Delta'' \Rightarrow \Delta'' = -\pi\rho_{\text{typ}} = \text{Im}\overline{G(\omega, \varepsilon_i)} \tag{4.15}$$

$$\Delta''(\omega) = -\exp\left\{\int d\varepsilon_i P(\varepsilon_i) \ln\left[\frac{-\Delta''}{(\omega - \varepsilon_i - \Delta') + \Delta''^2}\right]\right\} \tag{4.16}$$

$$\Delta'(\omega) = -H[\Delta''(\omega)]. \tag{4.17}$$

However as in previous section, we expect the critical exponent to be the same for *any* bare DOS.

To find the general critical behavior near the mobility edge, we need to expand Eq. (4.16) in powers of Δ''

$$\Delta'' = \Delta'' \exp\left\{-\int d\varepsilon_i P(\varepsilon_i) \ln\left[(\omega - \varepsilon_i - \Delta')^2 + \Delta''^2\right]\right\} \equiv \Delta'' f(\Delta''),$$
(4.18)

which cannot be done explicitly, since Δ' and Δ'' are related via Hilbert transform, which depends on the entire function $\Delta''(\omega)$, and not only on its form near $\omega = \omega_c$. Nevertheless, the quantity $\omega - \Delta'(\omega)$ assumes a well-defined W-dependent value at the mobility edge $\omega_c = \omega_c(W)$, making it possible for us to determine a *range* of values a critical exponent β may take.

After expanding $f(\Delta'') = 1$ defined by Eq. (4.18),

$$f(\Delta'') = a - b\Delta'' + O(\Delta''^2)$$

$$a = f(0) = \exp\left\{-2\int d\varepsilon P(\varepsilon) \ln|\omega - \varepsilon - \Delta'(\omega)|\right\}$$

$$b = \left.\frac{\partial f}{\partial \Delta''}\right|_{\Delta''=0} = a \lim_{\Delta''\to 0}\left[\int d\varepsilon P(\varepsilon)\frac{2\Delta''}{(\omega - \varepsilon - \Delta')^2 + \Delta''^2}\right]$$

$$= a\int d\varepsilon P(\varepsilon)2\pi\delta(\omega - \varepsilon - \Delta') = 2\pi a P(\omega - \Delta'),$$
(4.19)

we find that to the leading order Δ'' has the following ω dependence ($\delta a \equiv 1 - a$)

$$\Delta'' = \frac{1}{2\pi P(\omega - \Delta')}\left[\frac{1}{a} - 1\right]$$

$$\approx \frac{1}{2\pi P(\omega_c - \Delta'(\omega_c))}\delta a(\omega) \propto \delta\omega^\beta.$$
(4.20)

The functional form of $\delta a(\omega)$ is readily found

$$a = \exp\left\{-2\int d\varepsilon P(\varepsilon) \ln|\omega - \varepsilon - \Delta'(\omega)|\right\}$$
(4.21)

$$\delta a(\omega) = 2\int d\varepsilon P(\varepsilon)\frac{1}{\omega - \varepsilon - \Delta'(\omega)}(\delta\omega - \delta\Delta'(\omega)),$$
(4.22)

and combining Eqs. (4.20) and (4.22) we arrive to

$$\Delta'' = \Delta_0''(\delta\omega - \delta\Delta')$$
(4.23)

$$\Delta_0'' = \frac{1}{\pi P(\omega_c - \Delta'(\omega_c))}\int d\varepsilon\frac{P(\varepsilon)}{\omega_c - \varepsilon - \Delta'(\omega_c)}.$$

Note that $\delta\omega$ is negative, since in the range of interest $\omega < \omega_c$. In Eq. (4.23) $\int d\varepsilon \frac{P(\varepsilon)}{\omega_c - \varepsilon - \Delta'(\omega_c)}$ is the Hilbert transform of $P(\varepsilon)$, which is positive for $\omega_c - \Delta' > 0$ (right band edge), and it is negative for the left one, where $\delta\omega > 0$.

The lower bound on critical exponent β is 0, to insure that Δ'' is convergent and vanishing at $\omega = \omega_c$. Now, if we were to assume that the leading contribution to Δ'' comes from $\delta\omega$ (and $\delta\Delta'$ can be neglected), the conclusion would be that $\Delta'' \propto \delta\omega$, and the critical exponent $\beta = 1$, just like in $\omega = 0$ case. However, this value of β is unphysical, since the Kramers–Kroning predicts Δ' to be logarithmically divergent ($\Delta' \gg \delta\omega$) when $\Delta'' \propto \delta\omega$. This is in direct contradiction with our initial statement of $\delta\omega$ being a leading contribution in Eq. (4.23) ($|\delta\Delta'| \ll |\delta\phi|$), and we conclude that $\delta\Delta' \propto \delta\omega^\beta$ is the leading contribution

$$\Delta'' \approx -\Delta_0'' \delta\Delta' \propto -\delta\omega^\beta \qquad (4.24)$$

with $\beta \in (0, 1)$. This Δ'' being a negative definite quantity imposes a constraint $\delta\Delta' > 0$, which is only satisfied for $\beta > 1/2$ (see Fig. 19.9), thus our critical exponent can vary in the range $\beta \in (1/2, 1)$.

Although general arguments for second-order phase transitions[11] predict universality of exponent β, we find the exponent is non-universal, which is not uncommon in some special cases of mean field theories.[79] It is plausible that this critical exponent anomaly can be remedied if the MFT is extended to incorporate long range fluctuations effects beyond mean-field theory, but this remains an open problem for future work.

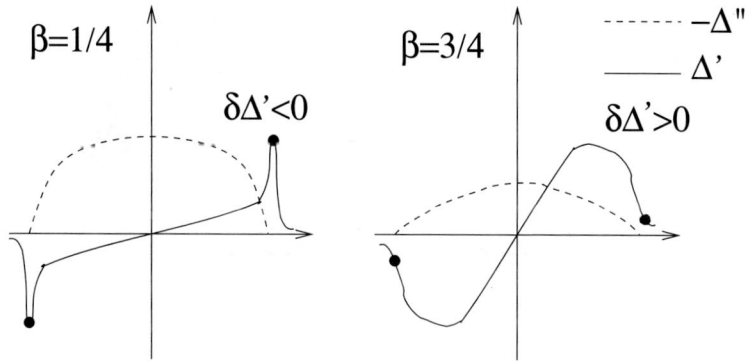

Fig. 19.9.   Requirement on $\delta\Delta'$ to be positive definite forces the value of critical exponent β to be larger than $1/2$.

4.2.3. *Scaling behavior near the critical disorder $W = W_c$.*

The complete analytical solution for TDOS is difficult to obtain for arbitrary ω and W. Still, the approach discussed in Sec. 4.2.1 can be extended to find a full frequency-dependent solution $\rho_{\text{typ}}(\omega, W)$ close to the critical value of disorder $W = W_c$ and which assumes a simple scaling form

$$\rho_{\text{typ}}(\omega, W) = \rho_o(W) f(\omega/\omega_o(W)). \tag{4.25}$$

Our numerical solution (see Figs. 19.10 and 19.11) has suggested that the corresponding scaling function assumes a simple *parabolic* form $f(x) = 1 - x^2$

$$\rho_{\text{typ}}(\omega) \approx \rho_0 \left(1 - \frac{\omega^2}{\omega_0^2} \right) \qquad |\omega| < |\omega_0|. \tag{4.26}$$

In the following, we analytically calculate the scaling parameters ρ_0 and ω_0 for semicircular bared DOS and box distribution of disorder

$$P(\varepsilon) = \begin{cases} \frac{1}{W} & \varepsilon \in [-\frac{W}{2}, \frac{W}{2}] \\ 0 & \varepsilon \notin [-\frac{W}{2}, \frac{W}{2}]. \end{cases} \tag{4.27}$$

$$\Delta'' = -\exp\left[\int d\varepsilon P(\varepsilon) \log\left(-\frac{\Delta''}{[(\omega - \varepsilon - \Delta')^2 + \Delta''^2]} \right) \right]$$

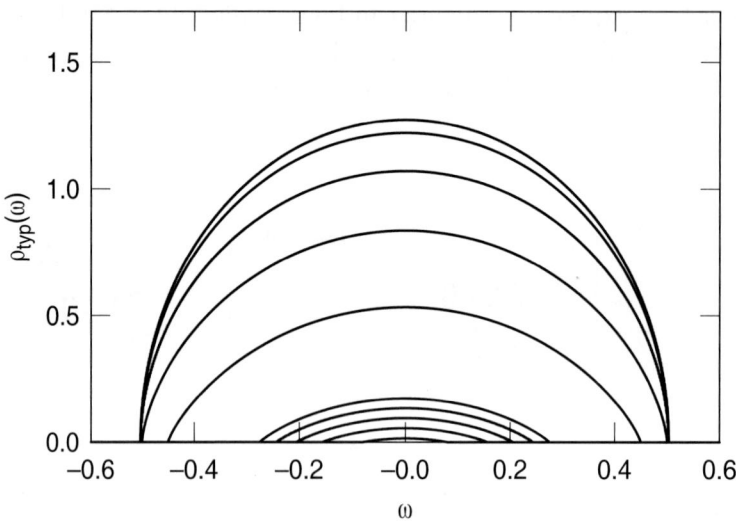

Fig. 19.10. Typical density of states for for the SC model, for disorder values $W = 0, 0.25, 0.5, 0.75, 1, 1.25, 1.275, 1.3, 1.325, 1.35$. The entire band localizes at $W = W_c = e/2 \approx 1.359$.

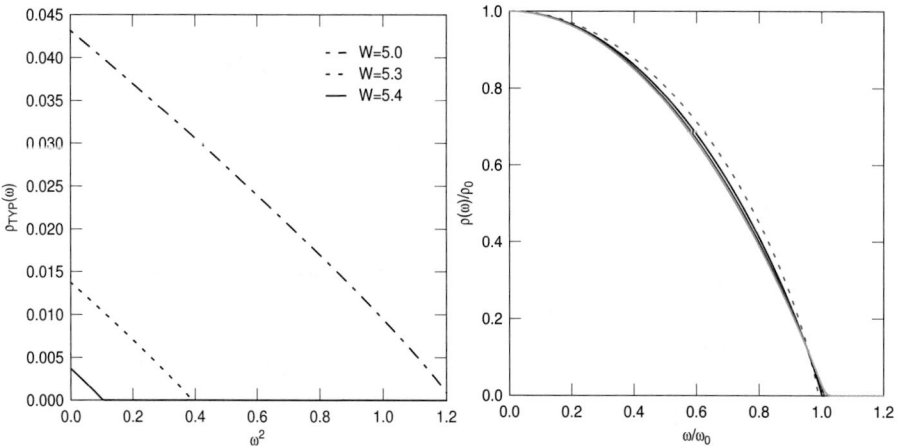

Fig. 19.11. Left: unscaled typical DOS for various disorder displays parabolic behavior near the MIT. Right: scaling behavior near the critical disorder. The range of disorders where parabolic behavior is observed is, in fact, quite broad — $W \in (1, W_c)$, $W_c = e/2$.

after averaging over disorder takes the following form

$$2W = a_- \log(\Delta''^2 + a_-^2) + a_+ \log(\Delta''^2 + a_+^2)$$
$$+ 2\Delta'' \left[\arctan\left(\frac{a_+}{\Delta''}\right) + \arctan\left(\frac{a_-}{\Delta''}\right) \right], \qquad (4.28)$$

where

$$a_\pm = \frac{W}{2} \pm (\Delta' - \omega). \qquad (4.29)$$

Exact expression for the real part of the cavity field Δ' is obtained by performing a Hilbert transformation of ansatz (4.26):

$$\Delta'' = -\pi\rho_0 \left(1 - \frac{\omega^2}{\omega_0^2}\right)$$
$$\qquad (4.30)$$
$$\Delta' = -H[\Delta''] = \rho_0 \left(2\frac{\omega_0\omega}{\omega_0^2} - \left(1 - \frac{\omega^2}{\omega_0^2}\right) \log\left|\frac{\omega - \omega_0}{\omega - \omega_0}\right|\right).$$

Expanding Eqs. (4.28) and (4.30) to the second order in small ω results (see Fig. 19.12) in a system of equations:

$$\frac{2\pi\rho_0}{W} \arctan\left(\frac{W}{2\pi\rho_0}\right) + \frac{1}{2}\log\left(\frac{W^2}{4} + \pi^2\rho_0^2\right) = 1$$

$$\frac{2\pi\rho_0}{W\omega_0^2} \arctan\left(\frac{W}{2\pi\rho_0}\right) = \frac{\left(\frac{4\rho_0}{\omega_0} - 1\right)^2}{2\left(\frac{W^2}{4} + \pi^2\rho_0^2\right)},$$

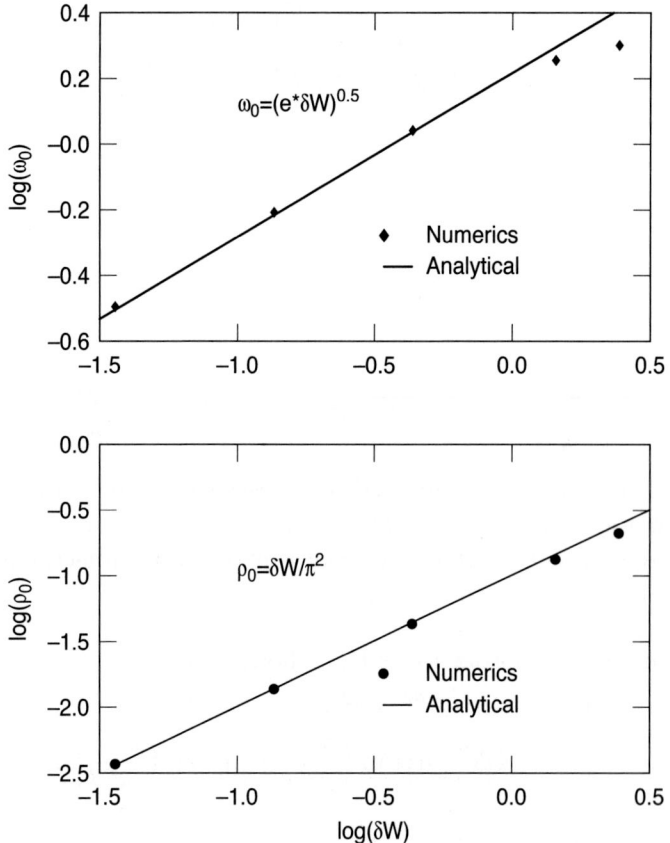

Fig. 19.12.   Scaling parameters ρ_0 and ω_0 as functions of the distance to the transition $\delta W = W_c - W$. Numerically obtained values ($\diamond$ and $\bullet$) are in good agreement with analytical predictions (full line).

which can be solved for scaling parameters used the in original ansatz, Eq. (4.26).

$$\rho_0 = \frac{4(W_c - W)}{\pi^2} \qquad (4.31)$$

$$\omega_0 = \sqrt{\frac{e}{2}} \sqrt{W_c - W}. \qquad (4.32)$$

4.3. Numerical test of TMT

In order to gauge the quantitative accuracy of our theory, we have carried out first-principles numerical calculations for a three-dimensional cubic lattice with random site energies, using exact Green functions for an open finite

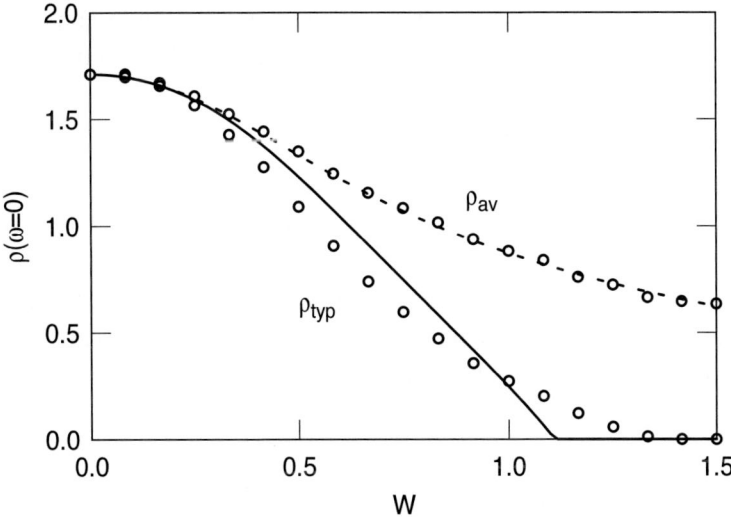

Fig. 19.13. Typical and average DOS for a three-dimensional cubic lattice at the band center ($\omega = 0$). Results from first-principle numerical calculations (circles) are compared to the predictions of TMT (for TDOS — full line) and CPA (for ADOS — dashed line).

sample attached to two semi-infinite clean leads.[41] We computed both the average and the typical DOS at the band center as a function of disorder, for cubes of sizes $L = 4$, 6, 8, 10, 12 and 16, and averages over 1000 sample realizations, in order to obtain reliable data by standard finite size scaling procedures. The TMT and CPA equations for the same model were also solved by using the appropriate bare DOS (as expressed in terms of elliptic integrals), and the results are presented in Fig. 19.13.

We find remarkable agreement between these numerical data[41] and the self-consistent CPA calculations for the ADOS, but also a surprisingly good agreement between the numerical data and the TMT predictions for the TDOS order parameter. For a cubic lattice, the exact value is $W_c \approx 1.375$,[80] whereas TMT predicts a 20% smaller value $W_c \approx 1.1$. The most significant discrepancies are found in the critical region, since TMT predicts the order parameter exponent $\beta = 1$, whereas the exact value is believed to be $\beta \approx 1.5$, consistent with our numerical data. Argument based on the multi-fractal scaling analysis,[76,77] together with numerical calculations[80] of the multi-fractal spectra of wavefunction amplitudes have suggested that in three dimensions, the TDOS order parameter exponent β should be equal to the conductivity exponent $\mu \approx 1.5$. The result $\beta = \mu = 1 + O(\varepsilon)$ is also found within the $2 + \epsilon$ approach.[76,77] Nevertheless, we conclude that TMT is as accurate as one can expect from a simple mean-field formulation.

4.4. Transport properties

Most previous conventional transport theories, while providing a wonderful description of good metals, fail to describe the transport properties of highly disordered materials.

In most metals, the temperature coefficient of resistivity (TCR) α is positive, because phonon scattering decreases the electronic mean free path as the temperature is raised. The sign of TCR

$$\alpha = \frac{d \ln \rho_{\mathrm{res}}(T)}{dT} \qquad (4.33)$$

can be deduced from Matthiessen's rule which asserts that the total resistivity in the presence of two or more scattering mechanisms is equal to the sum of the resistivities that would result if each mechanism were the only one operating, for example:

$$\rho_{\mathrm{res}} = \rho_{\mathrm{res}}^{(1)} + \rho_{\mathrm{res}}^{(2)}. \qquad (4.34)$$

Matthiessen's rule, as stated in Eq. (4.34), follows from the Boltzmann equation with the assumption of a wave-vector-independent relaxation time for each scattering mechanism, so if ρ_0 is the resistivity of a disordered metal at zero temperature and $\rho_{\mathrm{ph}}(T)$ is the resistivity of the ordered material due to electron–phonon scattering, then the total resistivity at finite temperature is

$$\rho(T) = \rho_0 + \rho_{\mathrm{ph}}(T) \geq \rho_0 \,, \qquad (4.35)$$

predicting that the TCR is positive ($\alpha > 0$).

Mooij[81] in 1973 has pointed out that there exist many highly disordered metals which are poor conductors and have $\rho(T) \leq \rho_0$ and negative TCRs, which clearly violate Matthiessen's rule. In fact in these materials, the Boltzmann equation formalism itself is breaking down. Apparently what is happening is that, because of the strong disorder and resulting multiple correlated scattering, the Boltzmann hypothesis of independent scattering events fails. The simple picture of temperature fluctuations impeding transport of electrons (positive TCR) is now replaced with the temperature fluctuations releasing the localized electrons and increasing the conductivity (negative TCR). When the transport properties are addressed within the TMT, the interplay of several localization mechanisms is considered, which is capable of producing the negative TCRs observed in highly disordered materials.

We start addressing the transport properties of our system within the TMT by pointing out that the escape rate from a given site can be rigorously defined in terms of the cavity field (see Eq. (4.2)), and using our solution of the TMT equations, we find $\tau_{\rm coc}^{-1} = -{\rm Im}\Delta(0) \sim \rho_{\rm TYP} \sim (W_c - W)$. To calculate the conductivity within our local approach, we follow a strategy introduced by Girvin and Jonson (GJ),[82] who pointed out that close to the localization transition, the conductivity can be expressed as $\sigma = \Lambda a_{12}$, where Λ is a vertex correction that represents hops to site outside of the initial pair i and j, and a_{12} is a two-site contribution to the conductivity, that can be expressed as

$$a_{12} = \langle A_{12}A_{21} - A_{11}A_{22}\rangle, \tag{4.36}$$

where $A_{ij} = -{\rm Im}G_{ij}$ is the spectral function corresponding to the nearest neighbor two-site cluster, $\langle \cdots \rangle$ represents the average over disorder.

We examine the temperature dependence of the conductivity as a function of W. Physically, the most important effect of finite temperatures is to introduce finite inelastic scattering due to interaction effects. At weak disorder, such inelastic scattering increases the resistance at higher temperatures, but in the localized phase it produces the opposite effect, since it suppresses interference processes and localization. To mimic these inelastic effects within our noninteracting calculation, we introduce by hand an additional scattering term in our self-energy, viz. $\Sigma \to \Sigma - i\eta$ or it can be treated as the imaginary part of $\omega \to \omega + i\eta$. The parameter η measures the inelastic scattering rate, and is generally expected to be a monotonically increasing function of temperature.

The relevant η-dependent Green's functions G_{ij}

$$G_{ii} = \frac{\omega + i\eta - \varepsilon_i - \Delta}{(\omega + i\eta - \varepsilon_i - \Delta)(\omega + i\eta - \varepsilon_j - t^2) - t^2}$$

$$\tag{4.37}$$

$$G_{ij} = \frac{t}{(\omega + i\eta - \varepsilon_i - \Delta)(\omega + i\eta - \varepsilon_j - t^2) - t^2},$$

reduce expression (4.36) to an integrable form

$$a_{12} = 4\frac{(\Delta'' - \eta)^2}{W^2} \int_{\omega - \Delta' - W/2}^{\omega - \Delta' + W/2} \frac{dx}{x^2 + (\Delta'' - \eta)^2} \arctan$$

$$\times \left[\frac{-x + y(x^2 + (\Delta'' - \eta)^2)}{(\Delta'' - \eta)(x^2 + (\Delta'' - \eta)^2 + 1)} \right] \Bigg|_{y=\omega - \Delta' - W/2}^{|y=\omega - \Delta' + W/2} \tag{4.38}$$

that can be solved numerically as a function of temperature η and disorder W.

We have computed a_{12} by examining two sites embedded in the effective medium defined by TMT (Δ_{TMT}), thus allowing for localization effects. The vertex function Λ remains *finite* at the localization transition,[82] and thus can be computed within. We have used the CPA approach to evaluate the vertex function as $\Lambda = \sigma_{\text{cpa}}/a_{12}^{\text{cpa}}$, where σ_{cpa} is the CPA conductivity calculated using approach described by Elliot[78]

$$\sigma(\omega) \propto \int_{-B/2}^{B/2} d\varepsilon \rho_0(\varepsilon) \text{Im}[G(\omega, \varepsilon)]^2 \qquad (4.39)$$

$$\rho_0(\varepsilon) = \left(\frac{B^2}{4} - \varepsilon^2 \right)^{3/2}, \qquad (4.40)$$

and a_{12}^{cpa} is the two-site correlation function embedded in the CPA effective medium (Δ_{CPA}). Since TMT reduces to CPA for weak disorder, our results reduce to the correct value at $W \ll W_c$, where the conductivity reduces to the Drude–Boltzmann form. The resulting critical behavior of the $T = 0$ conductivity follows that of the order parameter, $\sigma \sim \rho_{\text{TYP}} \sim (W_c - W)$, giving the conductivity exponent μ equal to the order parameter exponent β, consistent with what is expected.

The resulting dependence of the conductivity as a function of η and W is presented in Fig. 19.14. As η (i.e. temperature) is reduced, we find that the conductivity curves "fan out", as seen in many experiment close to the MIT.[7,68] Note the emergence of a "separatrix"[7,68] where the conductivity is temperature independent, which is found for $W \approx 1$, corresponding to $k_F \ell \sim 2$, consistent with some experiments.[7] At the MIT, we find $\sigma_c(\eta) \sim \rho_{\text{TYP}}(\eta) \sim \eta^{1/2}$.

5. Mott–Anderson Transitions

5.1. *Two-fluid picture of Mott*

A first glimpse of the basic effect of disorder on the Mott transition at half filling was outlined already by Mott,[1] who pointed out that important differences can be seen even in the strongly localized (atomic) limit.

For weak to moderate disorder $W < U$, the Mott insulator survives, and each localized orbital is singly occupied by an electron, forming a spin $1/2$ magnetic moment. For stronger disorder ($W > U$) the situation is different. Now, a fraction of electronic states are either doubly occupied or empty, as in an Anderson insulator. The Mott gap is now closed, although a finite fraction of the electrons still remain as localized magnetic moments. Such a state can be described[39,40] as an inhomogeneous mixture of a Mott and an

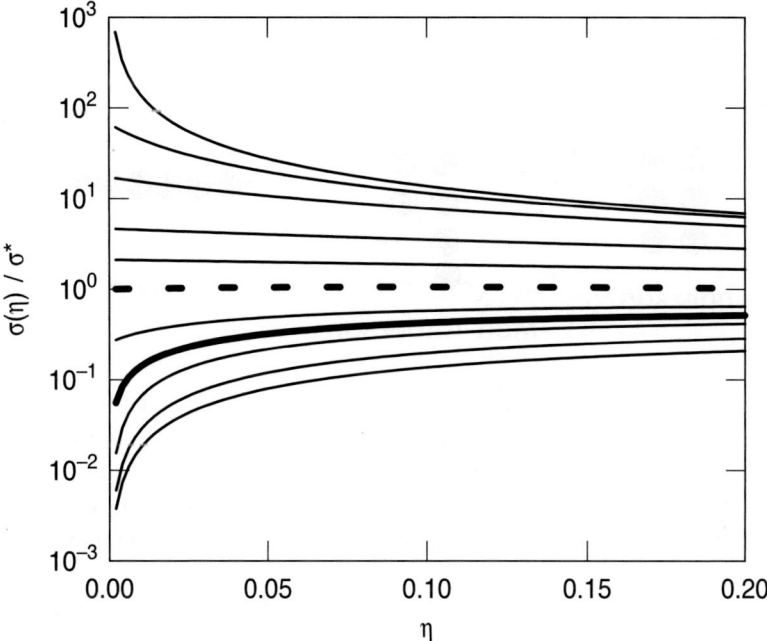

Fig. 19.14. Conductivity as a function of the inelastic scattering rate η for for the SC model at the band center and $W = 0$, 0.125, 0.25, 0.5, 0.75, 1, 1.25, 1.36, 1.5, 1.75, 2. The "separatrix" ($\sigma = \sigma^*$ independent of η, i.e. temperature) is found at $W = W^* \approx 1$ (dashed line). The critical conductivity $\sigma_c(\eta) \sim \eta^{1/2}$ corresponds to $W = W_c = 1.36$ (heavy full line).

Anderson insulator. A very similar **"two-fluid model"** — of coexisting local magnetic moments and conduction electrons — was proposed[83,84] some time ago on experimental grounds, as a model for doped semiconductors. Some theoretical basis of such behavior has been discussed,[37–40,73,85–87] but the corresponding critical behavior remains a puzzle.

This physical picture of Mott (see Fig. 19.15) is very transparent and intuitive. But how is this strongly localized (atomic) limit approached when one crosses the metal–insulator transition from the metallic side? To address this question one needs a more detailed theory for the metal–insulator transition region, which was not available when the questions posed by Mott and Anderson were first put forward.

5.2. *Mott or Anderson... or both?*

Which of the two mechanisms dominates criticality in a given material? This is the question often asked when interpreting experiments, but a convincing

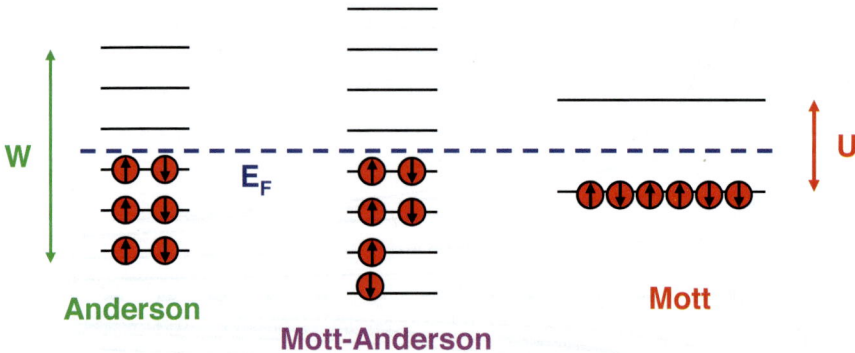

Fig. 19.15. Energy level occupation in the strongly localized (atomic) limit. In a Mott–Anderson insulator (center), the disorder strength W is larger then the Coulomb repulsion U, and a two-fluid behavior emerges. Here, a fraction of localized states are doubly occupied or empty as in an Anderson insulator. Coexisting with those, other states remain singly occupied forming local magnetic moments, as in a Mott insulator.

answer is seldom given. To answer it precisely, one must define the appropriate criteria — order parameters — characterizing each of the two routes. The conceptually simplest theoretical framework that introduces such order parameters is given by TMT-DMFT — which we introduced in the preceding section, and discussed in detail in the noninteracting limit. As in conventional DMFT, its self-consistent procedure formally sums up all possible Feynman diagrams providing local contributions to the electronic self-energy.[36] When the procedure is applied to systems with both interactions and disorder systems, the self energy is still local, but may display strong-site-to-site fluctuations. Its low-energy form

$$\Sigma_i(\omega_n) = (1 - Z_i^{-1})\,\omega_n + v_i - \varepsilon_i + \mu,$$

defines *local* Fermi liquid parameters[39,74]: the local quasi-particle (QP) weight Z_i, and the renormalized disorder potential v_i. This theory portrays a picture of a spatially inhomogeneous Fermi liquid, and is able to track its evolution as the critical point is approached.

In this language, Anderson localization, corresponding to the formation of bound electronic states, is identified by the emergence of discrete spectra[32] in the local density of states (LDOS). As we have seen above, this corresponds[39,41] to the vanishing of the *typical* (geometrically averaged) LDOS $\rho_{\mathrm{typ}} = \exp\langle \ln(\rho_i) \rangle$. In contrast, Mott localization of itinerant electrons into magnetic moments is identified by the vanishing of the local QP weights $(Z_i \to 0)$. It is interesting and important to note that a very similar physical picture was proposed as the key ingredient for "**local quantum critical-ity**",[88] or "**deconfined quantum criticality**"[89,90] at the $T = 0$ magnetic ordering in certain heavy fermion systems. A key feature in these theories is the possibility that Kondo screening is destroyed precisely at the quantum critical point. As a result, part of the electrons — those corresponding to tightly bound f-shells of rare earth elements — "drop out" from the Fermi surface and turn into localized magnetic moments. For this reason, it is argued, any weak-coupling approach must fail in describing the critical behavior. This is the mechanism several groups have attributed to the breakdown of the Hertz–Millis theory[91,92] of quantum criticality, which at present is believed to be incomplete.

Precisely the same fundamental problem clearly must be addressed for the Mott–Anderson transition. The transmutation of a **fraction** of electrons into local magnetic moments again can be viewed as the suppression of Kondo screening — clearly a **non-perturbative strong correlation effect** — that should be central to understanding the critical behavior. To properly characterize it, one must keep track of the evolution of the entire distribution $P(Z_i)$ of local quasi-particle weights — which can be directly obtained from TMT-DMFT approach[41] to the Mott–Anderson transition, which we outlined above. The first applications of this new method to correlated systems with disorder was carried out in recent studies by Vollhardt and collaborators,[93,94] who numerically obtained the phase diagram for the disordered Hubbard model at half-filling, and discussed the influence of Mott–Anderson localization on magnetically ordered phases. However, the qualitative nature of the critical behavior in the Mott–Anderson transition in this model has not been examined in these studies.

5.3. *Slave–Boson solution*

In the following we use complementary semi-analytical methods supplemented by Fermi liquid theorems, in order to clarify the precise form of criticality in this model.[95] By making use of scaling properties[96,97] of Anderson impurity models close to the MIT, we present a detailed analytic solution for this problem, which emphasizes the dependence of the system properties on its particle-hole symmetry. We consider a half-filled Hubbard model[40] with random site energies, as given by the Hamiltonian

$$H = -V \sum_{\langle ij \rangle \sigma} c_{i\sigma}^\dagger c_{j\sigma} + \sum_{i\sigma} \varepsilon_i n_{i\sigma} + U \sum_i n_{i\uparrow} n_{i\downarrow}. \qquad (5.1)$$

Here, $c_{i\sigma}^{\dagger}$ ($c_{i\sigma}$) creates (destroys) a conduction electron with spin σ on site i, $n_{i\sigma} = c_{i\sigma}^{\dagger}c_{i\sigma}$, V is the hopping amplitude, and U is the on-site repulsion. The random on-site energies ε_i follow a distribution $P(\varepsilon)$, which is assumed to be uniform and have width W.

TMT-DMFT[41,93] maps the lattice problem onto an ensemble of single-impurity problems, corresponding to sites with different values of the local energy ε_i, each being embedded in a typical effective medium which is self-consistently calculated. In contrast to standard DMFT,[31] TMT-DMFT determines this effective medium by replacing the spectrum of the environment ("cavity") for each site by its typical value, which is determined by the process of *geometric* averaging. For a simple semi-circular model density of states, the corresponding bath function is given by[41,93] $\Delta(\omega) = V^2 G_{\mathrm{typ}}(\omega)$, with $G_{\mathrm{typ}}(\omega) = \int_{-\infty}^{\infty} d\omega' \rho_{\mathrm{typ}}(\omega')/(\omega - \omega')$ being the Hilbert transform of the geometrically-averaged (typical) local density of states (LDOS) $\rho_{\mathrm{typ}}(\omega) = \exp\{\int d\varepsilon P(\varepsilon) \ln \rho(\omega, \varepsilon)\}$. Given the bath function $\Delta(\omega)$, one first needs to solve the local impurity models and compute the local spectra $\rho(\omega, \varepsilon) = -\pi^{-1}\mathrm{Im}G(\omega, \varepsilon)$, and the self-consistency loop is then closed by the the geometric averaging procedure.

To qualitatively understand the nature of the critical behavior, it is useful to concentrate on the low-energy form for the local Green's functions, which can be specified in terms of two Fermi liquid parameters as

$$G(\omega, \varepsilon_i) = \frac{Z_i}{\omega - \tilde{\varepsilon}_i - Z_i\Delta(\omega)}, \tag{5.2}$$

where Z_i is the local quasi-particle (QP) weight and $\tilde{\varepsilon}_i$ is the renormalized site energy.[31] The parameters Z_i and $\tilde{\varepsilon}_i$ can be obtained using any quantum impurity solver, but to gain analytical insight here we focus on the variational calculation provided by the "four-boson" technique (SB4) of Kotliar and Ruckenstein,[98] which is known to be quantitatively accurate at $T = 0$. The approach consists of determining the site-dependent parameters e_i, d_i and $\tilde{\varepsilon}_i$ by the following equations

$$-\frac{\partial Z_i}{\partial e_i}\frac{1}{\beta}\sum_{\omega_n}\Delta(\omega_n)G_i(\omega_n) = Z_i(\mu + \tilde{\varepsilon}_i - \varepsilon_i)e_i, \tag{5.3}$$

$$-\frac{\partial Z_i}{\partial d_i}\frac{1}{\beta}\sum_{\omega_n}\Delta(\omega_n)G_i(\omega_n) = Z_i(U - \mu - \tilde{\varepsilon}_i + \varepsilon_i)d_i, \tag{5.4}$$

$$\frac{1}{\beta}\sum_{\omega_n}G_i(\omega_n) = \frac{1}{2}Z_i(1 - e_i^2 + d_i^2), \tag{5.5}$$

where $Z_i = 2(e_i + d_i)^2[1 - (e_i^2 + d_i^2)]/[1 - (e_i^2 - d_i^2)^2]$ in terms of e_i and d_i and $\mu = U/2$. We should stress, though, that most of our analytical results rely only on Fermi liquid theorems constraining the qualitative behavior at low energy, and thus do not suffer from possible limitations of the SB4 method.

Within this formulation, the metal is identified by nonzero QP weights Z_i on *all* sites and, in addition, a nonzero value for both the typical and the average $[\rho_{av}(\omega) = \int d\varepsilon P(\varepsilon)\rho(\omega, \varepsilon)]$ LDOS. Mott localization (i.e. local moment formation) is signaled by $Z_i \longrightarrow 0$,[31] while Anderson localization corresponds to $Z_i \neq 0$ and $\rho_{av} \neq 0$, but $\rho_{typ} = 0$.[32,41] While Ref. 93 concentrated on ρ_{typ} and ρ_{av}, we find it useful to simultaneously examine the QP weights Z_i, in order to provide a complete and precise description of the critical behavior.

5.4. *Phase diagram*

Using our SB4 method, the TMT-DMFT equations can be numerically solved to very high accuracy, allowing very precise characterization of the critical behavior. In presenting all numerical results we use units such that the bandwidth $B = 4V = 1$. Figure 19.16 shows the resulting $T = 0$ phase diagram at half filling, which generally agrees with that of Ref. 93. By concentrating first on the critical behavior of the QP weights Z_i, we are able

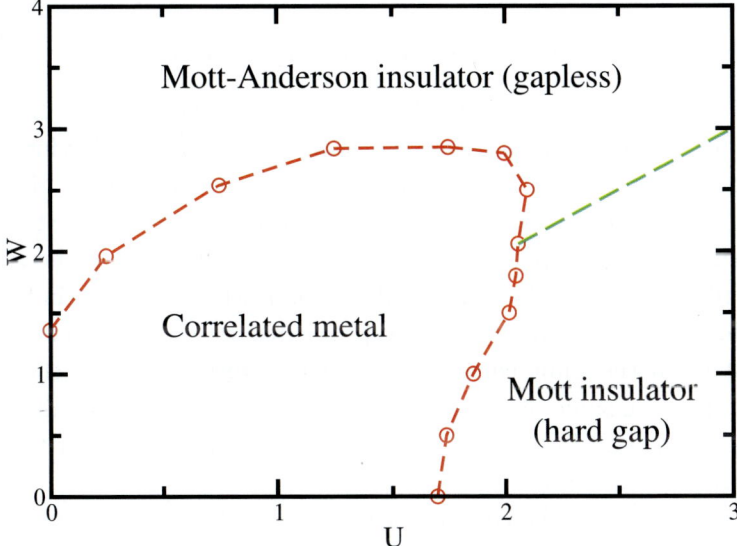

Fig. 19.16. $T = 0$ phase diagram for the disordered half filled Hubbard model, obtained from the numerical SB4 solution of TMT-DMFT.

to clearly and precisely distinguish the metal from the insulator. We find that at least some of the Z_i vanish all along the phase boundary. By taking a closer look, however, we can distinguish two types of critical behavior, as follows.

5.4.1. *Mott–Anderson vs. Mott-like transition*

For sufficiently strong disorder ($W > U$), the Mott–Anderson transition proves qualitatively different than the clean Mott transition, as seen by examining the critical behavior of the QP weights $Z_i = Z(\varepsilon_i)$. Here $Z_i \to 0$ only for $0 < |\varepsilon_i| < U/2$, indicating that only *a fraction* of the electrons turn into localized magnetic moments. The rest show $Z_i \to 1$ and undergo Anderson localization (see below). Physically, this regime corresponds to a spatially inhomogeneous system, with Mott fluid droplets interlaced with regions containing Anderson-localized quasiparticles. In contrast, for weaker disorder ($W < U$) the transition retains the conventional Mott character. In this regime $Z_i \to 0$ on all sites, corresponding to Mott localization of all electrons. We do not discuss the coexistence region found in Ref. 93, because we focus on criticality within the metallic phase. We do not find any "crossover" regime such as reported in Ref. 93, the existence of which we believe is inconsistent with the generally sharp distinction between a metal and an insulator at $T = 0$.

5.4.2. *Two-fluid behavior at the Mott–Anderson transition*

To get a closer look at the critical behavior of the QP weights $Z_i = Z(\varepsilon_i)$, we monitor their behavior near the transition. The behavior of these QP weights is essentially controlled by the spectral weight of our self-consistently-determined TMT bath, which we find to vanish at the transition. An appropriate parameter to measure the distance to the transition is the bandwidth t of the bath spectral function, which is shown in Fig. 19.17.

Considering many single-impurity problems, we observe a two-fluid picture, just as in the limit earlier analyzed by Mott.[1] Indeed, these results correspond to the same atomic limit discussed by Mott, since, although the hopping itself is still finite, the cavity field "seen" by the impurities goes to zero in the current case.

As in the atomic limit, the sites with $|\varepsilon_i| < U/2$ turn into local moments and have vanishing quasiparticle weight $Z_i \to 0$. The remaining sites show $Z_i \to 1$, as they are either doubly occupied, which corresponds to those with $\varepsilon_i < -U/2$, or empty, which is the case for those sites with $\varepsilon_i > U/2$.

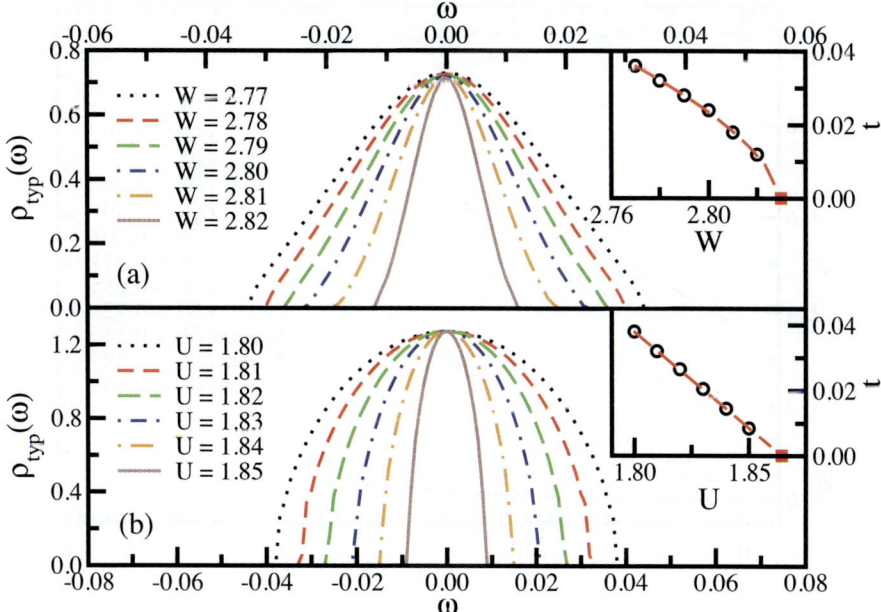

Fig. 19.17. Frequency dependence of the typical DOS very close to the metal–insulator transition for (a) the Mott–Anderson transition ($W > U$) at $U = 1.25$ and (b) the Mott-like transition ($W < U$) at $W = 1.0$. The insets show how, in both cases, the $\rho_{\text{typ}}(\omega)$ bandwidth $t \to 0$ at the transitions.

Consequently, as the transition is approached, the curves $Z(\varepsilon_i, t)$ "diverge" and approach either $Z = 0$ or $Z = 1$. These values of Z can thus be identified as two stable fixed points for the problem in question, as we discuss below.

Note that in Fig. 19.18 we restrict the results to positive energy values, as a similar behavior is observed for negative ε_i. In this case, there is precisely one value of the site energy $\varepsilon_i = \varepsilon^*$, for which $Z(\varepsilon^*, t) \to Z^*$. This corresponds to the value of ε_i below which Z "flows" to 0 and above which Z "flows" to 1. In other words, it corresponds to an unstable fixed point. Just as in the atomic limit, ε^* is equal to $U/2$ ($\varepsilon^*/W = 0.3125$ in Fig. 19.18, where $U = 1.75$ and $W = 2.8$).

5.4.3. *β-function formulation of scaling*

Our numerical solutions provide evidence that as a function of t the "charge" $Z(t)$ "flows" away from the unstable "fixed point" Z^*, and towards either stable "fixed points" $Z = 0$ or $Z = 1$. The structure of these flows show power-law scaling as the scale $t \to 0$; this suggests that it should be possible

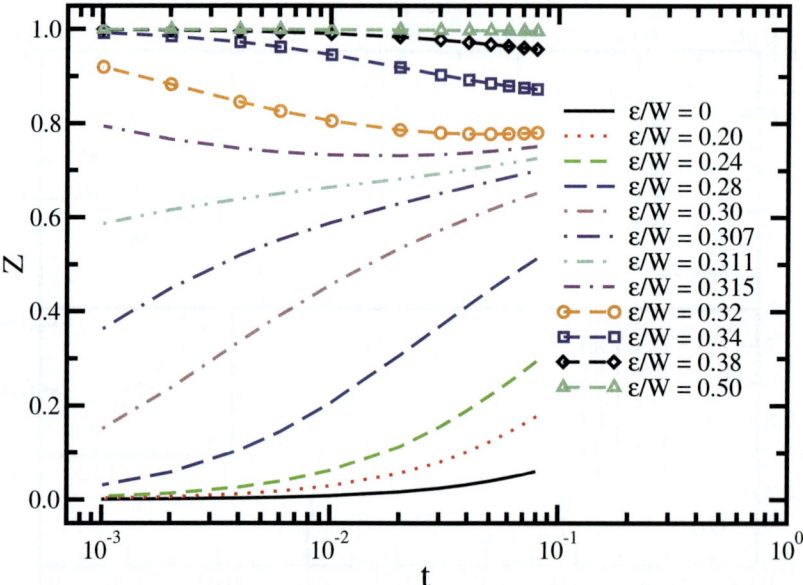

Fig. 19.18.   Quasiparticle weight Z plotted as a function of the distance to the Mott–Anderson transition t, for different values of the local site energy ϵ/W. We present the results only for positive site energies, as a similar behavior holds for negative ones.

to collapse the entire family of curves $Z(t, \delta\varepsilon)$ onto a single universal scaling function

$$Z(t, \delta\varepsilon) = f[t/t^*(\delta\varepsilon)], \tag{5.6}$$

where the crossover scale $t^*(\delta\varepsilon) = C^\pm |\delta\varepsilon|^\phi$ around the unstable fixed point. Remarkably, we have been able to scale the numerical data precisely in this fashion, see Fig. 19.19, and extract the form of $t^*(\delta\varepsilon)$. We find that $t^*(\delta\varepsilon)$ vanishes in a power law fashion at $\delta\varepsilon = 0$, with exponent $\phi = 2$ and the amplitudes $C^\pm$ differ by a factor close to two for $Z \gtrsim Z^*$.

As shown in Fig. 19.19, the scaling function $f(x)$ where $x = t/t^*(\delta\varepsilon)$ presents two branches: one for $\varepsilon_i < \varepsilon^*$ and other for $\varepsilon_i > \varepsilon^*$. We found that for $x \to 0$ both branches of $f(x)$ are linear in x, while for $x \gg 1$ they merge, i.e. $f(x) \to Z^* \pm A^\pm x^{-1/2}$. As can be seen in the first two panels, in the limit $t \to 0$, the curve corresponding to $\varepsilon_i < U/2$ has $Z(t) = B^- t$, while that for $\varepsilon_i > U/2$ follows $1 - Z(t) = B^+ t$. These results are for a flat cavity field but, as mentioned earlier, we checked that the same exponents are found also for other bath functions, meaning that they are independent of the exact form of the cavity field. The power-law behavior and the respective exponents

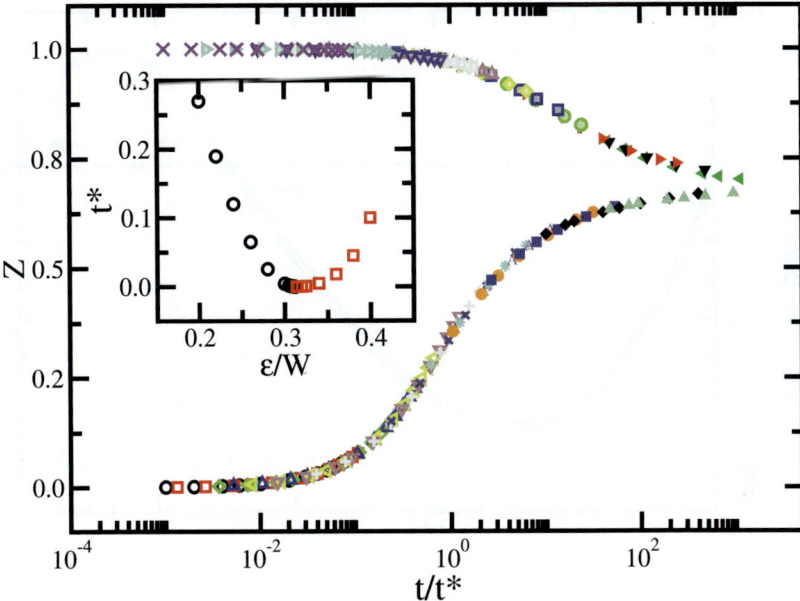

Fig. 19.19. Quasiparticle weight Z as a function of $t/t^*(\delta\varepsilon)$ showing that the results for different ε can be collapsed onto a single scaling function with two branches. The results for different ε correspond to different symbols. The inset shows the scaling parameter t^* as a function of ε/W for the upper (squares) and bottom (circles) branches.

observed numerically in the three limits above have also been confirmed by solving the SB equations analytically[97] close to the transition $(t \to 0)$.

In the following, we rationalize these findings by defining an appropriate β-function which describes all the fixed points and the corresponding crossover behavior. Let us assume that

$$\frac{dZ(t,\delta\varepsilon)}{d\ln t} = -\beta(Z) \qquad (5.7)$$

is an explicit function of Z only, but not of the parameters t or $\delta\varepsilon$. The desired structure of the flows would be obtained if the β-function had three zeros: at $Z = 0$ and $Z = 1$ with negative slope (stable fixed points) and one at $Z = Z^*$ with positive slope (unstable fixed point). The general structure of these flows can thus be described in a β-function language similar to that used in the context of a renormalization group approach; we outline the procedure to obtain $\beta(Z)$ from the numerical data.

The integration of Eq. (5.7) can be written in the form of Eq. (5.6) as

$$Z = f[t/t^*(Z_o)], \qquad (5.8)$$

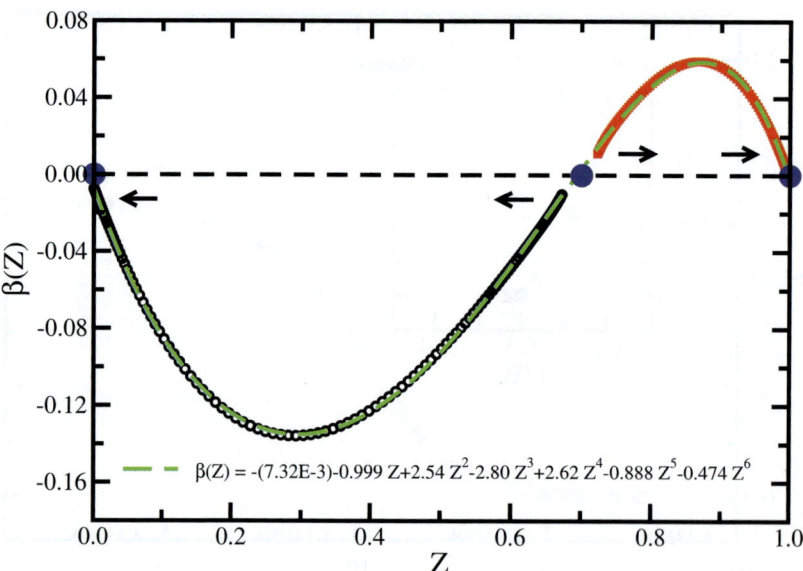

Fig. 19.20. β-function obtained as described in the text for the Anderson impurity models close to the Mott–Anderson transition. The filled circles indicate the three fixed points found for this problem. The arrows indicate how Z flows *to* the stable points ($Z = 0$ and $Z = 1$) and *from* the unstable one ($Z \approx 0.7$).

where Z_o is the initial condition (a function of $\delta\varepsilon$). With $x = t/t^*$ as before, Eq. (5.6) can be rewritten as

$$\beta(Z) = -xf'(x). \tag{5.9}$$

The numerical data for $Z = f(x)$ as a function of x is presented in Fig. 19.19. Thus, using Eq. (5.9), the β-function in terms of $x(Z)$ is determined, which can finally be rewritten in terms of Z. Carrying out this procedure, we obtain $\beta(Z)$ as shown in Fig. 19.20. In accordance with what was discussed above, we see that $\beta(Z)$ has three fixed points, as indicated in the figure by filled circles. $Z = 0$ and $Z = 1$ are stable, while $Z \approx 0.7$ is the unstable fixed point.

The scaling behavior and the associated β-function observed here reflect the fact these impurity models have two phases (singlet and doublet) when entering the insulator. The two stable fixed points describe these two phases, while the unstable fixed point Z^* describes the phase transition, which is reached by tuning the site energy.

Interestingly, the family of curves in Fig. 19.18 looks similar to those seen in some other examples of quantum critical phenomena. In fact, one can say that the crossover scale t plays the role of the reduced temperature, and the reduced site energy $\delta\varepsilon = (\varepsilon_i - \varepsilon^*)/\varepsilon^*$ that of the control parameter of the quantum critical point. As the site energy is tuned at $t = 0$, the impurity model undergoes a phase transition from a singlet to a doublet ground state. Quantum fluctuations associated with the metallic host introduce a cutoff and round this phase transition, which becomes sharp only in the $t \rightarrow 0$ limit.

5.5. *Wavefunction localization*

To more precisely characterize the critical behavior we now turn our attention to the spatial fluctuations of the quasiparticle wavefunctions, we compare the behavior of the typical (ρ_{typ}) and the average (ρ_{av}) LDOS. The approach to the Mott–Anderson transition ($W > U$) is illustrated by increasing disorder W for fixed $U = 1.25$ (Fig. 19.21 — top panels). Only those

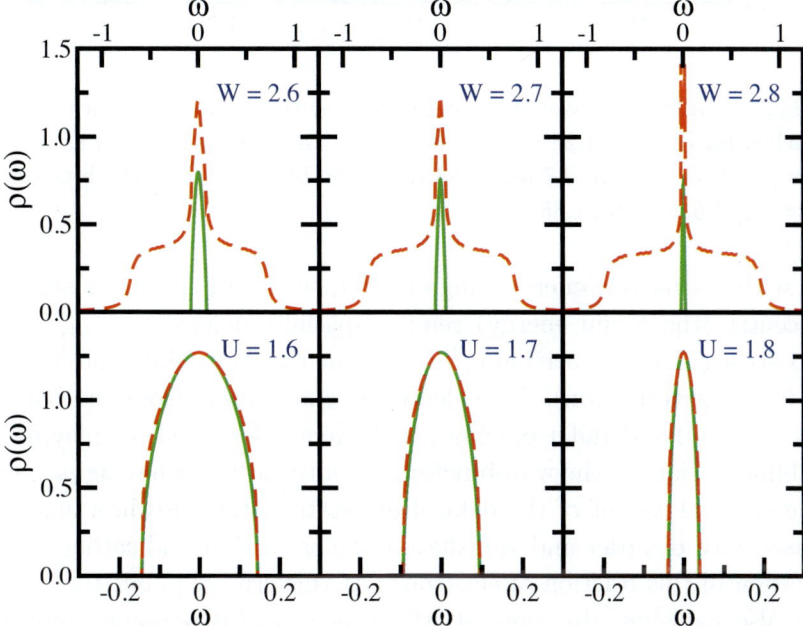

Fig. 19.21. Frequency dependence of ρ_{typ} (full line) and ρ_{av} (dashed line) in the critical region. Results in top panels illustrate the approach to the Mott–Anderson transition ($W > U$) at $U = 1.25$; the bottom panels correspond to the Mott-like transition ($W < U$) at $W = 1.0$. For the Mott–Anderson transition, only a narrow band of delocalized states remain near the Fermi energy, corresponding to $\rho_{\text{typ}} \neq 0$. In contrast, most electronic states remain delocalized $\rho_{\text{typ}} \approx \rho_{\text{av}}$ near the Mott-like transition.

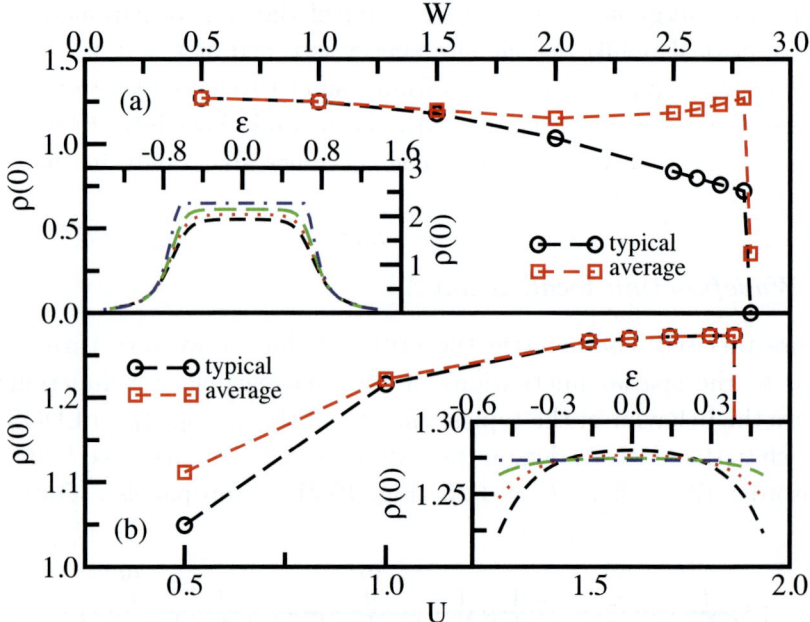

Fig. 19.22. Typical and average values of $\rho(0)$ as the metal–insulator transition is approached for (a) $U = 1.25$ and (b) $W = 1.0$. The insets show $\rho(0)$ as a function of ε for (a) $W = 2.5$, 2.6, 2.7 and 2.83 (from the black curve to the blue one) and (b) $U = 1.5$, 1.6, 1.7 and 1.86.

states within a narrow energy range ($\omega < t$, see also Fig. 19.17) around the band center (the Fermi energy) remain spatially delocalized ($\rho_{\text{typ}} \sim \rho_{\text{av}}$), due to strong disorder screening[31,97] within the Mott fluid (sites showing $Z_i \to 0$ at the transition). The electronic states away from the band center (i.e. in the band tails) quickly get Anderson-localized, displaying large spatial fluctuations of the wavefunction amplitudes[41] and having $\rho_{\text{typ}} \ll \rho_{\text{av}}$.

The spectral weight of the delocalized states (states in the range $\omega < t$) decreases with disorder and vanishes at the transition, indicating the Mott localization of this fraction of electrons. At this critical point, the crossover scale t also vanishes. In contrast, the *height* $\rho_{\text{typ}}(0)$ remains finite at the transition, albeit at a reduced W-dependent value, as compared to the clean limit. More precise evolution of $\rho_{\text{typ}}(0)$ is shown in Fig. 19.22(a), demonstrating its critical jump.

Behavior at the Mott-like transition ($W < U$) is dramatically different (Fig. 19.21 — bottom panel). Here $\rho_{\text{typ}} \approx \rho_{\text{av}}$ over the entire QP band, indicating the absence of Anderson localization. It proves essentially identical as that established for the disordered Hubbard model within standard

DMFT,[31] reflecting strong correlation-enhanced screening of disorder,[31,97] where both $\rho_{av}(\omega = 0)$ and $\rho_{typ}(\omega = 0)$ approach the bare $(W = 0)$ value (see also Fig. 19.22(b)). Similar results were found in Ref. 93, but an explanation was not provided.

The corresponding pinning[31,97] for $\rho(\omega = 0, \varepsilon)$ is shown in the insets of Fig. 19.3, both for the Mott–Anderson and the Mott-like transition. In the Mott–Anderson case, this mechanism applies only within the Mott fluid $(|\varepsilon| < U/2)$, while within the Anderson fluid $(|\varepsilon| > U/2)$ it assumes smaller values, explaining the reduction of $\rho_{typ}(0)$ in this case.

5.6. *Analytical solution*

Within our SB4 approach, the TMT-DMFT order-parameter function $\rho_{typ}(\omega)$ satisfies the following self-consistency condition

$$
\begin{aligned}
\rho_{typ}(\omega) = \exp \int d\varepsilon P(\varepsilon) \big\{ &\ln[V^2 Z^2(\varepsilon)\rho_{typ}(\omega)] \\
&- \ln[(\omega - \widetilde{\varepsilon}(\varepsilon) - V^2 Z(\varepsilon)\mathrm{Re}G_{typ}(\omega))^2 \\
&+ (\pi V^2 Z(\varepsilon)\rho_{typ}(\omega))^2] \big\}.
\end{aligned}
\tag{5.10}
$$

While the solution of this equation is in general difficult, it simplifies in the critical region, where the QP parameter functions $Z(\varepsilon)$ and $\widetilde{\varepsilon}(\varepsilon)$ assume scaling forms which we carefully studied in previous work.[97] This simplification allows, in principle, to obtain a closed solution for all quantities. In particular, the crossover scale t, which defines the $\rho_{typ}(\omega)$ mobility edge (see Fig. 19.17 and Ref. 97), is determined by setting $\rho_{typ}(\omega = t) = 0$.

Using this approach we obtain that, in the case of Mott-like transition $(W < U)$, the critical behavior of all quantities reduces to that found in standard DMFT,[31] including $t \sim U_c(W) - U$ (in agreement with the numerical results of Fig. 19.17(b)), perfect screening of site randomness,[31,97] and the approach of $\rho_{av}(\omega = 0)$ and $\rho_{typ}(\omega = 0)$ to the clean value. The precise form of the critical behavior for the crossover scale t is more complicated for the Mott–Anderson transition $(W > U)$ (as confirmed by our numerical results in Fig. 19.17(a), and this will not be discussed here.

Instead, we focus on elucidating the origin of the puzzling behavior of $\rho_c = \rho_{typ}(\omega = 0)$, which is known[41] to vanish linearly $\rho_c \sim (W_c - W)$ for $U = 0$, but which we numerically find to display a jump (i.e. a finite value) at criticality, as soon as interactions are turned on. For $\omega = 0$ our self-consistency condition reduces (for our model $\mathrm{Re}G_{typ}(0) = 0$ by particle–hole symmetry) to

$$\int d\varepsilon P(\varepsilon) \ln \frac{V^2 Z^2(\varepsilon)}{\tilde{\varepsilon}(\varepsilon)^2 + \pi^2 V^4 Z^2(\varepsilon)\rho_c^2} = 0, \qquad (5.11)$$

which further simplifies as we approach the critical point. Here, the QP parameters $Z(\varepsilon) \longrightarrow 0$ and $\tilde{\varepsilon}(\varepsilon) \sim Z^2(\varepsilon) \ll Z(\varepsilon)$ for the Mott fluid ($|\varepsilon| < U/2$), while $Z(\varepsilon) \longrightarrow 1$ and $|\tilde{\varepsilon}(\varepsilon)| \longrightarrow |\varepsilon - U/2|$ for the Anderson fluid ($|\varepsilon| > U/2$), and we can write

$$0 = \int_0^{U/2} d\varepsilon P(\varepsilon) \ln \frac{1}{(\pi V \rho_c)^2}$$
$$- \int_0^{(W-U)/2} d\varepsilon P(\varepsilon) \ln[(\varepsilon/V)^2 + (\pi V \rho_c)^2]. \qquad (5.12)$$

This expression becomes even simpler in the $U \ll W$ limit, giving

$$\frac{U}{W} \ln \frac{1}{\pi V \rho_c} + a - bV\rho_c + O[\rho_c^2] = 0, \qquad (5.13)$$

where $a(W, U) = (1 - U/W)\{1 - \ln[(W - U)/2V]\}$ and $b = \frac{2\pi^2 V}{W}$. This result reproduces the known result[41] $\rho_c \sim (W_c - W)$ at $U = 0$, but dramatically different behavior is found as soon as $U > 0$. Here, a *non-analytic* (singular) contribution emerges from the Mott fluid ($|\varepsilon| < U/2$), which assures that ρ_c must remain finite at the critical point, consistent with our numerical results (see Fig. 19.22). Note that the second term in Eq. (5.12), coming from the Anderson fluid ($|\varepsilon| > U/2$), vanishes in the case of a Mott-like transition ($U > W$), and our result reproduces the standard condition $\pi \rho_c V = 1$,[31] which corresponds to the clean limit.

A further glimpse on how the condition $\pi \rho_c V = 1$ is gradually violated as we cross on the Mott–Anderson side is provided by solving Eq. (5.12) for $U \lesssim W$ limit, giving

$$\rho_c \approx \frac{1}{\pi V} \left[1 - \frac{1}{24} \left(\frac{W}{V}\right)^2 \left(1 - \frac{U}{W}\right)^3 \right], \qquad (5.14)$$

again consistent with our numerical solution.

But what is the physical origin of the jump in ρ_c? To see it, note that the singular form of the first term in Eq. (5.12) comes from the Kondo pinning[31] $\tilde{\varepsilon}(\varepsilon) \sim Z^2(\varepsilon) \ll Z(\varepsilon)$ within the Mott fluid. This behavior reflects the particle–hole symmetry of our (geometrically averaged) $\rho_{\text{typ}}(\omega = 0)$ bath function, which neglects site-to-site cavity fluctuations present, for example, in more accurate statDMFT theories.[29,30,39,60,74,75] Indeed, in absence of particle–hole symmetry, one expects[31] $\widetilde{\varepsilon}(\varepsilon) \sim Z(\varepsilon)$, and the resulting ε-dependence should cut-off the log singularity responsible for the jump in μ_c. This observation provides a direct path to further refine the TMT-DMFT approach, reconciling the present results with previous statDMFT findings.[29,30,39,60,74,75] As a next step, one should apply the TMT ideas to appropriately chosen effective models,[62] in order to eliminate those features reflecting the unrealistic particle-hole symmetry built in the current theory. We emphasize that the two-fluid picture is a consequence of only a fraction of the sites showing $Z \to 0$ and is not dependent on either particle-hole symmetry or the consequent jump in the DOS. This interesting research direction is just one of many possible future applications of our TMT-DMFT formalism.

6. Conclusions and Outlook

This article described the conceptually simplest theoretical approach which is able to capture the interplay of strong correlation effects — the Mott physics — and the disorder effects associated with Anderson localization. It demonstrated that one can identify the signatures of both of these basic mechanisms for localization by introducing appropriate *local order parameters*, which are then self-consistently calculated within the proposed **Typical-Medium Theory**. We showed that key insight can be obtained by focusing on the evolution of the local quasiparticle weights Z_i as a *second order parameter* describing tendency to Mott localization, in addition to the Anderson-like TMT order parameter ρ_{typ}. Our main finding is that, for sufficiently strong disorder, the physical mechanism behind the Mott–Anderson transition is the formation of two fluids, a behavior that is surprisingly reminiscent of the phenomenology proposed for doped semiconductors.[99] Here, only a fraction of the electrons (sites) undergo Mott localization; the rest can be described as Anderson-localized quasiparticles. Physically, it describes spatially inhomogeneous situations, where the Fermi liquid quasiparticles are destroyed only in certain regions — the Mott droplets — but remain coherent elsewhere. Thus, in our picture the Mott–Anderson transition can be seen as reminiscent of the "orbitally selective" Mott localization.[100,101] To be more precise, here we have a "site selective" Mott transition, since it emerges in a spatially resolved fashion. Understanding the details of such "site selective" Mott transitions should be viewed as an indispensable first step in solving the long-standing problem of metal–insulator transitions in disordered correlated systems.

Acknowledgments

The author thanks Elihu Abrahams, Carol Aguiar, Eric Andrade, Gabi Kotliar, Eduardo Miranda, Andrei Pastor and Darko Tanasković for many years of exciting and fruitful collaboration. This work was supported by the NSF grant DMR-0542026.

References

1. N. F. Mott, *Metal–Insulator Transition* (Taylor & Francis, London, 1990).
2. Y. Kohsaka, C. Taylor, K. Fujita, A. Schmidt, C. Lupien, T. Hanaguri, M. Azuma, M. Takano, H. Eisaki, H. Takagi, S. Uchida and J. C. Davis, An intrinsic bond-centered electronic glass with unidirectional domains in underdoped cuprates, *Science* **315**, 1380 (2007).
3. W. D. Wise, K. Chatterjee, M. C. Boyer, T. Kondo, T. Takeuchi, H. Ikuta, Z. Xu, J. Wen, G. D. Gu, Y. Wang and E. W. Hudson, Imaging nanoscale Fermi surface variations in an inhomogeneous superconductor, *Nat. Phys.* **5**, 213 (2009).
4. A. N. Pasupathy, A. Pushp, K. K. Gomes, C. V. Parker, J. Wen, Z. Xu, G. Gu, S. Ono, Y. Ando and A. Yazdani, Electronic origin of the inhomogeneous pairing interaction in the high-t_c superconductor $Bi_2Sr_2CaCu_2O_{8+d}$, *Science* **320**, 196 (2008).
5. E. Dagotto, Complexity in strongly correlated electronic systems, *Science* **309**, 257 (2005).
6. E. Miranda and V. Dobrosavljević, Disorder-driven non-Fermi liquid behavior of correlated electrons, *Rep. Prog. Phys.* **68**, 2337 (2005).
7. P. A. Lee and T. V. Ramakrishnan, Disordered eletronic systems, *Rev. Mod. Phys.* **57**(2), 287 (1985).
8. E. Abrahams, P. W. Anderson, D. C. Licciardello and T. V. Ramakrishnan, Scaling theory of localization: Absence of quantum diffusion in two dimensions, *Phys. Rev. Lett.* **42**, 673 (1979).
9. F. Wegner, The mobility edge problem: continuous symmetry and a conjecture, *Phys. Rev. B* **35**, 783 (1979).
10. L. Schaffer and F. Wegner, Disordered system with n orbitals per site: Lagrange formulation, hyperbolic symmetry, and goldstone modes, *Phys. Rev. B* **38**, 113 (1980).
11. N. Goldenfeld, *Lectures on Phase Transitions and the Renormalization Group* (Addison-Wesley, New York, 1992).
12. K. B. Efetov, A. I. Larkin and D. E. Khmel'nitskii, Interaction between diffusion modes in localization theory, *Zh. Eksp. Teor. Fiz.* **79**, 1120 (1980).
13. A. M. Finkel'stein, Influence of Coulomb interaction on the properties of disordered metals, *Zh. Eksp. Teor. Fiz.* **84**, 168 (1983). [*Sov. Phys. JETP* **57**, 97 (1983)].
14. A. M. Finkel'stein, *Zh. Eksp. Teor. Fiz.* **86**, 367 (1984). [*Sov. Phys. JETP* **59**, 212 (1983)].

15. C. Castellani, C. D. Castro, P. A. Lee and M. Ma, Interaction-driven metal–insulator transitions in disordered fermion systems, *Phys. Rev. B* **30**, 527 (1984).
16. D. Belitz and T. R. Kirkpatrick, The Anderson–Mott transition, *Rev. Mod. Phys.* **66** (2), 261 (1994).
17. A. Punnoose and A. M. Finkel'stein, Dilute electron gas near the metal–insulator transition: Role of valleys in silicon inversion layers, *Phys. Rev. Lett.* **88**, 016802(4) (2002).
18. C. Castellani, B. G. Kotliar and P. A. Lee, Fermi-liquid theory of interacting disordered systems and the scaling theory of the metal–insulator transition, *Phys. Rev. Lett.* **56**, 1179 (1987).
19. R. B. Griffiths, Correlations in Ising ferromagnets. I, *J. Math. Phys.* **8**(3), 478 (1967).
20. R. B. Griffiths, Nonanalytic behavior above the critical point in a random Ising ferromagnet, *Phys. Rev. Lett.* **23**, 17 (1969).
21. H. V. Löhneysen, A. Rosch, M. Vojta and P. Wölfle, Fermi-liquid instabilities at magnetic quantum phase transitions, *Rev. Mod. Phys.* **79**(3), 1015 (2007).
22. S. Sachdev, *Quantum Phase Transitions* (Cambridge University Press, UK, 1999).
23. T. Vojta, Rare region effects at classical, quantum and nonequilibrium phase transitions, *J. Phys. A: Math. Gen.* **39**, R143 (2006).
24. D. S. Fisher, Critical behavior of random transverse-field Ising spin chains, *Phys. Rev. B* **51**(10), 6411 (1995).
25. T. Vojta, Disorder-induced rounding of certain quantum phase transitions, *Phys. Rev. Lett.* **90**, 107202 (2003).
26. J. A. Hoyos and T. Vojta, Theory of smeared quantum phase transitions, *Phys. Rev. Lett.* **100**(24), 240601 (2008).
27. V. Dobrosavljević and E. Miranda, Absence of conventional quantum phase transitions in itinerant systems with disorder, *Phys. Rev. Lett.* **94**, 187203 (2005).
28. M. J. Case and V. Dobrosavljević, Quantum critical behavior of the cluster glass phase, *Phys. Rev. Lett.* **99**(14), 147204 (2007).
29. E. Andrade, E. Miranda and V. Dobrosavljević, Energy-resolved spatial inhomogeneity of disordered mott systems, *Physica B* **404**(19), 3167 (2009).
30. E. C. Andrade, E. Miranda and V. Dobrosavljević, Electronic Griffiths phase of the $d = 2$ Mott transition, *Phys. Rev. Lett.* **102**, 206403 (2009).
31. D. Tanasković, V. Dobrosavljević, E. Abrahams and G. Kotliar, Disorder screening in strongly correlated systems, *Phys. Rev. Lett.* **91**, 066603 (2003).
32. P. W. Anderson, Absence of diffusion in certain random lattices, *Phys. Rev.* **109**, 1492 (1958).
33. N. E. Hussey, K. Takenaka and H. Takagi, Universality of the Mott–Ioff–Regel limit in metals, *Philos. Mag.* **84**, 2847 (2004).
34. M. M. Radonjić, D. Tanasković, V. Dobrosavljević and K. Haule, Influence of disorder on incoherent transport near the mott transition, *Phys. Rev. B.* **81**(7), 075118 (2010).

35. J. G. Analytis, A. Ardavan, S. J. Blundell, R. L. Owen, E. F. Garman, C. Jeynes and B. J. Powell, The effect of irradiation-induced disorder on the conductivity and critical temperature of the organic superconductor κ-(BEDT-TTF)$_2$Cu(SCN)$_2$, Phys. Rev. Lett. **96**, 177002 (2006).

36. A. Georges, G. Kotliar, W. Krauth and M. J. Rozenberg, Dynamical mean-field theory of strongly correlated fermion systems and the limit of infinite dimensions, Rev. Mod. Phys. **68**, 13 (1996).

37. V. Dobrosavljević, T. R. Kirkpatrick and B. G. Kotliar, Kondo effect in disordered systems, Phys. Rev. Lett. **69**, 1113 (1992).

38. V. Dobrosavljević and G. Kotliar, Strong correlations and disorder in $d = \infty$ and beyond, Phys. Rev. B **50**, 1430 (1994).

39. V. Dobrosavljević and G. Kotliar, Mean field theory of the Mott–Anderson transition, Phys. Rev. Lett. **78**, 3943 (1997).

40. V. Dobrosavljević and G. Kotliar, Dynamical mean-field studies of metal–insulator transitions, Philos. Trans. R. Soc. London, Ser. A **356**, 57 (1998).

41. V. Dobrosavljević, A. Pastor and B. K. Nikolić, Typical medium theory of Anderson localization: A local order parameter approach to strong disorder effects, Europhys. Lett. **62**, 76 (2003).

42. M. C. O. Aguiar, E. Miranda and V. Dobrosavljević, Localization effects and inelastic scattering in disordered heavy electrons, Phys. Rev. B **68**, 125104 (2003).

43. M. C. O. Aguiar, E. Miranda, V. Dobrosavljević, E. Abrahams and G. Kotliar, Temperature dependent transport of correlated disordered electrons: elastic vs. inelastic scattering, Europhys. Lett. **67**, 226 (2004).

44. E. Miranda and V. Dobrosavljević, Localization effects in disordered Kondo lattices, Physica B **259–261**, 359 (1999).

45. P. W. Anderson, Local moments and localized states, Rev. Mod. Phys. **50**(2), 191 (1978).

46. A. A. Pastor and V. Dobrosavljević, Melting of the electron glass, Phys. Rev. Lett. **83**, 4642 (1999).

47. V. Dobrosavljević, D. Tanasković and A. A. Pastor, Glassy behavior of electrons near metal–insulator transitions, Phys. Rev. Lett. **90**, 016402 (2003).

48. D. Dalidovich and V. Dobrosavljević, Landau theory of the Fermi-liquid to electron-glass transition, Phys. Rev. B **66**, 081107 (2002).

49. L. Arrachea, D. Dalidovich, V. Dobrosavljević and M. J. Rozenberg, Melting transition of an Ising glass driven by a magnetic field, Phys. Rev. B **69**(6), 064419 (2004).

50. S. Pankov and V. Dobrosavljević, Nonlinear screening theory of the Coulomb glass, Phys. Rev. Lett. **94**, 046402 (2005).

51. M. Muller and L. B. Ioffe, Glass transition and the Coulomb gap in electron glasses, Phys. Rev. Lett. **93**, 256403 (2004).

52. S. Sachdev and N. Read, Metallic spin glasses, J. Phys. Condens. Matter **8**, 9723 (1996).

53. A. M. Sengupta and A. Georges, Non-Fermi-liquid behavior near a $T = 0$ spin-glass transition, Phys. Rev. B **52**, 10295 (1995).

54. H. Westfahl, Jr., J. Schmalian and P. G. Wolynes, Dynamical mean field theory for self-generated quantum glasses, *Phys. Rev. B* **68**, 134203 (2003).
55. S. Wu, J. Schmalian, G. Kotliar and P. G. Wolynes, Solution of local-field equations for self-generated glasses, *Phys. Rev. B* **70**(2), 024207 (2004).
56. V. Dobrosavljević, T. R. Kirkpatrick and G. Kotliar, *Phys. Rev. Lett.* **69**, 1113 (1992).
57. E. Miranda, V. Dobrosavljević and G. Kotliar, Kondo disorder: a possible route towards non-Fermi liquid behavior, *J. Phys.: Condens. Matter* **8**, 9871 (1996).
58. E. Miranda, V. Dobrosavljević and G. Kotliar, Disorder-driven non-Fermi liquid behavior in Kondo alloys, *Phys. Rev. Lett.* **78**, 290 (1997).
59. E. Miranda, V. Dobrosavljević and G. Kotliar, Non-Fermi liquid behavior as a consequence of Kondo disorder, *Physica B* **230**, 569 (1997).
60. E. Miranda and V. Dobrosavljević, Localization induced Griffiths phase of disordered Anderson lattices, *Phys. Rev. Lett.* **86**, 264 (2001).
61. E. Miranda and V. Dobrosavljević, Griffiths phase of the Kondo insulator fixed point, *J. Magn. Magn. Mat.* **226–230**, 110 (2001).
62. D. Tanasković, E. Miranda and V. Dobrosavljević, Effective model of the electronic Griffiths phase, *Phys. Rev. B* **70**, 205108 (2004).
63. D. Tanaskovic, V. Dobrosavljevic and E. Miranda, Spin liquid behavior in electronic Griffiths phases, *Phys. Rev. Lett.* **95**, 167204 (2005).
64. B. Shklovskii and A. Efros, *Electronic Properties of Doped Semiconductors* (Springer-Verlag, 1984).
65. R. N. B. M. A. Paalanen, Transport and thermodynamic properties across the metal–insulator transition, *Physica B* **169**, 231 (1991).
66. S. V. Kravchenko, W. E. Mason, G. E. Bowker, J. E. Furneaux, V. M. Pudalov and M. D'Iorio, Scaling of an anomalous metal–insulator transition in a two-dimensional system in silicon at $b = 0$, *Phys. Rev. B* **51**, 7038 (1995).
67. D. Popović, A. B. Fowler and S. Washburn, Metal–insulator transition in two dimensions: effects of disorder and magnetic field, *Phys. Rev. Lett.* **79**, 1543 (1997).
68. E. Abrahams, S. V. Kravchenko and M. P. Sarachik, Colloquium: Metallic behavior and related phenomena in two dimensions, *Rev. Mod. Phys.* **73**, 251 (2001).
69. D. Simonian, S. V. Kravchenko and M. P. Sarachik, Reflection symmetry at a $b = 0$ metal insulator transition in two dimensions, *Phys. Rev. B* **55**(20), R13421 (1997).
70. V. Dobrosavljević, E. Abrahams, E. Miranda and S. Chakravarty, Scaling theory of two-dimensional metal–insulator transitions, *Phys. Rev. Lett.* **79**, 455 (1997).
71. M. Rozenberg, G. Kotliar and H. Kajueter, Transfer of spectral weight in spectroscopies of correlated electron systems, *Phys. Rev. B* **54**, 8542 (1996).
72. M. Mezard, G. Parisi and M. A. Virasoro, *Spin Glass Theory and Beyond* (World Scientific, Singapore, 1986).
73. V. Dobrosavljević and G. Kotliar, Hubbard models with random hopping in $d = \infty$, *Phys. Rev. Lett.* **71**, 3218 (1993).

74. V. Dobrosavljević and G. Kotliar, Dynamical mean-field studies of metal–insulator transitions, *Philos. Trans. R. Soc. London, Ser. A* **356**, 1 (1998).
75. E. Miranda and V. Dobrosavljević, Disorder-driven non-Fermi liquid behaviour of correlated electrons, *Rep. Prog. Phys.* **68**, 2337 (2005).
76. M. Janssen, Statistics and scaling in disordered mesoscopic electron systems, *Phys. Rep.* **295**, 1 (1998).
77. A. D. Mirlin, Statistics of energy levels and eigenfunctions in disordered systems, *Phys. Rep.* **326**, 259 (2000).
78. P. R. J. Elliot, J. A. Krumhansl and P. L. Leath, The theory and properties of randomly disordered crystals and related physical systems, *Rev. Mod. Phys.* **46**, 465 (1974).
79. H. Sompolinsky and A. Zippelius, Relaxational dynamics of the edwards-anderson model and the mean-field theory of spin-glasses, *Phys. Rev. B* **25**, 6860 (1982).
80. M. H. Grussbach, Determination of the mobility edge in the anderson model of localization in three dimensions by multifractal analysis, *Phys. Rev. B* **51**, 663 (1995).
81. J. Mooji, Electrical conduction in concentrated disordered transition metal alloys, *Phys. Status Solidi A Phys.* **17**, 521 (1973).
82. M. S. M. Girvin, Dynamical electron-phonon interaction and conductivity in strongly disordered metal alloy, *Phys. Rev. B* **22**, 3583 (1980).
83. J. D. Quirt and J. R. Marko, *Phys. Rev. Lett.* **26**, 318 (1971).
84. M. A. Paalanen, J. E. Graebner, R. N. Bhatt and S. Sachdev, *Phys. Rev. Lett.* **61**, 597 (1988).
85. M. Milovanović, S. Sachdev and R. N. Bhatt, Effective-field theory of local-moment formation in disordered metals, *Phys. Rev. Lett.* **63**, 82 (1989).
86. R. N. Bhatt and D. S. Fisher, Absence of spin diffusion in most random lattices, *Phys. Rev. Lett.* **68**, 3072 (1992).
87. M. Lakner, H. V. Löhneysen, A. Langenfeld and P. Wölfle, *Phys. Rev. B* **50**, 17064 (1994).
88. Q. Si, S. Rabello, K. Ingersent and J. L. Smith, Locally critical quantum phase transitions in strongly correlated metals, *Nature* **413**, 804 (2001).
89. T. Senthil, S. Sachdev and M. Vojta, Fractionalized Fermi liquids, *Phys. Rev. Lett.* **90**, 216403 (2003).
90. T. Senthil, M. Vojta and S. Sachdev, Weak magnetism and non-Fermi liquids near heavy-fermion critical points, *Phys. Rev. B* **69**, 035111 (2004).
91. J. A. Hertz, Quantum critical phenomena, *Phys. Rev. B* **14**, 1165 (1976).
92. A. J. Millis, Effect of a nonzero temperature on quantum critical points in itinerant fermion systems, *Phys. Rev. B* **48**(10), 7183 (1993).
93. K. Byczuk, W. Hofstetter and D. Vollhardt, Mott–Hubbard transition versus Anderson localization in correlated electron systems with disorder, *Phys. Rev. Lett.* **94**, 056404 (2005).
94. K. Byczuk, W. Hofstetter and D. Vollhardt, Competition between Anderson localization and antiferromagnetism in correlated lattice fermion systems with disorder, *Phys. Rev. Lett.* **102**, 146403 (2009).

95. M. C. O. Aguiar, V. Dobrosavljevic, E. Abrahams and G. Kotliar, Critical behavior at Mott–Anderson transition: a TMT-DMFT perspective, *Phys. Rev. Lett.* **102**, 156402 (2009).

96. M. C. O. Aguiar, V. Dobrosavljevic, E. Abrahams and G. Kotliar, Scaling behavior of an Anderson impurity close to the Mott–Anderson transition, *Phys. Rev. B* **73**, 115117 (2006).

97. M. C. O. Aguiar, V. Dobrosavljevic, E. Abrahams and G. Kotliar, Disorder screening near the Mott–Anderson transition, *Physica B* **403**, 1417 (2008).

98. G. Kotliar and A. E. Ruckenstein, New functional integral approach to strongly correlated Fermi systems: The Gutzwiller approximation as a saddle point, *Phys. Rev. Lett.* **57**(11), 1362 (1986).

99. M. A. Paalanen, J. E. Graebner, R. N. Bhatt and S. Sachdev, Thermodynamic behavior near a metal–insulator transition, *Phys. Rev. Lett.* **61**, 597 (1998).

100. L. De Leo, M. Civelli and G. Kotliar, $T = 0$ heavy fermion quantum critical point as an orbital selective Mott transition, *Phys. Rev. Lett.* **101**, 256404 (2008).

101. C. Pepin, Kondo breakdown as a selective Mott transition in the Anderson lattice, *Phys. Rev. Lett.* **98**, 206401 (2007).

Chapter 20

ANDERSON LOCALIZATION VS. MOTT–HUBBARD METAL–INSULATOR TRANSITION IN DISORDERED, INTERACTING LATTICE FERMION SYSTEMS

Krzysztof Byczuk,[*,†] Walter Hofstetter[‡] and Dieter Vollhardt[†]

*Institute of Theoretical Physics,
University of Warsaw, ul. Hoża 69,
PL-00-681 Warszawa, Poland

†Theoretical Physics III,
Center for Electronic Correlations and Magnetism,
Institute for Physics, University of Augsburg,
D-86135 Augsburg, Germany

‡Institut für Theoretische Physik,
Johann Wolfgang Goethe-Universität,
60438 Frankfurt/Main, Germany

We review recent progress in our theoretical understanding of strongly correlated fermion systems in the presence of disorder. Results were obtained by the application of a powerful nonperturbative approach, the dynamical mean-field theory (DMFT), to interacting disordered lattice fermions. In particular, we demonstrate that DMFT combined with geometric averaging over disorder can capture Anderson localization and Mott insulating phases on the level of one-particle correlation functions. Results are presented for the ground state phase diagram of the Anderson–Hubbard model at half-filling, both in the paramagnetic phase and in the presence of antiferromagnetic order. We find a new antiferromagnetic metal which is stabilized by disorder. Possible realizations of these quantum phases with ultracold fermions in optical lattices are discussed.

1. Introduction

In non-interacting quantum systems with disorder, e.g., in the presence of randomly distributed impurities, wavefunctions can either be spatially extended or localized. Until 1958 it was believed that a localized state corresponds to a bound state of an electron at the impurity. By contrast, in his landmark paper of 1958, Anderson[1] predicted that disorder can lead to quite a different type of localized state now referred to as "Anderson localized state". To understand its physical origin it should be noted that if a particle is inserted into a disordered system it will start to spread. As a consequence the wave is backscattered by the impurities, leading to characteristic "weak localization" effects.[2–4] The multiple scattering of the electronic wave can enhance these perturbative effects to such a degree that the electrons become spatially localized; for reviews see Refs. 4–6. In this case there is a finite probability for an electron to return to the point where it was inserted. If states are extended, this probability is zero. So, in contrast to localized states bound at an impurity, Anderson localized states are confined to a region of space due to coherent backscattering from randomly distributed impurities.

In the thermodynamic limit the excitation spectrum determined from the resolvent of the one-particle system or the one-particle Green functions is very different for extended and localized states. The one-particle Green function describing an extended state has a branch cut on the real axis, and the spectrum of the Hamiltonian is continuous. By contrast, the Green function for a localized state has discrete poles located infinitely close to the real axis, which implies a discrete point spectrum of the Hamiltonian. In particular, the point-like spectrum of an Anderson localized state is dense.

In the presence of interactions between the electrons the same classification of (approximate) eigenstates may, in principle, be used. Namely, if the one-particle Green function of the interacting system has a branch cut at some energies, the states at those energies are extended. If the Green function has discrete, separate poles the corresponding states are bound states, and if the poles are discrete and lie dense the states are Anderson localized. Since one-particle wave functions are not defined in a many-body system, they cannot be employed to describe the localization properties of the system. Instead the reduced one-particle *density matrix*, or the one-particle Green function $G(\mathbf{r} - \mathbf{r}')$ in position representation, may be employed. For localized states these quantities approach zero for $|\mathbf{r} - \mathbf{r}'| \to \infty$. For extended states, their amplitude only fluctuates very weakly, i.e., of the order $1/V$, where V is the volume of the system.

In the following, we are interested in the question how states of many-body systems change when the interaction and/or the disorder are varied. In general, the very notion of a metal or an insulator is related to the properties of two-particle Green functions, e.g., the current- and density-correlation functions. There exist different approaches to study the disappearance of a diffusion pole at the metal–insulator transition, and correspondingly, the vanishing of the DC conductivity in the thermodynamic limit.[4,5,7,8] On physical grounds it is very plausible to expect that the presence of Anderson localized states with dense, point-like spectrum at the Fermi level, discussed above in terms of one-particle Green function, implies zero conductivity. Mathematical proofs of this conjecture exist only for specific models and in limiting cases.[9] Indeed, it is usually assumed that the presence of Anderson localized states at the Fermi level implies the system to be an Anderson insulator, at least in the non-interacting case. This is also our line of approach which will be reviewed in this article.

The paper is structured as follows. In Sec. 2, we review general aspects of the interplay between interactions and disorder in lattice fermion systems. In particular, we discuss the important question concerning the appropriate average over the disorder, and describe the new developments in the field of cold atoms in optical lattices which will make it possible in the future to investigate disordered, interacting lattices fermions with unprecedented control over the parameters. The models of correlated fermions with disorder are introduced in Sec. 3, followed by an introduction into the dynamical mean-field theory (DMFT) (Sec. 4) and a more detailed discussion of arithmetic vs. geometric averaging over the disorder (Sec. 5). In Sec. 6, the DMFT self-consistency conditions for disordered systems are introduced. After having defined the characteristic quantities which help us to identify the different phases of the Anderson–Hubbard Hamiltonian (Sec. 7), the results for the ground state phase diagram at half-filling are reviewed (Sec. 8). In Sec. 9 the results are summarized.

2. Interplay between Interactions and Disorder in Lattice Fermion Systems

2.1. *Interactions vs. disorder*

The properties of solids are strongly influenced by the interaction between the electrons and the presence of disorder.[4,7,8] Namely, Coulomb correlations and randomness are both driving forces behind metal–insulator transitions (MITs) which involve the localization and delocalization of particles. While the electronic repulsion may lead to a Mott–Hubbard MIT,[7,10] the coherent backscattering of non-interacting particles from randomly distributed impurities can cause Anderson localization.[1,2]

Since electronic interactions and disorder can both (and separately) induce a MIT, one might expect their simultaneous presence to be even more effective in localizing electrons. However, this is not necessarily so. For example, weak disorder is able to weaken the effect of correlations since it redistributes states into the Mott gap and may thus turn an insulator into a (bad) metal. Furthermore, short-range interactions lead to a transfer of spectral weight into the Hubbard subbands whereby the total band-width and thus the critical disorder strength for the Anderson MIT increases, implying a reduction of the effective disorder strength. Hence the interplay between disorder and interactions leads to subtle many-body effects,[4,8,11–17] which pose fundamental challenges for theory and experiment not only in condensed matter physics,[4,7,8,18,19] but most recently also in the field of cold atoms in optical lattices.[20–27] Indeed, ultracold gases have quickly developed into a fascinating new laboratory for quantum many-body physics.[20,21,28–33] A major advantage of cold atoms in optical lattices is the high degree of controllability of the interaction and the disorder strength, thereby allowing a detailed verification of theoretical predictions. The concepts, models, and techniques for their solution to be discussed in this paper equally apply to electronic systems and cold fermionic atoms in optical lattices. In the following we will therefore refer generally to the investigation of "correlated lattice fermion systems".

2.1.1. *Average over disorder*

In general, the theoretical investigation of disordered systems requires the use of probability distribution functions (PDFs) for the random quantities of interest. Indeed, in physical or statistical problems one is usually interested in "typical" values of random quantities which are mathematically given by the most probable value of the PDF.[34] However, in many cases the complete PDF is not known, i.e., only limited information about the system provided by certain averages (moments or cumulants) is available. In this situation it is very important to choose the most informative average of a random variable. For example, if the PDF of a random variable has a single peak and fast decaying tails the typical value of the random quantity is well estimated by its first moment, known as the *arithmetic* average (or arithmetic mean). But there are many examples, e.g., from astronomy, the physics of glasses or networks, economy, sociology, biology or geology, where the knowledge of the arithmetic average is insufficient since the PDF is so broad that its characterization requires infinitely many moments.[35,36] Such systems are called non-self-averaging. One example is Anderson localization: when a disordered system is close to the Anderson MIT,[1] most electronic quantities fluctuate strongly and the corresponding PDFs possess long tails which can be described by a log-normal distribution.[37–42] This is well illustrated by the local density of states (LDOS) of the disordered system. Most recently it was shown for various lattices in dimensions $d = 2$ and 3 that the system-size dependence of the LDOS distribution is an unambigous sign of Anderson localization, and that the distribution of the LDOS of disordered electrons agrees with a log-normal distribution over up to ten orders of magnitude.[42] Therefore it is not surprising that the arithmetic mean of this random one-particle quantity does not resemble its typical value at all. In particular, it is non-critical at the Anderson transition[43-45] and hence cannot help to detect the localization transition. By contrast the *geometric* mean[35,36,46,47] of the LDOS, which represents the most probable ("typical") value of a log-normal distribution, is the appropriate average in this case. It vanishes at a critical strength of the disorder and hence provides an explicit criterion for Anderson localization in disordered systems,[1,39,42] even in the presence of interactions.[48,49]

2.1.2. *Dynamical mean-field approach to disordered systems*

In general, MITs occur at intermediate values of the interaction and/or disorder. Theories of MITs driven by interaction and disorder therefore need to be non-perturbative. Usually they cannot be solved analytically, and require numerical methods or self-consistent approximations. A reliable approximate method for the investigation of lattice fermions with a local interaction is provided by DMFT,[50-52] where the local single-particle Green function is determined self-consistently. If in this approach the effect of local disorder is taken into account through the arithmetic mean of the LDOS[53] one obtains, in the absence of interactions, the well-known coherent potential approximation (CPA).[54] However, the CPA does not describe the physics of Anderson localization since, as discussed above, the arithmetically averaged LDOS is non-critical at the Anderson transition.[45] To overcome this deficiency, Dobrosavljević and Kotliar[48] formulated a variant of the DMFT where the probability distributions (and not only the averages) of the local Green functions are determined self-consistently ("Statistical DMFT"). Employing a Slave–Boson mean-field theory as impurity solver, they investigated the disorder-driven MIT for infinitely strong repulsion off half-filling. This statistical approach was also employed in other studies of the Hubbard model[55] as well as in the case of electrons coupled to phonons[56] and the Falicov–Kimball model.[57] Subsequently, Dobrosavljević, Pastor and Nikolić[49] incorporated the geometrically averaged LDOS into the self–consistency cycle and thereby derived a mean-field theory of Anderson localization which reproduces many of the expected features of the disorder-driven MIT for non-interacting fermions. This scheme employs only one-particle quantities and is therefore easily incorporated into the DMFT for disordered electrons in the presence of phonons,[56] or Coulomb correlations.[58-61]

2.2. Cold atoms in optical lattices: a new realization of disordered, correlated lattice quantum gases

During the last few years, cold atoms in optical lattices have emerged as a unique tool-box for highly controlled investigations of quantum many-body systems. In recent years, the level of control in applying disordered potentials to ultracold quantum gases has greatly improved.[23,24] Anderson localization in its pure form has been demonstrated by the expansion of weakly interacting Bose–Einstein condensates in a disordered speckle light field, giving rise to characteristic localized condensate wave functions with exponentially decaying tails.[25,26] The additional influence of strong repulsive interactions has been investigated recently in the first full experimental realization of the 3d disordered Bose–Hubbard model, by using a fine-grained optical speckle field superimposed by an optical lattice.[27] In this experiment a strong reversible suppression of the condensate fraction due to disorder was observed, indicating the formation of a disorder-induced insulating state. Independent experimental evidence was obtained from interacting ^{87}Rb bosons in a quasi-random (bichromatic) optical lattice, where a strong reduction of the Mott gap was found and interpreted as possible evidence for a compressible Bose glass phase.[22] On the theoretical side, low-dimensional quasi-disordered Bose systems have been successfully described by DMRG simulations,[62] which extended previous weak-coupling calculations and found a direct transition from superfluid to Mott insulator. Regarding disordered bosons in higher dimensions, the status of theory is still more controversial, although significant insight was gained by a new stochastic mean-field theory,[63] which allows for an efficient description of the Bose glass phase and has already provided phase diagrams for realistic speckle-type disorder[64] such as used experimentally.[27] Under debate remains the issue of a direct transition between Mott insulator and superfluid, which was claimed to be ruled out in recent QMC simulations in three spatial dimensions, supported by general heuristic arguments.[65] Regarding disordered fermions, while no experiments in cold gases have been performed yet, theory has significantly advanced in recent years, mostly due to progress in the application of DMFT to disordered and inhomogeneous systems.[48,49,58,60,66] The phase diagram of spin-1/2 lattice fermions in a random potential has now been determined theoretically, both in the paramagnetic phase where Mott- and Anderson-insulator compete,[58] and in the low-temperature regime where antiferromagnetic ordering sets in and a new disorder-induced antiferromagnetic *metallic* phase was found.[60] In this way, predictions for single-particle spectral properties were also obtained, which are now becoming accessible experimentally via radio frequency spectroscopy measurements of strongly interacting fermionic quantum gases,[67] in analogy to photoemission spectroscopy of electronic solids. An alternative route towards single-particle spectroscopy based on stimulated Raman transitions has been discussed theoretically.[68] Very recently, also the dynamical structure factor of strongly interacting bosons in an optical lattice has been measured via two-photon Bragg scattering.[69,70] These new developments open the door towards controlled experimental realization and spectroscopy of strongly interacting and disordered fermions in optical lattices.

2.3. *Schematic phase diagram*

The Mott–Hubbard MIT is caused by short-range, repulsive interactions in the pure system and is characterized by the opening of a gap in the density of states at the Fermi level. By contrast, the Anderson MIT is due to the coherent backscattering of the quantum particles from randomly distributed impurities in a system without interactions; at the transition the character of the spectrum at the Fermi level changes from a continuous to a dense point spectrum. Already these two limits provide great challenges for theoretical investigations. It is an even greater challenge to explore the *simultaneous* presence of interactions and disorder in lattice fermions systems. In view of the construction of the dynamical mean-field approach employed here, the results which will be presented in the following are expected to provide a comprehensive description for systems in spatial dimensions $d = 3$ and larger, i.e., above the limiting dimension $d = 2$. Two particularly interesting questions are whether the metallic phase, which exists at weak enough disorder and/or interaction strength, will be reduced or enlarged, and whether the Mott and Anderson insulating phases are separated by a metallic phase. Corresponding schematic phase transition lines are shown in Fig. 20.1. It is plausible to assume that both MITs can be characterized by a single quantity, namely, the local density of states. Although the LDOS is not an order parameter associated with a symmetry breaking phase transition, it discriminates between a metal and an insulator which is driven by correlations and disorder.

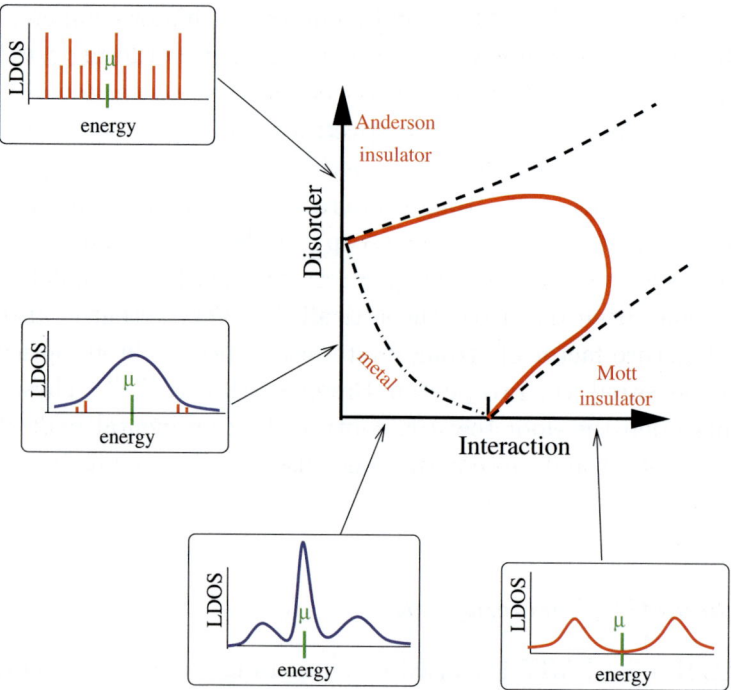

Fig. 20.1.   Schematic diagram of the possible phases and shapes of phase transition lines in disordered, interacting lattice fermion systems. In principle, the interplay between interactions and disorder could reduce the metallic regime (dash-dotted line), or enlarge it (full and dashed lines). As will be discussed below, investigations within DMFT find that the metallic phase increases when interactions and disorder are simultaneously present (full line), and that the two insulating phases are connected continuously, i.e., without critical behavior. Insets show the local density of states in the absence of disorder or interaction, respectively.

3. Models of Correlated Fermions with Disorder

Here we study models of correlated fermions on ionic crystals or optical lattices in the presence of diagonal (i.e., local) disorder within a tight-binding description. In general, these models have the form

$$H = \sum_{ij\sigma} t_{ij\sigma} c_{i\sigma}^{\dagger} c_{j\sigma} + \sum_{i\sigma} \epsilon_i \, n_{i\sigma} + U \sum_i n_{i\uparrow} n_{i\downarrow} \qquad (3.1)$$

where $c_{i\sigma}^{\dagger}$ and $c_{i\sigma}$ are the fermionic creation and annihilation operators of the particle with spin $\sigma = \pm 1/2$ at the lattice site i, $n_{i\sigma} = c_{i\sigma}^{\dagger} c_{i\sigma}$ is the particle number operator with eigenvalues 0 or 1, and $t_{ij\sigma}$ is the probability amplitude for hopping between lattice sites i and j. In the Hubbard model

$t_{ij\sigma} = t_{ij}$, i.e. the hopping amplitudes are the same for both spin projections. In the Falicov–Kimball model $t_{ij\sigma} = t_{ij}\delta_{\sigma\uparrow}$, i.e. only particles with one spin projection are mobile and the others are localized. The second term in (3.1) describes the additional external potential ϵ_i, which breaks the ideal lattice symmetry. For homogeneous systems we set $\epsilon_i = 0$, which defines the zero of the energy scale. The third term, a two-body term, describes the increase of the energy by $U > 0$ if two fermions with opposite spins occupy the same site. In Eq. (3.1) only a local part of the Coulomb interaction is included and other longer-range terms are neglected for simplicity. Note that this approximation is excellent in the case of cold gases in optical lattices, where the interaction between neutral atoms is essentially local.[28] The disorder affects the system through a local term $\sum_{i\sigma} \epsilon_i n_{i\sigma}$, where ϵ_i is a random variable drawn from a probability distribution function (PDF) $\mathcal{P}(\epsilon_1, \ldots, \epsilon_{N_L})$, where N_L is a number of lattice sites. Typically we consider uncorrelated, quenched disorder, where

$$\mathcal{P}(\epsilon_1, \ldots, \epsilon_{N_L}) = \prod_{i=1}^{N_L} P(\epsilon_i) \,. \tag{3.2}$$

Each of the $P(\epsilon_i)$ is the same, normalized PDF for the atomic energies ϵ_i. The quenched disorder means that $P(\epsilon_i)$ is time independent. In other words, the atomic energies ϵ_i are randomly distributed over the lattice and cannot fluctuate in time. This type of disorder is different from annealed disorder where the random atomic energies have thermal fluctuations.

In the following we use the continuous box-type PDF

$$P(\epsilon_i) = \frac{1}{\Delta} \Theta\left(\frac{\Delta}{2} - |\epsilon_i|\right) \,, \tag{3.3}$$

with $\Theta(x)$ as the Heaviside step function. The parameter Δ is therefore a measure of the disorder strength. The use of a different continuous, normalized function for the PDF would bring about only quantitative changes.

The Hubbard model and the Falicov–Kimball model defined by (3.1) are not only of interest for solid-state physics, but also in the case of ultracold atoms, where specific experimental realizations have been proposed.[20] By preparing a mixture of bosonic ^{87}Rb and fermionic ^{40}K in a 3d optical lattice, Ospelkaus *et al.* and Günter *et al.*[32] were able to create — to a first approximation — a version of the Falicov–Kimball model where the heavier bosonic species could be slowed down even further by using a species-dependent optical lattice and thus become "immobile" while the fermionic species remains mobile. Alternatively, if the heavy bosonic species could be frozen in a random configuration, this system would allow for a realization of the

Fermi–Hubbard model with quenched binary onsite disorder. A different approach towards quenched randomness in optical lattices was taken by White et al.[27] who implemented a fine-grained optical speckle potential superimposed onto a 3d optical lattice with interacting bosons and thus realized the bosonic version of the Anderson–Hubbard model (3.1) with continuous disorder. A third alternative approach to disordered cold gases is based on bichromatic optical lattices which are quasiperiodic, as implemented for the 3d Bose–Hubbard model by Fallani et al.[22] who observed a disorder-induced reduction of the Mott excitation gap, similar as discussed in the following for the fermionic case.

The Hamiltonian (3.1) is not solvable in general. Without disorder, i.e., for $\Delta = 0$, exact solutions on an arbitrary lattice and in arbitrary dimension exist only for $U = 0$ (non-interacting fermions), or $t_{ij\sigma} = 0$ (fermions in the atomic limit). In the $U = 0$ case the solution is obtained via discrete Fourier transform, i.e.,

$$H = \sum_{\mathbf{k}\sigma} \epsilon_{\mathbf{k}\sigma} c_{\mathbf{k}\sigma}^\dagger c_{\mathbf{k}\sigma}, \tag{3.4}$$

where $\epsilon_{\mathbf{k}\sigma} = \sum_{j(i)} t_{ij\sigma} e^{-i\mathbf{k}(\mathbf{R}_j - \mathbf{R}_i)}$ are free fermion dispersion relations. In the thermodynamic limit $N_L \to \infty$ the spectrum is continuous and eigenstates are extended. In the $t_{ij\sigma} = 0$ limit the lattice sites are uncorrelated and the exact partition function has the form $Z = \prod_i Z_i$, where

$$Z_i = 1 + 2e^{\beta\mu} + e^{-\beta U}, \tag{3.5}$$

where μ denotes the chemical potential within the grand canonical ensemble, and $\beta = 1/k_B T$ is the inverse temperature. In the thermodynamic limit the spectrum is point-like and the eigenstates are localized.

For finite disorder ($\Delta \neq 0$) an exact solution of the Hamiltonian (3.1) exists only for $t_{ij\sigma} = 0$. For a given realization of disorder, i.e., when all values of $\{\epsilon_1, \epsilon_2, \ldots, \epsilon_{N_L}\}$ are fixed, the partition function of the model (3.1) is given by

$$Z = \prod_i Z_i = \prod_i \left(1 + 2e^{-\beta(\epsilon_i - \mu)} + e^{-\beta U}\right). \tag{3.6}$$

As in the atomic limit discussed above ($t_{ij\sigma} = 0$) the spectrum is point-like in the thermodynamic limit and the eigenstates are localized.

The non-interacting limit ($U = 0$) of (3.1) with $t_{ij\sigma} \neq 0$ and disorder $\Delta \neq 0$ is not exactly solvable. In a seminal paper by Abou-Chacra, Thouless and Anderson[71] the model (3.1) with $U = 0$ and $t_{ij\sigma} = t$ between nearest neighbor sites was solved on the Bethe lattice, which is a tree-like graph without loops.[40,72] The solution is expressed by the one-particle Green function

$$G_{ii}(\omega) = \langle i| \frac{1}{\omega - H} |i\rangle = \frac{1}{\omega - \epsilon_i - \eta_i(\omega)}, \tag{3.7}$$

where the hybridization function

$$\eta_i(\omega) = \sum_{j \neq i} \frac{t^2}{\omega - \epsilon_j - \eta_j(\omega)} \tag{3.8}$$

describes a resonant coupling of site i with its neighbors. If in the thermodynamic limit the imaginary part of $\eta_i(z)$ is finite in some band of energies z, then the states with energies z are extended. Otherwise, if the imaginary part of $\eta_i(z)$ is finite at discrete energies z such states are localized. For bound states these energies z form a point spectrum, and for Anderson localized states the energies z form a dense point-like spectrum in the thermodynamic limit. The analysis of the self-consistent equations derived for $\eta_i(z)$ by Abou-Chacra, Thouless, Anderson[71] showed that, indeed, continuous and dense point spectra are separated by a mobility edge which depends on the value of the disorder Δ.

In the following we solve the full Hamiltonian (3.1) by applying a dynamical mean-field approximation to deal with the interaction and then discuss how to cope with disorder.

4. Dynamical mean-field theory (DMFT)

The dynamical mean-field theory (DMFT) started from the following observation[50]: if the hopping amplitudes are scaled with fractional powers of the space dimension d (or the coordination number Z), i.e., $t = t^*/\sqrt{2d} = t^*/\sqrt{Z}$ for nearest neighbour hopping on a hypercubic lattice, then in the limit $d \to \infty$ ($Z \to \infty$) the self-energy $\Sigma_{ij}(\omega)$ in the Dyson equation

$$G_{ij\sigma}(i\omega_n)^{-1} = G^0_{ij\sigma}(i\omega_n)^{-1} - \Sigma_{ij\sigma}(i\omega_n), \tag{4.1}$$

(here in a real-space representation) becomes diagonal[73]

$$\Sigma_{ij\sigma}(i\omega_n) = \Sigma_{i\sigma}(i\omega_n)\, \delta_{ij}, \tag{4.2}$$

where $\omega_n = (2n+1)\pi/\beta$ are fermionic Matsubara frequencies. In a homogeneous system the self-energy is site independent, i.e., $\Sigma_{ij\sigma}(i\omega_n) = \Sigma_\sigma(i\omega_n)\,\delta_{ij}$, and is only a function of the energy. The DMFT approximation when applied to finite dimensional systems neglects off-diagonal parts of the self-energy. In other words, the DMFT takes into account all temporal fluctuations but neglects spatial fluctuations between different lattice sites.[51,52]

Here we apply the DMFT to correlated fermion systems with disorder. Within DMFT we map a lattice site onto a single impurity, which is coupled to the dynamical mean-field bath. This coupling is represented by the hybridization function $\eta_{i\sigma}(\omega)$, which is determined self-consistently. The mapping is performed for all N_L lattice sites.

The partition function for a particular realization of disorder $\{\epsilon_1, \epsilon_2, \ldots, \epsilon_{N_L}\}$ is now expressed as a product of the partition functions which are determined for each impurity (representing lattice sites), i.e.,

$$Z = \prod_i Z_i = \prod_i \exp\left(\sum_{\sigma\omega_n} \ln[i\omega_n + \mu - \epsilon_i - \eta_{i\sigma}(\omega_n) - \Sigma_{i\sigma}(\omega_n)]\right). \quad (4.3)$$

The mean-field hybridization function $\eta_{i\sigma}(\omega_n)$ is formally a site- and time-dependent one-particle potential. In the interaction representation, the unitary time evolution due to this potential is described by the local, time-dependent evolution operator[74,75]

$$U[\eta_{i\sigma}] = T_\tau e^{-\int_0^\beta d\tau \int_0^\beta d\tau' c_{i\sigma}^\dagger(\tau)\eta_{i\sigma}(\tau-\tau')c_{i\sigma}(\tau')}, \quad (4.4)$$

where $c_{i\sigma}(\tau)$ evolves according to the atomic part H_i^{loc} of the Hamiltonian (3.1) in imaginary Matsubara time $\tau \in (0, \beta)$, and T_τ is the time ordering operator. We write the partition function (4.3) as a trace over the operator

$$Z = Z[\eta_{i\sigma}] = \prod_{i=1}^{N_L} \text{Tr}\left[e^{-\beta(H_i^{\text{loc}} - \mu N_i^{\text{loc}})} U[\eta_{i\sigma}]\right], \quad (4.5)$$

where N_i^{loc} is the local particle number operator.

Equation (4.5) allows us to determine the local one-particle Green function $G_{ii\sigma}(\omega_n)$ for a given dynamical mean-field $\eta_{i\sigma}(\omega_n)$. Indeed, the local Green function is obtained by taking a functional logarithmic derivative of the partition function (4.5) with respect to $\eta_{i\sigma}(\omega_n)$, i.e.,

$$G_{ii\sigma}(\omega_n) = -\frac{\partial \ln Z[\eta_{i\sigma}]}{\partial \eta_{i\sigma}(\omega_n)}. \quad (4.6)$$

Then we find the local Dyson equations

$$\Sigma_{i\sigma}(\omega_n) = i\omega_n + \mu - \epsilon_i - \eta_{i\sigma}(\omega_n) - \frac{1}{G_{ii\sigma}(\omega_n)}, \quad (4.7)$$

for each N_L lattice sites. For a single realization of disorder $\{\epsilon_1, \epsilon_2, \ldots, \epsilon_{N_L}\}$, Eqs. (4.1), (4.2), (4.5)–(4.7) constitute a closed set of equations. A solution of this set represents an approximate solution of the Hamiltonian (3.1).

5. Arithmetic vs. Geometric Averaging

A solution of Eqs. (4.1), (4.2), (4.5)–(4.7) is very difficult to obtain in practice. For each of the N_L impurities we need to determine the evolution operator (4.4) exactly. Using rigorous methods this can be done only for small N_L. However, Eqs. (4.1), (4.2), (4.5)–(4.7) should be solved in the thermodynamic limit, $N_L \to \infty$. This latter requirement might be overcome by performing a finite size scaling analysis. But such an analysis requires a large number of lattice sites N_L to reliably distinguish Anderson localized states from those belonging to the continuum. Here one faces a typical trade-off situation in computational physics. The computational problem is greatly reduced when the local interaction in (3.1) is factorized as in a Hartree–Fock approximation, whereby genuine correlations are eliminated.[15,76,77] Such approximate treatments can nevertheless provide valuable hints about the existence of particular phases. In our investigation[58–60] we employed the DMFT to include all local correlations as will be discussed in the next section.

If one could solve the DMFT equations exactly, one would obtain a set of local densities of states (LDOS)

$$A_{i\sigma}(\omega) = -\frac{1}{\pi} \mathrm{Im} G_{ii\sigma}(\omega_n \to \omega + i0^+)\,, \tag{5.1}$$

which are random quantities depending on the particular disorder realization $\{\epsilon_1,\ \epsilon_2,\ \ldots,\ \epsilon_{N_L}\}$. Usually one needs information about a system that does not depend on a particular disorder realization. Therefore one needs a statistical interpretation of the solutions of Eqs. (4.1), (4.2), (4.5)–(4.7).

When the system is large (cf., $N_L \to \infty$ in thermodynamic limit) one usually takes the arithmetic average of the LDOS $A_{i\sigma}(\omega)$ over many realizations of the disorder, i.e.,

$$\langle A_{i\sigma}(\omega) \rangle = \int \prod_{j=1}^{N_L} d\epsilon_j\, P(\epsilon_i)\, A_{i\sigma}(\omega; \{\epsilon_1, \ldots, \epsilon_{N_L}\})\,, \tag{5.2}$$

where the dependence on $\{\epsilon_1,\ \epsilon_2,\ \ldots,\ \epsilon_{N_L}\}$ is written explicitly. However, such a method holds only if the system is self-averaging. This means that sample-to-sample fluctuations

$$D_{N_L}(A_{i\sigma}(\omega)) = \frac{\langle A_{i\sigma}(\omega)^2 \rangle - \langle A_{i\sigma}(\omega) \rangle^2}{\langle A_{i\sigma}(\omega) \rangle^2} \tag{5.3}$$

vanish for $N_L \rightarrow \infty$, which is equivalent to the central limit theorem for independent random variables $A_{i\sigma}(\omega)$. By performing the arithmetic average, one restores the translational invariance in the description of the disordered system, i.e., $A_\sigma(\omega)_{\text{arith}} = \langle A_{i\sigma}(\omega) \rangle$ is the same for all lattice sites.

An example of a *non*-self-averaging system is a disordered system at the Anderson localization transition, or a system whose localization length is smaller than the diameter of the sample.[1] It implies that during the time evolution, a particle cannot explore the full phase space, i.e., cannot probe all possible random distributions. In such a case the arithmetic average (5.2) is inadequate. Here one is faced with the question concerning the proper statistical description of such a system.

The answer was given by Anderson[1]: one should investigate the full PDF for a given physical observable $P[A_{i\sigma}(\omega)]$ and find its most probable value, the "typical" value $A_\sigma(\omega)_{\text{typ}}$, for which the PDF $P[A_{i\sigma}(\omega)]$ has a global maximum. The typical value of the LDOS, $A_\sigma(\omega)_{\text{typ}}$, is the same for all lattice sites. By employing $A_\sigma(\omega)_{\text{typ}}$ one restores translational invariance in the description of a disordered system. This value will represent typical properties of the system. Using photoemission spectroscopy one could, in principle, probe the LDOS at a particular lattice site and measure its most probable value. We note that if sample-to-sample fluctuations are small, the typical value $A_\sigma(\omega)_{\text{typ}}$ would coincide with the arithmetic average $A_\sigma(\omega)_{\text{arith}}$. On the other hand, in a non-self-averaging system the PDF can be strongly asymmetric, with a long tail, in which case the typical value $A_\sigma(\omega)_{\text{typ}}$ would be very different from $A_\sigma(\omega)_{\text{arith}}$. The arithmetic mean is strongly biased by rare fluctuations and hence does not represent the typical property of such a system.

Statistical approaches based on the computation of the probability distribution functions would require the inclusion of very many (perhaps infinitely many) impurity sites. This is very hard to achieve in practice, in particular, in correlated electron systems discussed here, although there have been recent successful attempts in this direction.[66] Therefore, one should look for a generalized average which yields the best approximation to the typical value. Among different means the *geometric* mean turns out to be very convenient to describe Anderson localization. The geometric mean is defined by

$$A_\sigma(\omega)_{\text{geom}} = \exp\left[\langle \ln A_{i\sigma}(\omega) \rangle\right], \qquad (5.4)$$

where $\langle F(\epsilon_i) \rangle = \int \prod_i d\epsilon_i P(\epsilon_i) F(\epsilon_i)$ is the arithmetic mean of the function $F(\epsilon_i)$. The geometric mean is an approximation to the most probable, typical value of the LDOS

$$A_\sigma(\omega)_{\text{typ}} \approx A_\sigma(\omega)_{\text{geom}}. \tag{5.5}$$

It is easy to see that if $P[A_{i\sigma}(\omega)]$ is given by a log-normal PDF then $A_\sigma(\omega)_{\text{typ}} = A_\sigma(\omega)_{\text{geom}}$ holds exactly. It was shown that in the non-interacting case $A_\sigma(\omega)_{\text{geom}}$ vanishes at a critical strength of the disorder, hence providing an explicit criterion for Anderson localization.[1,39,48,49] We also note that by using the geometrically averaged LDOS we restore the translational invariance in our description of a disordered system. In addition, as we shall see in the next section, the restoration of translational invariance by averaging allows us to solve the DMFT equations in the thermodynamic limit. The problem of finite-size effects is then automatically absent.

6. DMFT Self-Consistency Conditions for Disordered Systems

According to the spectral theorem the geometrically averaged local Green function is given by

$$G_\sigma(\omega_n)_{\text{geom}} = \int d\omega \frac{A_\sigma(\omega)_{\text{geom}}}{i\omega_n - \omega}. \tag{6.1}$$

The DMFT self-consistency condition (4.6) is modified now to a translationally invariant form

$$\Sigma_\sigma(\omega_n) = i\omega_n + \mu - \eta_\sigma(\omega_n) - \frac{1}{G_\sigma(\omega_n)_{\text{geom}}}. \tag{6.2}$$

Here we assumed that $\langle \epsilon_i \rangle = 0$, which holds in particular for the box-shape PDF. We also used the translationally invariant hybridization function $\eta_\sigma(\omega_n)$. We can now perform a Fourier transform of the lattice Dyson equation (4.1) and obtain

$$G_\sigma(\omega_n)_{\text{geom}} = \int dz \frac{N_0(z)}{i\omega_n - z + \mu - \Sigma_\sigma(\omega_n)}, \tag{6.3}$$

where $N_0(z)$ is the density of states for a non-interacting and non-disordered lattice system.

Altogether the solution of the DMFT equations for interacting fermions with disorder requires the following steps:

(1) Select (i) N_L values of ϵ_i from a given PDF $P(\epsilon_i)$, (ii) an initial hybridization function $\eta_\sigma(\omega_n)$, and (iii) an initial self-energy $\Sigma_\sigma(\omega_n)$;

(2) for each ϵ_i solve the impurity problem defined by Eqs. (4.4)–(4.6);

(3) determine the LDOS $A_{i\sigma}(\omega)$ from the imaginary part of $G_{ii\sigma}(\omega)$, and $A_\sigma(\omega)_{\text{geom}}$ from Eq. (5.4);

(4) employ (6.1) to find $G_\sigma(\omega_n)_{\text{geom}}$;

(5) from Eqs. (6.2) and (6.3) find a new $\eta_\sigma(\omega_n)$ and $\Sigma_\sigma(\omega_n)$, then go to step (2) until convergence is reached.

It is clear that due to the averaging procedure we restore both translational invariance and the thermodynamic limit although N_L is finite. Therefore the method is superior to other stochastic methods which are affected by finite size effects.

In the presence of antiferromagnetic long-range order the self-consistency conditions are modified. In this case we introduce two sublattices $s =$A or B, and calculate two local Green functions $G_{ii\sigma s}(\omega_n)$. From this quantity we obtain the geometrically averaged LDOS $A_{\sigma s}(\omega)_{\text{geom}} = \exp\left[\langle \ln A_{i\sigma s}(\omega)\rangle\right]$, where $A_{i\sigma s}(\omega)$ is given as shown in Eq. (5.1). The local Green function is then obtained from the Hilbert transform (6.1). The local self-energy $\Sigma_{\sigma s}(\omega)$ is determined from Eq. (6.2). The self-consistent DMFT equations are closed by the Hilbert transform of the Green function on a bipartite lattice:

$$G_{\sigma s}(\omega_n)_{\text{geom}} = \int dz \, \frac{N_0(z)}{\left[i\omega_n - \Sigma_{\sigma s}(\omega_n) - \frac{z^2}{i\omega_n - \Sigma_{\sigma \bar{s}}(\omega_n)}\right]}. \tag{6.4}$$

Here $\bar{s}$ denotes the sublattice opposite to s.[51,53]

We note that if the geometric mean were replaced by the arithmetic mean one would obtain a theory where disorder effects are described only on the level of the CPA, which cannot detect Anderson localization. It should also be pointed out that in the presence of disorder the LDOS represented by $A_\sigma(\omega)_{\text{geom}}$ is not normalized to unity. This means that $A_\sigma(\omega)_{\text{geom}}$ only describes the extended states of the continuum part of the spectrum. Localized states, which have a dense point spectrum, are not included in the DMFT with geometric average. Therefore, this approach cannot describe the properties of the Anderson-insulator phase.

The accuracy of the DMFT approach with geometric average over disorder was checked against numerically exact results obtained for non-interacting fermions on a cubic lattice.[49,78] The critical disorder strengths at which Anderson localization occurs were found to agree within a factor of two[78] or better.[49] However, there exists a discrepancy regarding the shape of the mobility edge, which shows a pronounced reentrant behavior for non-interacting particles with box-type PDF of the disorder. This feature is not reproduced by our approach.[78] On the other hand, the re-entrant behavior is a non-universal feature. Namely, it is much less pronounced in the case of a Gaussian PDF for disorder, and does not occur at all for a Lorentzian PDF.[79]

It should be pointed out that the DMFT-based self-consistent approach to interacting lattice fermions with disorder discussed here, is not related to the self-consistent theory of Anderson localization by Vollhardt and Wölfle[5] and its generalizations.[80,81] Namely, the latter theory determines the frequency dependent diffusion coefficient $D(\omega)$ from arithmetically averaged two-particle correlation functions by considering diffuson and cooperon diagrams. The approach reviewed here does not make use of these coherent back-scattering contributions, but computes a one-particle correlation function, the LDOS, and thereby extracts information on Anderson localization. The fact that the DMFT is based on a local approximation through the limit of large spatial dimensions does not necessarily imply that back-scattering contributions are entirely absent in this approach. Indeed, contributions due to back-scattering are implicitly contained in the hybridization function, which describes the diffusion of one-particle excitations away from and back to a given lattice site.[1,71] Quite generally the relation between theoretical approaches based on one-particle and two-particle correlation functions, respectively, and their results for the critical disorder strength for Anderson localization, is still not sufficiently understood and will continue to be an important topic for future research. Perhaps the limit of high lattice dimensions will serve as a useful starting point.[54,82,83]

7. Identification of Different Phases

To characterize the ground state of the Hamiltonian (3.1) the following quantities are computed:

(1) the LDOS $A_{\sigma s}(\omega)_{\text{geom}}$ for a given sublattice s and spin direction σ;
(2) the total DOS for a given sublattice s at the Fermi level ($\omega = 0$) with
    $N_s(0)_{\text{geom}} \equiv \sum_\sigma A_{\sigma s}(\omega = 0)_{\text{geom}}$;
(3) the staggered magnetization $m_{\text{AF}}^{\text{geom}} = |n_{\uparrow A}^{\text{geom}} - n_{\uparrow B}^{\text{geom}}|$, where $n_{\sigma s}^{\text{geom}} = \int_{-\infty}^{0} d\omega\, A_{\sigma s}(\omega)_{\text{geom}}$ is the local particle density on sublattice s.[84]

For comparison we determine these quantities also with the arithmetic average.

The possible phases of the Anderson–Hubbard model can then be classified as follows: The systems is a

• paramagnetic metal if $N_s^{\text{geom}}(0) \neq 0$ and $m_{\text{AF}}^{\text{geom}} = 0$;

- AF metal if $N_s^{\text{geom}}(0) \neq 0$ and $m_{\text{AF}}^{\text{geom}} \neq 0$;
- AF insulator if $N_s^{\text{geom}}(0) = 0$ and $m_{\text{AF}}^{\text{geom}} \neq 0$ but $N_s^{\text{geom}}(\omega) \neq 0$ for some $\omega \neq 0$ (in fact, the last condition is already implied by $m_{\text{AF}}^{\text{geom}} \neq 0$);
- paramagnetic Anderson–Mott insulator if $N_s^{\text{geom}}(\omega) = 0$ for all ω.

Note, that we use the term "metal" also for neutral fermionic atoms if they fulfil the above conditions.

8. Ground State Phase Diagram of Interacting, Disordered Lattice Fermion Systems at Half-Filling

We now apply the formalism discussed above to the Anderson–Hubbard model at half-filling and compare the ground state properties in the paramagnetic and magnetic cases.[58,60]

In the following we choose a model DOS, $N_0(\epsilon) = 2\sqrt{D^2 - \epsilon^2}/\pi D^2$, with bandwidth $W = 2D$, and set $W = 1$. For this DOS and for a bipartite lattice, the local Green function and the hybridization function are connected by the simple algebraic relation $\eta_{\sigma s}(\omega)_{\text{geom}} = D^2 G_{\sigma \bar{s}}(\omega)_{\text{geom}}/4$.[51]

The DMFT equations are solved at zero temperature by the numerical renormalization group technique,[85] which allows us to calculate the geometric or arithmetic average of the local DOS in each iteration loop.

8.1. *Paramagnetic phase diagram*

The ground state phase diagram of the Anderson–Hubbard model at half-filling obtained within the DMFT approach discussed above is shown in Fig. 20.2.[58] Two different phase transitions are found to take place: a Mott–Hubbard MIT for weak disorder Δ, and an Anderson MIT for weak interaction U. The correlated, disordered metal is surrounded by two different insulating phases whose properties, as well as the transitions between them, will now be discussed. In this section, the spin index σ is omitted since all quantities are spin independent.

(*i*) *Disordered, metallic phase* — The correlated, disordered metal is characterized by a non-zero value of the spectral density at the Fermi level, $A(\omega = 0)_{\text{geom}} \neq 0$. In the absence of disorder, DMFT predicts this quantity to be given by the bare DOS $N_0(0)$, which is a consequence of the Luttinger theorem. This means that Landau quasiparticles are well-defined at the Fermi level. The situation changes completely when disorder is introduced since a subtle competition between disorder and electron interaction arises.

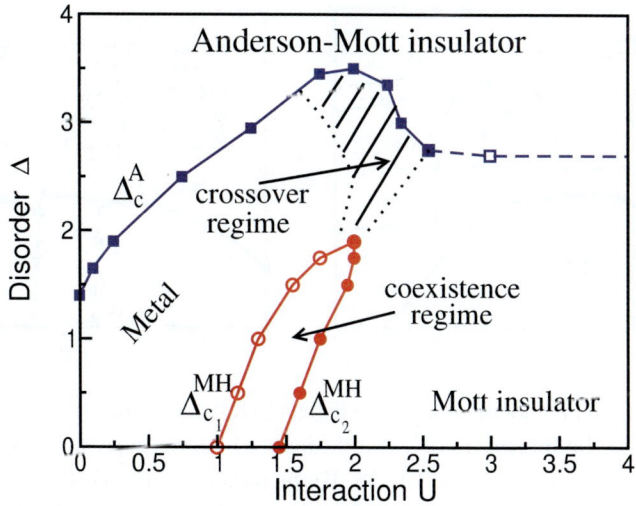

Fig. 20.2. Non-magnetic ground state phase diagram of the Anderson–Hubbard model at half-filling as calculated by DMFT with the typical local density of states; after Ref. 58.

Increasing the disorder strength at fixed U reduces $A(0)_{\text{geom}}$ and thereby decreases the metallicity as shown in the upper panel left of Fig. 20.3. The opposite behavior is found when the interaction is increased at fixed Δ (see right panel of Fig. 20.3 for $\Delta = 1$), i.e., in this case the metallicity improves. In the strongly interacting metallic regime the value of $A(0)_{\text{geom}}$ is restored, reaching again its maximal value $N_0(0)$. This implies that in the metallic phase sufficiently strong interactions protect the quasiparticles from decaying by impurity scattering. For weak disorder this interaction effect is almost independent of how the LDOS is averaged.

(*ii*) *Mott–Hubbard MIT* — For weak to intermediate disorder strength there is a sharp transition at a critical value of U between a correlated metal and a gapped Mott insulator. Two transition lines are found depending on whether the MIT is approached from the metallic side [$\Delta_{c2}^{MH}(U)$, full dots in Fig. 20.2] or from the insulating side [$\Delta_{c1}^{MH}(U)$, open dots in Fig. 20.2]. The hysteresis is clearly seen in right panel of Fig. 20.3 for $\Delta = 1$. The curves $\Delta_{c1}^{MH}(U)$ and $\Delta_{c2}^{MH}(U)$ in Fig. 20.2 are seen to have positive slope. This is a consequence of the disorder-induced increase of spectral weight at the Fermi level which in turn requires a stronger interaction to open the correlation gap. In the Mott insulating phase close to the hysteretic region an increase of disorder will therefore drive the system *back* into the metallic phase. The corresponding abrupt rise of $A(0)_{\text{geom}}$ is clearly seen in the left lower panel

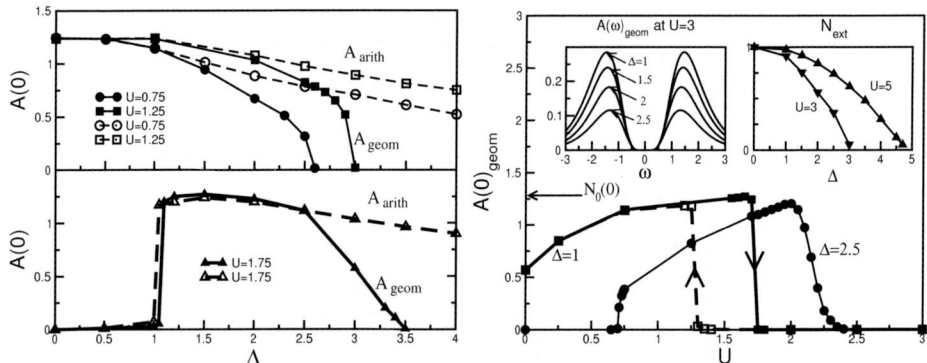

Fig. 20.3. Left panel: local density of states (LDOS) as a function of disorder Δ for various values of the interaction U. Solid (dashed) curves correspond to the geometrically (arithmetically) averaged LDOS. Right panel: geometrically averaged LDOS as a function of interaction U for different disorder strengths Δ. Solid (dashed) curves with closed (open) symbols are obtained with an initial metallic (insulating) hybridization function. Triangles: $\Delta = 1$; dots: $\Delta = 2.5$. Left inset: LDOS with Mott gap at $U = 3$ for different disorder strengths Δ. Right inset: integrated LDOS N_{geom} as a function of Δ at $U = 3$; after Ref. 58.

of Fig. 20.3. In this case the disorder protects the metal from becoming a Mott insulator.

Around $\Delta \approx 1.8$ the curves $\Delta_{c1}^{MH}(U)$ and $\Delta_{c2}^{MH}(U)$ terminate at a single critical point, cf. Fig. 20.2. For stronger disorder ($\Delta \gtrsim 1.8$) there appears to be a smooth crossover rather than a sharp transition from the metal to the insulator. This is illustrated by the U dependence of $A(0)_{\mathrm{geom}}$ shown in right panel of Fig. 20.3 for $\Delta = 2.5$. In this parameter regime the Luttinger theorem is not obeyed for any U. In the crossover regime, marked by the hatched area in Fig. 20.2, $A(0)_{\mathrm{geom}}$ vanishes gradually, so that the metallic and insulating phases can no longer be distinguished rigorously.[86]

Qualitatively, we find that the Mott–Hubbard MIT and the crossover region do not depend much on the choice of the average of the LDOS.[87] We also note the similarity between the Mott–Hubbard MIT scenario discussed here for disordered systems and that for a system without disorder at *finite* temperatures,[51,86] especially the presence of a coexistence region with hysteresis. However, while in the non-disordered case the interaction needed to trigger the Mott–Hubbard MIT decreases with increasing temperature, the opposite holds in the disordered case.

(*iii*) *Anderson MIT* — The metallic phase and the crossover regime are found to lie next to an Anderson insulator phase where the LDOS of the extended states vanishes completely (see Fig. 20.2). The critical disorder strength $\Delta_c^A(U)$ corresponding to the Anderson MIT is a non-monotonous function of the interaction: it increases in the metallic regime and decreases in the crossover regime. Where $\Delta_c^A(U)$ has a positive slope an increase of the interaction turns the Anderson insulator into a correlated metal. This is illustrated in Fig. 20.3 for $\Delta = 2.5$; at $U/W \approx 0.7$ a transition from a localized to a metallic phase occurs, i.e., the spectral weight at the Fermi level becomes finite. In this case the electronic correlations inhibit the localization of quasiparticles by scattering at the impurities.

Figure 20.3 shows that the Anderson MIT is a continuous transition. In the critical regime $A(0)_{\text{geom}} \sim [\Delta_c^A(U) - \Delta]^\beta$ for $U = \text{const}$. In the crossover regime a critical exponent $\beta = 1$ is found (see the case $U = 1.75$ in lower panel of Fig. 20.3); elsewhere $\beta \neq 1$. However, we cannot rule out a very narrow critical regime with $\beta = 1$ since it is difficult to determine β with high accuracy. It should be stressed that an Anderson transition with vanishing $A(0)_{\text{geom}}$ at finite $\Delta = \Delta_c^A(U)$ can only be detected within DMFT when the geometrically averaged LDOS is used (solid lines in Fig. 20.3). Indeed, using arithmetic averaging one finds a nonvanishing LDOS at any finite Δ (dashed lines in Fig. 20.3).

(*iv*) *Mott and Anderson insulators* — The Mott insulator (with a correlation gap) is rigorously defined only in the absence of disorder ($\Delta = 0$), and the gapless Anderson insulator only for non-interacting systems ($U = 0$) and $\Delta > \Delta_c^A(0)$. For finite interactions and disorder this distinction can no longer be made. On the other hand, as long as the LDOS shows the characteristic Hubbard subbands (left inset in Fig. 20.3) one may refer to a *disordered Mott insulator*. With increasing disorder Δ, the spectral weight of the Hubbard subbands vanishes (right inset in Fig. 20.3) and the system becomes a *correlated Anderson insulator*. The boundary between these two types of insulators is marked by a dashed line in Fig. 20.2. The results obtained here within DMFT show that the paramagnetic Mott and Anderson insulators are continuously connected. Hence, by changing U and Δ it is possible to move from one insulating state to another one without crossing a metallic phase.

8.2. *Magnetic phase diagram*

At half-filling and in the absence of frustration effects interacting fermions order antiferromagnetically. This raises several basic questions: (i) how is a non-interacting, Anderson localized system at half filling influenced by a local interaction between the particles? (ii) how does an antiferromagnetic

(AF) insulator at half filling respond to disorder which in the absence of interactions would lead to an Anderson localized state? (iii) do Slater and Heisenberg antiferromagnets behave differently in the presence of disorder? Here we provide answers to these questions by calculating the zero temperature, magnetic phase diagram of the disordered Hubbard model at half filling using DMFT together with a geometric average over the disorder and allowing for a spin-dependence of the DOS.[60]

The ground state phase diagram of the Anderson–Hubbard model (3.1) obtained by the above classification is shown in Fig. 20.4. Depending on whether the interaction U is weak or strong the response of the system to disorder is found to be very different. In particular, at strong interactions, $U/W \gtrsim 1$, there exist only two phases, an AF insulating phase at weak disorder, $\Delta/W \lesssim 2.5$, and a paramagnetic Anderson–Mott insulator at strong disorder, $\Delta/W \gtrsim 2.5$. The transition between these two phases is continuous. Namely, the local DOS and the staggered magnetization both decrease gradually as the disorder Δ increases and vanish at their mutual boundary (lower panel of Fig. 20.5). By contrast, the phase diagram for weak interactions, $U/W \lesssim 1$, has a much richer structure (Fig. 20.4). In particular, for weak disorder a *paramagnetic* metallic phase is stable. It is separated from the AF insulating phase at large U by a narrow region of *AF metallic* phase. The AF metallic phase is long-range ordered, but there is no gap since the disorder leads to a redistribution of spectral weight.[60]

To better understand the nature of the AF phases in the phase diagram we take a look at the staggered magnetization m_{AF}^{α}. The dependence of $m_{\mathrm{AF}}^{\mathrm{geom}}$ on U is shown in the upper panel of Fig. 20.5 for several values of the disorder Δ. In contrast to the non-disordered case a finite interaction strength $U > U_c(\Delta)$ is needed to stabilize the AF long-range order when disorder is present. The staggered magnetization saturates at large U for both averages; the maximal values depend on the disorder strength. In the lower panel of Fig. 20.5, the dependence of m_{AF}^{α} on the disorder Δ is shown for different interactions U. Only for small U do the two averages yield approximately the same results.

Another useful quantity is the polarization $P_{\mathrm{AF}}^{\alpha} = m_{\mathrm{AF}}^{\alpha}/I^{\alpha}$, where $I^{\alpha} = \int_{-\infty}^{+\infty} \sum_{\sigma s} \rho_{\sigma s}^{\alpha}(\omega)d\omega/2$ is the total spectral weight of $\rho_{\sigma s}^{\alpha}(\omega)$. It allows one to investigate the contribution of the point-like spectrum of the Anderson localized states to the magnetization. This provides important information about the spectrum since with increasing disorder more and more one-particle states of the many-body system are transferred from the continuous to the point-like spectrum. For weak interactions ($U = 0.5$)

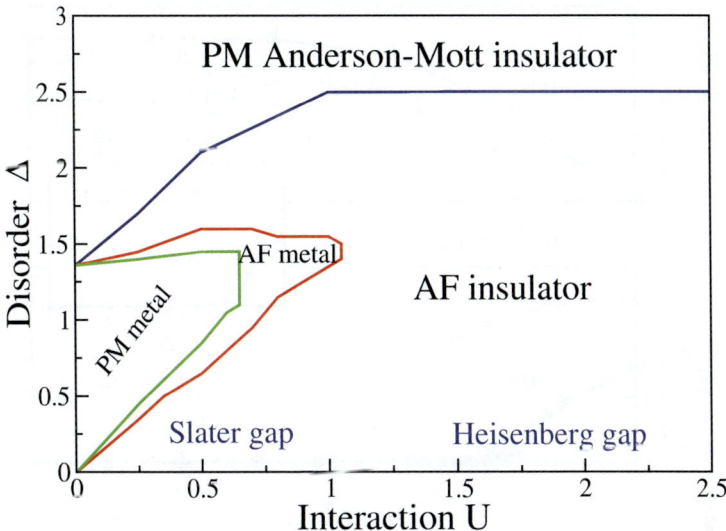

Fig. 20.4. Magnetic ground state phase diagram of the Anderson-Hubbard model at half-filling as calculated by DMFT with a spin resolved local DOS (see text); PM: paramagnetic, AF: antiferromagnetic; after Ref. 60.

the decrease of the polarization with increasing disorder Δ obtained with geometric or arithmetic averaging is the same (see inset in Fig. 20.5). Since within arithmetic averaging all states are extended, the decrease of m_{AF}^α (which is also the same for the two averages in the limit of weak interactions, see lower panel of Fig. 20.5) must be attributed to disorder effects involving only the continuous spectrum. At larger U, the polarization is constant up to the transition from the AF insulator to the paramagnetic Anderson–Mott insulator. In the latter phase the polarization is undefined, because the continuous spectrum does not contribute to I_{AF}^{geom}.

In the absence of disorder the AF insulating phase has a small ("Slater") gap at $U/W < 1$ and a large ("Heisenberg") gap at $U/W > 1$. These limits can be described by perturbation expansions in U and $1/U$ around the symmetry broken state of the Hubbard and the corresponding Heisenberg model, respectively. In agreement with earlier studies[88] our results for m_{AF} (upper panel of Fig. 20.5) show that there is no sharp transition between these limits, even when disorder is present. This may be attributed to the fact that both limits are described by the same order parameter. However, the phase diagram (Fig. 20.4) shows that the two limits *can* be distinguished by their overall response to disorder. Namely, the reentrance of the AF

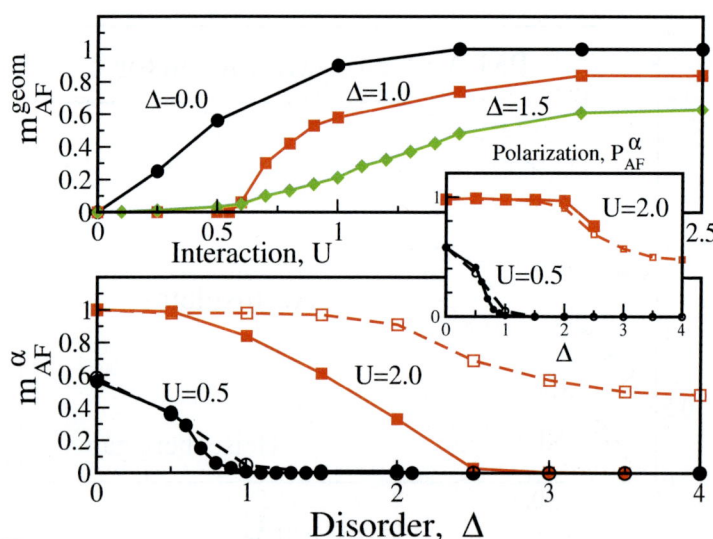

Fig. 20.5.   Upper panel: Staggered magnetization $m_{\mathrm{AF}}^{\mathrm{geom}}$ as a function of interaction U. Lower panel: m_{AF}^{α}, $\alpha = \mathrm{geom/arith}$, as a function of disorder Δ (full lines: arithmetic average, dashed lines: geometric average). Inset: Polarization P_{AF}^{α} as a function of disorder.[60] Reprinted with permission from *Phys. Rev. Lett.* **102**, 146403 (2009). © American Physical Society.

metallic phase at $\Delta/W \gtrsim 1$ occurs only within the Slater AF insulating phase.

The magnetic structure of the Anderson–Mott insulator cannot be determined by the method used here since it describes only the continuous part of the spectra and not the point spectrum. However, only the paramagnetic solution should be expected to be stable because the kinetic exchange interaction responsible for the formation of the AF metal is suppressed by the disorder. This does not exclude the possibility of Griffiths phase-like AF domains.[89,90]

It is interesting to note that even the DMFT with an arithmetic average finds a disordered AF metal.[53,91] However, the arithmetically averaged local DOS incorrectly predicts both the paramagnetic metal and the AF metal to remain stable for arbitrarily strong disorder. Only a computational method which is sensitive to Anderson localization, such as the DMFT with geometrically averaged local DOS employed here, is able to detect the suppression of the metallic phase for $\Delta/W \gtrsim 1.5$ and the appearance of the paramagnetic Anderson–Mott insulator at large disorder Δ already on the one-particle level.

9. Summary

In this article we reviewed the properties of low-temperature quantum phases of strongly correlated, disordered lattice fermion systems with application to correlated electronic systems and ultracold fermions in optical lattices. We discussed the Anderson–Hubbard model and a comprehensive nonperturbative theoretical method for its solution, the DMFT combined with geometrical averaging over disorder. This approach provides a unified description of Anderson- and Mott-localization in terms of one-particle correlation functions.

We presented low-temperature quantum phase diagrams for the Anderson–Hubbard model at half filling, both in the paramagnetic and the antiferromagnetic phase. In the paramagnet, we observed re-entrant metal–insulator transitions induced by disorder and interaction, where the corresponding Anderson- and Mott-insulating phases are continuously connected. In the presence of antiferromagnetic order, a new antiferromagnetic metallic phase was found, which is stabilized by the interplay between interaction and disorder.

It is expected that these new quantum states will be observable by using ultracold fermions in optical lattices where disorder and interactions are easily tunable in a wide range. While current experimental temperatures are still above those required for observing quantum antiferromagnetism, the paramagnetic Mott–Anderson insulator should be easily accessible within current setups.

Even after several decades of research into the complex properties of disordered, interacting quantum many-body systems many fundamental problems are still unsolved. Future investigations of the existing open questions, and of the new questions which are bound to arise, are therefore expected to provide fascinating new insights.

Acknowledgments

We thank R. Bulla and S. Kehrein for useful discussions. Financial support by the SFB 484, TTR 80, and FOR 801 of the Deutsche Forschungsgemeinschaft is gratefully acknowledged.

References

1. P. W. Anderson, *Phys. Rev.* **109**, 1492 (1958).
2. E. Abrahams, P. W. Anderson, D. C. Licciardello and T. V. Ramakrishnan, *Phys. Rev. Lett.* **42**, 673 (1979).

3. L. P. Gor'kov, A. I. Larkin and D. E. Khmel'nitskii, *Zh. Eksp. Teor. Fiz. Pis'ma Red.* **30**, 248 (1979) [*JETP Lett.* **30**, 248 (1979)].
4. P. A. Lee and T. V. Ramakrishnan, *Rev. Mod. Phys.* **57**, 287 (1985).
5. D. Vollhardt and P. Wölfle, in *Electronic Phase Transitions*, eds. W. Hanke and Ya. V. Kopaev (North-Holland, Amsterdam, 1992), Chapter 1, p. 1.
6. A. Lagendijk, B. van Tiggelen and D. S. Wiersma, *Phys. Today* **62**, 24 (2009).
7. N. F. Mott, *Metal–Insulator Transitions*, 2nd edn. (Taylor and Francis, London 1990).
8. D. Belitz and T. R. Kirkpatrick, *Rev. Mod. Phys.* **66**, 261 (1994).
9. For a recent review of exact results on Anderson localization see: M. Disertori, W. Kirsch, A. Klein, F. Klopp and V. Rivasseau, *Panoramas et Synthèses* **25**, 1 (2008).
10. N. F. Mott, *Proc. Phys. Soc. A* **62**, 416 (1949).
11. H. V. Löhneysen, *Adv. Solid State Phys.* **40**, 143 (2000).
12. S. V. Kravchenko and M. P. Sarachik, *Rep. Prog. Phys.* **67**, 1 (2004).
13. A. M. Finkelshtein, *Sov. Phys. JEPT* **75**, 97 (1983).
14. C. Castellani, C. Di Castro, P. A. Lee and M. Ma, *Phys. Rev. B* **30**, 527 (1984).
15. M. A. Tusch and D. E. Logan, *Phys. Rev. B* **48**, 14843 (1993); *ibid.* **51**, 11940 (1995).
16. D. L. Shepelyansky, *Phys. Rev. Lett.* **73**, 2607 (1994).
17. P. J. H. Denteneer, R. T. Scalettar and N. Trivedi, *Phys. Rev. Lett.* **87**, 146401 (2001).
18. B. L. Altshuler and A. G. Aronov, in *Electron–Electron Interactions in Disordered Systems*, eds. M. Pollak and A. L. Efros (North-Holland, Amsterdam, 1985), p. 1.
19. E. Abrahams, S. V. Kravchenko and M. P. Sarachik, *Rev. Mod. Phys.* **73**, 251 (2001).
20. M. Lewenstein, A. Sanpera, V. Ahufinger, B. Damski, A. Sen De and U. Sen, *Adv. Phys.* **56**, 243 (2007).
21. I. Bloch, J. Dalibard and W. Zwerger, *Rev. Mod. Phys.* **80**, 885 (2008).
22. L. Fallani, J. E. Lye, V. Guarrera, C. Fort and M. Inguscio, *Phys. Rev. Lett.* **98**, 130404 (2007).
23. A. Aspect and M. Inguscio, *Phys. Today* **62**, 30 (2009).
24. L. Sanchez-Palencia and M. Lewenstein, *Nature Phys.* **6**, 87 (2010).
25. J. Billy, V. Josse, Z. Zuo, A. Bernard, B. Hambrecht, P. Lugan, D. Clément, L. Sanchez-Palencia, P. Bouyer and A. Aspect, *Nature* **453**, 891 (2008).
26. G. Roati, C. D'Errico, L. Fallani, M. Fattori, C. Fort, M. Zaccanti, G. Modugno, M. Modugno and M. Inguscio, *Nature* **453**, 895 (2008).
27. M. White, M. Pasienski, D. McKay, S. Q. Zhou, D. Ceperley and B. DeMarco, *Phys. Rev. Lett.* **102**, 055301 (2009).
28. D. Jaksch, C. Bruder, J. I. Cirac, C. W. Gardiner and P. Zoller, *Phys. Rev. Lett.* **81**, 3108 (1998).
29. M. Greiner, O. Mandel, T. Esslinger, T. W. Hänsch and I. Bloch, *Nature* **415**, 39 (2002).
30. W. Hofstetter, J. I. Cirac, P. Zoller, E. Demler and M. D. Lukin, *Phys. Rev. Lett.* **89**, 220407 (2002).

31. M. Köhl, H. Moritz, T. Stöferle, K. Günter and T. Esslinger, *Phys. Rev. Lett.* **94**, 080403 (2005).
32. S. Ospelkaus, C. Ospelkaus, O. Wille, M. Succo, P. Ernst, K. Sengstock and K. Bongs, *Phys. Rev. Lett.* **96**, 180403 (2006); K. Günter, T. Stöferle, H. Moritz, M. Köhl and T. Esslinger, *Phys. Rev. Lett.* **96**, 180402 (2006).
33. R. Jördens, N. Strohmaier, K. Günter, H. Moritz and T. Esslinger, *Nature* **455**, 204 (2008); U. Schneider, L. Hackermüller, S. Will, T. Best, I. Bloch, T. A. Costi, R. W. Helmes, D. Rasch and A. Rosch, *Science* **322**, 1520 (2008).
34. The most probable value of a random quantity is defined as that value for which its PDF becomes maximal.
35. E. Limpert, W. A. Stahel and M. Abbt, *BioScience* **51**, 341 (2001).
36. E. L. Crow and K. Shimizu (eds.), *Log-Normal Distribution-Theory and Applications* (Marcel Dekker, Inc., New York, 1988).
37. A. D. Mirlin and Y. V. Fyodorov, *Phys. Rev. Lett.* **72**, 526 (1994); *J. Phys. I France* **4**, 655 (1994).
38. M. Janssen, *Phys. Rep.* **295**, 1 (1998).
39. G. Schubert, A. Weiße and H. Fehske in *High Performance Computing in Science and Engineering Garching 2004*, eds. A. Bode, F. Durst (Springer Verlag, 2005), pp. 237–250.
40. F. Evers and A. D. Mirlin, *Rev. Mod. Phys.* **80**, 1355 (2008).
41. A. Richardella, P. Roushan, S. Mack, B. Zhou, D. A. Huse, D. D. Awschalom and A. Yazdani, *Science* **327**, 665 (2010).
42. G. Schubert, J. Schleede, K. Byczuk, H. Fehske and D. Vollhardt, *Phys. Rev. B* **81**, 155106 (2010).
43. D. Lloyd, *J. Phys. C: Solid State Phys.* **2**, 1717 (1969).
44. D. Thouless, *Phys. Rep.* **13**, 93 (1974).
45. F. Wegner, *Z. Phys. B* **44**, 9 (1981).
46. E. W. Montroll and M. F. Schlesinger, *J. Stat. Phys.* **32**, 209 (1983).
47. M. Romeo, V. Da Costa and F. Bardou, *Eur. Phys. J. B* **32**, 513 (2003).
48. V. Dobrosavljević and G. Kotliar, *Phys. Rev. Lett.* **78**, 3943 (1997); *Philos. Trans. R. Soc. London, Ser. A* **356**, 57 (1998).
49. V. Dobrosavljević, A. A. Pastor and B. K. Nikolić, *Europhys. Lett.* **62**, 76 (2003).
50. W. Metzner and D. Vollhardt, *Phys. Rev. Lett.* **62**, 324 (1989).
51. A. Georges, G. Kotliar, W. Krauth and M. J. Rozenberg, *Rev. Mod. Phys.* **68**, 13 (1996).
52. G. Kotliar and D. Vollhardt, *Phys. Today* **57**, 53 (2004).
53. M Ulmke, V. Janiš and D. Vollhardt, *Phys. Rev. B* **51**, 10411 (1995).
54. R. Vlaming and D. Vollhardt, *Phys. Rev. B* **45**, 4637 (1992).
55. Y. Song, R. Wortis and W. A. Atkinson, *Phys. Rev. B* **77**, 054202 (2008).
56. F. X. Bronold, A. Alvermann and H. Fehske, *Philos. Mag.* **84**, 637 (2004).
57. M.-T. Tran, *Phys. Rev. B* **76**, 245122 (2007).
58. K. Byczuk, W. Hofstetter and D. Vollhardt, *Phys. Rev. Lett.* **94**, 056404 (2005).
59. K. Byczuk, *Phys. Rev. B* **71**, 205105 (2005).
60. K. Byczuk, W. Hofstetter and D. Vollhardt, *Phys. Rev. Lett.* **102**, 146403 (2009).

61. M. C. O. Aguiar, V. Dobrosavljević, E. Abrahams and G. Kotliar, *Phys. Rev. Lett.* **102**, 156402 (2009).
62. G. Roux, T. Barthel, I. P. McCulloch, C. Kollath, U. Schollwöck and T. Giamarchi, *Phys. Rev. A* **78**, 023628 (2008).
63. U. Bissbort and W. Hofstetter, *Europhys. Lett.* **86**, 50007 (2009).
64. U. Bissbort, R. Thomale and W. Hofstetter, preprint arXiv:0911.0923.
65. L. Pollet, N. V. Prokof'ev, B. V. Svistunov and M. Troyer, *Phys. Rev. Lett.* **103**, 140402 (2009).
66. D. Semmler, K. Byczuk and W. Hofstetter, *Phys. Rev. B* **81**, 115111 (2010).
67. J. T. Stewart, J. P. Gaebler and D. S. Jin, *Nature* **454**, 744 (2008).
68. T.-L. Dao, A. Georges, J. Dalibard, C. Salomon and I. Carusotto, *Phys. Rev. Lett.* **98**, 240402 (2007).
69. P. T. Ernst, S. Götze, J. S. Krauser, K. Pyka, D.-S. Lühmann, D. Pfannkuche and K. Sengstock, preprint arXiv:0908.4242.
70. N. Fabbri, D. Clément, L. Fallani, C. Fort, M. Modugno, K. M. R. van der Stam and M. Inguscio, *Phys. Rev. A* **79**, 043623 (2009).
71. R. Abou-Chacra, D. J. Thouless and P. W. Anderson, *J. Phys. C: Solid State Phys.* **6**, 1734 (1973).
72. K. Efetov, *Supersymmetry in Disorder and Chaos* (Cambridge University Press, Cambridge, 1997).
73. E. Müller–Hartmann, *Z. Phys. B* **74**, 507 (1989).
74. J. K. Freericks and V. Zlatić, *Rev. Mod. Phys.* **75**, 1333 (2003).
75. J. K. Freericks, *Transport in Multilayered Nanostructures — The Dynamical Mean-Field Approach* (Imperial College Press, London, 2006).
76. D. Heidarian and N. Trivedi, *Phys. Rev. Lett.* **93**, 126401 (2004).
77. H. Shinaoka and M. Imada, *Phys. Rev. Lett.* **102**, 016404 (2009).
78. A. Alvermann and H. Fehske, *Eur. Phys. J. B* **48**, 295 (2005).
79. B. Bulka, B. Kramer and A. MacKinnon, *Z. Phys. B* **60**, 13 (1985); B. Bulka, M. Schreiber and B. Kramer, *Z. Phys. B* **66**, 21 (1987).
80. E. Z. Kuchinskii, I. A. Nekrasov and M. V. Sadovskii, *JETP* **106**, 581, (2008); E. Z. Kuchinskii, N. A. Kuleeva, I. A. Nekrasov and M. V. Sadovskii, *J. Exp. and Theor. Phys.* **110**, 325 (2010).
81. P. Henseler, J. Kroha and B. Shapiro, *Phys. Rev. B* **77**, 075101 (2008); *Phys. Rev. B* **78**, 235116 (2008).
82. V. Janiš and D. Vollhardt, *Phys. Rev. B* **63**, 125112 (2001).
83. V. Janiš and J. Kolorenč, *Phys. Rev. B* **71**, 033103 (2005); *Phys. Rev. B* **71**, 245106 (2005).
84. The order parameter $m_{\mathrm{AF}}^{\mathrm{geom}}$ is related to the *number* of states in the continuous part of the spectrum and thus accounts for the existence of itinerant quasiparticles which then order magnetically. A finite value of $m_{\mathrm{AF}}^{\mathrm{geom}}$ corresponds to a spin density wave, which should be observable in neutron scattering experiments.
85. K. G. Wilson, *Rev. Mod. Phys. A* **47**, 773 (1975); W. Hofstetter, *Phys. Rev. Lett. A* **85**, 1508 (2000); R. Bulla, T. Costi and T. Pruschke, *Rev. Mod. Phys. A* **80**, 395 (2008).
86. R. Bulla, T. A. Costi and D. Vollhardt, *Phys. Rev. B* **64**, 045103 (2001).

87. K. Byczuk, W. Hofstetter and D. Vollhardt, *Physica B* **359–361**, 651 (2005).
88. T. Pruschke, *Prog. Theo. Phys. Suppl.* **160**, 274 (2005).
89. R. B. Griffiths, *Phys. Rev. Lett.* **23**, 17 (1969).
90. V. Dobrosavljević and E. Miranda, *Phys. Rev. Lett.* **94**, 187203 (2005).
91. A. Singh, M. Ulmke and D. Vollhardt, *Phys. Rev. B* **58**, 8683 (1998).

Chapter 21

TOPOLOGICAL PRINCIPLES IN THE THEORY OF ANDERSON LOCALIZATION

A. M. M. Pruisken

Institute for Theoretical Physics,
Valckenierstraat 65, 1018 XE Amsterdam, The Netherlands
pruisken@science.uva.nl

Scaling ideas in the theory of the quantum Hall effect are fundamentally based on topological principles in Anderson localization theory. These concepts have a very general significance and are not limited to replica field theory or disordered systems alone. In this chapter, we will discuss these ideas in several distinctly different physical contexts. We start with a brief overview that spans two and a half decades of experimental research on quantum criticality in strong magnetic fields. Secondly, we address the new understanding of universality that has emerged from the theory of Anderson localization and interaction phenomena. In the last part we show how the experimentally observed quantum phenomena fundamentally alter the way in which strong coupling problems in theoretical physics are perceived.

1. Quantum Hall Effect

1.1. *Introduction*

Following the seminal paper by Abrahams, Anderson, Liciardello and Ramakrishnan in 1979,[1] it has become conventional in the theory of the disordered electron gas to focus on the scaling properties of the dimensionless conductance σ_0 in units of e^2/h with the linear dimension L. The concept of "weak localization" that emerged, gave rise to the idea that the metal–insulator transition or "mobility edge" only appears in dimensions larger than two. In dimensions less than or equal to two the electronic wave functions are always Anderson localized with a finite localization length ξ. The two dimensional electron gas (2DEG) at high energies is merely quasi metallic at short distances

$$\sigma_0 = -\beta_1 \ln\left\{\frac{L}{\xi}\right\}, \qquad\qquad L < \xi \qquad\qquad (1.1)$$

with β_1 a universal constant. It generally becomes an insulator in the limit where L goes to infinity

$$\sigma_0 = \exp\left\{-\frac{L}{\xi}\right\}, \qquad\qquad L > \xi. \qquad\qquad (1.2)$$

With the advent of the (integral) quantum Hall effect[2] these famous predictions of the so-called scaling theory of Anderson localization ran into fundamental problems. Several scenarios have been proposed that explain the robust quantization of the Hall conductance based on the assumption that the states near the Landau band center are extended but localized elsewhere. The most popular amongst these are Laughlin's "gauge" argument,[3] the semi classical "percolation" picture[4] as well as Buttiker's "edge state" picture.[5] These scenarios highlight interesting aspects of electrons confined in Landau bands. They do not however, provide a microscopic theory of the quantum Hall effect nor do they reveal much about the mechanism of *de*-localization in strong magnetic fields (B).

In their pioneering work on the subject, Herbert Levine, Stephen Libby and the author[6,7] introduced several new ideas in the theory of Anderson localization that originally came as a complete surprise in physics. As is well known, the quantum Hall effect has turned out to be an interesting experimental realization of the ϑ angle concept, a topological issue that originally arose in QCD. Remarkably, this led to the idea of ϑ-renormalization by instantons.[8] This explains why not only the longitudinal conductance σ_0 but also the Hall conductance σ_H generally appears as a scaling parameter with varying linear dimension L of the system. This feature of the 2DEG is essentially non-perturbative and standardly illustrated by the scaling diagram in the σ_0–σ_H conductance plane, see Fig. 21.1.

The renormalization group flow lines are periodic in the dimensionless Hall conductance σ_H. These flow lines unify the semi classical Drude–Boltzmann theory valid at large values of σ_0 with quantum Hall effect. This robust quantization phenomenon primarily reveals itself in the regime of "bad" conductors $\sigma_0 \leq 1$ where one normally expects the 2DEG to be strongly Anderson localized. Besides the stable quantum Hall fixed points located at $\sigma_0 = 0$ and $\sigma_H = k$ with integer k, there are also the unstable *critical* fixed points located at $\sigma_0^* = \mathcal{O}(1)$ and $\sigma_H^* = k + \frac{1}{2}$. These critical fixed points are at the interface between "weak" and "strong" quantum interference and predict a *quantum phase transition* between adjacent quantum

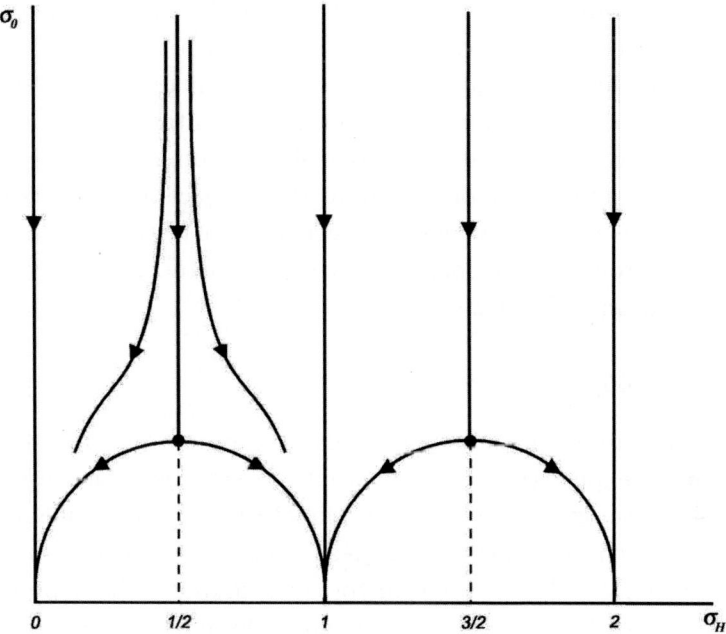

Fig. 21.1.   Scaling diagram of the conductances σ_0 and σ_H.

Hall plateaus. This transition is characterized by a continuously diverging localization length ξ and universal critical exponents.

The theory of scaling has had a lasting impact on how one thinks about the quantum Hall effect. For example, even long before the idea of ϑ-renormalization had been justified by detailed analytic work[9] it was already conjectured on phenomenological grounds that the scaling diagram of Fig. 21.1 should also include the fractional quantum Hall regime.[10] Serious attempts in this direction were possible only at a much later stage when, amongst many other things, the effects of the electron–electron interactions on Anderson localization had been clarified.[11–16] Nevertheless, the early ideas in the field already indicated that ϑ- or σ_H-renormalization is the fundamental principle that naturally unifies a whole range of entirely different physical phenomena.

Next, as was demonstrated in the original experiments of H. P. Wei *et al.* in 1988,[17] strongly disordered samples with a low mobility are a splendid laboratory for testing the unstable quantum *critical* fixed points in Fig. 21.1. It was assumed that the inelastic scattering processes affect the phase of the conducting electrons and provide a temperature dependent phase breaking length $L_\varphi \propto T^{-p/2}$ that serves as an effective sample size for scaling. So, by varying T the quantum critical behavior of the quantum Hall plateau transitions was observed in beautiful agreement with the theory.

Following the first experiments on scaling, the quantum Hall system became a popular subject for numerical simulations on the mobility edge problem that originally started with the work of MacKinnon and Kramer.[18] By this time several generations of computational physicists had worked on the problem. This has resulted in an impressive array of critical exponent values for the free electron gas that includes the multi fractal singularity spectrum of the electronic wave functions.[19,20]

1.2. *Physical objectives*

Given the original advances in the theory of the quantum Hall effect, one would normally expect that the field would rapidly be picked up by many and the results would extend in many different ways. Unfortunately, this is not what happened. The flow of new ideas mainly got stranded in the midst of biases that existed based upon historical thinking. The experiments in particular, have been misinterpreted by many. From the numerical side, the problem of Anderson localization in strong B has in general been mistaken for an ordinary statistical mechanics problem where the "exact" critical exponent values are the only thing that matters. However, the dilemma lies in the effects of the Coulomb interaction which invalidate any approach to the problem based on Fermi liquid type of ideas.

It is natural to assume that a conformal scheme for free electron approximations must exist and will eventually be found. However, it is also important to keep in mind that the numerous "attempts" in this direction already span more than two decades and have not as yet taught us anything about the experiments on the quantum Hall effect, nor have they revealed much about topological issues in quantum field theory in general. This is especially so because the most fundamental strong coupling aspect of the problem, the robust quantization of the Hall conductance, has so far remained unexplained.[21]

A profound consequence of the scaling theory of Anderson localization is the existence of universal scaling functions for the conductances that unify the concepts of "extended" and "localized" states.[22] It has slowly become evident over the years that these scaling functions are, in fact, the universal language in which the physics of the ϑ angle concept as well as the quantum Hall effect can in general be expressed. These scaling results, unlike the exponent values, always have exactly the same physical significance — termed *super universality* — which is independent of the details of the theory such as the number of field components or, for that matter, the replica method.[23] Super universality encompasses the statement of ordinary universality in critical phenomena phenomenology. This is a statement for only the critical exponent values.

Super universality dramatically alters the way the quantum Hall effect and Anderson *de*-localization are to be perceived. In this chapter, we will discuss this issue under three very different physical settings. These respective settings provide a nice illustration of the variety of physics that falls under the title of "Anderson localization."

First of all, from the point of view of the experiment, the main objective is obviously to establish the phenomenon of quantum criticality in the quantum Hall regime. This can be done, to a large extent, on the basis of the free particle theory which poses the most important experimental questions and directs us where to look.[22] It not only prescribes how the critical exponent values and universal scaling functions of the conductances can be extracted from the experimental data, but also defines the constraints on the experimental design such as the "range" of the potential fluctuations as well as the homogeneity of the sample. Inspite of these well defined objectives, it has taken over two decades to unravel the experimental situation. A brief summary is given in Sec. 2.

Next, from the condensed matter and general physics point of view, the central question that must be addressed is whether the infinitely ranged Coulomb interaction actually sustains the idea of Fermi liquid theory as proposed in Ref. 22. This is the main subject of Sec. 3 where we review the very different ideas that have emerged from the theory of Anderson localization and interaction phenomena. It turns out that Anderson localization in strong magnetic fields generally falls into two different universality classes. One is a *Fermi liquid* universality class and the other is a formerly unrecognized *F-invariant* universality class. Both these classes have a very different physical significance and distinctly different quantum critical behavior. These advances have altered the meaning of the experimentally measured critical exponent, popularly termed κ. Furthermore, they have important consequences for the unification of the integral and fractional quantum Hall physics, in particular the composite fermion ideas describing the abelian quantum Hall states.

Finally, from the point of view of the ϑ angle concept, one of the most compelling problems to address is the "arena for bloody controversies"[24] in quantum field theory that continues to haunt the subject. There are, in fact, entirely different views on the matter that have resulted from certain "exactly" solvable models of the ϑ angle such as the large N expansion of the CP^{N-1} model.[25,26] As is well known, the historical "large N picture" has set the stage for persistent conflicts with the semi classical theory of instantons and also the very existence of the quantum Hall effect. Following the "large N picture," for example, "the *mass gap* at $\vartheta = \pi$ remains *finite*",[27] "*no* critical exponents can be *defined*"[28] etc. These sorts of claims clearly upset the general arguments for *de*-localization, originally put forward by Levine, Libby and the author, which are independent of N.[7] These claims have nevertheless been taken for granted for many years. This is precisely why the topological concept of a ϑ vacuum is still not recognized as the fundamental theory of the quantum Hall effect.

It is important to keep in mind that unlike QCD or any other theory where the meaning of the ϑ parameter is rather obscure, the situation in Anderson localization theory is extremely clear. In Anderson localization theory the ϑ parameter becomes a physical observable and is in fact the Hall conductance. Macroscopic quantization phenomena such as the quantum Hall effect have previously never been associated with the ϑ angle, nor has the idea of ϑ renormalization ever been recognized before. The appropriate question to ask, therefore, is whether the physics of the quantum Hall effect can possibly be used in order to shed new light on the notorious strong coupling problems previously encountered in quantum field theory.

In Sec. 4 we review the Grassmannian $SU(M + N)/S(U(M) \times U(N))$ non-linear sigma model and in Sec. 5 we revisit the large N steepest descend methodology of the CP^{N-1} model. This methodology is standardly used for an infinite system. What is revealed is that the historical papers on the subject have overlooked a fundamental feature of the theory, the massless excitations that propagate along the "edges" of the ϑ vacuum. These "edge" excitations spontaneously break the $SU(N)$ symmetry and generally carry a fractional topological charge. These excitations are fundamentally different from the "bulk" excitations that always have a strictly integral topological charge.

By constructing an effective action for "edge" excitations it immediately becomes obvious that the large N expansion displays all the fundamental strong coupling features of the quantum Hall effect that earlier were concealed. Exact scaling results are obtained that complete the large N analysis recently reported by the author.[29] These demonstrate the robust quantization of the "Hall conductance" along with quantum criticality of the quantum Hall plateau transitions, in complete accordance with the theory of scaling[22] and the aforementioned statement of super universality.

In conclusion, more than 25 years after its introduction into the theory of the quantum Hall effect, the ϑ angle concept remains full of surprises and unforeseen complexity. The important lesson to be learned from all this is that even an exactly solvable theory can be totally misleading if the wrong questions are being posed and incorrect physical ideas are being pursued.

2. Experiments

2.1. *Mean field theory*

In order to establish the relationship between the theory and the experiments, it is necessary to first develop a general understanding of the semi classical theory of the conductances that defines the starting points for scaling. For strongly disordered systems with short ranged potential fluctuations which are of interest to us, this theory is provided by the self consistent Born approximation (SCBA).[29] For weak magnetic fields ($\omega_c \tau \leq 1$) one generally has the classical Boltzmann results

$$\sigma_0^0(B) = \frac{\sigma_0^0}{1 + (\omega_c \tau)^2} \quad , \quad \sigma_H^0(B) = \omega_c \tau \sigma_0^0(B) \tag{2.1}$$

describing a semi circle in the two dimensional conductance plane; see Fig. 21.2. Here, $\omega_c = eB/mc$ is the cyclotron radius and σ_0^0 is the $B = 0$ value of the longitudinal conductance that is usually of the order of a hundred units in e^2/h. Equation (2.1) describes the weakly localized phase and the logarithmic quantum corrections (i.e. Eq. (1.1)) are typically on the order of a few percent. The Hall conductance is unaffected by both the

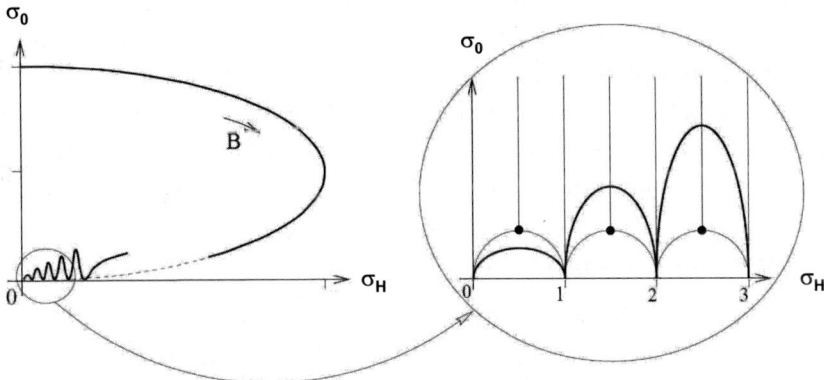

Fig. 21.2.   Mean field parameters $\sigma_0^0(B)$ vs. $\sigma_H^0(B)$ with varying B.

impurity and interaction diagrams and the renormalization group flow is therefore solely in the longitudinal direction.

For strong B ($\omega_c \tau \gg 1$) the Landau bands are no longer overlapping but separated by an energy gap or "Landau gap". Under these circumstances the mean field Hall conductance σ_H^0 in units of e^2/h is equal to the filling fraction ϕ of the Landau level system

$$\sigma_H^0(B) = \phi = \frac{n_0}{n_B} \tag{2.2}$$

with n_0 denoting the electron density and $n_B = eB/hc$ the density of a completely filled Landau band. The SCBA does not display the experimentally observed plateau features. These features are usually associated with the effects of strong Anderson localization in the tails of the Landau band $\phi \approx 0, 1, 2 \ldots$ where σ_0^0 is close to zero. In the scaling diagram of Fig. 21.1, these effects are described by the stable "quantum Hall" fixed points located at

$$\sigma_0 = 0 \quad , \quad \sigma_H = k \tag{2.3}$$

with integer k. On the other hand, near the half-integral filling fractions $\phi = \frac{1}{2}, \frac{3}{2}$ the electron gas becomes *quantum critical* and the transport is controlled by the critical fixed point with $\sigma_H^* = \mathcal{O}(1)$ in Fig. 21.2.

The maximum value of $\sigma_0^0(B)$ increases linearly with the Landau level index n. This means that the majority of the states in the higher Landau bands are *weakly localized* rather than quantum critical. This explains why at the time of the early experiments on the quantum Hall effect it was assumed that the "extended" states appear in energy bands, the width of which increases with increasing Landau level index n.[30] Similar conclusions were drawn from the early numerical work on the disordered free electron gas on the lattice (i.e. the Hofstadter model). Following the advent of the renormalization theory, these type of ideas were in general abandoned and the quantum Hall plateau transitions were then studied in accordance with the predictions of universality. This change of direction in both experimental and computational research has been one of the early victories of the ϑ angle approach to the quantum Hall effect.

2.2. Universal scaling functions

In accordance with the scaling theory[22] we express the macroscopic or measured conductances $\sigma_{0,H}(B, L)$ with varying B and length scale L in terms of universal scaling functions $F_{0,H}$ according to

$$\sigma_{0,H}(B, L) = F_{0,H}(X, Y) \tag{2.4}$$

where X, Y stand for the relevant, irrelevant scaling variable

$$X = \left(\frac{L}{L_0}\right)^{1/\nu} (\sigma_H^0(B) - \sigma_H^*) \;,\quad Y = \left(\frac{L}{L_0}\right)^{-y_0} (\sigma_0^0(B) - \sigma_0^*). \qquad (2.5)$$

Here, $\sigma_{0,H}^0(B)$ are the mean field conductances at a given microscopic length L_0 which we assume are close to the critical fixed point values $\sigma_0^* = \mathcal{O}(1)$ and $\sigma_H^* = \frac{1}{2}, \frac{3}{2}, \ldots$ in Fig. 21.2. The quantity ν in Eq. (2.5) is identified as the correlation length (or localization length) exponent

$$\xi \propto |\sigma_H^0(B) - \sigma_H^*|^{-\nu} \qquad (2.6)$$

and the exponent y_0 describes the leading corrections to scaling.

Specializing from now onwards to the lowest Landau bands in strong B we can substitute

$$\sigma_H^0(B) - \sigma_H^* = \phi - \phi^* \ll 1 \qquad (2.7)$$

in Eq. (2.5) with $\phi^* = \frac{1}{2}, \frac{3}{2} \ldots$ denoting the critical filling fraction. Next, the theory of Anderson localization only makes sense if one can rely on Fermi liquid principles such that the Coulomb interaction between the electrons can be ignored all together. For strongly disordered (spin polarized) particles it is generally assumed that the only effect of the Coulomb interaction is to produce an effective sample size L_φ for scaling which at finite temperature T is given by

$$L_\varphi \propto T^{-p/2} \ll L. \qquad (2.8)$$

Here, the inelastic scattering exponent p is taken as a phenomenological parameter. Once Eq. (2.8) is accepted one can substitute L_φ for L in Eq. (2.5)

$$X = \left(\frac{T}{T_0}\right)^{-\kappa} (\phi - \phi^*) \;,\quad Y = \left(\frac{T}{T_1}\right)^{\eta} \qquad (2.9)$$

such that the scaling of electron transport can be studied experimentally by varying B (or ϕ) and T. The exponents κ and η in Eq. (2.9) are equal to

$$\kappa = \frac{p}{2\nu} \;,\quad \eta = \frac{py_0}{2} \qquad (2.10)$$

and $T_{0,1}$ stand for fixed temperature scales that are determined by the microscopic details of the sample.

Finally, from the symmetries of the renormalization group flow diagram one can draw general conclusions about the actual form of the scaling functions $F_{0,H}$. For the transition between the $F_H = k$ and $k+1$ quantum Hall plateaus we generally can write

$$F_0(X,Y) = F_0(-X,Y) \quad , \quad F_H(X,Y) = 1 + 2k - F_H(-X,Y) \qquad (2.11)$$

which is also termed particle–hole (PH) symmetry.

2.3. *Early experiments (1985–2000)*

The experimentally measured quantities are the longitudinal resistance R_0 and Hall resistance R_H with varying B and T (see Fig. 21.3)

$$R_0(B,T) = \frac{F_0}{F_0^2 + F_H^2} \quad , \quad R_H(B,T) = \frac{F_H}{F_0^2 + F_H^2}. \qquad (2.12)$$

Like the conductances, these quantities are a function of X and Y only. An easy and popular way of extracting the critical exponent κ directly from the experimental data on R_0 and R_H is done by noticing the following

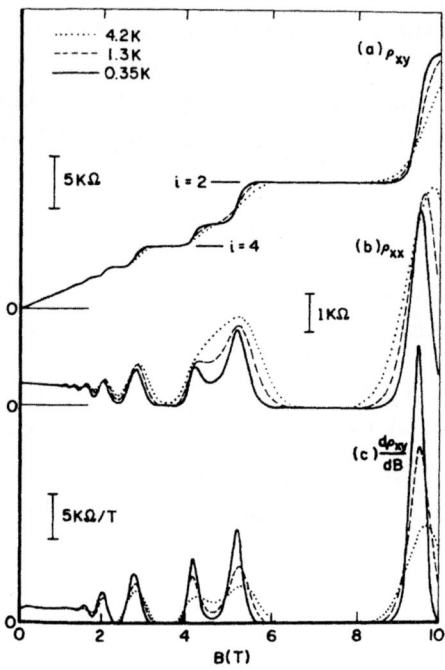

Fig. 21.3. (a) The Hall resistance $R_H = \rho_{xy}$, (b) the longitudinal resistance $R_0 = \rho_{xx}$ and (c) $\partial R_H / \partial B = d\rho_{xy}/dB$ with varying B taken from a low mobility InGaAs–InP heterostructure (H. P. Wei *et al.*, 1988). Reprinted (Fig. 1) with permission from *Phys. Rev. Lett.* **61**, 1294 (1998). © American Physical Society.

equalities[22]

$$\left[\frac{\partial R_H}{\partial B}\right]^{\max} = -\left(\frac{T}{T_0}\right)^{-\kappa}\left|\frac{\phi}{B}\frac{\partial R_H}{\partial X}\right|^{\max} \qquad (2.13)$$

$$\Delta B = \left(\frac{T}{T_0}\right)^{\kappa}\left(\frac{B}{\phi}\right). \qquad (2.14)$$

Assuming that the effects of the irrelevant scaling variable Y can be neglected then $\left|\frac{\phi}{B}\frac{\partial R_H}{\partial X}\right|^{\max}$ is a constant independent of B and T. Equation (2.13) then tells us that the slope of the steps between adjacent quantum Hall plateaus (also termed *plateau-plateau* or PP transition) diverges algebraically in T in the limit where T goes to absolute zero. On the other hand, ΔB in Eq. (2.14) can be identified with the half-width of the longitudinal resistance R_0 which in the transition regime roughly behaves like a simple Gaussian. This half-width vanishes algebraically in T with the same exponent κ.

The main advantage of Eqs. (2.13) and (2.14) is that the measurement of κ does not involve the precise critical value of ϕ^* or B^*, nor does it depend on an admixture of the R_0 and R_H data that are taken from different parts of the sample. However, this kind of measurement is utterly useless if one ignores any of the stringent constraints that are imposed on the experimental design. In particular, the type of quenched impurities in the sample[14] and the requirement of macroscopic homogeneity of the sample.[31] The experimental complications can be summarized as follows.

(1) To ensure that both the temperature scales T_0 and T_1 are within the experimental range $T > 10$ mK, it is imperative that the transport is dominated by an impurity potential that fluctuates randomly over distances on the order of the magnetic length or smaller. This is the opposite limit of the semi classical "percolation picture" which assumes a potential that varies very slowly over a magnetic length. Scattering is completely absent within this semi classical approach which means that T_0 has been reduced to zero.

Semi classical ideas on electron transport are usually associated with the GaAs–AlGaAs heterostructure where the remote ionized impurities cause a smoothly varying random potential relative to the magnetic length. Elastic scattering only takes place in an extremely small range of energies around the Landau band center. The scattering length L_0 can become arbitrarily large and the corresponding T_0 for scaling quickly falls outside the range of experimental T. The cross-over between "percolation" and "localization" generally competes with the cross-over between "weak" quantum interference and "strong" Anderson localization in the higher Landau bands. As mentioned earlier, the latter may result in unrealistically small values of T_1.

(2) Secondly, the sample must be sufficiently homogeneous. The degree of homogeneity is defined by Eq. (2.13) itself. For example, any spatial variation δn_0 in the electron density n_0 causes an uncertainty $\delta\phi = \delta n_0/n_B$ in the filling fraction ϕ. This uncertainty should be much smaller than the width $\Delta\phi \propto T^\kappa$ of the PP transition at any given T. The condition $\delta\phi \ll \Delta\phi$ therefore defines a characteristic temperature below which quantum criticality cannot be experimentally observed.

These constraints on the experiments on scaling are to a large extend self-evident even though they apply very specifically to transport measurements in the quantum Hall regime. However, only in recent years has the matter been studied in a systematic fashion. These studies have invalidated almost all the experiments on scaling that have been conducted over the past one and a half decades following the pioneering investigations by H. P. Wei *et al.* in 1986.[32]

The importance of the *range* of the impurity potential has already been understood and stressed from the very beginning.[33] In fact, a large set of different kinds of samples has been investigated before the final choice of H. P. Wei *et al.* fell on a few selected low mobility InGaAs–InP heterostructures where the transport is dominated by alloy scattering. These heterostructures were at that time the most likely candidate for samples with short ranged potential fluctuations.

The exponent value $\kappa = 0.42$ reported H. P. Wei *et al.* in 1988, has set the stage for the phenomenon of quantum criticality in the quantum Hall regime for many years. The data on Eqs. (2.13) and (2.14) taken from the lowest three PP transitions in the range 0.1–4.2 K provided a number of independent checks on the "universality" of the critical exponent κ, see Fig. 21.4. Experiments on certain strongly disordered GaAs–AlGaAs heterostructures clearly showed that the "range" of the impurity potential strongly affects observability of universality at finite T.[33] Only at a much later stage it became evident that these early measurements were complicated because of the problems with sample homogeneity (see Sec. 2.4).

H. P. Wei's original experiments have in the subsequent years been incorrectly interpreted by many. For example, the group of K. von Klitzing was unable to establish the universality of κ, simply because the samples were arbitrarily chosen and hence, inappropriate.[34,35] Very different exponent values were measured in the range $0.21 < \kappa < 0.85$. The different

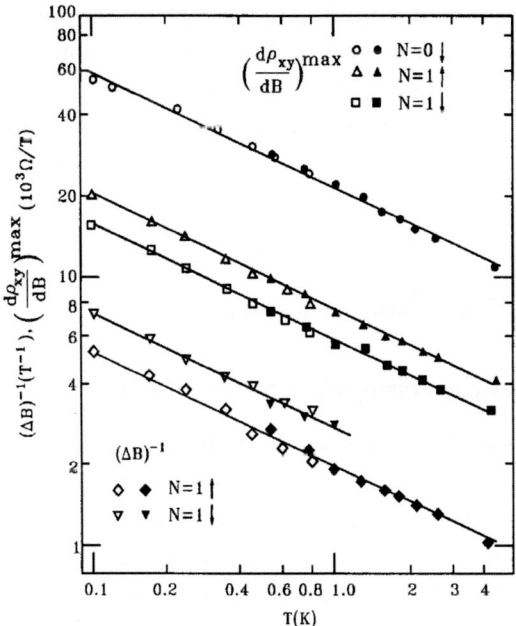

Fig. 21.4.   Inverse of the half-width $(\Delta B)^{-1}$ and $[\partial R_H/\partial B]^{\max} = (d\rho_{xy}/dB)^{\max}$ with varying T taken from the $N = 0 \downarrow$ and $N = 1 \uparrow\downarrow$ Landau bands of a low mobility InGaAs–InP heterostructure (H. P. Wei *et al.*, 1988). Reprinted (Fig. 2) with permission from *Phys. Rev. Lett.* **61**, 1294 (1998).   © American Physical Society.

values were dependent not only on the specific sample at hand but also on the PP transition of a given sample. Different values of κ were extracted even from the R_H and R_0 data taken from the same PP transition of the same sample. These experimental results merely showed that the "mobility edge" in Anderson localization theory is too elusive to catch using ordinary semiconductor technology.

The phenomenon of quantum criticality was in later years called into question even by the Princeton group. The transition in the lowest Landau level (also termed "plateau–insulator" or PI transition) was studied using a new series of strongly disordered GaAs–AlGaAs samples. The following results for the resistances have been found[36]

$$R_0(B, T) = c\, e^{-X} \ , \quad R_H(B, T) = 1 \tag{2.15}$$

where the geometrical factor c is close to unity when R_0 is normalized to the geometry of a square. Remarkably, the Hall resistance R_H at low T remains quantized throughout the PI transition and the dependence on T and B

only appears in the longitudinal component R_0. Equation (2.15) describes the transition between a quantum Hall "superconductor" and a quantum Hall "insulator" and these results have been verified for a variety of different samples. However, the variable X did not follow the predicted algebraic behavior in T. Instead of Eq. (2.9), the data were fitted to a kind of semi classical "linear law"

$$X = \{a\,T + b\}^{-1}\,(\phi - \phi^*) \tag{2.16}$$

with a and b sample dependent constants.[36]

2.4. *Amsterdam experiments (2000–2007)*

The aforementioned developments have been strongly refuted in Refs. 14 and 37. To explain the "linear law," the effects of the electron–electron interaction were studied based on a strongly disordered network consisting of sharp "edge states" and widely separated "saddle points." The results clearly indicated that the "linear law" is an artifact of the long-ranged potential fluctuations present in the samples of Ref. 36. The network model furthermore showed that Eq. (2.16) is controlled by a phase breaking length $L_\varphi \propto T^{p/2}$ with an inelastic scattering exponent p that lies in the range $1 < p < 2$ depending on the ramification of the network.

In order to re-establish the significance of short ranged disorder, the PI transition was investigated using the same InGaAs–InP samples that have originally been studied by H. P. Wei *et al.*[37–42] These new experiments clearly demonstrated that X depends algebraically on T rather than the "linear law" of Eq. (2.16). This time, however, an exponent value of $\kappa = 0.57$ was observed rather than the H. P. Wei value $\kappa = 0.42$ extracted earlier from the PP transitions of the same sample.

The resolution of this discrepancy was found in the effects of macroscopic sample inhomogeneities which previously had never been considered. Detailed studies have shown that the data on the PI transition are at least an order of magnitude less sensitive to experimental imperfections such as contact misalignment and density gradients as compared to the data on the PP transition.[31] The reason being that the PI transition does not display the usual sharp step in the Hall resistance R_H. Similarly, the longitudinal resistance R_0 with varying B does not display the usual sharp peak but varies smoothly throughout the transition. Extended numerical work on the effects density gradients and contact misalignment on the PP transition has furthermore shown that the experimental value of $\kappa = 0.42$ measured by H. P. Wei *et al.* is an *effective* exponent rather than the actual intrinsic one.[43]

Next, one can take complete advantage of the special features of the PI transition in order to extract the full scaling functions F_0 and F_H from the experiment. Notice that one can associate universal significance with Eq. (2.15) since the dependence on X apparently holds even for samples that do not display scaling at finite T. A more detailed study of the data taken from the InGaAs–InP sample has shown that the correct results for R_0 and R_H are as follows

$$R_0(X) = e^{-X-\gamma X^3} \quad , \quad R_H(X,Y) = 1 + Y R_0(X). \qquad (2.17)$$

Here, X and Y are given by Eq. (2.9). These expressions with $\gamma \approx 0.002$ and $Y R_0 \ll 1$ are a corrected version of Eq. (2.15) and valid in a large range $|X| < 5$.

To justify the Y dependence in Eq. (2.17) one has to show that the data on $(R_H - 1)/R_0$ are independent of B. The collapse of the data with varying B is plotted in Fig. 21.5. The results for $Y = Y(T)$ nicely expose the algebraic behavior of Eq. (2.9) with an exponent value $\eta = 2.4$ and a temperature scale $T_1 = 9.2$ K.

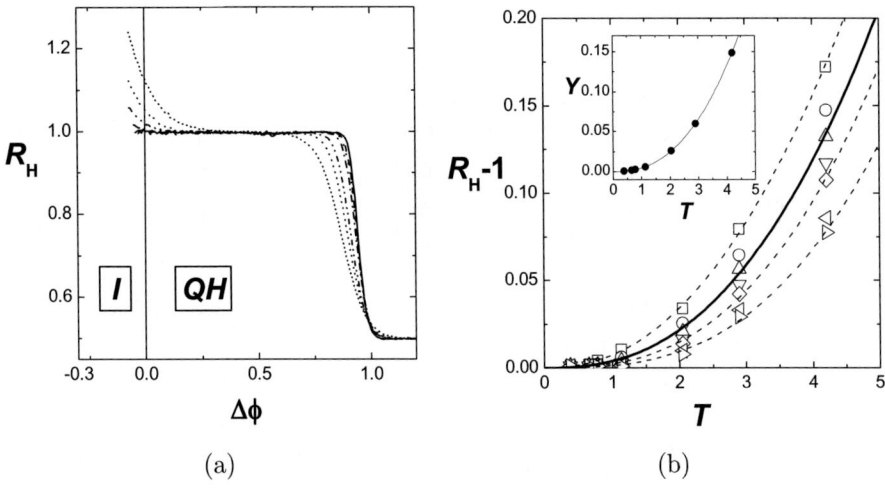

(a) (b)

Fig. 21.5. (a) The Hall resistance R_H with varying $\Delta\phi = \phi - \phi^*$ with $\phi^* \approx \frac{1}{2}$ in the Insulator (I) – Quantum Hall plateau (QH) regime of an InGaAs–InP heterostructure. The different curves are taken for different T in the range 0.38 K $< T < 4.2$ K. (b) $R_H - 1$ with varying T for different values of $\Delta\phi$ in the range $-0.025 < \Delta\phi < 0.05$. The inset: Collapse of the data for $Y = (R_H - 1)/R_0$ with varying T (de Lang et al., 2007; Pruisken et al., 2006). Figure 21.5(b) reprinted with permission from: (i) *Phys. Rev. B* **75**, 035315 (2007) (Fig. 1d) © American Physical Society, and (ii) *Solid State Commun.* **137**, 540 (2006) (Figs. 2a & 2b) © Elsevier.

Next, it is easy to see that the general form of Eq. (2.17) is in accordance with PH symmetry. In terms of conductances one can write

$$F_0(X,Y) = \frac{R_0(X)}{R_0^2(X) + 1 + 2Y R_0(X)} \qquad (2.18)$$

$$F_H(X,Y) = \frac{1 + Y R_0(X)}{R_0^2(X) + 1 + 2Y R_0(X)} \qquad (2.19)$$

which indeed satisfies Eq. (2.11) with $k = 0$.

The consequences for scaling are plotted in the diagram of Fig. 21.6. The T-driven flow lines for different values of B or ϕ are taken from Eqs. (2.18) and (2.19) using the experimental values $\kappa = 0.57$, $\eta = 2.4$, $T_0 = 188$ K and $T_1 = 9.2$ K. On the low T-side, the flow lines converge toward the semi circle described by $F_{0,H}(X,0)$. On the high-T side, however, the flow lines come together to form a new strong coupling fixed point located at $\sigma_0 = 0$ and

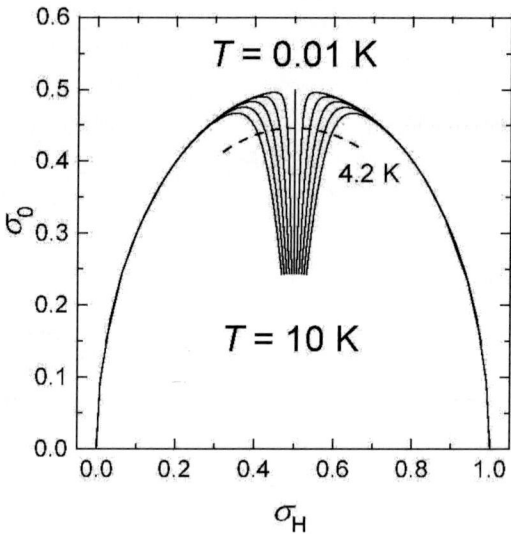

Fig. 21.6.   Experimental flow diagram of the longitudinal conductance (σ_0) and the Hall conductance (σ_H) in the quantum Hall regime. The scaling results for σ_0 and σ_H with varying T and magnetic field B have been taken from the lowest Landau level of an $InP - InGaAs$ heterostructure. The T-driven flow lines at constant B display a a quantum critical fixed point at $\sigma_0 = \sigma_H = 1/2$ and a perfect symmetry about the line $\sigma_H = 1/2$. (Pruisken $et.\ al.$, 2006; de Lang $et\ al.$, 2007). Reprinted (Fig. 3) with permission from $Solid\ State\ Commun.$ $\mathbf{137}$, 540 (2006). © Elsevier.

$\sigma_H = \frac{1}{2}$. This unstable fixed point is not an artifact of the extrapolation but physically describes a "Fermi liquid" of composite fermions. This issue will be revisited at a later stage.

It is natural to expect that Eqs. (2.18) and (2.19) are the universal scaling functions of the conductances in the integral quantum Hall regime. Based on the periodicity of scaling in the σ_0–σ_H conductance plane we perform the simple shift

$$F_0 \to F_0 \quad , \quad F_H \to F_H + k \qquad (2.20)$$

with integer k and obtain the correct scaling results for the PP transitions. This shift goes along with a redefinition of the parameters T_0, T_1 and ϕ^* as well as the sign of Y in Eq. (2.9). We conclude this section with several remarks.

(1) It is instructive to compare Eq. (2.20) and Fig. 21.6 for the $N = 0 \uparrow$ Landau band with the experimental flow lines that previously were taken directly from the measurements on the $N = 0 \downarrow$ Landau band of the same InGaAs/InP heterostructure by H. P. Wei et al.,[32] see Fig. 21.7. This comparison illustrates how the effects of sample inhomogeneity complicate the experiment on the PP transition. The earlier data on the PP transitions actually gave the wrong ideas about scaling, in particular the effects of the Fermi–Dirac distribution were interpreted incorrectly.

(2) It is important to stress that an improved sample technology alone does not necessarily reveal the correct behavior of the functions $F_{0,H}(X,Y)$. This is clear from the experimental papers on the PI transition that originally reported the results of Eqs. (2.15) and (2.16). Besides an

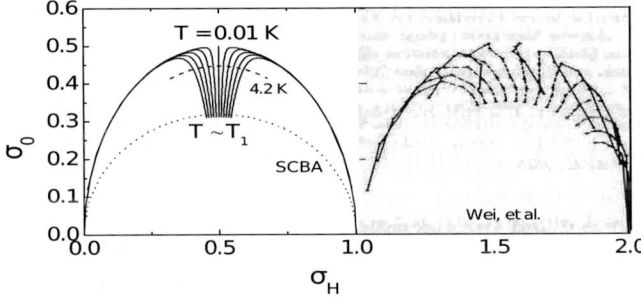

Fig. 21.7. Comparison of the experimental flow diagram of Fig. 21.6 and the bare data $1 < \sigma_H < 2$ taken from the $N = 0 \downarrow$ Landau band (H. P. Wei et al., 1986). Reprinted (Fig. 3) with permission from *Phys. Rev. B* **33**, 1488 (1986). © American Physical Society.

incorrect assessment of the "linear law," these studies ignored the Y dependence of the data and also mishandled the PH symmetry in quantum transport. This demonstrates that the experiments on Anderson *de*-localization can be extremely misleading if one is not guided by the virtues of a microscopic theory.

(3) By converting Eq. (2.20) back into resistances R_0 and R_H we obtain complex scaling functions, the details of which can hardly be extracted from the data on the PP transitions. By expanding the results to first order in Y we obtain

$$R_0(X,Y) = R_0^k(X) \left(1 - Y \frac{R_0^k(X)}{R_0^k(X_k)}\right) \ ,$$

$$R_H(X,Y) = R_H^k(X) - Y \frac{\partial R_0^k(X)}{\partial X} \tag{2.21}$$

where

$$R_0^k(X) = \frac{R_0(X)}{k^2 R_0^2(X) + (k+1)^2} \ , \quad R_H^k(X) = \frac{k R_0^2(X) + k + 1}{k^2 R_0^2(X) + (k+1)^2} . \tag{2.22}$$

The quantity $R_0^k(X_k)$ denotes the maximum of R_0^k that occurs for $X = X_k$ and the results are

$$R_0^k(X_k) = \frac{1}{2k(k+1)} \ , \quad X_k = \ln\left(\frac{k+1}{k}\right). \tag{2.23}$$

Therefore, the maximum in the longitudinal resistance R_0 with varying X or B depends weakly on T according to

$$R_0^{\max}(T) = R_0(X_k, Y) = \frac{1}{2k(k+1)} \left(1 \pm \left(\frac{T}{T_1}\right)^\eta\right). \tag{2.24}$$

Equation (2.24) in principle permits an extraction of the exponent η and the temperature T_1 from the measurement on the PP transition. The expression for X_k also implies that the position $\phi_{\max}$ (or $B_{\max}$) of the peak in R_0 varies with T according to

$$\phi_{\max}(T) = \phi^* + \ln\left(\frac{k+1}{k}\right) \left(\frac{T}{T_0}\right)^\kappa. \tag{2.25}$$

Provided the constraints of sample homogeneity are satisfied one can employ Eqs. (2.24) and (2.25) along with Eqs. (2.13) and (2.14) as an important check on the overall consistency of the experimental data on scaling.

2.5. *Princeton experiments (2005–2009)*

Following all the hard ground work, the importance of both short ranged potential fluctuations and sample homogeneity has finally been established. It is of course necessary to repeat exactly the same experiments but now with a series of specifically grown state-of-the-art samples with varying degrees of well defined disorder. There is in fact, a longstanding quest for more controlled experiments on scaling. The reason being that little microscopic knowledge exists on the alloy scattering present in the low mobility InP–InGaAs heterostructure. It is therefore not known whether or not long ranged potential fluctuations are present that would upset the "universality" of $\kappa = 0.57$ taken from the PI transition at finite T.

In more recent years the Princeton group re-investigated the PP transitions using a series of $Ga_x As_{1-x}–Al_{0.3}Ga_{0.7}As$ heterostructures with different Al concentration x that controls the alloy scattering. Due to the improved sample technology, the problem of macroscopic sample inhomogeneities can be ignored at least as far as the extraction of κ based on Eq. (2.13) is concerned.[44,45]

It turns out that by varying the Al concentration in the range $0 < x < 1.6\%$ one controls the admixture of short and long ranged scatters in the sample. Consequently, the value of κ taken at finite T depends on x and varies in the range $0.42 < \kappa < 0.58$. However, the new data give conclusive evidence for the fact that in the limit where T goes to absolute zero, the experimental κ is always the same and given by $\kappa = 0.42$.

This remarkable result is rather confusing. The universal value of κ is apparently identical to the old "H. P. Wei value," reported more than two decades ago, even though the latter is complicated by the effects macroscopic sample inhomogeneity. However, it is important to keep in mind that the new and old experiments have a totally different meaning and the numerical "agreement" is purely accidental, even magical.

It so happens that the long ranged potential fluctuations present in the InP–InGaAs heterostructure have a tendency to *increase* the intrinsic value of κ from 0.42 to 0.57. At the same time, the macroscopic inhomogeneities also present in the InP–InGaAs sample have a tendency to *reduce* the experimental κ taken from the PP transitions from 0.57 back to 0.42! So, the two different sample dependent effects cancel one another in the limited range of experimental T. This explains the magical "restoration" of "universality".

Unfortunately, since the lowest Landau level lies in the fractional quantum Hall regime, the new series of samples do not provide access to the universal scaling functions $F_{0,H}$ as studied in the Amsterdam experiments. In addition to this, there are complications with the R_0 measurements such that there is as yet no check on the overall consistency of the transport data. Establishing a complete experimental picture of the PP transition that includes the corrections to scaling is still awaited.

3. Electron–Electron Interactions

3.1. *The problem of disentangling p and ν*

With hindsight one can say that the experiments on scaling have in one respect been truly deceiving. The idea of Fermi liquid theory, in particular, has become so popular over the years, one tends to forget — or wishes to forget — that the Coulomb interaction in Anderson localization theory is an outstanding and notorious problem that has in general not been understood.

Following the first experiments on κ by H. P. Wei *et al.* in 1988, for example, there have been many attempts, both from the experimental and theoretical side, to demonstrate that the localization length exponent ν is identically the same for both interacting and free electrons in strong B.[45] The idea is to justify, in one way or the other, the equality

$$p = 2\nu\kappa = 2 \tag{3.1}$$

where the right hand side is usually taken to be an exact integer. Equation (3.1) is based on the *experimental* value of $\kappa = 0.42$ and the "commonly accepted" *free electron* value

$$\nu = 2.34 \pm 0.04 \tag{3.2}$$

as extracted from numerical simulations.[46] This sort of numerical agreement has already been pursued by many even though the real value of κ was until recently not known.

Given the complications in extracting critical exponents from semiconductor devices, Eqs. (3.1) and (3.2) have not really been established in any convincing manner.[47] Recently the numerical result of Eq. (3.2) has also been called into question. Slevin and Ohtsuki,[48] in particular, have reported a significantly different free electron value of $\nu = 2.6$ which, by the way, is much closer to the analytic result of $\nu = 2.8$ originally predicted by the dilute gas of instantons (Sec. 4.7.2).

3.2. *Unified renormalization theory*

It is important to keep in mind that the experiments conducted in the laboratory and those on the computer involve very different physical systems that in principle may have little to do with one another. The primary goal of the experimental advances on Anderson localization such as Fig. 21.6 is to understand the physics of both the integral and fractional quantum Hall effects. When viewed in the context of numerical work on free electron exponents, these advances clearly have a very different meaning. Our principle task is to unravel these differences and not conceal them in obscurity.

The only way in which one can possibly understand the experiments is through the so-called unified renormalization theory (URT) of the quantum Hall effect.[12–16] This theory essentially reconciles the topological issue of a ϑ angle with Finkelstein's generalized non-linear sigma model approach to Anderson localization and interaction phenomena.[49]

It is in many ways surprising that Finkelstein's radical ideas on the subject have hardly received the attention they deserve. They provide the much sought after answer to the problem of Anderson localization in the presence of the Coulomb interaction. However, to be able to study the generalized non-linear sigma approach as a field theory, several major advances have been necessary. Amongst these, the most important one is the electrodynamic $U(1)$ gauge invariance of the electron gas.

To demonstrate gauge invariance, a set of algebraic rules termed $\mathcal{F}$-algebra have been introduced.[12] By extending Finkelstein's theory to include both the ϑ angle concept and the Chern–Simons statistical gauge fields, one essentially paves the road towards a microscopic understanding of both the integral and fractional quantum Hall regimes. For example, the complete Luttinger liquid theory of "edge" excitations has been derived in this way.[15] This theory was previously proposed only on phenomenological grounds but now includes the effects of disorder as well as the coupling to external vector and scalar potentials.

What has remarkably emerged is a global symmetry, termed $\mathcal{F}$-*invariance*. This new symmetry explains why the Coulomb interaction fundamentally alters the theory of Anderson localization in strong B as a whole. Having completed the herculean task of developing different computational schemes, we now have what is necessary to address this theory. The schemes involve both perturbative and non-perturbative analyses and are a major advance in both the renormalization group and the instanton calculational technique.[13,16,50–52] We are rewarded however, by a physical simplicity that surfaces at the end of all the computations. For example, concepts such as ϑ or σ_H renormalization by instantons and also the idea of universal scaling functions $F_{0,H}$ for the conductances remain essentially unaltered although the quantum critical details are now clearly different. In particular, the critical exponents p and ν have a different physical meaning and become non-Fermi liquid like once the infinitely ranged Coulomb interaction is taken into account.

The strong similarity between the scaling theory of free electrons and the experiment cannot therefore be regarded as a manifestation of Fermi liquid theory. Instead topological principles in quantum field theory are at work, notably the discrete set of topological sectors that fundamentally explain how the quantum Hall effect reveals itself in the theory on the strong coupling side (Sec. 4).

We have already mentioned the fact that the ϑ angle concept in scale invariant theories generally displays the same scaling behavior and the same macroscopic quantization phenomena. The quantum Hall effect is, in fact, an interesting realization of this principle. *Super universality* is retained even though the infinitely ranged Coulomb interaction is an entirely different theory and gives rise to completely different physical phenomena.

3.3. *Scaling diagram*

The consequences of the URT for Anderson localization are encapsulated in the scaling diagram of Fig. 21.8. Besides the conductances σ_0 and σ_H, the renormalization of interacting, spin polarized or spinless electrons involves a third dimensionless parameter termed c which depends on the range of the (repulsive) pair potential $U(|\mathbf{r} - \mathbf{r}'|)$. The *free* electron gas is described by $c = 0$, the problem with *finite range* interactions is represented by $0 < c < 1$ and the case $c = 1$ corresponds to *infinite range* interactions, see Table 21.1.

Table 21.1.   The parameter c, see text.

c	Interaction potential	Terminology
$c = 0$	$U(r) = 0$	free electrons
$0 < c < 1$	$0 < \int_0^\infty rU(r)dr < \infty$	finite range interactions
$c = 1$	$\int_0^\infty rU(r)dr = \infty$	infinite range interactions

3.3.1. *β and γ functions*

The renormalization group β functions of the URT can in general be written as follows[51]

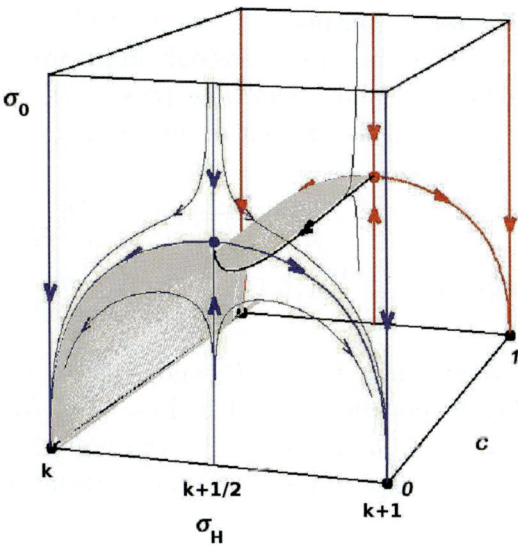

Fig. 21.8.   Three dimensional renormalization group flow diagram of Anderson localization in the presence of electron–electron interactions and strong B. The plane $c = 0$ describes the free 2DEG and $c = 1$ describes the 2DEG in the presence of the Coulomb interaction. The problem with finite ranged interactions $0 < c < 1$ lies in the domain of attraction of the free 2DEG. See text.

$$\frac{d\sigma_0}{d\ln b} = \beta_0(\sigma_0, \sigma_H, c) \tag{3.3}$$

$$\frac{d\sigma_H}{d\ln b} = \beta_H(\sigma_0, \sigma_H, c) \tag{3.4}$$

$$\frac{dc}{d\ln b} = c(1-c)\gamma(\sigma_0, \sigma_H, c) \tag{3.5}$$

where b denotes the length scale. The structure of Eqs. (3.3)–(3.5) is determined, to a large extend, by the quantization of the topological charge (see Sec. 4). This means that the functions β_0, β_H and γ can in general be expanded as an infinite trigonometric series according to

$$\beta_0 = \beta_0^0(\sigma_0, c) + \sum_{n=1}^{\infty} f_n(\sigma_0, c)\cos 2\pi n\sigma_H \tag{3.6}$$

$$\beta_H = \sum_{n=1}^{\infty} g_n(\sigma_0, c)\sin 2\pi n\sigma_H \tag{3.7}$$

$$\gamma = \gamma^0(\sigma_0, c) + \sum_{n=1}^{\infty} h_n(\sigma_0, c)\cos 2\pi n\sigma_H \tag{3.8}$$

with β_0^0, γ, f_n, g_n, and h_n unspecified functions of σ_0 and c. Generally speaking, one can evaluate these functions only in the weakly localized phase $\sigma_0 \gg 1$. The functions β_0^0 and γ can be expressed as usual in terms of a series in powers of $1/\sigma_0$ and the explicit results are known to two loop order

$$\beta_0^0(\sigma_0, c) = - \quad A_1(c) - \frac{1}{\sigma_0} A_2(c) + \mathcal{O}(\frac{1}{\sigma_0^2}) \tag{3.9}$$

$$\gamma(\sigma_0, c) = -\frac{1}{\sigma_0} B_1(c) - \frac{1}{\sigma_0^2} B_2(c) + \mathcal{O}(\frac{1}{\sigma_0^3}). \tag{3.10}$$

The non-negative coefficients A_1, A_2, B_1 and B_2 with varying c are listed in Table 21.2. The terms with $n > 0$ in Eqs. (3.6)–(3.8) are controlled by the instanton factors $f_n, g_n, h_n \propto e^{-2\pi n \sigma_0}$ which are invisible in ordinary perturbation theory. The explicit results for the single instanton sector $n = 1$ read as follows

$$f_1(\sigma_0, c) = -\mathcal{D}(c)\sigma_0^2 e^{-2\pi\sigma_0} \tag{3.11}$$

$$g_1(\sigma_0, c) = -\mathcal{D}(c)\sigma_0^2 e^{-2\pi\sigma_0} \tag{3.12}$$

$$h_1(\sigma_0, c) = -\mathcal{D}_\gamma(c)\sigma_0 e^{-2\pi\sigma_0} \tag{3.13}$$

with the coefficients $\mathcal{D}, \mathcal{D}_\gamma > 0$ listed in Table 21.2.

Table 21.2.   Coefficients of the renormalization group functions, see text. The expressions for $\mathcal{D}(c)$, $\mathcal{D}_\gamma(c)$ and $\mathcal{A} \approx 1.64$ are given in Refs. 16 and 51. A_2 and B_2 for $0 < c < 1$ are yet unknown and $\gamma_E \approx 0.577$ denotes the Euler constant.

c	A_1	A_2	B_1	B_2	$\mathcal{D}$	$\mathcal{D}_\gamma$
$c = 0$	0	$1/2\pi^2$	$1/\pi$	0	$16\pi/e$	$8\pi/e$
$0 < c < 1$	$2\left[1+\frac{1-c}{c}\ln(1-c)\right]/\pi$	-	$1/\pi$	-	$\mathcal{D}(c)$	$\mathcal{D}_\gamma(c)$
$c = 1$	$2/\pi$	$4\mathcal{A}/\pi^2$	$1/\pi$	$1/6 + 3/\pi^2$	$16\pi e^{1-4\gamma_E}$	$8\pi e^{1-4\gamma_E}$

Finally, the temperature dependence of quantum transport only appears through the combination zcT where zc stands for the *singlet interaction amplitude* also termed "Finkelstein z." The anomalous dimension of zc is given by

$$\frac{d\ln zc}{d\ln b} = \gamma(\sigma_0, \sigma_H, c) \tag{3.14}$$

where γ is the same function as in Eq. (3.5).

3.3.2. *Critical exponents*

As indicted by the flow lines in Fig. 21.8, the theory with finite range interactions $0 < c < 1$ is the domain of attraction of the *free electron* theory with $c = 0$. Fermi liquid ideas therefore apply only to the 2DEG with *finite range* interactions. We term this theory the *Fermi liquid* universality class. The exponent values are obtained in a standard manner[52]

$$\nu = \left(\frac{\partial \beta_H^*}{\partial \sigma_H}\right)^{-1} = 2.75 \quad , \quad y_\sigma = \frac{\partial \beta_0^*}{\partial \sigma_0} = -0.17 \quad , \quad p = \frac{2}{2 + \gamma^*} = 1.2 \quad (3.15)$$

where we employed the weak coupling expressions of Eqs. (3.9)–(3.14) with $c = 0$. The numerical results are remarkably close to those obtained from the computer. This numerical agreement endorses the victory of the instanton vacuum approach to the quantum Hall effect. It is not matched by the results of any alternative approach to the problem proposed over a period of more than two decades. The best numerical values for κ, η and p are in the range[52]

$$\kappa = p/2\nu = 0.27 \pm 0.04 \quad , \quad \eta = -py_\sigma/2 = 0.26 \pm 0.05 \quad , \quad p = 1.35 \pm 0.15.$$
$$(3.16)$$

These results clearly demonstrate that the Fermi liquid universality class is at odds with the experiment.

Let us next consider the case of the Coulomb potential $U(r) \propto \frac{1}{r}$ which is described by the theory with $c = 1$. This theory displays $\mathcal{F}$-invariance which is broken otherwise, i.e. in the theory with $c \neq 1$. As indicated by the flow lines in Fig. 21.8, the plane $c = 1$ represents a novel and separate universality class of Anderson localization phenomena. We term the case $c = 1$ the $\mathcal{F}$-*invariant* universality class.

Unfortunately, the weak coupling computational results of Eqs. (3.9)–(3.14) are insufficient to provide access to the quantum critical fixed points located in the $c = 1$ plane. We therefore do not as yet have reliable estimates for p and ν that can be compared with the experimental values. All we know about the Coulomb interaction problem is that p is bounded by $1 < p < 2$.[13,16,50] So, given the best experimental value $\kappa = 0.42$ it follows that the real value of the localization length exponent ν is in the range $1.2 < \nu < 2.4$. This is distinctly different from the aforementioned best free electron value $\nu = 2.6$.

3.4. *General remarks*

Besides exponent values there are of course many other ways in which one can *think* about the problem of the Coulomb interaction and *learn* about $\mathcal{F}$-invariant quantum criticality. This section briefly addresses some of the issues and examples that have had an important impact on the author's own view on the subject.

3.4.1. $2 + \epsilon$ dimension

An extremely important subject is of course the ordinary mobility edge problem in $2 + \epsilon$ spatial dimension.[13,16] This problem is defined by the perturbative results of Eqs. (3.9), (3.10) and (3.14) only whereas the function β_0^0 replaced by

$$\beta_0^0(\sigma_0, c) \rightarrow \beta_0^0(\sigma_0, c) + \epsilon\sigma_0. \tag{3.17}$$

As shown in Fig. 21.9, the meaning of the quantity c in $2 + \epsilon$ dimension is identically the same as in the strong magnetic field problem of Fig. 21.8.

The structure of Finkelstein's theory in $2 + \epsilon$ dimension is very similar to that of the classical Heisenberg ferromagnet. For $c = 1$ there is a conventional order parameter, notably the bosonic quasi particle density of states

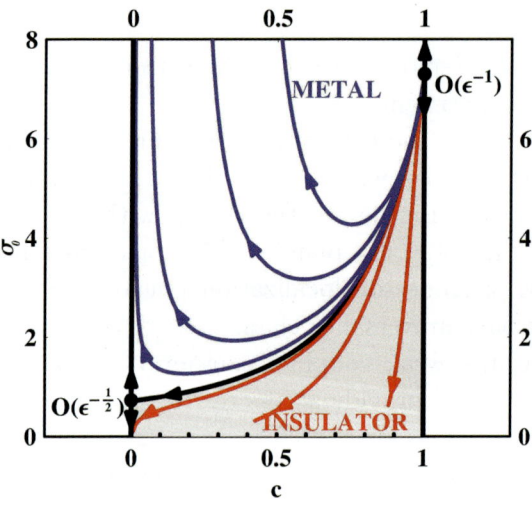

Fig. 21.9. The mobility edge problem of the interacting electron gas in $2+\epsilon$ dimension. The Fermi liquid fixed point σ_0^* is of order $\epsilon^{-1/2}$ whereas the $\mathcal{F}$-invariant fixed point is of order ϵ^{-1}. Like in Fig. 21.8, the problem with finite ranged interactions $0 < c < 1$ forms the domain of attraction of the free particle theory $c = 0$.

entering the expression for the specific heat c_v. At the mobility edge this quantity behaves according to

$$c_v \simeq zT^p \tag{3.18}$$

which is entirely non-Fermi liquid like. The dynamical aspects of scaling can be investigated in gory detail. Explicit scaling results for the longitudinal conductance σ_0 with varying temperature and/or frequency are obtained. The two independent critical exponents are given by

$$\frac{1}{\nu} = \frac{\partial \beta_0^*}{\partial \sigma_0} = \epsilon + \mathcal{A}\epsilon^2 + \mathcal{O}(\epsilon^3) \tag{3.19}$$

$$p = \frac{D}{D + \gamma^*} = 1 + \frac{\epsilon}{8} + \left(\frac{\pi^2 + 15 - 12\mathcal{A}}{192}\right)\epsilon^2 + \mathcal{O}(\epsilon^3) \tag{3.20}$$

which is one order in ϵ higher than what was originally obtained by Finkelstein[53] and others.[54] These quantities have the same physical significance as in the theory of the quantum Hall plateau transitions. In the theory with finite ranged interactions, Eq. (3.18) is replaced by

$$c_v \simeq zcT^p + z(1-c)T \tag{3.21}$$

and instead of Eqs. (3.19) and (3.20) we now have

$$\frac{1}{\nu} = 2\epsilon + 3\epsilon^2 + \mathcal{O}(\epsilon^3) \tag{3.22}$$

$$p = 1 + \sqrt{2\epsilon} + \mathcal{O}(\epsilon). \tag{3.23}$$

Notice that for free particles $c = 0$ the non-Fermi liquid piece in Eq. (3.21) disappears altogether and the Finkelstein z now stands for the free particle density of states. A standard feature of Anderson localization of free particles is the multi fractal singularity spectrum of the wave functions. As is well known, this feature does not exist in a conventional phase transition with an order parameter and hence, it is likely to disappear from the spin-polarized electron gas when the Coulomb interaction is taken into account.[16] As a final remark, it should be mentioned that the physical mechanism for dynamical scaling in the quantum critical phase is very different from that in the strongly localized phase. One expects, in particular, that the T dependence of the latter is controlled by the bare Coulomb interaction which acts like a dangerously irrelevant operator in the Finkelstein theory. This topic will not be addressed in the present chapter and the reader is referred to the literature.[50]

3.4.2. Integral quantum Hall regime

The most obvious way in which the $c = 1$ and $c = 0$ universality classes distinguish themselves is through the fractional quantum Hall effects. These strongly correlated phenomena only appear in the former but obviously do not exist in the latter. There is, however, very limited understanding of how these phenomena disappear as one increases the amount of disorder, nor is there any knowledge of how the fractional quantum Hall effect evolves when the range of the electron–electron interaction decreases from infinity to finite values. Nevertheless, based on our present understanding of both the integral and fractional quantum Hall effects, it is not difficult to establish the strong coupling features of the URT that are analytically inaccessible. More specifically, by combining the experimental results on scaling with composite fermion ideas as well as the aforementioned weak coupling results of the URT we obtain an extended scaling diagram in the σ_0–σ_H conductance plane as sketched in Fig. 21.10. This diagram consists of the following three pieces, each describing a totally different phase of the 2DEG.

(1) The flow lines for $\sigma_0 \gg 1$ indicated by FL. These flow lines emerge from the weakly localized phase and are obtained from Eqs. (3.9)–(3.13) with $c = 1$.

(2) The flow lines just above and below the semi circle indicated by EXPT. These flow lines are the experimentally observed scaling results of Fig. (21.6) and described by Eqs. (2.18) and (2.19). Plotted are the flow lines both positive and negative values of Y.

(3) The flow lines indicated by FL–CF which emerge from the high T fixed point at $\sigma_0 = 0$, $\sigma_H = \frac{1}{2}$. This fixed point describes the half-integer effect observed in the fractional quantum Hall regime. One generally thinks in terms of a "Fermi liquid" of composite fermions that is dual to the weakly localized phase of the 2DEG in weak B.[55] The corresponding flow lines are obtained directly from the renormalization group results of Eqs. (3.9)–(3.13) with $c = 1$ by performing a flux attachment transformation with two flux quanta per electron.[12]

Notice that the different parts of Fig. 21.10 are only observed in a piecewise fashion and under totally different experimental circumstances. They cannot possibly be seen simultaneously in a single experiment, the reason being that each of the different phases are separated by orders of magnitude in length scales and temperature scales. The combination of experimental and theoretical flow lines nevertheless provides a complete picture of the integral

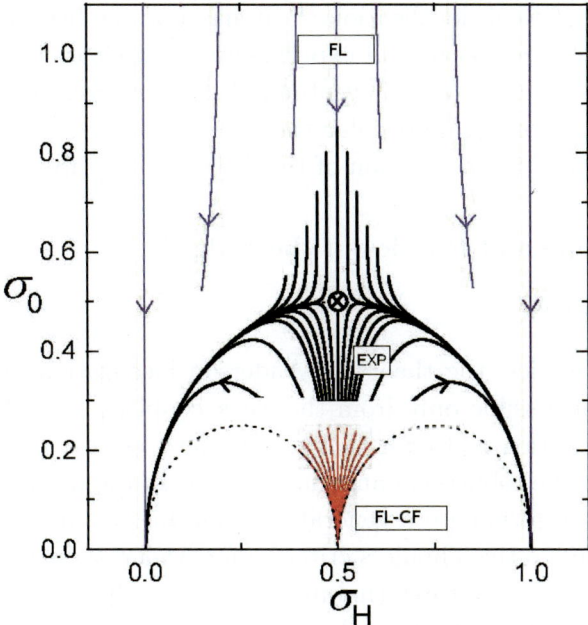

Fig. 21.10. The integral quantum Hall regime, see text.

quantum Hall regime. This regime is defined as the domain of attraction of the stable fixed points $\sigma_0 = 0$ and $\sigma_H = 0, 1$ located in the $c = 1$ plane of Fig. 21.8.

By comparing the flow lines of the $\mathcal{F}$-invariant universality class (plotted in Fig. 21.10) with those of the free particle theory (plotted in the $c = 0$ plane of Fig. 21.8) we conclude that the fixed point structure and the global features of scaling are identically the same in both cases. This remarkable feature of *super universality* prevails even though the flow lines stand for entirely different theories and physically have nothing to do with one another. For example, unlike Fig. 21.10, the strong coupling fixed point $\sigma_0 = 0$, $\sigma_H = \frac{1}{2}$ in the free particle theory can only be understood in terms of the aforementioned semi-classical picture of long ranged potential fluctuations.

3.4.3. *Dipole–dipole interaction*

Finally, it should be mentioned that an interesting way to experimentally distinguish between the $\mathcal{F}$-invariant and Fermi liquid universality classes is to place the 2DEG parallel to a metallic layer. The image charges in the metal then effectively alter the Coulomb pair potential into a dipole–dipole interaction which has a finite range. This idea has been proposed earlier in the context of weak localization.[11] It has not yet been pursued in the quantum Hall regime because of experimental complications. Nevertheless, progress in this direction would eventually help clarifying many of the issues raised in this section. In particular the different value of κ along with the possible differences in observation of the fractional quantum Hall effects.

4. The ϑ Angle and Physics of the "edge"

4.1. *Introduction*

As mentioned earlier, the theory of Anderson localization and interaction phenomena is accessible only from the weak coupling side. This raises the fundamental question as to whether or not the ϑ angle can in principle be used to explain the robustness and the precision of the quantum Hall effect.

The idea is to formulate a general criterion for the quantum Hall effect that is analogous to the Thouless criterion for Anderson localization. Recall that for problems without external magnetic fields the longitudinal conductance σ_0 at $T = 0$ can be regarded as a response parameter that probes the sensitivity of the system to changes in the boundary conditions. Provided there is a mass gap (or the states near the Fermi energy are Anderson localized) the system is insensitive to boundary conditions and hence σ_0 equals zero, apart from corrections that are exponentially small in the linear dimension of the system.

The early attempts to relate the quantum Hall effect to the appearance of a mass gap in the system were not really successful.[7,21] Unlike the usual situation where the choice of boundary conditions is immaterial, the problem dramatically changes in the presence of B. Suddenly the boundary conditions have an important topological significance and like in so many other situations, the experiment is needed in order to pursue the right physical ideas serving the correct mathematical objectives.

We have already mentioned the fact that the ϑ angle approach to the quantum Hall effect continues to be confronted with conflicting historical views in quantum field theory. The main purpose of the present section is to show that the idea of "massless chiral edge excitations" is the key for resolving some of the outstanding strong coupling problems that exist in both quantum field theory and condensed matter theory. From now on, we focus on the simplest case which is the replica field theory of free particles.

In Sec. 4.2, we briefly discuss the microscopic origins of the non-linear sigma model representation of Anderson localization as well as the problem of boundary conditions. In Secs. 4.3–4.5, we show how the massless edge excitations reveal themselves as a an unexpected universal feature of the ϑ angle concept in scale invariant theories. In Secs. 4.6 and 4.7, we show that the idea of massless chiral edge excitations quite naturally leads to the much sought-after criterion for Anderson localization in strong B. The computational results are summarized in Secs. 4.7.2 and 4.7.3.

4.2. *Problem of boundary conditions*

The non-linear sigma model representation of Anderson localization in a strong magnetic field involves the grassmannian field variable Q with $Q^2 = \mathbf{1}_{M+N}$ that can be written in a standard fashion as follows[6,21]

$$Q = T^{-1}\Lambda T. \tag{4.1}$$

Here, $T \in SU(M+N)$ and Λ denotes a diagonal matrix with M elements $+1$ and N elements -1

$$\Lambda = \begin{pmatrix} \mathbf{1}_M & 0 \\ 0 & -\mathbf{1}_N \end{pmatrix}. \tag{4.2}$$

The action is given by

$$S[Q] = S_\sigma[Q] + \pi\omega\rho \int d^2x \, \mathrm{tr}\, \Lambda Q \tag{4.3}$$

$$S_\sigma[Q] = -\frac{1}{8}\sigma_0^0 \int d^2x \, \mathrm{tr}\, \partial_\mu Q \partial_\mu Q + \frac{1}{8}\sigma_H^0 \int d^2x \, \mathrm{tr}\, \varepsilon_{\mu\nu} Q \partial_\mu Q \partial_\nu Q \tag{4.4}$$

and the number of field components (replicas) M, N are taken equal to zero in the end of all computations. The dimensionless quantities $\sigma_0^0 = \sigma_0^0(B)$ and $\sigma_H^0 = \sigma_H^0(B)$ are the earlier discussed mean field parameters for *longitudinal* conductance and *Hall* conductance respectively in units of e^2/h. The quantity $\rho = \rho(B)$ denotes the density of electronic levels and ω the external frequency.

In the original papers on the subject it was always assumed that Q is a constant matrix along the edge of the system.[7] This choice of boundary condition (which we term *spherical* boundary conditions) is equivalent to the mathematical statement that the two dimensional plane can be thought of as being compactified to a sphere S^2. The field configurations Q are then decomposed into a discrete set of topological sectors labeled by the topological invariant $\pi_2(SU(M+N)/S(U(M) \times U(N))) = Z$. The topological charge

$$\mathcal{C}[Q] = \frac{1}{16\pi i} \int d^2x \, \mathrm{tr}\, \epsilon_{\mu\nu} Q \partial_\mu Q \partial_\nu Q \tag{4.5}$$

is the integrated Jacobian of this mapping which, in fact, is equal to the integer Z.

The assumption of a quantized topological charge has far reaching consequences. It was immediately recognized, for example, that the quantum Hall effect is synonymous for $C[Q]$ being integer valued.[7,8,21] Most of the focus in the early investigations was on developing a consistent quantum theory of conductances that is based on the semi classical theory of instantons.[9] This theory originally led to the scaling diagram of Fig. 21.1 and the quantum critical behavior of the quantum Hall plateau transition as discussed in Sec. 2.2.

4.3. *Massless chiral edge excitations*

There is, however, the nagging and annoying problem that the spherical boundary conditions do not emerge from the microscopic origins of the non-linear sigma model in any obvious fashion. The effective action of Eq. (4.3) is actually defined for finite values of the symmetry breaking field ω but without any boundary conditions imposed on the Grassmannian field variable Q.[6,21] The resolution of this paradoxical situation is concealed in the peculiarities of the "edge" of the ϑ vacuum that historically have gone unnoticed.[14] There are, in fact, explicit contributions to Eq. (4.4) that originate from the diamagnetic "edge" currents of the system.[6,21] These contributions can be rewritten as "bulk" contributions by using the well known identity for the topological charge

$$\frac{1}{16\pi i} \int d^2x \, \mathrm{tr} \, \epsilon_{\mu\nu} Q \partial_\mu Q \partial_\nu Q = \frac{1}{4\pi i} \oint dx \, \mathrm{tr} \, T\partial_x T^{-1}\Lambda. \qquad (4.6)$$

To unravel the physics of the "edge" we first imagine the naive strong coupling limit obtained by placing the Fermi energy of the electron gas in a Landau gap. The mean field parameters $\sigma_0^0(B)$ and $\rho(B)$ in Eq. (4.4) are then equal to zero but $\sigma_H^0(B) = k$ with the integer k denoting the number of completely filled Landau bands. Under these circumstances one can write the action of Eq. (4.4) as a purely one dimensional theory of the "edge"

$$S[Q] \to S_{\mathrm{gap}}[Q] = \frac{k}{2} \oint dx \, \mathrm{tr} \, T\partial_x T^{-1}\Lambda + \pi\omega\rho_{\mathrm{edge}} \oint dx \, \mathrm{tr} \, \Lambda Q. \qquad (4.7)$$

We have added a term proportional to ρ_{edge} which physically stands for the density of current carrying "edge" states in the problem. Quite surprisingly, it turns out that Eq. (4.7) describes the "massless chiral edge excitations" which are a well known feature of free electrons confined in Landau bands.[14,15]

4.4. *Spontaneous symmetry breaking at the "edge"*

One can think of Eq. (4.7) in several distinctly different physical contexts. The action can be obtained as the effective replica field theory of disordered chiral electrons in one dimension described by the hamiltonian[14]

$$\mathcal{H}_{\text{edge}}^{jj'} = -iv_d\delta_{jj'}\partial_x + V_{jj'}(x). \tag{4.8}$$

The indices jj' run over k different orbitals or "edge" channels and $V = V^\dagger$ is a random hermitian potential with a Gaussian distribution. The quantity

$$v_d = \frac{k}{2\pi\rho_{\text{edge}}} \tag{4.9}$$

denotes the drift velocity of the chiral electrons.

A very different physical interpretation follows from the theory of dimerised $SU(M + N)$ quantum spin chains in $1 + 1$ space time dimension. By reading the imaginary time τ for the coordinate x then Eq. (4.7) stands for the bosonic path integral of a "dangling" edge spin with quantum number $s = k/2$ and in the presence an external magnetic field $B = v_d^{-1}$.[23,56]

In both cases it is relatively straightforward to deduce the exact expressions for all the multi point correlation functions of the matrix field Q. For example, by taking the length of the edge to infinity (or a dangling edge spin at zero temperature) then the non-vanishing one and two point correlations are evaluated to be

$$\langle Q(x)\rangle = \Lambda \tag{4.10}$$

$$\langle Q_{ab}^{+-}(x)Q_{ba}^{-+}(y)\rangle = \frac{4}{k}\theta_H(x - y - \epsilon)e^{-2\omega(x-y)/v_d}. \tag{4.11}$$

Here, the superscripts $+-$ and $-+$ denote the off-diagonal blocks of Q, θ_H is the Heaviside step function and the infinitesimal quantity ϵ defines the expectation at coinciding points x and y.

It is now clear why the "edge" has a fundamental significance in this problem: The $SU(M+N)$ symmetry is spontaneously broken and the "edge" excitations are massless or critical. Most remarkably, however, Eqs. (4.10) and (4.11) are completely independent of M and N which means that the massless edge excitations are universal. From now onward we identify the one dimensional action of Eq. (4.7) as the critical action of the quantum Hall state.

4.5. *Large k approximation*

Even though there are simple ways of deducing the exact expressions for the correlations, Eqs. (4.10) and (4.11) are nevertheless non-trivial. As an instructive example of the ambiguities inherent to the path integral formalism, we next address the semi classical theory obtained by taking s or k to infinity. This theory is a generalization of the Holstein–Primakoff representation of quantum spins. To start we rewrite the action of Eq. (4.7) as follows

$$S_{\text{gap}} = \frac{k}{2} \oint dx \ tr \ (\Lambda + \mathbf{1}_{M+N}) T \left(\partial_x + \omega v_d^{-1} \Lambda\right) T^{-1}. \tag{4.12}$$

The addition of the unit matrix implies that nothing depends on the overall phase of the matrix T. Equation (4.12) only depends on the the first M columns of T^{-1} and the first M rows of T. We represent these rows and columns as follows

$$\begin{bmatrix} W & V \end{bmatrix}, \qquad \begin{bmatrix} W^\dagger \\ V^\dagger \end{bmatrix}. \tag{4.13}$$

The blocks W and $W^\dagger$ are of size $M \times M$, the matrix $V^\dagger$ is of size $N \times M$ and V is of size $M \times N$. The matrices W and V are constrained by

$$WW^\dagger + VV^\dagger = \mathbf{1}_M. \tag{4.14}$$

It is convenient to introduce the change of variables $W = UP$ with $U \in U(M)$ and $P = P^\dagger$ given by

$$P = P^\dagger = \sqrt{\mathbf{1}_M - VV^\dagger}. \tag{4.15}$$

The matrix fields $Q = T^{-1} \Lambda T$ depend on the $M \times N$ independent complex fields V alone

$$Q = \begin{pmatrix} \mathbf{1}_m - 2VV^\dagger & 2\sqrt{\mathbf{1}_m - VV^\dagger}\ V \\ \\ 2V^\dagger \sqrt{\mathbf{1}_m - VV^\dagger} & -\mathbf{1}_n + 2V^\dagger V \end{pmatrix}. \tag{4.16}$$

By ignoring the constants we obtain the following simple form of the action

$$\begin{aligned} S_{\text{gap}} &= k \oint \ tr \ \left[UP(\partial_x + v_d^{-1})PU^\dagger + V(\partial_x - \omega v_d^{-1})V^\dagger \right] \\ &= k \oint \ tr \ U\partial_x U^\dagger + k \oint \ tr \ V(\partial_x - 2\omega v_d^{-1})V^\dagger. \end{aligned} \tag{4.17}$$

Notice that the new variables U and V represent the integral and fractional topological sectors respectively. The U produces a trivial phase factor and can be ignored. The massless "edge" excitations are solely described by the matrix field variable V which carries a fractional topological charge. The V integrals are simply gaussian in the limit $k \to \infty$. To define the correlations at a coinciding position x we introduce a slight redefinition of the mass term

$$S = k \oint dx \sum_{a=1}^{M} \sum_{b=1}^{N} \left[V_{ab}(x) \partial_x V_{ab}^*(x) - 2\omega v_d^{-1} V_{ab}(x + \epsilon) V_{ab}^*(x) \right] \quad (4.18)$$

with ϵ positive and infinitesimal. In the limit of large perimeters one can write the propagator as follows

$$\langle V_{ab}(x) V_{ab}^*(0) \rangle = \frac{1}{k} \theta_H(x - \epsilon) e^{-2\omega x/v_d}. \quad (4.19)$$

Based on Eqs. (4.16) and (4.19) we obtain Eqs. (4.10) and (4.11) directly from the $k = \infty$ limit. The corrections are zero, order by order in a systematic expansion in powers of $1/k$. This peculiarity of the large k or large s expansion was noticed first in Ref. 56. It was shown, in particular, that the Haldane mapping of dimerised quantum spin chains displays similar features. This mapping is standardly formulated for the theory with $s = \infty$ but the corrections are zero to all orders in $1/s$.

4.6. *Separating the "bulk" from the "edge"*

Since the massless edge excitations are physically very different from those in the "bulk" of the system they should be disentangled and studied separately. This can be done as follows. We introduce a fixed background field $t \in SU(M + N)$ and slightly redefine the general theory of Eq. (4.4) according to

$$Z(t) = \exp\{-\mathcal{F} + \mathcal{S}'_\sigma[t]\}$$
$$= \int_{\partial V} \mathcal{D}[Q_0] \exp \left\{ S_\sigma[t^{-1} Q_0 t] + \pi \omega \rho \int d^2 x \, \mathrm{tr} \, \Lambda Q_0 \right\}. \quad (4.20)$$

Here, the subscript ∂V indicates that the functional integral over Q_0 is performed with the boundary conditions $Q_0 = \Lambda$ at the edge. The matrix field t generally stands for the fluctuations about these special boundary conditions, i.e. $q = t^\dagger \Lambda t \neq \Lambda$ at the edge.

The theory of Eq. (4.20) has a distinctly different physical meaning dependent on the order of limits $\omega \to 0$ and $L \to \infty$ respectively with L denoting the linear dimension of the system.

(1) If one takes the limit $L \to \infty$ first then the boundary conditions on the matrix field Q_0 are immaterial. The background field t can now be

taken as the "source term" that generates the Kubo formula for the AC conductances σ_0 and σ_H. This aspect of replica field theory is extremely well known and has frequently been discussed earlier.[10,29]

(2) On the other hand, if $\omega \to 0$ is taken first then the boundary condition on the Q_0 is all important. The matrix field Q_0 now represents the "bulk" excitations which carry a strictly integral topological charge. By the same token one recognizes the matrix field t as the fractional topological sector describing the massless "edge" excitations in the problem. The action $S'_\sigma(t)$ in Eq. (4.20) is the effective action of "edge" excitations and it is understood that in the end one still has perform the integration over the "edge" field variable t.[29,57]

4.7. *Thouless criterion*

The explicit form of $S'_\sigma[t]$ in Eq. (4.20) is of primary physical significance since it contains all the information on the low energy dynamics of the system. Assuming that t obeys the classical equations of motion then one can write

$$S'_\sigma[t] = -\frac{\sigma_0}{8} \int d^2x \, \mathrm{tr} \, \partial_\mu q \partial_\mu q + \frac{\sigma_H}{8} \int d^2x \, \mathrm{tr} \, \epsilon_{\mu\nu} \partial_\mu q \partial_\nu q \qquad (4.21)$$

with $v = t^{-1}\Lambda t$. Equation (4.21) is the only possible local action with at most two derivatives that is compatible with the symmetries of the problem. The quantities of physical interest are $\sigma_{0,H} = \sigma_{0,H}(\omega, L)$ which in the replica limit $M, N \to 0$ precisely correspond to the Kubo expressions for the macroscopic conductances averaged over the impurity ensemble. However, the result of Eq. (4.21) has a general significance which is independent of M and N. One therefore expects that the "conductances" have the same physical meaning for all non-negative values of M and N with the replica limit only playing a role of secondary significance. Without going into much detail we briefly mention the following steps that lead directly to the Thouless criterion for Anderson localization in strong B.

4.7.1. *Quantum Hall effect*

In order to see in which way σ_0 and σ_H for $\omega = 0$ are a measure for the sensitivity of the system to changes in the boundary conditions we split the mean field parameter $\sigma_H^0 = \phi$ into an integral piece $k(\phi)$) and a fractional piece $-\pi < \theta_0(\phi) \leq \pi$ (see Fig. 21.11)[58]

$$\sigma_H^0 = \phi = k(\phi) + \frac{\theta_0(\phi)}{2\pi} . \qquad (4.22)$$

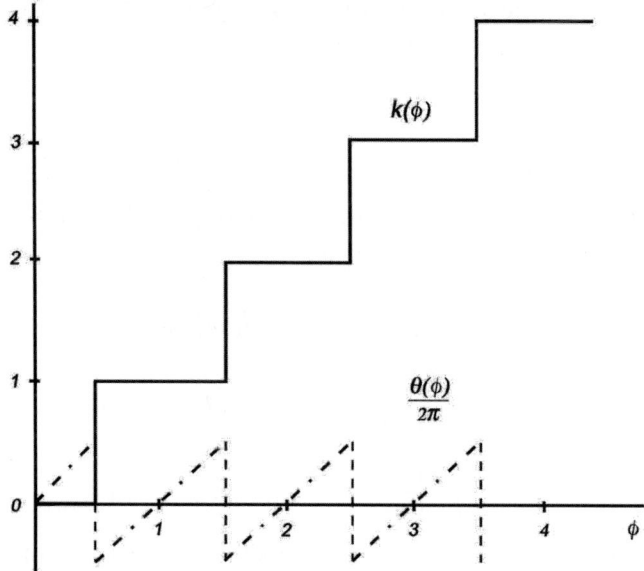

Fig. 21.11. The integral part $k(\phi)$ and the fractional part $\frac{\theta_0(\phi)}{2\pi}$ of the mean field Hall conductance $\sigma_H^0 = \phi$ with varying ϕ, see text.

Since the underlying theory of the Q_0 field is insensitive to the integer piece $k(\phi)$ we split Eq. (4.21) into an "edge" term and a "bulk" term according to

$$\mathcal{S}'_\sigma[t] = \frac{k(\phi)}{2} \oint \operatorname{tr} t \partial_x t^\dagger \Lambda - \frac{\sigma_0}{8} \int \operatorname{tr} \partial_\mu q \partial_\mu q + \frac{\theta}{16\pi} \int \operatorname{tr} \varepsilon_{\mu\nu} \partial_\mu q \partial_\nu q. \tag{4.23}$$

When written in this form it becomes obvious that the fractional piece θ is a probe for localization in the "bulk" of the system whereas the integer $k(\phi)$ describes *de*-localization or criticality at the "edge." Discarding corrections that are exponentially small in L, the criterion for Anderson localization in strong B is therefore

$$\sigma \to 0 \;, \quad \sigma_H = k(\phi) + \frac{\theta}{2\pi} \to k(\phi). \tag{4.24}$$

So provided there is a mass gap in the "bulk" of the system, the longitudinal conductance vanishes and the Hall conductance is robustly quantized. Under these circumstances, Eq. (4.23) stands for the critical action of the quantum Hall state describing massless chiral excitations along the edges. Notice that the criterion must break down at the points where $\sigma_H^0 = \phi$ is half odd-integral. At these points a transition takes place between adjacent quantum Hall plateaus and gapless excitations must therefore exist.

4.7.2. *Computational results*

For completeness we briefly summarize the results of explicit computations of σ_0 and θ on the weak coupling side. These quantities can be expressed in terms of the Noether current $J_\mu = Q_0 \partial_\mu Q_0$[9]

$$\sigma_0 = \sigma_0^0 \quad + \frac{(\sigma_0^0)^2}{8MN\Omega} \int_{x,x'} \langle \operatorname{tr} J_\mu(x) J_\mu(x') \rangle_0 \tag{4.25}$$

$$\theta = \theta_0(\phi) + \frac{\pi(\sigma_0^0)^2}{8MN\Omega} \int_{x,x'} \langle \operatorname{tr} \varepsilon_{\mu\nu} J_\mu(x) J_\nu(x') \Lambda \rangle_0$$

$$- \frac{\pi(M+N)\sigma_0^0}{4MN\Omega} \int_x \langle \operatorname{tr} \varepsilon_{\mu\nu} x_\mu J_\nu(x) \Lambda \rangle_0 \tag{4.26}$$

Here, the expectation $\langle \dots \rangle_0$ is with respect to the theory of Q_0 and the integrals are over an area of size Ω. The results for Eqs. (4.25) and (4.26) can be written in terms of β functions according to

$$\frac{d\sigma_0}{d\ln b} = \beta_0(\sigma_0, \theta) \ , \quad \frac{d\theta}{d\ln b} = \beta_\theta(\sigma_0, \theta). \tag{4.27}$$

Since the topological charge $\mathcal{C}[Q_0]$ is quantized we expand the β functions as an infinite trigonometric series in θ.[21] The explicit results for the lowest order terms are as follows[57]

$$\beta_0 = -\frac{M+N}{2\pi} - \frac{MN+1}{2\pi^2\sigma_0} - \mathcal{D}_{MN}(\sigma_0)\sigma_0^2 \exp(-2\pi\sigma_0) \cos\theta \tag{4.28}$$

$$\beta_\theta = -2\pi\mathcal{D}_{MN}(\sigma_0)\sigma_0^2 \exp(-2\pi\sigma_0) \sin\theta. \tag{4.29}$$

Here,

$$\mathcal{D}_{MN}(\sigma_0) = \left(\frac{16\pi}{e}\right) \frac{(2\pi\alpha\sigma_0)^{M+N}}{\Gamma(1+M)\Gamma(1+N)} \tag{4.30}$$

with $\alpha = 2\exp(-\gamma_E - 3/2)$ and $\gamma_E \approx 0.577$ the Euler constant.

4.7.3. *Discussion*

Equations (4.28)–(4.30) reveal fundamental features of the ϑ vacuum that were not obtainable from the ordinary theory of instantons.[60] In accordance with the criterion of Eq. (4.24) one expects that the system for all values of $M, N \geq 0$ will behave like an ordinary "metal" at short distances, and a quantum Hall "superconductor" (or a quantum Hall "insulator") in the limit where the scale size b goes to infinity.

The instanton pieces of Eqs. (4.28)–(4.30) have had a long and difficult history. They have originally been proposed in Ref. 8 and explained in a heuristic manner in Ref. 59. On the basis of the Kubo formula for the conductances they had later been computed explicitly in Ref. 9. They were re-evaluated again at a much later stage after it became clear that the massless edge excitations in the problem remove all the aforementioned ambiguities in the definition of the topological charge $\mathcal{C}[Q_0]$.

The concept of super universality that emerged sparked a stream of detailed investigations on both the weak and strong coupling side of the problem. The instanton methodology in particular, has been extended in several different ways. For example, by using the appropriate regularization scheme for both the perturbative and non-perturbative parts of the β functions, the correct numerical pre-factor in Eq. (4.30) was obtained which is in fact universal.[57] Moreover, as a major advance in the renormalization group computational technique, the methodology of "spatially varying masses" was introduced which provides the instanton corrections to the anomalous dimension of the composite operators of the matrix field Q_0. These advances have been instrumental in establishing the non-perturbative features of the much more ambitious URT as discussed earlier in this chapter.

The primary focus so far has been on the theory with small values of M and N. In accordance with the general ideas of the pioneering papers,[8,59] Eqs. (4.28) and (4.29) then display a quantum critical fixed point at $\theta = \pm\pi$ with σ_0^* of order unity. Figure 21.12 shows that the critical exponent ν varies continuously with varying values of $M = N$. As mentioned earlier, the value $\nu = 2.75$ obtained for $M = N = 0$ agrees very well with the results from numerical simulations. This remarkable numerical agreement includes the multi fractal singularity spectrum of the electron gas. As indicated in Fig. 21.12 we expect that the exact results for ν smoothly interpolate the free electron value $\nu \approx 2.6$ at $M = N = 0$ and the known value $\nu = \frac{2}{3}$ of the $O(3)$ model which corresponds to $N = M = 1$. Furthermore, the results show that the multi fractal aspects disappear from the theory in a continuous fashion as one moves away from the limit $N = M = 0$. We refer to the original papers for more detailed discussions of the replica limit[9] and a comparison of the analytic results with those of numerical simulations.[57]

Equations (4.28) and (4.29) do not display a critical fixed point with a finite σ_0^* for all values of M, N larger than unity. However, a major advantage of the grassmannian theory is that it has a tractable large N expansion. This means that the strong coupling features of the theory can be explored and investigated in great detail, in particular Eq. (4.24) and the physics of

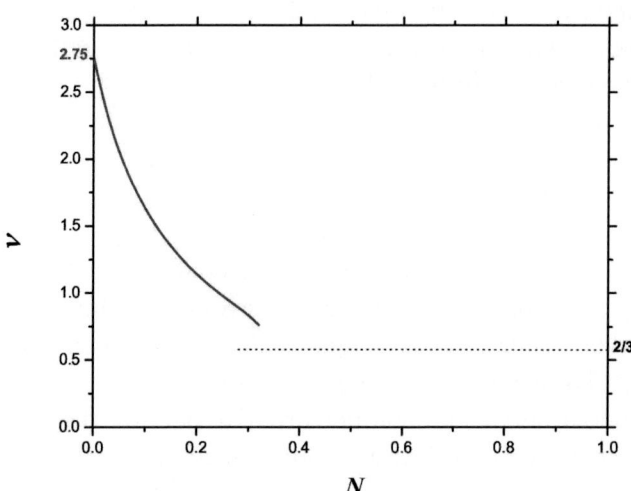

Fig. 21.12. The critical exponent ν with varying values of $M = N$. The solid curve with $0 \leq N \leq 0.3$ is obtained from the β functions of Eqs. (4.28) and (4.29) near the critical fixed point at $\theta = \pi$, see text.

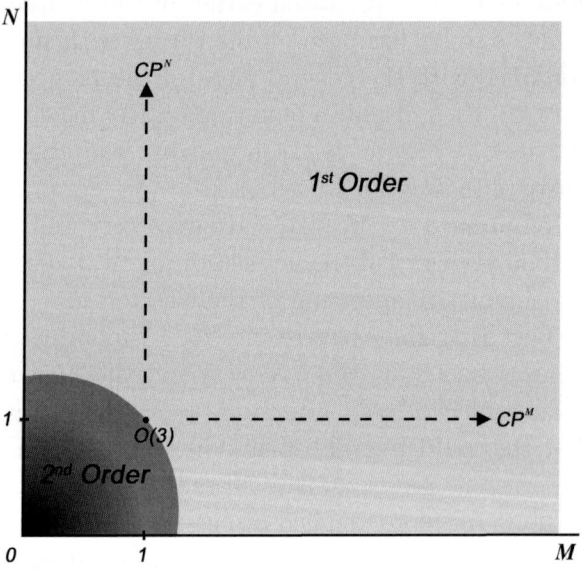

Fig. 21.13. The critical behavior at $\theta = \pi$ for different values of M and N. The theory displays a second order transition with varying exponents $\frac{2}{3} < \nu < 2.75$ for small values of M, N as well as multifractality in the limit where M and N go to zero. For larger values of M, N one expects a first order transition with $\nu = \frac{1}{2}$, see text.

de-localization at $\theta = \pi$. In anticipation of the results of Sec. 5 we expect that the critical fixed point for $M, N > 1$ lies in the strong coupling regime ($\sigma_0^* = 0$) describing a continuous quantum Hall plateau transition with a critical exponent $\nu = \frac{1}{2}$, even though the transition is formally a first order one. Figure 21.13 gives an overview of the expected critical behavior at $\theta = \pi$ for different values of M and N.

5. The CP^{N-1} Model with Large N

5.1. *Introduction*

Large N methods for grassmannian non-linear sigma models have been introduced in Ref. 61. Here we briefly review the large N expansion of the CP^{N-1} model which is obtained by substituting $M \to 1$ and $N \to N - 1$ in the grassmannian $SU(M + N)/S(U(M) \times U(N))$ theory. In terms of a complex vector field z_α

$$Q_{\alpha\beta} = 2z_\alpha^* z_\beta - \delta_{\alpha\beta} \quad , \quad \sum_{\alpha=1}^{N} z_\alpha^* z_\alpha = 1 \tag{5.1}$$

one can write the action as follows

$$S[Q] = \int d^2x \left[-\sigma_0^0 \left(\partial_\mu z_\alpha^* \partial_\mu z_\alpha + z_\alpha^* \partial_\mu z_\alpha z_\beta^* \partial_\mu z_\beta \right) + \sigma_H^0 \, \varepsilon_{\mu\nu} \partial_\mu z_\alpha^* \partial_\nu z_\alpha \right]. \tag{5.2}$$

This theory is usually taken in $1 + 1$ space-time dimension. By introducing a vector potential A_μ

$$S[Q, A_\mu] = \int d^2x \left[\sigma_0^0 z_\alpha^* D_\mu^2 z_\alpha + i\sigma_H^0 \varepsilon_{\mu\nu} \partial_\mu A_\nu \right] \tag{5.3}$$

with $D_\mu = \partial_\mu + iA_\mu$ denoting the covariant derivative, then the CP^{N-1} model becomes the finite temperature theory of N relativistic scalar particles in one spatial dimension, strongly interacting with a $U(1)$ gauge field and in the presence of a background electric field proportional to σ_H^0.

Within the large N steepest descend methodology the z_α particles acquire a mass $M_0 \propto e^{-2\pi/g_0}$ with $g_0 = N/\sigma_0^0$ a rescaled coupling constant. This leads to an effective action for the A_μ field alone[25]

$$S[A_\mu] = -\frac{N}{48\pi M_0^2} \int F_{\mu\nu}^2 + 2\pi i \sigma_H^0 \, C[A_\mu] \tag{5.4}$$

and the new expression for the topological charge is

$$C[A_\mu] = \frac{1}{2\pi} \int d^2x \, \varepsilon_{\mu\nu} \partial_\mu A_\nu. \tag{5.5}$$

So besides the dynamically generated mass M_0 there is also an emerging free electromagnetic field that was missing in the original theory of Eq. (5.3).

Without an explicit knowledge of the massless edge excitations in the problem, one would not anticipate that something has gone wrong in the interim steps. For example, in the original problem of Eq. (5.2) the $SU(N)$ symmetry is spontaneously broken at the "edge" of the system. However, in the final theory of Eq. (5.4) this phenomenon is completely absent. This setback has been totally overlooked in the original papers.[25,26] This has dramatic consequences. In particular, it was always assumed that the A_μ in Eq. (5.4) is a free field and $\mathcal{C}[A_\mu]$ unconstrained or *un*-quantized. For example, the "free energy" is evaluated to be

$$\mathcal{F} = -\ln \int \mathcal{D}[A_\mu] e^{S[A_\mu]} = \frac{12\pi\beta L M_0^2}{N} \left(\sigma_H^0\right)^2 \tag{5.6}$$

which lacks periodicity in σ_H. To restore the periodicity in σ_H^0 one has to argue in a heuristic manner that the theory is actually metastable for values of σ_H outside the range $-\frac{1}{2} < \sigma_H^0 < \frac{1}{2}$. Coleman's electrodynamics picture, in particular, leads us to believe that the system materializes a pair of static charges ("quark" and "antiquark") as σ_H^0 passes through $\pm 1/2$. These charges "move" to the opposite "edges" of the universe such as to maximally shield the background electric field σ_H^0.[63]

5.2. *Step 1: Retrieving the quantum Hall effect*

5.2.1. *Spin-charge separation*

Following the idea of massless edge excitations, however, one concludes that the $\mathcal{C}[A_\mu]$ in Eq. (5.4) is not unconstrained. Instead, one must physically distinguish between the integral and fractional pieces. The simplest way to do this by imposing the following constraint on the topological charge

$$\mathcal{C}[A_\mu] = n + \mathcal{C}[q]. \tag{5.7}$$

Here, n stands for the integral piece of the topological charge and $-\frac{1}{2} < \mathcal{C}[q] \leq \frac{1}{2}$ denotes the fractional piece with q standing for the matrix field variable of the "edge". Equation (5.7) leads to a slightly more complex theory that involves both the field variables A_μ and q. Introducing a sum over integers m one can write

$$Z[q] = \sum_m \int D[A_\mu] \, e^{S[A_\mu, q, m]} \tag{5.8}$$

where the new action is given by

$$S[A_\mu, q, m] = -\frac{N}{48\pi M_0^2} \int F_{\mu\nu}^2 + 2\pi i (\sigma_H^0 - m) \mathcal{C}[A_\mu] + 2\pi i m \mathcal{C}[q]. \quad (5.9)$$

The field A_μ is now free and can be eliminated

$$Z[q] = e^{-\mathcal{F}_0 + S_\sigma'[q]} \quad (5.10)$$

In the limit $\beta, L \to \infty$ one finds that the "bulk" free energy $\mathcal{F}_0$ is equal to

$$\mathcal{F}_0 = \frac{12\pi\beta L M_0^2}{N} \left(\frac{\theta_0(\phi)}{2\pi} \right)^2 = \frac{12\pi\beta L M_0^2}{N} \left(\sigma_H^0 - k(\phi) \right)^2 \quad (5.11)$$

with $\theta(\phi)$ and $k(\phi)$ defined by Eq. (4.22). The effective action of the q field is given by

$$S_\sigma'[q] = 2\pi i k(\phi) \mathcal{C}[q]. \quad (5.12)$$

So by separating the "bulk" degrees of freedom from "edge" excitations we obtain a lower free energy of the "metastable" phase $|\sigma_H^0| > \frac{1}{2}$ relative to the naive result of Eq. (5.6). Equation (5.11) is, in fact, precisely in accordance with Coleman's electrodynamics picture. We find, in particular, that Eq. (5.11) is periodic in $\sigma_H^0 = \phi$ with a sharp "cusp" or a first order phase transition at $\phi^* = k + \frac{1}{2}$. The quantity $k(\phi)$ in Eq. (5.11) is identified as the static electric field that, in Coleman's language, arises from the "quark" and "antiquark" located at the opposite sides of the universe.

However, Eq. (5.12) now shows that Coleman's particles also carry spin degrees of freedom that spontaneously break the $SU(N)$ symmetry at the "edges" of the universe. In the context of the electron gas we recognize Eq. (5.12) as the *critical* fixed point action of the quantum Hall state with $\sigma_0 = 0$ and $\sigma_H = k(\phi)$. The Hall conductance is quantized with a sharp plateau transition as ϕ passes through one of the critical values $\phi^* = k + 1/2$. This quantization phenomenon did not exist within the historical "large N picture" of the ϑ angle, simply because the topological charge was interpreted incorrectly.[29]

5.2.2. *Finite size scaling*

An important conclusion that one can draw from all this is that Coleman's scenario of "quarks" and "antiquarks" does not stand on its own as previously thought. Instead, Coleman's ideas are a direct consequence of the fact that the topological charge of "bulk" excitations is strictly quantized. This quantization is precisely in accordance with the expectations based on the semi classical theory of instantons. The cross-over between the weak coupling results of Eqs. (4.28)–(4.30) and the large N saddle point results has been discussed only recently.[29] It should be mentioned that similar ideas have been pursued a long time ago in the seminal work of Jevicki.[64]

We can carry the analysis of the previous section one step further. By taking the linear dimensions β and L to be finite then the sharp transitions at $\phi^* = k + \frac{1}{2}$ are smoothened out. Instead of Eq. (5.12) we find the following more general result

$$S'_\sigma[q] = 2\pi i \sigma_H \mathcal{C}[q] - \frac{1}{2}\zeta_2 \mathcal{C}^2[q] \tag{5.13}$$

Near the critical point the Hall conductance σ_H, the variance ζ_2 and the higher order cumulants can all be written as functions of the single scaling variable X

$$X = \frac{24\pi}{N}\beta L M_0^2 (\phi - \phi^*). \tag{5.14}$$

The explicit scaling functions are

$$\sigma_H = \phi^* + F_H(X) \quad , \quad \zeta_2 = \zeta_2(X) \tag{5.15}$$

with

$$F_H(X) = \frac{1}{2}\tanh X \quad , \quad \zeta_2(X) = \frac{\partial F_H}{\partial X} = \frac{1}{2} - 2F_H^2(X). \tag{5.16}$$

Equations (5.14)–(5.16) are an explicit demonstration of the scaling results discussed in Sec. 2.2. The main difference is that the *de*-localization now occurs at the critical value $\sigma_0^* = 0$ rather than $\sigma_0^* = \mathcal{O}(1)$. The large N theory has a diverging correlation length ξ according to

$$\xi \propto M_0^{-1}|\phi - \phi^*|^{-\nu} \tag{5.17}$$

and the critical exponent ν equals $\frac{1}{2}$. To appreciate the significance of these results in the context of the disordered free electron gas we recall that σ_H or F_H is the Hall conductance *averaged* over the impurity ensemble. The ζ_2 and the higher order cumulants play a role similar to the conductance *fluctuations* or the conductance *distribution* in disordered metals. The Hall conductance is therefore broadly distributed at the critical point. In the quantum Hall plateau regime both the corrections to exact quantization and the conductance fluctuations render exponentially small in the *area* βL. The large N theory is therefore the much sought after example where all the strong coupling features of the quantum Hall effect can be studied exactly.

The finite size scaling results are in accordance with the general ideas on Anderson localization in Sec. 4.7. However, based on exponential localization one expects that the corrections to exact quantization are exponentially small in the *linear dimension* of the system rather than the *area*. We will return to this issue at the end of this chapter.

5.3. *Step 2: Sine–Gordon model*

5.3.1. *Introduction*

To connect the known results with the theory on the weak coupling side one needs to include the "longitudinal conductance" σ_0 in the results of finite size scaling. Given the standard large N saddle point formalism there is no simple way of doing this.[29]

We now present the results of several formal steps necessary for a discussion of the large N steepest descend methodology for finite size systems. These steps are important because they reveal multiple ways of establishing the differences between the integral and fractional topological sectors of the ϑ vacuum. Furthermore, they show that the quantity σ_0 has an interesting geometrical significance and expose the leading corrections to exact quantization of the Hall conductance. The principal results of these steps are summarized in Sec. 5.4.3 where we give the scaling functions for the quantities σ_0, σ_H and ζ_2.

5.3.2. *Finite β*

As a first logical step toward finite size systems we consider finite temperature field theory in $1+1$ space-time dimension or, equivalently, an infinite cylindrical geometry in two spatial dimensions. This step has already been taken a long time ago[26] but unfortunately, the theory has been incorrectly treated and incorrect conclusions were drawn. The most important effect of finite temperatures on Eq. (5.4) is the appearance of a mass term in the static component Φ of the $A_0(x, \tau)$ field

$$\Phi(x) = \int_0^\beta d\tau\, A_0(x, \tau). \tag{5.18}$$

The effective action can be written as follows[65]

$$S[A_\mu] = -N \int_0^\beta d\tau \int_{-\infty}^\infty dx \left[\frac{1}{48\pi M_0^2} F_{\mu\nu}^2 + \frac{2z}{\beta^2}(1 - \cos \Phi) \right] + 2\pi i \sigma_H^0 \, \mathcal{C}[A_\mu]. \tag{5.19}$$

The dimensionless quantity z is exponentially small in β and computed to be

$$z = \sqrt{\frac{\beta M_0}{2\pi}} e^{-\beta M_0}. \tag{5.20}$$

The different frequency components of A_μ in Eq. (5.19) are all decoupled and only the zero frequency sector depends on z and σ_H^0. Therefore, the only field variable of interest is the static component Φ. The effective action is therefore the sine–Gordon action in one dimension

$$S[\Phi] = -N \int_{-\infty}^{\infty} dx \left[\frac{1}{48\pi\beta M_0^2} \partial_x \Phi \partial_x \Phi + 2\frac{z}{\beta}(1 - \cos\Phi) \right] + 2\pi i \sigma_H^0\ C[\Phi] \tag{5.21}$$

where the topological charge now reads

$$C[\Phi] = \frac{1}{2} \int_{-\infty}^{\infty} dx\ \partial_x \Phi. \tag{5.22}$$

5.3.3. Finite action

The sine–Gordon theory is interesting in and of itself. It is a toy model for instantons[66] (also termed "calorons" or "quantum instantons"[26]). The theory, as we shall further discuss below, permits exact solutions in the regime of physical interest.

The simplest way to proceed is by expanding the sine–Gordon theory as an infinite series in powers of z. This would standardly lead to the classical Coulomb gas representation in one dimension. However, in order to define the sine–Gordon theory in the limit $z = 0$ one must introduce a finite system size L and specify the boundary conditions on the field Φ. A natural way to do this is to fix the field Φ in Eq. (5.21) at one of its classical values ($\cos\Phi = 1$) everywhere *outside* a given area of size L and let it freely fluctuate *inside* the area, say $0 < x < L$. The sine–Gordon action is then non-zero only for $0 < x < L$ such that the limit $z = 0$ can be taken and is finite. Therefore, we define Eq. (5.21) as an integral over all paths $\Phi(x)$ that start from $\Phi = 2\pi n_l$ at $x = 0$ and end with $\Phi = 2\pi n_r$ at $x = L$. Since one can shift the integers n_l and n_r by an arbitrary integer, one can express the theory as a sum over all topological sectors n according to

$$Z = \sum_n Z(n) \tag{5.23}$$

where

$$Z(n) = e^{2\pi i \sigma_H^0 n} \int_{\substack{\Phi(\lambda) = 2\pi n \\ \Phi(0) = 0}} \mathcal{D}\Phi(x)\ e^{S_0[\Phi]}. \tag{5.24}$$

The action S_0 is given by

$$S_0[\Phi] = -N \int_0^\lambda dx \left[\partial_x \Phi \partial_x \Phi + \frac{2zc}{\lambda}(1 - \cos \Phi) \right]. \qquad (5.25)$$

We have introduced a scale factor λ and a geometrical factor c according to

$$\lambda = 48\pi \beta L M_0^2 \ , \quad c = \left(\frac{L}{\beta}\right). \qquad (5.26)$$

It is now clear why the infinitesimal sine–Gordon mass z has a fundamental significance: The requirement of finite action immediately implies that the topological charge of a finite size system is quantized.

Notice that Eqs. (5.24)–(5.25) and Eq. (5.9) define two different series expansions about the the same theory. Whereas in the first case we expanded in the fractional topological sectors, in the second case the expansion is in powers of z. The idea is to then combine these two different expansion procedures and find the finite size scaling behavior in the presence of z.

It can be shown that the expansion in powers of z leads to an elegant Coulomb gas representation in one dimension where both fractional and integral charges appear at the "edges".[67] This procedure is somewhat laborious, however, and we now focus on more effective ways of handling z.

5.3.4. *Hamiltonian formalism*

Path integrals like Eq. (5.24) are easiest evaluated in the hamiltonian formalism. If one reads the imaginary time τ for the coordinate x then the hamiltonian of the sine–Gordon theory is given by

$$\mathcal{H}(\sigma_H^0) = \beta M^2 \left(-i \frac{\partial}{\partial \Phi} - \sigma_H^0 \right)^2 - 2N \frac{z}{\beta} \cos \Phi \qquad (5.27)$$

where

$$M^2 = \frac{24\pi M_0^2}{N}. \qquad (5.28)$$

Like before, the $\cos \Phi$ term in Eq. (5.27) is all important. Following Bloch's theorem for periodic potentials we seek solutions of the form

$$\mathcal{H}(\sigma_H^0)\psi_m(\Phi) = E_m(\sigma_H^0)\psi_m(\Phi). \qquad (5.29)$$

The wave functions $\psi_m(\Phi) = \psi_m(\Phi + 2\pi)$ are periodic in Φ and the eigenstates are labeled by σ_H^0. One may think of Eq. (5.27) in terms of a particle on a circle with σ_H^0 standing for the magnetic flux threading through the circle.

The problem is readily solved for $z = 0$ and the result is

$$E_k^0(\sigma_H^0) = \beta M^2 (k - \sigma_H^0)^2 \quad, \quad \psi_k^0(\Phi) = e^{ik\Phi}/\sqrt{2\pi}. \tag{5.30}$$

Notice that the energy levels E_k^0 and E_{k+1}^0 cross one another at $\sigma_H^0 = k+1/2$. This level crossing explains why the "bulk" excitations at $\sigma_H^0 = k + 1/2$ are gapless. A band splitting occurs for finite values of z which can be dealt with using ordinary perturbation theory.

Let us first consider the quantum mechanical partition function Z which is easily computed to be

$$Z = \int_{-\pi}^{\pi} \frac{d\Phi}{2\pi} \, \langle \Phi | e^{-L\mathcal{H}} | \Phi \rangle = \sum_k e^{-LE_k(\sigma_H^0)}. \tag{5.31}$$

The Feynman path integral representation is standard and given by[66,68]

$$Z = \sum_n Z(n) \tag{5.32}$$

with

$$Z(n) = e^{2\pi i \sigma_H^0 n} \int_{\Phi(\lambda) = \Phi(0) + 2\pi n} \mathcal{D}\Phi(x) \, e^{S_0[\Phi]}. \tag{5.33}$$

The action S_0 is the same as in Eq. (5.25). Just like the finite action principle employed in the previous section, the hamiltonian formalism directly leads to a finite size system with a quantized topological charge.

For completeness we mention that the path integral of Eq. (5.24) can also be written in the hamiltonian form. The result is

$$Z = \sum_{k,k'} \langle \psi_k^0 | e^{-L\mathcal{H}} | \psi_{k'}^0 \rangle \tag{5.34}$$

where ψ_k^0 are the eigen states of the unperturbed problem. Equations (5.31) and (5.34) are identically the same for $z = 0$. The two theories are slightly different for $z \neq 0$, however.

5.3.5. *Edges or no edges ...*

To understand the subtle differences we express Eqs. (5.24) and (5.33) in terms of the $O(2)$ vector field variable $\mathbf{v} = (\cos \Phi, \sin \Phi)$ rather than Φ. We obtain the classical $X - Y$ model in one dimension

$$Z = \int_{\partial v} D[\mathbf{v}] \, e^{S[\mathbf{v}]} \tag{5.35}$$

where

$$S(\mathbf{v}) = -\int_0^\lambda \left\{ N \left(\partial_x \mathbf{v} \cdot \partial_x \mathbf{v} - \mathbf{h} \cdot \mathbf{v} \right) + i\sigma_H^0 \epsilon_{\mu\nu} v_\mu \partial_x v_\nu \right\}. \tag{5.36}$$

The "magnetic field" $\mathbf{h}$ equals $2\frac{zc}{\lambda}(1,0)$ and the subscript ∂v indicates the different boundary conditions imposed on the functional integral.

By choosing the boundary conditions

$$\mathbf{v}(0) = \mathbf{v}(\lambda) = (1,0) \tag{5.37}$$

then we obtain the theory of Eq. (5.24). The theory clearly has "edges" and the fluctuations about Eq. (5.37) should be identified with the dangling "edge" spins or massless chiral "edge" excitations at $x = 0$ and $x = L$ respectively. This situation physically applies to the dimerised quantum spin chain in $1 + 1$ dimension and the electron gas in $2 + 0$ dimension.

If, on the other hand, we choose periodic boundary conditions

$$\mathbf{v}(0) = \mathbf{v}(\lambda). \tag{5.38}$$

then we obtain the quantum mechanical partition function of Eq. (5.33). There are no physical "edges" in imaginary time quantum statistics and the fractional topological sectors now correspond to external fields that take the quantum system out of thermal equilibrium. This situation is physically realized in dimerised spins chains with periodic boundary conditions[23,56] and also the Ambegaokar–Eckern–Schön theory of the Coulomb blockade in $0+1$ dimension.[69,70]

We have thus found two very different physical situations in which the quantization of topological charge manifests itself. However, the theory displays the same macroscopic quantization phenomena in both cases. For example, it has recently been shown that the Ambegaokar–Eckern–Schön theory displays "macroscopic charge quantization" which in all respects is the same quantum phenomenon as the quantum Hall effect observed in completely different physical systems.[69,70]

5.4. Step 3: Fractional topological sectors

5.4.1. Introduction

We are now well equipped to embark on the third and final step which is to evaluate the effective theory of the fractionally charged excitations. Since the final answer does not strongly depend on which of the two "bulk" theories one considers we now present the results obtained from the hamiltonian formalism of Sec. 5.3.4.

By expanding in the generators of the Grassmannian q it follows after some elementary algebra that the general form of the partition function can be written as follows

$$Z[q] = Z e^{S'_\sigma[q]}. \tag{5.39}$$

Here, Z denotes the "bulk" theory of Eq. (5.31) and S'_σ is given by

$$S'_\sigma[q] = -\frac{1}{8} \int_0^L dx \int_0^\beta d\tau \, \mathrm{tr} \left[\sigma_{xx} \partial_x q \partial_x q + \sigma_{00} \partial_0 q \partial_0 q \right] + 2\pi i \sigma_H \, C[q] - \frac{1}{2} \zeta_2 \, C^2[q].$$
(5.40)

The response quantities σ_{xx}, σ_{00}, σ_H and ζ_2 can be expressed as ordinary thermodynamic derivatives of the "bulk" free energy $\mathcal{F}_0 = -\ln Z$ according to

$$\sigma_H = \sigma_H^0 - \lambda^{-1} \frac{\partial \mathcal{F}_0}{\partial \sigma_H^0}$$
(5.41)

$$\sigma_{xx} = 0 \quad , \quad \sigma_{00} = -\frac{1}{2Nc^2} \frac{\partial \mathcal{F}_0}{\partial \ln z} = z \langle \cos \Phi \rangle$$
(5.42)

$$\zeta_2 = \lambda^{-1} \frac{\partial \sigma'_H}{\partial \sigma_H^0}.$$
(5.43)

We see that the sine–Gordon mass z generates a finite value of the "longitudinal conductance" σ_{00} which is proportional to the magnetization $\langle \cos \Phi \rangle$ in the path integral language of Eqs. (5.31) and (5.33).

5.4.2. *Anisotropic case*

Let us next evaluate the quantum mechanical partition function Z as defined in Eq. (5.31). In the range $k < \sigma_H^0 = \phi < k+1$ one can express the result for small z as follows

$$Z = e^{-LM\epsilon_k^+(\phi)} + e^{-LM\epsilon_k^-(\phi)}.$$
(5.44)

Here, $M\epsilon_k^\pm$ are the lowest two energy levels obtained by projecting the corresponding eigenstates onto ψ_k^0 and ψ_{k+1}^0 of the unperturbed hamiltonian. The result is

$$\epsilon_k^\pm(\phi) = \beta M \left[(\phi - \phi^*)^2 + \frac{1}{4} \right] \pm \sqrt{\beta^2 M^2 (\phi - \phi^*)^2 + \left(\frac{Nz}{\beta M} \right)^2}.$$
(5.45)

Notice that in the limit $\beta \to \infty$ the theory is gapless at $\phi = \phi^* = k + \frac{1}{2}$ as it should be. Notice furthermore that the theory of Eq. (5.44) can be mapped onto the one dimensional Ising chain of length J and at low temperatures

$$Z = \sum_{\{s\}} e^{\sum_j (K s_j s_{j+1} + H s_j)}.$$
(5.46)

More specifically, if we assume periodic boundary conditions $s_j = s_{j+J}$ and under the identification

$$J = LM \quad , \quad H^2 = \beta^2 M^2 (\phi - \phi^*)^2 \ll 1 \quad , \quad e^{-2K} = \left(\frac{Nz}{\beta M} \right)^2 \ll 1 \quad (5.47)$$

then Eqs. (5.46) and (5.44) stand for the same theory with identically the same correlation functions. For example, one can express Eq. (5.41) in terms of the Ising model magnetization $m - \langle s_j \rangle$ according to

$$\sigma_H = k + \frac{1}{2} + \frac{1}{2}m. \qquad (5.48)$$

In the limit $J \to \infty$ we have the standard expression $(|H| \ll 1)$

$$m = \frac{H}{\sqrt{H^2 + e^{-4K}}} \qquad (5.49)$$

which is the correct result for the sine–Gordon theory with $L = \infty$. Notice that when $K \to \infty$ (or $z \to 0$) we recover the Hall conductance for infinite systems $\sigma_H = k(\phi)$. The discontinuity at $\phi = \phi^*$ is therefore none other than the spontaneous magnetization of the Ising model at zero temperature.

5.4.3. *Isotropic case*

We are primarily interested in a two dimensional geometry where the linear dimensions β and L are treated on an equal footing. Since the hamiltonian formalism corresponds to the geometry of a torus one would expect that the partition function of Eqs. (5.44)–(5.45) and also the effective theory of Eq. (5.40) are invariant under the interchange $\beta \leftrightarrow L$. However, this symmetry is broken by the small quantity z. This is so because we started out from the large N steepest descend methodology performed for a finite value of β but $L = \infty$. One can easily restore this symmetry by adding a term with $\bar{z}$ in Eq. (5.45)

$$\left(\frac{Nz}{\beta M} \right)^2 \to \left(\frac{Nz}{\beta M} \right)^2 + \left(\frac{\beta N \bar{z}}{L^2 M} \right)^2 \qquad (5.50)$$

where

$$\bar{z} = \sqrt{\frac{LM_0}{2\pi}} e^{-LM_0} \qquad (5.51)$$

is the same as z but with β replaced by L. Next, we redefine the quantity σ_{xx} in Eq. (5.42) as follows

$$\sigma_{xx} \to -\frac{c^2}{2N} \frac{\partial \mathcal{F}_0}{\partial \ln \bar{z}}. \qquad (5.52)$$

Finally, one can put $\beta = L$ in all the expressions such that the results are isotropic in the x and τ directions. This leads to the following form of the effective action

$$S'_\sigma[q] = -\frac{\sigma_0}{8} \int_0^L dx \int_0^L d\tau \, \mathrm{tr} \, \partial_\mu q \partial_\mu q + 2\pi i \sigma_H \, \mathcal{C}[q] - \frac{1}{2} \zeta_2 \, \mathcal{C}^2[q] \qquad (5.53)$$

where the response parameters σ_H, σ_0 and ζ_2 obey the scaling behavior

$$\sigma_H = \phi^* + F_H(X,Y) \quad , \quad \sigma_0 = F_0(X,Y) \quad , \quad \zeta_2 = F_2(X,Y) = \frac{\partial F_H}{\partial X} \quad (5.54)$$

with

$$F_H = \frac{1}{2}X\frac{\tanh\sqrt{X^2+Y^2}}{\sqrt{X^2+Y^2}} \quad , \quad F_0 = \frac{Y^2}{4N}\frac{\tanh\sqrt{X^2+Y^2}}{\sqrt{X^2+Y^2}}. \quad (5.55)$$

The two scaling variables X and Y are defined by

$$X = \frac{24\pi b^{1/\nu}}{N}(\phi - \phi^*) \quad ; \quad Y = N\sqrt{\frac{b}{\pi}}e^{-b} \quad ; \quad b = \beta M_0 = LM_0 \quad (5.56)$$

with the critical exponent ν equal to $\frac{1}{2}$.

5.4.4. Conclusion

Equations (5.54)–(5.56) are the principal results of this section. The crucial difference between the present results and the earlier expressions of Eqs. (5.15) and (5.16) is that the corrections to exact quantization are now exponential in b rather than b^2. More specifically, in the limit $X \to \pm\infty$ we obtain

$$\sigma_0 \approx \frac{Y^2}{4N^2|X|} \quad , \quad \sigma_H - k \approx \pm\frac{Y^2}{4X^2}. \quad (5.57)$$

So besides the quantization of topological charge, the sine–Gordon mass also leads to an agreement with the main expectations of exponential localization.

By combining the weak coupling results of Eqs. (4.28)–(4.30) with the strong coupling results of Eqs. (5.54)–(5.56) we obtain the scaling diagram of Fig. 21.14. The renormalization group flow lines clearly demonstrate how for large but fixed values of N the ϑ vacuum displays all the *super universal* features of the quantum Hall effect in the limit where the scale size λ goes to infinity. However, the limits $b \to \infty$ and $N \to \infty$ do not commute in this problem. The historical conflicts between the "large N picture" and the "instanton picture" of the ϑ angle,[25] which are in fact the result of the wrong order of limits, are now finally resolved.

The large N results of Eqs. (5.54)–(5.56) and the experimental scaling results of Eqs. (2.18)–(2.19) are strikingly similar and in fact describe exactly the same physical phenomena. The fact that we are comparing two completely different physical systems makes this truly remarkable.

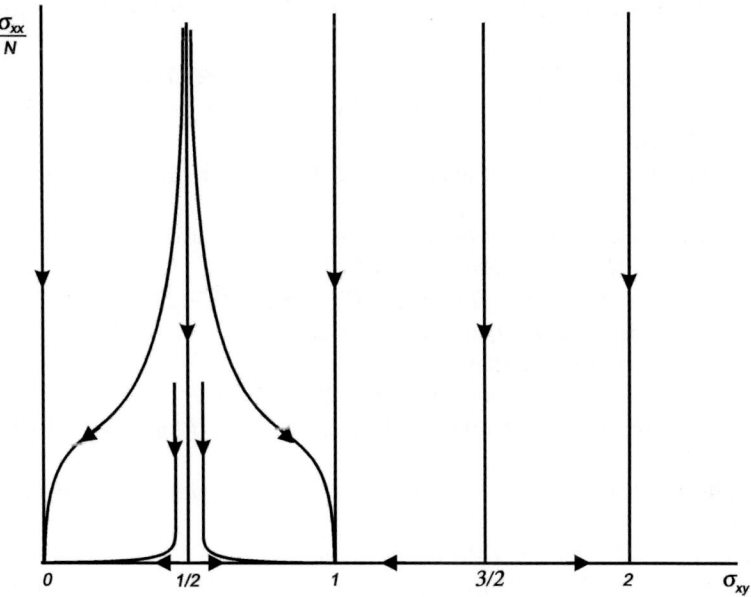

Fig. 21.14. Scaling diagram with $\sigma_0 = \sigma_{xx}$ and $\sigma_H = \sigma_{xy}$ of the CP^{N-1} model with large values of N, see text.

References

1. E. Abrahams, P. W. Anderson, D. C. Licciardello and T. V. Ramakrishnan, *Phys. Rev. Lett.* **42**, 673 (1979).
2. K. von Klitzing, G. Dorda and M. Pepper, *Phys. Rev. Lett.* **44**, 494 (1980).
3. R. B. Laughlin, *Phys. Rev. B* **23**, 5632 (1981).
4. See e.g. R. E. Prange, in *The Quantum Hall Effect*, eds. R. E. Prange and S. M. Girvin (Berlin, Springer, 1990).
5. M. Buttiker, *Phys. Rev. B* **38**, 9375 (1988).
6. A. M. M. Pruisken, *Nucl. Phys. B* **235**, 277 (1984).
7. H. Levine, S. Libby and A. M. M. Pruisken, *Phys. Rev. Lett.* **51**, 20 (1983); *Nucl. Phys. B* **240** [FS12] 30, 49, 71 (1984).
8. A. M. M. Pruisken, *Phys. Rev. B* **31**, 416 (1985).
9. A. M. M. Pruisken, *Nucl. Phys. B* **285**, 719 (1987); *Nucl. Phys. B* **290**, 61 (1987).
10. R. B. Laughlin, M. L. Cohen, J. M. Kosterlitz, H. Levine, S. B. Libby and A. M. M. Pruisken, *Phys. Rev. B* **32**, 1311 (1985).
11. A. M. M. Pruisken and M. A. Baranov, *Europhys. Lett.* **31**, 543 (1995).
12. A. M. M. Pruisken, M. A. Baranov and B. Škorić, *Phys. Rev. B* **60**, 16807 (1999).
13. M. A. Baranov, A. M. M. Pruisken and B. Škorić, *Phys. Rev. B* **60**, 16821 (1999).

14. A. M. M. Pruisken, B. Škorić and M. A. Baranov, *Phys. Rev. B* **60**, 16838 (1999).
15. B. Škorić and A. M. M. Pruisken, *Nucl. Phys. B* **559**, 637 (1999).
16. M. A. Baranov, I. S. Burmistrov and A. M. M. Pruisken, *Phys. Rev. B* **66**, 075317 (2002).
17. H. P. Wei, D. C. Tsui, M. Palaanen and A. M. M. Pruisken, *Phys. Rev. Lett.* **61**, 1294 (1988).
18. A. MacKinnon and B. Kramer, *Phys. Rev. Lett.* **47**, 1546 (1981).
19. B. Huckestein, *Rev. Mod. Phys.* **67**, 357 (1995).
20. F. Evers and A. D. Mirlin, *Rev. Mod. Phys.* **80**, 1355 (2008).
21. A. M. M. Pruisken, *The Quantum Hall Effect*, eds. R. E. Prange and S. M. Girvin (Springer, Berlin, 1990).
22. A. M. M. Pruisken, *Phys. Rev. Lett.* **61**, 1297 (1988).
23. A. M. M. Pruisken, R. Shankar and N. Surendran, *Phys. Rev. B* **72**, 035329 (2005).
24. This phrase is taken from S. Coleman in *Aspects of Symmetry* (University Press, Cambridge, 1989).
25. E. Witten, *Nucl. Phys. B* **149**, 285 (1979).
26. I. Affleck, *Phys. Lett. B* **92**, 149 (1980); *Nucl. Phys. B* **162**, 461 (1980); *Nucl. Phys. B* **171**, 420 (1980).
27. I. Affleck, *Nucl. Phys. B* **257**, 397 (1985).
28. I. Affleck, *Nucl. Phys. B* **305**, 582 (1988).
29. A. M. M. Pruisken, *Int. J. Theor. Phys.* **48**, 1736 (2009).
30. T. Ando, A. B. Fowler and F. Stern, *Rev. Mod. Phys.* **54**, 437 (1982).
31. B. Karmakar, M. R. Gokhale, A. P. Shah, B. M. Arora, D. T. N. de Lang, L. A. Ponomarenko, A. de Visser and A. M. M. Pruisken, *Physica E* **24**, 187 (2004).
32. H. P. Wei, D. C. Tsui and A. M. M. Pruisken, *Phys. Rev. B* **33**, 1488 (1986).
33. See e.g. H. P. Wei, S. Y. Lin, D. C. Tsui and A. M. M. Pruisken, *Phys. Rev. B* **45**, 3926 (1992).
34. S. Koch, R. J. Haug, K. von Klitzing and K. Ploog, *Phys. Rev. B* **43**, 6828 (1991).
35. S. Koch, R. J. Haug, K. von Klitzing and K. Ploog, *Phys. Rev. Lett.* **67**, 883 (1991).
36. D. Shahar, M. Hilke, C. C. Li, D. C. Tsui, S. L. Sondhi and M. Razeghi, *Solid State Commun.* **107**, 19 (1989).
37. T. F. van Schaijk, A. de Visser, S. M. Olsthoorn, H. P. Wei and A. M. M. Pruisken, *Phys. Rev. Lett.* **84**, 1567 (2000).
38. D. T. N. de Lang, L. A. Ponomarenko, A. de Visser, C. Possanzini, S. M. Olsthoorn and A. M. M. Pruisken, *Physica E* **12**, 666 (2002).
39. L. A. Ponomorenko, D. T. N. de Lang, A. de Visser, D. Maude, B. N. Zvonkov, R. A. Lunin and A. M. M. Pruisken, *Physica E* **22**, 236 (2004).
40. L. A. Ponomarenko, D. T. N. de Lang, A. de Visser, V. A. Kulbachinskii, G. B. Galiev, H. Künzel and A. M. M. Pruisken, *Solid State Commun.* **130**, 705 (2004).
41. A. M. M. Pruisken, D. T. N. de Lang, L. A. Ponomarenko and A. de Visser, *Solid State Commun.* **137**, 540 (2006).

42. D. T. N. de Lang, L. A. Ponomarenko, A. de Visser and A. M. M. Pruisken, *Phys. Rev. B* **75**, 035313 (2007).
43. L. A. Ponomarenko, Ph.D. thesis, University of Amsterdam (2005).
44. W. Li, G. A. Csthy, D. C. Tsui, L. N. Pfeiffer and K. W. West, *Phys. Rev. Lett.* **94**, 206807 (2005).
45. W. Li, C. L. Vicente, J. S. Xia, W. Pan, D. C. Tsui, L. N. Pfeiffer and K. W. West, *Phys. Rev. Lett.* **102**, 216801 (2009) and references therein.
46. B. Huckestein and B. Kramer, *Phys. Rev. Lett.* **64**, 1437 (1990).
47. A. M. M. Pruisken and I. S. Burmistrov, arXiv:0907.0356.
48. K. Slevin and T. Ohtsuki, *Phys. Rev. B* **80**, 041304 (2009).
49. A. M. Finkelstein in *Electron liquid in disordered conductors* (Harwood Academic Publishers, London, 1990).
50. A. M. M. Pruisken, M. A. Baranov and I. S. Burmistrov, *JETP Lett.* **82**, 150 (2005).
51. A. M. M. Pruisken and I. S. Burmistrov, *Annals Phys. (N.Y.)* **322**, 1265 (2007).
52. A. M. M. Pruisken and I. S. Burmistrov, *Pisma v ZhETF* **87**, 252 (2008).
53. A. M. Finkelstein, *JETP Lett.* **37**, 517 (1983); *Soviet Phys. JETP* **59**, 212 (1984).
54. C. Castellani, C. di Castro, P. A. Lee and M. Ma, *Phys. Rev. B* **30**, 527 (1984).
55. B. I. Halperin, P. A. Lee and N. Read, *Phys. Rev. B* **47**, 7312 (1993).
56. A. M. M. Pruisken, R. Shankar and N. Surendran, *Europhys. Lett.* **82**, 47005 (2008).
57. A. M. M. Pruisken and I. S. Burmistrov, *Annals Phys. (N.Y.)* **316**, 285 (2005).
58. For the microscopic origins of Eq. (4.22) see A. M. M. Pruisken, *Nucl. Phys. B* **295** [FS21], 653 (1988).
59. A. M. M. Pruisken, in *Localization, Interaction and Transport Phenomena*, Springer Series in Solid State Sciences, Vol. 61, eds. B. Kramer, G. Bergmann and Y. Bruynseread (Springer, Berlin, 1985).
60. See e.g. R. Rajaraman, *Instantons and Solitons* (North-Holland, Amsterdam, 1982).
61. E. Brézin, S. Hikami and J. Zinn-Justin, *Nucl. Phys. B* **165**, 528 (1980).
62. For an extensive review and numerical simulations see e.g. E. Vicari and H. Panagopoulos, *Phys. Rep.* **470**, 93 (2009).
63. S. Coleman, *Ann. Phys.* **101**, 239 (1976).
64. A. Jevicki, *Phys. Rev. D* **20**, 3331 (1979).
65. A. M. M. Pruisken, I. S. Burmistrov and R. Shankar, unpublished.
66. J. Zinn-Justin, *Quantum Field Theory and Critical Phenomena* (Oxford University Press, 2003).
67. A. M. M. Pruisken, I. S. Burmistrov and R. Shankar, arXiv:cond-mat/0602653.
68. M. Chaichian and A. Demichev, *Path Integrals in Physics*, Stochastic Processes and Quantum Mechanics, Vol. I (Institute of Physics Publishing, Bristol and Philadelphia, 2001).
69. A. M. M. Pruisken and I. S. Burmistrov, *Phys. Rev. Lett.* **101**, 056801 (2008).
70. A. M. M. Pruisken and I. S. Burmistrov, *Phys.Rev. B* **81**, 085428 (2010).

Chapter 22

SPECKLE STATISTICS
IN THE PHOTON LOCALIZATION TRANSITION

Azriel Z. Genack* and Jing Wang

*Department of Physics, Queens College of CUNY,
65-30 Kissena Boulevard, Flushing, NY 11367, USA*
*genack@qc.edu

We review the statistics of speckle in the Anderson localization transition for classical waves. Probability distributions of local and integrated transmission and of the evolution of the structure of the speckle pattern are related to their corresponding correlation functions. Steady state and pulse transport can be described in terms of modes whose speckle patterns are obtained by decomposing the frequency variation of the transmitted field. At the same time, transmission can be purposefully manipulated by adjusting the incident field and the eigenchannels of the transmission matrix can be found by analyzing sets of speckle patterns for different inputs. The many aspects of steady state propagation are reflected in diverse, but simply related, parameters so that a single localization parameter encapsulates the character of transport on both sides of the divide separating localized from diffusive waves.

1. Introduction

Anderson's prediction of electron localization in disordered lattices[1] has sparked interest in the localization of classical waves[2] associated with particles such as phonons[3,4] surface plasmons[5] and photons,[6] as well as cold atoms in random electromagnetic potentials.[7,8] Though the microscopic interactions of these waves are very different, the statistics of transport in random samples can be characterized in terms of a single localization parameter. The many aspects of propagation and localization surely may be reflected in different parameters, but these should be simply related and provide an index of the character of transport.

Anderson showed that electrons in random lattices that are not confined in a potential well may nevertheless be exponentially localized by disorder.

Subsequently, Thouless argued that the electronic conductance in bounded samples should depend only upon the dimensionless ratio of the average energy width and spacing of levels of the sample, which has come to be known as the Thouless number, $\delta = \delta E / \Delta E$.[9,10] Such levels correspond to resonances of an open system. These will often be referred to as quasimodes or simply as modes. The width of levels, which equals to the leakage rate of energy from the sample, is closely linked to the sensitivity of level energies to changes at the boundary since both are proportional to the ratio of the strengths of the mode at the boundary relative to the interior of the sample. When the wave is localized within the interior of a sample, the amplitude squared of the wave is exponentially small at the boundary and the mode is only weakly coupled to the surrounding medium. The mode lifetime is then long and its linewidth correspondingly narrow, so that, $\delta E < \Delta E$. On the other hand, when the wave is diffusive, modes extend throughout the medium; the wave then couples readily to its surroundings and the level is wide enough that its width overlaps several modes, $\delta E > \Delta E$. Thus, the localization threshold occurs at $\delta = 1$.

In many circumstances, it may be difficult to directly determine the characteristics of the modes of excitation of the medium. This is particularly true for diffusive waves when modes strongly overlap. But if δ characterizes propagation, it should be related to measurable transport quantities.[10–12] And, indeed, δ can be shown via the Einstein relation, which gives the conductivity as a product of the diffusion coefficient and the density of states, to equal the dimensionless conductance, $\delta = g = G/(e^2/h)$.[10] Here, G is the conductance, e is the electronic charge, and h is Planck's constant. Whereas δ is a property of the modes of a medium and is well-defined for any wave, g appears to relate specifically to electronic transport. But the link between δ and spatially averaged transport is maintained for classical waves and for cold atoms, as well, via the Landauer relation, in which g is expressed in terms of the transmittance, $g = T = \sum_{ab} T_{ab}$.[13,14] Here, T is the sum of transmission coefficients, T_{ab}, over all incoming and outgoing propagating transverse modes, a and b, respectively. The indices a and b may also refer to points on the input and output surfaces. For classical waves, it is natural to describe the width and spacing between resonances in terms of frequency rather than energy so that $\delta = \delta\nu/\Delta\nu = T$.

In strongly localized samples in which $\delta < 1$, sharp peaks are seen in the transmission spectrum as the frequency is tuned through resonances with modes of the medium.[15] In contrast, when modes overlap spectrally, so that, $\delta > 1$, fluctuations relative to average transmission are reduced. This is

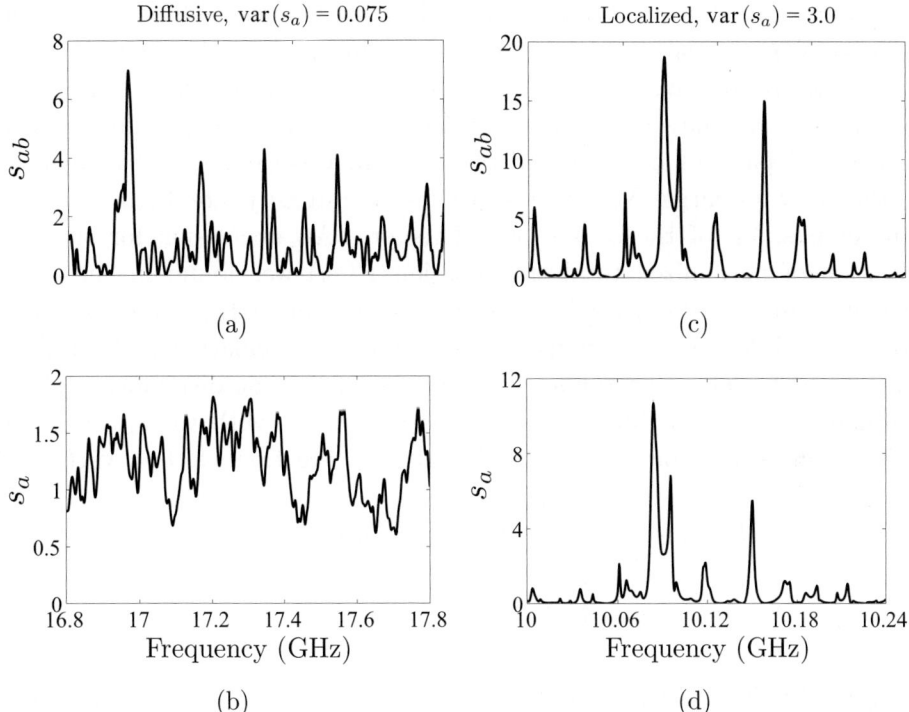

Fig. 22.1. Spectra of transmitted microwave intensity s_{ab}, and total transmission, s_a, relative to the ensemble average value for (a), (b) diffusive and (c), (d) localized waves. Radiation is transmitted through Q1D samples of randomly positioned dielectric spheres contained in a copper tube. Measurements of diffusive and localized waves are carried out in collections of polystyrene and alumina spheres, respectively.

seen in the spectra of total transmission for diffusive and localized microwave radiation transmitted through quasi-1D (Q1D) samples (Fig. 22.1). Peaks are observed in intensity spectra even for diffusive waves as a result of the changing speckle pattern with frequency shift, which may bring a null in intensity near the point of detection. Measurements of the field from which the intensity is obtained are made with a wire antenna at a point, while measurements of total transmission are made by translating the antenna over the output surface, as discussed below (Fig. 22.10).[16] The variance of total transmission normalized by its average over a random ensemble of samples with statistically equivalent disorder, $\mathrm{var}(s_a = T_a/\langle T_a \rangle)$, where, $T_a = \sum_b T_{ab}$, may then be expected to be inversely related to δ.

The simplest multichannel system is a Q1D sample with constant cross section, reflecting side walls, and length L much greater than the transverse dimensions. Examples of Q1D systems are wires and microwave waveguides. For diffusive Q1D samples, the conductance is Ohmic and given by, $g = N\ell/L$, where N is the number of incoming and outgoing transverse propagation channels and ℓ is the transport mean free path. Since the incident channels are completely mixed in such systems, the coupling between all pairs of incident and output transverse modes are equivalent in the limit of large N. The statistics of intensity are then independent of position of the source and detector. An expression for the distribution of normalized total transmission $P(s_a)$ was found in the diffusive limit in the absence of inelastic processes in Q1D samples by Van Rossum and Nieuwenhuizen using diagrammatic calculations,[17,18] and by Kogan and Kaveh using random matrix theory.[19] The distribution $P(s_a)$ is found to be a function of a single parameter, g, with $\mathrm{var}(s_a) = 2/3g$. It is therefore possible to express $P(s_a)$ as a function of $\beta = \mathrm{var}(s_a)$,[20]

$$P(s_a) = \frac{1}{2\pi i} \int_{-i\infty}^{i\infty} \exp(qs_a) F(3q\beta/2)\, dq,$$

$$F(q) = \exp\left[-\frac{2\ln^2(\sqrt{1+q}+\sqrt{q})}{3\beta}\right]. \tag{1.1}$$

Once the field in the speckle pattern is normalized by the square root of the average value of the intensity in the speckle pattern, it becomes a Gaussian random variable over the entire ensemble. The large fluctuations observed in intensity and total transmission are due to extended spatial intensity correlation which causes the brightness of the output speckle pattern as a whole to fluctuate as the frequency is tuned, as can be seen in Fig. 22.1. The distribution of a single polarization component of normalized intensity, $s_{ab} = T_{ab}/\langle T_{ab}\rangle$, in Q1D samples in which the wave is temporally coherent is thus obtained by mixing the distribution of normalized total transmission of Eq. (1.1) with the negative exponential distribution of intensity normalized by the average intensity within the speckle pattern, s_{ab}/s_a, found by Rayleigh for a Gaussian field pattern,[17,19,21]

$$P(s_{ab}) = \int_0^\infty P(s_a)\frac{\exp(-s_{ab}/s_a)}{s_a}\, ds_a. \tag{1.2}$$

This is equivalent to the relationship between the moments of s_a and s_{ab},[19]

$$\langle s_{ab}^n \rangle = n!\langle s_a^n \rangle. \tag{1.3}$$

Measurements of the first order statistics of relative total transmission and intensity for microwave radiation passing through samples of randomly mixed Polystyrene spheres contained in copper tubes of different length and diameter are shown in Fig. 22.2 and seen to be in excellent agreement with

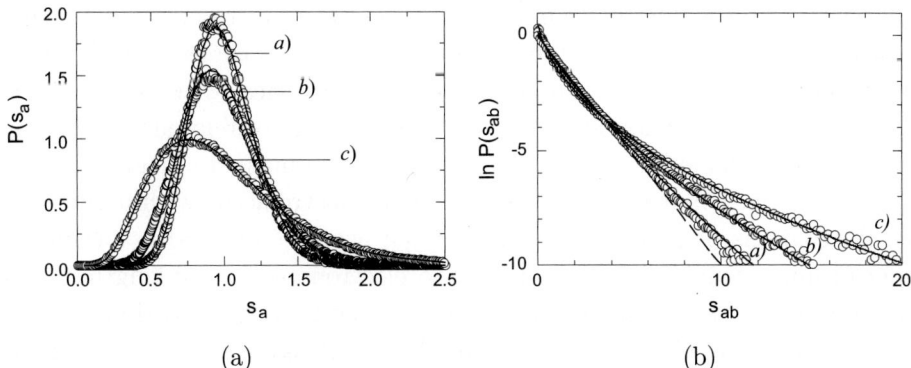

(a)   (b)

Fig. 22.2. Probability distribution functions of (a) normalized total transmission $P(s_a)$ and (b) normalized transmitted intensity $P(s_{ab})$, respectively, for three samples composed of polystyrene spheres in a copper tube with diameters and lengths: (a) $d = 7.5$ cm, $L = 66.7$ cm; (b) $d = 5.0$ cm, $L = 50$ cm; (c) $d = 5.0$ cm, $L = 200$ cm.[20] Solid lines are given by Eqs. (1.1) and (1.2) using measured values of $\mathrm{var}(s_a)$ of 0.50, 0.65 and 0.22 for samples (a), (b) and (c), respectively. The dashed line in (b) is a semi-logarithmic plot of the Rayleigh distribution, $P(s_{ab}) = \exp(-s_{ab})$.

Eqs. (1.1)–(1.2) using the measured value of $\mathrm{var}(s_a)$.[20] We will see in Secs. 4, 6, 7 and 10 that these formulas hold sway over a surprisingly wide range of physical phenomena, dimensionalities, scattering strengths, dissipation and time scales.

Enhanced fluctuations in transmission are a consequence of spatial correlation that extends beyond the short range of the field correlation length of the field, δx.[17–32] We will see below that for $N \gg 1$, the degree of correlation κ is given by, $\kappa \equiv \langle \delta s_{ab} \delta s_{ab'} \rangle = \mathrm{var}(s_a)$,[23,24,31] where $b \neq b'$ represent distinct transverse propagation modes or two points on the output surface at which field correlation vanishes, $F_E = \langle E(r)E(r \mid \Delta r) \rangle - 0$. The direct relation between propagation parameters characterizing different aspects of the statistics of transport in the absence of absorption in Q1D samples may thus be expressed as follows:

$$\delta = g = 2/3\mathrm{var}(s_a) = 2/3\kappa. \tag{1.4}$$

The relation between these parameters in geometries in which the wave is not confined laterally and the impact of absorption will be discussed in Sec. 4.

Since enhanced fluctuations of flux for temporally coherent waves in multiply scattering samples were first discovered for electronic conduction in micron-scale samples of size between the microscopic atomic scale of the wavelength, λ, and the mean free path, ℓ, and the macroscopic scale, they are termed mesoscopic.[33-35] In order for the electron wave to be coherent on scales much greater than ℓ, samples were cooled to ultralow temperatures to suppress electron–phonon scattering. Measurements of fluctuations in conductance were made by varying the voltage or the magnetic field threading a metallic ring. As a result of spatial current correlation within the sample, fluctuations of conductance did not self average and had a variance of order of unity, $\mathrm{var}(g) \approx 1$. In contrast to electrons with wavelength on the atomic-scale, classical waves have much larger wavelengths so that fluctuations of scattering elements relative to the wavelength scale are negligible. These samples can therefore be regarded as static even at room temperature. Waves may therefore be temporally coherent and exhibit mesoscopic fluctuations even in macroscopic samples. We further note that the term mesoscopic in the electronics community is often reserved for effects in conductors, $g > 1$. However, because the functional form of fluctuations and correlation is the same for diffusive and localized waves and depends upon a single parameter, the term mesoscopic is used in the optics community to describe phenomena arising from spatial correlation for localized as well as for diffusive waves.

The statistics of wave transport in random media reflect the spatial distribution of modes within the sample. For large samples, $L \gg \lambda, \ell$, the mobility edge separates samples with states extended throughout the sample from samples with exponentially peaked[1,3] states. The Thouless number represents the average overlap of modes in a random ensemble, but the number of modes contributing substantially to transmission at any frequency in a given configuration will vary. Multiply-peaked states form in space on both sides of the mobility edge whenever a small number of modes overlap spectrally with a number of peaks approximately equal to the number of overlapping modes.[36-39] This was explained by Mott[36] on the localized side as the hybridization of excitations due to the overlap of excitations peaked in neighboring localization centers. Such overlapping states, have been termed necklace states by Pendry, who showed that they dominate transmission for localized waves.[38,39] Necklace states are relatively short lived and contribute strongly to transmission since the distance to the sample boundary is typically shortened relative to localized states and the mode provides a path through the sample. Moreover, these short-lived modes impact transmission over a wide frequency range since the lines are broad. Finally, such short-lived states become even more pronounced in transmission in the presence of absorption since they are attenuated less by absorption than are long-lived localized states. In diffusive samples, frequencies at which the degree of overlap of resonant states is not as high as the average value, δ, within the random ensemble are unusually long-lived since the spatial spread of these modes is not as great as for typical modes. Such states[40] are of particular importance near the threshold of random lasers.[41]

The intensity distribution within the interior of a multiply scattering sample is generally inaccessible in three-dimensional samples, but can be examined in one-[42,43] and two-dimensional samples.[44] The presence of both isolated and overlapping modes within the same frequency range has been observed in measurements of field spectra carried out inside a single-mode waveguide containing randomly positioned dielectric elements.[43] The sample is composed of random binary elements with a number of single elements placed randomly in the sample. The binary element is composed of segments of equal length of high and low dielectric constant. This creates a pseudo-gap[45] in the frequency range of the stop band of the periodic structure with a low density of states. The waveguide containing the sample is slotted and covered with a movable copper strip to reduce leakage. Intensity spectra versus positions for isolated and overlapping waves in different configurations are shown in Fig. 22.3. Spectrally isolated lines are Lorentzians with the same width at all points within the sample and are strongly peaked in space. On the other hand, when modes overlap spectrally, the line shape varies with position within the sample and the spatial intensity distribution is multiply peaked. The field can be decomposed into a sum of modes at each position (See Eq. (8.1)). When this is done, the central frequencies and linewidths found for each of the modes in the superposition are the same at all positions. Thus the decomposition provides the shape of the mode along

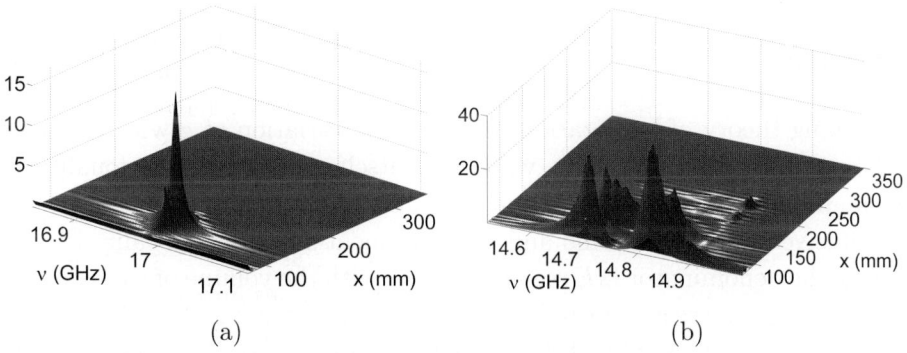

(a)   (b)

Fig. 22.3. Intensity spectra at positions within a random single-mode waveguide for (a) an isolated localized mode and (b) an overlapping necklace states.[43]

the length of the sample.[43] When a parameter of the sample configuration such as the spacing between two elements of the sample is changed, the frequency of one mode approach that of another mode and then stops while the frequency of the neighboring mode begins to move.[46] At the frequency of closest approach in this level anti-crossing, the spatial intensity distributions for the two modes are nearly identical. In cases in which the ratio of the linewidth to the strength of the spatial overlap of neighboring intensity peaks is large enough, the levels merge and a level crossing is observed. 1D localization has also been observed in optical measurements in single-mode optical fibers[47] and in single-mode channels that guide light within photonic crystals.[48,49]

In this article, we will consider the changing statistics of speckle in the Anderson localization transition and the relationships between key localization parameters. We will treat the scaling of the statistics of the field, intensity and integrated flux, as well as the structure of the transmitted speckle pattern and its evolution with frequency shift. The frequency variation of the field speckle pattern can be decomposed into a superposition of patterns associated with the modes of excitation of the sample. At the same time, analysis of a set of speckle patterns for different inputs can provide the eigenmodes of the transmission matrix. The subsequent sections will treat classical wave transport from the perspectives of scaling (Sec. 2); coherent backscattering and localization (Sec. 3); mesoscopic correlation and fluctuations (Sec. 4); generic speckle patterns (Sec. 5); intensity statistics beyond Q1D (Sec. 6); speckle evolution (Sec. 7); modes (Sec. 8); transmission eigenchannels (Sec. 9); and dynamics of localized waves (Sec. 10). A richer appreciation for the nature of transport can often be obtained by examining the problem from several perspectives.

2. Scaling

The scaling theory of localization provided the variation of g with the scale of the sample in terms of the value of g itself and the dimensionality of the sample, d.[11] The numerator in the ratio $g = \delta E / \Delta E$ scales as L^{-2} in the diffusive or Ohmic regime and falls exponentially in the localized limit, whereas the denominator ΔE scales inversely with the volume of the samples, as L^{-d}. Thus g increases with L for diffusive waves for $d > 2$, and decreases for localized waves. A critical fixed point marking the localization threshold exists at $g \sim 1$ for $d > 2$ for which g is invariant with sample size. For $d < 2$, g always falls with increasing sample size and multiply scattered coherent waves will always become localized in sufficiently large samples even when the scattering cross section is small.

Because fluctuations are often large, it is important to find the scaling of the full distribution of transmission quantities.[15,50] It is therefore crucial to differentiate between the ensemble average of the conductance, which we denote by, $g \equiv \langle T \rangle$, and the conductance or transmittance, T, in a specific sample at a particular energy or frequency.[12]

In 1D, the probability distribution of the conductance or transmittance, T, is essentially log-normal with a width that self averages in the limit of large sample length.[12] Expressing T in terms of the Lyapunov exponent, γ, $T = \exp(-2\gamma L)$, the single parameter scaling (SPS) hypothesis[12] predicts that for $\bar{\gamma}L \gg 1$, the probability density of $\gamma = -\ln T/2L$ is a Gaussian with var$(\gamma) = \bar{\gamma}/L$. Here, $\bar{\gamma} = 1/2\bar{\xi}$, is the average Lyapunov exponent and $\bar{\xi}$ is the average intensity localization length for $\bar{\gamma}L \gg 1$. For $\bar{\gamma}L < 1$, the distribution of T is very nearly a segment of a log-normal distribution with a maximum value of 1.[51,52] In a binary 1D sample, in which alternating elements with two different indices of refraction have random thickness, there is a minimum value of T, which occurs when the thicknesses of all layer satisfy the conditions, $2nd = (m+1/2)\lambda$. $P(T)$ falls to zero as this minimum value is approached. The probability density $P(T)$ must therefore fall below the log-normal distribution at low values of transmission. This can be seen in the simulation shown in Fig. 22.4 for optical transmission through a glass stack

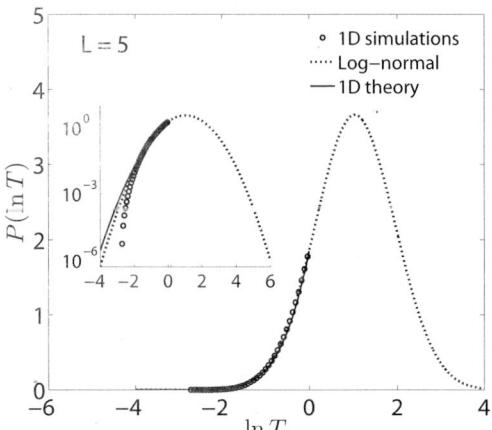

Fig. 22.4. Probability distribution of $\ln T$. 1D simulations, log-normal distribution, and 1D theory[52] are shown with open circles, dotted line and solid line, respectively. Simulations and theory are for a stack of parallel glass slides. The localization length is $\bar{\xi} = 11.4$.[53] The logarithm of $P(\ln T)$ is shown in the inset.

with five glass slides in which the index in the sample alternates between $n = 1$ for air and $n = 1.522$ for glass.[52] In this sample, the localization length is $\bar{\xi} = 11.4$ glass slides.[53] Remarkably, however, the bulk of the probability density of T fits a log-normal distribution even for $\bar{\gamma}L < 1$.

When the randomness in the elements of the 1D structure is restricted, a residual band structure can be observed and SPS no longer applies to the region of the pseudogap beyond the band center. In this case, scaling of T at a given frequency requires the introduction of second parameter which reflects the integrated density of states from the edge of the band.[54] Beyond 1D, a speckle pattern forms, so that points of vanishing intensity can be found at the output of the sample. We will see in Secs. 4–6 that this has a profound impact upon intensity statistics.

3. Coherent Backscattering and Localization

Interference of partial waves associated with different trajectories influences the average as well as correlation and fluctuations of transmitted flux in random ensembles of samples and also gives rise to a random volume speckle pattern within individual sample realizations. Whereas the random speckle pattern created by wave interference is washed out when averaged over a random ensemble, the suppression of average transport and the enhancement of fluctuations due to interference survives. Such interference involves combinations of paths within each sample realization in which the phase difference between paired paths is small. In this section, we outline the impact of interference on average transport and localization via coherent backscattering before reviewing mesoscopic fluctuations and correlation in the next section.

Interference is the source of weak localization, which suppresses diffusion and leads to Anderson localization. Average transport is reduced by constructive interference of pairs of partial waves that follow the same closed loop within the medium but in opposite senses. Because the amplitudes and phases associated with these partial waves are identical, the amplitude for return for paths following the loop for both senses is twice that for a single partial wave. The number of returns, N_{return}, which is proportional to the square of the amplitude is therefore four times as great as for a single path and twice the sum for the two paths separately. The enhancement of N_{return} leads to a suppression of transport away from a given point below the level of incoherent diffusion.

When $N_{\text{return}} \to 1$, wave transport is strongly suppressed by wave interference and the wave becomes exponentially localized within the medium. This coincides with the localization criterion, $g = 1$. Generally, $N_{\text{return}} = 1/g$ and the number of returns exceeds unity for localized waves. This can be seen by expressing N_{return} as the ratio of the volume of typical path transmitted through the medium with a cross sectional area for the trajectory equal to the coherence area, $A_c = (\lambda/2)^{d-1}$, of the speckle pattern, to the total volume, V, of the sample. The first volume can be written as A_c multiplied by the typical path length within the medium, $A_c v \tau_{\text{Th}}$, where τ_{Th} is the Thouless time, which is the average time in which waves are transmitted though a sample, and v is the velocity in the medium. Thus $N_{\text{return}} = A_c v \tau_{\text{Th}} / V = \tau_{\text{Th}} / [(V/A_c)/v] = \tau_{\text{Th}} / \tau_{\text{H}}$. Here, V/A_c is the length of the path required to sequentially visit each coherence volume of the sample, and the Heisenberg time, τ_{H}, is the corresponding time. τ_{Th} corresponds to L^2/D for diffusive waves, where D is the diffusion coefficient, and is equal to the inverse of the field correlation frequency, $\delta\nu$, which corresponds to the typical width of modes of the medium, $\tau_{\text{Th}} = 1/\delta\nu$. The inverse of τ_{H} is essentially the free spectral range in optics and so $\tau_{\text{H}} = 1/\Delta\nu$. Thus, $N_{\text{return}} = \tau_{\text{Th}}/\tau_{\text{H}} = \Delta\nu/\delta\nu = 1/\delta = 1/g$.

Weak localization can be observed directly in the enhanced backscattering of light from the surface of random samples,[55–57] as seen in Fig. 22.5.[58] The peak in the ensemble average of retroreflection is twice the background

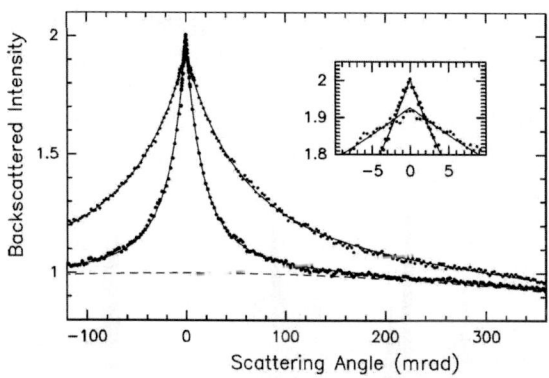

Fig. 22.5. Coherent backscattering of light measured in two samples with different values of the transport mean free path, ℓ.[58] The typical angular width varies as λ/ℓ. Narrow backscattering cone: a sample of $BaSO_4$ powder with $\ell/\lambda = 4$; broad cone: TiO_2 sample with $\ell/\lambda = 1$. The inset exhibits the triangular cusp predicted by diffusion theory, and also shows that the maximum enhancement factor is lowered for the sample with smaller value of ℓ/λ from the value of 2 in the diffusive limit.

level at large backscattering angles. The Fourier transform of the coherent part of the backscattered radiation, which is the scattered light above the background, is the Fourier transform of the point spread function of light on the incident surface.[59] When scattering in the sample is strong so that the transport mean free path is short, the point spread function is narrow resulting in a broad coherent backscattering peak. The angular width of the peak is of order $1/k\ell$, where $k = 2\pi/\lambda$ is the wave vector in air. The inset in Fig. 22.5 shows the triangular peak which is enhanced over the background by a factor of two in a weak scattering sample. For smaller values of $k\ell$, N_{return} increases and the enhancement of coherent backscattering over the background is lowered as a result of recurrent scattering in which the wave is scattered more than once by the same scatterer. Such recurrent scattering cannot be distinguished from the single scattering background in the coherent backscattering peak. In random 3D samples, $N_{\text{return}} \to 1$ when $k\ell \sim 1$. Propagation is then renormalized by coherent backscattering and this gives the Ioffe–Regel criterion for localization, $k\ell = 1$.[60] Wave propagation is unsustainable for $k\ell < 1$ since the spread in the wave vector is greater than the wave vector itself, $\Delta k \sim 1/\ell > k$.

Classical wave localization just beyond the localization threshold has been observed in 3D for ultrasound in a slab of aluminum beads weakly brazed together at a volume fraction of 0.55.[61] An example of the speckle pattern in this case is shown in Fig. 22.6. The intense narrow spikes in transmission[61] are associated with the multifractal nature of the wave function.[62,63] Localization in elastic networks is possible since scattering can be restricted to coupling between adjacent elements which can be made arbitrarily small.

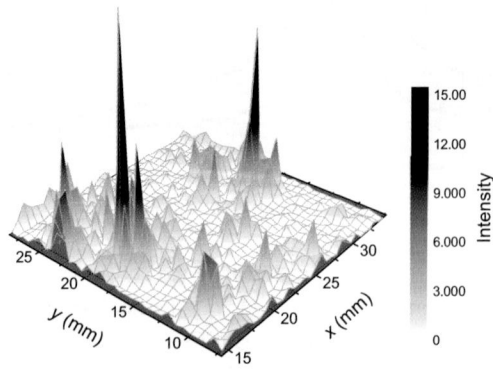

Fig. 22.6.   Near-field speckle pattern for localized acoustic waves.[61] The intensity pattern is multifractal.[62]

Achieving photon localization in random 3D dielectric structures without long-range structural correlation[45] has remained a particular challenge since the mean free path at a given volume fraction of scattering elements has a minimum when the wavelength is comparable to the diameter of the scattering element. But light can then only encounter a single scattering element in a wavelength and the cross section is not large enough to satisfy the Ioffe–Regel condition for localization. Photon localization can be achieved, however, in nearly periodic systems since the density of states can be arbitrarily low in the pseudogap. The level spacing is then large so that $\delta < 1$. Microwave localization was observed with $\text{var}(s_{ab}) > 7/3$ near the band edge of a 3D periodic copper wire mesh structure in which dielectric scatterers are floated.[15]

N_{return} increases in lower dimensions for a given value of $k\ell$ and localization can always be achieved in large enough samples for arbitrary values of $k\ell$ for $d < 2$.[11] Photon localization has been observed in 1D,[42,43,49,50,64] 2D,[44,65] Q1D,[15] layered samples,[53,66,67] and in the transverse dimensions[68,69] for samples which are uniform in the primary propagation direction.[70] Such transverse localization[70] is seen in Fig. 22.7.[68] A 2D periodic hexagonal lattice with superimposed random fluctuations is written into a photorefractive material. The random fluctuations are created by irradiation with random speckle patterns. The strength of disorder is adjusted by varying the strength of the speckle pattern. A transition from a diffusive (Fig. 22.7(b)) to a localized (Fig. 22.7(c)) wave is seen in the output plane as the ensemble

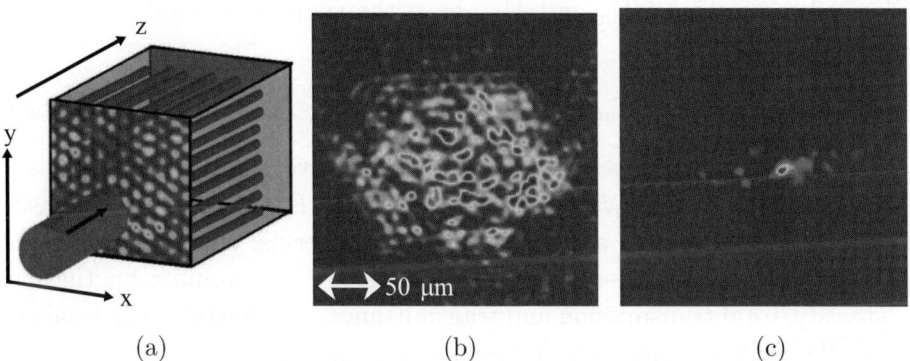

(a) (b) (c)

Fig. 22.7. Transverse localization[68]: (a) A probe beam enters a periodic 2D lattice in a sample which is uniform in the beam direction. The periodic lattice is prepared by interference of three laser beams in a photorefractive medium. (b) Transmission through 1 cm of a periodic lattice. (c) Transmission through 1 cm of a sample with 15% disorder in the lattice created by superimposing a random speckle pattern.

average of the spatial intensity distribution changes from being a Gaussian to an exponential function centered on the input beam.

4. Mesoscopic Correlation and Fluctuations in Quasi-1D

Enhanced mesoscopic correlation of intensity at two points on the sample output due to two sources at the input arises as a result of interference between the corresponding four partial waves with trajectories that cross within the sample and produce small overall phase for the four fields. Because the phase shift is small, the contributions of different possible pairings survive averaging over an ensemble of configurations. This gives rise to spatially extended correlation, which leads to enhanced fluctuations of integrated quantities such as T_a and T. The diagrammatic Green's function approach allows for direct calculations of correlation for diffusive waves due to wave interference for diffusive waves.[22,24]

The absence of self-averaging in mesoscopic systems, which leads to large fluctuations, is a consequence of extended spatial intensity correlation. Microwave measurements suggest, and diagrammatic calculations confirm, that the cumulant correlation function, C_I, of the normalized intensity with displacement of the source and detector can be expressed as the sum of three terms, with distinctive spatial dependences in Q1D samples.[23,24,31,32] Each term involves only the product or sum of the square of the correlation function on the input and output surfaces of the field normalized to the square root of the average intensity versus displacement and polarization, F_{in} or F_{out}, and a constant. F_{in} and F_{out} have the same functional form. This gives,

$$C_I = F_{\text{in}}F_{\text{out}} + A_2(F_{\text{in}} + F_{\text{out}}) + A_3(F_{\text{in}}F_{\text{out}} + F_{\text{in}} + F_{\text{out}} + 1). \quad (4.1)$$

C_I can be re-expressed as the sum of multiplicative, additive and constant terms, $C_I = (1 + A_3)F_{\text{in}}F_{\text{out}} + (A_2 + A_3)(F_{\text{in}} + F_{\text{out}}) + A_3$. For diffusive waves, the multiplicative, additive, and constant terms, which correspond to short-, long-, and infinite-range contributions to C_I, dominate fluctuations of intensity, total transmission and transmittance, respectively. For localized waves, the infinite-range term dominates correlation and fluctuations. For a fixed source, $F_{\text{in}} = 1$, and writing $F_{\text{out}} = F$, we have, $C_I = F + (A_2 + 2A_3)(1 + F)$, which can be written as,[31]

$$C_I = F + \kappa(1 + F). \quad (4.2)$$

where κ is the value of C_I at which $F = 0$.

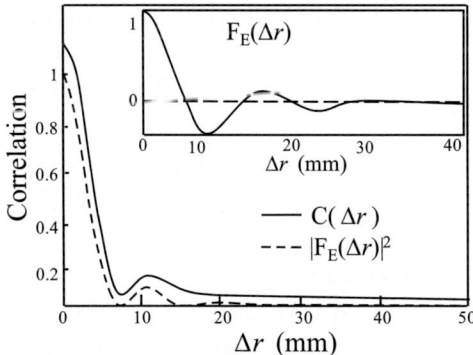

Fig. 22.8. Spatial intensity correlation and the square of the field correlation function within the speckle pattern on the output surface of Q1D diffusive samples of polystyrene spheres with $\kappa = 0.075$. The field correlation function is shown in the inset.[31]

In the interior of the sample, the normalized field correlation function versus displacement has the form, $F_E(\Delta r) = \exp(-r/2\ell)\sin k\Delta r/k\Delta r$, where ℓ is the scattering length.[71] The field correlation on the sample surface, $F_E(\Delta r)$ is the Fourier transform of the specific intensity, which is the ensemble average of the intensity per unit solid angle of the distribution of scattered intensity in the far field, $\langle I(k_\perp/k)\rangle$, where $k_\perp$ is the component of the k-vector of transmitted radiation in the plane of the surface.[72,73] The variations with displacement, Δr, of F_E, F, and C_I in a diffusive sample are shown in Fig. 22.8.[31] $C_I(\Delta r)$ is in accord with Eq. (4.2) with $\kappa = 0.075$. The structure of the correlation function at a fixed delay time from an exciting pulse is the same as in steady state measurement.[74,75] F_E is identical and C_I has the same form but with a time dependent value of the degree of correlation, $\kappa(t)$. In the far field, the variation with angle of polarization of the field correlation function is given by, $F_E(\Delta\theta) = \cos(\Delta\theta)$.[74,76] This simple form makes it particularly straightforward to distinguish the multiplicative, additive and constant contributions to C_I.[32] The infinite-range term, which is the origin of universal conductance fluctuations for diffusive waves, was observed in the time correlation function of light passed through a cylindrical pinhole separating two colloidal solutions.[28]

The variance of total transmission in Q1D samples with N transverse incident and outgoing propagating channels may be expressed as,

$$\text{var}(T_a) = \langle(\delta T_a)^2\rangle = \sum_{b,b'}\langle\delta T_{ab}\delta T_{ab'}\rangle = \sum_b\langle(\delta T_{ab})^2\rangle + \sum_{b,b'\neq b}\langle\delta T_{ab}\delta T_{ab'}\rangle$$

$$= N\text{var}(T_{ab}) + N(N-1)\langle\delta T_{ab}\delta T_{ab'\neq b}\rangle.$$

Normalizing T_a and T_{ab} by their respective ensemble averages, $\langle T_a \rangle = N\langle T_{ab} \rangle$, and $\langle T_{ab} \rangle$, and taking the limit $N \gg 1$ so that the second term with $N(N-1)$ terms dominates over the first term with N terms, gives, $\text{var}(s_a) = \langle \delta s_{ab} \delta s_{ab' \neq b} \rangle \equiv \kappa$. Similarly, expressing $\text{var}(T)$ as a sum of short-range terms, with $a = a'$, $b = b'$, intermediate-range terms with $a = a'$, $b \neq b'$ or $a \neq a'$, $b = b'$, and infinite-range terms, $a \neq a'$, $b \neq b'$, gives a result, which, for $N \gg 1$, is dominated by the sum with most numerous terms, the infinite-range terms. Thus for $N \gg 1$, $\text{var}(T) = \sum_{a,a',b,b'} \langle \delta T_{ab} \delta T_{a'b'} \rangle \sim \sum_{a,a' \neq a,b,b' \neq b} \langle \delta T_{ab} \delta T_{a'b'} \rangle \sim N^4 \langle \delta T_{ab} \delta T_{a'b'} \rangle = N^4 \langle T_{ab} \rangle^2 \kappa_\infty$, where $\kappa_\infty = \langle \delta T_{ab} \delta T_{a'b'} \rangle / \langle T_{ab} \rangle^2 = \langle \delta s_{ab} \delta s_{a'b'} \rangle$. In the diffusive limit, diagrammatic calculations give, $\kappa_\infty = 2/15g^2$.[23,24] Since, $\langle T \rangle = N^2 \langle T_{ab} \rangle$ in Q1D, $\text{var}(T) = \kappa_\infty \langle T \rangle^2 = 2/15$. The value of $\text{var}(T)$ is somewhat different in geometries other than Q1D but is independent of the scale of the system for diffusive waves.[33–35] This constant value of conductance fluctuations, know as universal conductance fluctuations, was discovered in electronic conductance measurements in samples cooled to suppress inelastic scattering process over the scale of the sample.[33]

Absorption affects different localization parameters differently. The equalities between fluctuations of transmission and corresponding measures of intensity correlation, $\text{var}(s_{ab}) = \kappa_0$, $\text{var}(s_a) = \kappa$, and $\text{var}(s) = \kappa_\infty$, where, $\kappa_0 \equiv \langle (\delta s_{ab})^2 \rangle$, are unaffected by absorption, though the values of these quantities are weakly affected by absorption. In the presence of absorption, $P(s_a)$ is no longer a function of g, however, $P(s_a)$ may still be expressed as a function only of $\beta = \text{var}(s_a)$, following Eq. (1.1).[20] Indeed, Eqs. (1.1)–(1.3) remain valid in Q1D even for absorbing samples and for localized waves.[15] These equations also describe fluctuations of intensity and total transmission at different delay times from an exciting pulse.[74] The dynamics of fluctuations reflects the changing renormalization of transport with time delay and the changing effective number of modes which contribute substantially to transmission following an exciting pulse. Equation (1.2) also gives the probability density of transmitted intensity for plane wave excitation even in 3D slabs or layered media with transverse disorder even though there is no well-defined area over which total transmission can be calculated, as would be the case in Q1D samples. Indeed, fluctuations of total transmission taken over an unbounded output surface vanish. Nonetheless, in these cases, $P(s_{ab})$ is still given by Eqs. (1.1)–(1.2) by treating s_a as a variable of integration. $P(s_{ab})$ then depends upon a single parameter, $g' = 2/3\text{var}(s_a)$, which we call the statistical conductance.[52] In nondissipative samples bounded in the transverse direction, in which g is well-defined, $g' = g$. g' represents the degree of renormalization of transport within the medium. The relationship in Q1D samples between the second moments of s_a and s_{ab} in Eq. (1.3) gives, $\mathrm{var}(s_{ab}) = 2\mathrm{var}(s_a) + 1$. This relation can be generalized to any dimension beyond 1D in which the speckle pattern is generic to, $\mathrm{var}(s_{ab}) = 4/3g' + 1$, so that measurements of intensity fluctuations give the degree of wave correlation and localization. The localization threshold is at $g' = 1$ or equivalently, $\mathrm{var}(s_{ab}) = 7/3$.[15,52,61]

Not only are the relationships between fluctuations and correlation not affected by absorption, but the value of these parameters is not significantly affected by moderate absorption. In addition, $\mathrm{var}(s_a) = 2/3g'$, scales linearly for $\mathrm{var}(s_a) < 2/3$ and exponentially for $\mathrm{var}(s_a) > 2/3$.[15] This is in contrast, to the quantities δ and g, which are strongly affected by absorption. The equality $\delta = g$ does not hold in the presence of absorption since the linewidth increases with absorption so that δ increases, while g, which is proportional to the transmitted flux, falls in dissipative samples. Since g scales as $\exp(-L/L_a)$ in absorbing diffusive samples for lengths greater than the absorption length, L_a, the exponential scaling of g does not by itself signify localization.[77–80] The exponential absorption length is, $L_a = \sqrt{D\tau_a}$, where D is the diffusion coefficient and $1/\tau_a$ is the absorption rate.

The usefulness of statistical measures, such as $\mathrm{var}(s_{ab})$, as localization parameters can be seen in measurements of microwave transmission in random Q1D samples of alumina spheres at low volume fraction shown in Fig. 22.9. Measurements of the spectra of the ensemble averages of transmission, $\langle I \rangle$, and the photon transit time, $\langle \tau \rangle$, in Fig. 22.9, show a series of dips and peaks, respectively, at each of the Mie resonances of the alumina spheres.[81] The smallest values of the mean free path are expected near sphere resonances, however, the relationship of the extrema in spectra of $\langle I \rangle$ and $\langle \tau \rangle$ to localization is not direct because of the possible impact of absorption. Since $\langle I \rangle = \langle T_a \rangle/N \sim \ell/L$, dips in $\langle I \rangle$ may indicate a low value of ℓ due to renormalization associated with localization. However, transmission may also be suppressed by absorption. The impact of absorption is particularly strong near sphere resonances because the dwell time is enhanced. The transit time or single-channel delay time is obtained from measurements of $\langle \tau_{ab} \rangle = \langle s_{ab} d\varphi_{ab}/d\omega \rangle$, where φ_{ab} is the phase accumulated by the field as it propagates through the sample, from incident channel a to outgoing channel b, and ω is the angular frequency.[82] In diffusive samples, the Thouless time is given by, $\tau_{\mathrm{Th}} = L^2/D$.[9,10] For classical waves in 3D, $D = v_E\ell/3$, where v_E is the transport velocity.[83] Thus dips in D can occur either because of the small value of ℓ, which might be renormalized by weak localization, or due

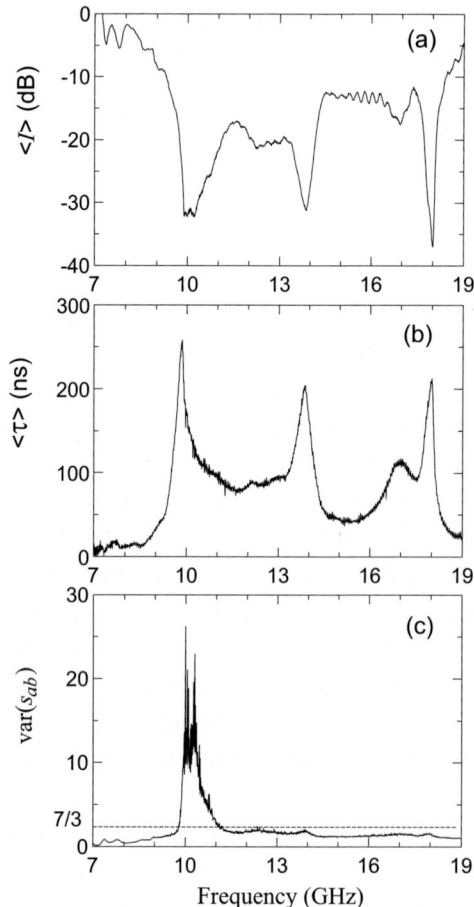

Fig. 22.9. (a) Average transmitted intensity, $\langle I \rangle$, (b) average photon transit time, $\langle \tau \rangle$ and (c) variance of normalized transmitted intensity in a Q1D alumina sample with $L = 80$ cm and alumina volume filling fraction $f = 0.068$. The dashed line indicates the localization threshold. A narrow window of localization in which, $\text{var}(s_{ab} = I/\langle I \rangle) > 7/3$, corresponding to, $\text{var}(s_a) > 2/3$, is found just above the first Mie resonance of the alumina spheres comprising the sample.[81]

to the suppression of v_E below the phase velocity. v_E is suppressed at resonance because of the high energy density of electromagnetic radiation within the dielectric sphere, which effectively slows down wave propagation.[83] Thus peaks in dwell time may be associated with resonant excitation of the sphere rather than with localization. Since dips in transmission and peaks in dwell time can be associated with absorption and microstructure resonances, the average transmission and dwell time do not provide definitive measures of the closeness to the localization threshold. This can be obtained, however, from the measurement of $\mathrm{var}(s_{ab})$, shown in Fig. 22.9(c). $\mathrm{var}(s_{ab})$ rises above the value at the localization threshold of $7/3$ indicated by the horizontal dashed line in a narrow window just above the first Mie resonance. Localization was not observed on resonance since the density of states is high on resonance so $\Delta\nu$ in the denominator of δ is small.

5. Generic Speckle

Wave interference is manifest in the random speckle pattern of scattered monochromatic radiation. The size of speckle spots is essentially the field correlation length, δx, which is the displacement for the first zero of the field correlation function. The field correlation function along the output surface of a scattering medium for monochromatic radiation is the Fourier transform of the specific intensity, $\langle I(k_\perp/k)\rangle$. As a result, $\delta x \sim 1/\delta k_\perp$, where $\delta k_\perp$ is the width of the specific intensity distribution. Within a locally 3D random medium, the scattered k-vector distribution is isotropic and gives, $\delta x = \lambda/2$. δx may be larger on the output surface of a sample when the angular spread of scattered light is narrower. In the case of anisotropic samples, the transmitted field may still be highly directional and δx may be much larger than λ.

Examples of microwave speckle patterns formed on the output surface of random Q1D sample are shown in Figs. 22.10–22.12.[16,82] A linear polarized component of the transmitted microwave field is detected using a wire antenna and vector network analyzer. Examples of the in- and out-of-phase spectra and the corresponding spectra of intensity and phase relative to the incident field at a single point are shown in Fig. 22.10. Full speckle patterns over the frequency range scanned are obtained from such measurements over a grid of points.

The intensity variation and equiphase lines[84,85] in a typical transmission speckle pattern at 10 GHz are shown in Fig. 22.11. The spatial variation of intensity is correlated with the variation in phase. The phase gradient is low near peaks in intensity, high in regions of low intensity, and singular at intensity nulls. At points of vanishing intensity, the phase cannot be defined since the in-phase component of the field vanishes. The speckle pattern is built upon a network of phase singularities towards which equiphase lines converge.[86–90] The phase jumps by π rad along any line passing though a phase singularity and changes by $\pm 2\pi$ rad in a circuit around single phase singularities. The most striking

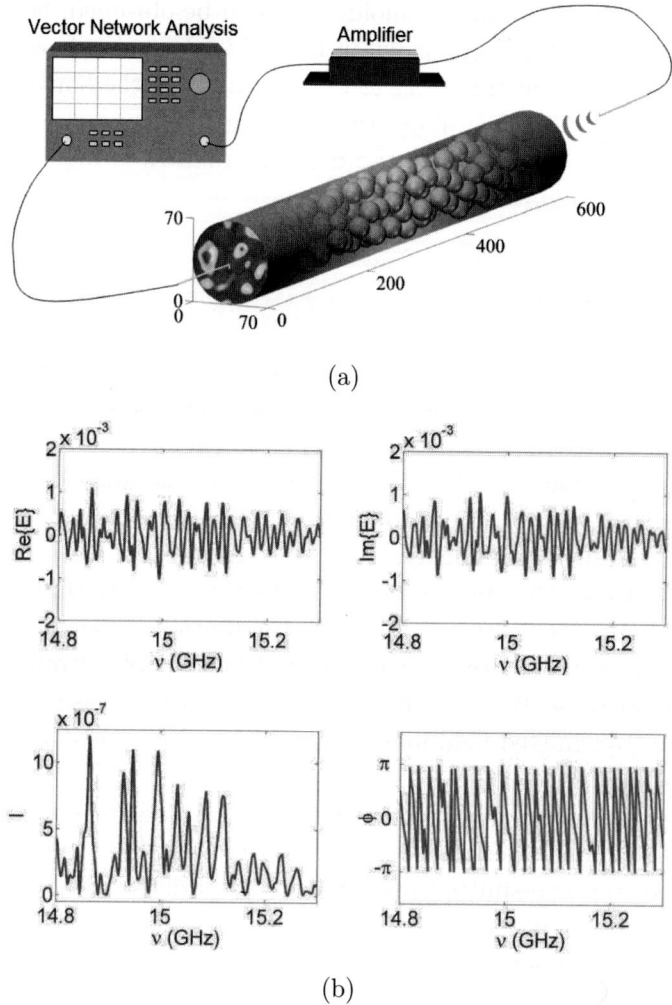

Fig. 22.10. (a) Experimental setup for measuring the field transmission coefficient and speckle pattern for microwave radiation in Q1D samples. Styrofoam shells containing alumina spheres are indicated inside the tube. (b) Examples of spectra of in- and out-of-phase components of the field and of the intensity and phase.[16]

feature of the evolving speckle pattern as the frequency is scanned is the creation and annihilation of phase singularities, which occurs in pairs of singularities with opposite senses of variation of the phase around the singularity.[86,87,89,90]

Measurements are carried out on a tight enough grid that the two-dimensional Whittaker–Shannon sampling theorem[91] can be applied to obtain the speckle pattern with arbitrary spatial resolution. Measurement

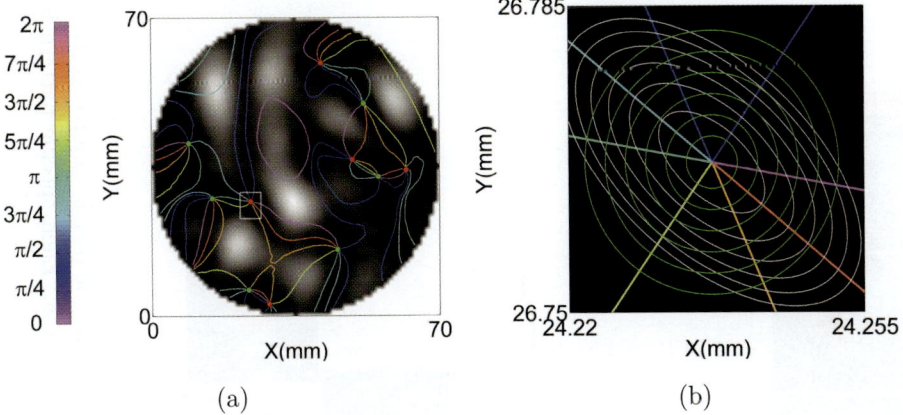

(a) (b)

Fig. 22.11. (a) An example of a speckle pattern is shown with equiphase lines with values of phase $\varphi = n\pi/4$ rad with $n = 0 \cdots 7$. (b) Magnification of the region near the singularity within the dashed rectangle in (a). Elliptical contours of intensity and circular contours of current magnitude are found.[16]

of the structure and statistics of the field structure at phase singularities can therefore be carried out. A micron-scale region surrounding the phase singularity in the rectangle in Fig. 22.11(a) for a speckle pattern created by radiation with a wavelength of 3 cm is shown in Fig. 22.11(b). Contours of intensity, I, and current magnitude, $|I\Delta\varphi|$, are ellipses and circles, respectively. The transverse electromagnetic flux circulating around the phase singularity forms a vortex in which the magnitude of flux increases linearly with displacement from the vortex center at a rate known as the vorticity, Ω.[90] The phase variation in the vortex core is determined by the orientation and eccentricity ϵ of the intensity ellipses in the vortex core.[84]

Measurements of statistics over an ensemble of equivalent random sample realizations can be made by rotating the sample tube to create new sample configurations. The distribution of eccentricity is a universal function, being the same for diffusive and localized waves. However, the statistics of vorticity are a mixture of the mesoscopic probability distribution of total transmission and the distribution of vorticity for Gaussian fields. Once the intensity in the pattern is normalized by the average intensity, however, the probability distribution of the normalized vorticity is universal.[84] This distribution is the same as for Gaussian fields found in random ensembles in the diffusive limit.[90]

Single speckle patterns formed in a sample irradiated at a specific wavelength are generic in that their structure and statistics are robust under perturbation and are governed by Gaussian field statistics. Sets of speckle

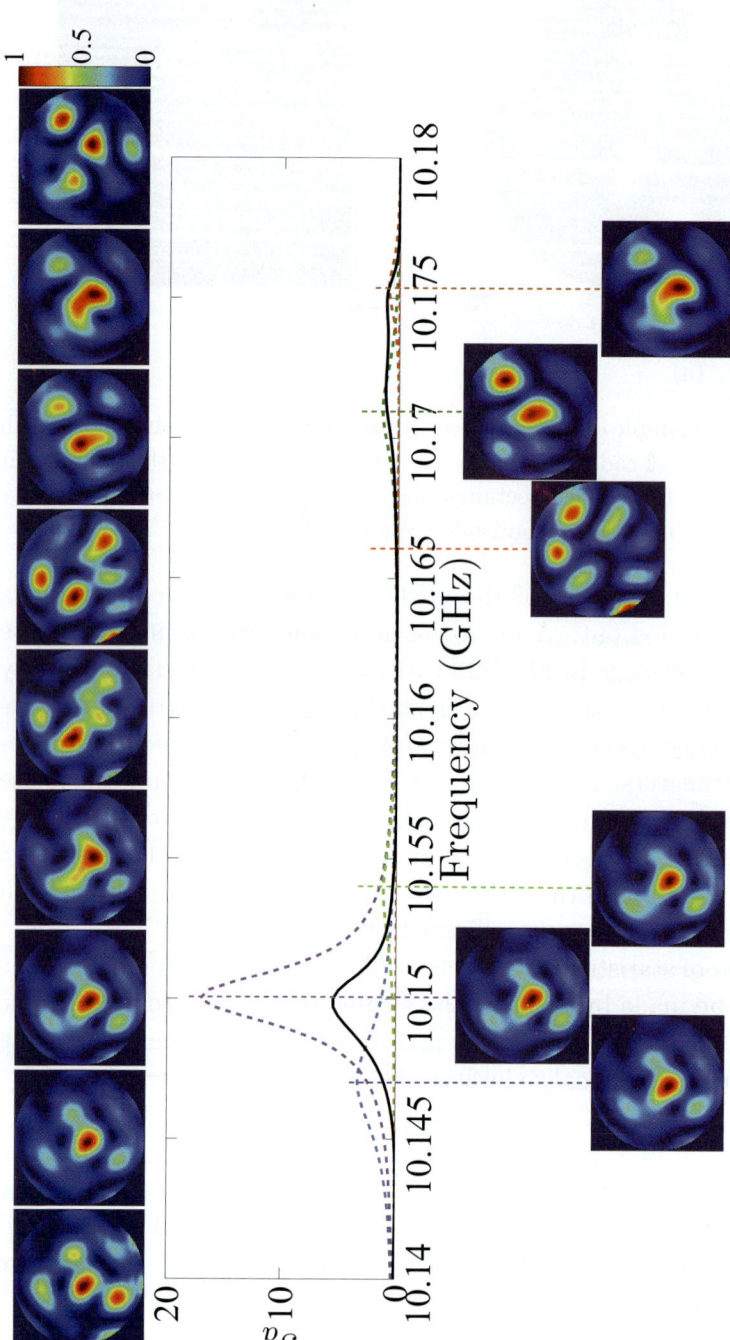

Fig. 22.12. Evolution of intensity speckle pattern with frequency shift is shown in the first row. Total transmission (black solid curve) and total transmission for individual modes (colored dashed curves) is shown in the middle panel. Colored vertical lines indicate the central frequencies of modes and connect to the corresponding speckle patterns. Each speckle pattern is normalized to the maximum value of intensity in the pattern.

patterns are also generic once the field at every point is divided by the square root of the average value of intensity within the speckle pattern. Thus mesoscopic correlation is not seen in the individual speckle patterns in Q1D but only in the average value of transmission and in fluctuations in relative total transmission. A sequence of normalized speckle patterns as the frequency is changed for transmission of localized waves in a sample drawn from an ensemble of configurations with $var(s_a) = 3.0$ is seen in the top row of Fig. 22.12. The structure of the patterns is independent of the total transmission and also of the value of $var(s_a)$ for a random ensemble of statistically equivalent samples. Large fluctuations seen in total transmission arise when the wave is on- or off-resonance with modes of the medium. These fluctuations correspond to a multiplicative factor of the speckle pattern as a whole and to correlation of intensity across the entire output surface.

6. Intensity Statistics Beyond Quasi-1D

We have seen that in 1D the probability density of intensity is a segment of a log-normal distribution while in Q1D it is given by Eqs. (1.1)–(1.2). We next consider intensity fluctuations in the slab geometry. In this geometry, the wave is not bounded laterally in a slab, so the total transmission and dimensionless conductance are not sharply defined.

Measurements of the probability distribution, $P(\hat{I} \equiv s_{ab})$, of normalized intensity of ultrasound radiation transmitted through a random network of brazed aluminum beads were made in samples with lateral dimensions greater than the sample thickness. Steady-state field spectra are obtained by Fourier transforming the pulsed response. A series of sharp dips in the spectra corresponding to alternating band gaps and pass bands associated with individual sphere resonances and possible ordering of the spheres in the sample. $P(\hat{I})$ at 2.4 MHz, well above the first band gap, is shown in Fig. 22.13.[61] A good fit of Eqs. (1.1)–(1.2) to $P(\hat{I})$ was obtained and gave a value for the statistical conductance of $g' = 4/3[var(\hat{I}) - 1] = 0.8$, indicating the waves was just below the localization threshold. This suggests that $P(\hat{I})$ is given by Eqs. (1.1)–(1.2) even though s_a is not strictly defined.

The applicability of Eqs. (1.1)–(1.2) to wave propagation in different dimensions can be explored by studying the variation of $P(\hat{I})$ in a transition from 1D to higher dimensions in samples with increasing numbers of layers. A crossover from localized to diffusive transmission associated with the change in dimensionality is seen in Fig. 22.14(a) in the scaling of average transmission, $\langle T(L) \rangle$, where L is the number of glass slides in a stack.[53]

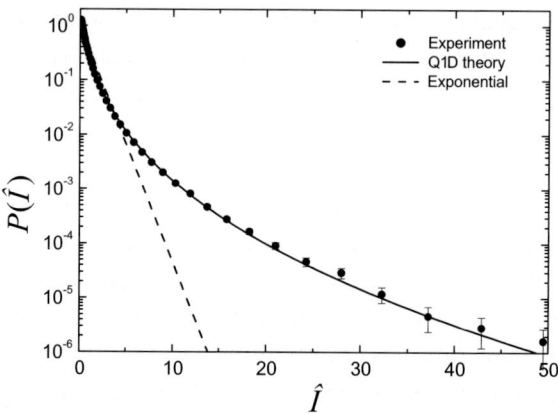

Fig. 22.13.   Probability distribution of normalized intensity $P(\hat{I})$ (solid circles) departs from the exponential distribution of the diffusive limit (dashed line).[61] The solid curve shows the Q1D theory of Eqs. (1.1)–(1.2).

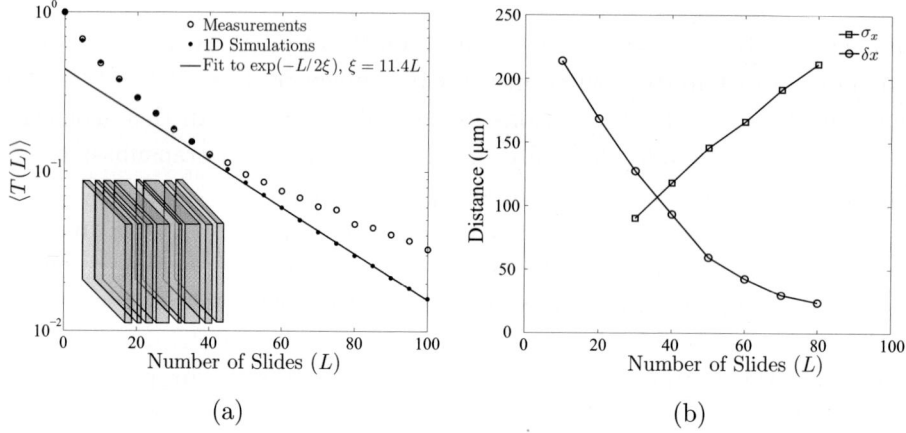

(a)	(b)

Fig. 22.14.   (a) Semilogarithmic plot of measurements (open circles) and 1D simulations (dots) of $\langle T \rangle$ versus number of glass cover slips, L. Schematic plot of random layer sample composed of glass cover slips is shown in the inset. (b) Measurement of beam spread, σ_x, and field correlation length, δx, versus L. Their crossing at $L = 35$ is consistent with the departures of $\langle T(L) \rangle$ from results of 1D simulations beginning at $L \sim 35$.[53]

At first, $\langle T(L) \rangle$ follows 1D simulations and approaches exponential scaling. But beyond a crossover thickness, $L_{co} \sim 35$, $\langle T(L) \rangle$ falls more slowly and transmission approaches an inverse falloff with L, characteristic of diffusion. As the sample thickness increases, the transverse spread of the wave σ_x increases while the coherence length, δx, drops (Fig. 22.14(b)). The coherence length is obtained from the Fourier transform of the specific intensity measured in the far field, which gives the field correlation function at the output surface of the sample. The curves giving the variation of σ_x and δx with thickness cross at $L \sim 35$. At this thickness, the impact of destructive interference, which results in wave localization in samples with uniform layers of random thickness, begins to be diminished as wave trajectories spread beyond a coherence length in the transmitted speckle pattern.

The transmitted intensity at the sample output is measured using a tapered optical fiber with small mode field diameter. The speckle pattern corresponding to plane wave excitation is found by translating the sample on a 10 μm square grid. Measurements of $P(s_{ab})$ for $L = 20$ are in excellent agreement with 1D simulations for $s_{ab} < 2.5$. Below a peak in $P(s_{ab})$ at a small value of s_{ab}, $P(s_{ab})$ drops sharply towards zero. The absence of nulls in transmission is a distinctive feature of 1D. $P(s_{ab})$ does not vanish above $s_{ab} = 3.4$, which corresponds to complete transmission, $T = 1$, as a result of interference on the output of the sample with waves that entered the sample at different points as a result of the transverse spread of light. With increasing thickness, the angular width of transmitted light and its transverse spread increase and a speckle pattern with a developed network of phase singularities develops. $P(s_{ab})$ then takes on the form predicted for Q1D. The mesoscopic function is the same as Eq. (1.1), and depends upon a single parameter, which may be expressed in terms of var(s_{ab}). Formally, this may also be expressed in terms of var(s_a) via the relationship between the moments in Eq. (1.3). However, Eqs. (1.1) and (1.3) only hold for Q1D in which all the transmitted energy passes through a defined area of the output over which the incident wave is thoroughly mixed. Thus, beyond Q1D, s_a is not related to the physical transmission but should be taken only as a variable of integration. It is convenient to express var(s_{ab}) in terms of g', since g' reduces to g in Q1D samples. The statistical conductance captures the extent to which wave trajectories cross within the sample, or equivalently the degree of mode overlap within the sample.

7. Speckle Evolution

Measuring the motion of phase singularities has proven to be an effective technique for tracking changes in a scattering medium.[92,93] It can also be used to determine the nature of waves within a random medium.[16] Examples of the changing speckle pattern with frequency shift are shown in the top row of Fig. 22.12. The overall change in the structure of the speckle

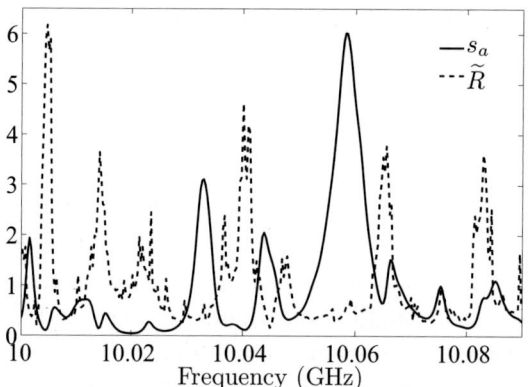

Fig. 22.15.   Spectra of normalized total transmission, s_a, and average singularity displacement, $\tilde{R}$, for localized waves for sample with var$(s_a) = 3.0$.[16]

pattern can be monitored using a variety of indicators. Among these are the average displacement of phase singularities, $R = \sum_i \Delta r_{si}/M$, where M is the number of singularities in a speckle pattern and Δr_{si} is the displacement of the ith singularity in a frequency shift $\Delta \nu$ which is much smaller than the correlation frequency. It is convenient to consider the average displacement, R, normalized to the ensemble average, $\tilde{R} = R/\langle R \rangle$. A comparison of the spectrum of $\tilde{R}$ and of s_a for localized waves in a single configuration drawn from measurements on the same ensemble as data in Fig. 22.12 is shown in Fig. 22.15. Peaks in $\tilde{R}$ occur most often between peaks of s_a. Peaks in s_a generally occur on resonance with modes of the medium. As the frequency is tuned though a resonance with a single mode strongly represented in transmission, the speckle pattern is dominated by the pattern of that mode and singularities are relatively stationary. We will see in the next section that when several modes overlap, their intensity patterns are similar so the speckle pattern is still quiescent as the frequency is tuned through a peak in transmission. The speckle pattern changes relatively rapidly, however, between strong peaks in total transmission over the frequency range in which the set of modes which dominates the speckle patterns changes.

The overall change of the speckle pattern can also be characterized by the standard deviations of phase changes, $\sigma_{\Delta\varphi}$, or the fractional intensity changes, $\sigma_{\Delta I^*}$, where the fractional intensity change is defined as: $\Delta I^* = [I(\nu + \Delta\nu) - I(\nu)]/[I(\nu + \Delta\nu) + I(\nu)]$.[16] When these very different measures of speckle change are normalized by their ensemble average values, which we indicate with a tilde, their spectra for the random ensemble with var$(s_a) = 3.0$ are strikingly similar. The corresponding probability

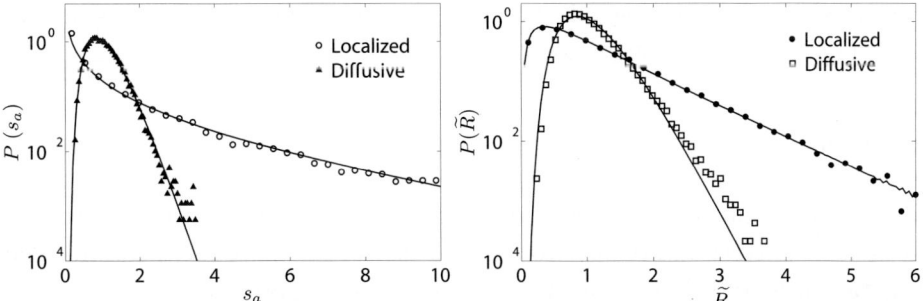

Fig. 22.16. Probability distributions of normalized total transmission, s_a, for diffusive (triangles) and localized (circles) waves (a) and normalized average displacement of singularities within one speckle pattern, $\widetilde{R}$, for diffusive (squares) and localized (filled circles) waves (b). The solid lines are calculated from Eq. (1.1) using the measured value of the corresponding variances.[16]

densities, $P(\widetilde{R})$, $P(\widetilde{\sigma}_{\Delta\varphi})$, and $P(\widetilde{\sigma}_{\Delta I^*})$, are also similar for both diffusive and localized waves and have precisely the same functional form as $P(s_a)$.[16] The probability densities of s_a and $\widetilde{R}$ in the ensemble with $\text{var}(s_a) = 3.0$ are shown in Fig. 22.16. Each of these probability density functions are described by Eq. (1.1) and are functions of the variances of the corresponding variable.

The surprising similarity in the probability densities of total transmission and various measures of speckle change in light of their different spectra is related to long-range correlation of the corresponding local variable.[21] This can be seen in a comparison of the spatial cumulant correlation functions of s_{ab} and the normalized velocity of phase singularities, $\widetilde{v} = v/\langle v \rangle$, where $v = dr/d\nu$ is the derivative of singularity position with frequency shift, denoted, $C_I(\Delta r)$ and $C_v(\Delta r)$, respectively.[21] Though v is only defined at singular points, while I is a continuous variable, the cumulant correlation functions of velocity has a similar form as for intensity, being a sum of short and long-rage contributions, as in Eq. (4.2). When the s_{ab} is normalized by s_a, $I' \equiv s_{ab}/s_a$, and v is normalized by a measure of speckle change, the standard deviation of fractional intensity change, $\sigma_{\Delta I^*}$, to give v', the corresponding cumulant correlation functions for both diffusive and localized waves reduce to the correlation function in the Gaussian limit, in which long-range correlation disappears. Thus the probability density for global fluctuations follows from the mixing of function for speckle change for Gaussian fields and a mesoscopic function representing the change in the speckle pattern as a whole. This is analogous to the probability density of intensity, which is a mixture of the intensity distribution for a Gaussian field and a mesoscopic function of total transmission. The mesoscopic function is associated with the presence of long-range correlation of the local variable.

8. Modes

We have seen that fluctuations in transmission and change in the structure of the speckle pattern are related to the underlying modes of the medium. The statistics of modes determine the statistics of all aspects of wave transport. The field inside an open random sample may be viewed as a superposition of modes, each of which corresponds to a volume speckle pattern of the field. The spatial and spectral variation of the polarization component j of the field for the nth quasimode is given by

$$A_{n,j}(r,\omega) = a_{n,j}(r)\frac{\Gamma_n/2}{\Gamma_n/2 + i(\omega - \omega_0)}. \tag{8.1}$$

Here, ω_n and Γ_n are the central frequency and linewidth, respectively, and $a_{n,j}(r)$ is the value of the complex amplitudes on resonance for the nth mode at position r. The modes of a random Q1D sample can be found by simultaneously fitting Eq. (8.1) to field spectra at 45 points in the speckle pattern with a single set of Γ_n and ω_n to obtain a set of complex mode amplitudes $a_{n,j}(r)$ at each of these points. Complex amplitudes are found for all other values of r by fitting field spectra at all points using the same set of Γ_n and ω_n. In contrast to modes of a closed system, quasimodes are eigenstates associated with open systems, in which energy is absorbed within the sample and leaks out through the boundaries. The decay rate of the local field amplitudes is $\Gamma_n/2$, corresponding to a decay rate of Γ_n for the mode. Although, in general, the eigenstates of a non-Hermitian operator do not form a complete basis, when the refractive index of materials in a leaky system varies discontinuously and approaches a constant asymptotic value sufficiently rapidly, the quasimodes are complete and orthogonal[94] so that an arbitrary state of the system can be expressed as a superposition of quasimodes.[95,96] In these cases the Green's function, which relates to the field at a point due to the interference of multiply scattered waves can be expanded in terms of these mutually orthogonal modes.

The decomposition of the transmitted field through a random sample of alumina spheres into a sum of field speckle patterns in a narrow frequency range is shown on the bottom of Fig. 22.12. The central frequency and frequency dependence of the total transmitted flux for each of the modes, as well as the relative total transmission for the individual modes are shown. The total transmission on resonance for individual modes can be much greater than the total transmission, as seen in the middle panel. This is because modes which are close in frequency have similar intensity speckle patterns but average phase differences that lead to destructive interference for the sum over the entire speckle pattern. The impact of destructive interference between neighboring modes can explain the dynamics of transmission, which will be discussed in Sec. 10. Key properties of modes that determine wave propagation are the joint statistics of linewidth, level spacing, integrated transmission, and average phase difference between modes as functions of δ. The study of modes allows for an investigation of microstatistics within a given ensemble in configurations and frequencies with the same degree of overlap.

An alternative way to approach modes of an open system is to consider the waves both inside and outside of the medium. In nondissipative systems, the eigenstates with constant flux for the system as a whole have real eigenvalues. Because these states are complete and orthorgonal,[96] the Green's functions can be expressed in terms of these constant flux modes. This has proved useful in the study of the spatial structure of modes in random lasers and nonlinear systems.[97]

9. Transmission Eigenchannels

Random matrix theory was developed in the 1950s to study the statistics of energy levels of heavy nuclei[98] and was later applied to other domains including quantum and classical wave transport[99,100] and chaotic system.[101] The field transmitted through a random medium can be expressed via the transmission matrix t as, $E_b = \sum_{a=1}^{N} t_{ba} E_a$, where a and b are, respectively, incoming and outgoing transverse propagation modes, or points on a grid over the input and output surfaces, and N is the total number of incident modes. Each configuration of a random medium can be related to a single realization of a large transmission matrix whose eigenvalues determine the the total transmission and the conductance. The multichannel Landauer formula, $g = \mathrm{Tr}(tt^{\dagger})$, implies that g is equal to the sum of the eigenvalues of the transmission matrix, $g = \sum_n T_n$. In diffusive samples, the distribution of T_n follows a bimodal distribution,[102–105] $\rho(T) = \frac{N\ell}{2L} \frac{1}{T\sqrt{1-T}}$, for, $T_{\min} \leq T < 1$, where the minimum transmission eigenvalue, $T_{\min} \approx 4e^{-2L/\ell}$, is determined from $\int_0^1 \rho(T)\, dT = N$.[100] Most of the eigenvalues are small so that the associated eigenchannels are referred to as "closed", while a number of channels, approximately equal to g, are "open" with transmission eigenvalues close to

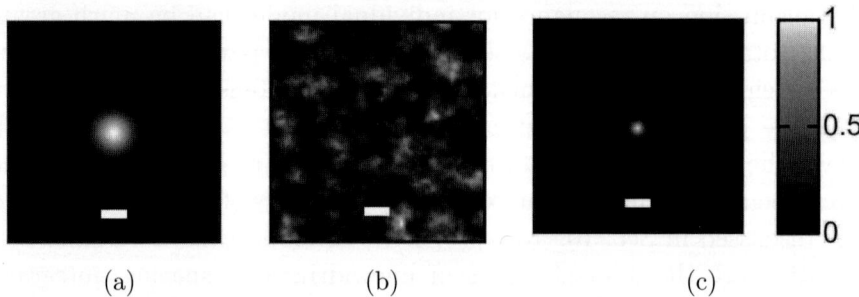

Fig. 22.17. (a) Unmodified incident wavefront. The focal width is of the order of the diffraction limit (62 μm, white bar). (b) Intensity transmission of the unmodified incident wavefront through a 6 μm layer of airbrush paint. (c) System with the sample present, and the wave shaped to achieve constructive interference at the target. A high-contrast, sharp focus is visible. The intensity plots are normalized to the brightest point in the image.[107]

unity.[103] Thus it is possible in principle to couple an incident wave to open channels and for a wave to be strongly transmitted through a strongly scattering opaque sample. Recently, a genetic algorithm was used to shape the wavefront of light reflected from a spatial light modulator to enhance total transmission by $\sim 44\%$[106] and to focus transmitted light on a detecting element in the transmitted speckle pattern with a thousand-fold enhancement of brightness over the average value within the original speckle pattern[107,108] (Fig. 22.17). The average value of total transmission through a random Q1D system is g/N. Thus the maximum value of the transmission eigenvalue of near unity represents an enhancement of N/g of transmission. Comparable enhancements have been observed recently in Q1D samples in microwave experiments in diffusive samples.[109] For localized waves, $g < 1$, and transmission in the strongest channel is typically much smaller than unity and is close to the value of the conductance.

The transmission matrix t can be written as, $t = U\Lambda V^\dagger$, using the singular value decomposition, in which U and V are unitary matrices transforming the input and output basis, respectively, and Λ is a diagonal matrix. The non-zero elements of Λ, λ_i, which are always real and positive, are called the singular values of t. Their distribution follows the quarter circle law,[110,111] $\rho(\lambda) = \sqrt{4 - \lambda^2}/\pi$. This law has been demonstrated in optical transmission[112](see Fig. 22.18) and ultrasound reflection.[113]

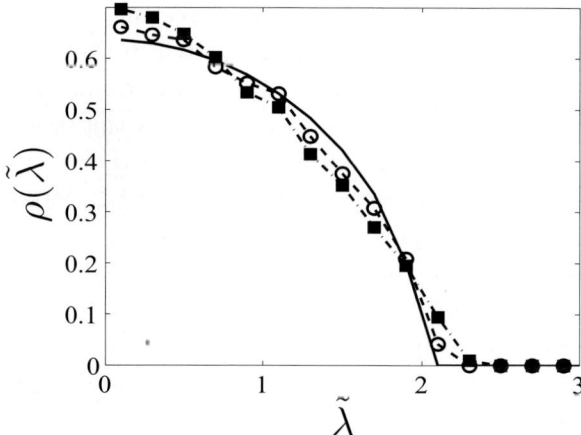

Fig. 22.18. Singular value distribution of the experimental transmission matrices obtained by averaging over 16 realizations of disorder. The solid line is the quarter-circle law predicted for random matrices. The solid squares and circles represent different filters to eliminate the contributions of the reference amplitude and inter-element correlations, respectively.[112]

10. Dynamics of Localized Waves

The dynamics of average transmission, as well as fluctuations and correlation of relative transmission can shed new light on wave transport in random media. Whereas steady-state measurements of the scaling of transmission can track the changing impact of weak localization on sets of samples of different size, measurements of pulsed transmission can reveal the time variation of the contributions of underlying modes with different lifetimes in samples of a particular scale.

In the diffusive limit, propagation can be described in terms of the diffusion of particles. For times longer than the diffusion time, τ_D, higher diffusion modes decay rapidly, so that energy becomes more concentrated in the lowest diffusion mode with a decay rate approaching the constant value, $\tau_D = L^2/\pi^2 D$. As δ decreases, however, a slowdown in the rate of decay of transmission with delay time from an exciting pulse is observed. A decreasing decay rate of transmission was observed in microwave[114] and optical[115] measurements in samples in which steady state measures of localization, $\text{var}(s_{ab})$ and $k\ell$, respectively, indicated the wave was diffusive. This slowdown can be explained in terms of a broadening distribution of mode lifetimes in the approach to localization,[114,116] in contrast to a sharply defined lifetime, equal to τ_D, in the diffusive limit. The slowdown of the rate of decay of transmission reflects the increasing weight of longer-lived modes with time delay. This has been calculated in the lowest order in $1/g$ using the supersymmetry approach.[116] The slowdown of the decay rate may also be calculated in terms of weak localization in the time domain[117] and in terms of the self consistent localization (SCLT) theory,[118–121] in which a renormalized diffusion coefficient as a function of depth into the sample and intensity modulation frequency, $D(z, \Omega)$,[120] is found.

A stronger slowdown of the decay rate is observed for localized waves. The time dependence of ultrasound transmission through the sample of aluminum sphere with $g' = 0.8$, for which a transmitted speckle pattern is shown in Fig. 22.4 and for which $P(\hat{I})$ is presented in Fig. 22.13, is shown in Fig. 22.19.[61] Measurements are well fit by SCLT. Measurements of pulsed microwave transmission of more deeply localized waves transmitted through a Q1D sample of random alumina spheres of length $L = 2\bar{\xi}$, in which $\mathrm{var}(s_a) = 3.0$, corresponding to $g' = 0.22$, are shown in Fig. 22.20.[122] Transmission spectra for samples within the same random ensemble are seen in Figs. 22.1 and 22.12. The impact of absorption[123] was removed statistically in the plot of Fig. 22.20 by multiplying the average measured intensity distribution by $\exp(t/\tau_a)$,[15] where the absorption time, τ_a, is determined from measurements of decay in a sample with reflecting end pieces so that leakage is negligible and the residual decay is dominated by absorption.[81] For

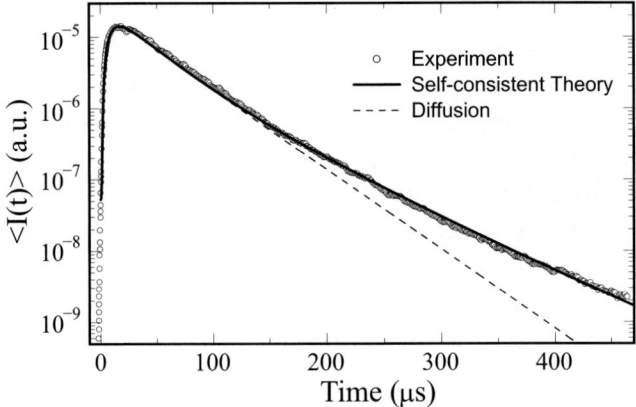

Fig. 22.19. Averaged time-dependent transmitted intensity $I(t)$, normalized so that the peak of the input pulse is unity and centered on $t = 0$, at representative frequencies in the localized regimes. The data are t by the self-consistent theory (solid curve). For comparison, the dashed line shows the long-time behavior predicted by diffusion theory.[61]

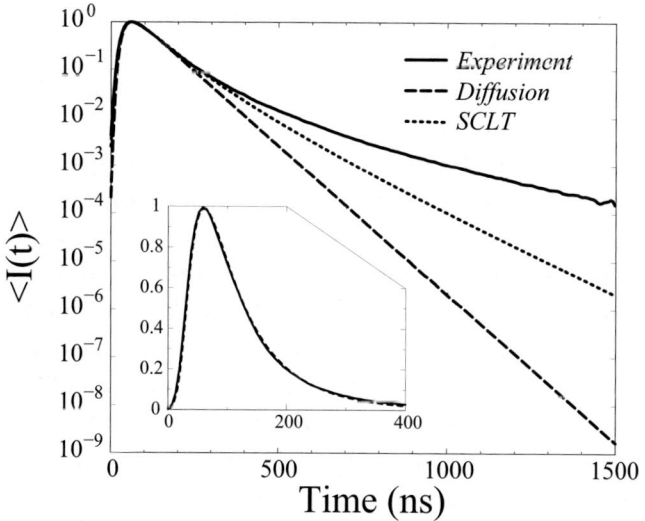

Fig. 22.20. Semilogarithmic plot of ensemble average of response, $\langle I(t) \rangle$, to a Gaussian incident pulse with standard deviation, $\sigma = 15$ MHz, in a sample of length, $L = 61$ cm. The dotted curve is the fit of SCLT at early times. The dashed curve gives the result of classical diffusion theory.[122]

times near the peak of the transmitted pulse, diffusion theory corresponds well with the measurements of $\langle I(t) \rangle$. For late times, however, pulsed transmission is higher than predicted by diffusion theory and SCLT.

Transmission associated with a single mode rises within the rise time of the exciting pulse. We have seen that, when many modes overlap, the intensity speckle patterns of neighboring modes are similar. Following an exciting pulse, transmission is suppressed by destructive interference between modes with similar intensity paterns but different phases. For times up to $4\tau_D$, pulsed transmission is in accord with SCLT, but transmission decays more slowly for longer times. This indicates the inability of this modified diffusion theory to capture the decay of long-lived localized states. Beyond the Heisenberg time, the decay rate approaches the predictions of the dynamic single parameter scaling (DSPS)[122] model, which reflects the increasing proportion of wave energy in long-lived localized states. The model is based upon the Gaussian distribution of Lyapunov exponents of the steady-state SPS model[12] and uniform random distribution of positions of localization centers within the medium. The mode decay rate is determined by its position and localization length. The model ignores the spectral and spatial overlap of modes that leads to the formation of more rapidly decaying necklace states. For this reason, the model cannot accurately represent transmission at early times, but it is able to represent aspects of decay at times long enough that dynamics is dominated by long-lived modes. At late times, the decay rates in 1D simulations and in the DSPS model converge towards the decay rates measured in the Q1D sample.[122]

The role of modes in dynamics can be further investigated by comparing measured transmission with the incoherent sum of modes found in the decomposition of spectra for a random ensemble of samples.[124] The time dependent total transmission for each mode is obtained by summing the square of the Fourier transform of the field in the form of Eq. (8.1) over all points in the output speckle pattern. To calculate the dynamics in the absence of absorption, the linewidth for each mode is the difference between the linewidth determined from the modal fit of the field spectrum and the intensity decay rate due to absorption, $(\Gamma_n - \Gamma_a)$, where $\Gamma_a = 1/\tau_a$. For $t > 250$ ns, the phasing of modes is randomized and the incoherent addition of decaying modes is in excellent agreement with measurements of dynamic transmission.

Once the phase between neighboring modes is randomized, the degree of correlation reflects the effective number of modes contributing to transmission.[125] Shortly after the ballistic pulse reaches the output of the sample, transmission is dominated by short-lived modes that release their energy quickly, while, at long times, short-lived modes have decayed and transmission is dominated by long-lived modes. At an intermediate time, both short and long-lived modes contribute to transmission. This explains the observation that, for the sample under discussion, $\kappa(t)$ has a minimum at an intermediate time after which correlation increases.[125]

A related reduction in transport has also been observed recently in the inhibition of the spread of matter waves due to Anderson localization of Bose–Einstein condensates[7,8] in a random 1D potential created by passing a laser though a diffusing plate. The spatial and temporal variation of the atomic density can be followed using fluorescence following atomic absorption of a probe beam. In a strong enough random potential, the tails of the spatial distribution fall exponentially and spread minimally in time.

11. Conclusion

The study of speckle from diverse perspectives has deepened our understanding of wave propagation, mesoscopic fluctuations and Anderson localization. The new approaches to the statistics of classical wave reviewed here indicate a quickening pace of discovery and hold promise for rapid progress in the near future. These approaches include studies of first and second order intensity statistics alongside the structure and evolution of the speckle pattern as a whole, steady state and pulsed measurements, and the decomposition of the transmitted speckle pattern into modes of the random medium and transmission eigenchannels.

Acknowledgments

We would like to thank our valued collaborators, Sheng Zhang, Jongchul Park, Bing Hu, Zhao-Qing Zhang, Andrey Chabanov, Patrick Sebbah, Zhou Shi, Yitzchak Lockerman, Victor Kopp, Boris Shapiro, Reuven Pnini, Valery Milner, Valentin Freilikher, Bart van Tiggelen and Jerome Klosner for discussions and for theoretical and experimental contributions. AZG would like to recall early support and encouragement of Narciso Garcia, Sajeev John and Philip Anderson. This work was supported by the NSF under grant DMR-0907285.

References

1. P. W. Anderson, *Phys. Rev.* **109**, 1492 (1958).
2. P. Sheng and Z. Q. Zhang, *Phys. Rev. Lett.* **57**, 1879 (1986).
3. M. Ya. Azbel, *Phys. Rev. B* **28**, 4106 (1983).
4. S. John, H. Sompolinsky and M. J. Stephen, *Phys. Rev. B* **27**, 5592 (1983).
5. S. Grésillon, L. Aigouy, A. C. Boccara, J. C. Rivoal, X. Quelin, C. Desmarest, P. Gadenne, V. A. Shubin, A. K. Sarychev and V. M. Shalev, *Phys. Rev. Lett.* **82**, 4520 (1999).
6. S. John, *Phys. Rev. Lett.* **53**, 2169 (1984).
7. J. Billy, V. Josse, Z. Zuo, A. Bernard, B. Hambrecht, P. Lugan, D. Clément, L. Sanchez-Palencia, P. Bouyer and A. Aspect, *Nature* **453**, 891 (2008).
8. G. Roati, C. D'Errico, L. Fallani, M. Fattori, C. Fort, M. Zaccanti, G. Modugno, M. Modugno and M. Inguscio, *Nature* **453**, 895 (2008).
9. J. T. Edwards and D. J. Thouless, *J. Phys. C* **5**, 807 (1972).
10. D. J. Thouless, *Phys. Rev. Lett.* **39**, 1167 (1977).
11. E. Abrahams, P. W. Anderson, D. C. Licciardello and T. V. Ramakrishnan, *Phys. Rev. Lett.* **42**, 673 (1979).
12. P. W. Anderson, D. J. Thouless, E. Abrahams and D. S. Fisher, *Phys. Rev. B* **22**, 3519 (1980).
13. R. Landauer, *Philos. Mag.* **21**, 863 (1970).
14. E. Economu and C. M. Soukoulis, *Phys. Rev. Lett.* **46**, 618(1981).
15. A. A. Chabanov, M. Stoytchev and A. Z. Genack, *Nature* **404**, 850 (2000).
16. S. Zhang, B. Hu, P. Sebbah and A. Z. Genack, *Phys. Rev. Lett.* **99**, 063902 (2007).
17. T. M. Nieuwenhuizen and M. C. van Rossum, *Phys. Rev. Lett.* **74**, 2674 (1995).

18. M. C. van Rossum and T. M. Nieuwenhuizen, *Rev. Mod. Phys.* **71**, 313 (1999).
19. E. Kogan and M. Kaveh, *Phys. Rev. B* **52**, R3813 (1995).
20. M. Stoytchev and A. Z. Genack, *Phys. Rev. Lett.* **79**, 309 (1997).
21. S. Zhang, Y. Lockerman and A. Z. Genack, arXiv:0904.1908.v1.
22. M. J. Stephen and G. Cwilich, *Phys. Rev. Lett.* **59**, 285 (1987).
23. P. A. Mello, E. Akkermans and B. Shapiro, *Phys. Rev. Lett.* **61**, 459 (1988).
24. S. Feng, C. Kane, P. A. Lee and A. D. Stone, *Phys. Rev. Lett.* **61**, 834 (1988).
25. A. Z. Genack, N. Garcia and W. Polkosnik, *Phys. Rev. Lett.* **65**, 2129 (1990).
26. J. F. de Boer, M. P. van Albada and A. Lagendijk, *Phys. Rev. Lett.* **64**, 2787 (1990).
27. E. Kogan and M. Kaveh, *Phys. Rev. B* **45**, 1049 (1992).
28. F. Scheffold and G. Maret, *Phys. Rev. Lett.* **81**, 5800 (1998).
29. P. W. Brouwer, *Phys. Rev. B* **57**, 10526 (1998).
30. P. Sebbah, R. Pnini and A. Z. Genack, *Phys. Rev. E* **62**, 7348 (2000).
31. P. Sebbah, B. Hu, A. Z. Genack, R. Pnini and B. Shapiro, *Phys. Rev. Lett.* **88**, 123901 (2002).
32. A. A. Chabanov, A. Z. Genack, N. Tregoures and B. A. van Tiggelen, *Phys. Rev. Lett.* **92**, 173901 (2004).
33. R. A. Webb, S. Washburn, C. P. Umbach and R. B. Laibowitz, *Phys. Rev. Lett.* **54**, 2696 (1985).
34. B. L. Altshuler and D. E. Khmelnitskii, *JETP Lett.* **42**, 359 (1985).
35. P. A. Lee and A. D. Stone, *Phys. Rev. Lett.* **55**, 1622 (1985).
36. N. F. Mott, *Philos. Mag.* **22**, 7 (1970).
37. I. M. Lifshits and V. Ya. Kirpichenkov, *Zh. Eksp. Teor. Fiz.* **77**, 989 (1979). [*Sov. Phys. JETP* **50**, 499 (1979).
38. J. B. Pendry, *J. Phys. C* **20**, 733 (1987).
39. J. B. Pendry, *Adv. Phys.* **43**, 461 (1994).
40. V. M. Apalkov, M. E. Raikh and B. Shapiro, *Phys. Rev. Lett.* **89**, 016802 (2002).
41. H. Cao, Y. Zhao, S. T. Ho, E. W. Seelig, Q. H. Wang and R. P. H. Chang, *Phys. Rev. Lett.* **82**, 2278 (1999).
42. S. He and J. D. Maynard, *Phys. Rev. Lett.* **57**, 3171 (1986).
43. P. Sebbah, B. Hu, J. Klosner and A. Z. Genack, *Phys. Rev. Lett.* **96**, 183902 (2006).
44. D. Laurent, O. Legrand, P. Sebbah, C. Vanneste and F. Mortessagne, *Phys. Rev. Lett.* **99**, 253902 (2007).
45. S. John, *Phys. Rev. Lett.* **58**, 2486 (1987).
46. K. Yu. Bliokh, Yu. P. Bliokh, V. Freilikher, A. Z. Genack and P. Sebbah, *Phys. Rev. Lett.* **101**, 133901 (2008).
47. O. Shapira and B. Fischer, *J. Opt. Soc. of Am. B* **22**, 2542 (2005).
48. J. Topolancik, B. Ilic and F. Vollmer, *Phys. Rev. Lett.* **99**, 253901 (2007).
49. L. Sapienza, H. Thyrrestrup, P. D. Garcia, S. Smolka and P. Lodahl, *Science* **327**, 1352 (2010).
50. A. Cohen, Y. Roth and B. Shapiro, *Phys. Rev. B* **38**, 12125 (1988).
51. A. A. Abrikosov, *Solid State Commun.* **37**, 997 (1981).
52. J. Park, S. Zhang and A. Z. Genack, private communication (2009).

53. S. Zhang, J. Park, V. Milner and A. Z. Genack, *Phys. Rev. Lett.* **101**, 183901 (2008).
54. L. I. Deych, A. A. Lisyansky and B. L. Altshuler, *Phys. Rev. Lett.* **84**, 2678 (2000).
55. E. Akkermans, P. E. Wolf and R. Maynard, *Phys. Rev. Lett.* **56**, 1471 (1986).
56. M. P. van Albada and A. Lagendijk, *Phys. Rev. Lett.* **55**, 2692 (1985).
57. P. E. Wolf and G. Maret, *Phys. Rev. Lett.* **55**, 2696 (1985).
58. D. S. Wiersma, M. P. van Albada, B. A. van Tiggelen and A. Lagendijk, *Phys. Rev. Lett.* **74**, 4193 (1995).
59. E. Akkermans, P. E. Wolf, R. Maynard and G. Maret, *J. de Physique (France)* **49**, 77 (1988).
60. A. Ioffe and A. R. Regel, *Prog. Semicond.* **4**, 237 (1960).
61. H. Hu, A. Strybulevych, J. H. Page, S. E. Skipetrov and B. A. van Tiggelen, *Nature Phys.* **4**, 945 (2008).
62. S. Faez, A. Strybulevych, J. H. Page, A. Lagendijk and B. A. van Tiggelen, *Phys. Rev. Lett.* **103**, 155703 (2009).
63. A. D. Mirlin, Y. V. Fyodorov, A. Mildenberger and F. Evers, *Phys. Rev. Lett.* **97**, 046803 (2006).
64. M. E. Gertsenshtein and V. B. Vasilév, *Theor. Probab. Appl.* **4**, 391 (1959).
65. R. Dalichaouch, J. P. Armstrong, S. Schultz, P. M. Platzman and S. L. McCall, *Nature* **354**, 53 (1991).
66. M. V. Berry and S. Klein, *Eur. J. Phys.* **18**, 222 (1997).
67. J. Bertolotti, S. Gottardo, D. S. Wiersma, M. Ghulinyan and L. Pavesi, *Phys. Rev. Lett.* **94**, 127401 (2005).
68. T. Schwartz, G. Bartal, S. Fishman and M. Segev, *Nature* **446**, 52 (2007).
69. Y. Lahini, A. Avidan, F. Pozzi, M. Sorel, R. Morandotti, D. N. Christodoulides and Y. Silberberg, *Phys. Rev. Lett.* **100**, 013906 (2008).
70. H. D. Raedt, A. Lagendijk and P. de Vries, *Phys. Rev. Lett.* **62**, 47 (1989).
71. B. Shapiro, *Phys. Rev. Lett.* **57**, 2168 (1986).
72. I. Freund and D. Eliyahu, *Phys. Rev. A* **45**, 6133 (1992).
73. J. H. Li and A. Z. Genack, *Phys. Rev. E* **49**, 4530 (1994).
74. A. A. Chabanov, B. Hu and A. Z. Genack, *Phys. Rev. Lett.* **93**, 123901 (2004).
75. N. Cherroret, A. Peña, A. A. Chabanov and S. E. Skipetrov, *Phys. Rev. B* **80**, 045118 (2009).
76. S. M. Cohen, D. Eliyahu, I. Freund and M. Kaveh, *Phys. Rev. A* **43**, 5748 (1991).
77. A. Z. Genack, *Phys. Rev. Lett.* **58**, 2043 (1987).
78. A. Z. Genack and N. Garcia, *Phys. Rev. Lett.* **66**, 2064 (1991).
79. D. S. Wiersma, P. Bartolini, A. Lagendijk and R. Righini, *Nature* **390**, 671 (1997).
80. F. Scheffold, R. Lenke, R. Tweer and G. Maret, *Nature* **389**, 206 (1999).
81. A. A. Chabanov and A. Z. Genack, *Phys. Rev. Lett.* **87**, 153901 (2001).
82. A. Z. Genack, P. Sebbah, M. Stoytchev and B. A. van Tiggelen, *Phys. Rev. Lett.* **82**, 715 (1999).
83. M. P. van Albada, B. A. van Tiggelen, A. Lagendijk and A. Tip, *Phys. Rev. Lett.* **66**, 3132 (1991).

84. S. Zhang and A. Z. Genack, *Phys. Rev. Lett.* **99**, 203901 (2007).
85. W. Wang, S. G. Hanson, Y. Miyamoto and M. Takeda, *Phys. Rev. Lett.* **94**, 103902 (2005).
86. J. F. Nye and M. V. Berry, *Proc. R. Soc. London, Ser A* **336**, 165 (1974).
87. M. V. Berry, *J. Phys. A* **11**, 27 (1978).
88. I. Freund, N. Shvartsman and V. Freilikher, *Opt. Commun.* **101**, 247 (1993).
89. N. Shvartsman and I. Freund, *Phys. Rev. Lett.* **72**, 1008 (1994)
90. M. V. Berry and M. R. Dennis, *Proc. R. Soc. London, Ser A* **456**, 2059 (2000).
91. J. W. Goodman, *Introduction to Fourier Optics* (McGraw-Hill, New York, 1968).
92. S. Sirohi, *Speckle Metrology* (Marcel Dekker, New York, 1993).
93. W. Wang, N. Ishii, S. G. Hanson, Y. Miyamoto and M. Takeda, *Opt. Commun.* **248**, 59 (2004).
94. P. T. Leung, S. Y. Liu and K. Young, *Phys. Rev. A* **49**, 3057 (1994).
95. E. S. C. Ching, P. T. Leung, A. Maassen van den Brink, W. M. Suen, S. S. Tong and K. Young, *Rev. Mod. Phys.* **70**, 1545 (1998).
96. H. E. Türeci, A. D. Stone and B. Collier, *Phys. Rev. A* **74**, 043822 (2006).
97. H. E. Türeci, A. D. Stone and L. Ge, *Phys. Rev. A* **76**, 013813 (2007).
98. M. L. Mehta, *Random Matrices* (Academic Press, New York, 1991).
99. C. W. J. Beenakker, *Rev. Mod. Phys.* **69**, 731 (1997).
100. C. W. J. Beenakker, arXiv:0904.1432v2
101. O. Bohigas, in *Chaos and Quantum Physics*, eds. M. J. Giannoni, A. Voros and J. Zinn-Justin (North-Holland, Amsterdam, 1990), p. 87.
102. O. N. Dorokhov, *Solid State Commun.* **51**, 381 (1984).
103. Y. Imry, *Europhys. Lett.* **1**, 249 (1986).
104. K. A. Muttalib, J.-L. Pichard and A. D. Stone, *Phys. Rev. Lett.* **59**, 2475 (1987).
105. J. B. Pendry, A. MacKinnon and P. J. Roberts, *Proc. R. Soc. London, Ser A* **437**, 67 (1992).
106. I. M. Vellekoop and A. P. Mosk, *Phys. Rev. Lett.* **101**, 120601 (2008).
107. I. M. Vellekoop, A. Lagendijk and A. P. Mosk, *Nature Photon.* **4**, 320 (2010).
108. I. M. Vellekoop and A. P. Mosk, *Opt. Lett.* **32**, 2309 (2007).
109. S. Zhou and A. Z. Genack, private communication (2010).
110. V. A. Marčenko and L. A. Pastur, *Math. USSR-Sbornik* **1**, 457 (1967).
111. E. P. Wigner, *SIAM Rev.* **9**, 1 (1967).
112. S. M. Popoff, G. Lerosey, R. Carminati, M. Fink, A. C. Boccara and S. Gigan, *Phys. Rev. Lett.* **104**, 100601 (2010).
113. A. Aubry and A. Derode, *Phys. Rev. Lett.* **102**, 084301 (2009).
114. A. A. Chabanov, A. Z. Genack and Z. Q. Zhang, *Phys. Rev. Lett.* **90**, 203903 (2003).
115. M. Störzer, P. Gross, C. M. Aegerter and G. Maret, *Phys. Rev. Lett.* **96**, 063904 (2006).
116. A. D. Mirlin, *Phys. Rep.* **326**, 259 (2000).
117. S. K. Cheung, X. Zhang, Z. Q. Zhang, A. A. Chabanov and A. Z. Genack, *Phys. Rev. Lett.* **92**, 173902 (2004).
118. D. Vollhardt and P. Wölfle, *Phys. Rev. B* **22**, 4666 (1980).

119. B. A. van Tiggelen, A. Lagendijk and D. Wiersma, *Phys. Rev. Lett.* **84**, 4333 (2000).
120. S. E. Skipetrov and B. A. van Tiggelen, *Phys. Rev. Lett.* **92**, 113901 (2004).
121. S. E. Skipetrov and B. A. van Tiggelen, *Phys. Rev. Lett.* **96**, 043902 (2006).
122. Z. Q. Zhang, A. A. Chabanov, S. K. Cheung, C. H. Wong and A. Z. Genack, *Phys. Rev. B* **79**, 144203 (2009).
123. R. Weaver, *Phys. Rev. B* **47**, 1077 (1993).
124. J. Wang, and A. Z. Genack, private communication (2010).
125. J. Wang, A. A. Chabanov, D. Y. Lu, Z. Q. Zhang and A. Z. Genack, to appear in *Phys. Rev. B* (2010).